Functional Biochemistry in Health and Disease

Eric A. Newsholme and **Tony R. Leech**

Merton College, Oxford and *Gresham's School, Norfolk*

Editorial Assistant
Mary Board

St Hilda's College, Oxford

WILEY-BLACKWELL

A John Wiley & Sons, Ltd., Publication

This edition first published 2009
© 2009 by John Wiley & Sons, Ltd

Wiley-Blackwell is an imprint of John Wiley & Sons, formed by the merger of Wiley's global Scientific, Technical and Medical business with Blackwell Publishing.

Registered office: John Wiley & Sons Ltd, The Atrium, Southern Gate, Chichester, West Sussex, PO19 8SQ, UK

Other Editorial Offices:
9600 Garsington Road, Oxford, OX4 2DQ, UK
111 River Street, Hoboken, NJ 07030-5774, USA

For details of our global editorial offices, for customer services and for information about how to apply for permission to reuse the copyright material in this book please see our website at www.wiley.com/wiley-blackwell

The right of the author to be identified as the author of this work has been asserted in accordance with the Copyright, Designs and Patents Act 1988.

Library of Congress Cataloguing-in-Publication Data

Newsholme, E. A.
    Functional biochemistry in health and disease / Eric Arthur Newsholme and Tony R. Leech.
        p. cm.
    Includes bibliographical references and index.
    ISBN 978-0-471-98820-5 (cloth) – ISBN 978-0-471-93165-2 (pbk.)   1. Biochemistry.   2. Metabolism.   I. Leech, A. R.   II. Title.
    QP514.2.N48 2009
    612′.015–dc22

                                                                                                    2009007437

ISBN: 978 0 471 98820 5 (HB) and 978 0 471 93165 2 (PB)

A catalogue record for this book is available from the British Library.

Set in 10/12 pt Times by SNP Best-set Typesetter Ltd., Hong Kong
Printed in Singapore by Markono Print Media Pte Ltd.

First printing   2010

To

Pauline Newsholme and Barbara Leech for patience,
encouragement and a willingness to share their husbands
with biochemistry.

Contents

Preface

This text attempts to provide a clear and straightforward account of the biochemistry underlying the physiological functions of different cells, tissues or organs essential for human life. The approach highlights the contribution that functional biochemistry makes to health and, when it is disturbed, to ill health and disease. It attempts to link biochemistry, medical education and clinical practice. Structural biochemistry and molecular biology are kept to a minimum in favour of a focus on the dynamic aspects of biochemistry and its immediate importance for health.

Unfortunately, in many medical or biomedical programmes, biochemical education has been reduced to the presentation of basic facts. It is the aim of this book to provide the link between these facts and medical practice to achieve a more complete biochemical education. Each topic is presented in one complete section or one chapter, which can be studies independently, unencumbered by extraneous material. Nonetheless, cross-references between topics are emphasised to help the student appreciate the common biochemical principles underlying health and disease.

Both authors and the editorial assistant have collaborated on each chapter; there have been no other contributors. Since the authors are not expert in all areas, much research in journals, reviews and books has been undertaken. The challenge faced has therefore been the same as that faced by every student entering a field for the first time and for this reason the authors hope to have been better placed to clarify the biochemistry which underlies physiology and pathology.

Background material necessary to tackle the core of each topic is provided in the introduction to each section or chapter. Students, including graduates, who are entering courses in medicine or the biomedical sciences, and who have limited knowledge of biochemistry, physiology or pathology, should find these introductions of particular value. Our approach should also help students who are taking courses which involve problem-based learning or who are preparing seminars. We hope, too, that it might be of use to physicians, surgeons and academics in understanding the biochemical background of illness and disease and help them to prepare lectures, reports and grant applications.

The manner in which biochemical and physiological knowledge and ideas are presented should help the student to develop the skills of critical analysis, debate issues and even challenge some of the dogmas in biomedicine. Such opportunities are not always available when there is over-dependence on electronic sources of information.

The topics covered in this book are presented in five Sections:

- Introduction

- Essential topics in dynamic biochemistry

- Essential metabolism

- Essential processes of life

- Serious diseases

Problems of ill health are discussed in most of these Sections, including fatigue, allergies, nutritional deficiencies, AIDS, chronic fatigue syndrome, Creutzfeldt-Jacob disease, malaria, neurological and eating disorders, tuberculosis, cancer and the deadly trio: atherosclerosis, hypertension and heart attack. The conditions of obesity, diabetes mellitus and disorders of fat metabolism are included in the on-line Appendices to allow full discussion and the clarification (and correction) of important aspects. The Appendices can be found on the companion website for the book – visit www.wiley.com/go/newsholme/biochemistry The reader will also find here a compilation of the book's figures and a full list of the References referred to throughout the text.

Eric A. Newsholme
Tony R. Leech

Acknowledgements

The authors wish to thank Professor Craig Sharp, Professor Philip Newsholme and Lindy Castell for their considerable help in providing much helpful information and for reading and constructively criticising the chapters relevant to their own fields. We also wish to thank Lindy Castell for carrying out literature searches and for introducing us to many of her clinical and physiological colleagues who were willing to answer specific queries.

Despite it being the age of the computer, the first drafts of all chapters were written by hand so we owe an enormous debt of gratitude to our typist, Judith Kirby, Fellows' Secretary at Merton College, Oxford. She miraculously transformed barely legible scribblings into immaculate electronic documents and then coped uncomplainingly with numerous further alterations.

We are exceedingly grateful for the way in which Ruth Swann was able to improve the proofs by detecting lack of clarity in places and by suggesting organisational improvements. She accepted what were sometimes quite substantial handwritten changes with good humour and executed them with deft professionalism.

We wish to identify three books which provided useful information on topics which were not otherwise easily available or accessible:

- *Modern Nutrition in Health and Disease* (2006) by Maurice Shils, Moshe Shike, Catharine Ross, Benjamin Caballero & Robert Cousins. Lippincott Williams & Wilkins.

- *The Encyclopedia of Molecular Biology* (1994) edited by John Kendrew. Blackwell Science.

- *The Fats of Life* (1998) by Caroline Pond. Cambridge University Press.

A further three books provided valuable information which extended the coverage of a number of topics which were not specifically biochemical:

- *The Faber Book of Science* (1995) edited by John Carey. Faber & Faber (London).

- *Drug Discovery: the Evolution of Modern Medicines* (1995) by Walter Sneader. John Wiley & Sons, Ltd. (Chichester).

- *The Oxford Medical Companion* (1994) edited by John Walton, Jeremiah Barondess and Stephen Lock, Oxford University Press.

Abbreviations

acetyl-CoA	acetyl-coenzyme-A
ACE	angiotensin-converting enzyme
ACP	acyl carrier protein
ACTH	adrenocorticotropic hormone
ADH	antidiuretic hormone
ARDS	adult respiratory distress syndrome
ALT	alanine aminotransferase
AMP	adenosine monophosphate
APC	antigen-presenting cell
APP	amyloid precursor protein
AST	aspartate aminotransferase
ATP	adenosine triphosphate
BMR	basal metabolic rate
BCAA	branched-chain amino acid
BMI	body mass index
cAMP	cyclic AMP (cyclic 3′,5′-adenosine monophosphate)
CHD	coronary heart disease
CJD	Creutzfeldt-Jacob Disease
CNS	central nervous system
CPS-I	carbamoyl phosphate synthetase
CoA	coenzyme A
CoASH	reduced coenzyme A
COX-1 and COX-2	cyclooxygenase enzyme
DAG	diacylglycerol
DTH	delayed-type hypersensitivity
DIC	disseminated intravascular coagulation
DNA	deoxyribonucleic acid
EFA	essential fatty acid
ELISA	enzyme-linked immunoabsorbent assay
FABP	fatty acid binding protein
FAD	flavin adenine dinucleotide
FFA	free fatty acid
FSH	follicle-stimulating hormone

G1P	glucose 1-phosphate
G6P	glucose 6-phosphate
GABA	gamma-aminobutyrate
GALT	gut-associated lymphoid system
GDP	guanosine diphosphate
GIP	glucose-dependent insulinotrophic polypeptide
GMP	guanosine monophosphate
GnRH	gonadotrophin-releasing hormone
GLP	glucagon-like peptide
GOT	glutamate-oxaloacetate transaminase
GPCR	G-protein coupled receptor
GPT	glutamate-pyruvate transaminase
GTP	guanosine triphosphate
HK	hexokinase
HRT	hormone replacement therapy
HDL	high density lipoprotein
IRS	insulin receptor substrate
IMP	inosine monophosophate
IP$_3$	inositol trisphosphate
IRS	insulin-receptor substrate
IRP	iron-regulating protein
K$_m$	Michaelis constant
kDa	kilodalton
LCAT	lecithin-cholesterol acyl transferase
LH	luteinising hormone
LO	lipoxygenase
LGIC	ligand-gated ion channel
LDL	low density lipoproteins
LSD	lysergic acid diethylamide
Mmolar	(mol/litre)
mM	millimolar (mmol/litre)
MDA	methylene dioxyamphetamine
MODY	maturity onset diabetes in the young
mol	mole

mmol	millimole		PIP$_2$	phosphatidylinositol bisphosphate
MOF	multiple organ failure		PKU	phenylketonuria
MSF	multiple systems failure		PPi	pyrophosphate ion
MW	molecular weight		PTH	parathyroid hormone
mV	millivolt(s)		PUFAs	polyunsaturated fatty acids
ME	myalgic encephalomyelitis			
MLCK	myosin light chain kinase		ROS	reactive oxygen species
mRNA	messenger RNA		REE	Resting energy expenditure
			RDA	recommended daily allowance
NAD$^+$	nicotinamide adenine dinucleotide			
NADH	reduced nicotinamide adenine dinucleotide		SAM	*S*-adenosyl methionine
NADP$^+$	nicotinamide adenine dinucleotide phosphate		SR	sarcoplasmic reticulum
			SIRS	systemic inflammatory response syndrome
NADPH	reduced nicotinamide adenine dinucleotide phosphate		TNF	tumour necrosis factor
NAGS	N-acetylglutamate synthetase		TAG	triacylglycerol
NEFA	non-esterified fatty acids			
NSAIDs	non-steroidal anti-inflammatory drugs		UDP	uridine diphosphate
			UDPG	UDP-glucose
OGDH	oxoglutarate dehydrogenase			
			VLDL	very low density lipoprotein
PAF	platelet activating factor		v-onc	viral oncogene
PAH	phenylalanine hydroxylase		V$_{max}$	maximal velocity
PEM	protein-energy deficiency		VFA	volatile fatty acids
Pi	phosphate ion			

I   INTRODUCTION

1
The Structural and Biochemical Hierarchy of a Cell and a Human

There is no magician's mantle to compare with skin in its diverse roles of waterproof overcoat, sunshade, suit of armor and refrigerator, sensitive to the touch of a feather, to temperature and to pain, withstanding wear and tear of three score years and ten, and executing its own running repairs.

(Lockhart *et al.*, 1959)

It is conventional to describe the structure of an organism in hierarchical terms: organelles make cells; cells make tissues; tissues make organs; organs make people. It is also possible to consider biochemistry in the same way. The biochemistry in the organelles contributes to that of the cell, which contributes to that of the tissues and organs and, eventually, they all contribute to the biochemistry and physiology that constitute the processes of life of a human. Indeed, the normal functioning of this hierarchy provides the basis for health and, when any component fails, the basis of disease.

In 1665, Robert Hooke examined thin slices of cork under his very simple microscope and discovered small, box-like spaces which he named cells. A few years later the Italian anatomist Marcello Malpighi described similar structures in animal tissues, which he called vesicles or utricles and, in 1672, the English botanist Nehemiah Grew published two extensively illustrated volumes greatly extending Hooke's findings. The concept of the cell as a unit of structure in the plant and animal kingdoms was launched, but it was two centuries later before scientists generally accepted the idea of the cell as the fundamental structural unit of living organisms. The 'cell theory' attributed to Matthias Schleiden & Theodor Schwann, was proposed in 1838: the cell is the smallest component of life that can exist independently. All living organisms are composed

---

Most human cells have diameters in the range 10–20 μm. A spherical cell of diameter 15 μm would have a volume of 1.8×10^{-9} mL and contain around 2.5×10^{9} molecules of protein and 4.5×10^{13} molecules of water. A single drop of blood (0.05 mL) contains around 2.25×10^{8} red blood cells.

---

of cells and, as concluded by Rudolf Virchow in 1855, cells can arise only from pre-existing cells. Multicellular organisms develop from one cell (a fertilised egg), which differentiates into many different cell types. The number of cells in a multicellular organism varies greatly from one organism to another. The average human comprises approximately 100 million million (10^{14}) cells.

Cell structure

The detailed knowledge of the structure and function of cells is largely due to the introduction of the electron microscope in the 1940s and the subsequent development of biochemical and cell biological techniques.

All cells are surrounded by a thin lipid membrane. This is a selective barrier, allowing some substances to pass across it and excluding others in order to maintain a relatively constant internal environment. Some of the different proteins that are embedded in the cell membrane transport compounds and ions across the membrane, whereas others act as receptors that respond to factors in the external environment and initiate responses within the cell, and still others provide a mechanism for cells to interact and communicate with each other.

Subcellular structures (organelles) are present in the cell. Each one has its own characteristic activities and properties that work together to maintain the cell and its functions. The remainder of the cell is the gel-like cytoplasm, known as cytosol (Figure 1.1). The largest organelle in the cell is the nucleus: it contains the genetic

---

Functional Biochemistry in Health and Disease by Eric Newsholme and Tony Leech
© 2010 John Wiley & Sons Ltd

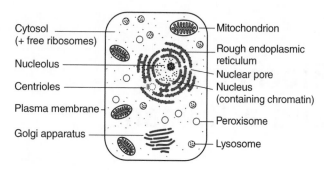

Cytosol (+ free ribosomes)
Nucleolus
Centrioles
Plasma membrane
Golgi apparatus
Mitochondrion
Rough endoplasmic reticulum
Nuclear pore
Nucleus (containing chromatin)
Peroxisome
Lysosome

Figure 1.1 *Diagrammatic representation of cell organelles as seen under an electron microscope.*

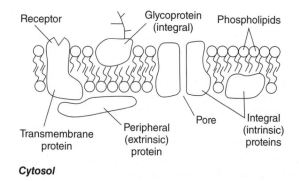

Extracellular fluid

Receptor
Glycoprotein (integral)
Phospholipids
Transmembrane protein
Peripheral (extrinsic) protein
Pore
Integral (intrinsic) proteins

Cytosol

Figure 1.2 *Diagrammatic representation of the plasma membrane. Arrangement of integral and peripheral membrane proteins with the molecular layer of phospholipids is shown.*

material, deoxyribonucleic acid (DNA). Through the information contained in a coded form within its chemical structure, DNA determines to a large extent, but not completely, the specific morphological and biochemical characteristics of each type of cell.

Plasma membrane

Every cell possesses a plasma (or cell) membrane which isolates its contents from its surroundings. This membrane consists of a double layer of phospholipid molecules with proteins attached or dispersed within. The uneven distribution of proteins and their ability to move in the plane of the membrane led to the description of this structure as a fluid mosaic (Figure 1.2) (Chapter 5). Some of these proteins facilitate the transport of molecules and ions through the membrane, while others are receptors for extracellular molecules which provide information about conditions in adjacent cells, blood and elsewhere in the body. Physical or chemical damage to these membranes can render them leaky so that, for example, Na^+ and Ca^{2+} ions, the concentrations of which are much higher in the extracellular fluid, can enter the cell causing damage. On the outer surface of

the membrane are short chains of sugars, which are attached to both phospholipids and proteins. They form a carbohydrate sheath known as the glycocalyx. This provides protection for the membrane, a site for attachment of proteins and an identity for the cell, enabling it to be identified as 'self' and not foreign by the immune system.

Mitochondria

All mitochondria have an outer membrane, which is permeable to small molecules, and an inner membrane, which is much less permeable and extensively folded to form cristae that extend into the matrix (Figure 1.3).

The main function of mitochondria is the generation of adenosine triphosphate (ATP) from adenosine diphosphate (ADP) and phosphate, which is achieved by energy transfer from oxidation of fuels. Some cells (e.g. hepatocytes) contain several thousand mitochondria, whereas others, such as erythrocytes, lack them entirely.

Endoplasmic reticulum

The endoplasmic reticulum, or ER, is an intricate system of membranes that spreads throughout the cytosol. Part of it is studded with ribosomes and is called the rough endoplasmic reticulum because of its appearance in the electron microscope (Figure 1.4). It has four main functions:

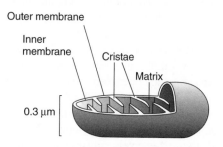

Outer membrane
Inner membrane
Cristae
Matrix
0.3 µm

Figure 1.3 *Diagrammatic representation of a mitochondrion.*

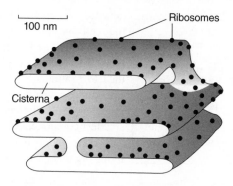

100 nm
Ribosomes
Cisterna

Figure 1.4 *Three-dimensional representation of a region of the rough endoplasmic reticulum.*

- Synthesis of those proteins that are destined for incorporation into cellular membranes or for export from the cell. Transport of those proteins that are destined for cell membranes or for release from the cell is achieved through vesicles that pinch off from the endoplasmic reticulum and fuse with membranes of the Golgi (see below).

- Synthesis of phospholipids and steroids.

- Hydroxylation (addition of an –OH group) of compounds that are toxic or waste products, which renders them more water soluble, hence they are more rapidly excreted. These are known as detoxification reactions.

- Storage of Ca^{2+} ions at a concentration 10 000 times greater than in the cytosol (i.e. similar to that in the extracellular fluid, about 10^{-3} mol/L). It is the release of some of these ions that acts as a signalling process in the cell. For example, stimulation of contraction of muscle by a nerve depends upon Ca^{2+} ion release from the reticulum into the cytosol of the muscle cell.

The Golgi complex

This organelle is variously known as the Golgi complex, the Golgi apparatus, the Golgi body or simply 'the Golgi'. It comprises a stack of smooth membranes that form flattened sacs (Figure 1.5). It 'directs' proteins that have been synthesised on the ribosomes and have then entered the endoplasmic reticulum to various parts of the cell. The *cis*-face of the Golgi complex faces towards the centre of the cell and proteins reach this face inside vesicles that bud off from the endoplasmic reticulum. Within the Golgi these proteins can be modified, e.g. by removal of some amino acids and addition of other compounds (e.g.

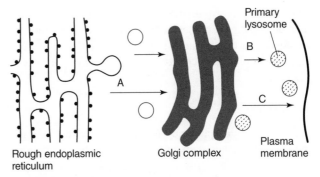

Figure 1.6 *Vesicular transport of proteins within the cell.* Vesicles from the endoplasmic reticulum [A] carry protein to the Golgi complex, they are repackaged in the Golgi from which they leave to form primary lysosomes [B] or fuse with the plasma membrane; this is to add proteins or to be secreted from the cell [C]. In the Golgi, new vesicles are formed to transport the proteins to the plasma membrane (e.g. transport proteins or proteins for export) or the lysosomes. This system transports, safely, 'dangerous' hydrolytic enzyme to the lysosomes and it also protects membrane proteins, or proteins for export, from degradation in the cytosol.

sequences of sugars). Vesicles, containing these modified proteins, bud from the *trans*-face of the Golgi and are then transported to other parts of the cell to form another organelle, e.g. the lysosome, or fuse with the plasma membrane where the proteins remain on are secreted into the extracellular space (Figure 1.6).

Lysosomes

Lysosomes are membrane-bound organelles that contain hydrolytic enzymes to break down macromolecules and other organelles taken up by the lysosomes. The pH within this organelle is very low (about 5.0) and the catalytic activities of the enzymes, within it, are highest at this pH. The pH in the cytosol is about 7.1, so that any enzymes released from the lysosome are not catalytically active in the cytosol.

The enzymes degrade a number of compounds:

- Proteins taken up from outside the cell or those damaged within the cell.

- Particles, including bacteria, taken up from the environment.

- Damaged or senescent organelles (e.g. mitochondria).

Any indigestible material within the lysosome is normally expelled through the plasma membrane, but as cells grow older, this process functions less effectively so that cells become loaded with unwanted lipid and protein, which is oxidised to produce a complex known as lipofuscin (an age pigment). Over many years, this can accumulate and impair

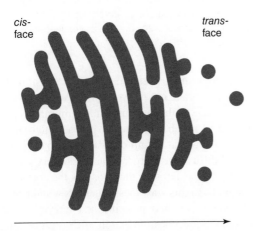

Figure 1.5 Cis *and* trans *faces of the Golgi.* The Golgi consist of four to six cisternae in a single stack but cells may contain more than one stack or, in the case of cells which have a major secretory function, larger stacks.

cell function. Accumulation occurs earlier in some cells than others: e.g. neurones in the spinal cord accommodate lipofuscin early so that the cytosol in neurones in very elderly people may consist of more than 70% of the pigment.

Peroxisomes

Peroxisomes are spherical vesicles bounded by a single membrane. They contain enzymes that catalyse oxidations that produce hydrogen peroxide which is degraded by the enzyme catalase. For example, very long or unusual fatty acids that are present in the diet but have no function are completely degraded in the peroxisomes.

Ribosomes

Unlike the organelles described above, ribosomes have no membrane but are aggregates of ribonucleic acid (RNA) and protein. Each ribosome consists of two subunits: a large and a smaller one (Figure 1.7).

Most of the protein synthesis in a cell takes place within or upon the ribosomes. They bring together messenger RNA (mRNA) and the components required for protein synthesis. This begins with the attachment of a ribosome to one end of the mRNA and continues with the progression of the ribosome along the messenger, generating the polypeptide chain as it moves (Figure 1.8) (Chapter 20).

The nucleus

All eukaryotic cells begin their existence with a nucleus and its loss or removal normally leads to death of the cell. The exception to this is the reticulocyte which, while within the bone marrow, extrudes its nucleus to form an erythrocyte (red blood cell).

Material within the nucleus (nucleoplasm) is separated from the cytoplasm by the nuclear membrane (also known as an envelope), a double membrane that is continuous

Figure 1.7 *Structure of a ribosome.* It is composed of two subunits: the large 60S* subunit has a mass of 2800 kDa and is composed of three RNA molecules and about 50 protein molecules. The smaller 40S subunit contains one RNA molecule plus around 30 protein molecules and has an aggregate mass of 1400 kDa. (*S is an abbreviation for a Svedberg, the unit of the sedimentation coefficient. This is measured in an analytical ultracentrifuge and is related to, but not simply proportional to, molecular mass.)

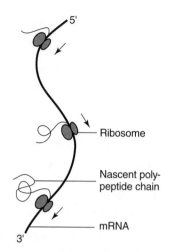

Figure 1.8 *Translation of messenger RNA.* The attachment of a ribosome to the mRNA involves protein initiation factors and the recognition of a particular base sequence, the start codon. A single mRNA can be simultaneously translated by more than one ribosome, forming a polyribosome. Synthesis occurs in the direction from the 5′ end of messenger RNA to the 3′ end. For further details of protein synthesis see Chapter 20.

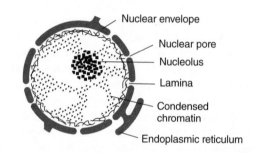

Figure 1.9 *Diagrammatic representation of the nucleus.* The nucleoplasm is not homogeneous but consists of darkly staining regions of heterochromatin, where the chromatin is more condensed, and paler regions of euchromatin. Transcription to form mRNA is restricted to the least condensed regions of euchromatin.

with the endoplasmic reticulum (Figure 1.9). Immediately inside this membrane is a network of protein filaments that defines the shape of the nucleus. The membrane is punctuated by a large number of nuclear pores, which are composed of proteins that permit diffusion of small molecules and limited diffusion of larger molecules. Very large molecules also diffuse across if they possess the correct 'identifying signal'. In this way, messenger RNA and ribosomal subunits can leave, and proteins and enzymes that are involved in DNA replication and in mRNA processing can enter the nucleus.

Within the nucleus resides DNA, which possesses the information required for the synthesis of almost all the proteins in the cell. The exception is the proteins synthesised within the mitochondria. Between cell divisions, this DNA exists as a complex of DNA with proteins (e.g. histones). This complex is called chromatin. The organisation of chromatin at the molecular level is not apparent when the nucleus is viewed under a light microscope, where it appears as a diffuse granular material unevenly distributed through the nucleoplasm. At cell division, however, the chromatin forms into clearly visible chromosomes, each of which contains one double-stranded molecule of DNA (Chapter 20).

A constant number of chromosomes is present in each cell. The somatic cells (i.e. not sperm or egg) are described as diploid because they contain two complete sets of chromosomes. There are 23 pairs of chromosomes in each cell, 22 pairs of somatic chromosomes (one of each pair derived from each parent) and one pair of sex chromosomes, either two Xs in the female or an X and Y in the male. Together, the 23 chromosomes contain about two metres of linear DNA or about three billion pairs of nucleotides. The linear structure of bases in DNA strands is called the primary structure of the chromosome. The secondary structure is the double helix, in which the two complementary strands of DNA twist about each other. One turn of the helix is called a pitch and consists of ten nucleotides.

A single chromosome, other than a sex chromosome, is called an autosome. The adjective, autosomal, describes a gene carried on an autosome. Autosomal dominant is a single gene trait that is encoded on an autosome and is expressed in a dominant fashion in heterozygotes, that is, the trait is expressed even if only one copy of the allele responsible is present. Examples of genetic disease with autosomal dominant inheritance are rare (Huntington's chorea is one). Autosomal recessive is a single gene trait that is encoded on an autosome and is expressed in a recessive fashion, that is, the trait is manifest only if two copies of the allele responsible are present. Many genetic diseases are of this type. The structure of a gene is described in Chapter 20.

Visible in the nuclei of most cells, especially those actively synthesising protein, is a nucleolus. It consists of a mass of incomplete ribosome particles and DNA molecules that code for ribosomal RNA: this is the site of synthesis of the ribosomal subunits.

New cells

The body continually needs to make new cells in order to grow or to replace those that have become senescent, lost or died. The replacement process goes on in many tissues, but the main sites are skin, bone marrow and the intestine. New cells are formed by division of existing cells, and it is estimated that about 100 000 000 000 (10^{11}) cells (about 0.002% of the total) divide each day. During cell division (mitosis) the genetic material in the nucleus, contained within the DNA, is duplicated so that identical sets of information are passed on to each of the two daughter cells.

The period between cell divisions is called interphase. The first phase of mitosis is known as prophase, which is signalled by the chromatin condensing to form chromosomes which now become visible for the first time. The centromere region of each chromosome then divides, separating the chromosome into two daughter chromatids. During early and late metaphase, which is the second phase, the membrane of the nucleus is lost and the chromatids attach to the centre of a framework of microtubules (the mitotic spindle) which extends across the cell. During anaphase the chromatids move outwards towards the ends of the spindle. Finally, during telophase, they complete their journey, the spindle breaks down and a nuclear membrane is formed around each new set of chromosomes, which begin to decondense into chromatin. The cytoplasm between the two nuclei then constricts, dividing the parent cell into two.

The daughter cells formed during mitosis have the same total number of chromosomes as the parent cell. In contrast, in meiosis, which occurs during the maturation of the germ cells, the number of chromosomes is halved (as the chromosomes divide once but the cells divide twice) so that the gametes – spermatozoa and ova – are haploid, i.e. they only have a single set of chromosomes. In addition, during the formation of the gametes, a process of crossing-over (exchanging of parts of each chromosome) occurs, ensuring a mixing of the genetic information that is passed on to the progeny. The processes of mitosis and meiosis are described and discussed in Chapter 20.

Cell differentiation

The different tissues of the body contain different types of cells, for example, neurones and glial cells in the brain, leucocytes and erythrocytes in the blood. All these cells are derived from a single fertilised egg by mitosis, and most contain exactly the same genetic information. Hence the development of differences between cells (differentiation) requires regulation of the expression of different genes at different times of development. At these times, there are some crucial cell divisions, during each of which at least one of the daughter cells becomes different from its parent. Over many cell divisions, divergences accumulate and the process of change becomes irreversible. For example, what has become a liver cell cannot regress to a common precursor cell that could redifferentiate as a different cell.

In addition to fully differentiated cells, many tissues of the body contain stem cells – precursor cells that are not fully differentiated. These can divide to produce more stem cells but can also give rise to progeny which can differentiate. Stem cells in the bone marrow, for example, give rise to all the different types of immune cells, erythrocytes and megakaryocytes, which give rise to platelets.

Tissues

Although cells are the basic unit of the body, it is their organisation into tissues that enables them to carry out their physiological roles. A tissue is an aggregate – a sheet or cluster – of similar cells that carry out one or more common functions. There are four basic types: *epithelial, connective, muscular* and *nervous*.

The existence of tissues implies that some cohesive force or structure holds the constituent cells together. In some cases, this also provides communication between cells to achieve coordination within the tissue. There are three classes of such *intercellular junctions*:

- anchoring junctions;

- tight, or occluding, junctions;

- gap junctions.

Anchoring junctions between cells involve cell adhesion proteins known as cadherins which are tissue specific, so that cells normally adhere only to similar cells. In tissues that require enhanced mechanical strength, special points of attachment exist where microfilaments in adjacent cells are linked to opposed arrays of cadherins.

In tight junctions, strands of protein form a sealing band around the entire circumference of a cell, greatly restricting the free movement of molecules between it and its neighbour. In the small intestine, tight junctions prevent microorganisms from getting through the epithelium. Their impermeability is, however, variable and this may play some part in controlling the rate at which water and other small molecules are absorbed.

Cells that possess gap junctions are able to exchange small molecules, up to around one kDa in molecular mass, with adjacent cells. Gap junctions are composed of clusters of proteins (connexins) in each cell membrane which form channels that align with each other.

Many cells secrete proteins and polysaccharides which remain associated with them to form features including:

- Extracellular matrix, often the major component of connective tissues.

- The basal lamina (basement membrane). This is a dense complex of macromolecules, forming a sheet that underlies epithelia and surrounds muscle cells.

Another family of cell adhesion molecules, the integrins, are the components of plasma membranes responsible for maintaining the link between cells and their extracellular associations. This link is not only mechanical but allows communication.

Epithelial tissue

Epithelium covers all surfaces of the body, internal and external. The simple epithelium consists of a single layer of cells that line the cardiovascular, respiratory, digestive and urogenital systems. The epithelium, consisting of endothelial cells, which lines the capillaries and arteries, is known as the endothelium. The epidermis, the outer layer of the skin, is a stratified epithelium consisting of several layers of epithelial cells (Figure 1.10). Epithelia form a barrier between the internal and external environments so that they fulfil an essentially protective role, especially as a first line of defence against invasion by pathogens. A second major role of epithelia is that of secretion, both from simple epithelial cells and from more differentiated cells in endocrine glands that are derived from epithelial tissue.

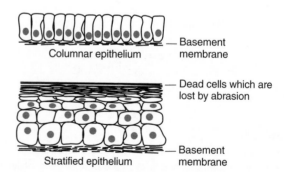

Figure 1.10 *Columnar and stratified epithelia.* Epithelial tissues are underlain by a basement 'membrane' which is a composite structure composed of an amorphous basal lamina, secreted by the epithelial cells, and a more fibrous reticular lamina derived from connective tissue.

The relative molecular mass of a molecule is the sum of the atomic masses of its constituent atoms. The term has replaced molecular weight because weight is a parameter that depends on the magnitude of gravitational attraction. Since relative molecular mass is a ratio (of the mass of the molecule to one-twelfth of the mass of the carbon-12 atom) no units are required. It has, however, become accepted practice to use daltons as a unit of molecular mass, commemorating John Dalton's atomic theory of matter. Relative molecular mass is an approximate indication of size; a spherical molecule of 5000 daltons (or 5 kDa) has a diameter of approximately 2.4 nm.

Connective tissue

The main role of connective tissue is the support of other tissues or structures. Its main characteristic is a large extra-cellular component that consists of two parts:

- Ground substance, which consists of modified sugars, in the form of polymers, and proteins which are associated with the polymers.

- Fibres of protein, including the inextensible collagen and the extensible elastic fibres.

It can be classified into different types.

Loose connective tissue   An example is subcutaneous, in which the ground substance is gel like and the fibres are irregularly disposed in sheets and bundles.

Dense connective tissue   This tissue has a large number of fibres. They may be irregularly arranged, as in the dermis, or in sheaths around tendons and nerves, or more regularly organised in the tendons, which link muscles to bones, and in ligaments which connect bone to bone.

Hard connective tissue   In these tissues, the ground substance is firm. In hyaline cartilage, there are few fibres: it is present in the articular cartilage in joints, as costal cartilage (allowing the ribs to move slightly relative to the sternum during breathing) and also in the trachea and larynx. In fibrocartilage there is a higher proportion of collagen fibres, giving it the firm mechanical properties appropriate to its role in, for example, intervertebral discs.

Bones start to be formed in the embryo. Cartilage becomes impregnated with crystals of hydroxyapatite, a calcium-containing compound of the approximate composition $Ca_3(PO_4)_2 \cdot Ca(OH)_2$. It is laid down by cells known as osteoblasts.

Adipose tissue   This is a specialised connective tissue that stores much of the energy reserve of the body in the form of fat (triacylglycerol) within the cells, known as adipocytes, which are close packed within the connective tissue. Adipose tissue is located in a number of anatomically distinct deposits in the body, e.g. below the skin, around major organs and between muscles.

Blood and lymph   These are the liquid tissues. It is, perhaps, surprising that they are classified as connective tissues, but their structure is the same as that of other connective tissues, except that the ground substance is fluid and the fibres are represented by the proteins such as fibrinogen and the strands of fibrin, which form when blood clots. The cells are the red and white cells (erythrocytes and leucocytes).

Figure 1.11 *Longitudinal section through a flight muscle of an insect as seen under an electron microscope.* The dark, approximately circular, objects are mitochondria packed between two myofibrils. Electron micrograph kindly provided by Professor David Smith, Oxford University. (For details, see Chapter 13).

Muscle tissue

In the cells of this tissue, which are known as fibres, the two major proteins, actin and myosin, are organised to form myofibrils. These are structural rods that can contract (Figure 1.11). This enables muscle cells to shorten, which provides for movement and locomotion (Figure 1.12).

There are three types:

- *Skeletal muscle* is attached to bones and brings about movement that is under conscious control. The cells are, in fact, long fibres, in which the microfilaments of adjacent myofibrils are in register to give the fibre a striated appearance (Chapter 13). In general, skeletal muscle exhibits a high power output and can contract rapidly.

- *Cardiac muscle* is present only in the heart. Its fibres are striated but, unlike skeletal muscle, they are short and branched. The force it produces is less than that of skeletal muscle but it is the endurance muscle *par excellence*.

- *Smooth muscle* is composed of spindle-shaped cells rather than fibres. It lacks striations since adjacent myofibrils are out of register. Power output is lower than that from skeletal muscle; it is responsible for the gentler, involuntary movements including, for example, those that cause movement of the intestine and those that change the diameter of blood vessels.

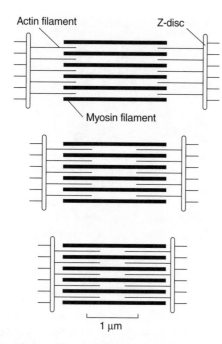

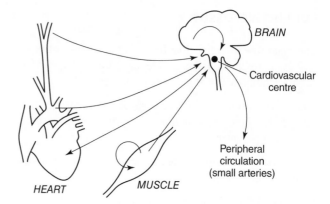

Figure 1.13 *Communication to and from the cardiovascular centre in the brain.* The cardiovascular centre controls changes in the output from the heart (cardiac output) and the flow of blood through peripheral tissues and organs. It is the efferent neurones that transfer information from the brain to the heart and peripheral vessels. The afferent neurones transfer information from the heart and other tissues, e.g. muscle, to the centre. Information transfers from the major arteries, the coronary arteries and peripheral muscles to the brain. There is also information transfer within the brain and within the muscle.

Figure 1.12 *Diagrammatic interpretation of contraction in a myofibril of skeletal muscle.* The diagram shows a single sarcomere, the basic contractile unit, limited at each end by a Z-disc. Muscle fibres are packed with hundreds of parallel myofibrils, each of which consists of many, often thousands, of sarcomeres arranged end to end. Contraction is the consequence of the thin actin filaments being pulled over the thick filaments to increase the region of overlap and telescope the sarcomere.

Nervous tissue

The nervous tissue provides a means of communication between different parts of the body, particularly to and from specific centres in the brain. This regulates and coordinates functions in the body (Figure 1.13).

Neurones are among the most specialised of cells in the body. Transient changes in electrical potential propagate along the axon and, through changes in the frequency of impulses, information is conveyed. Neurones communicate with other neurones; in some cases one neurone may communicate with as many as 1000 other neurones. Most neurones have one long process (the axon) and many shorter, extensively branched, processes, known as dendrites, through which they communicate with other neurones, receptors on various cells and muscle fibres (Figure 1.14). The communication between neurones is, however, not an electrical process. There is a small space between the two neurones, across which communication is made by chemicals, known as neurotransmitters. Although there are more than 50 different neurotransmitters, the biochemistry of the communication is the same. Most axons are surrounded by a sheath of a complex of lipid and protein,

known as myelin, which serves as a protective sheath and insulation, and provides for a mechanism to increase the rate of transmission of electrical activity along the axon.

In the central nervous system, cells other than neurones are present. Indeed, they outnumber the neurones. They constitute what is called, collectively, the neuroglia. These cells act as 'helper', 'nurse' or support cells for the neurones. There are at least four types of glial cells:

- Astrocytes, which transfer fuel and other material to the neurones and remove waste products from them; they perform the role of a 'nurse' or 'nanny'.

- Oligodendrocytes, which lay down the myelin sheath around the axons, so that they protect the nerve.

- Ependymal cells, which line the cavities of the central nervous system.

- Microglial cells, which are the phagocytes in the brain and protect against invasion by pathogens.

For discussion of these cells and the neurones, see Chapter 14.

The whole human

Systems in the body

Except in culture, where they are maintained by the artificial environment, tissues do not exist in isolation but are com-

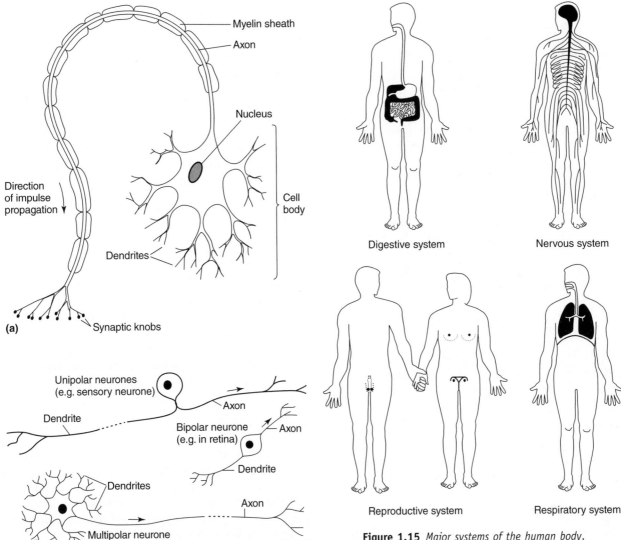

Myelin sheath

Axon

Nucleus

Cell body

Dendrites

Direction of impulse propagation ↓

(a) Synaptic knobs

Unipolar neurones (e.g. sensory neurone)

Axon

Dendrite

Bipolar neurone (e.g. in retina)

Axon

Dendrite

Dendrites

Axon

Multipolar neurone (the most abundant type in the brain; in peripheral motor neurones, the axon may be very long)

(b)

Digestive system

Nervous system

Reproductive system

Respiratory system

Figure 1.15 *Major systems of the human body.*

Figure 1.14 (a) *Basic structure of a neurone.* A motor neurone is shown, but the basic structure of all the neurones is the same. Dendrites transfer information from other nerves to the neurone, while the axon transfers information from the neurone to other neurones or tissues. The axon is particularly long in motor neurones (Chapter 14). **(b)** *Structures of unipolar, bipolar and multipolar neurones.* Unipolar neurones transfer information from tissues or organs to the brain. Multipolar neurones are the most abundant in the nervous system.

ponents of organs, which are anatomically distinct structures with defined functions. The systems of the body are as follows (Figure 1.15):

The nervous system

The nervous system is divided anatomically into two parts, each separately dignified by the term 'system' but, in reality, providing the body with a single communication and control network. The central nervous system (CNS) consists of the brain and spinal cord; all the other nerves constitute the peripheral nervous system.

Forming part of the peripheral nervous system is the autonomic nervous 'system' which controls the glands and non-skeletal muscles that are not under conscious control. This control is provided by two parts of this system: the sympathetic and parasympathetic divisions which, in general, bring about antagonistic responses.

The cardiovascular system

The circulatory system moves materials (and heat) from one organ to another. It is centred on the heart which pumps blood through arteries to capillaries, where exchange occurs before the blood returns to the heart via the veins (Figure 1.16). During its passage through the tissues, there

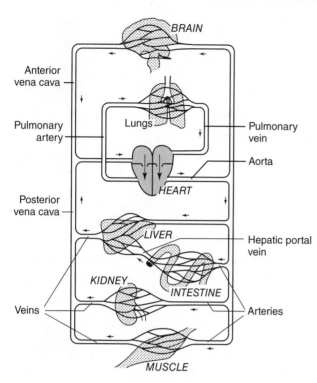

Figure 1.16 *A general plan of circulation of blood in a human.* For details of the hepatic portal vein, see Chapter 6.

is a net loss of plasma from the capillaries to form tissue fluid. This is returned to the circulation via the lymphatic system, the main role of which is as a component of the defence mechanisms of the body (Chapter 17).

Integumental system

Although often taken for granted, the skin is actually the largest organ in the body. It is responsible for protection (both physical and chemical) against attack by pathogens and prevents ultraviolet light from penetrating the tissues and most chemicals from entering the body. It also plays the major role in temperature regulation and bears a wide variety of receptors. It is a remarkable structure as indicated by the extract that introduces this chapter.

Respiratory system

It is the respiratory system that enables an adult human to absorb about 360 litres of oxygen in a typical day and excrete a slightly smaller volume of carbon dioxide. This is made possible by the branching system of airways in the lungs which services a vast surface area for gas exchange.

The urinary system

The urinary system is responsible for eliminating soluble waste products which are separated from the blood by the

kidney. In the male, it is anatomically combined with the reproductive system to form what is sometimes known as the urogenital system.

The digestive system

This is essentially a tube running from mouth to anus. It is, conventionally, considered to include the associated glands: liver, pancreas and gall bladder. It provides the enzymes required for digestion, the chambers in which this occurs and the mechanism for the absorption of the products of digestion. A major function of the liver is to chemically modify and store many of the products of digestion and detoxify or inactivate those that may be injurious to health.

The endocrine system

The endocrine system is an anatomically diverse assemblage of organs, united by the common function of secreting hormones. Organs, with quite different functions, including the kidney, liver and intestine, also secrete hormones and, in this sense, they too are part of the endocrine system. In a functional sense, the endocrine system also includes the blood which transports hormones to their target cells, tissues or organs.

The reproductive system

The reproductive system includes the ovaries in the female and testes in the male, together with the ducts and tubes in which the ovum or spermatozoa travel and meet in the fallopian tube, and the environment in which the foetus develops. Since both ovary and testis produce hormones, there is overlap with the endocrine system (Chapters 12 and 19).

The musculoskeletal system

Mechanically, the bones that constitute the skeleton provide protection, support and a framework of levers which enable attached muscles to develop the forces that make locomotion possible. Within cavities in the long bones of the limbs is the bone marrow, where erythrocytes (red cells) and immune cells (white cells) are produced for the blood and lymph. Adipocytes are also present in the bone marrow: indeed they outnumber the other cells (see Chapters 7 and 17).

The immune system

Unlike the other systems described above, the immune system has no organs as such and is anatomically diffuse. It consists of the cells that mediate the body's immune response and the tissues that produce and house them (also known as the lymphoreticular system) including bone marrow, lymph nodes, thymus and spleen.

The reticuloendothelial system

This is a diffuse and loosely defined system which encompasses those cells and tissues that are phagocytic. These include the endothelia of blood and lymphatic vessels (which are only weakly phagocytic), reticular cells of the spleen, endothelial cells, sinusoids in the liver and lymph nodes, macrophages and circulating phagocytes.

The biochemical hierarchy

A brief description of the biochemical function of organelles, cells or organs and systems in the body is provided above. A challenge for functional biochemistry is to explain how the biochemistry performed by organelles accounts for the biochemistry occurring in cells, which in turn explains the function of the tissue or organs in the body and, finally, how the hierarchy and the interaction between systems provide for many if not all the aspects of the life of humans. The information presented in this book responds to this challenge and, in so doing, helps to explain the biochemistry underlying good health and how disturbances can lead to ill-health and disease.

It is, in fact, possible to separate disease into classes that loosely correlate with the different systems in the human body, which are described above. These classes are listed in Appendix 1.1 together with an indication of the positions in the book where the diseases are described and discussed.

Two examples are provided here and expanded further in later chapters.

Mitochondria One characteristic of human life is movement and physical activity. It is the mitochondria in muscle that generate ATP to support the process of contraction of the myofilaments in the muscle fibre which results in contraction of the whole muscle.

Failure of mitochondria to generate sufficient ATP to power contraction of the muscles leads to fatigue, which if mild can be classified as ill-health but if severe can be classified as disease. Such failure could be caused by a deficiency of an enzyme in the mitochondria, damage to the membranes within the mitochondria or simply a reduction in number of mitochondria. Regular physical activity maintains or even increases the number of mitochondria in muscles and, hence, fatigue only occurs after intense or prolonged activity. This is an indication of good health. Lack of regular physical activity reduces the number of mitochondria in muscles so that even mild activity can result in fatigue, usually described as ill-health. One particular defect in mitochondrial ATP generation that occurs in adults, known as adult-onset mitochondrial disorder, became of considerable interest when the American cyclist Greg Lemond, who won the Tour de France cycle race three times, retired from competitive cycling in 1994. This decision was taken due to impaired mitochondrial ATP-generation (Chapter 13).

Failure of mitochondria to oxidise fatty acids sufficiently rapidly reduces the ability of the muscle fibres and other cells to utilise fat after a meal, so that fat levels in the blood and in the muscle increase to abnormally high levels. Over a period of time, this disturbance can lead to obesity or type 2 diabetes mellitus (Chapter 9).

B lymphocytes When stimulated during an infection, B lymphocytes differentiate to form effector lymphocytes, in which the rough endoplasmic reticulum together with the Golgi synthesise and secrete antibodies into the blood to fight invading pathogens. Detection of an invading pathogen by the B cells results in B cell proliferation and massive expansion of the rough endoplasmic reticulum to generate a vast number of the antibodies which will attack and defeat the pathogens. Hence, in B lymphocytes, the endoplasmic reticulum plays an essential role in maintaining health. However, infection of T lymphocytes with the human immunodeficiency virus (HIV) prevents these T cells from stimulating B cells to proliferate and differentiate, so that there is no massive expansion of the rough endoplasmic reticulum and hence the HIV and other pathogens are not attacked. The disease AIDS (acquired immune deficiency syndrome) is the result (Chapter 17).

II ESSENTIAL TOPICS IN DYNAMIC BIOCHEMISTRY

2
Energy: In the Body, Tissues and Biochemical Processes

Normal life may be defined as the conversion of energy to perform meaningful work at an acceptable metabolic cost. Illness and injury may be defined as energy conversion, work requirements or metabolic costs that have now become excessive. Therefore death may be defined as the irreversible loss of the ability to use energy to perform sufficient work in one or more vital organs.

(Kinney *et al.*, 1988)

The history of energy metabolism is a fascinating example of how progress depends alternatively on new concepts and, at other times, on new technology that allows elements of the problem to be revisited and objectively measured.

(Kinney *et al.*, 2006)

A problem with the term 'energy' is that it means different things to different people. In everyday life, it is associated with, for example, motion, speed, vitality, strength. Some people are described as having 'lots of energy' and others as having 'little energy'; some foods are described as 'high energy', others as 'low energy'. The physicist's definition of energy as the capacity to do work is too abstract to be useful here so probably the best way to understand it is to consider energy in its different forms and how it can be transformed from one form into another. The different forms include heat, electrical, light, mechanical and chemical, all of which are interconvertible: a dynamo converts mechanical into electrical energy and an internal combustion engine converts chemical into mechanical energy. Indeed, the transfer of chemical energy in one compound to another is essential to life and the conversion of light energy into chemical energy is the process upon which almost all life depends.

There are two laws of thermodynamics that govern the behaviour of energy. The first states that: *Energy can neither be created nor destroyed. It can only be converted from one form into another.* The second law can be stated in several ways but a simple one is: *Heat does not flow, spontaneously, from a cold object to a warmer one.* The relevance of these two laws in health and disease may not be immediately apparent but one aim of this chapter is to explain just this. The chapter is divided into three major sections:

- energy transformations in the whole body;
- energy transformations in tissues or organs;
- energy transformations in biochemical reactions or processes.

The first law applies to the material presented in the first two sections and both laws apply to the third section. One factor common to all these discussions is that the same unit, the joule, is used in all these energy transformations (Table 2.1).

Although details of the biochemistry of energy transformation have been established for over 50 years, it is only recently that this knowledge has been applied to key life processes in health and disease. This has been driven by several factors: appreciation of the importance of provision of chemical energy (i.e. food) for patients, in hospital, provision of chemical energy for physical activity of all kinds and the need to balance energy intake and expenditure for prevention and treatment of obesity. In discussions of the last point, even the mass media refer to the first law of thermodynamics.

Functional Biochemistry in Health and Disease by Eric Newsholme and Tony Leech
© 2010 John Wiley & Sons Ltd

Table 2.1 Conversions between units of energy

From	To	Factor
Calories	joules	×4.18
Joules	calories	×0.239
Kcal/day	MJ/day	×0.00418
Kcal/day	kcal/min	×0.000694
MJ/day	kJ/min	×0.694
MJ/day	kcal/day	×239

Unfortunately, in scientific literature energy is presented in various different units: the factors that are used to convert one unit into another are provided here.

The rate of energy expenditure is usually expressed as watts. One watt is equal to one joule per second. One kilowatt (kW) is equal to one kilojoule per second. A top class sprinter expends energy (i.e. converts chemical energy into heat) at a rate of about 3000 watts and a marathon rather about 1400 watts.

Energy transformations in the whole body

The most familiar contact with energy in everyday life is through eating and, because we eat intermittently, some of the energy is stored to provide for short or long periods of starvation and/or physical activity. Consequently, the first topic in this section presents the amount of the different fuels stored in the body.

Fuel stores

A fuel is a compound for which some of its chemical energy can be transformed into other forms when a chemi-

cal reaction takes place. Glycogen and triacylglycerol are the two major fuels stored in the body, the former in liver and muscle and the latter in adipose tissue. However, recently a third fuel has been identified that is also stored in skeletal muscle, the amino acid glutamine (Chapter 9). If the magnitude of these stores and the energy expenditure are known, it is possible to calculate how long they will last under different conditions (Table 2.2).

Glycogen

In the normal standard male (70 kg) the liver contains about 80 g of glycogen after the evening meal and the total content in muscle is about 350 g. Glycogen is always stored in association with water (1 g of glycogen with about 3 g of water), so complete repletion of the total glycogen stores increases body mass by about 2 kg, i.e. for each 4 kJ of energy stored as glycogen, body mass increases by approximately 1 g. A low energy diet, which results in depletion of the glycogen store in the body, produces a loss of about 2 kg of water. It is not a loss of fat!

The glycogen content can be measured by taking a small quantity of the muscle from a volunteer or patient (a biopsy), extracting it and measuring the amount of glycogen in the extract (Appendix 2.1). This is then multiplied by the total amount of muscle in the body. Since this is an invasive technique, it is not suitable for routine purposes. Although many biopsy samples of muscle from normal individuals have been taken, very few biopsies of the liver of normal volunteers have been taken. Since different muscles contain different amounts of glycogen, extrapolation from the amount in a biopsy of one muscle may produce inaccurate results. A nuclear magnetic resonance

Table 2.2 Approximate content of glycogen, triacylglycerol and glutamine in a normal adult male[e] and the estimated time for which they would last, if they were the only fuel used, during two forms of physical activity and during starvation

Fuel store	Approximate total fuel reserve		Estimated period for which fuel store would provide energy		
	g	MJ	Walking (days)[a]	Marathon running[b]	Starvation
Adipose tissue triacylglycerol	12 000	450	15	Several days	70 days[d]
Liver glycogen	80	1.3	0.04	15 min	24 hours
Muscle glycogen	350	5.8	0.5	70 min	–
Glutamine[c]	80	2.0	–	–	–

The size of fuel reserves, especially triacylglycerol, varies considerably in both male and female adults. It is assumed that the amount of adipose tissue in a 70 kg male is 15 kg, of which 75% is triacylglycerol. Data from Cahill (1970); Wahren (1979).

See Chapter 13 for a discussion of which fuels are actually used to support physical activity.

[a] Assuming that energy expenditure during walking at 4 miles/h is about 30 MJ/day (one day is 24 hours).

[b] Assuming that the energy expenditure is 84 kJ/min, or about 5 g glucose a minute, which is achieved by top class endurance runners (see below).

[c] Glutamine is stored and released from muscle but is not used. Glutamine is used by a large number of cells and tissues (Chapter 9). Glutamine is used in starvation but mainly to provide ammonia to buffer the hydrogen ions in the urine, and as a precursor for glucose but the amount of energy available from the glucose produced is not known (Chapters 6 and 8).

[d] See Chapter 16.

[e] Different amounts of fuels are stored in females due to different amounts of tissues (See Table 2.9).

spectrometer can measure the amount of glycogen from the content of ^{13}C-labelled glucose in glycogen in liver or muscle of a volunteer (for description of isotopes, see Appendix 2.2). This is a non-invasive technique, but it gives similar results to the biopsy method (Chapter 16) (Appendix 2.3).

Triacylglycerol

Most of the triacylglycerol is stored in adipose tissue depots. Smaller amounts are found in muscle and liver. Adipose tissue is connective tissue that contains adipocytes, blood vessels, collagen and lymphocytes. The last latter are present in lymph nodes, most of which are present within adipose tissue depots throughout the body. Indeed, there may be more lymphocytes in adipose tissue than there are adipocytes (see Chapter 17). The proportion of triacylglycerol in an adipose tissue depot is >80% and water is 14% of the wet weight. About 95% of an adipocyte is triacylglycerol.

> In the obese or overweight, the proportion of triacylglycerol in an adipose tissue depot is about 90%. In the newborn it is as low as 35%.

There are three ways, notionally, of compartmentalising the body, depending on whether the mass of adipose tissue or triacylglycerol itself is being measured:

- adipose tissue mass and lean body mass (LBM);

- fat mass (triacylglycerol) and fat-free mass (FFM);

- fat mass as a percentage of body weight.

Unfortunately, it is not always clear in the literature whether triacylglycerol or adipose tissue mass is being considered.

The range in the amount of fat in different humans is remarkable: from about 7% to more than 40% of body weight. The heaviest person recorded in the Guinness Book of Records was John Brower Minnoch (1943–83), who was more than 57 stone (362 kg) when he died. The maximum survivable is 500 kg.

There is no 'gold standard' for measuring the content of the triacylglycerol store, since adipose tissue is distributed in many different depots (see Appendix 2.4 for some of the methods that are available).

Glutamine

Glutamine is found in all cells in a combined form in peptides or proteins, but also in a free form. The highest free concentration of glutamine is found in muscle, where it acts as a store for use by other tissues. In fact, the total amount in all the skeletal muscle in the body is about 80 g, which is synthesised in the muscle from glucose and branched-chain amino acids (see Chapter 8). As with glycogen in the liver and triacylglycerol in adipose tissue,

glutamine is released during starvation and trauma (Chapters 16 and 17). The release requires a specific transport protein in the plasma membrane.

The content of glutamine in muscle is measured in a similar manner to that of glycogen. A biopsy of muscle is taken, extracted and the glutamine content measured by enzymes or by high pressure liquid chromatography (Appendix 2.1).

Energy intake and expenditure

The first law states that energy is neither created nor destroyed, which applies to the human body as to any other system. Thus the body mass represents a balance between energy intake (i.e. food) and that expended in various processes in the body, especially physical activity. Thus the law is particularly relevant in weight-reducing diets to overcome obesity or to maintain normal body weight. The subject of obesity is discussed in Appendix 1.5. The general principles of energy intake and expenditure are now discussed.

Energy intake

The only way that energy enters the body is in the form of food (i.e. chemical energy). An important question is, therefore, how can the amount of energy in food be measured? This question is answered by reference, once again, to the first law, which can be restated as: *the change in chemical energy in a reaction is equal to the heat produced by the reaction* (provided that no work is done by the reaction). The term enthalpy (H) is used to describe this chemical energy and its change is written as ΔH. In other words, chemical energy is measured when it is converted into another form, usually heat.

> The equation that governs this is
> $$\Delta H = q - w,$$
> where q is the heat released in the chemical reaction and w is the work done by the reaction. If no work is done, $\Delta H = q$. If some work is done, e.g. expansion of a gas, q is less than ΔH.

The energy in food is in the form of carbohydrate, fat and protein, and the oxidation of these compounds, in vitro, transfers the chemical energy into heat which can then be measured. This is done in what is known as a 'bomb calorimeter' (Figure 2.1). The heat released in the calorimeter when 1 g of carbohydrate, fat, protein or other fuel is fully oxidised (i.e. ΔH) is given in Table 2.3.

> It is important to note that the heat produced is a measure of a change in enthalpy, not its absolute amount, and that ΔH is independent of the chemical path followed by the reaction. For example, the value of ΔH is the same whether glucose is oxidised to CO_2 and H_2O in a laboratory or via metabolism in the body. The reaction is the same e.g.:
> $$glucose + 6O_2 \rightarrow 6CO_2 + 6H_2O$$

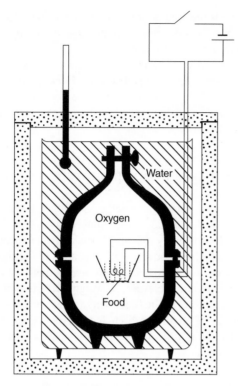

Figure 2.1 *A bomb calorimeter.* The food is ignited by an electric current within the inner compartment, which is known as the 'bomb' because the reaction within the box is generally so rapid as to be almost explosive. Insulation prevents heat loss and the thermometer measures the rise in temperature of the water that surrounds the bomb. From this increase, and the thermal capacity of the apparatus, the amount of heat released can be calculated.

These values can be used to calculate the energy content of a meal. The sequence of steps in the calculation is as follows.

• Weigh the food prior to eating and any left over at the end of the meal, so the difference gives the amount of food ingested.

• Calculate the amounts of the major components of the food (carbohydrate, fat, protein, including alcohol) ingested, from knowledge of the composition of the meal (obtained from information provided on the package of the food item or from food composition tables).

• Multiply each of these weights by the appropriate value of the heat released from oxidation of one gram (i.e. ΔH values given in Table 2.3) and add them together to give the total energy intake. Examples of the energy content of some meals are provided in Appendix 2.5.

Table 2.3 Energy released on oxidation of one gram of the major fuels

Constituent	ΔH (kJ/g)
Protein[a,b]	16.7
Fat[b]	37.6
Starch or glycogen[b]	16.7
Glucose[b]	15.7
Unavailable carbohydrate[c]	8.4
Ethanol[d]	29
Glutamine	13.2

[a] Provided that the end-products of oxidation in the body are the same as those in the bomb calorimeter, the information gained from the latter can be used for calculations on energy transformation in the body. This is the case for carbohydrate and fat but it is not the case for protein since, in a bomb calorimeter, nitrogen in protein is converted to its oxides whereas, in metabolism, nitrogen is incorporated into urea and excreted, and the conversion of the nitrogen to urea does not provide energy. Nonetheless, the effect of this difference can be calculated (it is 6.25 kJ/g protein) and, therefore, can be taken into account, which has been done for the value given in this table.

[b] Digestion and absorption of carbohydrate, fat or protein are not complete, making the values gained from the bomb calorimeter too high. This problem was first investigated by W. O. Atwater and the 'Atwater factors' have been used to produce the 'biological' values of ΔH, which are those given in this table. Digestion and absorption are very efficient and, in a mixed diet, 99% of the carbohydrate, 95% of the fat and 93% of the protein is actually absorbed, as monosaccharides, fatty acids and glycerol, and amino acids respectively. (Elia & Cumming S. 2007).

[c] Humans lack enzymes to hydrolyse cellulose, and some other carbohydrates in food. However, bacteria in the intestine can hydrolyse and ferment some of this carbohydrate to produce short-chain fatty acids, which are used by the colon and the liver. It is estimated that for each gram of unavailable carbohydrate in the diet, 8.4 kJ of energy is made available in this way, although this is influenced by factors such as ripeness of fruit or the way leguminous seeds are cooked. Nonetheless, these effects will be small and can be ignored unless the amount of such carbohydrate is high or very accurate results are required. (The subject of unavailable carbohydrate and fibre in the diet is discussed in Chapters 4, 6 and 15).

[d] In some individuals, ethanol accounts for more than 10% of the energy ingested.

There are, however, at least three factors that should be borne in mind if studies are being carried out and reasonably accurate results are required:

• It is usual for a subject to under-report the intake or to consume less than usual, if intake is being monitored by an observer.

• There is considerable day-to-day variation in food intake, so that studies should be carried out for a minimum of one week and under continual surveillance. The longest reported such study lasted 62 days.

• The heat released in the bomb calorimeter is not always the same as that released in the body (see footnotes to Table 2.3).

Energy expenditure

Four methods have been used to measure expenditure:

- use of formulae;

- direct calorimetry;

- indirect calorimetry;

- isotopic methods.

Use of formulae   Until recently, practical methods were not available for measuring energy expenditure in patients in hospital. Consequently, formulae, based on weight, height, age and gender of the patient were, and still are, in use. The earliest formulae were published in 1919, by Harris & Benedict, who studied 239 individuals aged 15–73. These were used for many years but the values were found to be too high. In 1985, a study on 7000 individuals provided new formulae known as Schofield standards (Schofield, 1985; Table 2.4). Whichever formula is used, it provides only approximate values, since metabolic rate varies markedly between individuals.

Direct calorimetry   Energy expenditure can be measured from the heat lost by an individual, which is the same as that produced. All the energy used in the daily activities of the body is released as heat. For example, the mechanical energy expended walking to work, sprinting for the train or running a marathon is converted into heat. The method used for measuring heat production is known as direct calorimetry (Box 2.1). The individual lives in a thermally insulated box, and water (or other fluid) is pumped around the calorimeter to absorb the heat produced. The rise in temperature of the water, multiplied by the thermal capacity of the water, gives the heat released for the period of time that the person is in the calorimeter. The main practical problem is that the calorimeter has to be large enough to allow the subject to eat, sleep and exercise, possibly for several days or more, and yet be extremely well insulated (Figure 2.2). Only a very few laboratories around the world have such facilities.

> The thermal capacity of a material is the amount of heat needed to raise its temperature by 1 °C. For 1 g of water, this is 4.2 joules.

Indirect calorimetry   This approach is so indirect that it does not involve measuring heat at all! All the energy expended by a person arises from the oxidation of carbohydrates, fats and/or proteins, which use oxygen and produce carbon dioxide (and urea), all of which can be measured.

With modern equipment it is now possible to measure oxygen consumption and carbon dioxide production breath by breath. Alternatively, for long-term studies, the volunteer can be enclosed in a room and the changes in content of the gases measured. For resting subjects and for patients, this is not necessary: a simple hood over the head may be sufficient to allow measurement of oxygen uptake and carbon dioxide production, if precision is not required.

Because similar amounts of heat are produced when a litre of oxygen is used to oxidise triacylglycerol or carbohydrate (19.28 kJ and 21.12 kJ, respectively) an

Table 2.4 Empirical formulae for the calculation of resting energy expenditure of men and women with worked examples

Harris & Benedict (1919) for adults
- Male: basal energy expenditure (in kJ/day) = $278 + 57.54\,W + 2093.4\,H - 28.3\,A$
- Female: basal energy expenditure (in kJ/day) = $2740.9 + 40.0\,W + 7.739\,H - 19.6\,A$

Where W is weight in kg, H is height in metres and A is age in years. To convert to kcal/day, multiply by 0.239.

Schofield (1985) for individuals 30–60 years old
- Male: basal energy expenditure (in kJ/day) = $48.0\,W - 11.0\,H + 3760$
- Female: basal energy expenditure (in kJ/day) = $34\,W + 6\,H + 3530$

Where W is weight in kg and H is height in metres.

Examples of calculations based on standard male and female adults:
Male 70 kg, 1.75 m, 45 years
Female 50 kg, 1.65 m, 45 years

Energy expenditure calculated from Schofield equation:

Male	3360 − 19 + 3760 = 7101 kJ/day (1699 kcal/day)
Female	1700 + 9.9 + 3560 = 5240 kJ/day (1253 kcal/day)

Box 2.1 History of the discoveries of the relationship between heat and metabolism

The first direct calorimeter worked on the principle that the heat produced by an animal could be measured by the amount of ice that it caused to melt. At 8:12 on the morning of 3 February 1783, Antoine Lavoisier and Pierre Laplace placed a guinea pig in an 'ice-machine' and initiated experiments that firmly established a link between the production of heat, the consumption of oxygen and the formation of carbon dioxide by animals. These experiments have been described as 'the most important group of experiments in the history of metabolic-heat studies'. In a paper published in 1777, Lavoisier had established that, during respiration, the content of the oxygen in the air is diminished, whereas that of carbon dioxide is increased and that of nitrogen remains the same. So began the realisation that physical and chemical processes were the source of body heat and that the use of the calorimeter enabled quantitative experiments to be carried out. Parallel investigations were taking place in Scotland at this time, although they focused more on the heat produced by animals than on gaseous exchanges. Before the studies of Lavoisier, it was widely accepted that the source of body heat was a fire in the left ventricle of the heart and that the role of the lungs was to cool this internal fire through ventilation.

Lavoisier's contribution to the study of animal energetics was not limited to his elucidation of the relationship between respiration and the production of heat. His studies with Seguin on the metabolism of humans, which involved the quantitative measurement of oxygen consumption and carbon dioxide production, uncovered hitherto unknown relationships. In a letter Lavoisier wrote to Joseph Black in 1790 he reported that oxygen consumption was increased by the ingestion of food, by the performance of muscular work and by exposure to cold. He also determined the minimal rate of metabolism in the resting, post-absorptive state and showed a proportionality between pulse frequency, ventilation frequency and metabolism.

Although the intuition and experiments and work of Lavoisier were remarkable, but discoveries over 100 years earlier had provided the basis for his work. In the latter part of the seventeenth century, William Harvey, personal physician to King Charles I of England, had attracted a talented group of scientists. Among this group of 'Oxford physiologists' was the chemist Robert Boyle, who elucidated the interrelationships of pressure, volume and temperature of a gas. With his colleague Robert Hooke he used a vacuum pump to show that a bird could not survive and a candle would not burn in the absence of air. They concluded that some gaseous element was needed to support both the flame and the life of the bird. This gaseous element was, of course, oxygen, discovered by Joseph Priestley in 1774 (Chapter 9), it was Lavoisier who appreciated the significance of Priestley's findings which eventually led to the development of chemistry. Lavoisier is therefore known as the 'father' of chemistry (and therefore biochemistry).

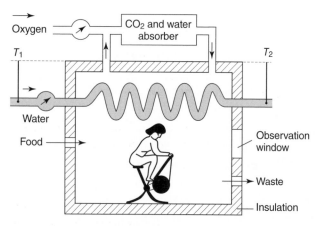

Figure 2.2 *A direct calorimeter.* The increase in temperature (T_2-T_1) of the water flowing through the tube is an indication of the amount of heat produced by the subject within the calorimeter, from which the precise amount of heat produced can be calculated.

average value of 20 kJ/L of O_2 can be used to calculate the approximate energy expenditure. The value is the same for production of CO_2 from carbohydrate. The advantage of indirect calorimetry is the greater accuracy when compared with the equations and the relative simplicity of the procedure, so that it is suitable for use in a hospital or laboratory.

This value of 20 kJ/L O_2 can be very useful: for example, it can be used to calculate that a top-class marathon runner requires oxidation of 5 g of carbohydrate every minute during the race, provided that no other fuel is used. The athlete takes up 4 litres of oxygen each minute, which is equivalent to 4×20, or 80 kJ each minute; as 1 g of glucose releases 16 kJ of energy, the glucose or glycogen used is 80/16 or 5 g each minute.

Isotopic methods   Indirect calorimetry is limited by the need to measure the gases, which means that it cannot be used for free-living studies. Consequently, methods involving the administration of isotopically labelled compounds have been devised. These are based on the principle of isotopic dilution, a widely employed method for estimating concentrations, particularly of hormones and other proteins in the blood (Appendix 2.6). The requirements for these methods are:

- The availability of the isotopically labelled substance.

- The assumption that, when administered, the substance will mix uniformly with the unlabelled substance in the body.

- The ability to measure the amount of isotope in a sample of the substance; that is, the amount of isotope per gram (or mole) of substance. Note that it is not necessary to know the size of this sample – a major advantage of the method.

In principle, isotope dilution could be used to measure the rate of production of water or carbon dioxide by measuring the rate at which the administered $^{14}CO_2$ or ^2H_2O was diluted. In practice, however, there are difficulties. In the case of water, the amount produced is small compared with the mass of body water, so that the changes in the content of isotope would be too small to measure accurately. The problem with carbon dioxide measurements is the need to take frequent samples because carbon dioxide is rapidly excreted. Two methods have been developed that overcome these problems (see Elia & Livesey 1992).

Dual isotope technique   The technique uses two heavy isotopes, oxygen (^{18}O) and deuterium (^2H). Water that contains these isotopes is prepared. The subject drinks a glass of this water, as part of a normal meal. Once equilibrated with body water, which occurs quickly, the content of ^2H in the water falls due to the production of unlabelled water from the oxidation of fuels. Similarly, the $^{18}O_2$ content in the water also falls but the rate is greater than that of ^2H since the $^{18}O_2$ equilibrates not only with the oxygen atoms in water but also with those in carbon dioxide. An equilibrium between water and carbon dioxide is rapidly established due to the activity of the enzyme carbonate dehydratase.

$$H_2O + CO_2 \rightleftarrows H_2CO_3$$

The greater the rate of production of carbon dioxide by the body, the greater the rate of fall of the $^{18}O/^2H$ ratio, which is measured in a sample of urine collected every day. From the difference in the slope of the decline of the two labels, the rate of CO_2 production is calculated (Figure 2.3).

The great advantage of the technique is that the measurements do not affect normal daily activities in any way, so it can be used to measure energy expenditure in a number of different activities or conditions (e.g. by cyclists in the Tour de France race, climbers on Mount Everest, members of a trans-Antarctic expedition, women during pregnancy or lactation, obese animals including humans carrying out their normal daily activities) (Prentice 1988).

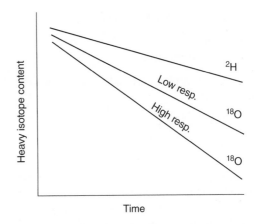

Figure 2.3 *Changes in the heavy isotope content of oxygen and hydrogen in the urine with time after administration of $^2H_2{}^{18}O$. The ^2H and ^{18}O are diluted by the formation of H_2O from oxidation of fuels and particularly that ingested in food and drink. The ^{18}O in water is diluted from the production of CO_2 via the oxidation of the fuels. The difference in the slopes indicates the rate of CO_2 production. The label 'heavy isotope content' is actually the isotopic enrichment. Physical activity, for example, increases fuel oxidation (i.e. respiration) and therefore greater dilution of ^{18}O.*

Inevitably, there are problems with the method:

- The respiratory exchange ratio (usually abbreviated, RQ or R) should be known but, in practice, it makes little difference unless very high accuracy is required.

- The method is not accurate for measurements over short periods (<10 days, or 5 days in children).

- It can underestimate the energy expenditure in children and in very obese individuals.

- The double-labelled water is expensive.

- A very sensitive mass spectrometer is required to measure ^{18}O and ^2H.

Isotopic dilution of urea   Urea is synthesised continually from carbon dioxide and ammonia in the body, in the ornithine cycle

$$CO_2 + NH_3 \leftrightarrows CO(NH_2)_2$$

After administration of ^{14}C-labelled hydrogencarbonate, the $^{14}CO_2$ rapidly equilibrates with the CO_2 in the whole body. Since the urea produced in the liver is in isotopic equilibrium with the CO_2 produced from the oxidation of fuels, changes in the radioactivity in urea in the urine can be used to measure the rate of CO_2 production and hence the rate of fuel oxidation. An advantage of the technique is that it is accurate over relatively short periods (e.g. 24 hours).

Components of energy expenditure

The amount of energy expended by an individual depends on three major components:

- resting energy expenditure (REE), also known as basal metabolic rate;

- thermic effect of food;

- physical activity.

Resting energy expenditure REE is defined as the energy expenditure at rest. It should be measured under conditions that minimise the effects of the other components, usually after the overnight fast or during a longer period of starva-

> The thermoneutral temperature is that at which there is no net gain or loss of heat from or to the environment. For humans, this temperature is not easy to ascertain but experiments suggest that the preferred temperature for subjects with light clothing is 23–26 °C. Mental rest (or complete relaxation) is not always an easy state to achieve under experimental conditions.

tion. The healthy individual should be at rest, both physically and mentally, and at a thermoneutral environmental temperature. If it is measured 2–4 hours after a light meal, the value includes a small contribution from the thermic effect of food. Basal metabolic rate (BMR) is an older term than resting energy expenditure. A problem with both is that there is a diurnal variation in energy expenditure: it is lowest after the overnight fast so that it should be measured at the same time each day.

A further problem is, on what basis should the REE be reported, e.g. per kilogram body weight or per kilogram fat-free mass? The former has the disadvantage of being influenced by the body composition, since adipose tissue has a lower metabolic activity than that of the fat-free component of the body. Per kilogram fat-free mass is now preferred because there is a linear increase in the value of the REE with the fat-free mass (Figure 2.4). There is a marked difference in the values of the REE for similar body weight in males and females, which is caused by their different fat-free masses: in females adipose tissue makes up a much higher percentage of body weight than in males.

Average values for the REE are 7.02 MJ (1680 kcal) per day for normal weight standard adult males (i.e. 70 kg) and 5.60 MJ (1340 kcal) per day for normal weight standard females (i.e. 58 kg). These values conceal as much as 30% variation between individuals, which is greater than changes in energy expenditure caused by normal levels of physical activity or by changes in ambient temperature. (Note difference for a female when calculated from schofield equations presented above.)

Thermic effect of food It has long been observed that eating a meal increases energy expenditure, a phenomenon also known as *diet-induced thermogenesis* or *specific dynamic action*. The increase depends on the type and quantity of food consumed. As an example, the effect of a carbohydrate-rich meal on the oxygen consumption of volunteers before and after exercise is shown in Figure 2.5.

The thermic effect of food has two main components:

> It is usually assumed that the total heat released when one mole of ATP is hydrolysed is equal to the enthalpy change (−21 kJ) but this is only in vitro. It ignores the fact that hydrolysis of ATP in vivo demands simultaneous generation of ATP to maintain its concurrence, which requires aerobic metabolism, which also expends energy. When this is taken into account, the hydrolysis of one mole of ATP is actually responsible for release of 90 kJ of heat (calculated on the basis that oxidation of one mole of glucose generates 30 moles ATP).

- *An obligatory component*, which results from the energy cost of digestion, absorption and assimilation of the food.

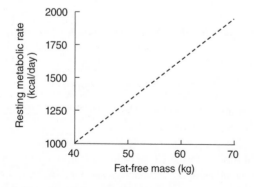

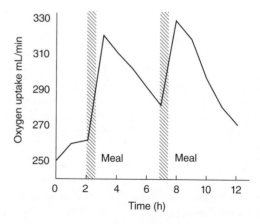

Figure 2.4 *A plot of resting metabolic rate against fat-free mass in men.* The line is a regression line based on 38 data points. The data are from Elia (1992). The largest contributor to fat free mass, is muscle.

Figure 2.5 *A plot of oxygen uptake after two meals in young adult males.* The vertical bars indicate the meal times. The measurements were carried out after physical activity which exaggerates the effect of the meal. The meal was high in carbohydrate.

• A *facultative component*, which results from the energy expended to regulate the rate of the various metabolic processes in relation to the metabolic load that is placed upon the body. For example, the amount of glucose absorbed into the bloodstream in a single normal meal is sufficient to increase the blood glucose level by about 20-fold (i.e. to about 100 mmol/L, well above the limit for the kidney to reabsorb glucose – about 10 mmol/L). To prevent this, regulatory mechanisms are required. Processes such as the secretion of several hormones, covalent modification of enzymes and substrate cycles (see Chapter 3) are activated to provide regulation and all require the expenditure of energy. Unfortunately, it is not possible to estimate accurately the contribution of these various processes to the facultative thermic effect.

Physical activity   Physical activity can increase energy expenditure markedly (Tables 2.5 and 2.6). Running 1 mile (1.5 km), at any speed, expends about 400 kJ (100 kcal) and a full marathon about 12 000 kJ (2900 kcal) for an average adult (about 4 kJ per kg per m). Thus, if the intensity is high and the duration long, dramatic increases are possible. Tour de France cyclists can increase their normal daily energy expenditure four- to five-fold (to over 30 MJ/day) during a particularly strenuous day in the mountains. The 'world best' for daily energy expenditure is 45–49 MJ (10.750–11.700 kcal), which is held by Mike Stroud and Ranulph Fiennes, achieved on particularly onerous days during their trek across Antarctica (see Chapters 13 and 15) weven, a similar value has been estimated for coalminers walking from the lift shaft to the

Table 2.5 Approximate energy expenditure during various activities

Work intensity	Approximate energy expenditure kJ/min
Complete rest	4–7
Sitting	6–8
Standing, light activity	9–13
Light work in the house	13–30
Light work in the garden	15–45
Coal miners heaving coal	42
Running[a]	
7 km/h	30–50
9 km/h	40–70
11 km/h	50–90
36 km/h (sprint)	200

[a] The faster the run, the greater the contribution from anaerobic glycolysis for ATP generation.

Data from Passmore & Durnin (1955); Ekblom (1992); Durnin & Passmore (1967); and Åstrand & Rodahl (1986). For the sprint see Margaria *et al.* (1966).

Table 2.6 Effects of prolonged and intense activity on the total daily energy expenditure

Subjects and activities	Total daily energy expenditure (MJ)
Soldiers	
Field training	14
Snow training	21
Athletes	
Cross-country skiing	30
Mountaineers climbing Mount Everest	15
Tour de France cyclists	34
Trans-Antarctic crossing[a]	49
Miners: coalminers walking to and from the coalface	40

[a] See also Chapter 13. The value for coal miners is close to the 'world best' set by Stroud and Fiennes (see Chapter 13).

Data obtained from Durnin & Passmore (1967).

coalface under difficult conditions, e.g. in a stooping position, which has been measured in volunteers performing similar activities (Table 2.6; Box 2.2).

Clinical conditions   Energy expenditure is increased by trauma, infections, burns or major surgery (Table 2.7). Energy expenditure can be very high in such patients even though they may be confined to bed. Indeed, before the advent of modern nursing, patients with severe burns almost doubled their resting energy expenditure. Cytokines are, in part, responsible for stimulation of thermogenesis which can result in fever (see Chapter 18 for the role of fever in such conditions).

In contrast, undernourished normal subjects have a lower rate of energy expenditure than expected on the basis of their lean body mass.

Low ambient temperature   A decrease in ambient temperature from 20 to 5 °C, for subjects wearing light clothes, can double the energy expenditure. Much of this extra energy is generated from shivering (see Chapter 9).

> A small decrease in ambient temperature, below the thermoneutral, results in a decrease in heat loss by reducing the flow of blood to the skin. If this is not sufficient to restore the normal body temperature, specific processes, including shivering, that result in heat generation, are stimulated.

Stress and stress hormones   Anxiety or stress can increase energy expenditure, although the effect is small. It is caused by increased sympathetic activity and hence increased levels of the stress hormones adrenaline and noradrenaline. Injection of these hormones increases oxygen consumption, as does caffeine, which

Box 2.2 The walk from the lift shaft to the coalface in a mine

In the mid-1930s, George Orwell visited working-class communities in northern England and wrote about the conditions that he witnessed. In his book *The Road to Wigan Pier.* He describes a visit to a coalmine: the section devoted to the walk from the mineshaft to the coalface is presented here. It makes it very clear why the energy expenditure involved in these activities is very high.

At the start to walk stooping is rather a joke, but it is a joke that soon wears off. I am handicapped by being exceptionally tall, but when the roof falls to four feet or less it is a tough job for anybody except a dwarf or a child. You have not only got to bend double, you have also got to keep your head up all the while so as to see the beams and girders and dodge them when they come. You have, therefore, a constant crick in the neck, but this is nothing to the pain in your knees and thighs. After a half a mile it becomes (I am not exaggerating) an unbearable agony. You begin to wonder whether you will ever get to the end – still more, how on earth you are going to get back. Your pace grows slower and slower. You come to a stretch of a couple of hundred yards where it is all exceptionally low and you have to work yourself along in a squatting position. Then suddenly the roof opens out to a mysterious height – scene of an old fall of rock, probably – and for twenty whole yards you can stand upright. The relief is overwhelming. But after this there is another low stretch of a hundred yards and then a succession of beams which you have to crawl under. You go down on all fours; even this is a relief after the squatting business. But when you come to the end of the beams and try to get up again, you find that your knees have temporarily struck work and refuse to lift you. You call a halt, ignominiously, and say that you would like to rest for a minute or two. Your guide (a miner) is sympathetic. He knows that your muscles are not the same as his. 'Only another four hundred yards,' he says encouragingly; you feel that he might as well say another four hundred miles. But finally you do somehow creep as far as the coalface. You have gone a mile and taken the best part of an hour; a miner would do it in not much more than twenty minutes.*

(Orwell, 1937)

Table 2.7 Percentage increase in resting energy expenditure in trauma, fracture, sepsis, burns or fever

Condition	Percentage increase in resting energy expenditure
Severe burns	25–60
Mild burns	10–25
Multiple trauma	20–50
Multiple fractures	10–25
Single fracture	0–10
Severe sepsis	20–50
Fever (increase of 2 °C)	10–25

raises the plasma level of catecholamines. The amount of caffeine in a cup of coffee increases the metabolic rate but the effect lasts for only about 20 minutes. For this reason, the drug ephedrine, which mimics some of the effects of the catecholamines, has been used to increase the metabolic rate with the intention of decreasing the amount of fat in the body but it has unacceptable side-effects including increased heart rate and tremor. Pharmaceutical firms attempted to develop specific thermogenic drugs, chemically related to the catecholamines, but they had limited success in humans although they were very effective in rats and mice!

Other hormones Thyroxine has long been known to increase metabolic rate, although the mechanism for this effect is not totally clear (Silvestri *et al.* 2005). More recently the hormone leptin, which is secreted by adipose tissue, has also been found to increase the metabolic rate. This effect of leptin is considered to play a role in controlling the amount of adipose tissue in the body, although this is a controversial subject (Chapter 12).

Energy transformations in tissues and organs

The REE of the body is the sum of the energy expenditure in the different tissues or organs. Organ-specific energy expenditure can be calculated from the arteriovenous difference in the concentration of oxygen across the organ and the blood flow. Data for various organs are presented in Table 2.8. The weights of organs (or tissues) for an adult male, female and a 6-month-old child are given in Table 2.9. They are particularly useful for studying the effect of different conditions on the energy expenditure by different organs or tissues.

Table 2.8 Metabolic rate of different organs or tissues in an adult human and contribution of the different organs and tissues to whole-body resting energy expenditure, (REE) in an adult male, female and a six month old child

| Organ or tissue | Organ or tissue metabolic rate kJ/kg per day* | Contribution to whole-body REE (% total) | | | Approx. oxygen utilisation of tissue for adult | | Equivalent rate glucose uptake[a] |
		Adult male	Adult female	Child (6 months)	mol/day	μmol/g tissue per min	μmol/g tissue per min
Liver	840	21	21	14	3.0	1.1	0.2
Brain	1000	20	21	44	2.7	1.2	0.2[b]
Heart	1850	9	8	4	1.2	2.5	0.4
Kidney	1850	8	9	6	1.1	2.3	0.4
Skeletal muscle (at rest)	54	22	16	6	2.9	0.1	0.02[c]
Adipose	16	4	6	2	–	–	–
Miscellaneous	50	16	20	24	–	–	–

*kJ per kg organ weight.

[a]The source of information is Elia (1992). For calculation and assumptions, see Newsholme & Leech (1983).

For calculation of rate of ATP generation from glucose, assuming complete oxidation of glucose, multiply rate of glucose utilisation by 30.

[b]On the basis of the data in this table, the calculated rate of glucose utilisation by the brain of an adult is about 3 g/hr. This is consistent with a rate of about 80 g in 24 hours (Chapter 14).

[c]The rate of glycolysis from glycogen during sprinting in a young adult male is about 50 μmol/g tissue per min, i.e. an increase of more than 1000-fold (see Chapters 3 and 13).

Table 2.9 Weights of different organs and tissues in a normal adult male, normal female and a child

| | Tissue or organ weight (kg) | | |
Organ	Man	Woman	Child (6 months)
Liver	1.80	1.40	0.26
Brain	1.40	1.20	0.71
Heart	0.33	0.24	0.04
Kidneys	0.31	0.27	0.05
Muscle	28.00	17.00	1.87
Adipose tissue	15.00	19.00	1.50
Miscellaneous	23.16	18.80	3.06
Total	70	58	7.5

Data from Elia (1992).

It should be noted that:

- The contribution of adipose tissue to the metabolic rate is small: in a subject with about 20% of body weight as fat, it contributes about 5% to the total expenditure.

- In the male, skeletal muscle (about 40% of body weight) contributes, at rest, about 15% to the total metabolic rate. In the female, skeletal muscle (about 30% of body weight) contributes about 10% to the total expenditure.

- In both male and female, the liver, brain, heart and kidney (about 50% of body weight) contribute about 50% to the total expenditure.

- The contribution of the lung is not known and is, therefore, excluded from Table 2.8. A guestimate is 20%.

- Per gram of tissue, the rate of oxygen consumption is highest in the heart and kidney, since they are doing the most work in the resting state. After a meal, it increases in the liver but it is not known by how much. An increase in physical activity increases the work of the heart and skeletal muscle. The increase in oxygen consumption by the active skeletal muscle and the heart during maximum prolonged physical activity is 20- to 50-fold in muscle and approximately 10-fold, in heart.

- One mole of glucose requires 6 moles of oxygen for complete oxidation, so that the uptake of glucose, on the basis of micromoles per gram tissue per minute, can be calculated from oxygen uptake provided no other fuel is being used. Such data are presented in the last column of (Table 2.8).

The metabolic rate per gram of each organ/tissue changes little during growth and development. The higher metabolic rate of infants, compared with that of adults, expressed in relation to body weight or fat-free mass, is explained by the presence of a larger proportion of metabolically active tissues. The brain of a child accounts for almost one-half of the total energy expenditure, so that the brain in a young child expends energy at about the same rate as that of an adult. This poses a potential clinical problem due to the smaller size of the liver and hence a smaller store of glycogen in comparison with the requirement of glucose by the brain in a child during starvation (Chapters 7 and 16).

Energy transformation in biochemical reactions and pathways

In early work on glycogen phosphorylase, in the early 1940s and 1950s, the activity was assayed by measuring the release of phosphate from glucose 1-phosphate in the reaction

glucose 1-phosphate + (glucose)$_n$ → (glucose)$_{n+1}$ + Pi

The symbols Pi and PPi are used to denote phosphate and pyrophosphate ions respectively. These widely used abbreviations originated from the designations inorganic phosphate and pyrophosphate. Under physiological conditions, the ions HPO_4^{2-} and $H_2PO_4^-$ will predominate.

where (glucose)$_n$ is glycogen and (glucose)$_{n+1}$ is the glycogen molecule extended by one glucose molecule.

The assay was chosen since phosphate was relatively easy to measure (the activity is now measured using purified coupling enzymes (Appendix 3.3). Since it was clear that phosphorylase could catalyse both chain extension and chain degradation of glycogen, it was considered, at that time, that the enzyme was involved in both synthesis and degradation of glycogen. It was only when an enzyme, glycogen synthase, was discovered, in 1957, that it was realised that phosphorylase did not catalyse the synthesis of glycogen in vivo. The discovery led to the identification of a specific pathway for glycogen synthesis, using uridine diphosphate glucose as the substrate for glycogen synthase. (Perhaps more surprising, even as late as the mid 1950s, the possibility that peptides (proteins) were synthesised by sequential catalysis by peptidase enzymes was still discussed, see Chapter 20).

That syntheses could be achieved by reversal of degradation processes was considered possible at that time, since, by definition, all chemical reactions are reversible. However, it gradually became apparent that synthetic and degradative processes in biochemistry in vivo were distinct, so that a process could only proceed in one direction.

The theoretical basis underlying the direction of a physical, chemical or biochemical process is derived from the second law of thermodynamics.

This law was formulated by Rudolph Clausius, a German physicist: *It is impossible for a machine, unaided by external agency, to convey heat from one body to another at a higher temperature that is, heat will not flow, of its own accord, from a cold place to a hot place.* This law is part of everyday life. For example, a refrigerator cannot lower the temperature within the box unless some external source of energy is used: this is usually electrical energy, which is well illustrated by the effect of a power cut, when the temperature within the box gradually increases towards the ambient temperature.

The factor that governs the direction of a reaction, which is central to the second law, is the change in entropy (ΔS). In formal terms, entropy is the heat (q) absorbed in a thermodynamically reversible reaction (at $T\,°K$) divided by the absolute temperature, T, thus $\Delta S = q/T$. A more qualitative representation of entropy is as the degree of disorder. The more disordered or random a system becomes, the more entropy it has, so that, in a spontaneous reaction, disorder must increase.

The degree of order can take a number of forms. Gases are less ordered than liquids and liquids less ordered than solids (because of the arrangement of molecules in the substance). In addition, some molecules have a greater degree of internal order than others, and so they have an inherently lower entropy. For example, proteins are highly ordered and possess a constrained conformation, but this changes to a much more random structure upon denaturation (see Chapter 3) and hence the increase in entropy during denaturation is considerable.

Gibbs free energy: combination of the first and second laws

It was an American scientist, Willard Gibbs, who realised that the change in enthalpy can be combined with a change in the entropy of the reactantion, to give the factor that defines whether a reaction can occur spontaneously. If, in a reaction, the change in enthalpy is smaller than the change in entropy ($T\Delta S$), i.e. the change in enthalpy minus the change in entropy is less than zero, the reaction can proceed spontaneously.

$$\Delta H - T\Delta S < 0$$

The quantity, $\Delta H - T\Delta S$, is known as the change in Gibbs free energy (ΔG). Thus, Gibbs free energy is derived from the first law (ΔH) and the second law (entropy), so that it expresses both laws in a single function. It should be noted that ΔG is not a form of energy in the conventional sense; it is not conserved in the way described by the first law. Indeed, use of the term 'energy' is unfortunate. It is better to consider ΔG as a factor that determines if a reaction can proceed in a given direction, that is, in which direction it will proceed. In fact, ΔG is much more akin to temperature than energy. As indicated above, temperature determines in which direction heat will flow, similarly ΔG determines in which direction a reaction will proceed.

Application of ΔG to biochemical reactions

It is the value of ΔG that determines the direction of a reaction or a biochemical pathway. If ΔG for a reaction is negative, the reaction can proceed in that direction. If it is zero or positive it cannot. If positive, the reverse reaction has a negative ΔG so that the reaction will occur in the opposite direction. If ΔG is zero, the reaction proceeds in neither direction and is said to be in a state of equilibrium.

Consider the reaction in which substance A is converted to substance B and the reaction is at equilibrium:

$$A \rightleftarrows B$$

If the concentration of either A or B changed, the equilibrium would be displaced and the reaction would proceed in the direction to maintain the equilibrium. For example, increasing the concentration of A will cause the reaction to move towards the right, producing substance B and lowering the concentration of A, until the equilibrium is re-established. Since ΔG determines the direction in which a reaction proceeds, it follows that the value ΔG must depend on reactant concentrations and the position of equilibrium (the equilibrium constant). It does so according to the equation:

$$\Delta G = \Delta G^\circ + R\Gamma \ln \frac{[B]}{[A]}$$

where R is the gas constant and Γ the absolute temperature, so that (at a fixed temperature) the value of ΔG depends on the constant ΔG° and the concentrations of the substrates and products (designated by square brackets). If there is more than one substrate or product, the concentrations of each are multiplied together. For explanation of derivation of this equation, see Bücher & Rüssman (1964) or Crabtree & Taylor (1979). In order to obtain the value of ΔG for a reaction in a living tissue, it is necessary to know the concentrations of substrate(s) and product(s) of the reaction in that tissue (i.e. A and B in the above example) and the value of ΔG°.

The ratio of the concentrations of product and substrate ([B]/[A]) for a reaction in a living organism is known as the mass action ratio and is given the symbol Γ. The constant, ΔG°, is known as the standard free energy change and is constant for a particular reaction at a given temperature. Its value depends upon the equilibrium constant, K_{eq}, which is the ratio of concentrations of product and substrate when the reaction is at equilibrium

$$\left(K_{eq} = \frac{[B_{eq}]}{[A_{eq}]} \right).$$

The relationship between the two constants, ΔG° and K_{eq}, is expressed in the equation:

$$\Delta G^\circ = -R\Gamma \ln K_{eq}$$

The value of ΔG° is therefore determined solely by the equilibrium constant (at a given temperature).

In summary, therefore, the sign and magnitude of ΔG (and hence the direction of a reaction) depends on two factors: the actual concentrations of the substrates and products and the value of the constant ΔG°.

The value of ΔG° alone is, therefore, not sufficient to determine the direction in which a reaction will proceed in a living cell. This explains why phosphorylase catalyses the breakdown of glycogen in vivo. Although ΔG° for the breakdown of glycogen is small and positive, the concentration ratio of phosphate to glucose 1-phosphate in vivo is >300 and it is this ratio that dominates the equation for the value of ΔG, and hence the direction of the reaction.

The equilibrium constant is measured in vitro as follows. The substrates and products are allowed to come to equilibrium in the presence of the enzyme that catalyses the reaction. At equilibrium, there is no further change in their concentrations, and the equilibrium constant, K_{eq}, is given by the ratio of these concentrations.

$$K_{eq} = \frac{\text{product concentration}}{\text{substrate concentration}}$$

From this the value of ΔG° is calculated.

Calculation of ΔG for a reaction in vivo

To calculate the value of the mass action ratio (Γ), the concentrations of the substrate(s) and product(s) of the reaction must be measured in the tissue/organ without perturbing the physiological state of the tissue. This is not a simple task. (The experimental details of how this can be done are given in Chapter 3.)

The ΔG value is then calculated from the equation

$$\Delta G = \Delta G^\circ + R\Gamma \ln \Gamma$$

(The value of R is 0.0082 kJ/mole and Γ is 310 °K, at 37 °C).

Two examples are presented: the hexokinase and phosphoglucoisomerase reaction in glycolysis (data taken from experiments with the isolated perfused rat heart).

Hexokinase catalyses the reaction, glucose + ATP $\rightarrow$ glucose 6-phosphate + ADP.

The value of ΔG° is -16.7 kJ/mole and the mass action ratio (Γ) is 0.08

(mass action ratio $= \dfrac{[\text{ADP}] \cdot [\text{glucose 6-phosphate}]}{[\text{ATP}] \cdot [\text{glucose}]}$)

Hence $\Delta G = -16.7 + (2.3 \times 0.0082 \times 310 \times \log 0.08)$
$\qquad = -23.1$ kJ/mole

The square brackets indicate concentration in tissue.

Phosphoglucoisomerase catalyses the reaction
$\qquad$ glucose 6-phosphate $\rightarrow$ fructose 6-phosphate.

The value of ΔG° is $+21$ and Γ is 0.24

$\Delta G = 21 + (2.3 \times 0.0082 \times 310 \times \log 0.24)$
$\qquad = 1.3$ kJ/mole

Values of ΔG for other reactions in glycolysis, calculated as above, are presented in Appendix 2.7.

Gibbs free energy and equilibrium and non-equilibrium reactions

The large and negative value for hexokinase indicates that this enzyme catalyses the reaction only in the direction of glucose 6-phosphate formation, i.e. it is a non-equilibrium reaction in vivo. In contrast, the low value for phosphoglucoisomerase indicates that it is a reaction that is close to equilibrium in vivo, that is, it can proceed in either direction. ΔG values for all the reactions of glycolysis indicate that those catalysed by hexokinase, phosphorylase, phosphofructokinase and pyruvate kinase are non-equilibrium; the others are near equilibrium (Figure 2.6).

It is these non-equilibrium reactions that provide for the direction of glycolysis, i.e. that glycolysis from glycogen or glucose always proceeds in the direction of pyruvate or lactate formation, i.e. it is these reactions that provide directionality in the pathway. The equilibrium nature of these reactions is discussed from a kinetic viewpoint in Chapter 3.

Gibbs free energy and the coupling of biochemical reactions

Two different processes are of fundamental importance in understanding the relationship between different reactions in biochemical pathways: these are *coupling-in-series* and *coupling-in-parallel*.

Coupling-in-series Reactions are said to be coupled-in-series when the product of one reaction is the substrate for the next reaction; the principle underlying this coupling is explained by reference to the Gibbs free energy equation. In the reaction A $\rightarrow$ B, in the hypothetical pathway S $\rightarrow$ P, reaction A to B is catalysed by enzyme E_2 as follows:

$$S \xrightarrow{E_1} A \xrightarrow{E_2} B \xrightarrow{E_3} P$$

ΔG° for this reaction could be either negative or positive, whereas ΔG must be negative. Consider the situation if the value for the constant ΔG° is positive. For the value of ΔG to be negative, the value of the term, $RT\ln([\text{B}]/[\text{A}])$, (in the equation $\Delta G = \Delta G^\circ + RT\ln([\text{B}]/[\text{A}])$ must be negative and larger than the positive value of ΔG° (thus the value of Γ must be less than that of K_{eq}). The question arises, what factors influence the concentrations of A and B? These are as follows.

If the value of the constant, K_{eq}, for reaction S $\rightarrow$ A is large and if the reaction is near equilibrium, this would produce a high concentration of the product, A. If K_{eq} for reaction B $\rightarrow$ P is large and the reaction is near equilibrium, the result would be a high concentration of P

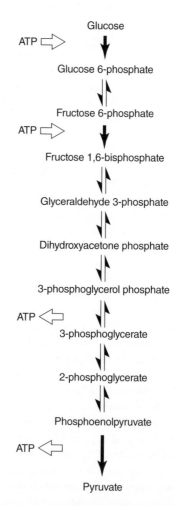

Figure 2.6 *The process of glycolysis illustrating the three non-equilibrium reactions. The reactions are catalysed by hexokinase, phosphofructokinase and pyruvate kinase which are indicated by the heavy unidirectional arrows. The reactions in which ATP is utilised and those in which it is produced are indicated (see Appendix 2.7).*

but a low concentration of B. A high concentration of A and a low concentration of B are the conditions necessary to produce a large and negative value of the term, $RT\ln([B]/[A])$ and hence a negative value for ΔG for the reaction catalysed by E_2.

It is important to appreciate that this principle of coupling-in-series underlies all biochemical pathways or processes, e.g. glycolysis, generation of ATP in the mitochondrion, protein synthesis from amino acids or a signal transduction pathway. Indeed, despite the fundamental importance of signalling pathways in biochemistry, a thermodynamic analysis of such a pathway has never been done, but the principles outlined above must apply even to signalling pathways.

Coupling-in-parallel For a two-substrate reaction, in which one reaction is inseparably linked to a second as follows:

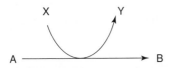

the linking of these two-part reactions is described as coupling-in-parallel, although it must be emphasised that the two reactions cannot occur separately, since a single mechanism is responsible for both parts, i.e. the overall reaction has two substrates (A and X) and two products B and Y. In general, the 'part' reaction A → B may be a component of a specific pathway or process. However, in many cases, X is regenerated (from Y) in a separate reaction:

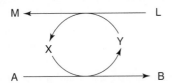

In this way, many biochemical systems can be linked together via a small number of pairs of compounds, with the functions of X and Y, i.e. such metabolic pairs play a major role in biochemistry. These compounds are known as coenzymes. Well-known examples include ADP/ATP, $NAD^+/NADH$ and $NADP^+/NADPH$. Such biochemical systems are discussed in Chapter 3.

The principle underlying this coupling is also explained by reference to the Gibbs free energy equation. An example, which illustrates the principle, is the phosphorylation of glucose in the reaction catalysed by the enzyme hexokinase:

The Gibbs free energy equation explains why glucose 6-phosphate cannot be formed, in vivo, from glucose and phosphate, as follows:

$$P_i + \text{glucose} \longrightarrow \text{glucose 6-phosphate} + H_2O$$

The reverse of this reaction does occur in some cells (e.g. hepatocytes), catalysed by the enzyme glucose-6-phosphatase, but the direction is always towards glucose formation, i.e. glucose 6-phosphate is never produced by this reaction in vivo. The question is, 'why not?'

This question is answered by calculation of the changes that would be necessary to produce a positive value of $\Delta G°$ for this reaction, i.e. a negative value for the reverse reaction of the

$$\text{glucose 6-phosphate} + H_2O \longrightarrow \text{glucose} + Pi$$

There are two situations, in which this could occur: (i) a large increase in the concentration of the substrates, glucose and/or phosphate; (ii) a large decrease in that of the product, glucose 6-phosphate.

$$\Delta G = \Delta G° - RT \ln \frac{[\textbf{glucose}].[P_i]}{[\text{glucose 6-phosphate}]}$$

indicating a high concentration of glucose or

$$\text{or} \quad \Delta G = \Delta G° - RT \ln \frac{[\text{glucose}].[P_i]}{[\text{glucose 6 - phosphate}]}$$

indicating a low concentration of glucose 6-phosphate.

- If the concentrations of glucose 6-phosphate and phosphate, which are normally present in the living cell, are inserted into this equation, the glucose concentration that is required for the reaction to proceed in the direction of glucose 6-phosphate must be greater than 1.6 mol/L. This concentration is more than 300-fold greater than that of blood glucose and about 3000-fold greater than the intracellular glucose concentration in, for example, muscle. This is physiologically unacceptable: a glucose concentration of 1.6 mol/L would cause major problems in a cell, including unwanted side reactions and a lethal osmotic effect.

- A positive ΔG for this reaction could also be achieved by a marked decrease in the concentration of glucose 6-phosphate. Why is this not feasible? Glucose 6-phosphate is an important metabolic intermediate and is involved in several metabolic pathways (e.g. glycogen synthesis, glycolysis, pentose phosphate pathway). Lowering its concentration by the two orders of magnitude, which would be necessary, would markedly decrease the rates at which these important pathways could proceed.

The important principle that can be derived from this discussion is that a reaction that has a large and positive $\Delta G°$ cannot proceed spontaneously unless it can be coupled, in parallel, to a reaction that has a large and negative $\Delta G°$, e.g. ATP hydrolysis. The overall reaction may then have a negative $\Delta G°$ and can proceed in the direction required at physiologically acceptable concentrations of substrates and products. (The values of $\Delta G°$ and ΔG for all the reactions in glycolysis are given in Appendix 2.7.)

The steady state condition

So far, discussion has focused on what determines the feasibility and direction of an individual reaction. This section extends the use of coupling-in-series to answer the question, how are the individual reactions organised to provide direction in the sequence of reactions that comprise a biochemical pathway or process? The answer is in two parts. First, the concept of closed and open systems is described.

Closed and open systems Consider the following hypothetical enzyme-catalysed reaction:

$$A \underset{v_f}{\overset{v_r}{\rightleftharpoons}} B$$

where the reactants are A and B and where v_f and v_r represent the actual rates of the forward and reverse components (not the rate constants).

If such a reaction is isolated from its surroundings, so that A and B neither enter nor leave the system, creating a closed system, these reactants will eventually reach equilibrium, i.e. the forward and reverse rates are the same. For a closed system, equilibrium is the only state in which the concentrations of A and B do not vary with time. Since $\Delta G = 0$, the reaction can do no useful work and has no direction. Hence a closed system is not a useful model of a living cell or, indeed, a living animal.

In an open system, there is a continual exchange of both matter and energy with the surroundings, yet in certain conditions it can result in constant concentrations of intermediates: this is known as the steady state:

$$\xrightarrow{x} S \xrightarrow{1} A \xrightarrow{2} B \xrightarrow{3} P \xrightarrow{y}$$

This system contains the same hypothetical reaction (A → B) but, in this case, there is a reaction (x) that continually supplies A (via S) from the surroundings (i.e. the environment) and a reaction (y) that continually removes B (via P) from the system into the environment. Conditions can be such that reaction (1) generates a constant flux (i.e. a constant flow of molecules from S, that is transmitted through reactions (2) and (3) by the concentrations of the substrates for these reactions (i.e. A and B). In

other words, the flux through all the reactions is the same; the flux through reaction (2) is a function of the concentration of A, and the flux through (3) is a function of concentration B. In this condition, there is a constant flux and the concentrations of S, A, B and P are constant, i.e. a *steady state* (see Chapter 3).

As an example, in glucose metabolism, blood glucose is S, which is the substrate for complete oxidation in several organs, e.g. the brain: glucose is provided, via food, from the environment and the products of the pathway, CO_2 and H_2O, are lost to the environment through the lungs.

$$\begin{array}{c} \qquad\qquad\qquad\text{uptake and oxidation} \\ environment \longrightarrow food \rightarrow glucose \longrightarrow \longrightarrow \longrightarrow \longrightarrow \\ CO_2 + H_2O \longrightarrow environment \end{array}$$

The concentrations of all the intermediates from intracellular glucose to carbon dioxide are constant, unless the flux through the pathway changes. A kinetic exploration of the steady state rather than a thermodynamic one is provided in Chapter 3.

Near-equilibrium and non-equilibrium reactions This steady-state condition is usually achieved by the presence, in the pathway, of two classes of reactions: those that are very close to equilibrium (near-equilibrium) and those that are far removed from equilibrium (non-equilibrium). The difference between these two types of reactions can be explained both thermodynamically (above) and kinetically (in Chapter 3).

It is the non-equilibrium reactions in which entropy is gained and, therefore, they provide the direction for the pathway, i.e. glucose is converted to lactate and not vice versa. (The new reactions in a pathway that have to be inserted to change the direction of a pathway, i.e. conversion of lactate to glucose, are discussed in Chapter 6.)

In comparison to glycolysis and a few other metabolic pathways, no systematic studies on the equilibrium nature of individual reactions in some important biochemical processes have been carried out (e.g. protein synthesis, signal transduction pathways see Chapters 12, 20 and 21).

Adenosine triphosphate: its role in the cell

The objective of this chapter is to place energy metabolism within the context of the biochemistry that takes place in the whole body, in the tissues and in biochemical pathways or processes. In all of these systems, the generation and utilisation of ATP are central. Consequently, the final dis-

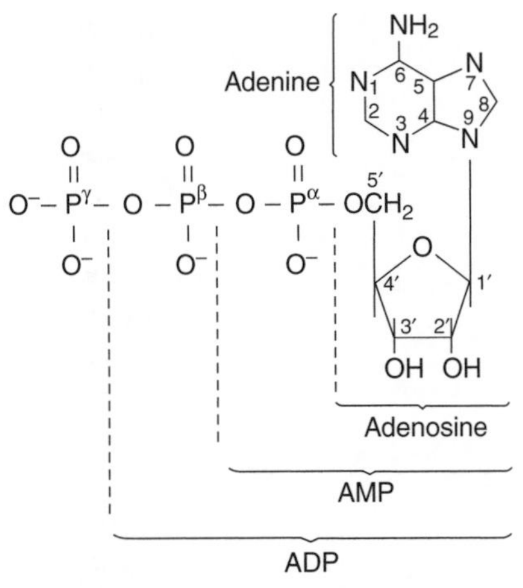

Figure 2.7 *Structure of adenosine triphosphate.* The molecule consists of adenine attached to ribose (forming adenosine) which is esterified to a triphosphate group with the three phosphorus atoms designated α, β and γ. The primes (') on the carbon atoms of ribose differentiate the numbers from those of the adenine ring.

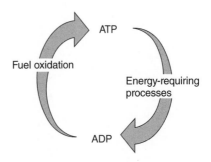

Figure 2.8 *The ATP/ADP cycle.* The major ATP-generating process from fuel oxidation is oxidative phosphorylation driven by electron transport in the mitochondria. In muscle, the major energy-requiring process is physical activity. The phosphate ion is omitted from the figure for the sake of simplicity.

cussion in this chapter relates to ATP. Surprisingly perhaps, ATP (adenosine triphosphate) is a rather ordinary molecule. It is a nucleotide (Chapter 10) in which a ribose sugar is connected to the base adenine, in its 1'-position and a triphosphate group in its 5'-position (Figure 2.7). Although it is one of five nucleoside triphosphates from which nucleic acids are synthesised, ATP justifies its pre-eminent position in biochemistry by its central role in energy transfer. It is the only compound that, either directly or indirectly, when hydrolysed, transfers chemical energy to all other processes that require energy. The generation of ATP, from ADP and phosphate, requires transfer of energy and this is from the oxidation of carbohydrates, fats and/or amino acids. It involves the transfer of chemical energy from the process of electron transport to the process involved in the generation of ATP from ADP and P_i. It therefore couples energy-requiring processes to energy-generating processes (Figure 2.8). What is more, it does this, so far as is known, in all organisms from bacteria to humans.

This role of ATP in the cell is analogous to the role of money in the economy of the world. Money can link any paid employment to the acquisition of any goods or services, a much more flexible system than bartering, in which specific goods or services are exchanged one for the other. Consequently, ATP has been called the *energy currency* of the cell. Its existence simplifies the coupling of energy-producing reactions to energy-requiring reactions, although the analogy with money must not be taken too far. Unlike cash, ATP cannot be accumulated, nor can it be transferred from cell to cell. Credit is not 'permitted'; once all the 'money' has been used, and no more can be generated, the cell is bankrupt and dies. Consequently, the cell possesses mechanisms to prevent this from happening, e.g. fatigue in muscle arrests ATP utilisation in physical activity; loss of electrical activity in the brain arrests ATP utilisation in neurones. A cardiomyocyte deprived of oxygen and hence of ATP generation stops beating in order to decrease the utilisation of ATP in an attempt to maintain the concentration of ATP that is essential for the life of the cardiomyocyte (see Chapter 22). Indeed, the maintenance of the ATP/ADP concentration ratio in all cells usually takes precedence over mechanisms involved in the physiological functions of the cells. The latter have sometimes been described as the 'luxury' reactions for the cell: for example, the cross-bridge cycle in muscle is not essential for the life of a skeletal muscle fibre, or for a cardiomyocyte; synthesis of a neurotransmitter by a neurone in the brain is not essential for the life of a neurone; and absorption of glucose from the lumen of the intestine is not essential for the life of an epithelial cell in the intestine. Yet they are, of course, not luxuries for the rest of the body: loss of the physiological function can sooner or later result in death of the whole body.

3
Enzymes: Activities, Properties, Regulation and Physiology

The primary and overriding interest of enzymes, however, is their connection with life. Of all the multitudinous chemical processes in the living cell on which its life depends, there is scarcely one which is not due to enzyme catalysis; there can be no life without enzymes.

(Dixon and Webb, 1979)

The thermodynamic principles described in Chapter 2 of this volume can be used to indicate whether or not a reaction can take place spontaneously. They do not, however, provide information about the *rate* at which a reaction will proceed. Most biochemical reactions proceed so slowly at physiological temperatures that catalysis is essential for the reactions to proceed at a satisfactory rate in the cell.

> Some RNA molecules are now known to possess catalytic activity and have been termed ribozymes.

The catalysts are enzymes, most of which are proteins. Not only are they catalysts, but their catalytic powers are enormous: they can increase the rate of a reaction by several orders of magnitude. Indeed, in the absence of enzymes, life as we know it would not be possible. One remarkable example of the use of the catalytic power of two enzymes in biology is the Bombardier beetle: it uses the enormous catalytic power of the enzymes catalase and a peroxidase to deter predators (Box 3.1).

Enzymology is therefore central to biochemistry and many books, chapters in books and review articles have been written on this topic. The aim of this chapter is to provide the reader with sufficient information to understand the key aspects of enzymology, how enzymes maintain the life of a cell and help it to perform its physiological functions that contribute to the life of a human and, finally, to show how individual enzymes can provide quantitative information on certain aspects of biochemistry and physiology in humans. Although the enormous catalytic potential is essential, it does pose a problem: if it was not regulated, all reactions in a cell would rapidly reach equilibrium and, once again, life as we know it would not be possible. Two questions, therefore, are: (i) What are the mechanisms that account for the enormous catalytic activity of enzymes? (ii) What are the properties of enzymes in the cell that allow this catalytic activity to be controlled? Answers to these questions provide an understanding of some of the basic principles of enzymology that apply in the cell.

There are only a limited number of chemical manipulations that a single enzyme can perform, so that for a major chemical change, more than one enzyme is required; for example, to convert glucose into ethanol or lactic acid, more than ten enzymes are required and they function as a coordinated sequence. Such a sequence is called a pathway. Since the first pathways to be elucidated were in the field of metabolism, the term *metabolic pathway* is a familiar term to almost all biochemists. However, the

> The early history of enzymes is associated with the process of brewing and the production of wine because of the economic importance of these processes in Europe in the nineteenth century. Following on from the work of Pasteur, it was Büchner and others who showed that an extract of yeast carried out fermentation (i.e. the conversion of glucose to alcohol) as well as the yeast cell itself. The agents that did this catalysis in the extract were simply described as 'in yeast' which, from the Greek *en*(in) and *zymē*, is the name *enzyme* was derived.

Functional Biochemistry in Health and Disease by Eric Newsholme and Tony Leech
© 2010 John Wiley & Sons Ltd

Box 3.1 Catalytic power and the Bombardier beetle

The Bombardier beetle possesses, at the end of its abdomen, a combustion chamber that contains a hydroquinone and hydrogen peroxide. When a predator approaches, the cells in the walls of the combustion chamber secrete two enzymes, catalase and peroxidase. Catalase causes decomposition of hydrogen peroxide to produce oxygen; peroxidase catalyses the oxidation of the hydroquinone to produce a quinone.

$$2H_2O_2 \rightarrow 2H_2O + O_2$$

$$\text{hydroquinone} + H_2O_2 \rightarrow \text{quinone} + H_2O$$

The catalytic power ensures that the reactions take place so rapidly that they result in an explosion.

The three things produced in the reaction are:

- heat;
- the gas, oxygen, which expands with the heat and acts as a propellant;
- the quinone, which acts as the deterrent.

The beetle directs the end of its abdomen towards the predator which is sprayed with a hot quinone, which deters further interest in the beetle.

It is unlikely that the terrorists who, in July 2005, attempted to cause explosions on the London Underground by using a mixture of hydrogen peroxide and flour, obtained the idea of this explosive mixture from a knowledge of the biology of the Bombardier beetle. The mixture failed to explode through failure of the detonators. The secretion of enzymes with their enormous catalytic potential ensures no such failure in the beetle.

term pathway also applies to other biochemical processes such as protein synthesis, gene expression or signal transduction and the same kinetic and energetic principles that apply to metabolic pathways also apply to these other pathways. Unfortunately, the depiction of pathways in textbooks, on wall charts or in handouts given to students conceals the fact that they have thermodynamic and kinetic structures that are based on the principles of physical chemistry and on the properties of enzymes. Indeed, it can be argued that understanding these aspects of enzymology is more important, for discussions in health and disease, than is knowledge of the three-dimensional structure of an enzyme (see below). A cell contains a remarkably large number of enzymes. There are approximately 2000 in a single bacterial cell and at least ten times more in a mammalian cell. The properties of the enzymes explain how this large number can be organised to allow the normal functioning of the cell and the physiological functions of the tissues that maintain the health of the individual.

Nomenclature and classification

The early elucidation of biochemical pathways resulted in the discovery of a large number of enzymes, which were given names according to their function. Those that catalysed degradation reactions were named simply by addition of the suffix -ase to the name of the substrate. For example, arginase catalyses the hydrolysis of arginine, and urease catalyses the hydrolysis of urea. This practice was extended to non-degradative enzymes by the addition of -ase to a term descriptive of the catalysed reaction; for example, kinases, isomerases, oxidases and synthases. Such terms were generally prefixed by the name of the substrate or, in the case of the synthases, that of the product. More etymologically inventive biochemists coined names such as cocoonase (for an enzyme secreted by some emerging silk moths to catalyse the hydrolysis of the silk of their cocoon) and nickase (for an enzyme catalysing the hydrolysis of a single phosphodiester bond in a strand of a DNA molecule). The natural growth of this unsystematic terminology resulted in single enzymes being given more than one name and, worse still, in more than one enzyme bearing the same name. To overcome these difficulties, in 1961 the International Union of Biochemistry adopted a systematic nomenclature and classification of enzymes, prepared by its Enzyme Commission (EC). For each well-characterised enzyme, the EC proposed a systematic name and a unique numerical designation (see Enzyme Nomenclature, 1992). Because the systematic names were frequently cumbersome and the numbers difficult to memorise, the EC also proposed that a single recommended (trivial) name should be retained (or invented) for each enzyme. For example, the enzyme catalysing the reaction:

$$ATP + AMP \rightleftharpoons 2\,ADP$$

bears the systematic name ATP:AMP phosphotransferase, the number EC 2.7.4.3 and the recommended name, adenylate kinase. The previously used name of myokinase, which erroneously suggested that the enzyme was restricted to muscle, was abandoned.

The EC system of classification is based on the division of enzymes into six major classes according to the reaction they catalyse (Table 3.1). Each class is subdivided. In class 1, oxidoreductases, this is done according to the nature of the electron donor: 1.1 with CHOH group as donor; 1.2 with aldehyde or oxo group as a donor; 1.3 with CH_2-CH_2 group as donor, etc. A further subdivision is achieved by consideration of the electron acceptor: 1.1.1 with NAD^+ as acceptor; 1.1.2 with cytochrome as acceptor; 1.1.3 with O_2 as acceptor, etc. Finally, each enzyme that catalyses a unique reaction is given a fourth number so that, for

Table 3.1 Major classes in the Enzyme Commission system for enzyme nomenclature

Number	Group	Reaction
1	Oxidoreductases	oxidation-reduction reactions.
2	Transferases	group transfer reactions.
3	Hydrolases	hydrolytic reactions.
4	Lyases	non-hydrolytic removal of groups to form double bonds.
5	Isomerases	isomerisations.
6	Ligases	bond formation with concomitant breakdown of nucleoside triphosphate.

example, enzyme 1.1.1.1 is alcohol dehydrogenase (or, to give it its systematic name, alcohol: NAD^+ oxidoreductase) and catalyses the reaction:

$$\text{alcohol} + NAD^+ \rightleftarrows \text{aldehyde (or a ketone)} + NADH + H^+$$

Even within a single organism, the same enzyme, as defined by the reaction it catalyses, can exist in more than one form, each of which is the product of a different gene. These are known as isoenzymes. The properties of the two enzymes are usually different and this may explain the different roles of the isoenzymes within either a single type of tissue or within different tissues, e.g. glucokinase and hexokinase in the hepatocyte (see below).

Basic facts

There are three characteristics of enzymes that form the basis of most of their properties: the active site, the enzyme-substrate complex and the transition state.

The active site

Of all the macromolecules in living organisms, only proteins have the diversity of structure necessary to create the immense number of catalysts required for the large number of reactions that occur in the cell. The smallest enzyme consists of just over 100 amino acids whereas the largest consists of several thousand. The amino acids are linked through peptide bonds to form one or more peptide chains, and an enzyme may comprise several chains. The flexibility of the peptide permits interaction between the side-chain groups of the amino acids that cause the peptide to fold into a three-dimensional structure that is unique to that protein. This folding brings together amino acids, most of which are not adjacent in the primary sequence, so that some amino acids (about ten) form a three-dimensional structure that binds with the substrate to form the enzyme-

substrate complex (Figure 3.1). This complex results in catalysis. The remainder of the amino acids in the enzyme are involved in maintenance of the three-dimensional structure of the enzyme, attaching the enzyme molecule to intracellular structures (e.g. membranes) or in binding molecules that regulate the activity of the enzyme. An introduction to the chemistry involved in reactions, and the bonds that are involved in the structures of small and large molecules is provided is Appendix 3.1. This should help students with material in this and other chapters.

A similar number of amino acids in other proteins are involved in binding specific molecules, e.g. an antigen to the antibody receptor protein on B lymphocytes, a neurotransmitter to the receptor in a postsynaptic nerve, a hormone or growth factor to a receptor on the surface of a cell.

The enzyme-substrate complex

Enzymes bind substrates to produce an enzyme-substrate complex as follows:

$$E + S \underset{k_2}{\overset{k_1}{\rightleftarrows}} ES$$

Weak bonds, generally non-covalent ones, are involved in formation of the complex, so that the reaction is readily reversed. The rate of the forward reaction is given by the concentration of substrate multiplied by the rate of constant k_1, and rate of the reverse reaction is given by the concentration of the product multiplied by the rate constant k_2. The dissociation constant for the ES complex is k_2/k_1. This is analogous to the formation of other complexes: for example receptor-hormone complex; receptor-neurotransmitter complex; antibody-antigen complex.

Formation of the enzyme-substrate complex can occur only if the substrate possesses groups that are in the correct three-dimensional orientation to interact with the binding groups in the active site. A 'lock and key' analogy has been widely used to explain specificity but it is inadequate because the formation of the enzyme-substrate complex involves more than a steric complementarity between enzyme and substrate. Chemical interactions occur between enzyme and substrate that have no counterpart in the mechanical analogy. Furthermore, the binding of substrate can modify the structure of the enzyme around the active site so that additional groups interact with the substrate, which further increases the effectiveness of its binding.

The transition state

In a chemical reaction, a substrate is converted to a product; that is, one stable arrangement of atoms (the substrate) is

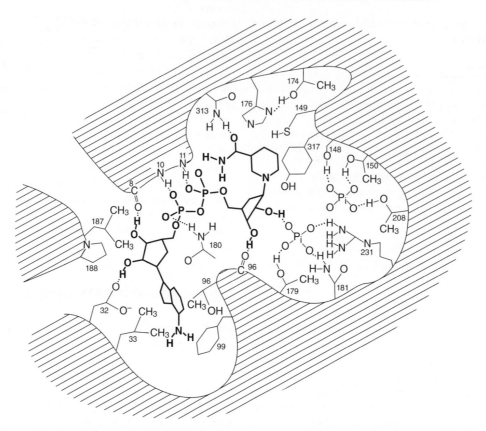

Figure 3.1 *Amino acid side-chain groups involved in binding NAD⁺ at the active site of an enzyme*. The enzyme is glyceraldehyde dehydrogenase. More than 20 amino acids, the position of which in the primary structure is indicated by the number, counting from the N-terminal amino acid, are involved in the binding. This emphasises the complexity of the binding that is responsible for the specificity of the enzyme for NAD⁺ (depicted in bold). The molecular structure of nicotinamide adenine dinucleotide (NAD⁺) provided in Appendix 3.3.

converted to another (the product). As this change proceeds, the atoms pass through an unstable arrangement, known as the transition state, which can be thought of as the 'halfway house' between the substrates and the products. For a reaction,

$$AB + C \rightarrow AC + B$$

the transition state is ACB, i.e. the reaction follows this sequence:

$$AB + C \rightarrow ACB \rightarrow AC + B$$

The relevance of the transition state to kinetics is that the rate of the overall reaction depends on the number of molecules in this state: the more molecules in the transition state, the greater is the rate. The role of an enzyme is to increase the number of molecules in this state, which is achieved by one or more of the mechanisms, described below. Detail of the kinetics and the thermodynamics of the formation of the transition state is given in Appendix 3.2.

Mechanisms by which an enzyme enhances the rate of a reaction

Examples of the enormous catalytic potential of enzymes are hexokinase, which increases the rate of glucose phosphorylation by a factor of $>10^{10}$, and urease, which increases the rate of hydrolysis of urea by $>10^{14}$. Just five mechanisms provide for these enormous catalytic effects: (i) general acid/base catalysis; (ii) covalent reaction with substrate; (iii) proximity and orientation; (iv) strain within the substrate; (v) nature of the environment within or surrounding the active site. The making and breaking of covalent bonds involves a redistribution of electrons. Indeed, reaction mechanisms in organic chemistry are usually depicted in terms of electron shifts. Since the transition state possesses the least stable electron distribution, an agent capable of supplying or withdrawing electrons to or from stable parts of the molecule in order to destabilise it, accelerates the rate of a reaction. All these mechanisms facilitate such electron shifts.

General acid/base catalysis

Addition of a proton (H^+) from an acid to a molecule can cause an electron to be withdrawn from one part of the molecule to the part which binds the proton. A base removes a proton from or within a molecule which will also cause electron shifts. If these shifts favour formation of the transition state, the rate of the reaction increases. Enzymes possess, in the active site, side-chain groups of the amino acids that act as acids or bases; that is, they can donate or withdraw electrons from the substrate, resulting in electron shifts that favour formation of the transition state (Figure 3.2).

The contribution of these groups is greatly enhanced if they act in a concerted manner so that, as an electron is withdrawn from one part of the substrate, another is donated to a different part. This is known as a 'push-pull' mechanism. It is only possible when the relevant groups in the active site are held in precisely the correct orientation so as to interact in this way with the substrate. One of the more versatile side-chains in this respect is the imid-

Table 3.2 Metal ions that are part of the active site of some enzymes

Metal ion	Enzyme	Pages in the text where these enzymes are discussed
Iron	Cytochromes	185–186
	Aconitase	183
	Catalase	185
Zinc	Carbonate dehydratase	89
	Superoxide dismutase	App. 9.6
Manganese	Pyruvate carboxylase	114
	Superoxide dismutase	App. 9.6
Molybdenum	Xanthine oxidase	218
Copper	Superoxide dismutase	App. 9.6
	Cytochrome oxidase	185

These metal ions perform a variety of functions and are attached to the proteins, either by a weak or a strong bond, in diverse ways (Williams & de Silva, 1991).

App. is abbreviation for Appendix.

azole group of the amino acid histidine. In one environment it can act as a donor (i.e. an acid) whereas, in another environment, the same group can act as an acceptor of electrons (i.e. a base). This occurs with two histidines in the same active site. Some enzymes contain metal ions within their active site, where they serve as centres of high positive charge, to withdraw electrons from substrates (Table 3.2).

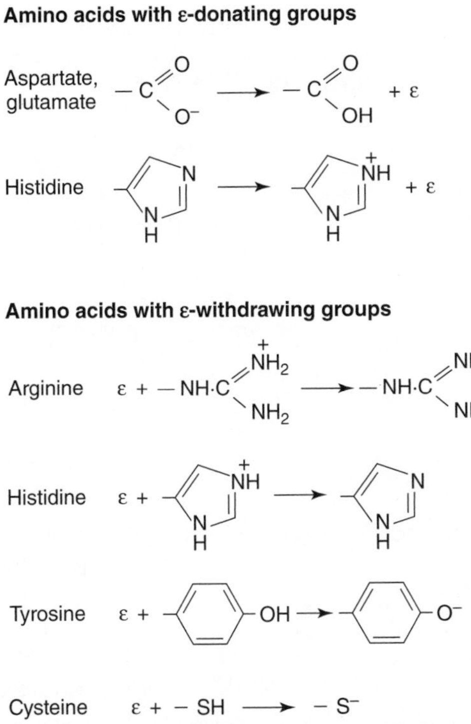

Covalent catalysis (formation of an intermediate)

Most enzymes bind their substrates in a non-covalent manner but, for those that do bind covalently, the intermediate must be less stable than either substrate or product. Many of the enzymes that involve covalent catalysis are hydrolytic enzymes; these include proteases, lipases, phosphatases and also acetylcholinesterase. Some of these enzymes possess a serine residue in the active site, which reacts with the substrate to form an acylenzyme intermediate that is attacked by water to complete the hydrolysis (Figure 3.3).

Proximity and orientation

The effects of proximity are easier to comprehend than the chemical mechanisms of rate enhancement. In a reaction

Amino acids with ε-donating groups

Amino acids with ε-withdrawing groups

Figure 3.2 *Examples of amino acids with electron-donating or electron-withdrawing groups.* An electron is represented by ε. Note that the imidazole group of histidine can either donate or withdraw electrons.

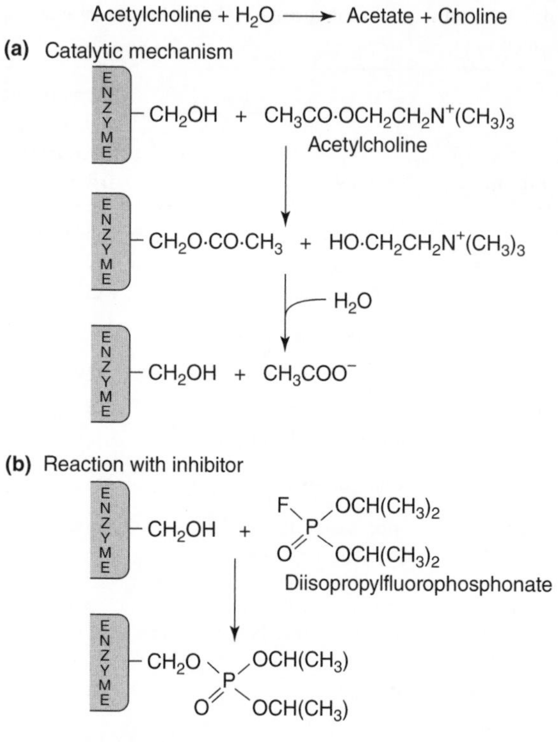

Acetylcholine + $H_2O \longrightarrow$ Acetate + Choline

(a) Catalytic mechanism

ENZYME—CH_2OH + $CH_3CO \cdot OCH_2CH_2N^+(CH_3)_3$
Acetylcholine

ENZYME—$CH_2O \cdot CO \cdot CH_3$ + $HO \cdot CH_2CH_2N^+(CH_3)_3$

H_2O

ENZYME—CH_2OH + CH_3COO^-

(b) Reaction with inhibitor

ENZYME—CH_2OH + Diisopropylfluorophosphonate

ENZYME—CH_2O

Figure 3.3 (a) *Covalent catalysis: the catalytic mechanism of a serine protease.* The enzyme acetylcholinesterase is chosen to illustrate the mechanism because it is an important enzyme in the nervous system. Catalysis occurs in three stages: (i) binding of acetyl choline (ii) release of choline (iii) hydrolysis of acetyl group from the enzyme to produce acetate. **(b)** *Mechanism of inhibition of serine proteases by diisopropylfluorophosphonate.* See text for details.

involving two substrates, the two must come together in order to react. The chance of them doing so depends upon their concentration in the solution: this is increased locally by providing adjacent binding sites for each substrate within the active site. This can increase the effective concentrations of the substrates about 1000-fold.

Even when a collision between two substrates occurs it is unlikely that they will both be in the correct orientation for a reaction to take place. Another property of the active site is that it binds the substrates in such a way that their orientation favours the reaction, i.e. it facilitates electron shifts that favour formation of the transition state.

Strain (distortion)

The structural precision between substrate and side-chain groups of the amino acids within the active site can be increased if the binding of the substrate changes the three-dimensional arrangement of the groups within the active site. This is known as 'induced fit'. Alternatively, the initial

binding in the active site can result in a slight change in structure of the substrate, known as strain or distortion, that favours formation of the transition state.

Environment of the active site

In many enzymes, the active site is present within a structural cleft in the protein structure. This can produce a chemical environment that favours a change in the structure of the substrates that enhances electron shifts within the substrate.

Cofactors and prosthetic groups

Despite the diverse chemical nature of side groups of amino acids in the active site, compounds other than amino acids may be co-opted to provide additional reactive groups. Such compounds play a part in the catalytic mechanism, so that they remain unchanged at the end of the catalytic process. Those that remain bound to the enzyme, even when catalysis is not occurring, are known as prosthetic groups (Figure 3.4). Compounds that have a similar

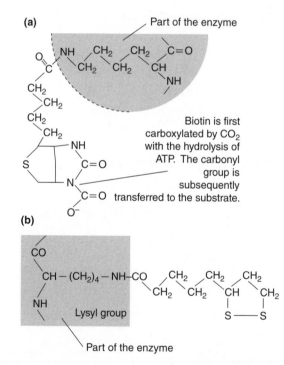

Biotin is first carboxylated by CO_2 with the hydrolysis of ATP. The carbonyl group is subsequently transferred to the substrate.

Figure 3.4 *Structure of two prosthetic groups* **(a)** *biotin* **(b)** *lipoate.* Biotin functions as a carboxyl group carrier, e.g. in acetyl-CoA carboxylase. Lipoate is presented in its oxidised form (-S-S-). It is a cofactor for pyruvate dehydrogenase and oxoglutarate dehydrogenase.

In assays of enzyme activities a cofactor, but not a prosthetic group, can be easily lost from the enzyme by dilution during extraction or purification, or removed by agents that will bind the cofactor. For these reasons, an excess of cofactor is routinely added to the assay medium (e.g. Mg^{2+} in kinase assays) for the measurement of enzyme activity.

role but only bind to the enzyme during catalysis are known as cofactors. Because the distinction between prosthetic group and cofactor depends only upon the strength of binding to the enzyme, it is not a very precise one and is of little theoretical significance.

Unfortunately, the term 'coenzyme' is sometimes confused with 'cofactor', but the difference is fundamentally important. Coenzymes do not function as part of the enzyme. They are, in fact, substrates (or products) in the reaction. The term coenzyme is usually used to describe compounds that link together different reactions or processes, so that they can transfer groups between biochemical pathways (Figure 3.5). To avoid the implication that 'coenzymes' have anything to do with the catalysis of the

An enzyme in a biochemical pathway may have two substrates, one of which is a cofactor. The other which is usually the substrate involved in the primary flux through the pathway is termed a pathway-substrate in this book.

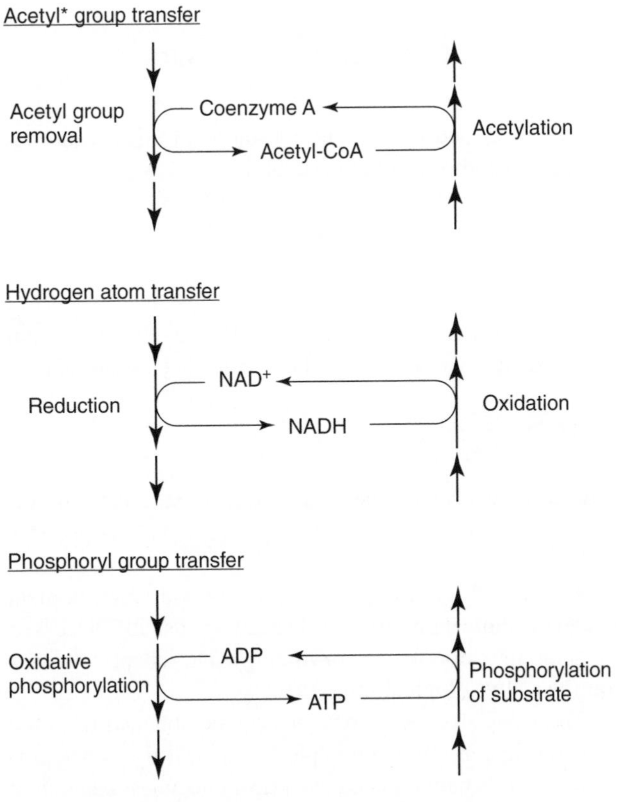

reaction, the term 'co-substrate' is introduced to provide a complementary term to 'pathway-substrate'.

Factors that change the activity of an enzyme

Many factors can change the catalytic activity of an enzyme. Those discussed in this book are the concentrations of substrates, protons (pH), temperature and inhibitors. The effects of these are discussed from a qualitative point of view, with more quantitative descriptions presented in Appendices 3.3, 3.4 and 3.5.

The effects of these factors and the means by which they are studied are usually described as *enzyme kinetics*. The basic information required to study enzyme kinetics is as follows.

Enzyme kinetics: the basics

- Conditions under which the catalytic activity is measured should ensure that the rate of the reaction is proportional to the concentration of the enzyme; that is, if the enzyme concentration is doubled, the catalytic activity must double.

- A satisfactory method for measuring the rate of the reaction catalysed by the enzyme is essential. The principles underlying some of the methods are described in Appendix 3.3, where methods for measuring the activities of two enzymes are given as examples.

- The rate of reaction that is measured is the initial rate, i.e. that which is measured over the first few seconds or minutes of the reaction, before a significant proportion of the substrate has been used or the product formed (Figure 3.6). The rate is known as the *enzyme activity*.

- The enzyme activity is usually reported in units of substrate consumed or product produced over a given period of time, at near-saturating concentration of substrate and at a given pH or temperature. For a pure enzyme, specific activity can be reported, for example, as μmol or mmol/min per mg enzyme-protein.

For an enzyme in an extract of a cell or tissue, enzyme activity is usually reported as μmol or mmol/min per mg protein or per mg DNA, in the extract, or per gram of fresh or dry weight of tissue from which the extract was

The most fundamental unit for reporting enzyme activity is its turnover number, the number of substrate molecules converted to product by one molecule of the enzyme in one minute. Its calculation, however, requires pure protein and knowledge of the enzyme's molecular mass.

Acetyl* group transfer

Acetyl group removal — Coenzyme A / Acetyl-CoA — Acetylation

Hydrogen atom transfer

Reduction — NAD^+ / NADH — Oxidation

Phosphoryl group transfer

Oxidative phosphorylation — ADP / ATP — Phosphorylation of substrate

Figure 3.5 *Group transfer by coenzymes.*

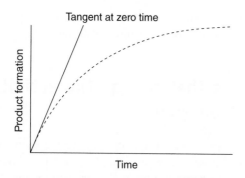

Figure 3.6 *The practical principle for measuring initial activity of an enzyme.* The catalytic rate must be linear with time: the precise activity can be obtained by plotting the amount of product formed against time, and drawing a tangent to zero time: it is known as the *initial activity* or simply 'enzyme activity'.

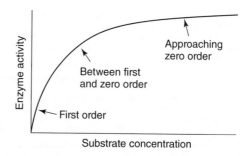

Figure 3.7 *First and zero order components of a hyperbolic curve of enzyme activity against substrate concentration.* For the first order, rate is directly proportional to concentration: at zero order, rate is independent or almost independent of substrate concentration.

made. For an enzyme in a biological fluid (e.g. blood), the activity is usually reported as μmol/min per litre.

Effect of substrate concentration on enzyme activity

The activity of an enzyme (v) varies according to the substrate concentration [S]. In most cases, the relationship is hyperbolic, that is:

- At very low substrate concentrations, the reaction rate is approximately first order (i.e. activity increases approximately linearly with increase in substrate concentration).

- At very high substrate concentrations, the rate of reaction approaches zero order (i.e. the increase in substrate concentration has very little effect on the rate of reaction).

- At intermediate substrate concentrations, the order is intermediate between zero and first order (Figure 3.7).

In this discussion, enzymes with a single substrate are considered but the same principles apply to enzymes with more than one substrate.

A hyperbolic curve is described by an equation of the form:

$$x = \frac{ay}{b+y}$$

In the case of an enzyme,

$$v = \frac{a[S]}{b+[S]}$$

A hyperbolic curve results when v is plotted against the concentration of S, where [S] indicates concentration of S and a and b are constants. Two questions arise:

1 What is the mechanism of catalysis that accounts for the hyperbolic relationship?

2 In relation to the mechanism, what are the constants a and b?

The answer to question 1 was provided by L. Michaelis and M.L. Menten in 1913. The mechanism is, in fact, very simple. The enzyme, E, binds the substrate S to produce an enzyme-substrate complex, which then breaks down to give rise to the product, P, and the free enzyme:

$$E + S \underset{k_2}{\overset{k_1}{\rightleftharpoons}} ES \overset{k_3}{\longrightarrow} E + P$$

where k_1, k_2 and k_3 are rate constants. From this mechanism, a hyperbolic equation is derived:

$$v = \frac{V_{max}[S]}{K_m + [S]}$$

where v is the enzyme activity, V_{max} is the maximum catalytic activity, and K_m is the Michaelis constant and is given by $\dfrac{k_2 + k_3}{k_1}$

Thus, the answer to question 2 is, $a = V_{max}$, and $b = K_m$. The derivation of the equation is provided in Appendix 3.4.

The values of both K_m and V_{max} are of both physiological and biochemical importance. The latter is usually discussed in detail in textbooks on enzymology. This chapter focuses on their physiological importance.

The values for these constants can be obtained from the hyperbolic curve when v is plotted against the concentration of S; this method is not very accurate unless calculated using a computer program but the most commonly used method is the double-reciprocal plot $\left(\dfrac{1}{v} \text{ vs. } \dfrac{1}{[S]}\right)$. Other methods for obtaining values are also described in Appendix 3.4.

Specificity

Enzymes exhibit a range of specificities. A few react with only one substrate. For example, the enzyme urease, which hydrolyses urea, only reacts with urea: this is known as absolute specificity. Others exhibit group specificity: for example the enzyme hexokinase phosphorylates several hexose sugars: glucose, fructose, mannose and 2-deoxyglucose. Stereospecificity, the property of enzymes to catalyse reactions with only one of a pair of stereoisomers, is a consequence of the three-dimensional nature of an active site and the fact that an enzyme interacts with its substrate at more than two points (Figure 3.1). For example, enzymes involved in the metabolism of unsaturated fatty acids are almost always specific for either the *cis* or the *trans* geometric isomer, and hexokinase will catalyse the phosphorylation of D-glucose, the naturally-occurring optical isomer, but not of L-glucose. Stereoisomerism is explained in Appendix 3.1.

Effect of hydrogen ion concentration (pH)

In living systems, the hydrogen ion concentration is usually very low, so that the negative logarithm is more frequently used to indicate the concentration. For an explanation of pH and its relationship to the hydrogen ion concentration, see Appendix 3.5.

The hydrogen ion concentration has an effect on enzyme activity because many of the amino acid groups in an enzyme bear ionisable groups. Changes in pH will alter the degree of ionisation of some of these groups and so affect the ionisation of the enzyme molecule as a whole, modifying enzyme activity in at least three ways:

- Extremes of pH (typically below pH 4.0 or above pH 10.0) change the degree of ionisation of so many groups that the three-dimensional structure is disrupted, resulting in irreversible loss of enzyme activity, which is known as denaturation.

> Practical use can be made of the denaturation by extremes of pH, since addition of excess acid or alkali to a solution containing an enzyme will immediately stop catalysis.

- Smaller changes of pH (within the range of approximately pH 4.5 to 9.5) affect a smaller number of ionisa-

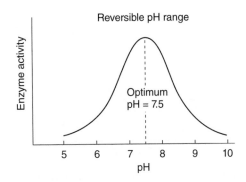

Figure 3.8 *Plot of enzyme activity against pH.*

ble groups. If these happen to be present in the active site, a marked change in enzyme activity can result, but the effect is reversible.

- Changes in pH may also cause changes in ionisation of the substrate and so affect catalytic activity by modifying the binding between substrate and enzyme.

For many enzymes, a plot of activity against pH gives a bell-shaped curve with a well-defined pH optimum (Figure 3.8). Most enzymes are adapted to function at the particular pH of their environment and, since the cytosolic pH is about 7.1, most intracellular enzymes have a pH optimum about 7.0. In contrast, the normal pH in the stomach is around 2.0, and most enzymes secreted by cells in the stomach have very low pH optima (e.g. pepsin, whose pH optimum is 2.0).

Effect of temperature

An increase in temperature increases the rate of chemical reactions. This is also the case for reactions that are catalysed by enzymes, except that above temperatures of 60–70 °C most of the bonds that are involved in maintenance of the three-dimensional structure are broken, the enzyme is denatured and the catalytic activity lost. The consequence of these opposing effects of an increase in temperature is that the graph of enzyme activity against temperature exhibits a maximum (Figure 3.9). Since thermal denaturation is a time-dependent process, the shape of the graph and the position of this 'optimal temperature' will depend upon the length of time the enzyme remains at the high temperature. All enzymes are moderately stable in vivo at temperatures that are normally experienced by the organism but the temperature at which denaturation occurs varies from enzyme to enzyme. Most are very rapidly denatured at 100 °C and show decreased activity after quite short exposures to temperatures above 60 °C. However, some bacteria live in hot springs in which the normal

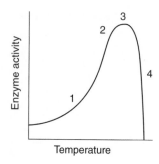

Figure 3.9 *Plot of enzyme activity against temperature.* The activity depends upon the time of incubation at a given temperature: 1 is low temperature (15–20 °C), 2 & 3 physiological temperature, 20–40 °C, and 4 >70 °C.

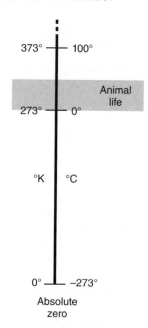

Figure 3.10 *Temperature scales: Celsius and Kelvin (absolute).* On the Celsius scale, absolute zero is −273 °C. The temperature at which water boils is, therefore, 373 °K. Most life on earth exists only between 277 and 293 °K.

temperature is greater than 90 °C: they are known as thermophilic bacteria. Enzymes in these bacteria remain active at such high temperatures, DNA polymerase is from such bacteria is purified for use in the polymerase chain reaction below).

It is not usually appreciated that the effect of an increase in temperature on the rate of a reaction is remarkable because the effect is concealed by the temperature scales that are usually used. The relevant temperature scale for assessing the effect of a temperature on a reaction rate is not the Celsius but the absolute or Kelvin scale (Figure 3.10). An increase from 4 to 20° Celsius is a fivefold increase on this scale yet, on the Kelvin scale, the increase is from 277 to 293°, only a 6% increase. Nonetheless, most life on earth exists within this temperature range. Of considerable clinical im-portance, even a very small decrease in the absolute temperature can have a profound effect on physiological processes (Box 3.2).

> When a gas is cooled by 1 °C, from 0 °C to −1 °C, it loses $\frac{1}{273}$ of its pressure. Since pressure is due to movement of the gas molecules, Lord Kelvin realised that a gas would have no pressure at −273 °C; that is, there would be no movement of the gas molecules: −273 °C is absolute zero. Since the rate of a reaction depends upon the movement and hence collision of molecules, the relevant temperature scale for chemical (and therefore biochemical) reactions is one whose zero is −273 °C. It is known as the Kelvin or absolute scale.

The reason for this remarkable effect of such a very small temperature change is that the relationship between the rate of a reaction and the temperature is exponential, as shown first by S. Arrhenius (Figure 3.11).

The Arrhenius equation is

$$k = Ae^{-\Delta G^{\ddagger}/RT}$$

where k is the rate constant for the reaction, A is a constant, R is the gas constant, T is the absolute temperature and $\Delta G^{\ddagger}$ is the free energy change for formation of the transition state.

Box 3.2 Temperature in heart surgery, in transfer of organs for transplantation surgery and in patients with trauma to the head

During surgery on the heart, it is cooled to about 32–33 °C by surrounding it with a 'slush' of ice or by infusion of the coronary arteries with a cold solution. Either treatment stops contractions of the heart. In addition, it lowers the metabolic rate so that glycolysis in cardiomyocytes is reduced and they are protected from damage that could be caused by lactic acid. For the same reason, an organ for transplantation is transferred between hospitals at a low temperature (4 °C). Some hospitals now use ice to cool the head in patients admitted with trauma to the head. These manipulations are successful despite the fact that a decrease in temperature from 37 °C to about 30 °C is only 2.3% on the Kelvin scale.

The equation demonstrates that the rate constant, which is the factor that determines the rate of the reaction, changes exponentially with temperature, so that even a small increase or decrease in the absolute temperature can change the rate of a biochemi-

> An exponential response is also important in regulation of enzyme activity. When a small percentage change in a regulator concentration is required to increase a flux by several- or many-fold, the mechanisms by which this is achieved may produce an exponential response. This is discussed below.

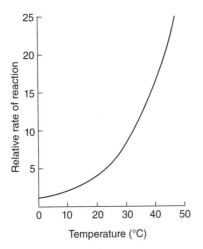

Figure 3.11 *The variation of the rate of a reaction with temperature.* The plot is based on the Arrhenius equation. Note that the reaction rate is not zero at 0 °C.

cal reaction quite markedly. The effect on a living organism can be quite dramatic: a cold-blooded animal may be totally inactive at 4 °C but fully active at 20 °C, yet the increase on the absolute scale is very small (Box 3.3).

Effect of inhibitors

A compound that reduces the activity of an enzyme is known as an inhibitor. Inhibitors are usually small molecules but some are peptides or proteins. For example, there are a number of proteolytic enzymes in the blood that have serine in their active site. If the activities of these enzymes are too high, they can cause problems. Consequently, inhibitor proteins, known as *serine proteinase inhibitors* (serpins), are present in blood: indeed, about 10% of all the plasma proteins are serpins (Box 3.4).

To help clarify the effect of inhibitors, it is useful to divide inhibition into four classes: reversible, allosteric, irreversible covalent, enzyme-catalysed covalent.

Reversible inhibition

Inhibition is termed reversible if, when the concentration of the inhibitor is reduced, the extent of inhibition is reduced (i.e. the dissociation constant for the binding of the inhibitor to the enzyme is not extremely high). Most inhibitors bind either to the enzyme or the enzyme substrate complex. In both cases, the binding is an equilibrium process, as follows:

$$E + I \underset{k_2}{\overset{k_1}{\rightleftharpoons}} EI$$

Box 3.3 Hibernation, migration or insulation in warm-blooded animals. Manipulations to overcome the effects of low temperatures

Most organisms are exposed to fluctuations in their environmental temperature which, according to the Arrhenius equation, will lead to considerable changes in the rates of their biochemical reactions and, therefore, considerable changes in their physiology. To compound the problem, not all rates would be similarly affected, so that precise coordination and control of biochemical processes would be difficult to achieve. Mammals and birds have overcome the problem by maintaining a constant internal temperature (homoiothermy) and one that is well above most normal environmental temperatures. The benefits are considerable but so is the cost; a greatly increased metabolic rate which needs to increase further as the temperature falls and the animal loses more heat. Hence, a fairly constant source of food, or a large store of fuel, is essential to survive a winter. However, if sufficient food is not available or fuel is not stored, when the environment temperature falls to low values, survival is only possible by changes in behaviour. These are (i) migration to a warmer climate; (ii) hibernation to reduce metabolic rate and hence the requirement for food; or (iii) increase in insulation. And biochemistry plays a role in the first two (for migration, see Chapter 7; for hibernation, see Chapter 11). Nonetheless, if some food is available, insulation improves the chance of survival. In addition, mechanisms exist for increasing heat formation (Chapter 9) so that by a combination of insulation and thermogenesis, survival in a cold environment or during the cold season is possible. All of these changes are necessary, it can be argued, because of the Arrhenius equation.

or

$$ES + I \underset{k_2}{\overset{k_1}{\rightleftharpoons}} ESI$$

where I is the inhibitor.

These equilibrium-binding relationships give rise to four different kinetic responses: competitive inhibition, uncompetitive inhibition, non-competitive inhibition, mixed inhibition. Details of the kinetics of these types of inhibition and how dissociation constants for the reactions can be measured are provided in Appendix 3.6.

Allosteric inhibition

Reversible inhibition in which the inhibitor binds to a specific site on the enzyme that is remote to the active site, i.e. a distinct binding site, is known as allosteric inhibition. It is of considerable physiological importance and is considered in detail below.

Irreversible covalent inhibition

Irreversible inhibition should be more correctly termed 'difficult-to-reverse inhibition' because any inhibition due to the binding of inhibitor molecules can be reversed theoretically by lowering the concentration of inhibitor to zero. The exceptions are some protease inhibitors (e.g. the serpins) since, once the inhibitor-enzyme complex is formed, it is destroyed by phagocytosis. It is, therefore, a specific form of irreversible inhibition (Box 3.4).

Some of the groups in the active centre of enzymes have greater reactivity than would be expected from the behaviour of the group when it is not in an active centre (this is known as 'enhanced reactivity') and is a consequence of the particular environment within the active centre. These groups, therefore, are particularly vulnerable to agents that will react with them to produce a strong bond, so that, once bound to the enzyme, the agent does not dissociate (i.e. the value of the dissociation constant approaches zero). The result is irreversible inhibition. In the cell, the concentration of most enzymes is very low (10^{-7} mol/L or lower), so that low concentrations of these agents can cause severe inhibition and they can be very toxic. The only way such inhibition can be reversed is by providing a compound to which the inhibitor binds more strongly than with the group in the active centre or by degradation of the enzyme-inhibitor complex, followed by its replacement by newly synthesised enzyme protein.

There are several such toxic agents that cause considerable medical, public and political concern. Two examples are discussed here: the heavy metal ions (e.g. lead, mercury, copper, cadmium) and the fluorophosphonates. Heavy metal ions readily form complexes with organic compounds which are lipid soluble so that they readily enter cells, where the ions bind to amino acid groups in the active site of enzymes. These two types of inhibitors are discussed in Boxes 3.5 and 3.6. There is also concern that some chemicals in the environment, (e.g. those found in industrial effluents, rubbish tips and agricultural sprays), although present at very low levels, can react with enhanced reactivity groups in enzymes. Consequently, only minute amounts concentrations are effective inhibitors and therefore can be toxic. It is suggested that they are responsible for some non-specific or even specific diseases (e.g. breast tumours).

> An irreversible enzyme inhibitor of clinical value is aspirin, which inhibits cyclooxygenase and therefore prostaglandin formation (Chapter 11).

Enzyme-catalysed covalent inhibition

Although not normally considered to be a form of enzyme inhibition, the enzyme-catalysed interconversion of active

---

Box 3.4 Serpins and pulmonary emphysema: chronic obstructive pulmonary disease (COPD)

Some irreversible covalent modifications of enzymes are physiologically important. One type of enzyme is inhibited by binding to another protein to form a complex (enzyme-inhibitor complex) that is taken up by liver cells or macrophages and degraded. Most proteolytic enzymes in the blood contain serine in the active centre and are inhibited by serine proteinase inhibitors (serpins). There are more than 40 serpins in blood and they play an important role in inhibition of some powerful proteolytic enzymes, e.g. those involved in blood clotting (see Chapter 17). Not surprisingly, they can be involved in pathology. Two examples are given.

One of the best-known examples of a serpin is α_1-antiproteinase (antitrypsin), which inhibits an enzyme known as elastase. The latter is a proteolytic enzyme that is secreted by macrophages: it is one of the many weapons used by these cells to kill or damage invading pathogens. The macrophages are particularly important in the lung, since it is a relatively easy point of entry for pathogens. However, once the pathogen has been killed, elastase activity must be inhibited and this is achieved by release of α_1-antitrypsin. Unfortunately, the free radicals in tobacco smoke cause chronic damage to the cells in the lung, one effect of which is to decrease their ability to secrete α_1-antitrypsin so that elastase is not inhibited. This results in further damage to the lung. The process is, therefore, a vicious circle, which eventually results in emphysema, a chronic inflammation of the airways. There is excessive growth of mucus-secreting cells and poor removal of mucus, which causes a permanent cough. Poor lung function results in wheezing, breathlessness and tightness in the chest.

Cytokines are peptides that are produced and secreted by cells of the immune system. They organise the immune response to invasion by a pathogen by communicating between the different cells. They are synthesised in the immune cells as precursor proteins (pro-proteins) from which a peptide is removed by a proteolytic enzyme to produce the active cytokine, prior to secretion. This enzyme is a serine protease. Perhaps surprisingly, some viruses are capable of synthesising serpins which inhibit this enzyme in the immune cells, so that secretion does not occur and communication and integration of the immune response to the viral infection is lost. This is one of many biochemical mechanisms by which pathogens can reduce or overcome the defence mechanisms of the host (Chapter 17).

Box 3.5 Heavy metals and inhibition of enzymes

There are a number of metals that are irreversible inhibitors of enzymes. Those that are found in increasing quantities in the environment and hence are causing concern include lead, mercury, cadmium and copper. Enzymes that possess the amino acid cysteine in their active site are most vulnerable, since the sulphydryl group (–SH) in this site readily reacts with the metal ions (Appendix 3.7).

Metal ions are particularly toxic in a complex with organic molecules, because the latter increase their lipid solubility so that they readily cross the plasma membrane to enter cells. They also cross the blood-brain barrier so that they very effectively gain access to nerve cells in the brain.

Organic heavy metal compounds are produced by the chemical industry (e.g. tetraethyl lead was produced and added to petrol as an 'anti knocking agent' for more than 30 years in the UK, until unleaded petrol became available in the mid-1980s). They can also be synthesised in some living organisms (e.g. bacteria).

Inorganic mercury compounds were discharged into the sea at Minamata Bay in Japan, from where they were taken up by bacteria and converted into methylmercury compounds. These bacteria entered the food chain to be taken up eventually by fish. Since fish are an important source of food in Japan, in a very short time many people died and many more were permanently affected by mercury poisoning.

The treatment of metal poisoning is to administer a compound that binds the metal ion more strongly than does the group in the active centre of the enzyme. These compounds are known as chelating agents. For lead, the compound ethylenediaminetetraacetic acid (EDTA) is used. For mercury, dimercaptopropanol (dimercaprol) is used.

There is now particular concern about the heavy metals that are present in most electronic machines. The constant upgrading of computers means that some are discarded without care so that the metal ions within them are gradually escaping into the environment. It is estimated that, in the USA alone, 60 million new computers are purchased each year and worldwide the number is about 130 million. Many of these are replaced within a few years, and disposal of old computers leads to large amounts of electronic waste (known as 'e-waste'). It is estimated that 5% of all solid waste in the USA is e-waste, accounting for about 70% of all the heavy metals in waste dumps. The problem of e-waste and human health is discussed in a book entitled *Challenging the Chip*, eds Smith, Sonnenfeld, *et al.* (2006).

Copper ions bind to and inhibit many enzymes. Of more importance, perhaps, is that in free solution or when bound to proteins, copper ions catalyse the Fenton reaction which produces the highly dangerous hydroxyl radical, OH^- (Appendix 9.6).

Box 3.6 Fluorophosphonates: nerve (war) gases and insecticides

The story began in the 1930s when the compounds known as dialkylphosphorofluoridates (now known as dialkylfluorophosphonates) were synthesised and their toxicity was recognised. Only much later was it realised that these organophosphorus compounds transferred dialkylphosphoryl groups to enzymes that possessed serine groups in their active site. One such enzyme, which is present at low concentrations and in specific sites in the body, is acetylcholinesterase (see page 40). This enzyme is localised in nervous tissue at synapses and at neuromuscular junctions. When a nervous impulse reaches a muscle, acetylcholine is released, which stimulates the muscle to contract. This effect is stopped by the activity of acetylcholinesterase (Chapter 13).

Symptoms of fluorophosphonate poisoning are difficulty in focusing the eye for near vision and difficulty in breathing, since the poison causes sustained contraction of the smooth muscles of the respiratory tract, which causes death by asphyxiation.

During the Second World War, even before the mechanism of toxicity had been established, related compounds including sarin and tabun were developed for use as so-called nerve gases. These compounds possess exceptionally high mammalian toxicity and are still stored as agents of chemical warfare. A plant disguised as a soap-making factory was opened near the German-Polish border in 1942 for production of tabun. By the end of the Second World War, 12 000 tons had been manufactured, although it was never used. As recently as May 1995, sarin was released into the Tokyo underground by terrorists: 12 people were killed and 5000 injured (Sneader, 1985).

Insects are very sensitive to fluorophosphonates, so that the compound parathion was synthesised and used as an insecticide soon after the Second World War. However, it entered the food chain and eventually found its way into mammals and caused death. An important breakthrough occurred with the synthesis of malathion, an insecticide which has high toxicity to insects, where it is converted to malaoxon, a potent acetylcholinesterase inhibitor. However, malathion is much less toxic to mammals, since it is readily detoxified (Appendix 3.8).

Treatment of fluorophosphonate poisoning involves the use of oximes, such as pralidoxime, which bind the phosphoryl group very strongly so that it is removed from the enzyme.

and inactive forms of an enzyme is a very widespread mechanism of enzyme activity regulation. There are three important points:

- Each covalent modification is achieved in an enzyme-catalysed reaction.

- It is termed reversible, since the covalent modification that is catalysed by one enzyme is reversed by another enzyme. However, each individual reaction is irreversible.

- In one reaction, the target enzyme is activated whereas, in the other, it is inactivated.

The most common form of such modifications is phosphorylation and dephosphorylation of the hydroxyl group of a serine, threonine or tyrosine amino acid in the protein. The reactions are catalysed by kinases (protein kinases) or phosphatases (protein phosphatases), respectively.

$$\text{Enzyme} + \text{ATP} \xrightarrow{\text{kinase}} \text{Enzyme-P} + \text{ADP}$$
$$\text{(inactive)} \qquad\qquad\qquad \text{(active)}$$

$$\text{Enzyme-P} + \text{H}_2\text{O} \xrightarrow{\text{phosphatase}} \text{Enzyme} + \text{P}_i$$
$$\text{(active)} \qquad\qquad\qquad \text{(inactive)}$$

A well-known example, indeed the first enzyme that was shown to be regulated by the phosphorylation/dephosphorylation mechanism, is glycogen phosphorylase, which catalyses the breakdown of glycogen (Box 3.7).

$$\text{glycogen}_{(n)} + \text{P}_i \rightarrow \text{glucose 1-phosphate} + \text{glycogen}_{(n-1)}$$

where $\text{glycogen}_{(n-1)}$ represents glycogen containing one less glucose molecule, and P_i represents the phosphate ion.

Since the change in rate of glycogen degradation, catalysed by glycogen phosphorylase, depends on a balance between the two activities, which constitute a cycle, the relationship is also known as an *interconversion cycle*. By convention, the active form is the *a* form and the inactive form is the *b* form (Figure 3.12).

Allosteric inhibition

The development of the concept of allosteric inhibition of enzymes began in the early 1950s but, surprisingly, not with studies on enzymes. It was discovered that addition of an amino acid to a culture of bacteria (*Escherichia coli*)

Box 3.7 Discovery of reversible phosphorylation in the regulation of enzyme activity

In the 1940s Carl and Gertrude Cori isolated and purified an active form (phosphorylase *a*) and an inactive form (phosphorylase *b*) of an enzyme from muscle. Phosphorylase *b* is activated by AMP (see page 64). In 1955, Fischer & Krebs found an enzyme that catalysed the conversion of phosphorylase *b* to phosphorylase *a*, together with hydrolysis of ATP to ADP. Thus it appeared to bring about phosphorylation of the enzyme. The enzyme was termed phosphorylase *b* kinase, was partially purified and the interconversion was established as

$$\text{phosphorylase } b + \text{ATP} \rightarrow \text{phosphorylase } a + \text{ADP}$$

This suggested that the inactivating enzyme was a phosphatase and a protein phosphatase was purified, which catalysed the following reaction:

$$\text{phosphorylase } a + \text{H}_2\text{O} \rightarrow \text{phosphorylase } b + \text{P}_i$$

The existence of two forms of phosphorylase with different catalytic activities, which were capable of being enzymatically interconverted, suggested that the two forms might be involved in regulation of glycogenolysis in muscle tissue. However, in the early studies, whenever phosphorylase was assayed in extracts of muscle it was always found to be in the *a* form. This was the case even in extracts prepared from resting muscle, in which it is known that glycogen is not degraded. At this stage, Fischer & Krebs realised that, if the distribution of the two forms of phosphorylase in muscle was to be accurately assessed, it was vital to inhibit the interconverting enzymes during extraction of the muscle tissue. Knowledge of the properties of these enzymes indicated suitable inhibitors: the phosphatase was inhibited by fluoride ions and the kinase was inhibited by EDTA, which chelates the essential Mg^{2+} ions. Under these conditions, extracts of resting muscles contained almost no phosphorylase *a* but this was increased by muscle activity. The theory developed that phosphorylase *b* was the inactive species of the enzyme and that, whenever glycogenolysis was necessary, phosphorylase *b* had to be converted to phosphorylase *a* via the protein kinase. This work by Krebs & Fischer was the first to show that phosphorylation of an enzyme by a protein kinase and dephosphorylation by a protein phosphatase was involved in changing the activity of an enzyme involved in regulation of a key biochemical process. For the discovery of the reversible phosphorylation of an enzyme, Edwin Krebs and Edmond Fischer were awarded the Nobel Prize for Chemistry in 1992. It is now known that approximately 1000 different enzymes or other proteins are regulated by this mechanism, many of which are involved in signal transduction pathways.

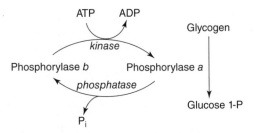

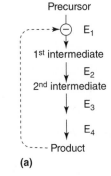

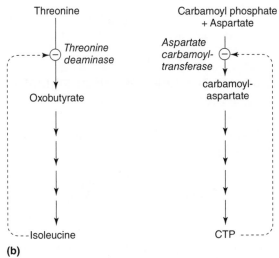

Figure 3.12 *The regulation of phosphorylase activity by reversible phosphorylation.* A reversible phosphorylation process is also known as an interconversion cycle: the latter term is preferred in this text, since the individual reactions *must* be irreversible, which can be confusing if the term 'reversible' is used to describe the overall process. In resting muscle, almost all phosphorylase is in the *b* form.

which were growing solely on glucose and ammonia, inhibited the synthesis of that particular amino acid. All other amino acids were synthesised normally. For example, addition of the amino acid isoleucine inhibited the synthesis of isoleucine but not that of any other amino acid.

It was subsequently discovered that the first enzyme in the pathway for isoleucine synthesis, which is threonine deaminase, was inhibited by isoleucine in an extract of *E. coli*. No other amino acid caused inhibition of the enzyme. Threonine deaminase is, in fact, the rate-limiting enzyme in the pathway for isoleucine synthesis, so that this was interpreted as a feedback control mechanism (Figure 3.13(a)). Similarly it was shown that the first enzyme in the pathway for cytidine triphosphate synthesis, which is aspartate transcarbamoylase, was inhibited by cytidine triphosphate (Figure 3.13(b)). Since the chemical structures of isoleucine and threonine, or cytidine triphosphate and aspartate, are completely different, the question arose, how does isoleucine or cytidine triphosphate inhibit its respective enzyme? The answer was provided in 1963, by Monod, Changeux & Jacob.

Figure 3.13 (a) *Feedback control of a hypothetical pathway.* **(b)** *Feedback control of threonine deaminase in the isoleucine synthetic pathway and of aspartate carbamoyltransferase in the cytidine triphosphate synthetic pathway in the bacterium* E. coli.

The 1963 allosteric model: separate catalytic and regulator sites

To provide a mechanism for the feedback inhibition of these enzymes, the allosteric model was put forward in 1963. It was proposed that the enzyme that regulates the flux through a pathway has two distinct binding sites, the active site and a separate site to which the regulator binds. This was termed the allosteric site. The word allosteric means 'different shape', which in the context of this mechanism means a different shape from the substrate. The theory further proposed that when the regulator binds to the allosteric site, it causes a conformational change in

the enzyme molecule that changes the three-dimensional structure of the active site, so that the enzyme activity is decreased. This conformational change can be considered as the means of transmission of information between the two sites (Figure 3.14). It should be noted that allosteric effects can also be positive; that is, binding of an allosteric regulator can increase rather than decrease the activity of an enzyme. That such a change could occur in an enzyme and produce a precise modification of the enzyme activity was a completely novel concept in 1963. Prior to this, it was unclear how small molecules could regulate the activity of an enzyme.

Not only does allosteric regulation have a profound influence on our understanding of biochemical regulation but the concept has considerable biological importance. These two sites within the enzyme, catalytic and allosteric, can evolve independently to provide the most efficient systems for both catalysis and regulation.

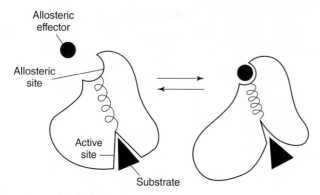

Figure 3.14 *Diagram illustrating regulation of enzyme activity by an allosteric regulator.* Note the representation of the conformational change.

The allosteric model of 1965: sigmoid responses and cooperativity

The allosteric model of 1963 was extended in 1965 by Monod, Wyman & Changeux. Unfortunately, this 1965 paper has sometimes caused confusion between the two models, possibly because the term 'allosteric' occurs in both models. In fact, the two models are totally different and they do two quite separate things. The 1963 paper put forward a simple but fundamental concept that an enzyme activity could be regulated by a factor that is structurally unrelated to either the substrate or the product and that binds at a specific regulator site on the enzyme molecule. The 1965 paper put forward a mechanism to account for the fact that the activity of many enzymes, which are involved in regulation, responds to an increase in substrate concentration in a sigmoid, rather than in a hyperbolic manner (Figure 3.15(a)) (Appendix 3.9). These enzymes are usually multi-subunit so that the enzyme binds more than one molecule of substrate and it is the interaction of these identical binding sites that provides the basis for the sigmoid response. The effect has been termed *cooperativity*. A sigmoid res-ponse can also occur in res-ponse to an allosteric inhibitor when it binds to more than one site on the enzyme (another cooperative effect) (Figure 3.15(a)). The physiological importance of this effect is that the sigmoid response increases the sensitivity of the change in enzyme activity to the change in concentration of the substrate, or to the change in concentration of the allosteric regulator, over a particular range of concentrations (i.e. the physiological range) of either substrate or regulator (Figure 3.15(b)) (also see below).

In fact, sigmoid binding of a small molecule to a protein was discovered long before that of enzymes. Bohr showed, in 1904, that a plot of the concentration of oxygen (i.e. partial pressure) against the extent of saturation of haemoglobin with oxygen was sigmoid.

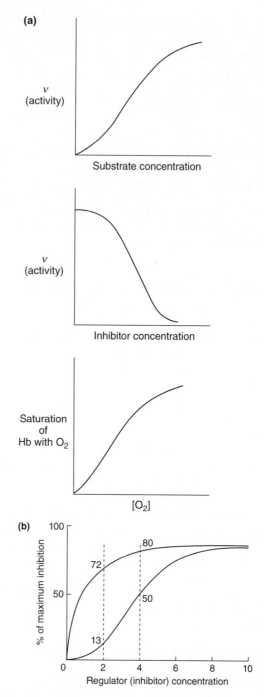

Figure 3.15 (a) *Three examples of sigmoid responses.* Enzyme activity against substrate concentration; enzyme activity against inhibitor concentration; saturation of haemoglobin (Hb) with oxygen against the concentration (i.e. the partial pressure) of oxygen. **(b)** *Hyperbolic and sigmoid responses of the percentage inhibition against changes in the inhibitor concentration.* Note the greater inhibitory effect provided by the sigmoid response to the same changes in the inhibitor concentration (i.e. a change that normally occurs in the living cell). The numbers represent percentage inhibition of enzyme activity caused by a doubling of the concentration of inhibitor.

The physiological significance of K_m and V_{max} values

The methods for measuring the value of these constants are described in Appendix 3.4. They are important to biochemists and enzymologists interested in kinetics and mechanisms of enzyme catalysis. It is not, however, always appreciated that they are also important in physiology and in the medical sciences.

Significance of the values of the maximum catalytic activity (V_{max})

The maximum catalytic activity (V_{max}) is measured at saturating concentrations of substrate and in the presence of activators and in the absence of inhibitors of the enzyme. It is the values of V_{max} of the enzymes that are obtained in extracts of cells, tissues or organs that are of importance in physiology: for example to establish or confirm (i) the role of an enzyme in a particular pathway, in vivo; (ii) the role of a particular pathway in a cell or tissue; or (iii) the maximum flux through a pathway in vivo.

In some cases it is necessary to isolate cell organelles before V_{max} is measured (Figure 3.16). A number of cells, a whole tissue or a sample of a tissue is taken, the cells are broken and the organelle in which the enzyme is present is separated, usually by centrifugation. The organelle is then extracted to provide a medium which can be used to measure the enzyme activity by a specific method.

Five examples of the physiological use of the V_{max} are given.

(i) The precursor for lactic acid formation in human muscle

The maximum rate of lactic acid formation in quadriceps muscle of an adult human during sprinting is about 50 units. The precursor for this process is either glucose or glycogen. Which precursor is used? The enzyme committing glucose to this process is hexokinase; the enzyme committing glycogen to this process is phosphorylase. The V_{max} for hexokinase, in an extract of the quadriceps muscle of an adult human, is about 2 units, whereas that for phosphorylase is about 50 units. It is, therefore, clear that glycogen, not glucose, is the fuel to generate ATP to provide the energy for sprinting (Figure 3.17). (For further discussion, see Chapter 13.)

> Units are μmol/min per g of fresh muscle at 37 °C.

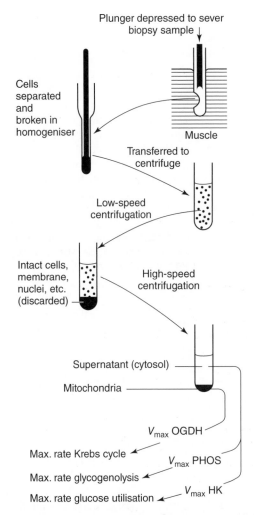

Figure 3.16 *Procedures for obtaining a sample of muscle, preparation of an extract of the sample and preparation of two compartments within the cell, for measurement of enzyme activity. A sample of muscle (or other tissue) is obtained by a biopsy needle. The fibres (cells) in the biopsy sample are damaged or broken (i.e. extracted) in an homogeniser (or other device), and then centrifuged to remove cellular 'debris' (e.g. unbroken cells, connective tissue) followed by high-speed centrifugation to prepare cytosol and mitochondria. Then the maximum activity of specific enzymes can be measured in these preparations, e.g. hexokinase (HK) or glycogen phosphorylase (Phos) in the cytosol, and oxoglutarate dehydrogenase (OGDH) in the mitochondria. These measurements indicate the maximum possible rates of glucose utilisation, glycogenolysis or Krebs cycle, in the tissue, respectively.*

(ii) Enzyme activity assessment of maximum flux through the Krebs cycle

The flux through the Krebs cycle can be estimated from the oxygen consumption of a cell or tissue. This has been done in a single human muscle during maximum physical

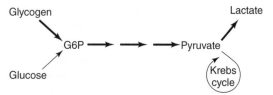

Figure 3.17 *Different capacities of glycolysis from glycogen or glucose in muscle.* In the pathway for the conversion of glycogen to pyruvate and then to lactate in skeletal muscle, the limiting (flux-generating) enzyme activity is phosphorylase. The large capacity of the pathway is indicated by the thick line: the low capacity from glucose, due to the low activity of hexokinase, is indicated by the thin line. However, in anaerobic muscle during exercise the pyruvate produced from glucose, via hexokinase, is not normally converted to lactate but enters the mitochondria for complete oxidation, in the Krebs cycle. At rest, some glucose is converted to lactate but at a low rate (Chapter 6).

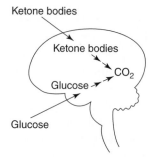

Figure 3.18 *Oxidation of glucose and ketone bodies by the brain.* Glucose is the sole fuel used by the brain, except in prolonged starvation in adults or relatively short-term starvation in children. In both cases, ketone bodies plus glucose are used.

Oxidation of one molecule of glucose requires six molecules of oxygen: two are required for conversion of glucose to acetyl-CoA (oxidation of four molecules of NADH) two are produced in glycolysis and two in conversion of pyruvate to acetyl-CoA) and four required for the two turns of the Krebs cycle. Hence one turn of the cycle consumes one-third of the total oxygen consumption of a cell, tissue or organ. i.e. in conditions of maximum rate of glucose/glycogen oxidation, one-third of oxygen consumption by the active muscle represents the maximum flux through one turn of the Krebs cycle (i.e. the maximum capacity of the cycle). (See also Chapter 9)

activity. The flux is similar to that calculated from the activity, in the extract of mitochondria of oxoglutarate dehydrogenase. This suggests that this activity can be used to estimate the maximum flux, through the Krebs cycle, in cells or tissues where oxygen consumption is impossible to measure. (Chapter 9).

(iii) Fuels for immune cells

For many years it was assumed that glucose was the only fuel used by cells of the immune system (lymphocytes, macrophages, neutrophils). However, studies in the 1980s on the activity of enzymes in these cells showed that the V_{max} of the enzyme glutaminase was similar to, if not higher than, that of hexokinase. As glutaminase is the first enzyme in the pathway for utilising glutamine, these results suggested that glutamine, as well as glucose, could be an important fuel for immune cells. This was confirmed from measurement of glucose and glutamine utilisation by the cells in culture or during incubation for a short period (e.g. one or two hours). It was, therefore, the V_{max} of an enzyme that first established that immune cells could use glutamine as well as glucose as a major fuel. This finding was and still is of considerable clinical importance. For example, glutamine is present in many commercial feeds for patients in intensive care units in hospitals to reduce the risk of infection (Chapters 17 and 18).

(iv) Ketone bodies and the brain

When experiments in the 1960s, using arteriovenous difference concentration across the human brain demonstrated that the brain could use the fat fuel, ketone bodies (acetoacetate and hydroxybutyrate) in addition to glucose, many physiologists and clinicians were sceptical. It was current dogma that the brain could use only glucose as a fuel. When it was shown that the V_{max} for the enzymes that are involved in the oxidation of ketone bodies were very high in samples of human brain taken at post mortem, the scepticism disappeared (Figure 3.18). Indeed, ketone bodies are now known to be quantitatively the most important fuel for the brain during prolonged starvation in adults and are particularly important during even short-term starvation in children (Chapter 16).

(v) Source of ammonia for buffering urine

During conditions of acidosis, hydrogen ions are excreted (secreted) by the kidney to maintain the normal pH of the blood. These ions are buffered in urine by ammonia, in the reaction

$$NH_3 + H^+ \rightarrow NH_4{}^+$$

and ammonium chloride is excreted

$$NH_4{}^+ + Cl^- \rightarrow NH_4Cl$$

For some years, the source of the ammonia was not known. The kidney does not take up ammonia from the blood, so it had to be produced within the kidney. The V_{max} values of two enzymes catalysing reactions that produce ammonia were shown to increase markedly in the kidney during acidosis, demonstrating their significance. The two enzymes are glutaminase and glutamate dehydrogenase and they function in sequence:

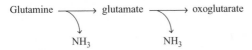

The increases in the V_{max} values are due to increased concentrations of the enzymes, as a result of acidosis increasing expression of the genes encoding these two enzymes (Chapter 10).

Significance of the values of the Michaelis constant (K_m)

It is necessary to know the K_m value of an enzyme in order to determine its maximal activity (V_{max}) which must be measured at saturating substrate concentrations. In practice this means that the substrate concentration must be at least ten times the K_m value (Appendix 3.3).

Knowledge of the K_m value also provides information about the role of enzymes in metabolic pathways. Two examples are given.

Identifying flux-generating enzymes

In a biochemical pathway there will be one reaction that is non-equilibrium and for which the enzyme approaches saturation with its substrate (pathway substrate). That is, the enzyme catalyses a zero order process (Figure 3.7).

In practical terms, that is the substrate concentration must be about or at least 10 times higher than the value of the K_m. This reaction is called the flux-generating reaction (Appendix 3.10) (Figure 3.19).

Glucose-phosphorylating isoenzymes

There are two different glucose-phosphorylating enzymes, hexokinase and glucokinase, which catalyse the reaction:

$$glucose + ATP \rightarrow glucose\ 6\text{-phosphate} + ADP$$

This reaction is important since it is the first glucose-utilising reaction in a cell.

These are isoenzymes, so that their properties are different, which is physiologically very important: (i) the K_m of hexokinase for glucose is 0.1 mmol/L, that for glucokinase is 100-fold higher (about 10 mmol/L) (Figure 3.20); (ii) hexokinase phosphorylates not only glucose but also fructose and mannose, whereas glucokinase is specific for glucose; (iii) hexokinase activity is inhibited by glucose 6-phosphate but glucokinase is not. These isoenzymes are present both in liver cells (hepatocytes), and in insulin-secreting cells (β-cells) in the pancreas, but it is

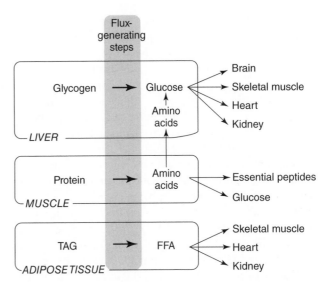

Figure 3.19 *Example of flux-generating reactions in carbohydrate, protein and fat metabolism.* The flux-generating reaction for provision of glucose for the body in conditions, such as starvation, is glycogen phosphorylase in the liver (Chapter 6). The flux-generating step for provision of amino acids for the formation of glucose in the liver (gluconeogenesis) during starvation, and other conditions (e.g. trauma) is the hydrolysis of muscle protein (Chapter 8). The flux-generating reaction for provision of long-chain fatty acids for oxidation during starvation and other conditions (e.g. physical activity, trauma) is hydrolysis of triacylglycerol (TAG) in adipose tissue (Chapter 7). FFA – long-chain fatty acids.

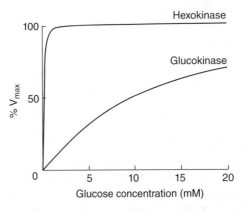

Figure 3.20 *A plot of the percentage of maximal activity of glucokinase and hexokinase against the glucose concentration.* Hexokinase is present in most if not all cells. Glucokinase is only present in hepatocytes and β-cells of the Islets of Langerhans in the pancreas (known as the endocrine pancreas). The enzyme is physiologically very important in both tissues; e.g. a low activity in the β-cells can be a cause of one type of diabetes mellitus.

glucokinase that is physiologically important in both. The transport of glucose across the plasma membrane in these cells is very rapid, that is the glucose transporter is very active, so that the process is near-equilibrium and the intracellular glucose concentration in these cells is identical to the concentration in the plasma. Although the concentration of glucose in the plasma in the peripheral circulation remains relatively constant, the liver and the pancreas receive blood from the hepatic portal vein, which drains the intestine, so that the concentration of glucose in this blood changes from about 4 mmol/L in short-term starvation (e.g. after the overnight fast), to 20 mmol/L or higher after a meal. Similar changes, therefore, occur within the hepatocytes and the β-cells. The K_m of hexokinase is such that its activity will not change over this range of glucose concentrations (i.e. even at 4 mmol/L it is already nearly saturated). In contrast, the activity of glucokinase increases proportionally with the increase in the plasma glucose concentration. These changes in activity of glucokinase have important and different roles in liver and β-cells.

Glucose phosphorylation in the liver

As the plasma glucose concentration increases after a meal, due to absorption from the intestine, the activity of liver glucokinase results in a higher rate of conversion of glucose to glucose 6-phosphate, which is subsequently converted to glycogen. This synthetic process is, in fact, stimulated by the increases in the glucose and glucose 6-phosphate concentrations in the liver cell, since they increase the activity of glycogen synthase (Chapter 6). Consequently, the enzyme glucokinase ensures that, after a meal, some of the increased amount of glucose in the plasma is converted to glycogen in the liver. This not only guarantees an increase in the store of glycogen in the liver but also restricts the increase in the plasma glucose level after the meal (Figure 3.21). In contrast, the role of hexokinase in the liver is to phosphorylate sugars other than glucose (e.g. mannose, fructose) and it is not involved in glycogen synthesis.

Glucose and the secretion of insulin by the β-cells

The role of the β-cells in the Islets of Langerhans is to secrete insulin in response to an increase in the glucose concentration in the plasma. It does this by assessing the increase in plasma glucose concentration: the higher the plasma glucose, the higher is the intracellular level of glucose in the β-cell, the greater is the rate of insulin secretion and hence the higher the level of insulin in the blood. It is the activity of glucokinase that senses the increase in the glucose concentration within the β-cell and gives rise to the metabolic changes in the cell that result in secretion of insulin (Figure 3.22).

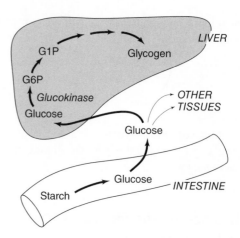

Figure 3.21 *The conversion of glucose to glycogen and the role of glucokinase in the liver*. Starch is hydrolysed to glucose in the intestine and the glucose is absorbed into the blood from where most of it can be taken up by the liver and converted to glycogen. It is glucokinase activity that responds to the increase in the plasma glucose level and plays a key role in conversion of some of this increased amount of glucose that is absorbed into the blood to glycogen. The broader arrows indicate the increased flux of glucose to glycogen in the liver.

Significance of glucokinase in regulation of the plasma glucose level

The primary role of glucokinase in liver cells and β-cells of the Islets of Langerhans is to restrict the increase in the peripheral plasma glucose level after a meal and also to restrict the decrease in the plasma level during short periods of starvation. In summary, the properties of glucokinase and its presence in the liver and in the β-cell play a key role in regulation of the plasma glucose concentration (Figure 3.23).

This is clinically important because a failure to control the blood glucose level adequately can lead to diabetes mellitus. Therefore, it can be argued that glucokinase is essential to prevent the development of diabetes in normal humans. Indeed, a deficiency of glucokinase in the β-cell gives rise to one type of diabetes, known as *maturity onset diabetes in the young* (MODY).

Enzymes as tools

Enzymes have been used as tools in many fields.

- For example, measurement of the concentrations of biochemical compounds that are of physiological or clinical value.

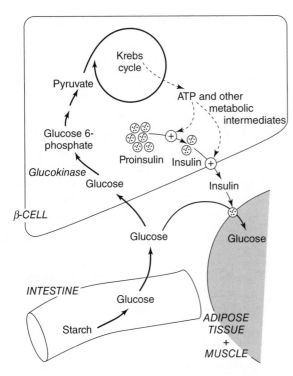

Figure 3.22 *Role of glucokinase in the control of insulin secreted by the β-cells in the endocrine pancreas.* Starch is hydrolysed to glucose in the intestine and the glucose is absorbed into the blood which increases the blood glucose level. The glucose level in the β-cells rises, which increases glucokinase activity. This stimulates the flux through glycolysis which increases the intra-cellular levels of ATP, intermediates of the Krebs cycle and some amino acids. These changes increase the conversion of proinsulin to insulin and the secretion of insulin. This increases the plasma level of insulin, which stimulates glucose uptake by muscle and helps to maintain the normal blood glucose level.

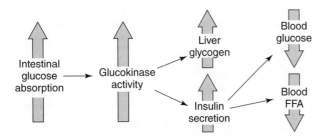

Figure 3.23 *A sequence of processes explaining the role of glucokinase in the liver and β-cells in regulation of the blood glucose concentration.* The increase in the plasma insulin increases glucose uptake by muscle and decreases fatty acid mobilisation from adipose tissue which lowers the plasma fatty acid level which also increases glucose uptake (Chapter 12).

- Measurement of the activities of some enzymes in blood: this is one factor involved in the diagnosis of disease.

- In forensic science and genetic investigations.

All these uses depend upon one or both of the two paramount properties of enzymes: *catalytic power* and *specificity*.

Use of enzymes in biochemical and clinical measurements

In biochemistry, physiology, pathology and medicine, there are usually two major problems in measuring the concentration of a compound:

(i) The compound is usually present at very low concentrations.

(ii) The compound is usually present in a solution that contains a large number of other similar compounds.

The immense catalytic power of enzymes means that a compound, even at a very low concentration, can rapidly be converted to a product whose concentration can then be measured by a sensitive technique. The specificity of enzymes ensures that they react with only one compound, even in the presence of many other similar compounds. This property can be used, for example, in measuring the concentration of glucose in body fluids.

One other class of biochemical proteins that offers a similar advantage in clinical biochemistry is antibodies, which are highly specific for interaction with other proteins (i.e. with antigens) and bind with a high affinity. The combination of antibodies with radioisotopes produces very sensitive and very specific assays. This system has been used particularly for measuring the concentrations of hormones (see below) peptides and cytokines.

Measurement of the concentration of glucose

The concentration of glucose in a blood sample, for example, can be measured using the enzyme, glucose oxidase, which catalyses, specifically, the oxidation of glucose:

$$\text{glucose} + O_2 + H_2O \rightarrow \text{gluconate} + H_2O_2 + H^+$$

The reaction proceeds almost totally to the right. Consequently, the amount of hydrogen peroxide (H_2O_2) produced is the same as that of glucose originally present in the sample. The amount of H_2O_2 is then measured using a second enzyme, peroxidase, which catalyses a reaction between the hydrogen peroxide and a reducing agent, which becomes oxidised:

$$H_2O_2 + AH_2 \rightarrow 2H_2O + A$$

In this hypothetical reaction, the compound AH_2 is the reducing agent. It is chosen because, when it is oxidised to form A, the colour changes, i.e. compound A absorbs light at a different wavelength from AH_2, so that its concentration can be measured at a specific wavelength in a spectrophotometer. The change in absorption of light is, therefore, proportional to the concentration of glucose in the original sample of blood. Because glucose is the only compound with which glucose oxidase reacts, the presence of even very similar compounds (e.g. fructose) is not a problem.

An extension of this assay has made it very flexible and easy to use. Absorbent plastic-backed strips are impregnated with the glucose oxidase and peroxidase, plus a compound (e.g. a dye) that reacts with hydrogen peroxide. When a drop of blood is placed on the strip or the strip is dipped into a sample of urine, the reactions outlined above proceed and the dye changes colour. By comparing the colour produced with a colour chart, the approximate concentration of glucose is determined. The approximate value may be all that is required for gaining information on the clinical relevance of any glucose in the urine.

A further extension is that the reaction catalysed by glucose oxidase can be modified in such a way that it generates a change in an electric signal, which is proportional to the glucose concentration, and the magnitude of the electric current can be displayed digitally. Such a device is generally known as a biosensor (Appendix 3.11).

Enzyme-linked immunoabsorbent assays

A method that combines the specificity of an antibody with the catalytic power of an enzyme, a marriage between enzymology and immunology, is known as an enzyme-linked immunoabsorbent assay (ELISA). To illustrate the method, measurement of the concentration of a hypothetical peptide X is described:

(i)   An antibody to peptide X is produced, which is the primary antibody.

(ii)   An enzyme (e.g. peroxidase) is attached to a second antibody to peptide X.

(iii)   The primary antibody is attached to a solid support.

(iv)   A solution (e.g. diluted blood) containing an unknown concentration of X is added to the immobilised primary antibody and, provided that the antibody is present in excess of X, the immobilised antibody will bind all of X in the sample. (The unbound material is washed away.)

(v)   The second peroxidase-linked antibody is now added: it binds to the already immobilised X. This procedure is carried out in the well of a small plate.

(vi)   This, too, is then washed to remove excess of the second antibody.

(vii)   The substrate for peroxidase is then added to the solid support (i.e. in the well) and the reaction allowed to proceed for a fixed period of time (e.g. 10 minutes).

(viii) The product of this reaction is a compound that can be readily measured, e.g. a compound that is coloured so that it absorbs light at a specific wavelength. The amount of product that is formed in the period of time of the incubation is proportional to the amount of enzyme which, in turn, is proportional to the amount of X present in the original sample.

(ix)   A standard curve is prepared with known quantities of X.

(x)   The advantage of the technique is its precision in measuring very low concentrations of peptides (in the nanomolar to the micromolar range).

Restriction enzymes and analysis of DNA

Bacteria contain enzymes that catalyse the breaking of phosphodiester links between nucleotides in DNA at specific sites, to which the enzyme is directed by a short sequence of bases. These are known as restriction enzymes and they have resulted in remarkable progress in analysing sequences of DNA fragments. They are endonucleases, i.e., they cleave DNA at the phosphodiester bonds within, rather than at the ends, of DNA chains. They cleave bonds such that sequences of nucleotides, typically 4–8 base pairs, are produced. These are the restriction sequences. There are more than 2000 restriction enzymes, each with their own specificity for restriction sites, so that several enzymes can result in many different short fragments of DNA. A reaction with one type of restriction enzyme results in a defined set of DNA fragments (known as restriction fragments) due to the DNA being cut at each copy of the recognition site and nowhere else along the chain. The different fragments can then be separated by chromatography or electrophoresis (Figure 3.24).

> Some bacteria are able to take up DNA from their surroundings and exchange regions with their own DNA. This recombination is the basis of a primitive sexual process known as transformation. However, recombination with DNA from species other than the same as the bacterium is unlikely to be beneficial and it is likely that restriction enzymes have evolved to destroy such 'foreign' DNA.

The sequences of some recognition sites for some restriction enzymes are given in Table 3.3.

Restriction enzymes are used in a number of areas, for example:

• In forensic science, such as helping to identify a suspect from a sample of tissue or body fluid obtained from the site of the crime.

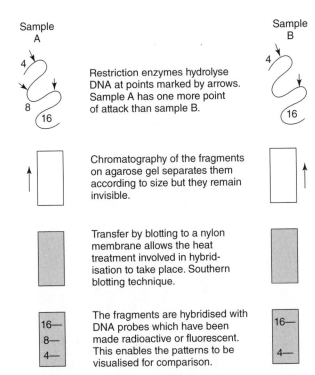

Restriction enzymes hydrolyse DNA at points marked by arrows. Sample A has one more point of attack than sample B.

Chromatography of the fragments on agarose gel separates them according to size but they remain invisible.

Transfer by blotting to a nylon membrane allows the heat treatment involved in hybridisation to take place. Southern blotting technique.

The fragments are hybridised with DNA probes which have been made radioactive or fluorescent. This enables the patterns to be visualised for comparison.

Figure 3.24 *Restriction fragments from digestion of two samples of DNA.* These are produced as a result of the activity of restriction enzymes on the samples of DNA and separation of the fragments by the Southern blotting technique (Appendix 3.12). Numbers refer to the length of the fragments. (i.e. number of nucleotides).

Table 3.3 Recognition sites for some restriction enzymes

Microorganism of origin	Identity of enzyme	Position of hydrolysis
Arthrobacter luteus	Alu1	AG↓CT TC↑GA
Escherichia coli	EcoR1	G↓AATTC CTTAA↑G
Klebsiella pneumoniae	Kpn1	GGTAC↓C C↑CATGG
Nocardia otitidis-caviarum	Not1	GC↓GGCCGC CGCCGG↑CG

The sequences are written 5′–3′ in the top strand and 3′–5′ in the bottom strand. The phosphodiester bonds cleaved by each enzyme are marked ↓ in the top strand and ↑ in the bottom strand. The shorthand used to identify specific restriction enzymes can be confusing to a student reading a scientific paper, in which the shorthand is not explained.

• The identification of the biological father in paternity cases.

Restriction enzyme techniques are frequently used in conjunction with the polymerase chain reaction.

Polymerase chain reaction (PCR)

The polymerase chain reaction uses the enzyme DNA polymerase to produce rapidly very many identical copies of a length of DNA. To catalyse the reaction, the polymerase has the following requirements:

• A length of DNA.

• The four deoxynucleotides (dATP, dGDP, dCMP, and dTPP) (Chapter 20).

• A primer for each strand of DNA: this comprises a small sequence of nucleotides that bind to the 5′ end of one strand and the 3′ end of the other strand of the DNA to initiate the polymerase reaction (Chapter 20).

• Mg^{2+} ions to bind the nucleotides to the polymerase.

The sequence of manipulations in the method is presented in Figure 3.25. An initial problem with the method was that, since the temperature used to separate the strands is about 90 °C, repetitive separation resulted in inactivation of the polymerase, so that fresh enzyme needed to be added for each cycle. The problem was solved by using a DNA polymerase extracted from the organism *Thermus aquaticus*, which lives in hot springs, so that the enzyme is stable at the high temperature needed to separate the strands.

DNA fingerprinting

DNA fingerprinting is a way of uniquely identifying a human. The technique depends on the fact that within the human genome there are multiple and highly variable repeats of small sequences of DNA. These sequences appear to have no function so that somatic mutations in them are not lost by natural selection. Consequently, each individual will have many such unique sequences, although there will be some similarity between individuals of the same family. The pattern of these sequences is revealed by hydrolysis of DNA with restriction enzymes. Since the pattern of restriction fragments is unique to each individual, it provides a 'DNA fingerprint'.

Provided a sample of DNA can be obtained, a restriction analysis can be carried out. A match between the restriction fragments from a sample of DNA left at the scene of a crime and that of a suspect is a valuable tool in forensic science. The usefulness of this technique is increased enormously by combining it with the polymerase chain reaction, since the amount of DNA extracted from a very small amount of tissue can be increased enormously, providing enough for a restriction analysis. Tissue samples as small as a single cell, a hair, a drop of saliva, a piece of dandruff or a smear of semen are sufficient to produce enough DNA. It has produced a revolution in forensic science. However, caution must be applied to interpretation of the results: for

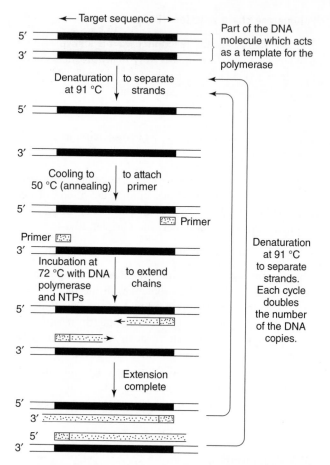

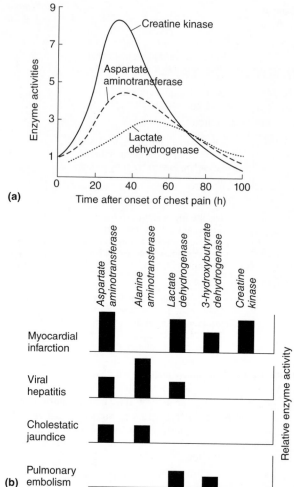

Figure 3.25 *Polymerase chain reaction.* The steps involved in the chain reaction are as follows: (i) Incubation of the DNA at a temperature above 90 °C in order to separate the two strands of the DNA duplex. (ii) Cooling of the solution to about 50 °C to allow annealing of the primers to the template (i.e. the nucleotides bind to the template DNA according to the base-pairing rules). (iii) Finally, addition of the polymerase and Mg^{2+} ions to extend the nucleotide primer and complete the synthesis of the complementary DNA, which takes place at about 70 °C. (iv) The sequence (i) to (iii) is repeated to allow another extension to occur; many repetitions can be carried out which results in enormous multiplication of the DNA strands. NTPs – deoxyribonucleoside triphosphates.

Figure 3.26 (a) *A plot of the increases in activities of creatine kinase, aspartate aminotransferase and lactate dehydrogenase in the blood against time after chest pain. These enzymes are assumed to be released from the heart and therefore indicate damage to heart muscle and support diagnosis of a myocardial infarction.* **(b)** *Relative activities of aspartate aminotransferase, alanine aminotransferase, lactate dehydrogenase and creatine kinase in the blood in the following conditions: a myocardial infarction, viral hepatitis, cholestatic jaundice and a pulmonary embolism. These activities can be used as an aid to diagnosis.*

example, a hair of a victim found on a suspect may have been passed on accidentally due to a common point of contact, independent of the crime scene.

Enzymes in diagnosis

Enzymes, especially those in the cytosol, may escape from a damaged or diseased cell and appear in the blood. Since different tissues contain different enzymes, or different amounts of isoenzymes, analysis of the enzymes in serum can identify which tissue is affected by the disease and hence can help with a diagnosis (Figure 3.26(a) and (b)).

The measurement of enzyme activities in tissues other than blood can also be carried out, for example, in extracts of tissue obtained by biopsy, from red or white blood cells or from cerebrospinal fluid. Such measurements are valuable in diagnosis of genetic diseases in which enzymes are either absent or present at very low activities.

Enzymes as therapeutic agents

In theory, it should be possible to alleviate the symptoms of an enzyme deficiency by administering the missing enzyme but this is made difficult because enzymes do not cross membranes and so fail to enter cells and they also induce an immune response. A solution is the use of liposomes – small lipoprotein vesicles in which enzymes are entrapped so that they fuse with the plasma membrane in vivo and introduce the enzymes into the cell. The liposome protects the introduced enzyme from both the host's immune system and from degradative enzymes. This approach has been used particularly in the treatment for deficiencies of enzymes in the lysosomes.

The heart attack

Many heart attacks are due to formation of a blood clot in one of the arteries supplying the muscles of the left ventricle. Clots in the coronary arteries can sometimes be removed by use of enzymes that dissolve them. The enzymes, which catalyse the hydrolysis of the fibrin fibres that support the clot, include streptokinase, which is obtained from bacteria of the genus *Streptococcus*, or urokinase and plasmin, which are present in human blood. These enzymes if infused sufficiently early, into blood can prevent serious damage. Other enzymes that have been used are plasminogen activators, which convert inactive plasminogen, already present in blood, to plasmin, the natural fibrinolytic enzyme (Chapter 17). Plasminogen activators are now produced in large quantities by genetic engineering.

Enzymes as targets for therapy

The roles of enzymes in biochemical pathways or specific reactions makes them ideal targets for drugs. There are, however, important constraints on this approach. Most enzymes are contained within cells and, indeed, within organelles, so that any drug must traverse one or more membranes to be effective. Although in principle drugs could act on enzymes either to increase or decrease their activity, the kinetic structure of biochemical pathways means that activation of an enzyme within the pathway may have no effect on the flux. Only if the enzyme catalyses the flux-generating step would an increase in activity increase the flux through the pathway. Even if this could

be achieved, the effect may be negated by feedback inhibition mechanisms, i.e. the normal control mechanism could overcome the effect of the drug. The same reservations apply even if the drug stimulated gene expression to increase the amount and hence the activity of the enzyme. Consequently, almost all drugs are inhibitors of enzymes. However, most drugs act on receptors which are involved in signalling, which are mostly located on the outer surface of the plasma membrane and can be inhibited on activated (antagonists or agonists).

Five examples of diseases that are influenced by drugs that inhibit enzymes are presented: depression, hypertension, bacterial infections, viral infections (retroviruses) and cancer.

Depression

This condition is caused by a deficiency of one or more of the monoamine neurotransmitters in the brain (e.g. noradrenaline, dopamine, 5-hydroxytryptamine). One means of increasing the concentration of the neurotransmitters is to inhibit one of the enzymes that degrade the neurotransmitter in the brain. For the monoamines, a key degradative enzyme is monoamine oxidase, which catalyses the reaction

$$R.CH_2NH_2 + \tfrac{1}{2}O_2 \rightarrow R.CHO + NH_3$$

Drugs that inhibit this enzyme result in an increase in the concentration of the monoamines and this can alleviate the depression, at least in some patients (Chapter 14).

Essential hypertension

The cause of this disease is not known, but one factor that increases blood pressure is a plasma protein, angiotensin-II, which is produced via an enzyme cascade in blood, as follows:

$$\text{angiotensinogen} \xrightarrow{(1)} \text{angiotensin-I} \xrightarrow{(2)}$$
$$\text{angiotensin-II} \xrightarrow{(3)} \text{amino acids}$$

The enzymes in this pathway are (1) renin, (2) angiotensin-converting enzyme (ACE) and (3) aminopeptidase. The key enzyme in this pathway is ACE. Drugs that inhibit this enzyme decrease the formation and hence the concentration of angiotensin-II, which lowers blood pressure. Compounds that inhibit this enzyme are very successful antihypertensive drugs (discussed in detail in Chapter 22).

Box 3.8 The first drugs to inhibit an enzyme in bacteria

In 1904 Paul Ehrlich showed that a dye, Trypan Red, killed trypanosomes, which cause sleeping sickness. In 1927, IG Farbenindustrie opened a laboratory for experimental pathology. Its first director, Gerhard Domagk, screened azo-dyes for their ability to cure mice infected with *Streptococcus pyogenes*, a major cause of fatal pneumonia at the time. It turned out that the azo-dyes, although toxic to bacteria in culture, were ineffective in the mice. However, when diazotised sulphonamides (originally developed to improve the binding of azo dyes to wool) were used, the situation was reversed; the drug was now more effective in the mouse than in culture. Furthermore, it was remarkably non-toxic. A patent for 'prontosil rubrum' and other sulphonamides was submitted on 25 December 1932. By 1935, prontosil had proved its clinical value in the treatment of puerperal fever in Queen Charlotte's Maternity Hospital, London, and Domagk was awarded the Nobel Prize for Medicine in 1939. On Hitler's instructions he was persuaded to reject it. He had to wait until 1947 to receive the medal in Stockholm (Sneader 1985).

It was discovered that the active part of the prontosil molecule was not the azo group but the 4-aminobenzenesulphamide (sulphanilamide) which was released from the parent compound within the patient. Other sulphanilamide drugs were developed, including sulphapyridine produced by May & Baker Laboratories, in 1937. It was this drug, generally known as M&B, that cured Winston Churchill (the British Prime Minister during the Second World War) of pneumonia during his visit to North Africa in December 1943. [An extract from *The Second World War*, by Winston Churchill (p.89) (Heron Books) emphasises the benefit he obtained from M&B. 'The admirable M and B, from which I did not suffer any inconvenience, was used at the earliest moment, and after a week's fever the intruders were repulsed . . . There is no doubt that pneumonia is a very different illness from what it was before this marvellous drug was discovered.']

Donald Woods discovered that sulphonamides exerted their action by inhibiting an enzyme used by bacteria to synthesise folic acid. The compound 4-aminobenzoic acid is the precursor for folic acid, and is structurally similar to sulphonamide. Bacteria that were unable to synthesise folic acid were unable to achieve de novo synthesis of purines for their DNA and RNA synthesis and hence could not proliferate. Such competitive inhibitors, which mimicked normal metabolites, became known as antimetabolites (many are used in cancer chemotherapy, Chapter 21).

Bacterial infection

To be an effective antibacterial agent, a drug must inhibit an enzyme that is present in the bacteria but not in the host. One well-known example is a transpeptidase involved in cell wall synthesis in some bacteria. Inhibition prevents bacteria from synthesising their cell wall so that proliferation stops. A drug that inhibits this enzyme is the antibiotic, penicillin first used in 1941 (see Chapter 17). However, the first durg to inhibit bacterial growth was developed from a dye (Box 3.8).

Anti-retroviral drugs and AIDS

The genetic material in a retrovirus is RNA not DNA. The best known retrovirus is the human immunodeficiency virus (HIV) which infects lymphocytes and hence interferes with the immune system, giving rise to the disease AIDS. Once the virus infects its host, it converts its RNA into DNA by an enzyme known as reverse transcriptase and the DNA is then inserted into the genome of the host cell (in this case the lymphocyte):

viral RNA →→ viral DNA →→ host cell genome

Reverse transcriptase is a retroviral-specific enzyme and is essential to the virus. Drugs that inhibit this enzyme are used to treat the infection and hence the disease AIDS. As well as reverse transcriptase, the virus needs other proteins (e.g. an enzyme that catalyses insertion of the viral DNA into the genome of the host cell). These enzymes are first formed as pro-enzymes (precursor proteins) so that they require activation by a proteolytic enzyme, which is also specific to the virus. Inhibitors of this enzyme, protease inhibitors, have had some success in improving life expectancy of AIDs patients, presumably because they kill the virus (Chapter 17).

Cancer

Drugs that can be used to control tumour cell proliferation inhibit a variety of enzymes, including thymidylate synthase and topoisomerase (Chapter 20). The enzyme aromatase converts a ring in a steroid to an aromatic ring. It converts, for example, adrenal steroid hormones into female sex hormones, which bind to oestrogenic receptors in the ovary or breast and increase the risk of ovarian or breast cancer. Aromatase inhibitors are used to treat patients with breast or ovarian cancers that are sensitive to oestrogen. Unfortunately, none of the inhibitors is specific for enzymes in tumour cells and they can therefore have severe side-effects (Chapter 21).

Kinetic structure of a biochemical pathway

Many enzymes in the cell are organised into sequences, so that the reactions they catalyse are integrated into pathways or processes. In these pathways, a precursor or substrate is converted to a product, e.g. glucose is converted to lactic acid; amino acids are polymerised to form protein; glutamine is converted to aspartate. These pathways have both a thermodynamic and a kinetic structure. The thermodynamic structure is presented in Chapter 2. The kinetic structure is described here. There are three basic facts that must be appreciated before the kinetic structure is explained.

- Reactions in a pathway can be divided into two classes: those that are very close to equilibrium (near-equilibrium) and those that are far removed from equilibrium (non-equilibrium). This is discussed in Chapter 2 but is summarised here using kinetic principles to explain how enzyme catalysis can give rise to two separate types of reaction in one pathway.

- One of the enzymes that catalyses a non-equilibrium reaction approaches saturation with substrate, so that it is the flux-generating step, (i.e. the beginning of the pathway).

- The kinetic and thermodynamic structure of a pathway or process in a cell or a tissue can only be maintained because living systems are open: that is, they exchange matter and energy with the environment (Chapter 2).

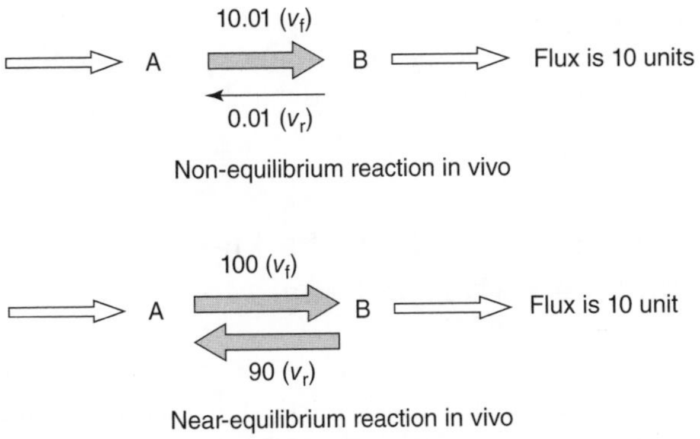

Equilibrium and non-equilibrium reactions: a kinetic explanation

A reaction in a metabolic pathway is likely to be non-equilibrium if the maximum catalytic activity of the enzyme that catalyses the reaction is low in comparison with those of other enzymes in the pathway. In consequence, the concentration of substrate of this reaction is likely to be high whereas that of the product is likely to be low, since the next enzyme in the sequence readily catalyses its removal. Because the concentration of this product is low, the rate of the reverse component of the reaction is very much less than the rate of the forward component. This situation characterises a non-equilibrium process. Conversely, a reaction is near-equilibrium if the maximum catalytic activity of the enzyme is high in relation to those of other enzymes in the pathway; in this case, the rates of the forward and the reverse components of the reaction are similar and both are much greater than the overall flux

Figure 3.27 *Representation of the rates of the forward and reverse reactions for non- and near-equilibrium reactions in one reaction in a hypothetical pathway.* The values represent actual rates, not rate constants. The net flux through the pathway is given by (V_f-V_r). In the non-equilibrium reaction, the rate of the forward reaction dominates, so that the net flux is almost identical to this rate. In the near-equilibrium reaction, both forward and reverse rates are almost identical but considerably in excess of the flux.

through the pathway. A quantitative explanation should help to clarify and is presented in Figure 3.27. From this explanation, it should be clear why the terms *reversible* and *irreversible* are sometimes used in place of *equilibrium* and *non-equilibrium*. The latter terminology is used in this text. The concept of the flux-generating reaction is now discussed.

Flux-generating reactions

Consider an enzyme at the beginning of a pathway whose pathway-substrate concentration is much less than that required to saturate the enzyme (see Figure 3.7), e.g. similar to or lower than that of the K_m. As the catalysis proceeds, the concentration of substrate falls so that the activity of the enzyme decreases more and more. Consequently, the activity of such an enzyme cannot maintain a constant flux through a pathway, so that a steady state cannot be achieved.

In contrast, if the enzyme is saturated with its pathway-substrate, (i.e. zero order) a decrease in the concentration will not decrease its activity, so that it could generate a constant flux through the reaction and hence through the pathway. It is, therefore, an enzyme that can generate a constant flux: a non-equilibrium reaction that is saturated with pathway-substrate is termed a *flux-generating reaction*. Examples of some flux-generating reactions are given in Figure 3.19. The physiological significance of such enzymes is discussed for specific pathways in later chapters. The kinetic significance in establishing a pathway and defining how the flux is transmitted through a pathway is now discussed.

Transmission of flux: the kinetic structure of the pathway

If a biochemical pathway possesses a flux-generating reaction and, by definition, it should, it follows that the flux through all reactions in the pathway must conform to that of the flux-generating step. To see how this works, consider an increase in the activity of enzyme E_1, the flux-generating reaction of the hypothetical pathway depicted in Figure 3.28. The immediate consequence will be a rise in the concentration of A, which will increase the rate of E_2, since it is not saturated with substrate. This will raise the concentration of B, and hence the activity of E_3, and so on along the pathway, so that the rates of all the reactions will increase in parallel with the rate of the reaction catalysed by E_1 and, in time, a new steady-state will be established. This mechanism of regulation of the activities of E_2 and E_3 etc. by changes in the concentrations of their substrates can be described as *internal regulation*, i.e. regulation is internal to the pathway. The activity of E_1

results, therefore, in a steady-state flux through the whole pathway which is described as a *transmission sequence*.

Regulation of flux through a pathway

If a compound (e.g. X, an allosteric activator) increases the activity of E_1, the flux through the transmission sequence will increase. This type of regulation is termed *external regulation* (i.e. it is achieved by a factor external to the pathway). However, a change in the activity of any other enzyme would not change the flux. For example, a change in flux through the pathway would not occur if only the activity of E_4 increased: it would result only in a decrease in the concentration of C, until the activity of E_4 decreased to its previous value. However, if E_4 communicated with E_1, such a change could modify the flux. Appropriate communication could come about if compound C is an allosteric inhibitor of E_1. Thus an increase in activity of E_4, via an effect of an allosteric regulator, would lower the concentration of C, which would then increase the activity of E_1, so that the flux through the transmission sequence would increase (Figure 3.28(c)). Such inhibition, from a final product or a precursor of the product of the pathway is common in the control of biochemical pathways, and it is known as *feedback inhibition* (see Figure 3.13).

From this discussion it should be clear that if the activity of any enzyme in the sequence was decreased, for any reason, to such an extent that its maximum activity fell below that of E_1, the concentration of its substrate would rise sufficiently to saturate the enzyme, when the activity could increase no longer. If there were no feedback regulation, i.e. no control-structure to the pathway, this increase would continue until the substrate was removed by a side reaction or escaped from the cell to be modified and excreted in the urine. Such a loss of enzyme activity is exceedingly rare. When it occurs, it is usually due to a genetic deficiency; that is, the enzyme is inactive due to an 'inborn error' or an *enzyme deficiency disease*. Since the enzyme is not part of a coordinated regulatory mechanism, nothing prevents the substrate concentration from increasing excessively, even to pathological levels (Figure 3.28(d)).

Well-known examples of enzyme deficiency diseases include:

- Phenylalanine hydroxylase deficiency, giving rise to phenylketonuria (PKU) (Box 3.9).

- Pyruvate dehydrogenase deficiency giving rise to lactic acidosis (Chapter 9).

- Glycogen phosphorylase deficiency in muscle gives rise to muscle weakness, frequent cramp and ease of fatigue (McArdle's syndrome). It also gives rise to hypoglycaemia if the liver enzyme is deficient (Chapter 6).

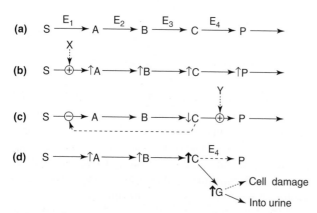

Figure 3.28 *A hypothetical pathway and modes of regulation*
(a) The hypothetical pathway in which E_1 is the flux-generating step.
(b) Factor X activates E_1, which results in increased concentrations of all the intermediates in the pathway (i.e. the transmission sequence).
(c) Factor Y activates E_4, concentration of C decreases, which stimulates E_1 (because it is an inhibitor).
(d) Enzyme E_4 is absent, so that C accumulates and it is then converted to G by a side reaction and G is excreted in urine, and can cause damage to the all or other tissues.

Box 3.9 Phenylketonuria

Phenylketonuria (PKU) is a group of inherited disorders caused by a deficiency of the enzyme phenylalanine hydroxylase (PAH) that catalyses the conversion of phenylalanine to tyrosine, the first step in the pathway for catabolism of this amino acid. As a result, the concentration of phenylalanine in the liver and the blood increases. This high concentration in the liver increases the rate of a side reaction in which phenylalanine is converted to phenylpyruvic acid and phenylethylamine, which accumulate in the blood and are excreted in the urine.

The disease develops at 3 to 6 months of age and it is characterised by developmental delay, eczema, hyperactivity and mental retardation. Newborn babies are routinely screened for PKU in many countries. Treatment is a phenylalanine-restricted diet and supplementation with tyrosine.

$$\text{Diet} \rightarrow \text{phenylalanine} \dashrightarrow \text{tyrosine} \dashrightarrow$$
$$\downarrow$$
$$\text{phenylpyruvate}$$

Regulation of enzyme activity

Investigating the regulation of enzyme activity requires identification of the external regulators and how regulation of the enzyme activity affects the flux through a pathway. There are four important questions that must be answered before mechanisms of regulation can be usefully discussed. These are:

(i) Which enzyme(s) in a pathway is subject to external regulation?

(ii) What is the biochemical mechanism by which the enzyme activity is regulated by an external regulator?

(iii) How does regulation of the activity of an enzyme regulate the flux through a pathway?

(iv) What is meant by sensitivity in regulation?

These questions are considered below.

(i) The enzyme that catalyses the flux-generating step must be regulated to change the flux through the pathway. Enzymes that catalyse non-equilibrium reactions are more likely to be regulated by external factors than those that catalyse near-equilibrium reactions, so it is these enzymes that are studied in (ii).

(ii) The biochemical mechanisms by which enzyme activity is regulated are suggested by studying the properties of the enzyme in vitro. The proposed mechanism must then be investigated in vivo. (This approach is used to establish a mechanism for different pathways or processes, in many chapters in this book.)

(iii) Once the mechanism(s) has been identified, the means by which it can change the flux requires information on how each enzyme is involved in the transmission sequence.

(iv) In many biochemical or physiological processes, a weak stimulus produces a large response. For example, the increase in flux through glycolysis in the leg muscle of a sprinter leaving the blocks, to achieve a maximum power output, is approximately 1000-fold. Yet the factors that regulate glycolysis change nothing like 1000-fold. Similarly, the increase in the Krebs cycle from rest to maximum aerobic physical activity in muscle is approximately 50-fold. To understand how such marked changes in activity can be produced by small changes in the concentration of a regulator, the concept of sensitivity in regulation must be addressed.

Sensitivity in regulation is defined as the quantitative relationship between the relative change in enzyme activity and the relative change in concentration of the regulator. For example, if an enzyme activity needs to increase 100-fold to produce the necessary change in flux through the pathway, how large an increase in concentration of regulator is required? The greater the change in response of enzyme activity to a given change in regulator concentration, the greater is the sensitivity. This is defined mathematically as follows.

The concentration of a regulator (x) changes by Δx, so that the relative change in concentration is $\Delta x/x$. This results in a change in flux, J, by ΔJ, so that the relative change in flux is $\Delta J/J$. The sensitivity of the flux to the change in concentration of x is given by the ratio $\Delta J/J$ to $\Delta x/x$, i.e. $S = \dfrac{\Delta J/J}{\Delta x/x}$ where S is the sensitivity.

The next quantitative problem is to understand the basic mechanism of interaction between the regulator (in this case x) and its binding to the target enzyme (i.e. the enzyme that regulates the flux through the pathway).

Equilibrium-binding of a regulator to an enzyme

To modify the activity of an enzyme, or any protein, the regulator must bind to the protein and, in almost all cases, the binding is reversible. Such binding is described as equilibrium-binding.

$$E + X \rightleftharpoons E * X$$

where E is the enzyme, X is the regulator and E* is the altered form of the enzyme. The asterisk indicates that the binding of X has changed the conformation of the enzyme so that the structure of the catalytic site has changed to increase or decrease the catalytic activity.

The normal response of enzyme activity to the binding of the regulator (or the binding of the substrate) is hyperbolic, as described above. Unfortunately, this response is relatively inefficient for sensitivity in regulation of the activity of the enzyme. The maximum sensitivity, as defined quantitatively above, is unity. This is the part of the response that is first order (see Figure 3.7). For example, a twofold change in regulator concentration will change the enzyme activity by no more than twofold (i.e. the value of S, in the above equation, is unity). This interpretation may be difficult to accept from simply viewing the initial part of a hyperbolic curve. However, it must be appreciated that sensitivity is *not* the slope of the plot of activity versus concentration of substrate or regulator; it is the relationship between the *relative* change in activity and the *relative* change in concentration of the regulator.

Since the hyperbolic response is the simplest relationship between protein and regulator, it can be considered as the basic response with which any mechanism for improving sensitivity can be compared. Four such mechanisms are now examined.

Mechanisms for improving sensitivity

Multiplicity of regulators

It is possible for an enzyme to be regulated by several different external regulators that all bind at different allosteric sites on the enzyme. In this case, if the concentrations of all the regulators change in directions to change the activity of the enzyme in the same direction, the effect of all external regulators could be cumulative (Figure 3.29).

Cooperativity

For many enzymes that play a role in regulation, the response of their activity to the substrate or regulator concentration is sigmoid, not hyperbolic. This phenomenon is known as *cooperativity* (see above). For part of the concentration range of the substrate or regulator, the effect on the enzyme activity is greater than that provided by the hyberbolic response, i.e. the sensitivity is greater than unity (see Figure 3.15(b)).

Substrate cycles

A totally different mechanism for improving sensitivity is known as the substrate cycle. It is possible for a reaction that is non-equilibrium in the forward direction of a pathway (i.e. A→B, see below) to be opposed by a reaction

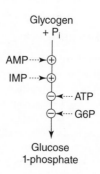

Figure 3.29 *Control of an enzyme activity by multiple allosteric regulators.* The enzyme glycogen phosphorylase *b* in muscle is regulated by changes in the concentrations of AMP and inosine monophosphate (IMP) (which are activators) and ATP and glucose 6-phosphate (G6P), which are inhibitors.

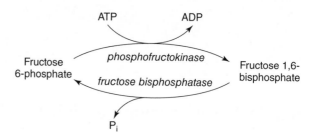

Figure 3.30 *The fructose 6-phosphate/fructose 1,6-bisphosphate cycle.* The forward reaction is catalysed by the enzyme phosphofructokinase, the reverse reaction by fructose bisphosphatase.

that is non-equilibrium in the reverse direction of the pathway (i.e. B→A). For example,

$$S \xrightarrow{\text{E}_1} A \underset{\text{E}_5}{\overset{\text{E}_2}{\rightleftharpoons}} B \xrightarrow{\text{E}_3} C \xrightarrow{\text{E}_4} P$$

The substrate cycle between A and B, is catalysed by enzymes E_2 and E_5 in the pathway S → P. The reactions must be chemically distinct and non-equilibrium and catalysed by different enzymes (i.e. E_2 and E_5, above). It is possible that these two opposing reactions are components of two separate pathways that function under different conditions (e.g. glycolysis and gluconeogenesis in the liver – see Chapter 6) but the reverse reaction (E_5 in the above example) may not be part of any other pathway but only present in the cell to provide a cycle for regulation of flux, through that reaction; that is, for improving sensitivity in regulation. An example is the fructose 6-phosphate/ fructose bisphosphate cycle in muscle (Figure 3.30).

If the two enzymes are simultaneously active, A will be converted to B and the latter will be converted back to A, thus constituting the cycle. There are, thus, two fluxes: a linear flux through the cycle, A to B, as part of the pathway by which S is converted to P, and a cyclical flux between A and B. Both fluxes are to a large extent independent and calculations show that the improvement in sensitivity is greatest when the cyclical flux is high but the linear flux is low, i.e. the ratio, cycling rate/flux, is high (Table 3.4, Figure 3.31).

In some conditions, to achieve satisfactory regulation of flux, an enzyme activity may have to be reduced to values approaching zero. Even with a sigmoid response, this would require that the concentration of an activator be reduced to almost zero or that of an inhibitor be increased to an almost infinite level. Such enormous changes in concentration never occur in living organisms, because they would cause osmotic and ionic problems and unwanted side reactions; that is, they are physiologically unacceptable. However, the net flux through a reaction can be reduced to very low values (approaching zero) via a substrate cycle (Figure 3.32).

It is possible that, via a cycle, the direction of a flux can be completely reversed. An example is glucose metabolism in the liver: at a low blood glucose level the liver releases glucose, whereas at a high concentration of blood glucose the liver takes up glucose. This is the result of a substrate cycle between glucose and glucose 6-phosphate in the liver (Figure 3.32) (discussed in detail in Chapter 6).

Since the net result of a cycle, in addition to an increase in sensitivity, is the hydrolysis of ATP, it is unlikely that high rates of cycling will be maintained for any prolonged periods of time. One means of providing high sensitivity, but low rates of cycling transiently, is to increase the rate of cycling only when increased sensitivity. Chronically, is required. For example, a stressful condition increases the release of the stress hormones, adrenaline and noradrenaline. These hormones could increase the activity of both enzymes, i.e. those that catalyse the forward and reverse reactions in the cycle (e.g. by a change in a specific messenger, e.g. cyclic AMF). The role of these hormones is to prepare the body for 'fight or flight', i.e. increased physical activity. An increase in the rate of cycling and hence an increase in sensitivity in preparation for increased ATP generation, would be an advantage if fight or flight had to take place (Figure 3.31).

In some circumstances, substrate cycles may operate not only to regulate flux through biochemical pathways but to achieve the controlled conversion of chemical energy (i.e. ATP) into heat. This occurs in two conditions.

Table 3.4 Effect of an increase in the concentration of a regulator on net flux through a reaction that is regulated by a direct effect of the regulator on the activity of an enzyme. The hypothetical pathway is

$$S \xrightarrow{E_1} A \underset{E_5}{\overset{E_2}{\rightleftharpoons}} B \xrightarrow{E_3} C \xrightarrow{E_4} P$$

The quantitive effect is examined when there is no substrate cycle and when there is substrate cycle.

Concentration of regulator (x)	Enzyme activities[a] (units/min)		Net flux A to B (J)	Relative fold increase in flux	Sensitivity (S)
	E_2	E_5			
No cycling (i.e. enzyme E_5 is inactive)					
Basal	10	zero	10		
Fourfold above basal	40	zero	40	4	1.0
Cycling (i.e. enzymes E_2 and E_5 are active)					
Basal	10	9.8	0.2		
Fourfold above basal	40	1.0	39	195	approx. 50.0

[a] The units are arbitrary.

E represents the enzymes catalysing the reactions in the pathway. Simultaneous activities of E_2 and E_5 produce a substrate cycle between A and B.

In the *no cycling* condition, enzyme E_5 is absent (or inactive).

In the *cycling* condition, the regulator not only increases the activity of E_2 but decreases that of E_5. However, the improvement in the relative increase is not much greater if E_5 activity does not change. The relative change in the concentration of regulation (i.e $\Delta x/x$) is 4.0 in both conditions; in the *no cycling* condition $\Delta J/J$ (the relative change in flux) is 4.0 but it is approx. 200 in the *cycling* condition. Consequently the values for $\Delta J/J/\Delta x/x$ (i.e. sensitivity, S) are unity and about 50, in the no cycling and cycling contitions, respectively.

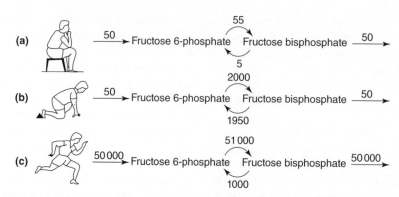

Figure 3.31 *Representation of the role of a substrate cycle improving the sensitivity of the regulation of the flux through the reaction in which fructose 6-phosphate is converted to fructose bisphosphate in muscle during sprinting.* The upper arrow represents phosphofructokinase activity and the lower arrow represents fructose bisphosphatase activity. **(a)** Resting before the sprint, when cycling rate is low, and flux is low; **(b)** on the starting block, when stress hormones increase the cycling rate markedly (i.e. 'preparation for flight or fight'); **(c)** about six seconds after the start of the sprint, when allosteric regulators have increased the activity of phosphofructokinase and decreased that of fructose bisphosphatase. The enzyme activities represent (a) the relaxed sprinter, (b) the stressed sprinter immediately before the race and (c) the sprinter at maximum speed. The activities are hypothetical. From this it can be seen that a 25-fold increase in activity of phosphofructokinase and a 50% decrease in that of fructose bisphosphatase, both caused by changes in allosteric regulator concentrations, at the beginning of the sprint, increase the glycolytic flux 1000-fold: a well-established biochemical fact. Indeed, this increase in sensitivity must be required in most sporting activities (Chapter 13). It must be noted that the activity of glycogen phosphorylase must also increase by a thousand fold. This is achieved by an inter-convention cycle.

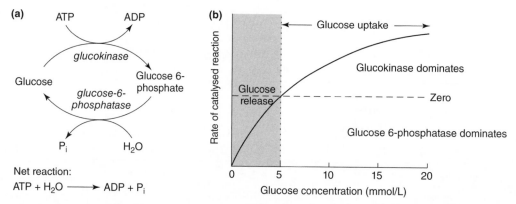

Figure 3.32 *Generation of a threshold response of a reaction in a pathway to a change in regulator concentration. The example is that is provided by the glucose/glucose-6-phosphate cycle.* **(a)** The glucose/glucose-6-phosphate cycle. **(b)** The net reaction is achieved by subtracting the activity of the enzyme that catalyses the reverse reaction from that of the enzyme catalysing the forward reaction. A threshold (vertical dotted line) is achieved since there is zero net flux, when both activities are identical. The example given is the cycle between glucokinase and glucose 6-phosphatase in liver. When glucose 6-phosphatase activity exceeds glucokinase activity, glucose is released from the liver; when glucokinase activity exceeds glucose 6-phosphatase activity, glucose is taken up from the blood. This remarkable effect is achieved solely by changing the activity of glucokinase, but not that of glucose 6-phosphatase. It depends solely on changes in the concentration of glucose, the substrate for glucokinase (see above and Chapter 6).

- To produce heat so as to maintain body temperature (known as non-shivering thermogenesis) (Chapter 9).

- To reduce body mass by 'burning off' stored fuel. This is put forward as one mechanism by which the amount of triacylglycerol stored in adipose tissue can be reduced (Chapters 7, 12 and 15).

Interconversion cycles

The topic of interconversion cycles in providing inhibition or activation of a target enzyme, the activity of which regulates the flux through a pathway, is discussed above. In brief, an enzyme exists in two forms, conventionally designated *a* and *b*, one being a covalent modification of the other. This is brought about, for example, by phosphorylation with ATP, so that one form is a phos-phorylated modification of the other. Since only one of the two forms, *a*, has significant catalytic activity, the flux can be regulated by altering the amount of the target enzyme in this form (Box 3.7). The basis for improving sensitivity by this mechanism is discussed above and indicated in Figure 3.12. It is also discussed in Chapter 20 where enzyme interconversion cycles are compared with control of the concentration of a regulatory protein by protein synthesis and protein degradation.

It is important to point out that these four mechanisms are not mutually exclusive. Indeed, it is probable that, for some reactions, all four mechanisms play a role in regulation of flux and this combination could provide an enormous increase in sensitivity. An example is the regulation of the enzyme phosphorylase in muscle and liver, and hence the process of glycogenolysis (Chapters 6 and 12).

4
Transport into the Body: The Gastrointestinal Tract, Digestion and Absorption

October 21ˢᵗ 1779, for 5 people, dinner (between 1 and 3pm) consisted of leg of pork (boiled), a roasted turkey and a couple of ducks.

For supper (probably 6–7.30pm) the same people had a couple of fowls boiled, a fine pheasant roasted and some cold things, pears, melons, apples and walnuts.

(Woodforde, 1935)

225 g sugar, 50 g tea, 1,800 ml milk, 50 g butter, 100 g margarine, 50–100 g cheese, 1 shelled egg (or 1 packet dried egg), 100 g bacon or ham, meat to the value of 6p, 75 g sweets.

(Information provided by Second World War Coastal Defence Battery,
Battery Gardens, Brixham, South Devon, England.)

The capacity of the human intestine to adjust digestion and absorption to accommodate a huge range of quantity and content of food, is remarkable. A comparison can be made between the content and composition of dinner and supper given by Parson James Woodforde in England in the 18th century on a single day and the weekly ration for one adult in wartime Britain in 1944.

William Beaumont (1785–1853) experimented with the gastric juices of a patient with a permanent hole in his stomach. He published his results in 1833, and was the first to explain how the stomach could digest food. Extracts from this publication are fascinating:

> *I think I am warranted, from the results of all the experiments in saying that the gastric juice, so far from being 'inert as water', as some authors assert, is the most general solvent in nature . . . even the hardest bone cannot withstand its action. We must, I think, regard this fluid as a chemical agent, and its operation as a chemical action . . . Its taste, when applied to the tongue is similar to thin mucilaginous water, slightly acidulated with muriatic acid [hydrochloric acid].*

Not only did Beaumont explain the basis of digestion, but also the bactericidal nature of gastric juice:

> *a powerful antiseptic, cheating the putrefaction of meat; and effectually restorative of healthy action, when applied to old foetid sores, and foul, ulcerating surfaces.*

(Beaumont, 1833)

The macronutrients – carbohydrate, fat and protein – are the major components of food. To be utilised by the body, they must be broken down to compounds that are small enough to be absorbed through the cells of the intestine and transported by the blood to appropriate tissues and organs. The gastrointestinal tract is responsible for both the breakdown and the absorption of the products of breakdown. The structure of the tract, and biochemical events within it, both contribute to the following: transport of material from mouth to anus; secretion of enzymes, other proteins and ions; hydrolysis of large molecules (digestion); absorption of the products of digestion; transport of these products into blood or lymph; and generation of ATP to support these processes.

The control of gut function involves interplay between neurones and peptide hormones. Information from a variety of receptors along the digestive tract is processed by a network of nerves, the *enteric nervous system*, which also receives input from the brain.

Functional Biochemistry in Health and Disease by Eric Newsholme and Tony Leech
© 2010 John Wiley & Sons Ltd

Gross structure of the gastrointestinal tract

The gastrointestinal tract is about 4.5 metres long, running from mouth to anus. Its division into a number of clearly defined regions, each of which possesses a variety of differently specialised cells, enables it to carry out its diverse functions (Figures 4.1(a)&(b)). In general, the substructure of the wall of each region is the same (Figure 4.2).

Buccal cavity

Food is taken into the buccal cavity, where it is masticated by the teeth and mixed with saliva from three pairs of salivary glands. It moistens the food and dissolves some molecules enabling them to interact with the taste receptors on the tongue. Saliva contains Na^+, K^+, Cl^- and HCO_3^- ions and a protein, mucin, which is a component of mucus that lubricates the chewed food on its way down the oesophagus. The pH of saliva is about 7.8, which neutralises acid formed by bacteria in the mouth: this protects tooth enamel from acid attack. The enzyme amylase is also present in saliva and it initiates digestion of the starch in the food.

Stomach

The stomach receives food from the buccal cavity, it partially digests protein, fat and carbohydrate and it then delivers the resulting mixture (chyme) into the small intestine. The inner surface of the stomach is folded into ridges, to allow for distension after a meal, they contain gastric pits into which several gastric glands discharge their secretions (Table 4.1).

Gastric secretions

The stomach secretes pepsinogens, which are inactive proteolytic enzymes, and protons – the high concentration of which initiates hydrolysis of the pepsinogens to form active pepsins, which then continue their own activation, via an autocatalytic, hydrolysis (Appendix 4.1).

$$\text{pepsinogen} \xrightarrow{\text{H}^+} \text{pepsin} + \text{peptide}$$

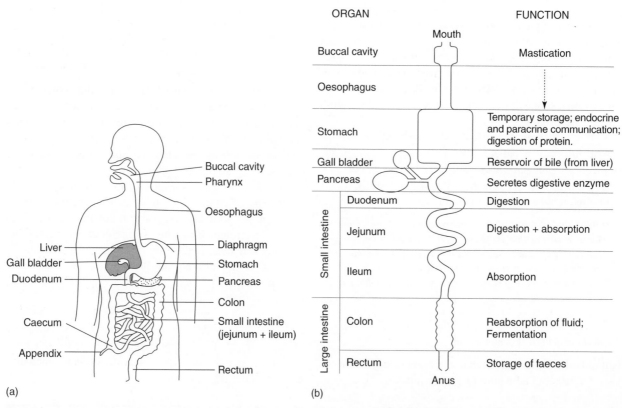

Figure 4.1 **(a)** *The gastrointestinal tract and its position in the body.* **(b)** *General organisation and functions of the organs of the gastrointestinal tract.* (Not to scale)

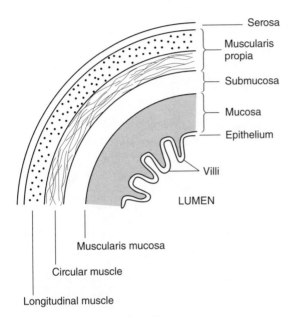

Figure 4.2 *Substructure of the wall in the gastrointestinal tract.* (Not to scale). Epithelial cells of the villi possess micro villi which extend into the lumen (Not shown). The epithelium consists of absorptive enterocytes, endocrine cells that secrete peptide and goblet cells that secrete mucus.

Table 4.1 Main secretions into the lumen by the stomach

Compounds secreted	Cells responsible
Acid (i.e. protons)	Parietal
Mucus	Mucous cells
Hydrogencarbonate ion	Parietal
Pepsinogens	Chief (zymogen) cells
Gastrin	G-cells
Gastric lipase	Chief and mucous cells
Ghrelin	(see Chapter 12)

The protons are secreted by the parietal cells in exchange for potassium ions (Figure 4.3). This process is unique to the parietal cells (Box 4.1).

As well as functioning as a lubricant, the mucus secreted by the cells serves to protect the epithelial cells from the damaging effects of pepsin and the protons. The mucous layer itself is protected from the protons, since they are secreted by the gastric gland with sufficient hydrostatic pressure

> Pathogens need to adhere to the cell surface of the epithelial cells in order to enter the cells, so the mucous layer protects against attack by pathogens.

that they form narrow channels known as 'viscous fingers' that do not mix well with the mucus. Despite this protection, the lifespan of a gastric epithelial cell is only about 48 hours.

None of these functions are vital; provided food is ingested, a little at a time, and in a suitable form, it is

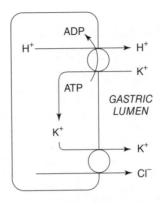

Figure 4.3 *The H^+/K^+-ATPase: transport of H^+ and K^+ ions in the stomach.* The H^+ ions are derived from carbonic acid and the resultant hydrogencarbonate ion is transported out of the cell into the interstitial space and hence into the blood. The H^+ ions are transported into the lumen of the stomach in exchange for K^+ ions, which requires ATP hydrolysis. K^+ ions are transported back into the lumen of the stomach along with Cl^- ions.

possible to survive partial or total gastrectomy. However, vitamin B_{12} must be provided otherwise the patient will become anaemic. This is because an additional function of the parietal cell is secretion of a protein which is required for the absorption of this vitamin. The protein is called the intrinsic factor. The complex formed between B_{12} and intrinsic factor binds to

> One means of losing weight (i.e. adipose tissue), when all others have failed, is by reducing the volume of the stomach by a surgical process known as *stomach stapling*: this restricts the amount of food that can be eaten at one time.

> Deficiency of vitamin B_{12} interferes with production of red blood cells so that anaemia (megaloblastic anaemia) develops (Chapter 15).

a specific receptor protein present on cells in ileum, from where it is absorbed. This vitamin is essential for some of the reactions that synthesise nucleotides, and hence for RNA and DNA synthesis.

Control of acid secretion

The basal rate of proton secretion is around 10% of maximal but the perception of food (smell, taste, sight or even just the thought of it) increases secretion. This is the cephalic effect of food. Nervous signals from the brain cause release of acetylcholine, histamine and gastrin to stimulate acid secretion from the parietal cells. When food actually reaches the stomach, distension, proteins, peptides and amino acids further stimulate the release of gastrin.

Small intestine

The small intestine is divided into three regions: the duodenum, jejunum and ileum. The duodenum, which is the

Box 4.1 Acid, bacteria and drugs

The acid and proteolytic enzymes in the stomach would damage epithelial cells if they were not protected by a layer of mucus. However, if secretion of mucus is impaired or acid is over-produced, damage can result. Slight penetration of the defences causes dyspepsia but, if the mucus-secreting cells are damaged, a vicious circle develops causing damage to epithelial cells and development of an ulcer. If excess acidic chyme leaves the stomach, a duodenal ulcer can also develop. The condition is exacerbated by agents which increase acid secretion, such as alcohol and caffeine, and especially aspirin or other non-steroidal anti-inflammatory drugs (Chapter 11). Some relief is obtained by the use of antacids: sodium hydrogencarbonate, magnesium hydroxide or aluminium hydroxide, which neutralise the acid. A pharmacological breakthrough occurred, however, when James Black developed compounds which blocked the H_2 receptor for histamine on the surface of parietal cells. The receptor blocker, ranitidine, became a best-selling drug. Alternatively, drugs that inhibit the H^+/K^+-ATPase in the parietal cells also reduce acid secretion.

A surprising discovery was that a bacterium, *Helicobacter pylori*, survives and proliferates even in the acid conditions of the stomach; its presence can contribute to development of stomach ulcers. The infection is difficult to treat since antibiotics are not very effective in acidic conditions.

Anti-inflammatory drugs (e.g. aspirin) are beneficial in reducing the pain of arthritis but the side-effect of excess acid secretion is a major problem. Since the drugs are required chronically, there is a risk of development of an ulcer. Unfortunately, a drug that reduced pain of arthritis but minimised acid secretion had severe side-effects and was withdrawn (Chapter 11).

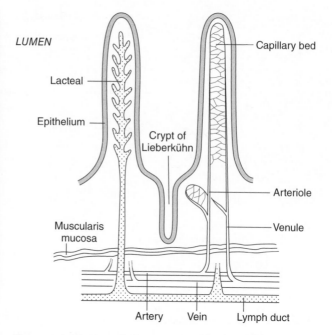

Figure 4.4 *Diagram of the structure of a villus.* Most of the absorbed materials enter the blood vessel, but chylomicrons enter the lymph in the lacteals.

Table 4.2 Some regulatory peptides secreted by the gastrointestinal tract and main effects

Compound	Main effects
Cholecystokinin (CKK)	Stimulates enzyme secretion from pancreas and contraction of gall bladder.
Gastric inhibitory peptide (GIP)	Enhances insulin secretion by the endocrine cells in the pancreas.
Glucagon-like peptide (GLP)	Enhances insulin secretion and suppresses glucagon secretion.
Gastrin	Stimulates acid secretion by stomach.
Ghrelin	Stimulates appetite.
Motilin	Increases motility of intestine.
Neurotensin	Inhibits emptying of stomach and acid secretion.
Opioid peptides	Decrease motility of intestine.
Secretin	Stimulates secretion of HCO_3^- from pancreas.
Vasoactive intestinal polypeptide (VIP)	Relaxes smooth muscle in blood vessels to increase blood flow.

shortest (~30 cm), is most clearly defined. Digestion starts in the duodenum, which receives an enzyme-rich secretion from the pancreas and bile from the gall bladder. Absorption is normally completed before the end of the jejunum is reached. The structure of all parts of the small intestine is similar: an inner ridged surface with each fold bearing numerous villi, slender extensions of the absorptive surface that greatly increase its surface area (Figure 4.4). Various cells in the duodenum secrete the following: bicarbonate to neutralise the acid entering from the stomach; about three litres of mucus each day; peptides that control the secretion of other substances (Table 4.2) and opioid peptides that inhibit the contraction of the smooth muscles, which slows the movement of food along the intestine. Most of the epithelial cells lining the

Consequently the opioid drugs (e.g. codeine, morphine), which are normally used to control pain, can cause constipation.

villi are enterocytes, which absorb the products of digestion. Numerous microvilli (each about 0.1 μm in diameter) produce a luminal border for these cells, known as the brush border. Altogether, the folds, villi and microvilli increase the surface area by about 600-fold. The total area is estimated at 200 m^2, which is larger than a doubles tennis court!

Most cells have a superficial layer of polysaccharide, known as the glycocalyx, which is attached to the cell surface. This is particularly well developed in enterocytes. Some of the digestive enzymes from the lumen are adsorbed onto the glycocalyx. The bulk of digestion occurs in the lumen of the intestine, but the enzymes on the glycocalyx catalyse the final stages of some processes.

Colon

The colon is the main part of the large intestine. It is wider but shorter (100–150 cm) than the small intestine and it contains wide tubular crypts. The colon is divided into ascending, transverse and descending sections (Figure 4.1). The epithelium contains three types of cells: colonocytes, mucus-secreting and peptide-secreting cells. An important function of the colonocytes is reabsorption of water, sodium ions and chloride from the lumen while secreting potassium and hydrogencarbonate ions into it. This results in the formation of more solid faeces. The lumen contains a massive number (approx. 10^{14}) of microorganisms (microflora) which degrade those carbohydrates that are not digested in the small intestine. The resultant sugars are fermented by the microflora to produce short-chain fatty acids (Chapter 6), which are important fuels for the colonocytes. In addition, butyrate reduces proliferation of colonocytes, which might reduce the risk of cancer of the colon (Box 4.2). Burkitt (1975).

> Ingestion of commercial preparations that contain digestion-resistant starches and bacteria (e.g. homolactic lactobacilli) increases volatile fatty acid formation in the colon. This provides more fuel for colonocytes: it is claimed regular intake of these preparations improves intestinal function and hence mood, known sometimes as the 'feel good' factor.

Dietary fibre

Dietary fibre was defined by Hugh Trowell as 'the plant polysaccharides and lignin which are resistant to hydrolysis by the digestive enzymes of humans'. This definition lacks chemical precision, because non-fibrous pectins and gums are also present. The term non-starch polysaccharide (NSP) is often preferred, although the term dietary fibre still persists. Unfortunately, NSP is also not satisfactory since some starch, known as 'resistant' or 'partially resistant starch', does not undergo hydrolysis, or complete hydrolysis, in the small intestine, due to its physical form in food, which depends upon processing and cooking of the food and even on the ripeness of fruit. For example, 90% of the starch in unripe bananas is resistant. Since resistant starch is degraded to sugar by the microflora, and is fermented, it could be included in the definition of dietary fibre. The term dietary fibre, as used in this text, includes NSP. Fibre is most abundant in cereals, bread made from whole grain flour, leafy and root vegetables, mature leguminous seeds, nuts and fruits. One component of fibre is cellulose, which is not digested by humans but can be degraded to glucose by some microorganisms, which then ferment the glucose. It is not known how much cellulose can be digested by microorganisms in the human intestine and it probably varies from one human to another. However, it is well developed in ruminants (Box 4.3).

> Adults in developed countries may consume about 40 g of such starch and between 10 and 20 g of fibre each day. Certain types of dietary fibre increase the faecal loss of bile salts. The loss can be increased artificially by the administration of ion exchange resins that bind the bile salts. This is one means of lowering the liver and blood levels of cholesterol (Box 4.2).

Urea salvage

Approximately 30% of the urea produced by the liver diffuses into the colon where it is degraded by some of the microorganisms, i.e. those that possess the enzyme urease, to form ammonia

$$\text{urea} \rightarrow NH_3 + CO_2$$

The microorganisms use the ammonia to synthesise essential and non-essential amino acids and hence protein. When they die, their protein is degraded and some of the amino acids are absorbed into the bloodstream, to be used by the host. This process, urea salvage, is quantitatively significant in provision of amino acids for the host, especially for individuals on a protein deficient diet (Chapter 8).

Biochemistry of cooking and food preparation

Chewing of food and salivation in the buccal cavity is often considered to be the first stage of digestion but, for humans, it can be considered that cooking and preparation of the food can be included. An excellent discussion of the chemistry and biochemistry of cooking is provided by Pond (1998). Plant tissues that are rich in starch are not palatable and are difficult to chew and digest. This is because starch consists of two glucose polymers, one of which is branched (amylopectin) and the other linear (amylose). The amylose lies within the branches of amylopectin producing a tightly packed structure (similar to that of reinforced concrete). The heat of cooking (boiling, baking or frying) breaks down this structure and promotes the uptake of water or fat, which loosens the carbohydrate structure. For example, pasta when boiled takes up so much water that its volume

Box 4.2 Dietary fibre and health

The health benefits of dietary fibre were first brought to general attention by Denis Burkitt, Peter Cleave and Hugh Trowell in 1960s (Box 4.4). As much as 50% of the contents of the colon are accounted for by bacteria and between 400 and 500 different species are present. Much of the increase in stool bulk arising from a high-fibre diet is actually due to an increase in the number of bacteria. A high intake of fibre decreases transit time through the gut and increases the bulk of the stool. It is a traditional remedy for constipation. A high-fibre diet has also been used in the successful treatment of diverticular disease, where pockets form in the wall of the colon. However, the greatest impact on the general public has come from claims that cardiovascular disease, diabetes, obesity and, particularly, colon cancer may be caused by a lack of fibre in the diet. In response, manufacturers of some foods provide information on fibre content of the food.

It is not clear how the presence of fibre on partly fermented fibre in the gastrointestinal tract could exert such protective effects but a number of hypotheses have been put forward:

(i)   Fibre binds carcinogenic substances; in the intestine this should prevent their absorption into the colonocytes. Alternatively, the increase in faecal mass due to the presence of fibre could simply 'dilute' a potential carcinogen.

(ii) Fibre binds bile salts so that they are lost in the faeces which decreases their re-uptake from the intestine, causing the liver to make more at the expense of blood cholesterol, the concentration of which decreases, with possible cardiovascular benefits.

(iii) The volatile fatty acid, butyric acid, formed by fermentation of carbohydrate in the fibre by the microorganisms, inhibits proliferation of colonocytes and facilitates DNA repair.

(iv) Fibre slows the absorption of glucose from the small intestine and hence it reduces the increase in the blood glucose after a meal. In particular, a chronic decrease in the peak concentration of glucose could reduce glycosylation of low density lipoprotein (LDL) in the blood: glycosylated ('damaged') LDL is involved in development of atherosclerosis (Chapter 22) (Trowell & Burkitt 1981).

The bacteria in the human colon also produce a considerable volume of gas which includes carbon dioxide, hydrogen and methane. Indeed, the extent of fermentation in the gastrointestinal tract can be estimated from the amounts of hydrogen and methane that appear in exhaled air. If too much gas is produced, flatulence can result which can be relieved by taking charcoal-containing tablets to absorb the gas. Since the two gases are flammable, diets that reduce their formation are important for long-term space flights.

Box 4.3 Digestion of cellulose

The largest amount of renewable energy on the planet is contained in cellulose, a linear polymer of glucose. There is, however, only one enzyme that degrades cellulose to produce glucose: this is cellulase. Although animals do not produce this enzyme, some bacteria and fungi do and are made use of by animals, a remarkable example of symbiosis. For example, ruminants (e.g. cows, sheep) have developed a large and compartmentalised stomach to provide an environment for the bacteria. The largest of the compartments is the rumen, a vast fermentation chamber (100 litre capacity in the cow) in which bacteria thrive. Some of these possess cellulase and hydrolyse the cellulose ingested to produce glucose. The glucose is fermented by the microflora in the rumen, as in the colon of other mammals, to produce a mixture of acetic, propionic, lactic and butyric acids (the volatile on short chain fatty acids). A prodigious volume of methane is also generated – cows need to belch up to 80 litres of gas each day and, if this is prevented, a distressing condition known as 'bloat' can develop. It is believed that the quantities of methane produced by ruminants throughout the world contributes to global warming!

The contents of the rumen pass steadily into the third chamber of the ruminant 'stomach', the omasum, where the fatty acids, together with water and salts, are absorbed. These fatty acids provide much of the energy for the ruminant but a price is paid: virtually all the carbohydrate in the diet is fermented and almost none enters the body. Consequently, glucose must be synthesised to provide lactose in the lactating animals. The lactic and propionic acids are the precursors for the glucose (Chapter 6). It is unclear if some of this glucose is used by the brain of the ruminant.

Finally, the contents of the omasum, now a thick slurry of microorganisms, pass into the abomasum into which are secreted acid and proteinases to produce an environment corresponding to that of the human stomach. Some of the microflora passing from the rumen to the omasum die and are digested by the acid and the enzymes. This provides the ruminant not only with an additional energy source but with vitamins and essential amino acids that its own tissues cannot synthesise.

Box 4.4 Peter Cleave and fibre

Peter Cleave was a surgeon in the Royal Navy, with no scientific training. He noticed that certain diseases were more prevalent in developed countries than in underdeveloped countries. These are sometimes known as the 'Western Diseases'. On the basis of these observations he came up with the concept of 'the saccharine disease' (Cleave 1974).

The contribution of Peter Cleave to the appreciation of the importance of fibre and refined carbohydrate has been admirably summarised by Sir Richard Doll:

Interest in dietary fibre is . . . so recent that we can still recognise its origin in Peter Cleave's concept of 'the saccharine disease'. Under this title Cleave brought together a variety of conditions characteristic of industrial society which, he thought, were due to overconsumption of carbohydrates, made easy to absorb and unsatisfying to the appetite by the refinement that they had undergone in the course of their preparation for the Western market. Stimulating though this idea was,

it did not attract much support, because it failed to provide a comprehensible explanation for the pathogenesis of many of the diseases concerned. When, however, Trowell and Burkitt inverted the idea by suggesting that the dietary fibre that had been removed in the course of the refinement of carbohydrate was a specific nutrient and that many pathological effects could be attributed to a deficiency of it, a whole new vista of possible mechanisms was revealed . . . and the corollary that we should return to a more natural diet corresponded so well with the ecological spirit of the times, that the idea was widely accepted and national diets began to be modified, while the scientific evidence lagged behind. Now, however, the biochemical and physiological facts are beginning to emerge.

Sir Richard Doll, foreword to *Dietary Fibre, Fibre-depleted Foods and Disease* (Trowell *et al.*, 1985)

---

increases threefold. This makes it softer, easier to chew and improves access for the digestive enzymes. Similarly, an uncooked potato is difficult to eat but, when cut into pieces and fried, it becomes easier to eat and very palatable.

Fats Fat boils at higher temperatures than water, thus speeding up the cooking process. Unfortunately, the high temperatures involved in frying or baking can cause oxidation of unsaturated fatty acids to form toxic products, so that mixing and boiling, which increase access to oxygen, should be avoided. Slow cooking at a lower temperature reduces the risk of oxidation. Another important role of fat is to dissolve spices, many of which are not soluble in water. Most sauces that contain spices are based on oil or fat.

The high temperatures used in cooking convert solid fat in the food into liquid fat, which more readily forms an emulsion, especially in combination with bile acids in the small intestine. In this form, the fat is more accessible to digestive enzymes (see below).

Protein Except in vegetarians, most of the dietary protein is from muscle of other animals, which is fibrous. These fibres must be broken down for palatability, ease of eating and digestion. Breakdown is more readily achieved if the meat is cut into thin strips and heated to a high temperature by baking, frying or grilling. If the meat is boiled, it must be cooked for a longer time relative to baking as the temperature is much lower. One method of preparation, now not commonly used in developed countries, is 'hanging'

the meat for several days before cooking, during which time the proteolytic enzymes within the fibres begin to degrade the myofibrils, so that the meat becomes tender. A similar result can be achieved by injection of commercially available proteolytic enzymes into the meat.

Digestion and absorption

The nutrients discussed in this section are carbohydrate, fat, protein and nucleic acids. The nucleic acids, upon digestion, provide phosphate, bases and nucleosides (Chapters 10 and 20). Each is discussed under four separate headings:

- Digestion.

- Transport of the products of digestion from the lumen into the enterocyte.

- Metabolism of some of the absorbed compounds within the enterocyte.

- Transport of the compounds across the basolateral membrane into the interstitial space and then into the blood.

Carbohydrate

Carbohydrate accounts for about 40% of the energy content of a normal meal. In the UK, the daily intake is typically 60 g starch, 120 g sucrose, 30 g lactose, 10 g of glucose and fructose. Starch, sucrose and lactose are digested to glucose, fructose and galactose, which are the carbohydrate molecules absorbed by the enterocytes.

Digestion

Digestion of starch involves the hydrolysis of the bonds between the glucose molecules. Two classes of hydrolytic enzymes are required: amylases and oligo- and disaccharidases (Figure 4.5). The disaccharidases are also involved in hydrolysis of sucrose and lactose.

Amylases

The enzyme amylase is present in saliva and in the pancreatic secretions. It hydrolyses bonds both in amylose and amylopectin. The products are maltose (a disaccharide of two glucose molecules), maltotriose and oligosaccharides. Digestion of starch begins in the mouth, continues in the stomach and is completed in the lumen of the duodenum, where the oligosaccharidases and disaccharidases are also present. Some of these two enzymes are attached to the glycocalyx, so that they are close to the enterocytes, which increases the local concentration of the monosaccharides to facilitate their absorption. In addition, it decreases the availability of glucose for pathogens that may be present in the lumen.

The most abundant sugar in the human diet is sucrose, which is hydrolysed to glucose plus fructose by a disaccharidase, sucrase

$$sucrose \rightarrow glucose + fructose$$

Lactose is the sugar present in milk, which is hydrolysed by the enzyme lactase

$$lactose \rightarrow glucose + galactose$$

The galactose and fructose produced in these reactions are absorbed and transported to the liver, where they are converted to glucose.

Transport of monosaccharides into the enterocyte

Monosaccharides are transported across the luminal surface of the enterocytes that line the microvilli. There are two transporter systems: one for glucose and galactose, which is dependent upon Na^+ ions, and the other for fructose, which does not require Na^+ ions: a molecule of glucose is transported into the cell against its concentration gradient and the energy for this is obtained from the transport of a Na^+ ion down its concentration gradient (Chapter 5). This ensures that the glucose concentration in the enterocyte is higher than that in the bloodstream, so that glucose is transported across the contralateral membrane into the interstitial space (and then into the blood) by a glucose transporter that does not require Na^+ ions. In contrast, the concentration of fructose in the lumen of the intestine will be higher than in the enterocyte, so that it enters the enterocyte down its concentration gradient and the transporter does not require Na^+ ions (Figure 4.6).

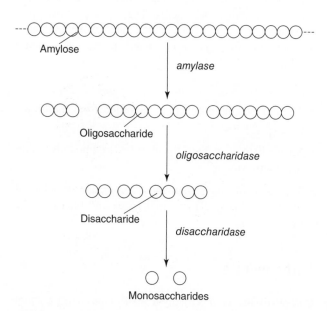

Figure 4.5 *Digestion of starch and saccharides.* The three enzymes hydrolyse the bonds between the glucose molecules.

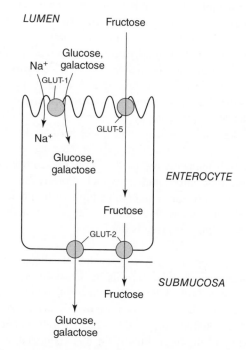

Figure 4.6 *Transport of glucose, galactose and fructose in and out of the enterocyte.* The transporter GLUT-5 is specific for fructose: GLUT-2 is non-specific (see Chapter 6).

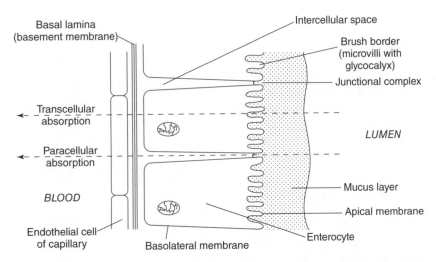

Figure 4.7 *Entry into the blood via transcellular and paracellular routes.*

Another means of transport across the intestine is via the paracellular route, that is between the adjacent enterocytes. Water can enter the intestinal space through this route and take with it small molecules including glucose, amino acids and small peptides. This is known as 'solvent drag' (Figure 4.7). Unfortunately, the quantitative importance of this route is not known.

Metabolism of monosaccharide within the enterocyte

Not all of the glucose or fructose taken up by the enterocyte passes directly through and into the blood. Hexokinase and the other glycolytic enzymes are present in these cells and some glucose and fructose is converted into lactic acid. How much glucose is metabolised in this way is not known, since the lactic acid is taken up by the liver and converted back to glucose. It may be as high as 50% of the glucose absorbed (Chapter 6).

The ATP generated from glycolysis will provide energy to maintain the Na^+ ion gradient for the transport of glucose and amino acids and for the formation of chylomicrons in the enterocyte.

Fat

Adults in developed countries ingest about 100 g of fat each day (about 40% of the total energy intake) most of which is from the adipose tissue of other animals oily fish, where fat is present in the muscle (e.g. salmon, mackerel, sardine) and the seeds or fruits of a number of plants (sunflower, peanut, safflower, palm, olive). The best place to see the fat that is eaten by humans is in a butcher's shop or the meat counter in a supermarket: the adipose tissue is in separate depots, which can be readily observed in animal carcasses but there is also a considerable quantity attached to the meat, especially in commercially reared animals, which are largely the ruminants (Figure 4.8). The fat consists of triacyglycerol, in which the three hydroxyl groups of glycerol are esterified with long-chain fatty acids (Chapter 7).

The digestion and absorption of fat is considerably more complex than that of carbohydrate or protein because it is insoluble in water, whereas almost all enzymes catalyse reactions in an aqueous medium. In such media, fat can form small droplets, an emulsion, which is stable in this medium. Formation of an emulsion is aided by the presence of detergents: these possess hydrophobic and hydrophilic groups, so that they associate with both the fat and the aqueous phases. Such compounds are known as emulsifying agents and those involved in digestion are mainly the bile salts and phospholipids.

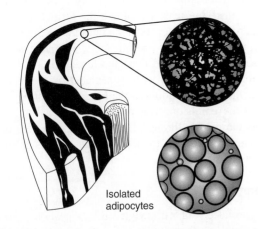

Figure 4.8 *Fat in a lamb or pork chop.* Adipose tissue consists of connective tissue in which adipocytes are dispersed. Adipocytes are small spherical cells comprising 90% triacylglycerol (Chapter 7).

Digestion

The hydrolysis of triacylglycerol is catalysed by lipases, two of which are present in the stomach. These are lingual lipase, which is secreted by the soft palate, and gastric lipase, which is secreted by the gastric glands of the stomach. Gastric lipase is particularly important in the newborn since, at this stage of life, pancreatic secretions contain relatively little lipase.

Lingual and gastric lipases catalyse only partial lipolysis to produce fatty acids and diacylglycerol.

$$\text{triacylglycerol} \rightarrow \text{fatty acid} + \text{diacylglycerol}$$

This lipolysis is important for four reasons:

- The long-chain fatty acids stimulate the secretion of cholecystokinin, which increases secretion of gastric lipase (i.e. positive feedback).

- The long-chain fatty acids (containing mainly 16 or 18 carbon atoms) have antibacterial activity.

- Diacylglycerol is a better substrate than triacylglycerol for the lipolytic enzymes that are secreted by the pancreas.

$$\text{diacylglycerol} \xrightarrow{\text{pancreatic lipase}} \text{monoacylglycerol} + \text{FA}$$
$$\text{monoacylglycerol} \xrightarrow{\text{pancreatic esterase}} \text{fatty acids}$$

- The formation of monoacylglycerol aids fat absorption (see below).

Emulsifying agents

Since triacylglycerol is immiscible with the aqueous phase, lipase activity is restricted to the surface of lipid droplets. The smaller the droplets, the greater their surface area and the more readily the lipase can hydrolyse the triacylglycerol. Emulsifying agents facilitate the formation of very small droplets; such agents in the intestine are: monoacylglycerol, phospholipids and bile salts. The latter are synthesised in the liver and enter the lumen via the bile. In addition, colipase, a protein, is secreted by the pancreas and facilitates the binding of the lipase to the triacylglycerol in the emulsion. The process of peristalsis, which moves the chyme through the intestine, also churns the contents, which helps to maintain the small size of the droplets in the emulsion.

Absorption

The products of triacylglycerol digestion, mainly monoacylglycerol and long-chain fatty acids, interact with bile salts to form micelles, which comprise bile salts/ monoacylglycerols/phospholipids and fatty acids. The micelle aids the absorption of monoacylglycerol and fatty acids by effectively increasing the solubility of these compounds and hence their diffusion through the aqueous contents of the lumen to the epithelial cell surface. At this surface, the monoacylglycerols, fatty acids and phospholipids are released from the micelle to be absorbed into the enterocyte (Figure 4.9). The bile salts remain in the lumen to continue their role in the absorption of the fat. They are eventually absorbed from the ileum by a specific transport system (see below). The uptake of fatty acids into the enterocytes is facilitated by the presence of a fatty acid binding protein (FABP) in the cytosol. FABP is a small protein with a very high affinity for long-chain fatty acids, so that it lowers their concentration in the cytosol, promoting diffusion into the cell (Chapter 7). The concentration of FABP increases in enterocytes in response to a fat-rich diet.

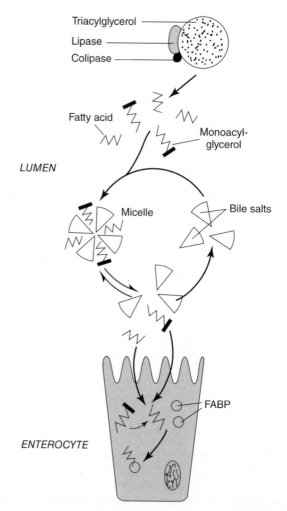

Figure 4.9 *Formation of a micelle and its role in the uptake of fatty acids and monoacylglycerol into enterocytes. The micelle is stable in the aqueous environment of the intestinal lumen and is necessary for movement of the monoacylglycerol and fatty acids in the lumen.*

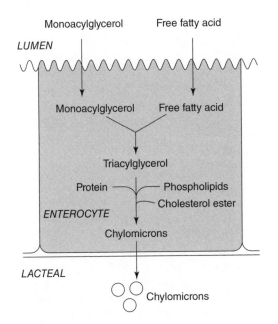

Figure 4.10 *Monoacylglycerol pathway for synthesis of triacylglycerol and formation of chylomicrons within the enterocyte.*

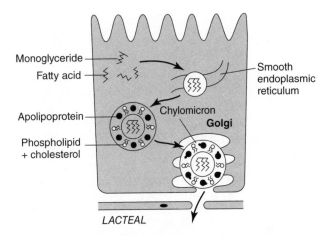

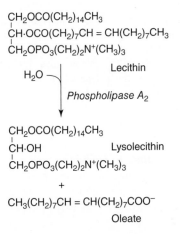

Figure 4.11 *Details of the formation of a chylomicron and its structure.* Triacylglycerol is synthesised upon the smooth endoplasmic reticulum, chylomicrons are synthesised in the cytosol and then secreted into the lacteal via the Golgi.

Metabolism of the monoacylglycerol and fatty acids in the enterocyte: formation of chylomicrons

After their entry into the enterocyte, fatty acids and monoacylglycerols are re-esterified to form triacylglycerol (Figure 4.10). The triacylglycerol so formed then associates with phospholipids and cholesterol ester, together with a specific protein (apolipoprotein B-48) to form complexes known as chylomicrons (Figure 4.11). These are secreted into the lymphatic vessels, or lacteals, within the villi (see Figure 4.4). The lymph then flows through the lacteals into the large thoracic duct of the lymphatic system and the chylomicrons enter the blood via the subclavian vein. Subsequently, the triacylglycerol in the chylomicrons is hydrolysed by a lipoprotein lipase in adipose tissue and muscle and the fatty acids are taken up by these tissues. Chylomicrons are secreted into the lymphatic system rather than into the hepatic portal vein in order to bypass the liver. They are large particles which might become trapped within the capillaries or sinusoids of the liver and hence block the blood supply.

Phospholipids, cholesterol, bile salts

Phospholipids are digested and absorbed in a similar manner to that of triacylglycerol. Pancreatic lipase has some hydrolytic activity towards phospholipids and removes the fatty acid from the 1-position. The product is a lysophospholipid such as lysolecithin (Figure 4.12). It also acts as a detergent and contributes to the stability of the mixed micelles.

$$CH_2OCO(CH_2)_{14}CH_3$$
$$CH{\cdot}OCO(CH_2)_7CH = CH(CH_2)_7CH_3$$
$$CH_2OPO_3(CH_2)_2N^+(CH_3)_3$$

Lecithin

$$H_2O$$

Phospholipase A$_2$

$$CH_2OCO(CH_2)_{14}CH_3$$
$$CH{\cdot}OH$$ Lysolecithin
$$CH_2OPO_3(CH_2)_2N^+(CH_3)_3$$

+

$$CH_3(CH_2)_7CH = CH(CH_2)_7COO^-$$

Oleate

Figure 4.12 *Hydrolysis of a phospholipid (lecithin) in the lumen by a phospholipase. Lysolecithin is a lysophospholipid and is a detergent. At high concentrations it can damage membranes. It is also produced during repair of damaged phospholipids (Chapter 11)*

About 1 g of cholesterol is ingested by adults each day in developed countries. A similar amount enters the lumen via the bile, synthesised from acetyl-CoA in the liver, is also released from sloughed epithelial cells. Absorption of cholesterol also occurs from the mixed micelles. Within the enterocyte, it is esterified and the cholesterol ester is incorporated into the chylomicrons.

Bile salts In contrast to cholesterol, bile salts are absorbed mainly in the jejunum. They are returned to the liver through the hepatic portal vein (in association with proteins) and can thence be re-secreted into the bile. The transport of bile salts between liver and intestine is known

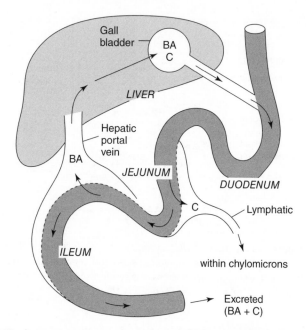

Figure 4.13 *Uptake of bile acids in the jejunum.* Bile acids (BA) and cholesterol (C) are secreted from the liver, via the bile, into the duodenum. Cholesterol is transported back into the blood, from the enterocyte, within chylomicrons. The latter enter the lymphatic system (i.e. the lacteals). Bile acids are absorbed from the jejunum into the hepatic portal vein for re-uptake into the liver.

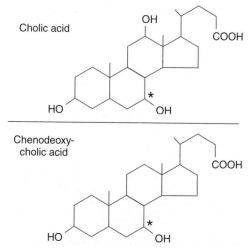

Figure 4.14 *Structures of cholic and chenodeoxycholic acids which are the acids that form the bile salt.* The asterisk indicates the position at which an ester bond is formed with taurine or glycine so that bile salts are taurocholate, chenodeoxytaurocholate, glycocholate, and glycochenodeoxycholate are formed. The structure of taurine is $H_2NCH_2CH_2SO_3$ and glycine is H_2NCH_2COOH.

as the *enterohepatic cycle* of bile salts (Figure 4.13). Some of these salts and cholesterol are, however, lost in the faeces, necessitating the synthesis of new bile salts from cholesterol in the liver (Figure 4.14) (Appendix 4.2).

Protein

About 90 g of protein is ingested each day by an adult on a typical Western diet and a similar amount enters the intestine from the secretion of enzymes, proteins and from sloughed epithelial cells (Chapter 8). The proteins are hydrolysed by proteolytic enzymes.

Digestion

Protein digestion occurs in two stages: endopeptidases catalyse the hydrolysis of peptide bonds within the protein molecule to form peptides, and the peptides are hydrolysed to form the amino acids by exopeptidases and dipeptidases. Enteropeptidase initiates pro-enzyme activation in the small intestine by catalysing the conversion of trypsinogen into trypsin. Trypsin is able to achieve further activation of trypsinogen, i.e. an autocatalytic process, and also activates chymotrypsinogen and pro-elastase, by the selective hydro-

lysis of a small number of peptide linkages (Appendix 4.1). The processes can be summarised by the equations:

$$\text{trypsinogen} \xrightarrow{\text{enteropeptidase}} \text{trypsin} + \text{peptide}$$

$$\text{chymotrypsinogen} \xrightarrow{\text{trypsin}} \text{chymotrypsin} + \text{peptide}$$

$$\text{pro-elastase} \xrightarrow{\text{trypsin}} \text{elastase} + \text{peptide}$$

These proteolytic enzymes are all endopeptidases, which hydrolyse links in the middle of polypeptide chains. The products of the action of these proteolytic enzymes are a series of peptides of various sizes. These are degraded further by the action of several peptidases (exopeptidases) that remove terminal amino acids. Carboxypeptidases hydrolyse amino acids sequentially from the carboxyl end of peptides. They are secreted by the pancreas in pro-enzyme form and are each activated by the hydrolysis of one peptide bond, catalysed by trypsin. Aminopeptidases, which are secreted by the absorptive cells of the small intestine, hydrolyse amino acids sequentially from the amino end of peptides. In addition, dipeptidases, which are structurally associated with the glycocalyx of the enterocytes, hydrolyse dipeptides into their component amino acids.

Although amino acids are the major products of protein digestion that are absorbed, some dipeptides and tripeptides are produced and are absorbed without further digestion.

> Peptides are now included in some parenteral feeds for patients because they have less effect than amino acids on the osmotic pressure of the solution. Osmotic pressure can be a problem since the feed enters the blood directly.

Absorption of amino acids and small peptides

Amino acids and some small peptides are absorbed into the enterocytes in the jejunum. The transport of amino acids from the lumen into the cell is an active process, coupled to the transport of Na^+ ions down a concentration gradient. There are at least six carrier systems with different amino acid specificities: neutral amino acids (i.e. those with no net charge, e.g. branched-chain amino acids); neutral plus basic amino acids; imino acids (proline, hydroxyproline) and glycine; basic amino acids (e.g. arginine and lysine); β-amino acids and taurine; acidic amino acids (glutamic and aspartic acids).

Additional 'back door' carriers are needed to transport amino acids across the basolateral membrane of enterocytes into the interstitial space and then into the blood. These 'back door' carriers are similar to those present in other cells (see Chapter 5).

Metabolism of amino acids within the enterocyte

Some metabolism of amino acids occurs in the enterocyte:

- Several peptides are synthesised, for example, glutathione (which is a tripeptide consisting of glutamate, glycine and cysteine. Chapter 8), and peptides that control activities within the intestine (see Table 4.2).

- Arginine is converted to citrulline, which is released into the blood. This protects arginine from uptake and degradation in the urea cycle in the liver. Citrulline is released into the blood and transported to the kidney, where it is converted back to arginine (Chapter 8).

- Some of the glutamine that is absorbed is metabolised in the enterocytes. It is used, along with glucose, as a fuel to generate ATP (Chapter 8). The ammonia and the alanine that are produced enter the blood for uptake by the liver.

- Glutamate and aspartate are transaminated to their respective oxoacids (oxoglutarate and oxaloacetate) which are oxidised to generate ATP. Therefore, they do not enter the bloodstream, so that their concentrations in the blood are very low (Chapter 8). This is important because these amino acids act as neurotransmitters in the brain, where their concentrations are strictly controlled. The transaminase enzymes in the intestine can, therefore, be considered to act as detoxifying enzymes. A health problem can arise if a large amount of glutamate is ingested in one meal (Chapter 8).

Most of the amino acids that enter the portal blood are taken up and metabolised by the liver. Hence, some synthesis of peptides taken place in the enterocytes where there is a plentiful supply of amino acids for these processes.

Nucleic acids

In common with the digestion of other macromolecules, nucleic acids are hydrolysed in a stepwise manner, by pancreatic nuclease (diesterase enzymes) which hydrolyse the bonds between two adjacent phosphate groups in RNA and DNA. The resultant oligoribonucleotides and oligodeoxy ribonucleotides are hydrolysed to form nucleoside monophosphates, which lose their phosphate to form nucleosides, by the action of pancreatic phosphatase. In brief, the process is:

$$nucleic\ acid \rightarrow oligonucleotides \rightarrow nucleoside$$
$$monophosphates \rightarrow nucleosides$$

(For details of these reactions, see Chapter 10).

Absorption

Nucleosides are absorbed into the enterocyte by nucleoside transporters and are then transferred to the interstitial space and into the blood. They are taken up by cells in various tissues and converted back to nucleoside triphosphates, which can then be re-incorporated into nucleic acids (Figure 4.15).

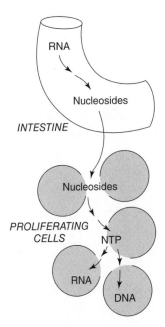

Figure 4.15 *A summary of the fate of nucleosides that are produced from RNA digestion in the lumen of the intestine. The nucleosides produced from RNA in the lumen are absorbed by the enterocytes and then transported from the intestine into the blood from where they are taken up by cells (especially proliferating cells, e.g. in the bone marrow) to form nucleotides for DNA and RNA synthesis. (See Chapter 10) NTP is nucleoside triphosphate.*

Immunonutril is an enteral feed produced by Sandoz that contains RNA. The nucleosides that enter the blood after digestion of the RNA can be taken up by the immune cells to provide nucleotides required for their proliferation to improve the immune response. Hence the name of the feed.

The nucleosides are converted back to nucleosides is reactions known as salvage pathways (Chapter 20). RNA supplements in the diet are also hydrolysed to form nucleosides, which can be taken up by cells in the body to form nucleotides. Although proliferating cells can synthesise nucleotides *de novo*, provision of nucleosides bypasses a considerable amount of biochemistry. This is important, for example, in lymph nodes for proliferating immune cells in the bone marrow and also for stem cells in the crypts of the villi (Chapter 20).

The gastrointestinal tract and disease

Considering the importance of the gastrointestinal tract for the entry of essential compounds into the body, it is surprising that its function is so rarely impaired by the massive variations in quantity and composition of the food consumed by humans. Nonetheless, when it is impaired the consequences can be severe. Examples are:

- Sensitivity to the protein gliadin, present in gluten, results in coeliac disease.

- Trauma, major surgery, burns or haemorrhagic shock can result in damage to the small and large intestine. Damage to the colon in particular can increase the translocation of bacteria into the peritoneal cavity leading to sepsis (see below and Chapter 18).

- Crohn's disease, in which segments of the intestine suffer chronic inflammation. The cause is unknown but it may be an autoimmune disease.

- Infection of the intestine is common if the immune system is impaired by malnutrition.

- Patients suffering from cystic fibrosis have problems digesting fat.

- Patients suffering from obstruction of the bile duct fail to digest fat, due to a deficiency of bile salts to maintain an effective emulsion of fat.

Coeliac disease

Coeliac disease probably arose around 10000 BC when humans switched from the hunter-gatherer way of life to cultivation of cereals (barley, wheat and oats). An illness resembling the disease was described by Aretaeus of Cappadocia as early as the first century (he also described diabetes mellitus).

The condition results in an immune response in the intestine to the protein gliadin, from gluten, a constituent of the germ of wheat, barley, oats and rye. Sensitivity to gliadin arises through stimulation of B-lymphocytes to produce antibodies of class E which activate mast cells. These mast cells release their contents of granules that contain many toxic compounds (see Chapter 17 for a full discussion). This results in loss of villi, especially in the proximal part of the small intestine. The symptoms include pale, bulky, loose, offensive stools, abdominal distortion and discomfort presumably caused by failure of the small intestine to digest a considerable proportion of the food, which then passes into the colon where it is fermented. Treatment is a gluten-free diet for life but damage to the mucosa may be such that it can take up to six months for symptoms to disappear.

Sprue (post-infective malabsorption)

The healthy small intestine contains only a small bacterial population, unlike the colon. However, an acute infection of the mucosa by a virus, bacterium or other parasite can reduce its motility, allowing a huge proliferation of the resident bacteria. Absorption of both macro- and micronutrients is impaired, resulting in the disorder known as sprue. Folic acid is particularly poorly absorbed, causing reduced rates of repair of mucosal cells. Hence, the damage persists and worsens to create a vicious circle. Treatment involves administration of an antibiotic to kill the bacteria and folic acid to allow damaged tissue to recover. The clinical presentation includes bulky stools, steatorrhoea (fatty faeces) and weight loss.

The intestine usually constitutes an effective immunological barrier to an invasion by pathogens but this protection can be overwhelmed by frequent repeat attacks. This may explain the prevalence of sprue in the tropics (hence the name, tropical sprue) especially for visitors who are unaccustomed to contaminated food and water and hence have no immune protection against the bacteria specific to that part of the tropics.

Enzyme deficiencies

Huge amounts of digestive enzymes can be produced by the gastrointestinal tract in order to digest vast quantities of food that may be ingested. Nonetheless, low activities of three enzymes can occur: sucrase, lactase or pancreatic amylase, which can lead to problems.

Sucrase A deficiency of sucrase results in failure to hydrolyse sucrose, which passes unchanged into the large intestine. Here it raises the osmotic pressure in the colon, reducing water reabsorption and promoting diarrhoea. In addition, bacteria in the colon ferment the sugar, producing short-chain fatty acids and the gases methane and hydrogen, which distend the colon and cause discomfort.

Lactase The disaccharide lactose is the only carbohydrate present in milk, which is essential for survival of an infant. Consequently, the enzyme lactase is essential for babies. Caucasians retain lactase activity into adulthood, whereas many Asian or African groups progressively lose its activity in adult life. This could, therefore, be described as an adult deficiency disease. Ingestion of milk in these individuals causes nausea, diarrhoea and stomach cramps. Symptoms disappear if milk is excluded from the diet or if a source of lactase is ingested along with or before ingestion of milk. The bacteria that are involved in the production of yoghurt contain the enzyme lactase.

Pancreatic amylase Infants with this deficiency begin life unable to digest starch. However, after a few months, the pancreas starts to produce sufficient amylase. Adults produce such an excess of this enzyme that even patients with severe pancreatitis (who are unable to produce sufficient lipolytic or proteolytic enzymes) can produce sufficient amylase to cope with a normal amount of starch in the diet.

An inhibitor of α-amylase, acarbose, obtained from the red kidney bean, prevents breakdown of starch which, consequently, passes into the colon unchanged, where it is fermented. It is sometimes used to reduce the peak levels of blood glucose in diabetic patients after a meal, as it slows the absorption of glucose.

Intestinal damage in response to shock or trauma

Haemorrhagic shock, heat stress or trauma can lead to loss of blood or diversion of blood away from the small intestine. As a result, the tips of the villi receive much less blood than the crypts. If prolonged, this leads to loss of the microvilli and glycocalyx and damage to the tight junctions. Less mucus may be secreted making it easier for bacteria to bind to and enter epithelial cells. Furthermore, when normal blood flow returns to the intestine, there can be a marked increase in the formation of oxygen free radicals, resulting in damage (reperfusion damage, Appendix 9.6).

The health of the small intestine can also be compromised by poor nutrition or starvation, which can regularly occur in hospital. Patients are normally starved overnight, prior to operation, and starvation may continue well after completion of the surgery because of the nature of the operation or because of the anorexia that can result from surgery or anaesthesia. This will deny adequate nutrition to epithelial cells in the intestine, which can result in slow recovery after surgery and, in addition, may compromise the immune system. Provision of food by the enteral route, as soon as possible after injury, surgery, sepsis or burns, is therefore highly desirable (Chapter 18).

5
Transport into the Cell: Particles, Molecules and Ions

We are proposing that the transfer of glucose into the cell is similarly not a free and unlimited process but rather an important limiting factor upon which insulin acts to promote entry. The subsequent fate of the glucose will depend on the presence and state of activity of the enzyme apparatus . . . A hypothesis of insulin action is proposed which attributes to insulin the role of facilitating the rate of transfer of some hexoses into the cell.

(Levine et al., 1950)

A paper published six decades ago was the first to draw attention to the possibility that a change in the rate of transport of a molecule across the plasma membrane could play a role in regulation of both intra- and extracellular metabolism; that is, the regulation of the blood glucose level and the intracellular metabolism of glucose. The paper was entitled 'A hypothesis of insulin action is proposed which attributes to insulin the role of facilitating the rate of transport of some hexoses into the cell as opposed to a direct effect on intracellular metabolism' (Levine *et al.* 1950).

The plasma membrane forms a boundary between the extra- and intracellular environments whereas membranes within a cell form boundaries between the organelles and the cytosol. These are discussed in other chapters, whereas the material in this chapter focuses on the plasma membrane. A primary function of this membrane is to serve as a barrier to prevent the entry of some molecules and ions into the cell and to retain others within the cell (Table 5.1). The plasma membrane has other roles, which are related to the presence of proteins within or attached to the membrane. These are:

- Transport of fuels into the cells.

- Regulation of the rate of transport of fuels, according to the requirements of the cell or extracellular signals (e.g. hormones).

- To maintain a concentration difference of ions across the membrane, which can generate a potential difference. This can be used in a number of ways, for example to transmit electrical activity along the plasma membrane of a muscle or a nerve.

- Communication between the extracellular and intracellular environments; the response within the latter is usually an increase or decrease in the rates of specific biochemical activities or processes.

- To provide a surface for attachment of proteins on the outside or inside of the cell, including receptors, enzymes and identifying molecules.

- To provide a surface for attachment of other cells, either cells in the same tissue or different cells, e.g. immune cells during a local infection.

- To provide a source of polyunsaturated fatty acids and other compounds that can be released to act as intracellular messengers or as precursors for such messengers. These are usually part of a communication system from the extracellular to the intracellular environments.

Structure of the plasma membrane

The composition of the membrane is about half phospholipid and half protein. In an aqueous medium, phospholipid molecules associate in such a way that their hydrophobic tails exclude water molecules, whereas the hydrophilic heads orient towards the water molecules (Figure 5.1). The result is two layers of phospholipid molecules: the outer

---

Functional Biochemistry in Health and Disease by Eric Newsholme and Tony Leech
© 2010 John Wiley & Sons Ltd

Table 5.1 The concentrations of some ions in the extracellular and intracellular compartments

Ion	Concentration (mmol/L)	
	Extracellular	**Intracellular**
Calcium	1	1.5
Chloride	110	10
Hydrogen carbonate	24	10
Magnesium	1.5	12
Phosphate	2	40
Potassium	4	150
Sodium	145	15

The intracellular concentrations are total – the *free* concentration of some ions, such as Ca^{2+} and phosphate, is much lower in the cytosol of the cell due to binding to proteins or compartmentation. For example, the free concentration of Ca^{2+} ions in the cytosol is approximately 0.1 μmol/L. The concentration of phosphate in the cytosol is approximately 1 mmol/L. Most of the Ca^{2+} ions in the cell are present in the endoplasmic reticulum (see Chapter 13).

and inner leaflets. In 1972 S.J. Singer and G. Nicolson proposed a fluid mosaic model for membrane structure which was based on the two-dimensional arrangement of the self-assembled phospholipid molecules: fluid describes the movement of phospholipid molecules in the plane of the membrane; mosaic describes the arrangement of the proteins within the membrane (Figure 5.2). Mobility is such that a phospholipid molecule can diffuse once round an erythrocyte in little over ten seconds, whereas movement from one side of the bilayer to the other (transverse diffusion or 'flip-flop') is very low. The proteins can either be embedded within the membrane (intrinsic proteins, which traverse the whole membrane), or attached to the inner or outer leaflet (extrinsic proteins).

> Changes in phospholipid composition between the inner and outer leaflets are brought about by the activity of a membrane enzyme known as a phospholipid translocase (or 'flipase').

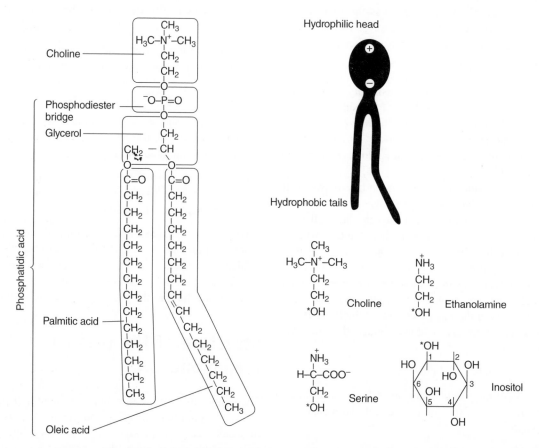

Figure 5.1 *The structure of a glycerophospholipid. A simple diagram showing the charges on the head group.* In this struction, palmitic and oleic acids, provide the hydrophobic component of the phospholipids and choline (and four bases) and the phosphate group provide the hydrophilic head. The unsaturated fatty acid, oleic acid, provides a 'kink' in the structure and therefore some flexibility in the membrane structure which allows for fluidity. The more unsaturated fatty acid, the larger is the 'kink' and hence more fluidity in the membrane. Cholesterol molecules can fill the gaps left by the kink and hence reduce flexibility. Hydroxyl groups on the bases marked * are those that form phosphoester links. Choline and inositol may sometimes be deficient in the diet so that they are, possibly, essential micronutrients (Chapter 15).

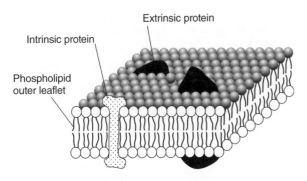

Figure 5.2 *The fluid mosaic model of the plasma membrane.*

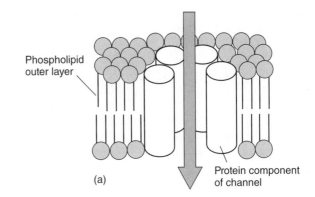

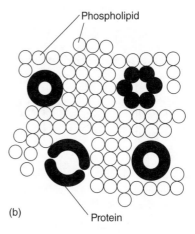

Figure 5.3 **(a)** *Diagram of a section through a membrane.* It shows a portion of the bilayer and some of the protein components of the ion channel. The arrow shows the path of an ion. **(b)** *View of ion channels from above the bilayer.* This shows the possible arrangements of proteins that make up the ion channel.

Diffusion through membranes

Molecules in a gas or liquid phase move by virtue of their kinetic energy. At body temperature, the movement of small molecules is very rapid, although movement in any one direction is hindered by collisions with other molecules. Nonetheless, such motion permits molecules to reach uniform concentration throughout a solution. This is diffusion, as a result of which ions and molecules move without the application of a force. This means that an input of external energy is not required. It is, however, only effective over very small distances: it requires about 3.5 seconds for a glucose molecule to diffuse 10 μm through an aqueous solution but if the distance were increased to 10 metres, it would take 11 years. Since the diffusion distance between the nearest capillary and the centre of a muscle cell is around 20 μm, glucose, oxygen and carbon dioxide, at physiological concentrations, can diffuse at a sufficient rate to satisfy the requirements of the muscle. The problem is, however, more complex when the molecule or ion has to diffuse through a membrane. For small molecules, such as oxygen or carbon dioxide, or for slightly larger fat-soluble molecules, such as acetate or ethanol, the membrane does not restrict diffusion sufficiently to interfere with normal physiological functions of the cell, whereas diffusion of ions or larger molecules is far too slow. There are two things that speed up diffusion across a membrane: the presence of a hole or the presence of a transporter within the membrane. Ions (e.g. Na^+, K^+, Ca^{2+} ions) enter or exit the cell through pores, which are composed of protein molecules and are known as ion channels. Molecules cross a membrane via transporters (or carriers) which are also proteins.

Water enters cells not only via channels but also by diffusion through the plasma membrane, although the quantities of water in the latter case are likely to be small. It also enters via endocytosis (see below).

Ion channels

Channels allow ions and water molecules to diffuse through the membrane without coming into contact with the phospholipid molecules (Figures 5.3(a) and (b)). The diameter of the channel prevents entry of any molecules, or ions above a particular size. In addition, charges on the surface of the proteins within the channels may provide further restriction. Of particular biochemical and physiological significance, it is possible to open or close an ion channel. This is achieved by a conformational change in one or more of the proteins that constitute the channel (Figure 5.4). This means that, through the conformational change, the open/closed state of the channel can be regulated which, in turn, can regulate the rate of diffusion. There are three means by which the conformation of a protein in a channel, and hence the open/closed state of the channel, can be changed:

The rate of entry of ions can be large (about 10^6 ions per second per channel).

(i) By a change in membrane potential (i.e. a change in voltage): that is, the conformation of the protein is voltage sensitive;

(ii) By a change in the concentration of a molecule that binds to the protein (i.e. a ligand for the protein);

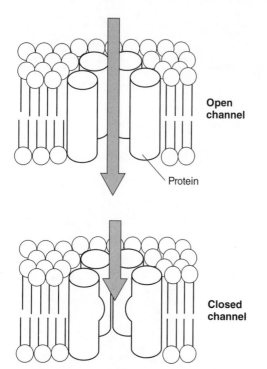

Figure 5.4 *Diagram of an open and a closed ion channel.* The diagram of the closed channel attempts to portray the conformational change as a closed gate.

(iii) By phosphorylation of a protein in the ion channel, via a protein kinase, or dephosphorylation via a protein phosphatase.

The investigations of membrane potential and ion channel activities were originally carried out by physiologists rather than biochemists, so that a different terminology has developed. Since opening and closing a channel is reminiscent of opening or closing a gate, physiologists have used the term 'gate' in descriptions of the open or closed state. A channel that is opened or closed by a change in voltage is known as a voltage-gated channel, whereas if the binding of a ligand changes the open/closed state, the channel is known as a ligand-gated channel. For a biochemist, the opening or closing of a gate would be described as regulation of the channel. Consequently, these physiology terms are synonymous with the terms voltage-regulated and ligand-regulated channels. Indeed, the latter mechanism of regulation of an ion channel is likely to be similar, if not identical, to the allosteric mechanism for regulation of enzyme activity (Chapter 3).

Transporters

Molecules such as glucose and amino acids are transported across a membrane by specific protein transporters. The

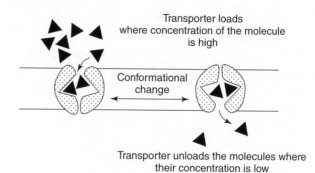

Figure 5.5 *Model of a possible mechanism for a transporter.* The number of molecules on each side of the membrane is an attempt to represent transport from a higher to a lower concentration of the molecule. Note that the conformational change is reversible, so that the transporter can transport the molecules in either direction, according to the concentration difference.

name suggests that the protein physically carries the molecule across the membrane. This is not the case. One proposed mechanism is that the transporter binds the molecule at one side of the membrane and, after a conformation change, releases the molecule on the other side (Figure 5.5). If there is no energy input into the process, the molecule is only transported from a region of high concentration to a region of lower concentration. In addition, the transporter only speeds up transport across the membrane, in the same way that an enzyme only speeds up the rate of a reaction. The process is, therefore, termed transporter-mediated diffusion or facilitated diffusion. Indeed, there is considerable similarity between enzyme catalysis and transporter-mediated transport. To illustrate this, a comparison is made between the properties of the glucose transporter and the enzymes that react with glucose: hexokinase or glucokinase.

- The transporter binds glucose at a specific site, then changes its conformation, which results in transport of glucose across the membrane to be released on the other side. The enzyme hexokinase (or glucokinase) binds glucose at a specific site, then catalyses its phosphorylation and releases the product, glucose 6-phosphate.

- There are several different glucose transporters. There are two glucose phosphorylating enzymes glucokinase and hexokinase (Chapter 6).

- *Hyperbolic response* The relationship between the rate of glucose transport and glucose concentration is hyperbolic; that is, the transporter has a limited number of glucose-binding sites and there is V_{max} for transport. The concentration of glucose at half the V_{max} is known as the K_m (Figure 5.6). This is exactly the same as an enzyme (Chapter 3).

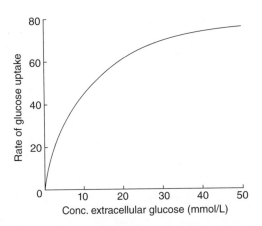

Figure 5.6 *Plot of the rate of glucose uptake (i.e. activity of the carrier) against glucose concentration by a skeletal muscle cell. The curve obeys a hyperbolic equation. The K_m is about 5 mM, which is similar for the enzyme glucokinase for glucose when its activity is plotted against the glucose concentration (see Chapter 3).*

- *Specificity* The transporter has a similar specificity to hexokinase: it transports several hexoses, glucose, mannose, 2-deoxyglucose. Similarly, the optical isomer, D-glucose, but not L-glucose, is transported (Chapter 3).

- *Competition* The rate of transport of one sugar is reduced by the transport of another sugar: the kinetics are identical to that of competitive inhibition of hexokinase by the presence of other sugars.

- *Specific activation or inhibition* Transport of glucose can be increased or decreased by specific compounds: insulin increases the transport whereas phloridzin, a plant glycoside, inhibits glucose transport by muscle. Insulin increases glucokinase activity in liver, whereas a plant sugar, mannoheptulose, inhibits glucokinase activity. Hexokinase is inhibited by its product, glucose 6-phosphate.

Co-transport and counter-transport

Co-transport, also known as symport, is the transport of one compound that is obligatorily linked to that of another. An example is the transport of glucose and Na^+ ions (see below). In counter-transport, also known as antiport, the transporter transports one compound in one direction and this is obligatorily linked to transport of another compound in the opposite direction. An example is the transport of HCO_3^- and Cl^- ions into and out of erythrocytes which it is

> The 'chloride shift' was originally detected as a markedly lower concentration of chloride in plasma from venous blood compared with that from arterial blood.

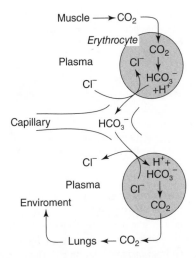

Figure 5.7 *An example of counter-transport in the erythrocyte.* The transport of CO_2 from peripheral tissues to the lungs for excretion is more complex than simple solution of CO_2 in the plasma and transport in the blood. The CO_2 produced by the muscle (or any other tissue) enters the blood and then enters an erythrocyte where it reacts with water to produce hydrogencarbonate, catalysed by the enzyme carbonate dehydratase:

$$H_2O + CO_2 \rightarrow H_2CO_3 \rightarrow H^+ + HCO_3^-$$

The HCO_3^- is transported out of the red cell via a transporter that transports the chloride ion into the erythrocyte. The reverse process occurs in the lungs: the hydrogencarbonate is transported into the erythrocyte in exchange for the chloride ion. Within the cell, the hydrogencarbonate ion is converted to CO_2, via the same enzyme,

$$HCO_3^- + H^+ \rightarrow H_2CO_3 \rightarrow H_2O + CO_2.$$

The CO_2 diffuses from the erythrocyte into the blood and across the membrane of the alveoli into the air in the alveoli, for loss to the enviroment.

essential for transport of carbon dioxide from tissues to the lung (Figure 5.7). It is known as the 'chloride shift'.

Active transport

The transport of molecules and ions across a membrane from a low to a high concentration requires an input of energy: it is called active transport. There are two ways in which this can be achieved:

- The transport is directly coupled to adenosine triphosphate (ATP) hydrolysis, which is known as primary active transport.

- The transport is coupled to that of another compound which is transported down its concentration gradient. This is called secondary active transport, since it is the

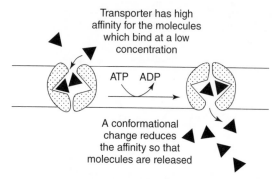

Figure 5.8 *Active transport is achieved by phosphorylation and dephosphorylation of the transporter.* The dephosphorylated form has a high affinity for the molecules whereas the phosphorylated form has a low affinity. This is achieved by a conformation change resulting from phosphorylation by ATP and dephosphorylation via phosphate release.

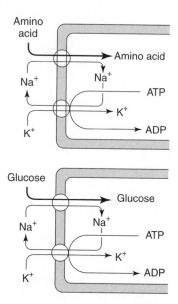

Figure 5.9 *The sodium ion/glucose transporter and sodium ion/ amino acid transporter.* The biochemistry of the two processes is identical. To maintain electroneutral transport K^+ ion replaces Na^+ ion, via Na^+/K^+ ATPase. The broader arrow indicates overall effect (i.e. unidirectional) transport.

maintenance of a concentration gradient of the second compound that requires energy, usually ATP hydrolysis. This mechanism is equivalent to *coupling-in-parallel*, discussed in Chapter 2.

Primary active transport

The majority of active transport systems involve phosphorylation of the carrier by ATP. The phosphorylated form of the transporter has a higher affinity for the transported molecule than the dephosphorylated form (Figure 5.8).

Secondary active transport

An example is the coupling of the transport of glucose into a cell, against its concentration gradient, with that of Na^+ ions, which move down their concentration gradient from outside to inside the cell. A similar mechanism transports amino acids (Figure 5.9). This occurs in the absorption of glucose by the epithelial cells in the small intestine (Box 5.1). It also occurs in the absorption of glucose from the glomerular filtrate by the epithelial cells in the kidney tubules. The glomerulus in the kidney simply filters blood, so that small molecules in blood, including glucose, enter the kidney in what is known as the glomerular filtrate. To prevent loss of glucose in the urine, glucose must be re-absorbed by the tubules, which is achieved by active transport.

The K_m of the glucose transporter in the kidney tubules is relatively low, so that the transporter approaches saturation at glucose concentrations of about 10 mmol/L. If the

Box 5.1 Rehydration drinks

Diarrhoea resulting from intestinal infections or malnutrition is prevalent especially in children in underdeveloped countries. A serious complication of the diarrhoea is dehydration and electrolyte imbalance. (An electrolyte is a substance which, when dissolved, dissociates into ions making it possible for the solution to conduct an electric current. The term is used clinically in a more restricted sense to refer to the inorganic constituents of body fluids, particularly Na^+ and K^+.) This condition can be treated with rehydration drinks which contain both Na^+ ions and glucose. Due to the active transport of glucose associated with the Na^+ ion gradient and the Na^+/K^+ ATPase, the absorption of glucose from the intestine results in the absorption of Na^+ ions. Restoration of intra-cellular Na^+ levels increases the osmotic pressure, which withdraws water from the intestinal lumen. The solution contains sodium chloride (or sodium lactate) and glucose. More recently, some solutions may contain potassium chloride. The drink restores hydration which also can reduce diarrhoea.

concentration of glucose in the blood is above about 10 mmol/L, that in the filtrate will also be above 10 mmol/ L, so that it might exceed the capacity of the transporter to remove all of the glucose and some will be lost in the urine. This explains why some diabetic patients, particularly if

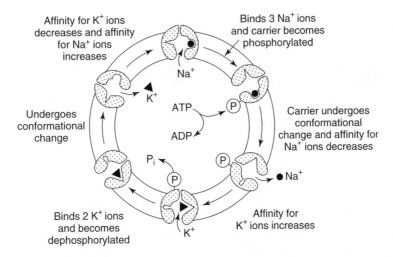

Figure 5.10 *The Na⁺/K⁺ ATPase.* ATP hydrolysis results in phosphorylation of the carrier when the Na^+ ion is bound. This changes the conformation of the transporter so that the Na^+ ion dissociates into the extracellular space. Dephosphorylation of the transporter changes the conformation so that the K^+ ion dissociates into the cytosol.

not well controlled with insulin, excrete glucose in the urine.

In addition, the presence of glucose in the tubules increases the osmotic pressure which 'pulls' water from the blood into the tubule. Consequently, both glucose and water are lost from the body. This results in polyuria, dehydration, thirst and, hence, polydipsia. Indeed, it is these two symptoms that may persuade an unsuspecting diabetic to seek medical advice.

Na⁺/K⁺ ATPase

The Na^+ ion concentration within the cell is maintained low, due to activity of an enzyme known as the Na^+/K^+ ATPase. This enzyme/carrier is present in the plasma membrane. It is an antiport system that transports three Na^+ ions out of the cell and two K^+ ions into the cell, for each molecule of ATP that is hydrolysed (Figure 5.10). It is responsible for maintaining a low Na^+ ion concentration but a high K^+ ion concentration within the cell. Its constant activity in many it not all cells requires constant ATP hydrolysis, which accounts for more than 10% of the resting energy expenditure of an adult.

One additional benefit is that by continuously pumping more Na^+ ions out than K^+ ions into the cell, it results in a higher intracellular K^+ ion concentration than might be expected. Since the K^+ ion channel is open, the K^+ ions diffuse out to an extect dependent upon the concentration difference, so that the Na^+/K^+ ATPase contributes to the magnitude of the membrane potential across a cell.

Endocytosis and exocytosis

Endocytosis

When some cells are observed under a microscope, regions of the plasma membrane can be seen to fold into the cell, forming pockets that pinch off to produce membrane-bound vesicles within the cell. This is the means by which

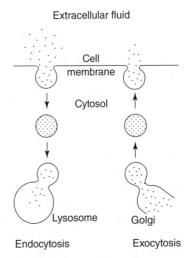

Figure 5.11 *A diagrammatic representation of endocytosis and exocytosis.*

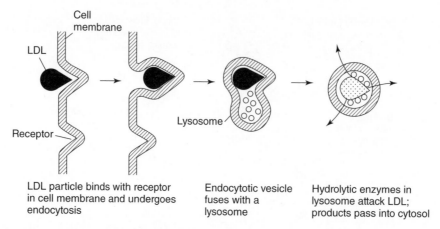

LDL particle binds with receptor
in cell membrane and undergoes
endocytosis

Endocytotic vesicle
fuses with a
lysosome

Hydrolytic enzymes in
lysosome attack LDL;
products pass into cytosol

Figure 5.12 *Receptor-mediated endocytosis of the LDL particle.* The specificity of binding depends upon two proteins that are components of the LDL particle, apolipoprotein E and apolipoprotein B (Chapter 11).

cells can take up extracellular fluid, macromolecules, particles or microorganisms. This process is known as endocytosis (Figure 5.11), of which there are three types: fluid phase, receptor-mediated and phagocytosis.

(i) *Fluid phase endocytosis (pinocytosis)* This is responsible for uptake of extracellular fluid and solutes. Within the cell, the vesicle is known as an endosome and releases its contents into the cytosol, whereas the membrane component is recycled back to the membrane. In cells in which endocytosis is frequent, it takes less than 30 minutes to recycle the whole of the plasma membrane in this way. The process is a little more complex than indicated in Figure 15.11. Thus the endocytotic vesicles form from the plasma membrane due to a protein, clathrin, which is present in small indentations or pits known as 'clathrin-coated pits'. Clathrins self-assemble to form polyhedral cular cages which distort the membrane into invaginations that break away to give rise to the endosome.

(ii) *Receptor-mediated endocytosis* This is a mechanism for uptake of macromolecules or particles. They bind to a specific receptor on the plasma membrane which induces the uptake of the receptor plus the macromolecule. An example is the uptake of cholesterol into a cell. Cholesterol is transported in the blood as cholesterol ester within a lipoprotein particle, the low-density lipoprotein (LDL). A protein component of LDL binds to the receptor and, together with part of the cell membrane, the LDL and its receptor are taken into the cell by endocytosis to form an endosome. This fuses with a lysosome and enzymes in the latter hydrolyse protein to form amino acids and cholesterol ester to form free cholesterol, which is released into the cytosol to be used within the cell (Figure

5.12). This process is important for proliferating cells, in which cholesterol is required for formation of new membranes, and for some endocrine cells (in the testis, ovary and adrenal cortex) for the synthesis of steroid hormones.

There is a mutant gene that produces an inactive receptor and patients with this mutant suffer from high levels of LDL-cholesterol in their blood. They have a markedly increased risk of developing atherosclerosis and coronary artery disease.

Another example is uptake of the iron-containing protein, transferrin, which circulates in the blood. It binds to its receptor to form a complex that enters the cell via endocytosis. The iron is then released from the endosome for use in the cell (e.g. haemoglobin formation for erythrocyte production or cytochrome production in proliferating cells). The number of transferrin receptors in the plasma membrane increases in proliferating cells and the number in the liver is increased by cytokines during infection. This results in a lower concentration of iron in the blood which decreases the proliferation of invading pathogens (Chapters 15 and 18).

(iii) *Phagocytosis* This is another form of endocytosis, carried out by phagocytes (e.g. macrophages, neutrophils). The fate of the particle that is phagocytosed depends on its nature. For solid particles and dead or dying cells, the endosome fuses with the lysosome and is then digested by the enzymes in the lysosome. If a pathogen is phagocytosed, the endosome, now called a phagosome, fuses with a lysosome to produce a phagolysosome. The membrane of the latter contains a small electron transfer chain that transfers electrons from NADPH to oxygen, producing the superoxide radical that kills pathogens (see Chapter

17). Macrophages circulate in the blood and lymph but also reside in some tissues (e.g. lung, liver spleen and lymph nodes). It should be noted that the role of macrophages in the lung is to phagocytose not only pathogens that enter in each breath but also dust particles that otherwise would collect in the lung and interfere with the transfer of oxygen and carbon dioxide across the membranes of the alveoli.

Exocytosis

A process similar to endocytosis occurs in the reverse direction when it is known as exocytosis (Figure 5.11). Membrane-bound vesicles in the cytosol fuse with the plasma membrane and release their contents to the outside of the cell. Both endocytosis and exocytosis are manifestations of the widespread phenomenon of vesicular transport, which not only ferries materials in and out of cells but also between organelles, e.g. from the endoplasmic reticulum to the Golgi and then to the lysosomes or to the plasma membrane for secretion (Chapter 1). Many hormones are also secreted in this way, as are neurotransmitters from one nerve into a synaptic junction that joins two nerves (Chapters 12 and 14).

Physiological importance of some transport systems

Transport into the cell is vitally important in the physiology of the cell. Some examples are given.

Transport of fuels into the cell

There are three major fuels for cells: glucose, fatty acids and glutamine (Chapter 9). Transporter molecules are present in the membrane for each of these fuels.

Glucose   There are several different glucose transporters with different properties, the distribution of which relates to the function of the cell (Chapter 6). The transporter can be sufficiently active to bring the concentrations of glucose on both sides of the membrane to the same level, i.e. the transport process is at equilibrium or near-equilibrium; the equilibrium constant for transport of glucose is unity. This is the case in the liver and the insulin-secreting cells in the islets of Langerhans in the pancreas (β-cells). The physiological importance of this, in both cells, is that a change in plasma glucose level changes the intracellular glucose concentration to the same extent. In hepatocytes, an increase in the intracellular glucose level stimulates glycogen synthesis and in the β-cells it stimulates glycolysis which results in an increase in the secretion of insulin. Both of these effects are important in controlling blood glucose level after a meal, discussed in Chapters 3 and 6.

In skeletal muscle, glucose transport is non-equilibrium, so that an increase in activity of the transporter increases glucose utilisation. Factors that increase the activity of the transporter (e.g. the number of transporter molecules) in the membrane are insulin and sustained physical activity. In contrast, the hormone cortisol decreases the number of transporters in the membrane. This decreases glucose uptake and is one of the effects of cortisol that helps to maintain the normal blood glucose level (Chapter 12).

Fatty acids   Despite the fact that fatty acids are lipid soluble, so that they will diffuse across membranes without a transporter, one is present in the plasma membrane to speed up entry into the cells, so that it is sufficient to meet the demand for fatty acid oxidation. Triacylglycerol transport into cells also depends on the fatty acid transporter. Since it is too large to be transported *per se*, it is hydrolysed within the lumen of the capillaries in these tissues and the resultant fatty acids are taken up by the local cells via the fatty acid transporter (Chapter 7). Hence the fatty acid transporter molecule is essential for the uptake of triacylglycerol.

Glutamine   Glutamine is an important fuel for many cells, including immune cells, proliferating cells in the bone marrow and cells in the small and large intestine. It is transported into the cell via a specific transporter (Chapter 9). Tumour cells also use glutamine as a fuel and the transporter is more active in tumour cells compared with normal proliferating cells. The transporter is also important to transport glutamine out of cells that produce it (e.g. muscle, lung, adipocytes).

Transport of amino acids

Several transporters that are specific for transport of groups of amino acids are present in all tissues. They are of particular physiological importance in some tissues.

Liver   Liver is the major organ for metabolism of almost all amino acids; the exception is the branched-chain amino acids which are metabolised in muscle and adipose tissue (see Chapter 8). Of the amino acids absorbed from the lumen of the intestine in the adult, at least 70% are metabolised in the liver. Transporters for almost all amino acids are, therefore, very active in the membranes of hepatocytes.

Muscle All the amino acids are required for protein synthesis so that the transporters are important particularly during growth. Consistent with this, the major anabolic hormone, insulin, increases the transport of some amino acids into the muscle, which contributes to the stimulation of protein synthesis by insulin.

Brain Some amino acids are precursors for neurotransmitters so that the carriers for these amino acids are of vital importance in neural activity: these are glutamine, tyrosine and tryptophan. Indeed, changes in the activity of one or more of these carriers can change the concentration of neurotransmitters in the brain, which can modify the behaviour of the human (Chapters 13 and 14).

Ions

The transport of ions into some cells is of considerable biochemical and physiological importance. These include Na^+, Ca^{2+} and Fe^{2+}.

Sodium ions A marked and rapid increase in the Na^+ ion transport across the plasma membrane into a nerve or muscle cell, via the Na^+ ion channel, causes a depolarisation of the membrane that initiates a transient flux of electrical activity along the nerve or muscle (that is, an action potential) (Chapters 13 and 14).

Calcium ions Movement of Ca^{2+} ions from the extracellular environment into the cytosol is achieved via calcium ion channels. An increase in the number of Ca^{2+} ion channels that are open in cells of smooth or cardiac muscles stimulates contraction. Excessive rates of entry can, however, cause problems. For example, increased entry of Ca^{2+} ions into vascular smooth muscle increases contraction which reduces the diameter of blood vessels which can lead to hypertension (Chapter 22).

Iron ion The protein transferrin binds ferric ions and transports them in the serum around the body. The ions are taken up by cells via a transferrin receptor which is present in the plasma membrane. The receptor binds transferrin and the complex enters the cell where transferrin releases the ion. Iron is extremely important in proliferating cells for the synthesis of mitochondrial proteins involved in electron transfer. The number of transferrin receptors in the membrane of a cell is increased when it is stimulated to proliferate (e.g. lymphocytes in response to an infection) (Chapters 17 and 18).

Channelopathies

As with other proteins, mutations in genes that encode for ion channels can give rise to diseases. These are known as channelopathies. A number of diseases in both the nervous system and in skeletal muscle are known to be due to defects in the structure of ion channels. However, given the large number of ion channels in these tissues, the channelopathies currently known probably represent the tip of an iceberg. For many channels, total loss of function is probably not compatible with life so that it is only subtle changes in structure and function that lead to diseases. In general, chronic defects in channels in the nervous system give rise to seizures whereas those in muscle give rise to some form of paralysis. (For review, see Ashcroft 2000).

Cystic fibrosis in a channelopathy. It is a chloride channel that is defective. It is, present in epithelial cells of several issue, including the intestine, lung, pancreatic ducts and sweat glunts. Impaired movement of chloride ions art of the cells interferces with movement of fluid act of the cells. In the lungs, this increases the viscosity of mucous in the bronchioles which is normally flows upwards due to ciliary action. A more viscous mucous cannot be moved, so that it clogs the bronchioles which impairs breathing and clogs of mucous are ideal for bacterial growth and Lence infection of the lung. There are several different classes of mulations of the channel. One mulation result in a change in structure of the channel which causes it is become stuck in the membranes of the golgi on its way to the plusma yembrone. If the channel can be freed, its friction in the plasma membrane is normal suggestions have been made that changes in the fatty acid composition of the membranes in the golgi may allow the channel to escape from the vesicles. Whether an increase in the content of polyunsaturated fatty acids in the phospholipid in these membranes could increase fluility with a greater chance of escape of the channel has, to the authors' knowledge, never been tested.

III ESSENTIAL METABOLISM

6
Carbohydrate Metabolism

Colin Blakemore* describes the effects of a low blood glucose concentration as follows:

As blood glucose falls, there is a window of experience between sanity and coma in which self-control is lost and that precious feeling of conscious choice disappears.

He then recalls the feelings of one person suffering from hypoglycaemia:

I was totally convinced I was in charge of everything and that all I had to do was think and the thing would happen. I was being extremely tempted by the knives about me . . . and especially the thin knife which it seemed would be very easy to stick into somebody . . .

(Blakemore, 1994)

The liver plays a major role in glucose homeostasis by releasing into the systemic circulation the exact amount of glucose required to match extrahepatic glucose utilisation and maintain plasma glucose concentrations within tight normal limits. It performs this task by mobilising glucose stored within hepatocytes as glycogen and/or by converting lactate, glycerol, and amino acids into glucose (gluconeogenesis).

(Jéquier and Schneiter, 2002)

The extract from Colin Blakemore's book provides a vivid account of the behavioural consequences of a low blood glucose level. This is because the brain uses glucose as the only fuel, except in prolonged starvation. Problems also arise if the blood glucose level increases well above the normal. An increase of only about twofold above the normal can result in loss of glucose in the urine, since not all the glucose in the glomerular filtrate can be absorbed by the kidney, a condition known as glycosuria. Furthermore, a chronic increase above the normal level can contribute to several debilitating diseases (e.g. atherosclerosis, kidney failure, blindness. These problems arise because glucose reacts spontaneously with some side-chain groups of amino acids in proteins, a process known as glycosylation. Indeed, glucose can be regarded as a toxic molecule and glucotoxicity is a pathological condition (see Appendix 6.1, Chapter 22). Consequently, maintenance of the normal blood glucose concentration is vitally important and much of carbohydrate metabolism is designed to do just that. The second extract above, by Jéquier & Schneiter, emphasises the importance of the liver in regulation of the blood glucose level (discussed below). Not only is it important to avoid hypoglycaemia for the brain, but for many cells and tissues that also depend upon glucose as a major fuel (e.g. erythrocytes, immune cells, proliferating cells in the bone marrow, tissues in the eye and the kidney medulla, see below). In addition, glucose is the starting point for a number of important metabolic pathways (Figure 6.1) and some of the metabolites of glucose are precursors for the synthesis of various key compounds (Figure 6.2).

The topics discussed in this chapter are glycolysis, glycogen synthesis, fructose synthesis, pentose phosphate pathway, gluconeogenesis, regulation of blood glucose concentration, regulation of glycolysis and gluconeogenesis and the causes of hypoglycaemia.

*Emeritus Professor of Physiology, University of Oxford. Former Chief Executive of Medical Research Council, London.

Functional Biochemistry in Health and Disease by Eric Newsholme and Tony Leech
© 2010 John Wiley & Sons Ltd

Glycolysis

Although glycolysis is a single process, the term is used to describe two processes:

(i) Glucose or glycogen breakdown to form a three carbon compound, lactate.

(ii) The first part of an overall process that produces pyruvate and then acetyl-CoA, for complete oxidation in the mitochondria (Figure 6.3(a)&(b)).

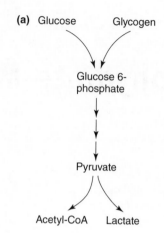

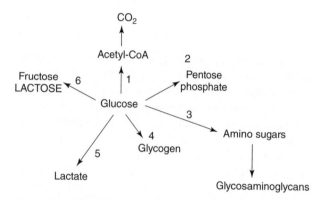

Figure 6.1 *Glucose is a substrate for several metabolic pathways.*
1. *Aerobic* glycolysis to form acetyl-CoA.
2. Pentose phosphate pathway.
3. Amino sugars to glycosaminoglycans (Appendix 6.2).
4. Glycogen synthesis.
5. *Anaerobic* glycolysis.
6. Formation of fructose and lactose.

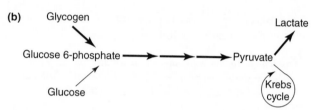

Figure 6.3 (a) *Glucose and glycogen as substrates for* aerobic *and* anaerobic *glycolysis.* For *aerobic* glycolysis, pyruvate is converted to acetyl-CoA; for *anaerobic* glycolysis pyruvate is converted to lactate.

(b) *An indication of maximum capacities of aerobic and anaerobic glycolysis in muscle.* The maximum capacity of anaerobic glycolysis from glycogen is at least 10-fold greater than aerobic glycolysis from glucose. This is indicated by the thickness of the arrows. When glycogen is the fuel for oxidation, the rate of glycolysis is limited by the lower capacity of the Krebs cycle.

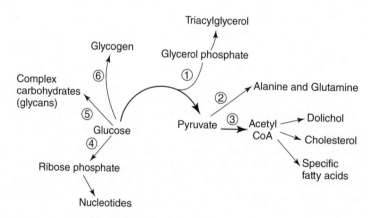

Figure 6.2 *Glucose and glycolytic intermediates as precursors for synthetic pathways.*
1. Glycerol phosphate for phospholipid and triacylglycerol formation.
2. Pyruvate for alanine and glutamine formation.
3. Pyruvate for acetyl-CoA which is precursor for several compounds (cholesterol, specific fatty acids, dolichol).
4. Glucose for ribose 5-phosphate (for formation of nucleotides).
5. Glucose for complex carbohydrate formation (e.g. glycans, glycoproteins, glycosaminoglycans, Appendix 6.2).
6. Glucose for glycogen formation.

It is important to distinguish between them since they are described in different ways in different books. In this book, process (i) is called *anaerobic* glycolysis and process (ii) is called *aerobic* glycolysis. The adjectives are in italics to denote that they do not describe accurately the actual conditions in which they occur. For example, process (i) occurs in erythrocytes, which is hardly an anaerobic environment. The difference, however, should be clear from the context. Another term that is used in relation to glycolysis, is glycogenolysis, which is the breakdown of glycogen to glucose 1-phosphate, which then feeds into glycolysis at the position of glucose 6-phosphate.

The pathway of glycolysis

For most of the lay population, carbohydrate is understood as starch and sugar as sucrose. In the body, however, the main carbohydrate is glycogen and the sugar is glucose. A point to re-emphasise is that either glucose or glycogen can be substrates for glycolysis (Figure 6.3(a)&(b)). There are ten reactions that convert glucose into pyruvate, eleven reactions that convert it to lactate, and 13 reactions that convert glycogen to lactate. The process from glucose can be described in a series of five stages, as follows:

- *Initial phosphorylations:* The formation of fructose bisphosphate occurs via two phosphorylation reactions that involve ATP hydrolysis, which are catalysed by hexokinase and phosphofructokinase:

(i)
$$\text{glucose} \xrightarrow[\text{ATP}]{\quad\quad\quad \text{ADP}} \text{glucose 6-phosphate}$$

Glucose 6-phosphate is then isomerised to form fructose 6-phosphate.

$$\text{fructose 6-phosphate} \xrightarrow[\text{ATP}]{\quad\quad\quad \text{ADP}} \text{fructose bisphosphate}$$

(ii) *Splitting of hexose:* This stage gives glycolysis its name, literally 'sugar splitting', since it involves the cleavage of fructose bisphosphate into two triose phosphates, glyceraldehyde 3-phosphate and dihydroxyacetone phosphate, which are interconvertible, catalysed by an isomerase enzyme.

(iii) *Oxidation:* Glyceraldehyde is oxidised to a carboxylic acid, NAD$^+$ is reduced to NADH and inorganic phosphate is incorporated to form 3-phosphoglycerol phosphate.

(iv) *ATP generation:* In the next series of reactions, which form pyruvate, ATP is generated.

(v) *Formation of lactate or acetyl-CoA from pyruvate.*

These stages are presented in Figure 6.4, and the complete pathway is presented in Figure 6.5.

Transport of glucose into the cell

Before glycolysis from glucose can begin, glucose has to be transported into the cell. This is achieved by a transporter protein, in the plasma membrane (Chapter 5). There are five different types of glucose transporter, all encoded by separate genes. The proteins have slightly different properties, different tissue distribution and somewhat different roles in these tissues. Their roles are briefly described in Table 6.1. A sixth transporter is specific for fructose: it is

> This transporter is also present in the plasma membrane of the sperm, since fructose, not glucose, is the fuel provided in semen (Chapter 19).

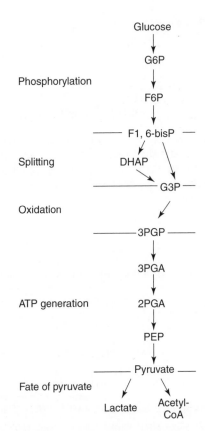

Figure 6.4 *Glycolysis divided into five stages.* It is not clear whether the reaction in which pyruvate is converted to acetyl-CoA, catalysed by pyruvate dehydrogenase, should be classified as a glycolytic enzyme or an enzyme of the Krebs cycle. Since the enzyme 'presents' acetyl-CoA to the cycle, it is considered in this book as an enzyme of glycolysis (i.e. *aerobic* glycolysis).

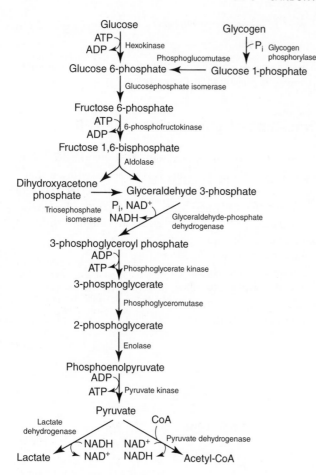

Figure 6.5 *Details of the glycolytic pathway.* All the enzymes except pyruvate dehydrogenase are present in the cytosol. The pathway, presently the molecular structured of the intermediates, is described in Appendix 6.3.

important for fructose absorption in the intestine. As with enzymes, transporters only speed up the rate of transport.

Glycogen is a branched polymer of glucose consisting of linear chains of glucose molecules linked by glycosidic bonds and branches in which two glucose molecules are linked by a different glycosidic bond. The linear parts are hydrolysed by the enzyme glycogen phosphorylase in a reaction involving phosphate, whereas the branches are hydrolysed by a debranching enzyme (Figure 6.6). The whole glycogen molecule can be broken down by the combined actions of both enzymes. A summary of the process is:

In the initial stages of breakdown of glycogen, phosphorylase hydrolyses all the glucosidic bonds external to the branch points. This represents about 50% of the glycogen. Debranching enzyme must catalyse hydrolysis of the branch points to allow phosphorylase activity to continue. The hydrolysis of the branch point is such that glucose rather than glucose 1-phosphate is produced. Indeed, as a result of this activity of debranching enzyme, about 10% of all glucose residues in glycogen are released as glucose. In liver, this glucose can be transported out of the ven into the blood. Since anoxic conditions prevail after death, glycogen breakdown would occur in liver and muscle, which may result in hyperglycaemia after death – thus concealing hypoglycaemia prior to death. This might pose a problem in deciding on a cause of death.

$$(\text{glucose})_n + \text{P}_i \rightarrow (\text{glucose})_{n-1} + \text{glucose 1-phosphate}$$

where $(\text{glucose})_n$ represents glycogen.

Details of the breakdown are presented in Appendix 6.4. The glucose 1-phosphate is converted to glucose 6-phosphate by the action of the enzyme phosphoglucomutase. The glucose 6-phosphate is then metabolised by glycolysis.

Table 6.1 Glucose (and the fructose) transporters in the plasma membrane of human cells and tissues

Type	K_m for hexose uptake (mmol/L)	Location of transporter
GLUT-1	1–2	Red blood cells, placenta, brain, kidney
GLUT-2	15–20	Liver, β-cells[a], kidney, small intestine
GLUT-3	1–2	Brain, intestine
GLUT-4	5	Skeletal muscle, cardiac muscle, adipose tissue
GLUT-5	6–11 (fructose)	Small intestine, sperm

The hexose transporters are numbered GLUT-1 to GLUT-5 in order of their discovery.

GLUT-1 is present in large numbers of tissues but is not present in muscle or liver. Its activity is particularly high in endothelial cells in capillaries in the brain to ensure sufficient entry of glucose from the blood into the brain, where it is taken up by glial cells (Chapter 14).

GLUT-2 has high activity in liver to maintain transport close to equilibrium for both uptake and release of glucose (see below). It can also transport galactose, mannose and fructose – all of which can be converted to glucose in the liver.

GLUT-5 is a fructose transporter.

[a]β-cells are those present in the Islets of Langerhans in the pancreas that secrete insulin in response to an increase in the plasma glucose concentration.

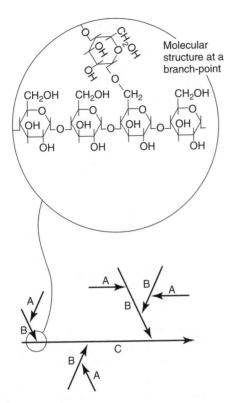

Figure 6.6 *Diagrammatic representation of a small part of the glycogen molecule.* Each line represents a chain of glucose molecules linked between the 1 and 4 positions of the glucose units. The arrowhead represents a branch-point with structure of branch-point indicated in the insert. This link is between the one and six positions.

Fate of the NADH produced in glycolysis

In glycolysis, NAD^+ is converted to NADH (Figure 6.5). Since the NAD^+ content of tissues is very small, for glycolysis to continue NAD^+ must be regenerated from NADH. In aerobic tissues and in the presence of oxygen, the NADH is oxidised to NAD^+ within the mitochondria, but not directly. It requires transport of hydrogen atoms, via metabolic shuttles which are described in Chapter 9. Under these conditions, the pyruvate that is produced is transported into the mitochondria for complete oxidation by the Krebs cycle. In this case, glycolysis is the first part of the process by which glucose is completely oxidised, i.e. *aerobic* glycolysis.

In cells that lack mitochondria, or during hypoxia in aerobic tissues, NADH is oxidised to NAD^+ in a reaction in which pyruvate is reduced to lactate, catalysed by lactate dehydrogenase (i.e. *anaerobic* glycolysis).

$$pyruvate + NADH \rightarrow lactate + NAD^+$$

Lactate and proton production in glycolysis

An important point to note is that this the above reaction produces lactate, not lactic acid. Nonetheless, protons are produced in glycolysis but in another reaction (Appendix 6.5). Consequently, the two end-products are lactate plus protons, which can be described as lactic acid. Despite this discussion, it can be argued that lactate dehydrogenase is not the terminal reaction of glycolysis, since the lactate plus protons have to be transported out of the cell into the interstitial space. This requires a transporter protein, which transports both lactate and protons across the plasma membrane and out of the cell.

Physiological importance of transport of lactate and protons out of the cell

Under some conditions, the rate of glycolysis from glycogen to lactate plus protons can be greater than the capacity of the transporter to transport all the lactate and protons out of the muscle. In this case, they accumulate in the muscle and the pH falls to about 6.5 or even lower (Chapter 13). This can occur, for example, when a muscle is working close to maximum, or when there is a poor blood supply to a working muscle. If this continued for any length of time, the increase in proton concentration could be sufficient to damage some of the proteins in the muscle. Indeed, if this decrease in pH occurred in the blood, it would be very serious and could rapidly lead to death. This is prevented in two ways:

- Under the conditions when protons accumulate in muscle, the physical activity of muscle is decreased by fatigue, hence the rate of ATP utilisation and, therefore, ATP generation via glycolysis, are decreased. Consequently the rate of lactic acid formation is decreased. It is unclear it protons are directly involved in fatigue. It is discussed in Chapter 13.

- The increase in proton concentration in the muscle decreases the activities of two key enzymes, which regulate the flux through glycolysis: phosphorylase and phosphofructokinase (Chapter 13) (Figure 6.7).

Although the condition is hypothetical, it can be calculated that if all of the skeletal muscles in the body degraded glycogen to lactic acid at the maximum rate and, if all of the protons that were produced were transported into the blood, it would exceed the buffering capacity of the blood and the pH would fall dramatically (see below). This could soon result in death. Hence the inhibitory effect of protons

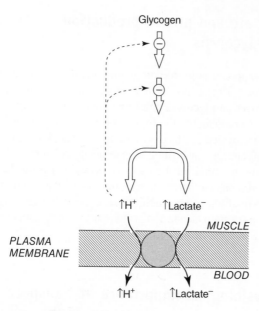

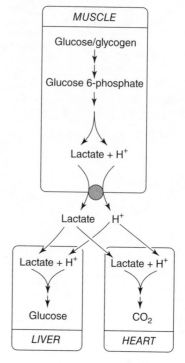

Figure 6.7 *Lactate and proton transporter in the plasma membrane and the effect of proton accumulation.* It should be noted that lactate and protons are formed in two different reactions in the glycolytic pathway (Appendix 6.5). The arrows adjacent to H⁺ and lactate indicate the increase in concentrations lactate during physical activity. The two key enzymes inhibited by protons are phosphorylase and phosphofructokinase.

Figure 6.8 *Fate of lactate and protons produced by muscle and other tissues.* The heart and liver play major roles in removing both lactate and protons from blood.

It is not physiologically possible to exercise all skeletal muscles in the body maximally and aerobically at the same time because the output of blood by the heart is not sufficient to transport enough oxygen to all the muscles (i.e. cardiac output is limiting).

on the two key glycolytic enzymes and the phenomenon of physical fatigue are of considerable physiological importance: they restrict the magnitude of the increase in lactic acid concentration not only in the muscle but also in the blood.

Removal of lactate and protons from the blood

The lactate and proton concentrations, and hence the pH in the blood, depend not only on the rate of release from muscle, and other tissues, but also on the rate of their removal from the blood. The transporter that transports lactate and protons out of cells also transports them into cells. There are two tissues that are important in their removal from the blood: liver and heart (Figure 6.8).

Removal by the liver   Lactate and protons are transported into the cells and then converted to glucose via gluconeogenesis. The overall process is

$$2 \text{ lactate}^- + 2\text{H}^+ \rightarrow \text{glucose}$$

This is, quantitatively, extremely important. One condition in which the liver fails to take up protons and lactate is when a considerable amount of blood has been lost (e.g. during or after an accident). To compensate for this, blood flow to the splanchnic area, including the liver, is severely reduced so that the liver becomes hypoxic, and not enough ATP can be generated to maintain gluconeogenesis. Consequently, lactate and protons accumulate in the blood and the pH falls: a condition known as lactic acidosis. It is important, therefore, to maintain blood volume after an accident, which is usually done by infusion of fluid.

Removal by the heart   Lactate and protons are transported into the cardiomyocytes and lactate is fully oxidised to CO_2:

$$\text{lactate}^- + \text{H}^+ + 3\text{O}_2 \rightarrow 3\text{CO}_2 + 3\text{H}_2\text{O}$$

Note that oxidation of lactate utilises a proton so that, essentially, lactic acid is oxidised.

All aerobic tissues are able to oxidise lactate to generate ATP but heart is especially important, since lactate is readily transported into cardiomyocytes and there is always a demand for ATP.

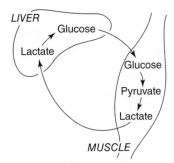

Figure 6.9 *The Cori cycle.*

The protons are removed so effectively by these two processes that, even during intense physical activity, the blood pH is maintained at or very close to normal (pH 7.4).

The Cori cycle

Even at rest, a low rate of lactic acid is formed from glucose in muscle and released. Most is transported to the liver where it is removed and converted to glucose. This glucose is released back into the blood, where it is, once again, taken up by muscle and converted to lactic acid. This is known as the Cori cycle, named after Carl Cori who first recognised the process (Figure 6.9). However, it is now known that other cells and tissues continuously convert glucose into lactic acid which is released into the blood (e.g. erythrocytes, proliferating cells in the bone marrow, immune cells in lymph nodes and epithelial cells in the skin). The Cori cycle can, therefore, be expanded to take these into account (Figure 6.10). During trauma, sepsis or after major surgery, proliferation of cells in the wound and in the bone marrow increases, as does the number of immune cells: these increases provide new cells for repair of damaged tissue and more immune cells to fight an infection. This results in a greater rate of formation of lactic acid. Hence, flux through the Cori cycle is increased (Chapter 18). Since Cori cycle activity involves hydrolysis of ATP, and hence metabolism is increased to generate more ATP, when the flux through this cycle increases, energy expenditure increases, which is characteristic of patients in Intensive Care Units. Nutrition provided for these patients must take this increase in expenditure into account (Chapters 2, 15 and 18).

The physiological pathway of glycolysis

In the pathway, glycolysis-from-glycogen, the flux-generating step is catalysed by phosphorylase, the activity of which is controlled by the physical activity of the muscle

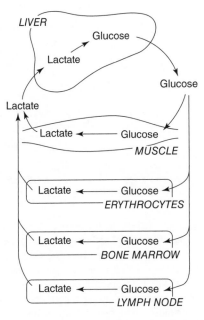

Figure 6.10 *The extended Cori cycle.* Tumour cells convert glucose to lactate at a high rate and could be incorporated into this figure (Chapter 21).

and by several hormones (Chapter 3). In contrast, the pathway of glycolysis-from-glucose in muscle does not contain a flux-generating step. This step is either the absorption of glucose from the intestine or the breakdown of glycogen in the liver (see Figure 6.11). That is, the pathway spans more than one tissue. This can give rise to two interesting physiological conditions.

- During prolonged physical activity, glycogen in muscle is used first as a fuel but, as it becomes depleted, liver glycogen is broken down to provide glucose. The flux-generating step is the conversion of glycogen to glucose 1-phosphate in the liver, that blood glucose is an intermediate in the pathway between liver and muscle. The blood glucose concentration remains constant provided that the rate of degradation of glycogen is identical to that taken up by the muscle. The blood glucose level can fall, markedly, if the glycogen in the liver is depleted, so that there is no longer a flux-generating step in the pathway (see section on hypoglycaemia, below).

- After a carbohydrate-containing meal, the blood glucose level increases which stimulates glucose uptake by muscle. The rate of uptake is enhanced by secretion of insulin. However, if insulin is secreted when glucose is not being absorbed into the body, i.e. there is no flux-generating step, the stimulated rate of glucose uptake by muscle will result in hypoglycaemia, even severe

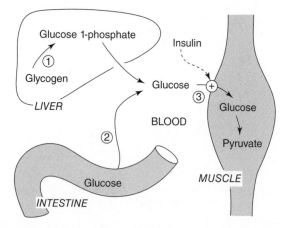

Figure 6.11 *The physiological pathway for aerobic glycolysis from glucose.* The flux-generating step(s) for the pathway are glyco-gen breakdown ① in the liver or glucose absorption from intes-tine ②. When there is no absorption (i.e. process ② is zero) then the pathway starts with glycogen breakdown in liver. When there is glucose absorption from intestine, and process ① is zero, glucose absorption is the start of the physiological pathway. The uptake of glucose by muscle is stimulated by insulin ③. If excess insulin is present in blood, glycogen breakdown is inhib-ited and, if the subject has not eaten, hypoglycaemia can result, due to a high rate of ③. The hypoglycaemia can be sufficiently severe to result in death.

hypoglycaemia. Similarly tumour cells in the Islets of Langerhans may secrete, at times, excessive amounts of insulin and hypoglycaemia can develop rapidly. Indeed, a fear of every type 1 diabetic patient is injection of too high a dose of insulin and consequent hypoglycaemia. In view of this, injection of a high dose of insulin has been used in attempts to commit suicide and as a murder weapon! The first recorded case of murder was in 1957, when Kenneth Barlow murdered his wife by injecting insulin into her buttocks. Even though the methods for measuring insulin concentration were somewhat impre-cise at that time, sufficient was detected in the buttock of the body to provide enough evidence for conviction of murder (Burkinshaw *et al.*, 1958).

The yield of ATP from glycolysis

The number of ATP molecules generated by glycolysis depends upon the substrate for the process and the eventual fate of the pyruvate. *Anaerobic* glycolysis from glucose generates two molecules of ATP but, from glycogen, three molecules of ATP are generated for each glucose molecule in glycogen that is converted to lactic acid. However, con-siderably more ATP is generated if the pyruvate produced

is completely oxidised to CO_2 (see Table 9.5). Despite the small number of ATP molecules generated in *anaerobic* glycolysis, from glycogen the capacity of the process may be so large that, if fully stimulated, it can produce as many if not more ATP molecules than *aerobic* glycolysis, when the pyruvate is completely oxidised. This is the case since more molecules of glucose-in-glycogen are broken down. For example, even in aerobically trained volunteers, the maximum rate of *anaerobic* glycolysis can generate more ATP molecules compared with that from the maximum rate of the oxidation of pyruvate, i.e. the Krebs cycle (see Table 9.7).

The biochemical and physiological importance of *anaerobic* glycolysis

Glucose or glycogen can be the substrate for anaerobic glycolysis and, although there is not a complete division, some tissues use primarily glucose whereas others use primarily glycogen. The significance of which substrate is used to generate ATP, in various cells, is as follows.

Glycogen to lactic acid

When a human muscle, which comprises exclusively anaerobic (i.e. type IIß) fibres is physically active, glyco-gen conversion to lactate generates all the ATP that is required to support the activity. Type I or IIa fibres use this process only when the demand for ATP exceeds that which can be generated from aerobic metabolism, e.g. during hypoxia. The significance of these processes for generation of ATP by muscle during various athletic events is dis-cussed in Chapter 13.

One condition in which all tissues or organs may use glycogen is the mammalian foetus during birth. This is especially the case if the placenta separates from the uterus prior to birth, which may occur if parturition is slow. In this situation, prior to birth and the first breath, the tissues of the foetus must generate ATP from conversion of gly-cogen to lactic acid. Survival depends on this process! To this end, the glycogen level in the tissues of the foetus increases just prior to birth.

Cells that use glucose conversion to lactic acid to generate ATP

Those that depend upon this process are cells in the tissues of the eye, the kidney medulla, the epithelial cells in the skin, the enterocytes in the small intestine, erythrocytes, immune cells and cells during proliferation (Chapters 9, 17 and 18).

Tissues of the eye   In the tissues of the eye, the presence of too many blood vessels, or too many mitochondria within cells, could interfere with the transmission of light. Consequently, much of the ATP is generated from glucose conversion to lactic acid. The tissues involved are the cornea, lens and those parts of the retina that require high visual acuity (e.g. the fovea). Hence, although blood vessels overlie the retina, very few are present over the fovea; and the capacity of anaerobic glycolysis in the retina, as a whole, is high indicating that *anaerobic* glycolysis contributes significantly to ATP generation. The glucose is provided and lactic acid removed by the fluids in the eye, the aqueous and vitreous humours, so that the flow of these fluids in the eye is important.

Kidney medulla   From the metabolic point of view the kidney is virtually two organs, the cortex and the medulla. The cortex contains the glomeruli, through which the blood is filtered, the proximal tubules and part of the distal tubules, from which ions and molecules are reabsorbed. The cortex is well supplied with blood so that ATP is generated by the oxidation of fuels. The medulla is metabolically quite different. Here the ATP is required for the reabsorption of ions from the loop of Henle. Some ATP is generated by *anaerobic* glycolysis, since the supply of blood, and therefore of oxygen, to the medulla is much poorer than to the cortex. This reflects control of the uptake of water and Na^+ ions into the blood by the counter current mechanism. This depends on a slow flow of the blood in the capillaries.

Roles of glycolysis other than ATP generation

One role of glycolysis is to provide intermediates which are substrates for other pathways (Figure 6.2). Roles in specific cells or tissues are now presented.

> Mosquitoes detect lactic acid, which is volatile, as a signal for the presence of an animal surface for feeding. A receptor site for lactic acid is present on the antennae of the mosquito.

Epithelial cells in the skin   Some of the lactic acid produced by glycolysis is secreted in perspiration. It lowers the pH of the surface of the skin, where it has an antibacterial effect. The concentration on the surface may be as high as 15 mmol/L.

Glial cells in the brain   Some glial cells convert glucose to lactic acid, which is the fuel used by the neurones (Chapter 14).

Enterocytes in the small intestine   The conversion of glucose to lactic acid in the enterocytes provides another means of absorbing glucose into the blood. They absorb glucose from the lumen of the intestine and then convert it to lactic acid, which enters the hepatic portal blood from where it is removed by the liver, which then converts the lactate back to glucose. The glucose is then released into the blood. An advantage of this process is that it slows the entry of glucose into the blood which minimises the peak blood sugar level after a meal. It is possible that as much as 50% of the glucose absorbed into the enterocytes enters the blood in this manner.

Erythrocytes   There are several important roles of glycolysis in these cells, in addition to generation of ATP.

(i)   Not only is phosphoglyceroyl phosphate an intermediate in glycolysis, but it is a precursor for formation of 2,3-bisphosphoglycerate (BPG) which has a fundamental role in transport of oxygen within the blood (Figure 6.12). It binds to haemoglobin and, in so doing, modifies the binding of oxygen to haemoglobin. In the absence of BPG, the graph of the saturation of haemoglobin with oxygen is hyperbolic whereas, in its presence, it is sigmoid. This phenomenon permits a significant amount of oxygen to dissociate from

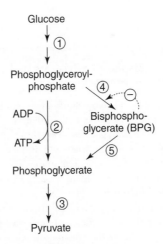

Figure 6.12 *The bisphosphoglycerate shunt.*
The concentration of BPG is the same as that of haemoglobin and the position where the BPG is bound to the haemoglobin molecule has been identified from the 3-dimensional structure.
① represents the first part of glycolysis, ② is phosphoglycerate kinase, ③ is the latter part in glycolysis, ④ is bisphosphoglycerate mutase and ⑤ is bisphosphoglycerate phosphatase.
Bisphosphoglycerate mutase is inhibited by BPG, which is a feedback inhibitory mechanism.

haemoglobin at physiological concentrations of oxygen in the capillaries (Figure 6.13). The graph indicates that, in the absence of BPG, life as we know it would not be possible. Almost no oxygen would dissociate from haemoglobin in the capillaries. Even a small change in the concentration of BPG can make a marked difference to the extent of dissociation at physiological concentrations of oxygen. Deficiencies of the enzymes hexokinase or pyruvate kinase can influence the concentration of BPG in the erythrocyte and hence erythrocyte physiology (Box 6.1).

(ii) In order for the iron ion in haemoglobin to bind oxygen, it must be in the ferrous (Fe^{2+}) form. However, the high concentration of oxygen in the erythrocyte causes a continuous oxidation of ferrous to ferric iron to form a protein known as methaemoglobin (equation a). However, methaemoglobin is continuously converted back to haemoglobin, in a reaction catalysed by methaemoglobin reductase (equation b) Consequently, under normal conditions, only 1% of haemoglobin is in the form of methaemoglobin.

This is important since not only does it affect oxygen transport but, if it accumulates, the cells are then damaged, which can result in anaemia.

The reactions are

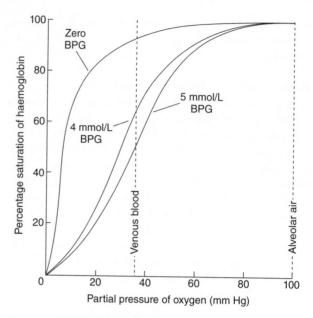

Figure 6.13 *The effect of different concentrations of 2,3-bisphosphoglycerate on the oxyhaemoglobin dissociation curve.* The increase in the concentration of BPG from 4 to 5 mmol/L results in an increase in the amount of oxygen released in the capillaries by more than 20%. The concentration of BPG decreases on storage of erythrocytes, so that cells from the blood bank have a higher affinity for oxygen and hence discharge less oxygen in the tissue.

Box 6.1 2,3-Bisphosphoglycerate and the dissociation of oxygen from oxyhaemoglobin: in patients and athletes

Both haemoglobin and myoglobin bind reversibly with oxygen. The dissociation curve of oxyhaemoglobin is sigmoid whereas that of myoglobin is hyperbolic. This difference ensures that unloading of oxygen from haemoglobin is significant at the oxygen concentrations found in capillaries in tissues other than the lung. In contrast, myoglobin releases its oxygen close to mitochondria within muscle fibres, where the partial pressure of oxygen is very low. Mammalian life as we know it would not be possible if erythrocytes contained myoglobin in place of haemoglobin; the partial pressure of oxygen (i.e. the concentration) in the capillaries would be too low to maintain an adequate diffusion rate from capillary into cells. Early studies indicated that the sigmoid curve was considerably reduced if the haemoglobin was dialysed, to remove all small molecules from the solution. In 1921, Adair postulated that some additional factor was involved. Only in 1967 was this identified as 2,3-bisphosphoglycerate. The advantage of this effect is that the extent of the sigmoidicity of the dissociation curve for

haemoglobin can be modified within the erythrocyte by changing the concentration of 2,3-BPG. The normal concentration of 2,3-BPG in erythrocytes is around 4 mmol/L but this increases in individuals exposed to low partial pressures of oxygen such as found at high altitudes and in patients with chronic hypoxia, due to heart or lung problems. This enables more effective unloading of the smaller amount of oxygen that will be carried under these circumstances.

The concentration of 2,3-BPG in erythrocytes is regulated through negative feedback. Deoxygenated haemoglobin binds more 2,3-BPG than oxygenated haemoglobin and, as the concentration of the deoxygenated form rises during hypoxia, less free 2,3-BPG will be present. This reduces the inhibition of bisphosphoglycerate mutase so that more 2,3-bisphosphoglycerate is produced and its concentration rises, thus affecting the shape of the dissociation curve. It is tempting to speculate that the remarkable performances in endurance runs by Kenyan athletes are due to a high level of 2,3-BPG in their erythrocytes.

(a) haemoglobin + O_2 → methaemoglobin + $O_2^{-\bullet}$

(b)
$$\text{methaemoglobin} \xrightarrow[\text{NADH}]{\quad\quad} \text{haemoglobin}$$
NADH   NAD$^+$

It is important to note that the oxidation produces the superoxide free radical. Since it is toxic, the radical produced in reaction (a) must be removed. This is done in a reaction catalysed by superoxide dismutase, which produces hydrogen peroxide. However, this also must be removed (see Appendix 9.6 for discussion of free radicals). Removal of hydrogen peroxide is achieved in a reaction with reduced glutathione, catalysed by glutathionine peroxidase.

$$2GSH + H_2O_2 \rightarrow GS\text{--}SG + H_2O$$

Since this reaction is required continuously, the oxidised glutathione must be reduced continuously and this is achieved with NADPH, as follows

$$G\text{--}S\text{--}S\text{--}G + 2NADPH \rightarrow 2GSH + 2NADP^+$$

The NADPH is produced from glucose 6-phosphate in the first three reactions in the pentose phosphate pathway (see below). Hence the pentose phosphate pathway is essential in the erythrocyte; and glycolysis provides the substrate glucose 6-phosphate. Individuals with a reduced amount of glucose 6-phosphate dehydrogenase can suffer from oxidative damage to their cells and hence haemolysis.

It is estimated that over 400 million people, principally in the tropical Mediterranean regions, have a deficiency of this dehydrogenase. It is, in fact, the most common inborn error of metabolism. These individuals are more sensitive to oxidative agents which increase the formation of methaemoglobin and hence the superoxide radical. Therefore, formation of hydrogen peroxide is increased: failure to reduce glutathione results in accumulation of hydrogen peroxide and damage to the cells. Such oxidative agents include the antimalarial drugs primaquine and paraquine. Another oxidative chemical is present in the fava bean, a vegetable consumed in the Mediterranean region (*Vincia faba*). It induces haemolysis in individuals deficient in the dehydrogenase. It is a sufficient problem to be given a name, *favism*.

Regulation of the flux through glycolysis

The substrate for glycolysis can be either glucose or glycogen (Figure 6.3) and control of the first steps in each process is different.

Glucose transport and the control of glycolysis in muscle

There are two biochemical situations in which to consider glucose transport and regulation of glycolysis:

1 Glucose transport is non-equilibrium In this case, it is rate-limiting for glycolysis. For example, an increase in the rate of glucose transport increases the intracellular glucose concentration which increases hexokinase activity, which then increases the flux through the remainder of glycolysis by 'internal regulation' (see Chapter 3 for an explanation of 'internal regulation'). One factor that stimulates transport and glycolysis is insulin (Fig. 6.14).

It is the transporter protein, GLUT-4, that is involved in regulation of the rate of glucose transport, when it is in non-equilibrium. The rate is regulated by changing the number of transporters in the plasma membrane. The number is decreased chronically by cortisol but is increased acutely by insulin, physical activity and factors that lower the ATP/ADP concentration ratio. The transporters are stored within the cell and they are translocated to the plasma membrane and back again to the store. Translocation can change the number of transporters in the plasma membrane and hence the rate of transport (Figure 6.15).

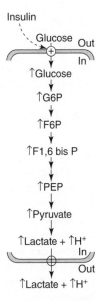

Figure 6.14 *An increase in the rate of glucose transport, in response to insulin, which increases the rate of glycolysis.* This is achieved by increasing the concentrations of all the intermediates in the pathway, indicated by the arrows adjacent to the intermediates. Insulin, physical activity or a decrease in the ATP/ADP concentration ratio all result in increased rates of glucose transport in skeletal muscle. Insulin increases the rate about fivefold, physical activity about 50-fold.

Intracellular   Extracellular

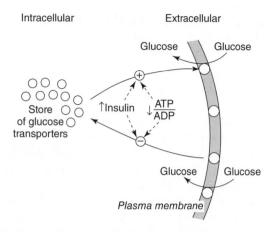

Figure 6.15 *Regulation of the number of glucose transporters in the plasma membrane.* The transporter affected is GLUT-4. It is unclear which translocation process is affected by insulin, physical activity or a change in the ATP/ADP concentration ratio. Effects on the translocation from within the cell to the membrane or vice versa are indicated here.

2 Glucose transport is near-equilibrium   In this condition in the muscle, hexokinase is rate-limiting for glycolysis. Its activity is regulated by phosphofructokinase activity (see below). If the enzyme glucokinase is present in the tissue, changes in the blood glucose concentration regulate the rate of glycolysis. This is important in (i) the β-cells of the Islets of Langerhans in the pancreas, i.e. the insulin-secreting cells, since glycolytic flux plays a role in regulation of insulin secretion (Chapter 3 and see below); (ii) the liver, in which changes in the blood glucose level can increase glycogen synthesis (Chapter 3 and see below).

Regulation of the entry of glycogen into the glycolytic pathway

The enzyme that controls the entry of glucose molecules in glycogen into the glycolytic pathway is phosphorylase (specifically glycogen phosphorylase). The enzyme exists in two forms, *a* and *b* and, in muscle, is regulated by two mechanisms. (i) Phosphorylase *b* is normally inactive but it can be activated by an allosteric mechanism. (ii) The proportion of *a* and *b*, at any one time, is regulated by an interconversion cycle of phosphorylation and dephosphorylation, which are regulated by physical activity (Ca^{2+} ions) and the stress hormone adrenaline. In resting muscle, all the phosphorylase is in the *b* form and, in this condition, *b* is inactive, so that glycogen breakdown does not occur. However, when muscle is physically active phosphorylase *b* is activated allosterically and some is converted to the *a* form. Both mechanisms of control are discussed in Chapter 3 and see also Chapter 12 for discussion of the effect of adrenaline.

Regulation of flux through glycolysis, from glucose to pyruvate and then to acetyl-CoA in muscle

Although the rate of glucose transport can regulate the flux through glycolysis, as indicated above, several enzymes also play a role.

- Hexokinase, the activity of which is inhibited by its product, glucose 6-phosphate, which is relieved by phosphate (i.e. phosphate can activate the enzyme when it is inhibited by glucose 6-phosphate).

- Phosphofructokinase, the activity of which is inhibited by ATP and phosphocreatine, but this inhibition is relieved by AMP and phosphate.

- Pyruvate kinase, the activity of which is inhibited by ATP and phosphocreatine but is activated by ADP, which is its substrate (Figure 6.16(a)).

- Pyruvate dehydrogenase, the activity of which is affected by a number of factors: the ATP/ADP, the NADH/NAD$^+$ and the acetyl-CoA/CoASH concentration ratios. It is also activated by Ca^{2+} ions. These factors change the enzyme activity via an interconversion cycle (Figure 6.16(b)).

From this description of the properties of these enzymes, it is concluded that the major factors involved in the regulation of glycolytic flux are changes in the ATP/ADP concentration ratio and changes in the concentrations of AMP, the phosphate ion and phosphocreatine. The integration of these factors in the regulation of glycolysis and the Krebs cycle in muscle during physical activity and in other conditions is described in Chapters 9, 13 and 16.

Regulation of the flux through glycolysis by insulin

Insulin increases glucose uptake by muscle through stimulation of the rate of glucose transport. The fate of the glucose is either conversion to lactate or conversion to glycogen. The stimulation of glycolysis is achieved by increasing the intracellular concentration of glucose and the concentrations of all the glycolytic intermediates which increases the activities of all the enzymes, since they are not saturated with their substrates (Figure 6.14). This is an excellent example of control via a 'transmission sequence' as described in Chapter 3.

Glycogen synthesis

Glycogen is a polymer of glucose and, although most tissues contain some glycogen, quantitatively important

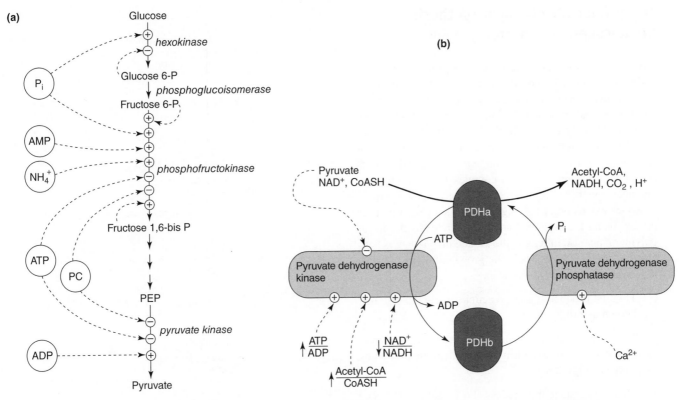

Figure 6.16 **(a)** *Factors involved in regulation of key steps in glycolysis.* PC is phosphocreatine; P_i is the phosphate ion.
(b) *Factors involved in the regulation of the interconversion cycle enzymes for pyruvate dehydrogenase.* Phosphorylation of pyruvate dehydrogenase (PDH) inhibits the enzyme (PDHb) whereas dephosphorylation activates the enzyme: i.e. PDHa is the active form and PDHb the inactive form form PDHa. Note that none of the regulators affect the dehydrogenase enzyme directly. Note also that since pyruvate dehydrogenase is an intramitochondrial enzyme, changes in Ca^{2+} ion concentration also activate isocitrate dehydrogenase and oxoglutarate dehydrogenase but by direct effects on the enzyme molecule. Consequently, stimulation of pyruvate dehydrogenase occurs simultaneously with an increase in flux through the Krebs cycle (Chapter 9). This increases formation of acetyl CoA for the increased flux through the cycle. Note that the two factors responsible for the increased rate of glycolysis and pyruvate oxidation are the ATP/ADP concentration ratio and the Ca^{2+} ions (see below). The role of Ca^{2+} ions in control the cycle is discussed in Chapter 9.

stores are found only in muscle and liver (Chapter 2). The pathway of synthesis is usually considered to begin with phosphorylation of glucose to form glucose 6-phosphate. In the liver, this reaction is catalysed by glucokinase whereas in the muscle, hexokinase is responsible. The significance of glucokinase, and the fact that it has a high K_m for glucose, in the regulation of blood glucose level and glycogen synthesis is discussed in Chapter 3.

There are five enzymes involved in the synthesis of glycogen, hexokinase (or glucokinase), phosphoglucomutase, uridylyltransferase, glycogen synthase and a branching enzyme. A summary of the pathway is shown in Figure 6.17. Two additional systems are required: a glycogen primer and a small intermediate form of glycogen which acts as a building platform for the large polymer.

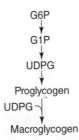

Figure 6.17 *A summary of the reactions in the pathway for glycogen synthesis.* Full details of the pathway are given in Figure 6.18. UDPG is the abbreviation for uridine diphosphate glucose.

The primer for glycogen synthesis, proglycogen and macroglycogen

Glycogen can only be synthesised if a primer is present, which is a small branched oligosaccharide, comprising eight to ten glucose molecules. The oligomer is synthesised by a small protein called glycogenin. In fact, glycogenin itself catalyses formation of the primer from glucose and the primer remains attached to the glycogenin, i.e. it catalyses its own glucosylation. Once glucosylation is complete, glycogenin forms a complex with the enzyme, glycogen synthase. In this complex, glycogen synthase catalyses extension of the oligosaccharide chain, upon the primer, using UDP-glucose as the substrate. As the linear polymers increase in length, the branching enzyme is activated to produce branches in the growing polysaccharide. However, it proceeds in two stages:

(1) The first product of these reactions is a small polymer known as proglycogen.

(2) The proglycogen is extended to produce macroglycogen by action of glycogen synthase and the branching enzyme (Figure 6.18).

To help to understand this process, it is useful to appreciate that there is a degree of similarity in the sequences of the three major synthetic processes, glycogen, fatty acid and peptide syntheses. The similarities are presented in Table 6.2.

Control of glycogen synthesis

Glycogen synthesis is important in two tissues, muscle and liver. In muscle the major factors regulating the rate of synthesis are insulin and the amount of glycogen already present in the muscle. In liver, the major factor is the intracellular concentration of glucose (see below) (Figure 6.19).

Synthesis of fructose and lactose

Fructose is synthesised in the prostate gland. It is secreted into the seminal fluid as the fuel for sperm. It is synthesised from glucose as follows

$$\text{Glucose} \rightarrow \text{sorbitol} \rightarrow \text{fructose}$$

Reactions are catalysed by aldose reductase and iditol dehydrogenase. (For details, see Appendix 6.6).

Lactose is a disaccharide of glucose and galactose linked by a 1,4 glycositic bond. It is the only disaccharide synthesised in large quantities in mammals where its sole

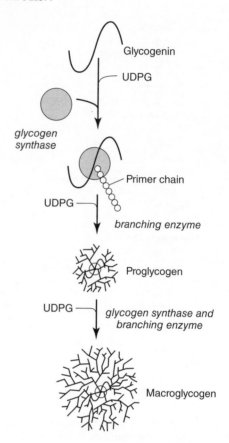

Figure 6.18 *The pathway for synthesis of pro- and macroglycogen. Note that the glycogenin–primer complex is formed prior to complexing with the glycogen synthase.*

function is to provide the sugar for milk formation. The synthesis occurs by transfer of a galactosyl residue from UDP galactose to a glucose molecule, catalysed by the enzyme lactose synthase. The enzyme is, of course, present in the mammary glands.

$$\text{UDP-galatose} + \text{glucose} \rightarrow \text{lactose} + \text{UDP}$$

The synthesis of UDP galactose is described in Appendix 6.7.

The pentose phosphate pathway

This pathway is variously known as the pentose phosphate, hexose monophosphate or phosphogluconate pathway, cycle or shunt. Although the pentose phosphate pathway achieves oxidation of glucose, this is not its function, as indicated by the distribution of the pathway in different tissues. Only one of the carbons is released as CO_2, the key products are NADPH and ribose 5-phosphate, both of which are important for nucleotide phosphate formation and hence for synthesis of nucleic acids (Chapter 20). The

Table 6.2 Similarities in the processes involved in synthesis of glycogen, fatty acid and peptides

Process	Activation	Large binding system	Binding sites	Condensation	Elongation and movement	Termination	Post-synthetic modifications
Glycogen synthesis	Formation of UDP-glucose	Glycogenin	Binding of UDP-glucose to glycogenin–glycogen synthase complex	Condensation between glucose in oligosaccharide chain and glucose attached to UDP	?Movement of complex as oligosaccharide chain extends and movement of branching enzyme	Not known, possibly loss of activity of glycogen synthase	Hydration of glycogen to form granules[a]
Fatty acid synthesis	Formation of malonyl-CoA	Acyl carrier protein (ACP)	(i) acetyl site on ACP (ii) malonyl site on ACP	Condensation between acetyl (acyl) group and malonyl-CoA	Transfer of extending chain to 'acetyl site' on ACP	Release of acyl chain (palmitate) from ACP, catalysed by thioesterase	(i) Formation of triacylglycerol or phospholipids from fatty acid (ii) Specific desaturation and elongation reactions
Peptide synthesis	Formation of tRNA-amino acid complexes	Ribosome	(i) amino site on ribosome (ii) peptide site on ribosome	Condensation between amino acyl group and peptide	Transfer of amino acyl group to peptide site on ribosome at a termination site	Release of peptide from ribosome at a termination site	(i) Formation of protein from peptides (ii) Specific methylation, hydroxylation etc. reactions

[a]It is unclear if hydration is a specific process or simply cytosolic water spontaneously filling the spaces between the polysaccharide chains. Nonetheless it has considerable physiological significance.

APC – acyl carrier protein.

For details of fatty acid synthesis, see Chapter 11; for peptide synthesis, see Chapter 20.

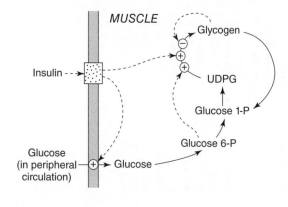

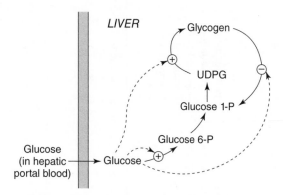

Figure 6.19 *Regulation of the synthesis of glycogen from glucose in liver and muscle.* Insulin is the major factor stimulating glycogen synthesis in muscle: it increases glucose transport into the muscle and the activity of glycogen synthase, activity which is also activated by glucose 6-phosphate but inhibited by glycogen. The latter represents a feedback mechanism and the former a feedforward. The mechanism by which glycogen inhibits the activity is not known. The mechanism for the insulin effect is discussed in Chapter 12.

The glucose concentration is the major factor regulating glycogen synthesis in liver. Glucose activates glucokinase directly as a substrate and indirectly via an increase in the concentration of fructose 6-phosphate. It also activates glycogen synthase but it inhibits glycogen phosphorylase (see text).

pathway can most clearly be envisaged as consisting of two parts (Figure 6.20). The first consists of three reactions that produce NADPH and a pentose phosphate:

(i)

$$\text{glucose 6-phosphate} \xrightarrow[\text{}]{\text{NADP}^+ \quad \text{NADPH}} \text{phosphogluconolactone}$$

(ii)   phosphogluconolactone + $H_2O \rightarrow$ phosphogluconate

(iii)

$$\text{phosphogluconate} \xrightarrow[\text{}]{\text{NADP}^+ \quad \text{NADPH} + CO_2} \text{ribulose 5-phosphate}$$

The three reactions are catalysed by glucose 6-phosphate dehydrogenase, gluconolactonase and phosphogluconate

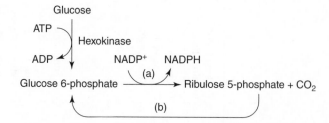

Figure 6.20 *An outline of the pentose phosphate pathway.*
(a) represents the three reactions described in the text.
(b) represents the reactions catalysed by transketolases, transaldolases, aldolase and epimerase enzymes by which ribulose 5-phosphate is converted back to glucose 6-phosphate (Appendix 6.8).

Table 6.3 Some roles of NADPH in various cells

Tissue	Role in the cell
Testes (Leydiy cells) Ovary (Follicles and corpus luteum) (Chapter 19) Adrenal cortex	Steroid hormone synthesis
Liver	Fatty acid synthesis Elongation of fatty acids (Chapter 11) Desaturation of fatty acids Detoxification reactions
Adipose	Fatty acid synthesis (Chapter 11)
Erythrocytes and many other cells	Reduction of oxidised glutathione (see above)
Macrophages Neutrophils	Production of free radicals for killing pathogens (Chapter 17)
Proliferating cells	Synthesis of nucleotides (Chapter 20)

dehydrogenase, respectively. These three reactions result in the conversion of a hexose phosphate into a pentose phosphate.

The second part consists of a series of reactions in which ribulose 5-phosphate is reconverted back to glucose 6-phosphate (i.e. '5C' $\rightarrow \rightarrow$ '6C') (Figure 6.20) (Appendix 6.8). Some key processes that depend on NADPH, and therefore the pentose phosphate pathway, are identified in Table 6.3 and presented in Figure 6.21.

Gluconeogenesis: glucose formation from non-carbohydrate sources

Glucose enters the blood from three different sources depending on the conditions.

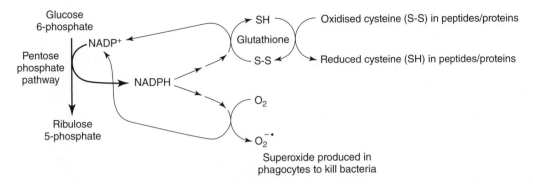

Figure 6.21 *Two examples of the roles of NADPH generated from the pentose phosphate pathway.* It reduces the oxidised glutathione to reduced glutathione, which can reduce the S-S bridges formed by oxidation of SH groups in proteins. This is particularly important in maintaining the SH group in the active site of many enzymes (e.g. repair of oxidative damage of proteins). NADPH is substrate for the NADPH oxidase complex, which results in conversion of the oxygen molecule to the superoxide radical ($O_2^{-\bullet}$) in phagocytes, which is one of the weapons used to kill bacteria that are phagocytosed by neutrophils or macrophages (Chapter 17). Reduced glutathione peroxide catalysed by the enzyme glutathione peroxidase (see text above).

(i) After a meal, glucose enters the blood from carbohydrate that is digested in the intestine and is absorbed by the enterocytes in the small intestine.

(ii) Between meals, particularly during the overnight fast, it enters the blood from the breakdown of glycogen stored in the liver.

(iii) During starvation, even short-term starvation (<24 hours), some of the glucose released into the blood is produced from non-carbohydrate sources, a process known as gluconeogenesis which occurs mainly in the liver but with some contribution from the kidney cortex. Gluconeogenesis in starvation is not usually important in adults in developed countries, since starvation for longer than a few hours is uncommon, unless breakfast and lunch are missed and diner supper is late. In contrast, it is important in developing countries where starvation for longer periods is not rare. It is, however, important in young children, even in developed countries, during short periods of starvation, since the amount of glycogen stored in the liver is much less than in an adult. It may also be important in the homeless or poor in developed countries since chronic undernutrition can be prevalent which will restrict glycogen synthesis in the liver, so that the store of glycogen in the liver may be small (Chapter 16).

Gluconeogenesis

Gluconeogenesis is the de novo synthesis of glucose from none carbohydrate sources. These sources (precursors) are lactic acid, glycerol and the amino acids, especially alanine, glutamine and aspartic acid (Figure 6.22).

Lactic acid   The normal concentration of lactate in the blood is about 1 mmol/L which is the balance between production and utilisation. Various tissues produce lactic acid and release it into the blood, from where it is taken up by liver and converted to glucose (or glycogen) (the Cori cycle, see above).

$$2\,\text{lactate} + 2\text{H}^+ \rightarrow \text{glucose}$$

Glycerol   Glycerol is produced from lipolysis in adipose tissue and, since the process is continuous, glycerol is released continuously into the blood from where it is also taken up by the liver and converted to glucose or glycogen. Within the liver cell, glycerol is phosphorylated to produce glycerol 3-phosphate in a reaction catalysed by the enzyme glycerol kinase:

$$\text{glycerol} + \text{ATP} \rightarrow \text{glycerol 3-phosphate} + \text{ADP}$$

The glycerol 3-phosphate is oxidised by glycerol 3-phosphate dehydrogenase to produce dihydroxyacetone phosphate

$$\text{glycerol 3-P} + \text{NAD}^+ \rightarrow \text{dihydroxyacetone-P} + \text{NADH} + \text{H}^+$$

This is interconverted to form glyceraldehyde 3-phosphate and both combine, via the enzyme aldolase, to produce fructose bisphosphate, en route to form glucose or glycogen.

Oxoacids derived from amino acids   After a meal containing protein, the amino acids that are absorbed into the blood are largely metabolised in the liver (>70%) and muscle. Most of these are converted to oxoacids which have two main fates:

- complete oxidation to CO_2

- conversion to glucose or glycogen, via gluconeogenesis.

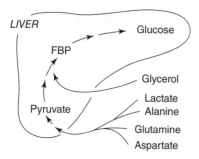

Figure 6.22 *Major precursors for gluconeogenesis.* FBP is fructose 1,6-bisphosphate.

In fact, these two processes are metabolically linked. The oxidation generates ATP whereas gluconeogenesis utilises this ATP. Consequently, in the well-fed human, gluconeogenesis is essential for oxidation of amino acids, otherwise oxidation is limited by the need to utilise the ATP (Chapter 8). The reactions in which amino acids are converted to compounds that can enter the gluconeogenic pathway are described in Chapter 8. The position in the gluconeogenic pathway where amino acids, via their metabolism (Chapter 8), enter the pathway is indicated in Figure 6.23.

The pathway of gluconeogenesis

The pathway for gluconeogenesis is shown in Figures 6.23 and 6.24. Some of the reactions are calalysed by the glycolytic enzymes; i.e. they are the near-equilibrium. The non-equilibrium reactions of glycolysis are those catalysed by hexokinase (or glucokinase, in the liver), phosphofructokinase and pyruvate kinase and, in order to reverse these steps, separate and distinct non-equilibrium reactions are required in the gluconeogenic pathway. These reactions are:

• The conversion of glucose 6-phosphate to glucose occurs in a reaction catalysed by the enzyme glucose-6-phosphatase:

    glucose 6-phosphate + H$_2$O → glucose + P$_i$

This bypasses the hexokinase reaction.

• The conversion of fructose 1,6-bisphosphate to fructose 6-phosphate in a reaction catalysed by the enzyme fructose-bisphosphatase:

    fructose bisphosphate → fructose 6-phosphate + P$_i$

This bypasses the phosphofructokinase reaction.

• The conversion of pyruvate into phosphoenolpyruvate, which bypasses the pyruvate kinase reaction, requires two separate reactions: carboxylation of pyruvate to

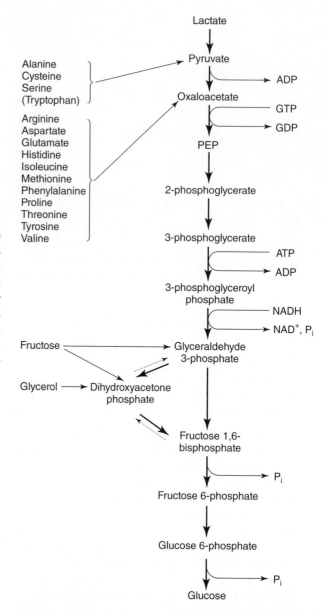

Figure 6.23 *Positions in the gluconeogenic pathway where amino acids, fructose and glycerol enter the pathway.* For details of the metabolism that provides the intermediates that actually enter the pathway from the amino acids, see Chapter 8. Not all of the carbon in some of the amino acids is incorporated into glucose (e.g. tryptophan). Two amino acids, leucine and lysine, do not give rise to glucose.

form oxaloacetate, catalysed by pyruvate carboxylase, and conversion of oxaloacetate to phosphoenolpyruvate, in a reaction catalysed by phosphoenolpyruvate carboxykinase.

    pyruvate + CO$_2$ + ATP → oxaloacetate + ADP + P$_i$

    oxaloacetate + GTP → phosphoenolpyruvate + CO$_2$ + GDP

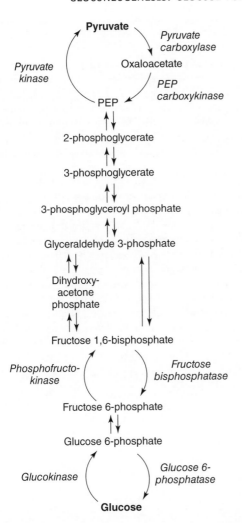

Figure 6.24 *The gluconeogenic pathway indicating the glycolytic and gluconeogenic non-equilibrium reactions.* The non-equilibrium reactions provide for the substrate cycles. (See Chapter 3 for a discussion of substrate cycles and their role in regulation.)

The mitochondrial barrier: mitochondrial and cytosolic distribution of enzymes

A problem for gluconeogenesis is that pyruvate carboxylase, which produces oxaloacetate from pyruvate, is present in the mitochondria but phosphoenolpyruvate carboxylase, at least in human liver, is present in the cytosol. For reasons given in Chapter 9, oxaloacetate cannot cross the mitochondrial membrane and so a transporter is not present in any cells. Hence, oxaloacetate is converted to phosphoenolpyruvate which is transported across the membrane (Figure 6.25).

Gluconeogenesis is relatively 'energy expensive': six molecules of ATP are hydrolysed for every two molecules of lactate converted to one of glucose, but, in addition, since substrate cycles are involved in three steps in the

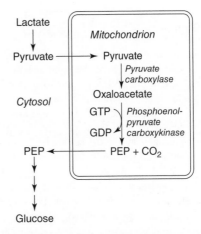

Figure 6.25 *The intracellular location of the gluconeogenic enzymes.* The gluconeogenic enzymes are located in the cytosol, except for pyruvate carboxylase which is always present within the mitochondria: phosphoenolpyruvate carboxykinase is cytoplasmic in some species including humans. Consequently phosphoenolpyruvate must be transported across the inner mitochondrial enzyme by a transporter molecule in order for gluconeogenesis to take place.

process, even more energy must be expended. It is, however, not possible to quantify this expenditure since the rates of cycling are not known and are probably very different between different conditions. In starvation, the ATP required for gluconeogenesis is generated from the oxidation of fatty acids. This has physiological and pathological significance (Chapter 16, see Box 6.2).

The physiological pathway of gluconeogenesis

The reactions catalysed by pyruvate carboxylase, phosphoenolpyruvate carboxykinase, fructose bisphosphatase and glucose 6-phosphatase are non-equilibrium. Comparison of the concentrations of substrates for these non-equilibrium reactions with the K_m of the enzymes catalysing them, in both the liver and kidney, indicates that none of these enzymes approaches saturation with pathway substrate. Consequently, to find the flux-generating step, and hence the start of the physiological pathway, it is necessary to consider the precursors of the pathway. The gluconeogenic pathway can, in fact, be treated as starting with the formation of oxoacids in liver, lactate in muscle (and other tissues) and glycerol in adipose tissue. Gluconeogenesis has to be seen, therefore, as a complex, branched pathway that spans more than one tissue (Figure 6.26). This is consistent with the fact that increasing the concentrations of any of these precursors, in the blood, increases the rate of gluconeogenesis.

Box 6.2 Gluconeogenesis and death

The aim of this box is to emphasise the physiological significance of gluconeogenesis and a major problem that can arise if the process is severely inhibited under some conditions. The biochemical basis for the problem is that once the store of glycogen the liver is depleted, the only source of blood glucose is either from digestion and absorption of carbohydrate within the intestine or gluconeogenesis. Consequently, during starvation, gluconeogenesis is the only source of glucose.

A well-established inhibitor of gluconeogenesis is ethyl alcohol. This can cause problems in at least two situations. When alcoholic patients enter an alcoholic 'binge', they do not eat, so that liver glycogen is soon depleted. Since gluconeogenesis is inhibited, both hypoglycaemia and, as indicated below, lactic acidosis can develop. Indeed they may be the two most important factors that precipitate coma and collapse in the alcoholic patient.

Another, perhaps less dramatic condition, is the effect of social drinking after a very short period of starvation. A period of eating a high protein, low carbohydrate meal for several days, followed by a missed breakfast and a late lunch might result in a low level of liver glycogen. If now, lunch is preceded by alcoholic drinks, hypoglycaemia could readily develop, and, if severe, could lead to coma with possible serious consequences. Even social drinking can be a problem since gluconeogenesis is inhibited by low concentration of ethanol (1 mmol/L): a small sherry or whisky, especially on an empty stomach, can increase the blood alcohol concentration to 2–4 mmol/L. Although the concentration in the liver is not known, ethanol is lipid soluble so it might reach a similar concentration in the liver. This condition is sometimes known as the 'business executive's lunch syndrome'. To the author's knowledge there have been no studies on the condition so it remains hypothetical, although there have been several reports of hypoglycaemic incidents on long-haul flights, when alcoholic drinks are servil before lunch.

In this book it is suggested that one possible cause of death in prolonged starvation is severe hypoglycaemia. This may be due to a lack of amino acid precursors since almost all the body protein has been broken down. Alternatively, the fat store in the body has been totally depleted, so that the plasma fatty acid level will be close to zero. Consequently, there will be no fatty acid oxidation in the liver and therefore little or no ATP generation to support gluconeogenesis. Post-mortem studies on individuals who have died of starvation show that the fat stores are totally depleted. This topic is discussed further in Chapter 16.

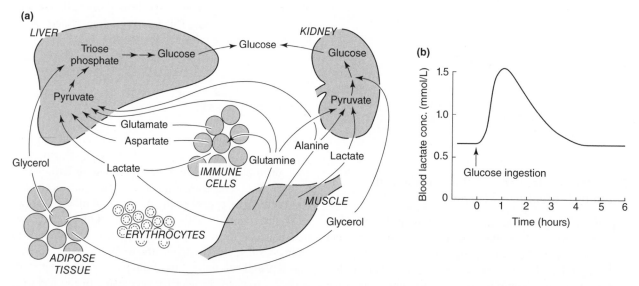

Figure 6.26 (a) *The branched nature of an extended (physiological) gluconeogenic pathway*. Note, the immune cells produce lactate from glucose and also glutamate and aspartate from glutamine. Since there are more than 10^{12} lymphocytes plus other immune cells, the rate of formation of these gluconeogenic precursors is large. These will, therefore, be released into the lymphatic system as well as the blood. It has not been considered previously that lymph may transport precursors for the process of gluconeogenesis and that immune cells provide gluconeogenic precursor. Note that the kidney is also an important gluconeogenic organ, although the quantitative importance of each organ is not known. It may vary in different conditions. **(b)** *The effect of ingestion of a solution containing 75 g of glucose on the level of blood lactate*. Note that ingestion of glucose results in almost a threefold increase in the blood lactate concentration. This is an important precursor not so much for gluconeogenesis, but for glyconeogenesis (i.e. synthesis of glycogen from lactate in the liver).

Source of glucose 6-phosphate for glycogen synthesis in the liver

There is a marked increase in the content of glycogen in the liver after a meal that contains carbohydrate. The precursor for the synthesis of glycogen is glucose 6-phosphate, which is converted to glycogen in the pathway shown above. However, there are two sources of glucose 6-phosphate in liver after a meal.

- One is the glucose that is absorbed from the intestine and enters the blood in the hepatic portal vein from where some of it is taken up by the liver and phosphorylated to form glucose 6-phosphate, which then stimulates the formation of glycogen (for discussion of regulation of this process, see below). This is known as the *direct pathway* for glycogen synthesis.

- Another source of glucose 6-phosphate is gluconeogenesis. After a carbohydrate meal, the increase in the concentration of insulin in the blood stimulates glucose uptake in muscle and adipose tissue, which results in an increase in the rate of glycolysis and an increase in lactate formation, which is released into the bloodstream and the concentration in the blood increases (Figure 6.26b). This lactate is taken up by the liver and converted to glucose 6-phosphate. Similarly, amino acids, absorbed after digestion of a meal, are metabolised to oxoacids in the liver, some of which are converted, via gluconeogenesis, to glucose 6-phosphate. The glucose 6-phosphate formed via this process, in the fed condition, is not released as glucose but is converted to glycogen. The conversion of such precursors to glycogen is known as glyconeogenesis. It is also known as the *indirect pathway* for synthesis of glycogen.

It is estimated that each process, the direct and indirect pathways, contributes equally to glycogen synthesis after a meal (Figure 6.27). It is, therefore, important to appreciate that the process of gluconeogenesis is of considerable importance in synthesis of liver glycogen during absorption of a meal; that is, all three macronutrients can contribute to the restoration of normal liver glycogen levels after a meal: carbohydrate, via glucose and lactate, fat, via glycerol, and protein, via oxoacids. Although amino acids via oxoacids can contribute to glycogen synthesis, carbohydrate, via the monosaccharides, is the most effective macromolecule.

Role of the liver in the regulation of the blood glucose concentration

The liver is the only organ that can either add or remove glucose to or from the blood. It removes glucose when the concentration in the hepatic portal blood is above normal. It releases it when the concentration is below normal. Furthermore, the rate of addition or removal is proportional to the size of the deviation from the normal glucose concentration: the higher the plasma concentration, the higher is the rate of uptake and vice versa. The liver must, therefore, be able to assess the concentration of glucose in the blood and respond to any changes. Remarkably, it can do this without any exogenous controls since this ability is apparent in an isolated perfused liver. The characteristics of the liver that are relevant to this biochemical feat are as follows.

- The liver receives most of its blood via the hepatic portal vein. The significance of this is that the blood passes through the absorptive area of the small intestine before it passes through the liver (Figure 6.28). After a meal containing carbohydrate, the concentration of glucose in this vein will be high (possibly 15 mmol/L or higher). In contrast, between meals, especially if the interval is long (e.g. the overnight fast), the concentration will be below normal (as low as 4 mmol/L or less). The liver, therefore, is exposed to much greater variations in the concentration of glucose in the blood than any other tissue or organ.

- As discussed in Chapter 5, the transport of glucose into the liver cell is near-equilibrium, so that the intracellular glucose concentration is similar to and follows precisely the changes in that of the hepatic portal blood. Furthermore, since it is near-equilibrium glucose can be transported into or out of the cell.

- The enzyme glucokinase, which is the dominant glucose phosphorylating enzyme in the liver, has a K_m for glucose about 10 mmol/L. The high K_m of glucokinase, and the near-equilibrium nature of glucose transport, means that the activity of glucokinase varies according to the

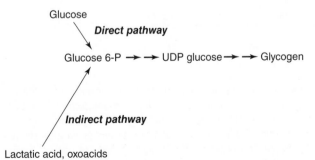

Figure 6.27 *The direct and indirect pathways for glycogen synthesis in the liver in the fed condition.*

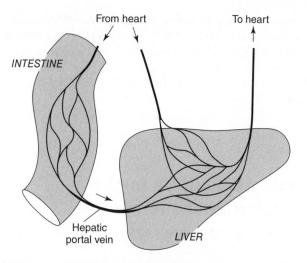

Figure 6.28 *The vascular link between the absorptive area of the intestine and the liver: the hepatic portal vein.* This vein is the only one that links capillaries in two organs and is fundamentally important in ensuring that the liver has access to all the nutrients, including the micronutrients, absorbed by the intestine. Note that the liver has two sources of blood.

changes in the concentration of glucose in the hepatic portal blood (Chapter 3). The higher the concentration, the higher is the activity of glucokinase.

• The enzyme glucose 6-phosphatase, which catalyses the hydrolysis of glucose 6-phosphate to form glucose, is also present in the liver cell.

• Both glucose 6-phosphatase and glucokinase are simultaneously active, the result of which is a substrate cycle, the glucose/glucose 6-phosphate cycle (Figure 6.29(a)).

The mechanism by which the glucose/ glucose 6-phosphate cycle regulates glucose uptake and release

At the normal blood glucose concentration in the hepatic portal vein, the activities of glucokinase and glucose 6-phosphatase are identical, so that there is neither uptake nor release of glucose. When the blood glucose concentration increases (e.g. in the fed condition), the intracellular concentration increases, so that the activity of glucokinase increases above that of glucose 6-phosphatase. Hence, glucose is converted to glucose 6-phosphate. The result is that glucose is taken up by the liver. However, as the blood glucose concentration decreases below normal (e.g. during starvation), the intracellular glucose concentration falls, so that the activity of glucokinase falls and, when it falls below that of glucose 6-phosphatase, the rate of hydrolysis of glucose 6-phosphate exceeds that of glucose phosphorylation and glucose is released from the liver (Figure 6.29b). Not only does this explain how the direction of glucose metabolism can be changed, it also explains how the liver responds precisely to the extent of the change in glucose concentration (Figure 6.30).

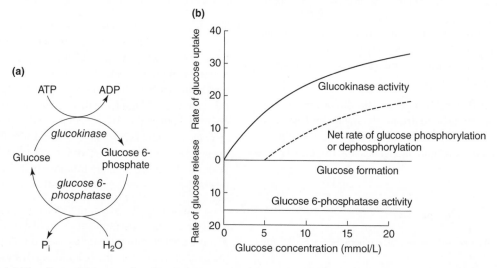

Figure 6.29 **(a)** *The glucose/glucose 6-phosphate substrate cycle in the liver.* The reactions are catalysed by glucokinase and glucose 6-phosphatase in the liver. **(b)** *A hypothetical graph of the effect of changes in the glucose concentration on glucokinase activity and the net rate of glucose phosphorylation and hence glucose uptake or release.* The net rate of glucose phosphorylation and therefore glucose uptake is obtained from the activity of glucokinase minus that of glucose 6-phosphatase, at any given glucose concentration. At the normal blood concentration of about 5 mmol/L, the two activities are equal, so that glucose uptake is zero. Below about 5 mmol/L, glucose 6-phosphatase activity is greater than that of glucokinase, so that the liver releases glucose, and above 5 mmol/L the liver removes glucose from the blood, since glucokinase activity exceeds that of glucose 6-phosphatase. This phenomenon is presented as the dotted carve in this figure.

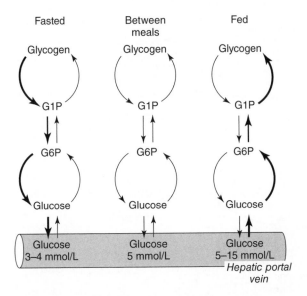

Figure 6.30 *The net direction of glucose flux in the liver via the cycle in the fed, fasted and in the normoglycaemic conditions.* In the normoglycaemic condition, the activities of glucokinase and glucose 6-phosphatase are identical, so that there is no net flux. In the fasted condition, the net flux is in the direction of glucose release. In the fed condition, the net flux is in the direction of glucose uptake. Glycogen degradation provides the glucose in the fasted condition. Glycogen synthesis is the fate of the glucose in the fed condition. The thickness of the line indicates the net direction of flux.

Although the cycle provides the basis of the mechanism, other processes must contribute, as follows.

- For the uptake of glucose to continue, the glucose 6-phosphate that is formed from the glucose must be converted to glycogen (Figure 6.30).

- For the release of glucose to continue, glycogen must be degraded to glucose 6-phosphate (Figure 6.30). Alternatively, glucose 6-phosphate can be formed from lactate or other compounds, via gluconeogenesis. This will be the case after about 24 hours of starvation in an adult.

How these processes are regulated in response to changes in the blood glucose concentration is now considered.

(i) The enzymes involved in regulation of glycogen synthesis in the liver

Since the enzyme glycogen synthase catalyses the rate-limiting step in glycogen synthesis, it is the activity of this enzyme that must be increased as the blood glucose concentration increases. This is achieved via an interconversion cycle (i.e. reversible phosphorylation). A protein kinase phosphorylates it, which inactivates the enzyme, whereas a protein phosphatase dephosphorylates it, which activates the enzyme. It is the protein phosphatase that plays a major role in regulation (see below for further discussion).

(ii) The enzymes involved in the regulation of glycogen breakdown in the liver

Phosphorylase is the enzyme that regulates the breakdown of glycogen and, as in muscle, it is regulated by reversible phosphorylation. However, unlike the muscle enzyme, phosphorylase *b* is totally inactive so that the enzyme can only be activated by conversion to the *a* form: i.e. there is no allosteric regulation of phosphorylase *b* in liver. Furthermore, an increase in the concentration of glucose actually inhibits prevents conversion of phosphorylase *b* to the *a* form, so that it decreases phosphorylase activity. This makes 'physiological sense' since a decrease in the hepatic concentration of glucose should lead to an activation of phosphorylase, which should increase glycogen breakdown to provide glucose 6-phosphate for conversion to glucose (Figure 6.31a). However, a mechanism exists, by which inhibition of phosphorylase activity can, remarkably, lead to activation of glycogen synthase and hence glycogen synthesis. The mechanisms for the regulation of both synthesis and breakdown of glycogen by glucose are described (Figure 6.31a).

Mechanism of regulation of glycogenolysis by glucose

This mechanism explains how a decrease in the glucose concentration stimulates glycogen degradation in the liver: it is an intriguing mechanism as follows.

At a high concentration, glucose binds to the enzyme, phosphorylase *a*, at a specific binding site and, when glucose is bound, phosphorylase *a* is a better substrate for protein phosphatase, so that the *a* form is converted to the *b* form (i.e. the inactive form) which decreases the rate of glycogenolysis. Glucose is, therefore, an allosteric regulator: its binding to phosphorylase *a* changes the conformation of the protein, which is the preferred substrate for the phosphatase. Consequently, an increase in glucose concentration in the blood and, therefore, in the liver, above the normal level, leads to a decrease in the rate of glycogenolysis. However, when the blood glucose concentration falls, the reverse process occurs: that is, glucose dissociates from phosphorylase *a* so that it is a less effective substrate for the protein phosphatase and it is not inactivated (i.e. not converted to phosphorylase *b*). Hence, glycogenolysis can now take place to provide glucose 6-phosphate for the glucose 6-phosphatase enzyme and formation of glucose (Figure 6.31(b)).

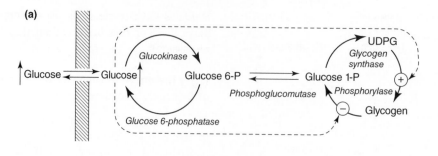

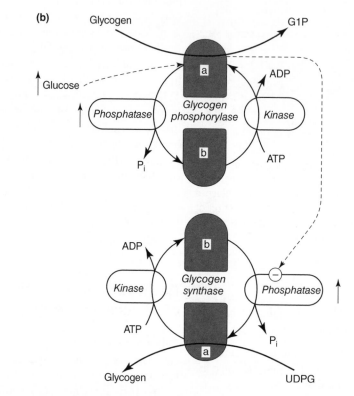

Figure 6.31 **(a)** *An increase in the intracellular concentration of glucose in the liver results in an increased activity of glycogen synthase and a decreased activity of glycogen phosphorylase.* The mechanisms for these effects are shown in (b).

(b) *The mechanism by which an increase in the hepatic concentration of glucose increases glycogen synthase activity but decreases glycogen phosphorylase activity.* The primary effect of glucose is binding to a specific site on phosphorylase, which changes its conformation, so that it is a better substrate for the phosphatase, the activity of which, therefore, increases, resulting in conversion of phosphorylase *a* to phosphorylase *b*. This lowers the concentration of phosphorylase *a* which that activates glycogen synthase inhibits the phosphatase. Consequently, lowering of concentrations of phosphorylase *a* leads to an increase in the proportion of the synthase in the active form, and hence glycogen synthesis is stimulated.

Integration of regulation of glycogen synthesis and glycogen breakdown by glucose

In addition to inhibition of glycogenolysis, glucose also activates glycogen synthase and hence stimulates glycogen synthesis; and this is achieved by an even more intriguing mechanism. The protein phosphatase, which activates glycogen synthase, is inhibited by phosphorylase *a*, which is the result of a direct interaction between the two proteins. Hence, a decrease in the proportion of phosphorylase in the *a* form, which is caused by an increase in glucose concentration in the liver, as described above, leads to an *activation* of the enzyme glycogen synthase and hence an increase in glycogen synthesis. This ensures that some of the glucose 6-phosphate formed via glucokinase, from the increase in the plasma glucose concentration after a meal, is converted to glycogen.

In summary, an increase in the glucose concentration in the portal blood results in an increase in the hepatic concentration of glucose which leads to three changes:

- An increase in the activity of glucokinase, which converts glucose to glucose 6-phosphate.

- A decrease in the activity of phosphorylase, so that glycogen breakdown is decreased.

- An increase in the activity of glycogen synthase, so that glucose 6-phosphate is converted to glycogen (Figure 6.31(a)&(b)).

This is the mechanism to explan the changes shown in (Figure 6.30).

This discussion illustrates that glucose is not only a molecule of importance as a fuel or precursor but it is a remarkable signalling molecule, the effect of which is to maintain the blood concentration as constant as possible. For example, when glucose is absorbed, then a meal, the changes in the plasma glucose concentration are small (Table 6.4). The total amount of glucose in the extracellular fluid in a normal standard male is 10–12 g, that in the plasma is 3–5 g. Yet, rapid ingestion of a solution containing 75 g of glucose, on an empty stomach, increases the peripheral plasma glucose level by little more than 50% (Figure 6.32). This is a remarkably efficient dynamic buffer system for maintenance of the blood glucose concentration. Its dynamic nature is similar to the buffer system that maintains the proton concentration in the blood as constant as possible (Chapter 13).

Regulation of gluconeogenesis

Gluconeogenesis is the only source of glucose when starvation exceeds 24 hours or less and hence the factors that regulate the flux through gluconeogenesis are very important. It is a complex process since it has many substrates (precursors), spans several organs/tissues, is involved in synthesis of glycogen after a meal, removes lactate plus protons from the blood and plays a major role in the regulation of the blood glucose level. It is not surprising, therefore, that the mechanism of regulation is complex. In this

Table 6.4 Plasma peripheral concentration of glucose over 24 h after breakfast, lunch and dinner in normal subjects

Time of sample	Plasma concentration of glucose	
	mg/100 mL	mmol/L
08:00	78	4.3
09:00	122	6.8
09:30	89	5.0
10:30	73	4.1
13:00	73	4.1
14:30	101	5.6
15:00	93	5.2
17:00	85	4.7
19:00	79	4.4
20:30	91	5.1
21:00	103	5.7
24:00	77	4.3
02:00	75	4.2
06:00	78	4.3
08:00	79	4.4

Breakfast was taken at 08:30, lunch at 14:00 and dinner at 20:00.

The meals were high in carbohydrate. Note the largest increase in plasma glucose concentration occurs after breakfast but is only a little 50%.

Data kindly provided by Professor G. Dimitriadis. The National Diabetes Centre, Athens.

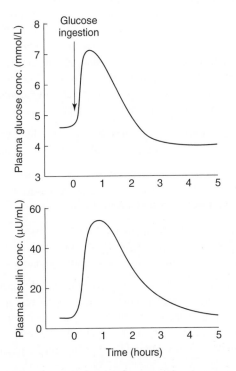

Figure 6.32 *The effect of ingestion of 75 g of glucose on the plasma concentrations of glucose and insulin.* Normal volunteers drank a solution containing 75 g of glucose on an empty stomach. Note that the insulin concentration increases by about 10-fold whereas the glucose concentration increases by about 50%. Data kindly provided by Professor G. Dimitriadis.

section, an account of the biochemical mechanisms and the physiological importance of the regulation of gluconeogenesis are presented.

Biochemical factors involved in the regulation of gluconeogenesis

There are only five key biochemical factors that provide for the regulation: (i) substrate cycles; (ii) regulation of phosphofructo-2-kinase (PFK-2); (iii) phosphorylation/dephosphorylation interconversion cycles (i.e. reversible phosphorylation); (iv) gene expression of gluconeogenic enzymes; (v) concentrations of precursors in the blood.

Substrate cycles

Three steps in gluconeogenesis are non-equilibrium and these are, therefore, targets for regulation. At each step, however, a substrate cycle exists which comprises the gluconeogenic and their corresponding glycolytic enzymes. It is suggested that the main role of the glycolytic enzymes in the liver cells that carry out gluconeogenesis (i.e. the periportal cells) is to provide for the regulation of the rate of gluconeogenesis, via the substrate cycles. (The glycolytic enzymes that convert glucose to pyruvate are present mainly in the perivenous cells.) Evidence from isotopic studies indicates that cycling does in fact occur at each step, supporting the suggestion that cycling occurs in the periportal cells due to the presence of the specific glycolytic enzyme. The three cycles involved in gluconeogenesis are shown in Figure 6.24. They are:

- The glucose/glucose 6-phosphate cycle, which involves the enzymes glucokinase and glucose 6-phosphatase.

- The fructose 6-phosphate/fructose 1,6-bisphosphate cycle, which involves the enzymes phosphofructokinase and fructose 1,6-bisphosphatase.

- The pyruvate/phosphoenolpyruvate cycle, which involves the enzymes pyruvate kinase, pyruvate carboxylase and phosphoenolpyruvate carboxykinase.

Cycles arise because both the glycolytic and gluconeogenic enzymes are simultaneously active. Consequently, an increase in gluconeogenic flux can be achieved either by a decrease in the activity of the glycolytic enzyme(s) and/or an increase in the activity of the gluconeogenic enzyme(s).

Glucose/glucose 6-phosphate cycle

This cycle plays a major role in controlling the rates of both glucose uptake and release by the liver. It is the changes in concentration of glucose in the liver that determine the direction and rate of glucose metabolism (described above).

Fructose 6-phosphate/fructose 1,6-bisphosphate cycle

The cycle plays a role in the regulation of gluconeogenesis but this is achieved via another enzyme, which is involved only in regulation of gluconeogenesis, phosphofructo-2-kinase.

Phosphofructo-2-kinase

The enzyme catalyses the phosphorylation of fructose 6-phosphate to form fructose 2,6-bisphosphate.

$$\text{fructose 6-phosphate} \xrightarrow[\text{ATP} \quad \text{ADP}]{} \text{fructose 2,6-bisphosphate}$$

To distinguish between the two enzymes, the glycolytic enzyme is identified as phosphofructo-1-kinase, abbreviated to PFK-1. The regulatory enzyme, phosphofructo-2-kinase, is abbreviated to PFK-2.

Fructose 2,6-bisphosphate is not a metabolic intermediate but an allosteric regulator. It has two important roles: it increases the activity of PFK-1 but decreases the activity of fructose 1,6-bisphosphatase (FBPase). Consequently an increase in the concentration of fructose 2,6-bisphosphate favours glycolysis but restricts gluconeogenesis.

The activity of PFK-2, and hence the concentration of fructose 2,6-bisphosphate, is regulated in two ways:

(i) by reversible phosphorylation (see below);

(ii) by a change in the concentration of glucose, as follows.

An increase in the hepatic concentration of glucose increases glucokinase activity, which results in an increase in the concentration of glucose 6-phosphate which is in equilibrium with fructose 6-phosphate via an isomerase. An increase in fructose 6-phosphate stimulates the activity of PFK-2 (simply as an increase in substrate concentration), which increases concentration of fructose 2,6-bisphosphate. This results in stimulation of the activity of PFK-1 and inhibition of that of fructose 1,6-bisphosphatase. Both of these changes result in a decrease in the rate of gluconeogenesis through this cycle. This can, therefore, be seen as a feedback inhibition mechanism: an increase in the blood glucose concentration decreases the rate of gluconeogenesis.

Summary of the enzymes affected by an increase in the hepatic portal blood concentration and hence in the intracellular concentration of glucose in the liver

The information discussed above presents the mechanisms by which changes in the concentration of glucose affect several different enzyme activities that are involved in the

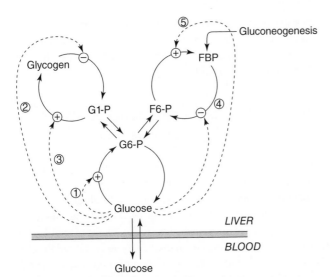

Figure 6.33 *The effects of an increase in the intracellular concentration of glucose that minimise the increase in the blood glucose concentration and ensure glycogen synthesis in the liver after a meal. The effects of glucose are listed* ① *to* ⑤. ① *An increase in glucokinase activity due to the increased concentration of glucose (its substrate).* ② *Binding of glucose to phosphorylase which enhances conversion of phosphorylase* a *to* b *(i.e. inactivation)* ③. *Activation of glycogen synthase due to the decrease in amount of phosphorylase* a*, which inactivates the enzyme that activates glycogen synthase.* ④ *Glucose leads to activation of phosphofructo-2-kinase, due to an increase in its substrate fructose 6-phosphate: this activation increases the level of fructose 2-6-bisphosphate which inhibits the gluconeogenic enzyme, fructose 1,6 bisphosphatase.* ⑤ *It also activates the glycolytic enzyme, phosphofructo-1-kinase. The information in this figure illustrates the remarkable amount of biochemical regulation is one small metabolic area, that even a simple carbohydrate meal initiates.*

early stages of glucose and glycogen metabolism in the liver. An increase in the blood concentration of glucose in the hepatic portal vein results in increased rates of glucose uptake, increased activity of glucokinase and synthesis of glycogen, but also inhibition of both glycogen breakdown and gluconeogenesis. The effect of these changes is to minimise the increase in the blood glucose concentration after a meal and ensure that as much of the increased amount of glucose in the blood is stored as glycogen in the liver. These effects are brought together in Figure 6.33.

Reversible phosphorylation (interconversion cycles)

Two key enzymes in the pathway are regulated by interconversion cycles: they are the regulatory enzyme PFK-2, and the glycolytic enzyme pyruvate kinase. There are two separate protein kinases that phosphorylate these enzymes and they both result in activation of these enzymes. Dephosphorylation inactivates them.

- Cyclic AMP-dependent protein kinase (protein kinase-A).

- Ca^{2+}-calmodulin protein kinase.

These two kinase enzymes are involved in regulation of gluconeogenesis by hormones.

Hormones and control of gluconeogenesis

The effects of glucose as a signal are acute but hormones have more long-term effects which are achieved either by phosphorylation/dephosphorylation or by changes in the concentration of enzymes. Four hormones can individually regulate the flux through gluconeogenesis: they are insulin, glucagon, adrenaline and cortisol. They can also act in concert. The overall effects of these hormones on glucose metabolism are described in Chapter 12. In this section, mechanisms underlying these effects are described. Gluconeogenesis is inhibited by insulin but stimulated by other hormones: glucagon, cortisol and adrenaline.

- Insulin and glucagon regulate gluconeogenesis via changes in cyclic AMP concentration.

- Cortisol regulates gluconeogenesis by activation of genes which express some gluconeogenic enzymes so that the concentration of these enzymes is increased.

- Adrenaline regulates gluconeogenesis via changes in the Ca^{2+} ion concentration.

Cyclic 3′,5′-AMP

Glucagon stimulates adenylate cyclase activity and this increases the concentration of cyclic AMP. Insulin antagonises this effect via an increase in the activity of cyclic AMP phosphodiesterase, which hydrolyses cyclic AMP to AMP, which results in a decrease in the concentration of cyclic AMP (Figure 6.34).

An increase in cyclic AMP concentration activates protein kinase-A. The latter phosphorylates the following enzymes, which leads to an increase in the rate of gluconeogenesis.

(i) Phosphorylation of phosphofructo 2-kinase (PFK-2), which leads to a decrease in activity so that the concentration of fructose 2,6-bisphosphate is decreased which leads to a decrease in PFK-1 activity but an increase in that of fructose 1,6-bisphosphatase, which stimulates gluconeogenesis, at this substrate cycle. These effects of glucagon are antagonised by insulin.

(ii) The phosphorylation of pyruvate kinase results in the conversion of the active form of the enzyme (pyruvate

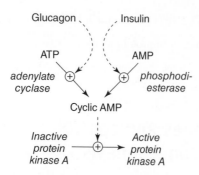

Figure 6.34 *Effects of glucagon and insulin on the cyclic AMP level.* Glucagon increases the activity of adenylate cyclase, which increases the concentration of cyclic AMP whereas insulin activates the phosphodiesterase which hydrolyses cyclic AMP to form AMP. Cyclic AMP activates protein kinase A.

kinase *a*) to the inactive form (pyruvate kinase *b*). Inhibition of this glycolytic enzyme results in an increase in the net flux in the direction of gluconeogenesis at this substrate cycle.

The role of glucagon and insulin in the regulation of gluconeogenesis, along with other factors, is to maintain the blood glucose concentration in starvation. This is discussed in Chapter 12.

Ca²⁺-calmodulin-dependent protein kinase

Adrenaline increases the rate of gluconeogenesis: it binds to the α-receptor on the surface of the liver cell, which results in an increase in cytosolic concentration of Ca^{2+} ions (Chapter 12). This increases the activity of the Ca^{2+}-calmodulin-dependent protein kinase which phosphorylates and causes similar changes in the activities of the enzymes PFK-2 and pyruvate kinase to those resulting from activation of cyclic-AMP-dependent protein kinase. Hence Ca^{2+} ions increase the rate of gluconeogenesis.

Effect of hormones on gene expression

The hormones glucagon, cortisol and insulin regulate the concentrations of some enzymes and hence their activities. These include glucokinase, pyruvate kinase and phosphoenolpyruvate carboxykinase. Most work has been carried out on the carboxykinase enzyme, for which it is known that glucagon and cortisol increase the concentration whereas insulin decreases it. These changes are brought about at the transcriptional level by changing the activity of transcription factors (Chapter 20). Since the hormones act rapidly on transcription, and since the half-life of the enzyme is short, the concentration of phosphoenolpyruvate carboxykinase changes within an hour, in response to changes in the levels of these hormones.

The effect of precursor concentrations on the rate of gluconeogenesis

Hormones can modify the concentration of precursors, particularly the 'lipolytic' hormones (growth hormone, glucagon, adrenaline) and cortisol. The lipolytic hormones stimulate lipolysis in adipose tissue so that they increase glycerol release and the glycerol is then available for gluconeogenesis. Cortisol increases protein degradation in muscle, which increases the release of amino acids (especially glutamine and alanine) from muscle (Chapter 18).

The plasma concentrations of lactate and amino acids increase after a meal; they are taken up by the liver and converted to glucose 6-phosphate, via gluconeogenesis. Under these conditions, the glucose 6-phosphate is converted to glycogen not glucose. The mechanism by which glucose 6-phosphate is directed to glycogen rather than to glucose is not known.

Regulation of glycolysis and gluconeogenesis by ATP/ADP concentration ratio in the liver

The demand for glucose may be sufficiently high in some conditions (e.g. starvation) that the rate of gluconeogenesis can utilise about 50% of the maximum rate of ATP generation in the liver. It is vital, therefore, for the survival of the liver, that, if sufficient ATP cannot be generated, the rate of gluconeogenesis is decreased. The properties of the key enzymes involved in the regulation of gluconeogenesis indicate that a decrease in the ATP/ADP concentration ratio will decrease their activities and hence decrease the rate of gluconeogenesis. In addition, the properties of the glycolytic enzymes are similar to those in muscle, so that a decrease in the ratio will increase their activities and, via the substrate cycles, the rate of gluconeogenesis will be further decreased.

It is unlikely that the rate of ATP generation in the liver under normal conditions will be decreased to such an extent that ATP/ADP concentration will be seriously affected. However, some extreme physiological conditions, such as intense and prolonged physical activity, prolonged starvation and haemorrhagic shock may result in reduced perfusion of the liver and lack of oxygen that could result in a marked decrease in the rate of ATP generation. The effect of a decreased supply of oxygen will not only decrease the rate of gluconeogenesis but will stimulation that of glycolysis to generate as much ATP as possible. Decreased gluconeogenesis prevents removal of lactic acid from the blood whereas increased glycolysis adds lactic acid to the blood. A low blood supply to the liver can therefore lead to lactic acidosis, with a severe decrease in blood pH (Chapter 13).

Hypoglycaemia

Hypoglycaemia is defined as a plasma concentration of glucose of 2.5 mmol/L or below. Some of the symptoms are caused not by hypoglycaemia per se but by high levels of catecholamines that are secreted to stimulate breakdown of liver glycogen to maintain the blood glucose concentration. These symptoms include pallor, sweating, tremor, palpitations, nausea, vomiting and anxiety. Other symptoms, which include dizziness, confusion, tiredness, headache and inability to concentrate, are caused by failure to provide enough glucose to satisfy energy requirements of some parts of the brain. These are usually known by the rather cumbersome term neuroglycopaenia (i.e. shortage of glucose for the brain).

An excess of insulin can cause hypoglycaemia and the hormones that respond to this condition to restore normal glucose levels are known as the counter-regulatory hormones. They are adrenaline, glucagon, growth hormone and cortisol. An increase in the blood levels of these hormones can sometimes be used to confirm a diagnosis of hypoglycaemia.

The concentration of glucose in the blood is maintained as a balance between rates of glucose utilisation and glucose supply and changes in one or both of these can lead to hypoglycaemia. Three situations are considered.

Hypoglycaemia caused by stimulation of the rate of glucose utilisation and inhibition of the rate of release of glucose by the liver

This situation arises mainly when the blood level of insulin is high – abnormally high for the given condition. Insulin not only stimulates peripheral utilisation but also inhibits glucose output by the liver so that hypoglycaemia can develop rapidly. Four examples are given.

(i) Diabetic mothers who are not well-controlled may suffer from hyperglycaemia and if this occurs just prior to birth it can result in a high blood glucose concentration in the foetus and hence high levels of insulin, due to secretion by the insulin-secreting cells in the pancreas of the foetus. If these high levels are maintained after birth, it can result in hypoglycaemia in the baby. This can also occur, although in a less severe form, if a normal mother is given an intravenous infusion of glucose during labour.

(ii) A tumour of the beta cells in the Islets of Langerhans in the pancreas (an insulinoma) can result in excessive rates of secretion of insulin and hence hypoglycaemia which may be intermittent but can be severe.

(iii) Hypoglycaemia can sometimes occur, surprisingly, in response to a meal. This is known as *reactive hypoglycaemia*. The cause is not known but it may be due to an excessive secretion of insulin in a patient who is suffering from the early stages of damage to the beta cells (i.e. the prediabetic condition).

(iv) It can occur in a well trained athlete who eats a small quantity of carbohydrate immediately after physical activity. The sensitivity of the peripheral tissues (especially muscle) to insulin is high in well-trained subjects and can be increased further by a single bout of physical activity. Secretion of even a small amount of insulin can result in such a marked increase in uptake by muscle that hypoglycaemia results. Research workers who infuse insulin for experimental purposes in well-trained volunteers should be aware that even a low dose of insulin can result in hypoglycaemia unless glucose is provided.

Hypoglycaemia that arises when an increased rate of utilisation exceeds that of glucose release by the liver

This usually occurs during prolonged and intense physical activity. In this condition, the glycogen stores in the liver and muscle can become depleted due to the high demands for ATP generation by the muscles that are physically active. Once they are depleted, maintenance of a high intensity can result in hypoglycaemia, even severe hypoglycaemia, which can have serious consequences. Usually, hypoglycaemia results in a decrease in the rate of ATP generation in some parts of the brain that leads to fatigue, known as central fatigue, which forces a reduction in the intensity of the activity (Chapter 13). It can be considered, therefore, that fatigue is a safety mechanism to prevent dangerous hypoglycaemia. Overriding the safety mechanism by the desire to win a race or beat a record can be dangerous. In the women's Olympic marathon in Los Angeles in 1984 the Swiss runner Gabriella Anderson-Scheiss completed the last few hundred metres in a state of clouded consciousness as she staggered from side to side around the athletic stadium. She was suffering from severe hypoglycaemia but responded rapidly to intravenous glucose after finishing the race. In the 1982 Boston marathon, Alberto Salazar and Dick Beardsley battled out the last 10 miles and Salazar won in a record time of 2 h 8 min 5 s, only 5 s ahead of Beardsley. After completion of the race, Salazar became severely hypoglycaemic but after receiving an intravenous infusion of glucose, he rapidly recovered.

Similar situation, during which hypoglycaemia can occur, is when intense physical activity takes place in an extremely low environmental temperature without adequate clothing, so that hypothermia develops. Hypothermia appears to increase the risk of hypoglycaemia. This can lead to mental confusion, which can be dangerous in some situations (e.g. walking or running in mountainous regions) (Chapter 13).

Hypoglycaemia caused by inappropriate low rates of gluconeogenesis

The specific gluconeogenic enzymes in the liver of the foetus develop late in pregnancy, so that premature babies can develop hypoglycaemia soon after birth and provision of glucose is essential for their survival.

Some types of modern dancing, particularly when it takes place in clubs and at parties, can be described as intense physical activity although the authors are not aware of any experimental studies. The mechanism in the brain that results in central fatigue may be disturbed by recreational drugs, which may be more available at parties and clubs. Consequently, these activities could readily lead to depletion of liver glycogen. This would not normally cause a problem, since gluconeogenesis would be increased, by mechanisms described above. Problems can arise, however, if alcohol is consumed by the dancers since this inhibits gluconeogenesis (see above). The result could be a rapid onset of hypoglycaemia. Perhaps this might be one cause of the aggressive and socially disruptive behaviour observed in teenagers or in young adults after a 'night out'. It is usual to blame alcohol per se for this bad behaviour but a low blood sugar level could also be a factor. Indeed the extract from Blakemore's book presented at the beginning of this chapter would support this suggestion.

Hypoglycaemia in pregnancy

In view of the essential compounds that are synthesised from glucose or its metabolites, particularly those that are essential for proliferation of cells, failure of the mother to provide enough glucose for the developing foetus could give rise to inadequate proliferation of cells or abnormal structure of cells which could be responsible for congenital malformations or slow development giving rise to abortion or low birth weight. Careful studies carried out in Western Holland during the period of undernutrition in the winter of 1944/45 reported increased incidence of low birth weight babies and infant mortality (Chapter 16). It is unclear, however, if this was due to a chronic low blood glucose level. Other deficiencies could also be responsible. Similarly, chronic or frequent acute episodes of hypoglycaemia might be one factor to account for the correlation between low birth weight and diseases in adult life (Chapters 15 and 19).

Hyperglycaemia

Just as an insufficient rate of gluconeogenesis can lead to hypoglycaemia, too high a rate of gluconeogenesis can lead to hyperglycaemia. And just as hypoglycaemia can be pathological, so can hyperglycaemia. The difference, however, is that hypoglycaemia can lead to both acute and chronic problems whereas hyperglycaemia leads to chronic problems, so that its effects can be insidious.

The blood glucose level is normally maintained, at least in part, by a balance between the effects of two hormones, insulin and glucagon, on metabolism in the liver. In type 1 diabetic patients, who do not secrete sufficient insulin, the liver is exposed to glucagon without the balancing effect of insulin so that the rate of gluconeogenesis is high. In type 2 diabetics, in whom the liver is insulin-resistant, the glucagon effect on the liver dominates. Many of the afflictions that decrease the quality of life for diabetic patients (damage to the retina and lens, to the kidneys and to peripheral nerves) are the result of chronic hyperglycaemia. In addition, it can also result in damage to the beta cells of the pancreas, causing problems with insulin secretion in type 2 diabetic patients and therefore exacerbating the hyperglycaemia. This is known as the toxic effect of glucose. The pathological subject of diabetes mellitus is discussed in Appendix 6.9.

7
Fat Metabolism

The basic structure of biological fats was elucidated by a Frenchman, Michel-Eugène Chevreul. . . . Napoleon recognised the importance of science in military operations, civilian prosperity and national prestige, and befriended and financed its practitioners. Armies need good rations and hunger fuels discontent among civilians, so the newly established French Republic vigorously promoted research into what we would now call food technology, enlisting the help of its most promising scientists. . . . Starting in 1811, Chevreul devoted himself to identifying the 'immediate principles' in mutton fat . . . he heated fats with alkalis and purified the resulting mixtures. He named the clear, syrupy, sweet-tasting liquid that he extracted 'glycérine' (from the Greek word γλυκυς, sweet). In spite of its sweet taste, glycerine, now called glycerol, is an alcohol, not a sugar.

Chevreul named the other major 'principles' he found in animal fat 'fatty acids', and showed that they occurred in the proportions of three fatty acids to each glycerol. When separated from the glycerol, fatty acids dissolve in alcohol and, by repeated extraction and precipitation with salts, could be purified sufficiently to form crystals.

Pond (1998)

The term lipid includes all compounds that release fatty acids on digestion, so that this definition is more extensive than that of fat. Compounds which are insoluble in water, soluble in organic solvents but not derived from fatty acids have been termed non-saponifiable lipids. They include steroids, terpenes and ubiquinone.

Fats are solid triacylglycerols, oils are liquid triacylglycerols: it is the fatty acid composition of the triacylglycerol that determines its physical properties. Triacylglycerols with a high proportion of unsaturated fatty acids have lower melting points than those rich in saturated fatty acids, so the former are oils – that is, they are liquid at room temperature. The unsaturated fatty acids are identified by the number of carbon atoms and the position of the double bonds (see below).

Fats in the diet are an important source of energy not only for immediate use but particularly for storage. Unfortunately, ingestion of too much fat and hence too much energy contributes to obesity, which is now a major problem in developed countries and of much concern. Two early findings and two recent ones illustrate the importance of fat metabolism:

- The discovery in 1929 that some fatty acids required by humans were not synthesised in the body and were, therefore, essential components of the diet.

- The discovery in the 1950s that the amount and type of fat consumed by humans could be related to the incidence of coronary heart disease.

- In addition to fuel storage, fats and fatty acids are now known to have several key roles in the body, e.g. as messengers, precursors of molecular messengers, gene regulators, components of phospholipids (which form the major part of membranes) and possible modifiers of the immune response.

- Deficiencies of polyunsaturated fatty acids may be a factor in the development of neurological and behavioural disorders, which has considerable clinical potential and significance in public health.

These topics are discussed in Chapters 11, 14 and 22. Discussion in this chapter focuses primarily on the metabolism of fats that are fuels and that provide energy for various tissues under different conditions, many of which are common in everyday life. It also provides basic knowledge for the discussion in Chapters 11, 14 and 22.

Functional Biochemistry in Health and Disease by Eric Newsholme and Tony Leech
© 2010 John Wiley & Sons Ltd

Fats in nutrition

This subject is discussed in Chapter 15, but a brief summary is given here, as an introduction to fat metabolism. The three major components of a human diet are meat, fish and plants, and the fats associated with each component can be different. Meat usually comprises the muscle of mammals and birds but it is always associated with fat. Even apparently lean meat may have a high fat content: a 250 g beef steak may contain 60 g fat in addition to its 80 g protein. Some of this will be adipose tissue between muscles; the remainder is triacylglycerol (TAG) within the fibre. In addition, milk and other dairy products contain significant amounts of fat: about a quarter of the fat in the average UK diet comes from dairy produce.

Fish are the only cold-blooded animals eaten by humans in any quantity. The fat is mainly in the form of oil, that is, triacylglycerol containing a high proportion of unsaturated acids. As foods, fish are classified either as lean, cod) which store their TAG in the liver, or oily (e.g. sardines, salmon, mackerel) which store TAG in muscle. Many of the unsaturated fatty acids in fish oil are the omega 3 acids. The structures, systematic and common names of the naturally occurring fatty acids are provided in Appendix 7.1. Biochemical and pathological significance of omega-3 oils containing omega-3 acids are discussed in Chapters 11 and 22.

Large amounts of fat are stored in the fruits or seeds of some plants. Those that grow in the tropics store saturated fatty acids in the TAG, which is liquid at the high environmental temperatures (e.g. coconut, cocoa). In contrast, those that grow in cooler climates store unsaturated fats to ensure they remain liquid at lower environmental temperatures (e.g. sunflower, safflower and soybean).

The seeds of some plants (wheat, oats, rye, acorns, hazelnuts) store mainly carbohydrate.

Fat fuels

There are five classes of fat or fat-derived fuels:

- triacylglycerol;
- long-chain fatty acids;
- medium-chain fatty acids;
- short-chain fatty acids;
- ketone bodies (acetoacetate and 3-hydroxybutyrate).

A summary of the sources and fates of fatty acids and ketone bodies is presented in Figure 7.1 and Table 7.1. A major problem with long-chain fatty acids and TAGs is their lack of solubility in the aqueous medium of the blood and interstitial fluid. How this is overcome for fatty acids is discussed in this chapter, and for triacylglycerol in Chapter 11. Unfortunately, the need to transport relatively large quantities of triacylglycerol in the blood can lead to pathological problems (Chapter 11).

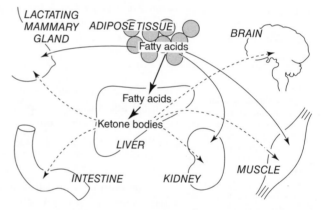

Figure 7.1 *The sources and fates of fat fuels.* The fat fuels considered in this figure are the long-chain fatty acids and ketone bodies that are transported in the blood.

Table 7.1 Fat fuels: sites of origin, their form in blood and sites of uptake

Fuel	Form in bloodstream	Origin	Major site of uptake
Triacylglycerol	(i) chylomicrons	intestine (diet)	adipose tissue, muscle, lactating mammary gland
	(ii) very low density lipoproteins	liver	
Long-chain fatty acid	albumin bound	adipose tissue	liver, skeletal muscle, kidney, cardiac muscle, liver
Medium-chain fatty acid	albumin bound	diet (especially dairy produce)	
Short-chain fatty acids (butyrate, propionate, acetate)	free (unbound)	microorganisms in colon	colonocytes, liver
Ketone bodies (acetoacetate, hydroxybutyrate)	free	liver	cardiac muscle, brain, kidney, skeletal muscle, small intestine

Triacylglycerol

Triacylglycerol occurs in three separate locations in the body: blood, adipose tissue and some other tissues.

Triacylglycerol in blood

Triacylglycerol in blood occurs as an emulsion either in the form of chylomicrons or very low density lipoproteins (VLDLs). A full description of the metabolism of triacylglycerol in the blood is given in Chapters 4 and 11. A brief summary is as follows. Triacylglycerol in food is digested in the small intestine and the resultant fatty acids and monoacylglycerol are absorbed by the enterocytes, in which they are esterified back to triacylglycerol. This triacylglycerol, along with proteins and phospholipids, is 'packaged' into chylomicrons which are secreted into the lymph in the intestine (the lacteals) then into the thoracic duct and then the lymph enters the blood. The VLDL is formed in the liver: fatty acids that are taken up from the blood are

> Chylomicrons in emulsion can adsorb pathogens onto the surface which facilitates attack by antibodies present in the lymph.

esterified and, along with protein and phospholipid, are packaged to form VLDL, which is then secreted into the blood (Chapter 11). The emulsions of both chylomicrons and VLDL are stabilised by the protein, phospholipids and cholesterol (see Figure 7.2(a)&(b)).

The particles that form the emulsions are too large to cross the capillaries, so the triacylglycerol is hydrolysed to fatty acid in the capillaries by the enzyme, lipoprotein lipase, which is attached to the luminal surface of the capillaries. Most of the fatty acids that are released are taken up by the surrounding tissue but about 30% escape into the general circulation (Figure 7.3). In the fed state, lipoprotein lipase is active primarily in the capillaries of adipose tissue. After sustained exercise, it is active in the muscles that have been involved in the exercise, and, during lactation, it is active in the mammary gland. Hence, according to the conditions, the fatty acids in the triacylglycerol in chylomicrons or VLDL end up in adipose tissue, muscle or mammary gland (Figure 7.4) (see below for more details).

Triacylglycerol in adipose tissue

The largest store of fuel in the body occurs in adipose tissue. Approximately 80% of adipose tissue is triacylglycerol (the remainder is connective tissue, water, proteins and DNA). Approximately 90% of an individual adipocyte is triacylglycerol (Figure 7.5). Despite this, triacylglycerol is not released from the adipose tissue. Instead hydrolysis (lipolysis) of the triacylglycerol within adipose tissue

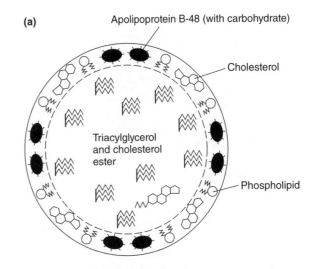

(a)

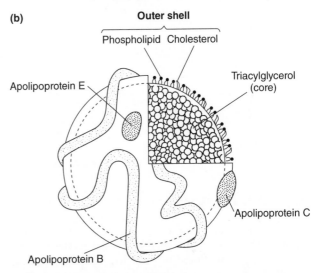

(b)

Figure 7.2 (a) *Diagrammatic structure of a chylomicron.* An outer coat consists of proteins (apolipoproteins), cholesterol and phospholipid. The core consists of triacylglycerol and cholesterol ester. **(b)** *Detailed structure of VLDL particle.* The core is mainly triacylglycerol. The other shell, which stabilises the particle, comprises phospholipids and cholesterol. Apolipoprotein B (Apo B100) helps to maintain the structure. Apolipoprotein E is a ligand for binding to specific receptors on cells. Apolipoprotein C activates lipoprotein lipase and is therefore essential for hydrolysis of triacylglycerol. Diagram from Gibbons & Wiggins (1995) with permission.

results in formation of long-chain fatty acids which are released into the blood. The fate of these fatty acids is oxidation to CO_2 to provide energy to generate ATP in skeletal muscle, heart or kidney (Figure 7.6).

> Adipose tissue is not the only tissue that releases long-chain fatty acids. Macrophages also release long-chain fatty acids which increases their concentration in the lymph, where the fatty acids can damage bacterial membranes and kill the bacteria (Chapter 17).

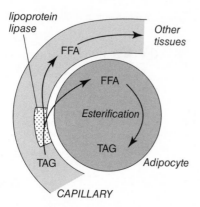

Figure 7.3 *The action of lipoprotein lipase in the hydrolysis of triacylglycerol in the blood and the fate of the fatty acids produced.* Lipoprotein lipase is attached to the luminal surface of the capillaries in the tissues that are responsible for removal of triacylglycerol from the bloodstream (e.g. adipose tissue, muscle, lactating mammary gland).

Figure 7.5 *An interference contrast photograph of a white adipose tissue cell. Note the spherical lipid droplet fills most of the cell.*

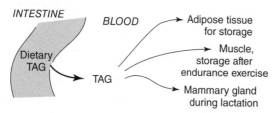

Figure 7.4 *Fate of triacylglycerol that is present in fuel blood after secretion by the intestine.* The dietary triacylglycerol in the intestine is hydrolysed to long-chain fatty acids and monoacylglycerol, both of which are taken up by the enterocytes in which they are then re-esterified. The triacylglycerol is released in the form of chylomicrons into the blood, from where it is hydrolysed to fatty acids and glycerol by the enzyme lipoprotein lipase in specific tissues (Figure 7.3). The fatty acids are taken up by adipocytes, muscle fibres and secretory cells in the mammary gland.

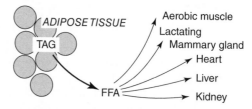

Figure 7.6 *Release of fatty acids from the triacylglycerol in adipose tissue and their utilisation by other tissues.* Fatty acids are long-chain fatty acids, abbreviated to FFA (see below). Hydrolysis (lipolysis) of triacylglycerol in adipose tissue produces the long-chain fatty acids that are released from the adipocytes into the blood for oxidation by various tissues by β-oxidation (see below).

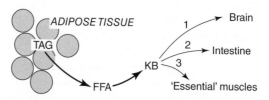

During starvation or hypoglycaemia, the liver partially oxidises fatty acids to form ketone bodies, which are released and oxidised by the brain, intestine and the essential muscles (see below) (Figure 7.7).

Fatty acids

Although emphasis is usually placed on long-chain fatty acids (i.e. containing more than 14 carbon atoms) as a fat fuel, three others are also important: short-chain (<6 carbon atoms), medium-chain (6–14 carbon atoms) and ketone bodies. Unless otherwise specified, the term fatty acids

Figure 7.7 *Release of fatty acids from triacylglycerol and conversion to ketone bodies in the liver and their fate.* The formation of ketone bodies (KB) from fatty acids takes place only in the liver and they are oxidised in brain, intestine or essential muscles. (1) The oxidation by brain takes place during prolonged starvation in adults and during a shorter period in children; and also during hypoglycaemic conditions. (2) About 50% of energy required by the small intestinal cells during starvation is produced by oxidation of ketone bodies. (3) 'Essential' skeletal muscles are those which are essential to life, e.g. heart, diaphragm, intercostal (for breathing), masseter muscles (for chewing food). They are all aerobic muscles, i.e. they obtain most of their energy from the mitochondrial oxidation processes.

refers to long-chain fatty acids. One group not included above, since they are much less important as fuels, are those containing more than 20 carbon atoms, known as very long-chain fatty acids. These are oxidised within the peroxisomes (see below).

Short-chain fatty acids (butyric, propionic and lactic acids)

Short-chain fatty acids are not important components in most human diets although small amounts are present in bovine milk and therefore in dairy produce. Nonetheless, they are an important fuel for cells of the colon (colonocytes), since they are formed endogenously. The lumen of the colon contains a very large number of microorganisms, some of which can break down the polysaccharides that are not digested in the small intestine. The sugars that are produced are fermented to form the short-chain fatty acids, known collectively as the volatile fatty acids (VFA). In developed countries, the average daily consumption of digestion-resistant polysaccharide produces about 15 g of VFA, which has an energy content of around 600 kJ.

> As well as providing fuel, butyrate (which contains four carbon atoms) can reduce the proliferation of colonocytes, which may reduce the risk of tumour development. This is one suggestion to explain how high-fibre diets protect against colon cancer (Chapter 21).

Medium-chain fatty acids

Medium-chain fatty acids are also present in bovine milk and some plant oils (e.g. coconut). After digestion of the triacylglycerol, they are taken up by the enterocytes in the small intestine but are not esterified. Instead they pass directly into the hepatic portal blood, from where they are taken up by the liver for complete oxidation or conversion to ketone bodies.

Long-chain fatty acids

Long-chain fatty acids in the human diet are mainly palmitic (16C), oleic (18C) and stearic acids (18C) as well as the polyunsaturated fatty acids (18C, 20C and 22C). Almost all of the fatty acids in the diet have an even number of carbon atoms and are unbranched. Many are unsaturated. Indeed, in human adipose tissue, more than half of the total fatty acids are unsaturated (Chapter 11).

Long-chain fatty acid binding to albumin: a problem of transport, pathology and terminology

To distinguish between those fatty acids that are esterified in either triacylglycerol or phospholipids and those present in their free form in plasma, fatty acids are known either as non-esterified fatty acids (NEFA) or free fatty acids (FFA). The latter term is now more frequently used in the scientific literature and is used in this text. Unfortunately, the term 'free fatty acid' is ambiguous. Long-chain fatty acids are almost insoluble in plasma at physiological pH, the maximum concentration being about 10^{-6} mol/L. Yet the normal plasma concentration is much higher (e.g. 0.5 to 2.0 mmol/L). This is achieved by binding to a soluble protein in the blood, albumin, according to the reaction

$$albumin + fatty\ acid \rightleftarrows (\ albumin-fatty\ acid\ )$$

Thus the fatty acid not bound to albumin, which constitutes about 10% of the total, is also known as free fatty acid. The different uses of the term 'free fatty acid' should be clear from the context.

A molecule of albumin possesses three high-affinity binding sites for fatty acids but, under normal physiological conditions, only one or two of these are filled. However, as the plasma FFA concentration increases and reaches a value of about 2 mmol/L, all three sites become filled. Above 2 mmol/L, there is a rise in the concentration of fatty acids that are not bound to albumin. As the concentration increases, they begin to form fatty acid micelles (Chapter 4) which act as detergents and can disrupt the conformation of proteins and the organisation of membranes, causing damage to tissues. Consequently, even in the most extreme conditions (e.g. diabetic acidosis) the plasma concentration of fatty acid rarely exceeds 2 mmol/L (Table 7.2).

The FFA released by the adipocytes is collected by albumin and is transported to the various tissues in the blood. The albumin–FFA complex is able to cross the endothelial barrier in the capillaries and enter the interstitial space and so deliver this important fuel to the plasma membrane of the cell. To facilitate the transport of free fatty acids across the plasma membrane and within the cell, other transport proteins are present: these are known as fatty acid binding proteins (FABP).

> The FABP isolated from mammary gland, a tissue with a high level of fatty acid metabolism, has the ability to inhibit proliferation of tumour cells in vitro. The concentration of FABP increases in the cytosol of the mammary gland during lactation. Could this explain the protection against breast cancer afforded by an early pregnancy and hence lactation at an early age (Chapter 21)?

To help clarify this process an analogy can be drawn between the transport of fatty acids and that of oxygen since, as with fatty acids, the solubility of oxygen in an aqueous medium is very low.

Table 7.2 Concentrations of long-chain fatty acids in plasma during different conditions in humans

Condition	Approx. plasma fatty acid concentration (mmol/L)
Normal fed	0.30
Stress (racing driver)	1.7
Stress (public speaking)	1.0
Starvation (8 days)	1.9
Three hour post-surgery	0.80
Diabetic acidosis	1.6
Severe burns	1.3
Acute trauma	0.75
Prolonged exercise (60% of $\dot{V}O_{2max}$)	
Rest	0.30
30 min	0.35
60 min	0.45
120 min	0.80
180 min	1.4
Post-exercise	
5 min	1.8
15 min	2.0
30 min	1.5
Fasting	
12 hr[a]	0.60
36 hr	1.1
72 hr	1.3
Obesity (moderate)	0.80

Data taken from various sources (Newsholme & Leech, 1983).

Blood for measurement of the effect of exercise was taken via an indwelling catheter.

For the racing driver, the blood sample was taken as soon as possible after the race.

It appears from the data in this table that a level of 2 mmol/L plasma fatty acid is about the highest that can be achieved in any condition so far studied. A similar message comes from studies on exercise in domestic animals. Horses that were ridden over 50 miles at an average speed of 8 mph had a plasma fatty acid concentration of 1.4 mmol/L. Dogs that had been run on a treadmill for 4 hours had a plasma concentration of 2.1 mmol/L.

[a] In pregnancy, there is a sharp increase in the plasma fatty acid level after about 12 hours of fasting, much sooner than in the non-pregnant woman. This may be important in maintaining the plasma glucose level not only for the mother but also the foetus. This maintenance is achieved via the glucose fatty acid cycle (Chapter 16).

In the blood

$$albumin + FFA \rightleftarrows (albumin–FFA)$$

$$haemoglobin + O_2 \rightleftarrows (haemoglobin–O_2)$$

In the cell

$$binding\ protein + FFA \rightleftarrows (binding\ protein–FFA)$$

$$myoglobin + O_2 \rightleftarrows (myoglobin–O_2)$$

The analogy is described in more detail in Figure 7.8.

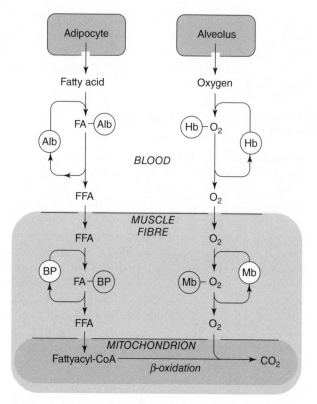

Figure 7.8 *Comparison of oxygen transport from lung to a cell and then into a mitochondrion with fatty acid transport from an adipocyte to a cell and then into the mitochondria in various tissues/organs. Fatty acid is transported in blood bound to albumin, oxygen is transported in blood bound to haemoglobin. Fatty acid is transported within the cell attached to the fatty acid-binding protein (BP), oxygen is transported within a cell attached to myoglobin (Mb). Alb represents albumin, Hb haemoglobin.*

Ketone bodies

Acetoacetate and 3-hydroxybutyrate are known as ketone bodies. They are classified as fat fuels since they arise from the partial oxidation of fatty acids in the liver, from where they are released into the circulation and can be used by most if not all aerobic tissues (e.g. muscle, brain, kidney, mammary gland, small intestine) (Figure 7.7, Table 7.1). There are two important points: (i) ketone bodies are used as fuel by the brain and small intestine, neither of which can use fatty acids; (ii) ketone bodies are soluble in plasma so that they do not require albumin for transport in the blood.

The plasma concentration of ketone bodies can increase to much higher values than those of fatty acids. They can increase to levels similar to or higher than those of glucose (Table 7.3). Hence they can compete more effectively with glucose as a fuel. Ketone bodies provide an excellent alternative fuel to glucose during prolonged starvation or severe hypoglycaemia (see below and Chapter 16).

Table 7.3 Concentrations of total ketone bodies in plasma under different conditions

Condition	Total plasma ketone body concentration (mmol/L)
normal fed	<0.1
8 hours without food after prolonged exercise[a]:	
normal diet prior to exercise;	0.3
low carbohydrate diet prior to exercise[b].	2.8
Prolonged starvation:	
8 days (lean male);	5.3
8 days (obese female).	7.0
Diabetic ketoacidosis	~25.0

For sources of data see Newsholme & Leech (1983).

Total ketone body concentration is acetoacetate plus hydroxybutyrate.

[a]Exercise was at high percentage $\dot{V}O_{2max}$ for 90 min.

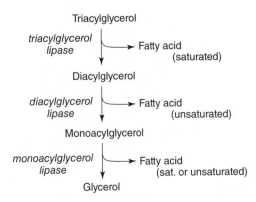

Figure 7.9 *The degradation of triacylglycerol in adipose tissue to fatty acids and glycerol.* The figure indicates the progressive release of fatty acids and the types of fatty acid that are usually present at each position and, therefore, released from each position as the triacylglycerol molecule. Sat. – Saturated. A lipase that is not regulated by hormones is also present is adipose tissue. It is continually active. Its role is described below.

Ketone bodies are misnamed, for they are not bodies and hydroxybutyrate is not a ketone. The appellation 'ketone body' was given to them towards the end of the nineteenth century by German physicians who found that the urine of diabetic patients gave a positive reaction when tested for ketones. They were described as 'ketones *of* the body' but sloppy translation from the German reduced this to 'ketone body'. Not only acetoacetate but also acetone was detected by this reaction. Acetone is formed by the spontaneous decarboxylation of acetoacetate, so that it is only detectable when the concentration of the latter is abnormally high. Some of the acetone is metabolised to acetate but most is excreted through the kidneys and lungs, accounting for the characteristic sweet smell on the breath of diabetic patients in a ketotic coma.

> This is the basis of a story, probably apocryphal. A physician had been summoned to attend a family member who had collapsed. As soon as the physician entered the house, the family were stunned by the immediate diagnosis of diabetes mellitus by the physician, without even seeing the patient. The physician recognised the sweet smell of acetone.

Pathways of oxidation of fat fuels

One important fate of all the fat fuels is complete oxidation to CO_2 but long-chain fatty acids can also be esterified to produce triacylglycerol or phospholipid.

Long-chain fatty acids

Most of the long-chain fatty acids that are oxidised in the body are released from triacylglycerol in adipose tissue. The first step is hydrolysis of triacylglycerol within the adipocyte. It is a three-stage process catalysed by three separate lipases acting, consecutively, on triacylglycerol, diacylglycerol and monoacylglycerol, releasing the fatty acids at each stage, which are then transported into the blood (Figure 7.9). Of these lipases, the triacylglycerol lipase is the rate-limiting step and, consequently, its activity regulates the overall rate of lipolysis and hence the supply of fatty acids to other tissues. Not surprisingly, its activity is regulated by a variety of hormones, which has engendered its name, 'hormone-sensitive lipase' (Figure 7.10). Other tissues (e.g. aerobic muscle and liver) contain intracellular lipases but the resultant fatty acids are not released but are oxidised within the tissue.

After release from the adipocyte, the fatty acids are transported in the blood as a complex with albumin, as described above. They are then taken up by cells for oxidation This involves transport through the plasma membrane, the cytosol and finally the inner mitochondrial membrane of the cell for oxidation of the fatty acids within the mitochondria.

Transport across the plasma membrane

Although fatty acids are lipid soluble and might be expected to diffuse through the plasma membrane sufficiently rapidly to satisfy the required rates of oxidation, this is not the case and a transporter protein is present in the plasma membrane.

Transport within the cell

The very low solubility of fatty acids poses a problem for their transport within the cell, as it does in the blood. The problem is solved by the presence of the fatty acid binding protein (FABP). It binds fatty acids at the inner surface of

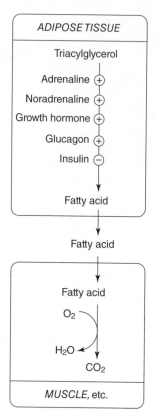

Figure 7.10 *Hormones that regulate the activity of the hormone-sensitive lipase in adipose tissue.* Each hormone binds to a receptor on the outside of the plasma membrane and changes the activity of the lipase within the adipocyte, via a messenger molecule (Chapter 12). A hormone – independent lipase is also present with provides a low rate of release of fatty acid when the former is inactive.

the plasma membrane, lowering the local concentration of fatty acids so that the transport across the plasma membrane is facilitated, just as myoglobin encourages the entry of oxygen into the cell. Within the cytosol, the complex dissociates to release the fatty acid where it is then 'activated', i.e. converted to the acyl-CoA derivative, which is catalysed by the enzyme, acyl-CoA synthetase. The FABP also binds the fatty acyl-CoA, which it transports to the mitochondria. This binding keeps the free concentration of this metabolite very low, since it is a powerful detergent, and a high concentration of fatty acyl-CoA can damage membranes, enzymes and proteins in signalling pathways (Chapter 9, Figure 9.31).

Acyl-CoA synthetases

These synthetases are responsible for formation of acyl-CoAs:

$$R.COOH + CoASH + ATP \rightarrow R.CO.SCoA + AMP + PP_i$$

There are three acyl-CoA synthetases, which are specific for fatty acids of different chain lengths:

- Short-chain acyl-CoA synthetase activates short-chain fatty acids, acetic, butyric and propionic acid. The enzyme is present in both the cytosol and in the mitochondrial matrix of most tissues: the activity is especially high in the liver and the colon.

- Medium-chain acyl-CoA synthetase, which is present within the mitochondrial matrix of the liver, activates fatty acids containing from four to ten carbon atoms. Medium-chain length fatty acids are obtained mainly from triacylglycerols in dairy products. However, unlike long-chain fatty acids, they are not esterified in the epithelial cells of the intestine but enter the hepatic portal vein as fatty acids to be transported to the liver. Within the liver, they enter the mitochondria directly, where they are converted to acyl-CoA, which can be fully oxidised and/or converted into ketone bodies. The latter are released and can be taken up and oxidised by tissues.

This latter situation is particularly beneficial for patients who are being fed intravenously because, if triacylglycerols containing medium-chain fatty acids are included in parenteral feeds, they are readily converted into ketone bodies so that a soluble fat fuel is rapidly made available in the blood that can be oxidised by most tissues.

- Long-chain acyl-CoA synthetase activates fatty acids containing from 10 to 18 carbon atoms.

A separate very long-chain-acyl-CoA synthetase is present in peroxisomes for the 'activation' of very long-chain fatty acids, such as arachidonate (20 carbon atoms). These fatty acids are degraded exclusively in the peroxisomes.

Transport across the mitochondrial membrane

Mitochondria contain all the enzymes necessary for oxidation of fatty acids but, before this can take place, the fatty acids have to be transported into the mitochondria. Transport requires the formation of an ester of the fatty acid with a compound, carnitine, in a reaction catalysed by the enzyme carnitine palmitoyltransferase:

> The structure of carnitine is $(CH_3)_3N^+.CH_2CH(OH).$ CH_2COO^-. It is synthesised in the liver from γ-butyrobetaine but possibly not rapidly enough to support the high levels needed in some conditions, e.g. during intensive physical training, lactation or trauma. For such groups, carnitine may be a 'conditionally essential nutrient', equivalent to conditionally essential amino acids (Chapters 8 and 9). It has also been used as a dietary supplement for babies fed on soy-protein infant formulas (in such infants, casein produces an allergic response) as soy-protein lacks carnitine.

fatty acyl-CoA + carnitine $\rightarrow$ fatty acyl-carnitine + CoASH

It is the fatty acyl-carnitine that is transported across the inner mitochondrial membrane from the cytosol to the matrix so that two different enzymes are required for the transport. The first enzyme, carnitine palmitoyltransferase-I, is located on the outer surface of this membrane and the second enzyme, carnitine palmitoyltransferase-II, is located on the inner side of this membrane (Figure 7.11). Carnitine may have this role since it is smaller than CoASH and has no net charge.

The activity of carnitine palmitoyltransferase-I plays an important role in the regulation of fatty acid oxidation; malonyl-CoA is an allosteric exhibitor of the enzyme. Malonyl-CoA is a key intermediate in fatty acid synthesis, which ensures that fatty acid oxidation is decreased when synthesis is taking place. Nonetheless, malonyl-CoA has a major role in the control of fatty acid oxidation in all tissues in which fatty acid oxidation occurs, even if no synthesis takes place.

Oxidation of fatty acids

The oxidation of long-chain fatty acids is known as β-oxidation as it is the second carbon atom from the carboxyl end that is 'attacked' in the process. In summary, the molecules of long-chain fatty acyl-CoA are shortened by the step-wise removal of two-carbon fragments to form acetyl-CoA, which is oxidised by the Krebs cycle. Since virtually all naturally occurring fatty acids have an even number of carbon atoms, β-oxidation results in their complete degradation to acetyl-CoA (Appendix 7.2).

In liver cells, two β-oxidation pathways are present: one in mitochondria and a slightly different pathway in peroxisomes.

Mitochondrial β-oxidation

Each turn of the β-oxidation spiral splits off a molecule of acetyl-CoA. The process involves four enzymes catalysing, in turn, an oxidation (to form a double bond), a hydration, another oxidation (forming a ketone from a secondary alcohol) and the transfer of an acetyl group to coenzyme A (Figure 7.12). The process of β-oxidation operates as a multienzyme complex in which the intermediates are passed from one enzyme to the next, i.e. there are no free intermediates. The number of molecules of ATP generated from the oxidation of one molecule of the long-chain fatty acid palmitate (C18) is given in Table 7.4. Unsaturated fatty acids are also oxidised by the β-oxidation process but require modification before they enter the process (Appendix 7.3).

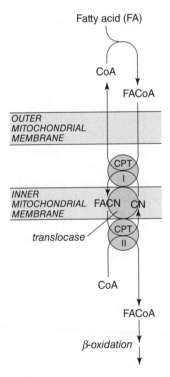

Figure 7.11 *Mechanism of transport of long-chain fatty acids across the inner mitochondrial membrane as fatty acyl carnitine.* CPT is the abbreviation for carnitine palmitoyl transferase. CPT-I resides on the outer surface of the inner membrane, whereas CPT-II resides on the inner side of the inner membrane of the mitochondria. Transport across the inner membrane is achieved by a carrier protein known as a translocase. FACN – fatty acyl carnitine, CN – carnitine. Despite the name, CPT reacts with long-chain fatty acids other than palmitate. CN is transported out of the mitochondria by the same translocase.

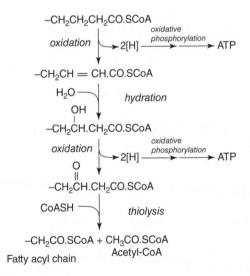

Figure 7.12 *The process of β-oxidation of fatty acyl-CoA in mitochondria.* This cycle of reactions converts all the carbon atoms in fatty acyl-CoA to acetyl-CoA which is oxidised in the Krebs cycle. In the first oxidation, hydrogen atoms are transferred to FAD and in the second to NAD⁺. Both reduced coenzymes are oxidised via the electron transfer pathway.

Table 7.4 Products of β-oxidation and the number of molecules of ATP generated from the oxidation of one molecule of palmitate

Reaction/Process	No. of molecules produced	Molecules ATP per molecule produced	ATP yield
Synthesis of palmitoyl-CoA thioester from palmitate	–	–2[a]	–2
FADH$_2$ oxidation via electron transfer chain	7	1.5	10.5
NADH oxidation via electron transfer chain	7	2.5	17.5
Acetyl-CoA oxidation via Krebs cycle	8	10	80
Total			**106**

[a]Molecules of ATP *consumed* per molecule palmitate activated.

The oxidation is separated into four processes to help clarify the positions in the actions that generate ATP (see Chapter 9, Table 9.5).

The physiological pathway of fatty acid oxidation

The physiological pathway for oxidation of fatty acids in organs or tissues starts with the enzyme triacylglycerol lipase within adipose tissue, that is, the hormone-sensitive lipase. This enzyme, plus the other two lipases, results in complete hydrolysis of the triacylglycerol to fatty acids, which are transported to various tissues that take them up and oxidise them by β-oxidation to acetyl-CoA. This provides a further example of a metabolic pathway that spans more than one tissue (Figure 7.13) (Box 7.1).

Regulation of the rate of fatty acid oxidation

There are several conditions when an increase in fatty acid oxidation is necessary:

• When there is an increase in demand for ATP generation that cannot be met by an increase in glucose oxidation, e.g. skeletal and cardiac muscle during sustained physical activity.

• To provide an alternative fuel to glucose during starvation. Indeed, fatty acid oxidation restricts the rate of glucose utilisation, which maintains the blood glucose level, via a regulatory mechanism known as the glucose/fatty acid cycle (Chapter 16).

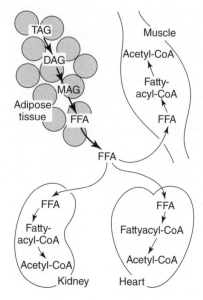

Figure 7.13 *Physiological pathway for fatty acid oxidation.* The pathway starts with the hormone-sensitive lipase in adipose tissue (the flux-generating step) and ends with the formation of acetyl-CoA in the various tissues. Acetyl-CoA is the substrate for the flux-generating enzyme, citrate synthase, for the Krebs cycle (Chapter 9). Heart, kidney and skeletal muscle are the major tissues for fatty acid oxidation but other tissues also oxidise them.

• During trauma, when fatty acids are once again required to provide an alternative fuel to glucose, so that the blood glucose level is maintained. This is important since proliferating cells (e.g. cells in the bone marrow, in lymph nodes and in a wound) require glucose. If the patient is not eating or not being fed, glucose is also required for the brain (Chapter 18).

• During stress, fatty acids are released to provide energy for the 'fight or flight' activity that might follow the stress condition.

At least two mechanisms, which work in concert, regulate their rate of oxidation (Figure 7.14).

(i) The *plasma concentration of fatty acids.* An increase in the plasma concentration increases the rate of fatty acid oxidation. The plasma concentrations of fatty acids depend primarily on the activity of the hormone-sensitive lipase in adipose tissue: the activity is stimulated by the hormones, adrenaline/noradrenaline, growth hormone and glucagon and is inhibited by insulin (Figure 7.10) which therefore play a role in regulation of the rate of fatty acid oxidation.

(ii) The *intracellular concentration of malonyl-CoA.* This compound inhibits the activity of the key enzyme in the oxidation of fatty acids, carnitine palmitoyltrans-

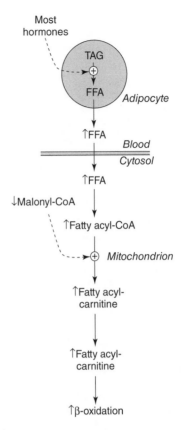

Box 7.1 A short history of experiments leading to the conclusion that long-chain fatty acids are the important lipid fuel

It was appreciated by the beginning of the twentieth century that glucose was the major fuel of carbohydrate oxidation. However, it was not until 1956 that it was established that long-chain fatty acids were the major fuel by which fat is oxidised.

One problem was that a precise and sensitive technique to measure the concentrations of fat fuels was not available. Another problem was that the concentration of long-chain fatty acids in the blood is very low, so that it comprises only a small proportion of the total concentration of the fat in the blood and was therefore considered to be unimportant. However, when a sensitive method for their assay became available, it was shown that the plasma fatty acid concentration changed in various conditions: for example it decreased after a meal but was elevated during starvation and the high plasma fatty acid concentration during starvation was rapidly decreased following the administration of glucose or insulin. It was Gordon & Cherkes (1956) who realised the full significance of such findings, namely that fatty acids were the form in which fat was liberated from adipose tissue and made available as an oxidisable substrate for other tissues. They also realised that if fatty acids were to be an important fat fuel, they required a high turnover rate. In fact, in the same year Havel & Fredrickson reported a very short half-life (about 2 minutes) for plasma fatty acids.

From this it was calculated that the oxidation of fatty acids could provide a significant proportion of the total energy required, for example in starvation. The rise in plasma fatty acid concentration during starvation is at least one mmol/L, and the turnover is approximately 25% per minute. Hence, the total increase in fatty acid in the 3 litres of plasma in the body is 3×1.0, that is 3.0 mmoles, equivalent to approximately 0.75 g FFA. The daily turnover is, therefore, $0.75 \times 0.25 \times 60 \times 24 = 270$ g, the oxidation of which generates $270 \times 37 \approx 1000$ kJ – sufficient to satisfy a large proportion of the resting energy expenditure during starvation. It was from such a calculation that long-chain fatty acids were proposed as the important fat fuel in the blood. The physiological and clinical significance of fatty acids as a fuel, however, was not generally appreciated until the mid 1960s. This is not exceptional, transpeptidation was still considered as a mechanism for protein synthesis in the late 1950s (Chapter 20).

Figure 7.14 *Regulation of rate of fatty acid oxidation in tissues.* Arrows indicate direction of change (i) Changes in the concentrations of various hormones control the activity of hormone-sensitive lipase in adipose tissue (see Figure 7.10). (ii) Changes in the blood level of fatty acid govern the uptake and oxidation of fatty acid. (iii) The activity of the enzyme CPT-I is controlled by changes in the intracellular level of malonyl-CoA, the formation of which is controlled by the hormones insulin and glucagon. Insulin increases malonyl-CoA concentration, glucagon decrease it. Three factors are important: TAG-lipase, plasma fatty acid concentration and the intracellular malonyl-CoA concentration.

(i) Insulin decreases the activity of 'hormone-sensitive' lipase in adipose tissue, which decreases the rate of fatty acid release. This decreases the plasma fatty acid level, which decreases their rate of oxidation.

(ii) Insulin increases the intracellular level of malonyl-CoA and hence decreases the activity of carnitine palmitoyltransferase and, consequently, fatty acid oxidation (Chapters 9 and 12). Glucagon has the opposite effect.

These contribute to a marked decrease in oxidation. This is physiologically important since insulin is the only inhibitory hormone. The major hormone regulating ketogenesis is the other important hormone regulating

ferase (see below): thus a decrease in the intracellular concentration of malonyl-CoA stimulates fatty acid oxidation.

The hormone insulin decreases the rate of fatty acid oxidation via two mechanisms.

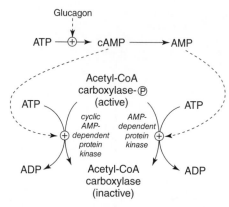

Figure 7.15 *Inhibition of acetyl-CoA carboxylase by cyclic AMP dependent protein kinase and AMP dependent protein kinase: the dual effect of glucagon.* Phosphorylation of acetyl-CoA carboxylase by either or both enzymes inactivates the enzyme which leads to a decrease in concentration of malonyl-CoA, and hence an increase in activity of carnitine palmitoyltransferase-I and hence an increase in fatty acid oxidation. Insulin decreases the cyclic AMP concentration maintaining an active carboxylase and a high level of malonyl-CoA to inhibit fatty acid oxidation.

fatty acid oxidation, which is glucagon. This controls the formation of malonyl-CoA by regulating the activity of acetyl-CoA carboxylase, the enzyme that catalyses the formation of malonyl-CoA. There are two enzymes that result in phosphorylation, cyclic AMP dependent protein kinase and AMP dependent protein kinase (Figure 7.15). Phosphorylation inactivates the enzyme, so that malonyl-CoA formation is inhibited, which will lead to stimulation of fatty acid oxidation. Dephosphorylation, catalysed by a phosphatase, activates the enzyme. Glucagon increases cyclic AMP dependent protein kinase activity by stimulating adenylate kinase, which increases the AMP concentration. The latter is degraded to AMP so that glucagon stimulates both protein kinase enzymes leading to inactivation of the carboxylase, and hence decreasing the malonyl-CoA concentration.

Peroxisomal β-oxidation

Peroxisomes are organelles which are bounded by a single membrane. They are present in the liver where very long-chain fatty acids are oxidised by β-oxidation in peroxisomes, which is different from mitochondrial oxidation.

- A different transporter protein is required to transport the acyl-CoA across the peroxisomal membrane.

- The initial oxidation reaction is unusual since it uses molecular oxygen. It is catalysed by the enzyme acyl-CoA oxidase.

- β-oxidation continues only as far as a medium-chain CoA derivative, which is transported out of the peroxisome as its carnitine ester. It then enters the mitochondria where β-oxidation continues to produce acetyl-CoA. The acetyl-CoA, which is produced in the peroxisome, is also transported out for oxidation in the mitochondria (Appendix 7.4).

The capacity of β-oxidation in the peroxisomes is only about 10% of that in the mitochondria but it plays an important role in oxidising unusual fatty acids: for example, very long-chain fatty acids, polyunsaturated fatty acids, dicarboxylic fatty acids.

> The capacity of the oxidation process in peroxisomes is increased by activation of a transcription factor. A diet high in polyunsaturated fatty acids increases the capacity of the process.

Short-chain fatty acids

These short-chain fatty acids are acetic, butyric, lactic and propionic acids, also known as volatile fatty acids, VFA. They are produced from fermentation of carbohydrate by microorganisms in the colon and oxidised by colonocytes or hepatocytes (see above and Chapter 4). Butyric acid is activated to produce butyryl-CoA, which is then degraded to acetyl-CoA by β-oxidation; acetic acid is converted to acetyl-CoA for complete oxidation. Propionic acid is activated to form propionyl-CoA, which is then converted to succinate (Chapter 8). The fate of the latter is either oxidation or, conversion to glucose, via gluconeogenesis in the liver.

> In ruminants, lactic and propionic acids are the major precursors of glucose. This is particularly important during lactation, since all the carbohydrate in the food is fermented by the bacteria in the rumen, so that no glucose enters the body but glucose is required for the formation of lactose for the milk (Chapter 6).

Ketone bodies

The fatty acids that are taken up by the liver have three fates:

- Esterification to form triacylglycerol or phospholipids, which are secreted by the liver in the form of VLDL (see below).

- Complete oxidation to CO_2, via β-oxidation and the Krebs cycle, to generate ATP.

- Oxidation to acetyl-CoA, which is then converted to ketone bodies, as follows.

Pathway of ketone body formation

Acetyl-CoA, from β-oxidation, is the immediate substrate for synthesis of ketone bodies, which is achieved in four reactions as follows:

(i) 2 acetyl-CoA → acetoacetyl-CoA
   (2C)   (4C)

(ii) acetoacetyl-CoA + acetyl-CoA →
    (4C)   (2C)
                 hydroxymethylglutaryl-CoA
                              (6C)

(iii) hydroxymethylglutaryl-CoA → acetoacetate +
     (6C)   (4C)
                              acetyl-CoA
                                (2C)

(iv) acetoacetate ⇌ hydroxybutyrate
     (4C)   (4C)

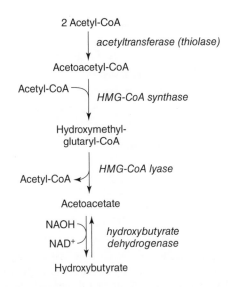

Figure 7.16 *Pathway for the synthesis of ketone bodies: the HMG-CoA pathway.*

The Cs indicate the number of carbons in each intermediate and in the final products. Details of the pathway are presented in Figure 7.16, where the names of the enzymes are given. There are five important points, as follows:

- The pathway is only present in liver.

- It produces two ketone bodies, acetoacetate and 3-hydroxybutyrate.

- Acetoacetate is formed first and then converted to hydroxybutyrate but both are released from the liver.

- Approximately five times more 3-hydroxybutyrate than acetoacetate is produced and released.

Once equilibrium is established for the 3-hydroxybutyrate dehydrogenase reaction, the following equation applies:

$$\frac{[\text{hydroxybutyrate}]}{[\text{acetoacetate}]} = K_{eq} \times \frac{[\text{NADH}]}{[\text{NAD}^+]}$$

In some poorly controlled diabetic patients the high rate of fatty acid oxidation decreases the mitochondrial NAD^+/NADH concentration ratio so that the 3-hydroxybutyrate/acetoacetate concentration ratio can rise to as high as 15 in the blood. Since a test for ketone bodies in the urine (using Clinistix or similar material) detects only acetoacetate this can result in a serious underestimate of the concentration of ketone bodies in the urine.

Regulation of rate of ketone body formation

Ketone bodies are produced in the liver by the partial oxidation of fatty acids.

Two conditions in which the rate of ketone body formation is increased are hypoglycaemia and prolonged starvation in adults or short-term starvation in children. What is the mechanism for increasing the rate? Although there are several fates for fatty acids in the liver, triacylglycerol, phospholipid and cholesterol formation and oxidation via the Krebs cycle, the dominant pathway is ketone body formation (Figure 7.20). Three factor regulate the rate of ketone body formation (i) hormone sensitive lipase activity; (ii) activity of carnitine palmitoyltransferase and (iii) activity of HMG-CoA synthase. Regulation of factors (i) and (ii) are discussed above. HMG-CoA synthase in regulated by an interconversion cycle which involves and desuccinylation (Appendix 7.5).

Oxidation of ketone bodies

Ketone bodies are oxidised by most aerobic tissues including skeletal muscle, heart, kidney, lung, intestine and brain. Since the last two cannot oxidise fatty acids, their ability to oxidise ketone bodies is very important, because they provide another fuel in addition to, or as an alternative to, glucose. Hence, they can be used to replace some of the glucose to maintain the blood glucose concentration (e.g. in prolonged starvation or hypoglycaemia).

The oxidation pathway for ketone bodies occurs within the mitochondria and comprises three reactions:

(i)   Hydroxybutyrate is converted to acetoacetate in a reaction catalysed by hydroxybutyrate dehydrogenase.

(ii) The conversion of acetoacetate to acetoacetyl-CoA. This is in an unusual reaction: it uses the Krebs cycle intermediate, succinyl-CoA, to provide the CoASH for transfer to acetoacetate. The reaction is catalysed by the enzyme 3-oxoacid CoA-transferase:

succinyl-CoA + acetoacetate →

   succinate + acetoacetyl-CoA

(Such activation reactions for the carboxylic acids, usually involve CoASH and ATP.)

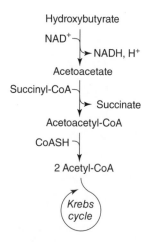

Figure 7.17 *The pathway of ketone body oxidation: hydroxybutyrate to acetyl-CoA.* Hydroxybutyrate is converted to acetoacetate catalysed by hydroxybutyrate dehydrogenase; acetoacetate is converted to acetoacetyl-CoA catalysed by 3-oxoacid transferase and finally acetoacetyl-CoA is converted to acetyl-CoA catalysed by acetyl-CoA acetyltransferase, which is the same enzyme involved in synthesis of acetoacetyl-CoA.

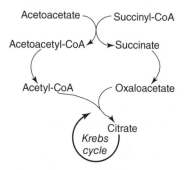

Figure 7.18 *Oxoacid transferase is the key reaction in acetoacetate oxidation.* The reaction produces acetoacetyl-CoA and succinate: the former produces the substrate acetyl-CoA, and the latter produces the co-substrate, oxaloacetate, for the first reaction in the Krebs cycle.

(iii) Acetoacetyl-CoA is split into two molecules of acetyl-CoA in a reaction catalysed by the enzyme, acetyl-CoA acetyltransferase (thiolase), the same enzyme that is used in acetoacetyl-CoA synthesis (Figure 7.17).

It is possible to put forward a biochemical advantage for this unusual reaction in the activation of acetoacetate. The succinate that is produced in the reaction is further metabolised to oxaloacetate, via the reactions of the Krebs cycle. Since acetyl-CoA requires oxaloacetate for formation of citrate and oxidation in the cycle, one product of

Box 7.2 A deficiency of the hepatic glycogen store: hypoglycaemia and ketosis in a child

A child presented to a physician with dizzy spells prior to breakfast but no other problems. An examination of the metabolic changes over 24 hours could only be explained by a deficiency of glycogen synthase in the liver. During the overnight fast, the blood glucose concentration fell to about 1 mmol/L or less, whereas it increased to 10 mmol/L after breakfast and remained elevated for the remainder of the day. As the concentration of glucose decreased during the night, that of ketone bodies increased but, as the blood glucose concentration increased during the day, that of ketone bodies fell. In fact, the plasma ketone body concentration increased in six hours to a similar concentration (7–8 mmol/L) that is achieved by an adult during a fast of 6–8 days. Apart from dizzy spells first thing in the morning, the child had no other symptoms. It suggests that function of the brain during the overnight fast was maintained by ketone body oxidation (Aynsley-Green *et al.* 1977).

A similar suggestion is made for maintenance of the mental activities of Mike Stroud and Ranulph Fiennes during their long trans-Antarctic expedition when, at times, their blood glucose fell to about 1 mmol/L (Chapter 13). Unfortunately the blood level of ketone bodies was not measured during this trek.

this reaction, succinate, facilitates the oxidation of the other, acetoacetate (Figure 7.18). This is important because, under some conditions, ketone bodies must replace glucose as a fuel for essential organs, such as the brain, kidney, heart and some skeletal muscles. When ketone bodies are taken up by a cell under conditions, such as hypoglycaemia and starvation, they must be oxidised immediately in order to generate ATP so that the tissue or organ can maintain its essential function. Any interruption in fuel oxidation could inpair physiological function with serious consequences for health (Box 7.2).

Physiological pathway of ketone body oxidation

The physiological pathway for oxidation of ketone bodies starts with the hydrolysis of triacylglycerol in adipose tissue, which provides fatty acids that are taken up by the liver, oxidised to acetyl-CoA by β-oxidation and the acetyl-CoA is converted to ketone bodies, via the synthetic part of the pathway. Both hydroxybutyrate and acetoacetate are taken up by the tissues, which can oxidise them to generate ATP (Figure 7.19).

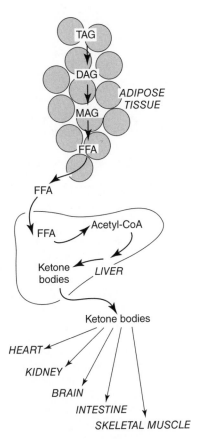

Figure 7.19 *The physiological pathway for ketone body oxidation: from triacylglycerol in adipose tissue to their oxidation in a variety of tissues/organs.* The pathway spans three tissues/organs. The flux-generating step is the triacylglycerol lipase and ends with CO_2 in one or more of the tissues/organs.

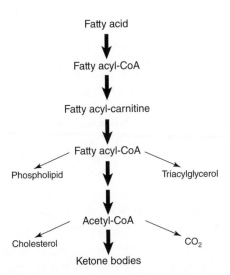

Figure 7.20 *The major quantitative pathway for fatty acid metabolism in the liver is ketone body formation.* This is another indication of the importance of ketone bodies as a fuel.

Regulation of the oxidation of ketone bodies

The immediate signal for increased ketone body uptake by a cell is the increase in their plasma concentration. This can be substantial: up to 50-fold in some conditions (Table 7.3, Figure 7.21) and will result in a marked increase in the oxidation of ketone bodies. At there high concentrations of ketone bodies the uptake and oxidation by the brain can provide more than 50% of its ATP requirement so that glucose utilisation is decreased by almost 50% (Chapter 16).

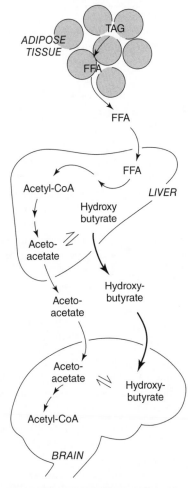

Figure 7.21 *Provision of the fat fuels for the brain during hypoglycaemia.* During hypoglycaemia it is essential that ketone bodies are available for the brain to provide a fat fuel for ATP generation to maintain mental functions. This sequence of processes from adipose tissue to the brain is therefore, a survival pathway especially for children during short-term starvation or hypoglycaemia. (Box 7.2) (Chapter 16).

Physiological importance of fat fuels

The various roles of fats were summarised in the introduction to this chapter. Two questions arise from this: what is the advantage of storing lipid as the major energy reserve? Why are there three lipid fuels in the blood while there is only one carbohydrate fuel?

The answers to these questions demonstrate the physiological importance of fat fuels. There are two parts to the answer for question one:

(a) Fat is more reduced than carbohydrate, so that the amount of energy available to generate ATP from the same quantity of fuel is almost 2.5 times greater from triacylglycerol (see Table 2.3).

(b) Since lipids are hydrophobic, they are stored dry (more than 90% of the mass of a typical adipocyte is pure triacylglycerol) (Figure 7.5). In contrast, glycogen is stored in an hydrated form: for every gram of glycogen stored, 3 grams of water is structurally associated with it. Thus, as much as 75% of the weight of glycogen stored in a tissue is water, so that the amount of energy liberated on oxidation of one gram of the glycogen in a tissue is only about 4.2 kJ/g. When comparing the energy stored, on the basis of total wet weight of fuel, this makes fat approximately nine times

> Examples are the ruby-throated hummingbird which can fly non-stop across the Gulf of Mexico, a distance of approximately 1300 miles, and the migratory locust which can fly 200 miles or more non-stop in search of food. Both animals store only fat to provide energy for such long-distance flights.

better than carbohydrate, as a store of energy (see Chapter 2). The difference is especially important in terrestrial animals, in which an increase in mass interferes with mobility, and in birds and insects that fly, since mass of the animal is critically important.

Two further interesting questions are raised as an extension of this topic. A normal young adult woman stores considerably more fat than a normal young adult man. About 25% of the weight of an average young woman is due to adipose tissue whereas the proportion for an adult male is about 15%. It is rarely appreciated that this is a unique human characteristic, i.e. no other animal so far investigated exhibits such a gender difference. What is the explanation for this difference and why is it unique to humans? Why is the distribution of adipose tissue so different between adult men and women (Figure 7.22). Possible answers to these questions are given by Pond (1998).

To answer the question raised above, 'why are there three fat fuels in the blood but only one carbohydrate fuel', the three fat fuels are discussed separately.

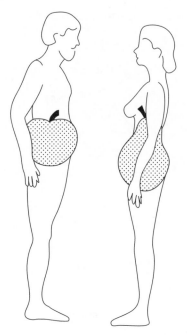

Figure 7.22 *A common difference in the distribution of adipose tissue between adult men and women.* Men usually expand their adipose tissue around the waist to give an 'apple' appearance, whereas women usually expand in the region of the hips to give a 'pear'-shaped appearance (a difference readily observable on any beach). The differences are exaggerated in obese men and women. The biological, biochemical or physiological significance of this difference is not known, but see Pond, 1998.

Triacylglycerol in blood

The packaging of triacylglycerol into chylomicrons or VLDL provides an effective mass-transport system for fat. On a normal Western diet, approximately 400 g of triacylglycerol is transported through the blood each day. Since these two particles cannot cross the capillaries, their triacylglycerol is hydrolysed by lipoprotein lipase on the luminal surface of the capillaries (see above). Most of the fatty acids released by the lipase are taken up by the cells in which the lipase is catalytically active. Thus the fate of the fatty acid in the triacylglycerol in the blood depends upon which tissue possesses a catalytically active lipoprotein lipase. Three conditions are described (Figure 7.23):

- In the normal fed state, the enzyme is active in adipose tissue, so that the fatty acids in the triacylglycerol are taken up by adipocytes and then esterified for storage.

- During lactation, the enzyme is active in mammary gland and the released fatty acids are taken up and used to synthesise triacylglycerol for the milk. A few days before parturition, in preparation for lactation, the activity of lipoprotein lipase is increased in the mammary gland,

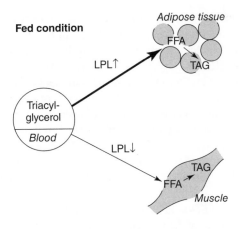

Fed condition

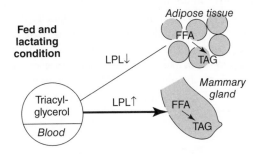

Fed and lactating condition

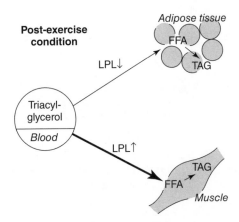

Post-exercise condition

Figure 7.23 *Fate of blood triacylglycerol (in the chylomicrons and VLDL) in three conditions: role of changes in activity of lipoprotein lipase in directing the uptake of fatty acids.* It is primarily the activity of lipoprotein lipase that directs which tissue/organ takes up the fatty acids from the blood triacylglycerol. The abbreviation LPL↑ indicates a change to a higher activity of lipoprotein lipase; LPL↓ indicates a change to a lower activity of lipoprotein lipase. The broadness of the arrow indicates the dominant direction of the fate of the fatty acid.

whereas that in the adipose tissue is decreased, so that the uptake of fatty acids from the blood triacylglycerol is directed to the mammary gland.

- After sustained physical activity, the enzyme is active in the muscles that have been exercising, which results in uptake of the fatty acids from the triacylglycerol in blood by these muscles. The fatty acids are esterified within the muscle and the triacylglycerol so formed replenishes that which was used in the physical activity.

An interesting additional point is that during trauma the cytokine, tumour necrosis factor, results in a decrease in lipase activity in adipose and other tissues, so that there is an increase in the level of VLDL and chylomicrons in the blood. The significance of this is unclear but it may be that pathogens in the blood are adsorbed onto the emulsion of VLDL or chylomicrons which reduces the risk of adsorption of the pathogen onto the surface of a cell, which is necessary for the pathogen to enter the cell. This localisation also aids attack by antibodies (Chapter 17).

These changes give rise to the concept of 'fuel direction', which is an important advantage of triacylglycerol as a fuel. The physiological significance is as follows:

- The changes in lipase activity in lactation direct the fuel away from adipose tissue to the mammary gland, i.e. the change in the fate of the triacylglycerol in blood is due to opposite changes in the activity of only one enzyme in the two tissues.

- After prolonged physical activity, the uptake of fatty acids is directed to the muscles involved in the activity and away from adipose tissue and the inactive muscles, due to a change in the activity of one enzyme. This allows restoration of the triacylglycerol store within the muscle.

In summary, the advantage of changes in the activity of the lipase is that it directs the uptake of fatty acid into a tissue that requires the fat (see Figure 7.23). No such direction occurs with the fatty acids that are released from adipose tissue.

Long-chain fatty acids

Fatty acids are released from adipose tissue into the bloodstream, from where they can be taken up and used by 'aerobic' tissues, with the exception of brain and the intestine. In addition, an increase in the plasma fatty acid concentration is one factor that increases the rate of fatty acid oxidation by tissues. Hence, an increase in the mobilisation of fatty acid from adipose tissue is an immediate signal for tissues such as muscle, heart and kidney cortex to increase their rate of fatty acid oxidation. This not only generates ATP for the tissue but reduces the rate of glucose oxidation, which is part of the mechanism that helps to maintain blood glucose concentration. Thus, in contrast to triacylglycerol, fatty acid uptake by tissues depends upon the plasma level of fatty acid, the activity of the rate-limiting enzyme (carnitine palmitoyltransferase) and the demand for ATP generation in the tissue.

Ketone bodies

Since their discovery in the urine of diabetic patients in the latter part of the nineteenth century, ketone bodies have had a varied history, being considered at different times either as products of disturbed metabolism or as important fuels. Their association with a disease branded them, initially, as undesirable metabolic products but the demonstration in the 1930s that various tissues could oxidise ketone bodies led to the suggestion that they could be important as a lipid fuel. Thus, in a review in 1943, MacKay wrote: 'Ketone bodies may no longer be looked upon as noxious substances which are necessarily deleterious to the organism.' Despite this, in 1956 evidence was presented that long-chain fatty acids were the important fat fuel in starvation, so that ketone bodies were further neglected and once more associated with pathological conditions. In 1968, Greville & Tubbs wrote: 'Clearly it is not obvious in what ways ketogenesis in fasting is a good thing for the whole animal; should the liver be regarded as providing manna for the extrahepatic tissues or does it simply leave them to eat up its garbage?'

In the 1970s the physiological importance of ketone bodies was revived, quite dramatically, when George Cahill and his colleagues demonstrated that ketone bodies could be used by the human brain during prolonged starvation and, furthermore, contributed significantly to ATP generation. Not only did they generate ATP, they decreased the rate of glucose utilisation by the brain by about 50%, which contributes significantly to survival in prolonged starvation (Chapter 16). This important observation was made directly on the subjects by measurement of the arteriovenous differences across the brain, (this is a procedure that would no longer gain ethical permission Table 7.5).

The plasma concentration of ketone bodies in fed, healthy humans is very low (about 0.1 mmol/L) so that the rate of utilisation is very low. However, it is elevated in several conditions, e.g. starvation, hypoglycaemia, affer physical activity. In starvation in normal adults, it increases to about 3 mmol/L after three days and to 5–6 mmol/L after several more days (Figure 7.24). Nevertheless, it can increase to 3 mmol/L or higher within a few hours of completing a prolonged period of physical activity if food, particularly carbohydrate, is not eaten (known as 'accelerated starvation') (Table 7.3). Ketone bodies are particularly important in children, since starvation can quickly result in severe hypoglycaemia. This is due to the fact that the amount of glycogen stored in the liver of a child is

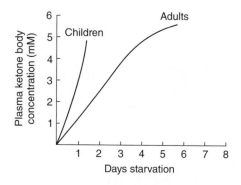

Figure 7.24 *The increase in the plasma ketone body concentration during starvation in adults and children.* The ketone body comprises both acetoacetate and hydroxybutyrate.

Table 7.5 Arteriovenous differences of oxygen, glucose, fatty acids, acetoacetate and hydroxybutyrate across the brain of obese females during prolonged fasting

	Arteriovenous concentration difference (mmol/L)				
	Oxygen	**Glucose**	**Acetoacetate**	**Hydroxybutyrate**[a]	**Fatty acid**
Fed	−3.3	−0.51	–	–	–
5–6 weeks' fasting	−3.0	−0.14	−0.06	−0.34	−0.02

In the fed state, the glucose uptake by the brain is 0.32 mmol/min and the substrate is completely oxidised. This accounts for 94% of the oxygen consumption (2.1 mmol/min) (Reimuth *et al.*, 1965). In the fasted state, glucose oxidation (0.09 mmol/min) accounts for 30% of oxygen uptake (1.9 μmol/min). Ketone body oxidation accounts for the remainder (Owen *et al.*, 1967). The negative sign indicates uptake of the fuel by the brain. Measurements were made by insertion of a catheter into the brachial artery and the vein draining blood from the brain. The negative sign indicates uptake.

[a] Note that considerably more hydroxybutyrate than acetoacetate is taken up by the brain.

low in comparison with that in an adult, yet the rate of glucose utilisation by the brain is similar (see Chapters 6 and 16). Consequently, the blood ketone body level increases to about 5 mmol/L in children during a short period of starvation (Figure 7.24). This fact, and its clinical significance, appear to have been ignored by health professionals, physicians and paediatricians (Chapters 6 and 16).

Ketone bodies are produced in the liver by partial oxidation of long-chain fatty acids arising from the triacylglycerol stored in adipose tissue, so that the question arises, why should one lipid fuel be converted into another? There are several reasons.

- There are two tissues that cannot use long-chain fatty acids, the small intestine and the brain. Both can, however, oxidise ketone bodies and therefore can restrict glucose utilisation. It is not known why these tissues do not oxidise fatty acids: possibly the activity of the enzymes in oxidation is very low.

- Some muscles carry out essential physiological functions, for example, heart, respiratory muscle (e.g. diaphragm and intercostal muscle), postural muscles, muscles used for chewing food. Severe hypoglycaemia could result in such a low uptake of glucose that the rate of ATP generation would not sufficient to meet the energy be needs for these essential functions. In such a situation, the formation of ketone bodies and their uptake by the muscles would replace glucose as a fuel to generate ATP for these essential functions.

- Ketone bodies provide a fuel for the brain which can significantly reduce the rate of glucose utilisation by this organ. During starvation, this permits a reduction in the loss of muscle protein, which otherwise breaks down to provide glucose, via gluconeogenesis from amino acids, for the brain. This reduction in the rate of breakdown of muscle protein is one factor that increases the length of time of survival during starvation (Chapter 16).

- The rate of utilisation of ketone bodies by the brain of the neonate is considerably higher than that of an adult (about fourfold). This high rate of ketone body oxidation may restrict that of glucose, so that it is available for synthesis of compounds such as amino sugars, glycosylation of proteins and provision of glycerol 3-phosphate for phospholipid synthesis, which is especially important for the developing brain (Chapters 14 and 15). In addition, the utilisation of glucose by the pentose phosphate pathway, rather than oxidation, provides NADPH, which is necessary for several reduction reactions, and for ribose 5-phosphate formation, which is required for the synthesis of nucleotides for RNA and DNA synthesis (Chapters 6 and 20) (Box 7.3).

Box 7.3 Ketone bodies in the neonate

The major fuel for the foetus is glucose, provided by the mother via the placenta. However, any glycogen stored in the foetus prior to birth is used during parturition since hypoxia can readily develop as the placenta separates from the mother (Chapter 6). The major fuel in the milk of the mother is fat, so that in the neonate the nature of the major fuel changes from carbohydrate to fat. Since the brain is an organ that demands most of the fuel in the neonate and the brain cannot use fatty acids, the fat fuel that is provided is ketone bodies. A high rate of ketogenesis, occurs therefore, in the neonate. This is achieved as follows:

- There is a marked increase in the activity of the key enzymes that convert fatty acids into ketone bodies: carnitine palmitoyltransferase and HMG-CoA synthase.
- The carnitine palmitoyltransferase is insensitive to the inhibitor malonyl-CoA, so that it cannot be inhibited by this allosteric effector (see above).
- The activity of the enzyme that forms malonyl-CoA, acetyl-CoA carboxylase, is very low, so that fatty acid oxidation is high.
- HMG-CoA synthase is not succinylated, i.e. the interconversion cycle involving succinylation does not occur (Appendix 7.5).

These changes provide further biochemical support for the mechanisms proposed for regulation of ketone body synthesis that are discussed above.

- The plasma ketone body level increases rapidly in starvation in pregnancy, which is important in preventing hypoglycaemia in the mother and thus maintaining a supply of glucose for the developing foetus (Chapter 19).

Limitations or drawbacks of fats as a fuel

Although there are advantages in the provision of three lipid fuels in the body, there are a number of drawbacks to the storage of all the fuel reserves as triacylglycerol and the use of fat fuels:

- Fat fuels can only generate ATP via the Krebs cycle and electron transport, so that generation of ATP requires molecular oxygen. Consequently, ATP cannot be generated under anoxic or hypoxic conditions from fat fuels.

- They cannot be oxidised if the tissue lacks mitochondria, e.g. red blood cells, kidney medulla, lens of the eye, type IIB muscle fibres (anaerobic fibres) – see Chapter 6.

- During sustained exercise, fat oxidation alone cannot provide sufficient ATP to maintain a high power output (see Chapter 13).

- Fat is stored in the body in separate depots, most of which need to be remote from muscles, so as not to interfere with their function. A considerable amount is stored around the waist or hip where it does not affect the centre of gravity of the body, which is important in locomotion. Consequently, the fat fuel requires transport from the site of storage to the tissue of utilisation, via the blood, which requires specific transport mechanisms (see above).

- Fatty acids require about 7% more oxygen than does carbohydrate to generate the same amount of ATP. Under circumstances where oxygen supply is limiting, for example in parts of the myocardium after an occlusion in one of the arteries, glucose is the preferred fuel and attempts are made to increase blood glucose levels and decrease mobilisation of fatty acids in this condition (see Chapter 22).

Genetic defects in fatty acid oxidation

The study of genetic defects in the oxidation of fat fuels is a relatively new field compared with the study of such defects in carbohydrate and amino acid metabolism. The first genetic defect was reported in 1970 and the first enzyme deficiency in 1973. The probable reasons for the late discovery of these defects are of some interest:

- In an adult, it is only after about 24 hours of fasting or during prolonged physical activity that fatty acid oxidation plays a major role in energy provision. Neither condition is common in developed countries, so that an inability to generate ATP from fat oxidation is not normally apparent.

- Except for measurement of ketone bodies and triacylglycerol or cholesterol, clinical chemistry laboratories seldom measure the plasma concentrations of fatty acids, which could provide some information about fatty acid oxidation.

- Abnormal metabolites of fatty acids in urine, which could indicate defects in fat metabolism, can only be detected by gas chromatography or mass spectrometry. These techniques are not routinely available in clinical laboratories.

Defects in several proteins involved in fatty acid oxidation are known. These are carnitine palmitoyltransferases, any of the three acyl-CoA dehydrogenases, or the protein that transports carnitine into cells. All are inherited in an autosomal recessive manner and between them give rise to the following signs. *If the liver is involved*, it presents, in infancy or early childhood, with recurrent episodes of hypoglycaemia and possibly coma, since ketone bodies cannot be produced. *If muscle is involved*, it usually presents with pain, cramps, myoglobinuria and chronic muscle weakness.

Carnitine palmitoyltransferase deficiency

Patients with this deficiency present with myopathy, recurrent aching muscles and myoglobinuria after prolonged exercise or starvation. It is interesting to note that there are more cases of a deficiency of this enzyme in muscle than there are cases of a deficiency of any of the glycolytic enzymes (including phosphorylase, see Chapter 6).

Carnitine transporter deficiency

Children with a primary deficiency of the carnitine transporter present with acute episodes of hypoglycaemia leading to loss of consciousness during even a short fast. (See Chapter 9 for a role of carnitine) in the Krebs cycle.

Acyl-CoA dehydrogenase deficiencies

The commonest form of this deficiency is absence of medium-chain acyl-CoA dehydrogenase activity but, if the long-chain acyl-CoA dehydrogenase activity is absent, the symptoms, which appear in childhood, are more severe. These include vomiting and severe drowsiness. Problems arise during starvation when neither fatty acids nor ketone bodies can spare the demand for glucose so that hypoglycaemia develops rapidly. A precise diagnosis is difficult unless methods are available for measurement of glycine-conjugates of fatty acids in the urine.

Pathological concentrations of fat fuels

The three fat fuels and their metabolism are involved directly or indirectly in diseases such as diabetes mellitus, syndrome X, obesity, atherosclerosis and coronary heart disease, which are discussed in other chapters in this book. This section considers the problems associated with high blood levels of ketone bodies and long-chain fatty acids.

Ketoacidosis

An elevated plasma ketone body concentration, of about 7–8 mmol/L, occurs in a number of conditions but it causes no problems (Table 7.3). In contrast, in type 1 diabetic patients the plasma ketone body level can, on occasion, exceed 20 mmol/L. Although this is not common in well-controlled diabetic patients, it can arise if the patient has failed to inject sufficient insulin to control the blood glucose level and inhibit fatty acid mobilisation and ketogenesis in the liver (see above). The problem is not the concentration of ketone bodies but the fact that they are acidic and, if the concentration is high, blood pH can fall to values less than 7.0. A low pH together with hyperglycaemia can lead to a coma. Indeed, this may be the first indication, especially in a young child, of severe insulin deficiency. Since the patient can develop this extreme degree of ketosis gradually over a considerable period of time, diabetes mellitus may not always be considered as a cause. Such ketoacidosis can also occur in a well-controlled diabetic patient when another illness, especially gastroenteritis, increases the requirement for insulin for which the patient is unaware.

Long-chain fatty acids

The plasma level of fatty acids in a fed subject is between 0.3 and 0.5 mmol/L. As discussed above, the maximal safe level is about 2 mmol/L. This is not usually exceeded in any physiological condition since, above this concentration, that of the free (not complexed with albumin) fatty acids in the blood increases markedly. This can then lead to the formation of fatty acid micelles which can damage cell membranes: the damage can cause aggregation of platelets and interfere with electrical conduction in heart muscle (Chapter 22). The cells particularly at risk are the endothelial cells of arteries and arterioles, since they are directly exposed to the micelles, possibly for long periods of time. Two important roles of endothelial cells are control of the diameter of arterioles of the vascular system and control of blood clotting (Chapter 22). Damage to endothelial cells could be sufficiently severe to interfere with these functions; i.e. the arterioles could constrict, and the risk of thrombosis increases. Both of these could contribute to the development of a heart attack (Chapter 22) (Box 7.4).

In addition, a high intracellular concentration of fatty acids raises that of long-chain acyl-CoA, which can damage

mitochondrial membranes, uncoupling oxidative phosphorylation and impairing ATP generation. This could seriously interfere with the function of a tissue. Unfortunately, since the plasma level of fatty acids is not normally measured, unlike glucose or cholesterol, high concentrations of fatty acids are not considered to be a problem by most clinicians.

Triacylglycerol in the blood

Triacylglycerol in the forms of chylomicrons or very low density lipoproteins constitutes the mass transport system of fat in the blood. Excessive levels, particularly of VLDL, can give rise to various pathological problems which are grouped together under the title lipoproteinaemias and are discussed in Chapter 11 (Appendix 11.9).

Box 7.4 Stress and long-chain fatty acids

Stress, whether caused by anxiety or aggression, elevates plasma levels of the stress hormones adrenaline, noradrenaline and cortisol. These hormones stimulate the hormone sensitive lipase in adipose tissue which elevates the fatty acid concentration in the blood. The hormones increase the levels of fuels in the blood in preparation for 'fight or flight' but, especially in humans living in developed countries, the stress-inducing situations are very different from those of early humans. For example, driving a car in crowded traffic, discussion of a contentious issue at a committee meeting or watching exciting sporting events can induce stress in some individuals. In these situations, blood fatty acid concentrations could be raised to 'toxic' levels.

After the Athens earthquake in 1981 there was a marked increase in deaths due to heart attacks over a period of five days after the quake; on the day of the earthquake in Los Angeles, there was an increase in heart attacks. In the first few days after the Iraqi missile attacks on Israel in 1991, there was a sharp increase in heart attacks in Israel compared with control periods. The psychological stress caused by these sudden disasters could have raised the blood fatty acid levels above the 'safe' level in some subjects which could have increased the risk of a heart attack.

8
Amino Acid and Protein Metabolism

A consequence of the oxidation of amino acids is that the amino group is lost to the system, mainly through the formation and excretion of urea. Of the urea produced, the proportion which is excreted in the urine varies depending on the . . . metabolic activity of the colonic microflora. From 25 to 90% of the urea nitrogen produced may be retained in the system following hydrolysis by colonic microflora, thereby playing a fundamental role in achieving nitrogen balance and making a significant contribution to the nitrogen economy of the body.

(Jackson, 1998)

At least 30 different amino acids are found in Nature. It was considered for some time that only 20 amino acids were present in mammalian proteins but now it is known that there are 21; the 21st is selenocysteine (Table 8.1). The metabolism of each is different, making amino acid metabolism considerably more complex than that of either glucose or lipid. An unfortunate consequence of this is that a series of pages describing the metabolism of each individual amino acid is daunting, to say the least, and may have little value except for reference purposes. Of greater utility is an understanding of the general principles of amino acid metabolism, which are described in this chapter. Specific details of the roles played by individual amino acids in the immune system, behaviour, physical fatigue, cancer and trauma are discussed in later chapters, but the background for these discussions is provided in this chapter.

Introduction

Protein represents the second largest store of chemical energy in the body, but it is not used for generating ATP except in some diseases and in some extreme conditions (e.g. prolonged starvation, very sustained exercise). The largest deposit of protein in the body is in skeletal muscle (about 40% of body weight). The synthesis of proteins requires amino acids whereas degradation of proteins produces amino acids and the two processes occur simultaneously. Hence, there is a continual turnover of protein, which accounts for at least 20% of the resting energy expenditure.

The intra- and extracellular concentrations of amino acids in humans

The intracellular concentrations of amino acids in two tissues and those in plasma are given in Table 8.2. Some general comments are:

- In muscle, the concentrations of alanine, aspartate, glutamate, glutamine, leucine, serine and valine are high; that of glutamine is the highest (*c.* 20 mmol/L). The lowest are those of methionine, tryptophan and tyrosine.

- In plasma, the glutamine concentration is also the highest (*c.* 0.6 mmol/L).

- The total intracellular concentration of all the free amino acids is about 15 times higher than those in the plasma.

- The *total* nitrogen content in the whole body is about 24 g per kg body wt.

- Hence the *free* amino acids comprise only 1% of the total amino acid N: i.e. 99% of all the amino acids in the body is bound in proteins.

Functional Biochemistry in Health and Disease by Eric Newsholme and Tony Leech
© 2010 John Wiley & Sons Ltd

Table 8.1 The 21 amino acids present in mammalian proteins

Amino acids in proteins	Standard abbreviation		Molecular mass[b]
	3-Letter	1-Letter[a]	
Essential (indispensable)[c]			
Histidine	His	H	155
Isoleucine	Ile	I	131
Leucine	Leu	L	131
Lysine	Lys	K	146
Methionine	Met	M	149
Phenylalanine	Phe	F	165
Threonine	Thr	T	119
Tryptophan	Trp	W	204
Valine	Val	V	117
Non-essential (dispensable)			
Alanine	Ala	A	89
Arginine	Arg	R	174
Aspartate	Asp	D	133
Cysteine[d]	Cys	C	121
Asparagine	Asn	N	132
Glutamate	Glu	E	147
Glutamine	Gln	Q	146
Glycine	Gly	G	75
Proline	Pro	P	115
Serine	Ser	S	105
Tyrosine[d]	Tyr	Y	181
Selenocysteine[e]			168

Data from Matthews (2006).

[a] The single-letter abbreviations are often used to indicate amino acid sequences in proteins.

[b] Molecular mass is rounded to the nearest whole number and represents the number of grams per mole of amino acid.

[c] For difference between essential and indispensable, and for histidine, see text.

[d] Cysteine and tyrosine are described as *conditionally* essential (see text).

[e] Selenocysteine is not always included in lists of amino acids but it obeys the definition of an amino acid, it is present in the diet, is present in some mammalian proteins, and it is incorporated directly into these proteins as selenocysteine during the normal process of translation: there is a specific tRNA for this amino acid: the anticodon is AGU.

Structure and classification of amino acids

Amino acids bear both carboxyl (–COOH) and amino (–NH$_2$) groups, which become ionised at pH 7.0:

$$
\begin{array}{c}
H \\
| \\
R - C - COO^- \\
| \\
NH_3^+
\end{array}
$$

Since both groups are attached to the α-carbon, they are known as α-amino acids. The identity of an amino acid depends on the structure of its side chain (R group), which

Table 8.2 Concentrations of free amino acids in plasma, liver and muscle of humans

Amino acid	Concentrations (mmol/L)		
	Plasma	Liver	Muscle
Alanine	0.36	3.2	3.2
Arginine	0.06	0.03	0.57
Aspartic acid	0.01	18.7	1.0
Asparagine	0.05	0.32	0.40
Cysteine	0.09	–	0.16
Glutamic acid	0.02	4.1	3.8
Glutamine	0.60	5.1	20.0
Glycine	0.20	3.7	1.5
Histidine	0.07	0.77	0.40
Isoleucine	0.05	0.10	0.10
Leucine	0.11	0.30	0.24
Lysine	0.16	0.25	1.2
Methionine	0.02	0.05	0.1
Phenylalanine	0.06	0.10	0.09
Proline	0.22	–	1.6
Serine	0.10	1.0	0.71
Taurine[a]	0.07	8.5	25.0
Threonine	0.11	0.55	0.67
Tryptophan	0.04	0.03	0.1
Tyrosine	0.05	0.15	0.14
Valine	0.21	0.32	0.30

For muscle, data from Blomstrand *et al.* (1995); Matthews (2006); Barle *et al.* (1996). From liver, data from

[a] Taurine is a sulphur amino acid, which is not present in protein (see text).

For the calculation from the amount in fresh (wet) tissue, e.g. µmol/g fresh tissue, it is assumed that the intracellular water makes up 40% of the weight.

Concentrations reported by different authors vary slightly. The concentrations of selenocysteine have not been measured. The total concentration of all the essential amino acids (without cysteine) is the same in muscle and liver. Similarly the total concentration of all the non-essential amino acids (excluding taurine) is similar.

is a single H in the simplest amino acid (glycine) but in tryptophan is complex.

Amino acids can be divided into six classes according to the nature of their side chains (R groups) (Figure 8.1). The complete structure of each amino acid can be determined from the R group. All the amino acids that occur in proteins are the L-enantiomers (see Appendix 3.1) except glycine, which is not optically active. The D-enantiomers of a few amino acids do occur in Nature but are found not in proteins but in small peptides in some bacterial cell walls.

The words peptide and polypeptide are both used to describe chains of amino acids linked by peptide bonds. Peptides are short chains of amino acids that do not form part of a protein. The number of amino acids is indicated by a prefix (e.g. dipeptide, hexapeptide, oligopeptide). Polypeptides (or polypeptide chains) are longer sequences which can form part of a protein, which may consist of several polypeptides (Appendix 3.1).

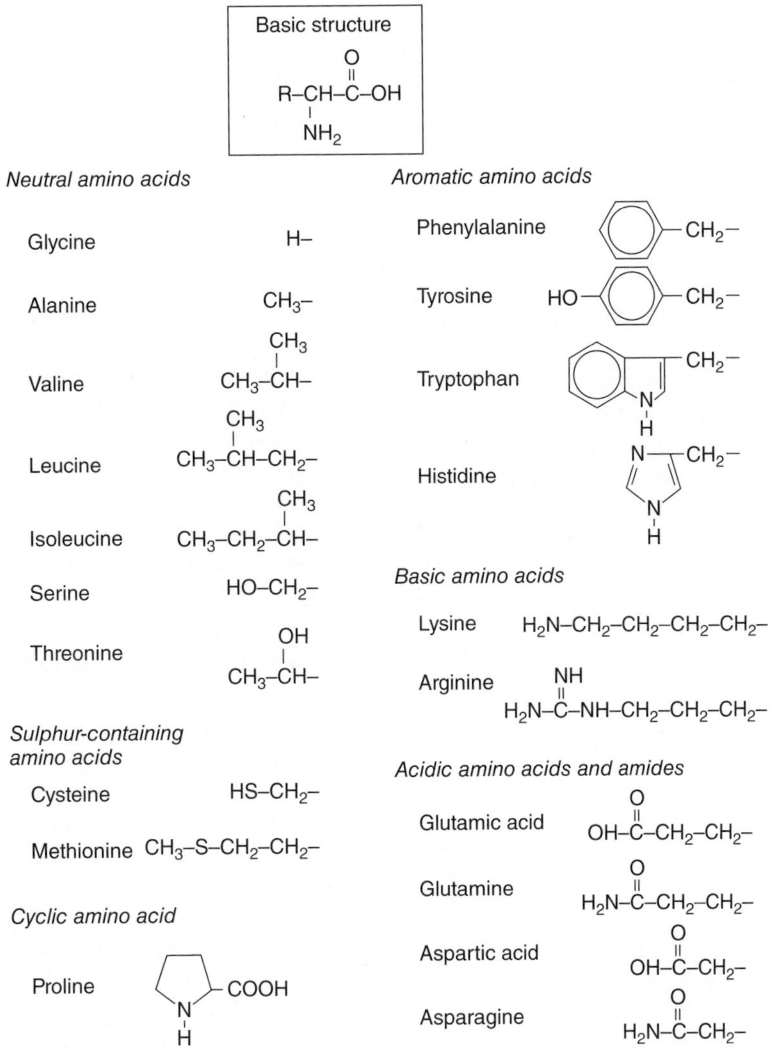

Basic structure

$$\underset{\underset{NH_2}{|}}{R-CH}-\overset{\overset{O}{\|}}{C}-OH$$

Neutral amino acids

Glycine H–

Alanine CH_3–

Valine $CH_3-\overset{\overset{CH_3}{|}}{CH}-$

Leucine $CH_3-\overset{\overset{CH_3}{|}}{CH}-CH_2-$

Isoleucine $CH_3-CH_2-\overset{\overset{CH_3}{|}}{CH}-$

Serine $HO-CH_2-$

Threonine $CH_3-\overset{\overset{OH}{|}}{CH}-$

Sulphur-containing amino acids

Cysteine $HS-CH_2-$

Methionine $CH_3-S-CH_2-CH_2-$

Cyclic amino acid

Proline

Aromatic amino acids

Phenylalanine

Tyrosine

Tryptophan

Histidine

Basic amino acids

Lysine $H_2N-CH_2-CH_2-CH_2-CH_2-$

Arginine $H_2N-\overset{\overset{NH}{\|}}{C}-NH-CH_2-CH_2-CH_2-$

Acidic amino acids and amides

Glutamic acid $OH-\overset{\overset{O}{\|}}{C}-CH_2-CH_2-$

Glutamine $H_2N-\overset{\overset{O}{\|}}{C}-CH_2-CH_2-$

Aspartic acid $OH-\overset{\overset{O}{\|}}{C}-CH_2-$

Asparagine $H_2N-\overset{\overset{O}{\|}}{C}-CH_2-$

Figure 8.1 *Structure of side-chain groups (R) of amino acids except for proline, for which the entire structure is presented.*

Sources of amino acids

There are four sources of amino acids that enter the free amino acid pool in the body: proteins in food; proteins secreted into the stomach and intestine by the digestive glands; endogenous proteins; and microorganisms that die and release their protein in the colon.

(i) *Food* The average intake of protein in developed countries is about 90 g a day. During digestion, protein is hydrolysed to release the amino acids that are absorbed into the enterocytes of the small intestine and then enter the blood, from where they are taken up by the tissues for peptide and protein synthesis or to enter the pathways of metabolism, which are described in this and subsequent chapters. (Free amino acids are present in food but amounts are small, unless the diet is supplemented with amino acids.)

(ii) *Small intestine and pancreas* About 70 g of protein enters the lumen of the intestine every day from the secretory cells in the form of digestive enzymes and mucus and from desquamated epithelial cells.

(iii) *Endogenous protein* The process of protein turnover involves hydrolysis of cellular protein, with release of free amino acids into the intracellular compartment.

(iv) *Bacterial and other microorganisms in the intestine* These are present mainly in the colon. Death of the microorganisms is followed by their digestion and the release of amino acids into the lumen. The amino acids are then available for use by other microorganisms, by the colonocytes or the liver, after their uptake from the lumen. The use by the liver is quantitatively significant in some conditions (see below).

A summary of the contributions of various processes to protein intake, protein loss and whole-body protein turnover is presented in Table 8.3.

Protein turnover

Protein turnover in an adult is about 4 to 5 g per kg body wt, equivalent to about 250 to 350 g of protein hydrolysed and resynthesised every day in the tissues of an adult human. This represents considerably more protein than is ingested in food. The rates of protein turnover vary enormously, depending on the nature of the protein, the condition of the subject and the tissue (Table 8.3). Proteins (mainly enzymes) in the liver are replaced every few hours or days whereas structural proteins (e.g. collagen, contractile proteins) are stable for several months. Contractile proteins can be degraded relatively rapidly in some conditions (see below).

Table 8.3 Approximate values for (a) protein intake, (b) protein loss, (c) protein turnover, each day in a normal adult human

Protein intake	
Process	**Amount (g)**
Dietary intake	90
Proteins secreted into GI tract	70
Total absorbed	150[a]

Protein loss	
Process	**Amount (g)**
In faeces	10
In urine[a]	75
Sloughing of skin	5
Total lost	90

Protein turnover	
Tissue	**Amount (g)**
Muscle	75
Liver, gut, lung	130
White blood cells	20
Red blood cells	8
Albumin	12
Other tissues	8

[a]Calculated from the nitrogen in urea and ammonia which represents primarily protein metabolism (the nitrogen in nucleic acid is released as ammonia and converted to urea).

Data from Matthews (2006).

Four questions arise:

(i)   Why should this turnover occur at all?

(ii)   Why should its rate be so variable?

(iii)   What are the processes responsible for intracellular protein degradation?

(iv)   How is it controlled?

There are at least two answers to question (i). First, abnormal proteins can arise in cells due to spontaneous denaturation, errors in protein synthesis, errors in post-translational processing, failure of the correct folding of the protein or damage by free radicals. They are then degraded and replaced by newly synthesised proteins. Secondly, turnover helps to maintain concentrations of free amino acids both within cells and in the blood. This is important to satisfy the requirements for synthesis of essential proteins and peptides (e.g. hormones) and some small nitrogen-containing compounds that play key roles in metabolism (see Table 8.4).

An answer to question (ii) is as follows. Turnover can be considered as a large-scale substrate cycle (Chapter 3), one role of which is to regulate, precisely, the concentrations of specific proteins. This explanation predicts that the turnover rate of enzymes that control metabolism should be highest, since their activity and therefore their concentration is of key importance in the regulation of metabolism. The greatest sensitivity of control, and therefore the greatest precision provided by a cycle, is achieved when there is a high rate of cycling compared with the net rate of synthesis or degradation (Chapter 3). This is indeed the case (Table 8.5). It is also consistent with the fact that the rate of protein turnover is proportional to the rate of energy expenditure: i.e. a high rate of protein turnover results in a high rate of energy expenditure: i.e. energy expenditure depends in part, on rates of protein turnover.

In answer to question (iii), there are three pathways for protein degradation:

- the lysosomal-autophagic system;

- the ubiquitin-proteasome system;

- the calpain-calpastatin system.

The quantitative importance of each pathway varies from one tissue to another and from one protein to another. Although hydrolysis of the peptide bonds does not involve ATP, the various processes of protein degradation require considerable expenditure of energy, possibly more than is required for protein synthesis. It is not suprising, therefore, that protein turnover contributes at least 20% to resting energy expenditure (basal metabolic rate).

The answer to question (iv) is that although much is known about the control of protein synthesis, very little is known about the processes of degradation.

Table 8.4 Some small nitrogen-containing molecules synthesised from amino acids

Amino acid precursor	Molecule synthesised
Arginine	Creatine, nitric oxide
Aspartate	Purines and pyrimidine nucleotides
Cysteine	Glutathione
	Taurine
Glutamate	Glutathione, N-acetylglutamate, γ-aminobutyrate
Glutamine	Purine and pyrimidine nucleotides, amino sugars
Glycine	Creatine
	Porphyrins (for the synthesis of haemoglobin and cytochromes)
	Purine nucleotides
Histidine	Histamine
Lysine	Carnitine
Methionine	Creatine
	Choline
	Ornithine, putrescine
Proline	–
Serine	Ethanolamine and choline, sphingosine
Tyrosine	Adrenaline
	Noradrenaline
	Dopamine
	Melanin, thyroxine
Tryptophan	5-Hydroxytryptamine (serotonin), melatonin

Table 8.5 Half-lives of enzymes and serum proteins

Source	Enzyme/Protein	Approximate half-life (h)
Cellular	Ornithine decarboxylase[a]	0.2
	Hydroxymethylglutaryl CoA reductase[a]	2
	Glucokinase[a]	12
	Lactate dehydrogenase	140
	Cytochrome *c*	150
Serum	Albumin	480
	Transferrin	210
	Pre-albumin	72
	Retinol-binding protein[b]	12

The half-life is a precise indication of the rate of turnover.

[a] Key regulatory enzymes.

[b] Retinol-binding protein is crucial for binding vitamin A in the serum.

Intracellular protein degradation systems

Lysosomal-autophagic system

Lysosomes are vesicles which contain approximately 40 enzymes capable of hydrolysing all major cell components. A variety of proteases (cathepsins) and peptidases exist so that proteins can be hydrolysed completely to amino acids. The pH within the lysosome is 4.5–5.0 and all lysosomal enzymes exhibit low pH optima. This ensures that, if they leak into the cytosol, their activity is very low and little damage is done. This low pH within the organelle is maintained by a proton pump, driven by the hydrolysis of ATP, thus contributing to the energy requirement of degradation.

Proteins enter the lysosome by three mechanisms:

- Vesicles transport extracellular particles and membrane-proteins into the cell, where they fuse with the lysosomes (endocytosis, see Chapter 5).

- The endoplasmic reticulum engulfs some cytosolic proteins to form vesicles which fuse with the lysosomes.

- The direct uptake of cytosolic proteins by the lysosomes.

> Enzymes hydrolysing peptide bonds within the peptide chain are referred to as proteases (known also as proteinases) while those catalysing the removal of terminal amino acids are known as peptidases. Proteases are grouped into four classes according to the group in the active site: serine proteases (e.g. pancreatic enzymes, complement, and clotting factor enzymes); cysteine proteases (e.g. lysosomal enzymes, calpains); metalloproteases (e.g. collagenase, angiotensin-converting enzyme); and acidic proteases (renin, HIV protease).

The first two are quantitatively the most important. Proteins that are taken up directly by the lysosomes have specific short amino acid sequences which are bound by a 'recognition' protein that transports them to the lysosome. The concentration of the recognition protein increases in starvation and other conditions in which the concentrations of anabolic hormones are low. Thus, protein degradation is stimulated under these conditions.

Ubiquitin-proteasome system

This system is quantitatively the most important process for protein breakdown in mammalian cells. It is so named because it involves the proteolytic enzyme, the proteasome, and the peptide ubiquitin.

The proteasome is a very large complex of at least 50 subunits. It is present in a wide variety of tissues and can constitute up to 1% of soluble protein in a cell. The catalysis occurs within the central core of the molecule and ATP hydrolysis is required to 'drive' the protein into the core. Before the complex can break down proteins, the latter must first have been 'tagged' by complexing with *ubiquitin*, a peptide of molecular mass 8.5 kDa. This attachment requires three enzymes and the hydrolysis of ATP and it results in a peptide link between the carboxylic group at the C-terminus of ubiquitin and the NH_2 of a lysine side-chain in the condemned protein. Several ubiquitin molecules are attached to a single lysine so that a chain of four or more ubiquitin molecules is formed (Figure 8.2). It is,

of course, vitally important that the correct molecules receive this 'kiss of death'. Some of the signals that identify these proteins are known: e.g. those with a predetermined short life have specific amino acids at their N-terminus (e.g. lysine, tryptophan, aspartate) or they have a short specific sequence of amino acids within the overall peptide sequence. Proteins are also ubiquitinated if they become damaged or denatured, since this reveals more lysine side-chains for attachment. In addition, some proteins must be phosphorylated before they can be ubiquitinated. Immediately before the proteolysis by the proteasome, the ubiquitin is removed from the target protein, which also requires hydrolysis of ATP.

> The concentration of ubiquitin increases in injured cells, and any damaged proteins are removed by reaction with ubiquitin. It also increases in trauma, which may be a protective mechanism responding to the increased likelihood of damage to proteins.

Some key intracellular processes, which involve degradation of protein, utilise this system:

- *The cell cycle* The concentrations of specific proteins regulate the cell cycle by activation (known as cyclins) is achieved of cell-division cycle kinases, during key steps in the cycle. The concentration of these proteins (cyclins) is regulated by synthesis and degradation. The latter is this proteolytic system (Chapter 20).

- *Transcription factors* These factors activate the expression of genes. In order to carry out their regulatory function, they must have short half lives. Their degradation is carried out by this system (Chapter 20).

- *Formation of antigens from the intracellular degradation of pathogens* The proteolytic system hydrolyses proteins of pathogens that are present within the host cell (e.g. a virus), to produce a short peptide which forms a complex with a specific protein, known as the major histocompatibility complex (MHC) protein. The peptide is, in fact, the antigen. At the plasma membrane, the MHC protein locates within the membrane and the small peptide sits on the outside of the membrane, where it can interact with the receptor on a cytotoxic T-lymphocyte to kill the host cell and the virus (Chapter 17).

- *Protein processing in the endoplasmic reticulum makes mistakes.* All membrane-associated proteins and proteins that are secreted by the cell are synthesised on membrane-bound ribosomes and pass into the lumen of the reticulum, where they are modified by post-translational processes, so that much biochemical manipulation of the proteins takes place. Consequentially mistakes are often made. Such abnormal proteins are exported from the lumen into the cytosol for ubiquitination and degradation in the proteasome.

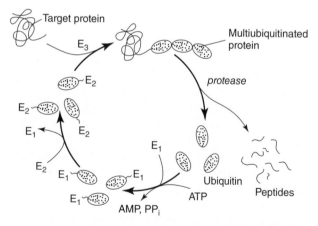

Figure 8.2 *The tagging of a protein molecule with ubiquitin to identify it for degradation.* Three enzymes are involved: (i) E_1 is activated by binding to ubiquitin (ii) E_2 displaces E_1 (iii) E_3 catalyses the transfer of ubiquitin from E_2 to the 'doomed' protein: more than one molecule is transferred to produce multi-ubiquitinated protein, which is the ideal substrate for the proteolytic enzyme.

The calpain-calpastatin system

Calpains are enzymes that consist of a proteolytic subunit and a calcium binding subunit. In the cytosol, these enzymes are inactive due to binding of the inhibitory protein, calpastatin. Attachment to the cell membrane removes this inhibition and activation occurs at low concentrations of Ca^{2+} ions. The enzymes hydrolyse proteins as far as peptides: complete hydrolysis requires peptidases, which are also present in the cytosol.

Regulation of degradation

Very little is known of the mechanism(s) by which the rate of degradation is controlled. The current view is that the concentrations of ubiquitin, together with changes in the activity of the proteasome complex control the rate of proteolysis by this system. Lysosomal degradation may be controlled by the number of particles transported into the cell. The calpains might be controlled by the Ca^{2+} ion concentration.

Defects in protein degradation and diseases

If a protein is lost it must be replaced if health is to be maintained: excessive loss contributes to death, for example during prolonged starvation or cachexia. The cause of death is usually infection, due to an impaired immune system, or heart failure (Chapters 16 and 22). In children, poor nutrition and infections are factors that result in the nutritional disorder known as protein-energy malnutrition. In the elderly, trauma, major surgery or poor nutrition can result in loss of so much skeletal muscle that normal daily activities (e.g. walking, climbing stairs, dressing and, in severe cases, even getting out of a chair or up from a toilet seat) are impaired (Chapters 15 and 18).

The importance of the proteasomal-ubiquitin system in the degradation of cellular proteins or proteins of pathogens suggests that any defects in this system could result in disease. Indeed, one possible cause of some neurodegenerative diseases, such as Alzheimer's, Parkinson's, motor neurone disease and spongiform encephalopathies, is the accumulation of insoluble proteins which form aggregates in neurones. This could be caused by impaired proteolytic digestion of the proteins prior to aggregation (Chapter 14).

Failure to control the rate of degradation of cyclins could lead to their over-expression, increasing the risk of tumour development.

Muscular dystrophies are characterised by variable degrees of muscle weakness and degeneration. The most common forms are Duchenne (severe) and Becker (benign). Both are caused by a mutation in a gene on the X chromosome.

In Duchenne, the first symptom is muscle weakness in the early years of life which gradually worsens so that patients are unable to walk by the age of 10 years. Death from cardiac or respiratory insufficiency usually occurs before the age of 25. In Becker type, weakness and wasting becomes apparent between 5 and 25 years but, although severely disabled, patients can survive to a normal age.

In 1987 the gene responsible for muscular dystrophy was identified, leading to the isolation of a protein, known as dystrophin, which is either totally absent in Duchenne, or partially absent in the Becker type. The protein is located on the inside of the plasma membrane of all muscles (and some neurones). Although its precise function is not known, the mutant form results in structural abnormalities of the plasma member which results in degradation of myofibrils, but the link between the abnormalities of the membrane and degradation is not known. One theory is that it leads to an increase in the activity of a Ca^{2+} ion channel in the membrane and, therefore, a marked increase in the Ca^{2+} ion concentration in the cytosol. This chronic elevation results in the activation of calpain, which leads to protein breakdown and the degeneration within the fibre (Chapter 13).

Protein and amino acid requirements

The presence of proteins in the diet is essential for health. An important question, therefore, is what is the minimal amount of protein that must be provided to maintain health? It is not an easy question to answer. Even when no protein or amino acid is consumed, in an otherwise adequate diet, urea is lost from the body due to body protein break down. The daily loss of protein is about 0.34 g per kg or about 24 g protein each day for a 70 kg person (i.e. when no protein is consumed). However, this amount does not represent the minimal intake required, since other factors, (such as the amount of energy consumed, other components in the diet, and trauma physical activity can affect this amount.) The recommended dietary allowance (RDA) for a young adult is 0.8 g per kg per day (Table 8.6).

Essential and non-essential (indispensable and dispensable) amino acids

For protein synthesis to take place, *all* the amino acids must be available within the cell. To this extent, all amino

Table 8.6 Recommended daily intake of high-quality protein[a] for infants, children, teenagers and adults

Age (years)	Body weight (kg)	Recommended Dietary Allowance of protein (g/kg per day)	
0–0.5	6	2.2	
0.5–1	9	1.6	
1–3	13	1.2	
4–6	20	1.1	
7–10	36	1.0	
		Males	Females
11–14	–	1.0	1.0
15–18	–	0.9	0.8
19+	–	0.8	0.8

Data from Food and Nutrition Board, Institute of Medicine. *Dietary proteins and amino acids. Reference Intakes for Energy, etc.* Washington, DC: National Academy Press (2002) **10:** 1–143 (see also Matthews (2006)).

[a]The term 'high quality' has a number of different interpretations but for human nutrition it usually means protein of animal origin, especially milk or whole egg (see Appendix 8.1).

Vegetarians need to be aware of the amino acids present in their diet, since most animal proteins contain amino acids approximately in proportion to those required by humans but this is not true for all vegetable proteins. This problem is particularly severe for those dependent on a single source of plant protein, for example corn or rice, as is frequently the case in poorer parts of the world. In general, legumes are low in methionine while cereals are low in lysine. Some strains of corn now contain lysine, but the best advice to vegetarians is to include as wide a variety of plants as possible in their diet (Chapter 15).

acids are essential. However, studies in which animals have been fed proteins of known amino acid composition or pure amino acids, have demonstrated that some amino acids are essential for maintaining nitrogen balance and growth, whereas others are not essential. Such work led to the simple division of the amino acids into essential and non-essential. The terms indispensable and dispensable amino acids are also used, but there is now an important difference in their meaning. In biochemical terms, non-essential amino acids are defined as those for which a synthetic pathway is present in the body, which is of sufficient capacity to satisfy the normal requirement. Essential amino acids are defined as those for which a synthetic pathway is not present. In nutritional terms, dispensable amino acids are defined as those that can be excluded from the diet without affecting nitrogen balance, whereas indispensable amino acids are defined as those which, when excluded from the diet, result in negative nitrogen balance. Difficulties arise, however, since some amino acids that are normally dispensable can, under certain conditions, become indispensable (see below).

There are several means of determining whether an amino acid is indispensable, although results vary according to the method used. The nitrogen balance method has been used most frequently. In a normal healthy adult, if the intake of nitrogen in the diet is equal to the loss from the body (in faeces, urine and skin), this is termed nitrogen balance. If more nitrogen is ingested than lost (e.g. during periods of growth or tissue repair) this is termed positive nitrogen balance. In contrast, during malnutrition, starvation or in some diseases, less nitrogen is ingested than lost: this is termed negative nitrogen balance.

To determine whether an amino acid is indispensable, it is omitted from the diet while all the other amino acids are included. If the omission results in negative nitrogen balance, the amino acid is deemed indispensable. If, because of the absence of this single amino acid, the body has been unable to synthesise proteins, the nitrogen which would have been used in this synthesis is excreted. By this method, the following amino acids are considered to be indispensable for humans: isoleucine, leucine, lysine, methionine, phenylalanine, threonine, tryptophan and valine. Histidine occupies a grey area. To obtain accurate data for this amino acid, nitrogen balance studies would have to be carried out over long periods and such extended studies are not ethical. More sensitive experiments to determine the essential nature of amino acids in humans are performed by omitting particular amino acids from the diet for short periods of time and determining how this affects the concentrations of plasma proteins that turn over rapidly (Table 8.5). These include albumin, prealbumin, retinol-binding protein and transferrin: their concentrations drop rapidly if the rate of synthesis is decreased. Such studies have shown that histidine can be added to the list of essential amino acids. The daily requirements of all these amino acids are given in Table 8.7.

The non-essential amino acids are alanine, arginine, aspartate, asparagine, cysteine, glutamate, glutamine, glycine, proline, serine and tyrosine. A summary of the reactions involved in their synthesis is given in Figure 8.3 and full details of these pathways are provided in Appendix 8.2.

Conditionally essential amino acids

The distinction between essential and non-essential amino acids is not always clear-cut; some amino acids become essential only under certain conditions; otherwise they are classified as non-essential. Examples are as follows:

- The rate of synthesis of some amino acids that are normally considered to be indispensable is not sufficient under conditions when the demand for them is increased (e.g. after severe trauma, major surgery or during sepsis). These amino acids include glutamine, cysteine and,

Table 8.7 Recommended daily intake of essential amino acids for adults, children and infants

	Daily requirement of amino acids (mg/kg per day)			
Amino acid	Adults	Child (1–3 y)	Child (4–13 y)	Infant (7–12 months)
Histidine	14[a]	21	16[a]	32
Isoleucine	19 (23)	28	22	43
Leucine	42 (40)	62	48	96
Lysine	38 (30)	58	45	89
Methionine plus cysteine	19 (13)	28	22	43
Phenylalanine plus tyrosine	33 (39)	54	41	84
Threonine	20 (15)	32	24	49
Tryptophan	5 (6)	8	6	13
Valine	24 (20)	37	28	58

Data from Matthews (2006).

In adults, data are also presented in parentheses: these are reassessments of the requirements by Young & Borgonha (2000).

[a] Although histidine may not be needed beyond infancy, histidine has been recommended for children and adults based on a calculation of the histidine content of protein and the recommended dietary allowance for protein, for each of these age groups.

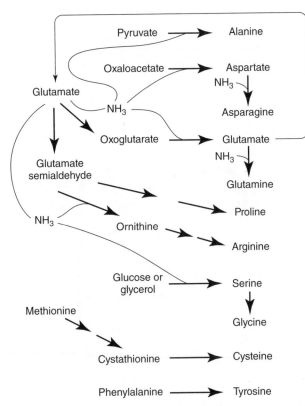

Figure 8.3 *A summary of pathways involved in the synthesis of non-essential amino acids.* Glutamate is produced from ammonia and oxoglutarate. Glutamate is the source of nitrogen for synthesis of most of the amino acids. Cysteine and tyrosine are different because they require the essential amino acids (methionine and phenylyalanine) for their synthesis. These two amino acids are, therefore, conditionally essential, i.e. when there is not sufficient methionine or phenylyalanine for their synthesis, they are essential (Details are in Appendix 8.2).

possibly, glycine and arginine. The essential nature of these amino acids is, therefore, *conditional*. The provision of nutrients for patients recovering from trauma, etc., should take into account this increased requirement.

- A problem with tyrosine and cysteine is that they can be synthesised from other amino acids, phenylalanine and methionine, respectively. Since both phenylalanine and methionine are indispensable amino acids, if they are not present in the diet at or below minimal requirement, then there is not sufficient to synthesise tyrosine or cysteine and, therefore, these amino acids become indispensable, i.e. *conditionally* essential.

Fate of amino acids

There are three major fates of amino acids:

- *Synthesis of new proteins for growth or repair.* This is discussed in Chapter 20. The rate of protein synthesis is a major factor determining the overall rate of amino acid metabolism: the higher the rate of synthesis, the lower is the amino acid concentration which reduces the rate of catabolism.

- *Synthesis of a range of nitrogen-containing small compounds.* These compounds are identified in Table 8.4.

- *Catabolism.* This results, eventually, in formation of ammonia and small carbon-containing compounds. The carbon skeletons are used for the synthesis of glucose and triacylglycerol or for complete oxidation to CO_2,

---

Box 8.1 Taurine: bile, brain, cats, heart, milk, sharks, shrimps and skeletal and cardiac muscle

Taurine ($H_3N^+CH_2CH_2SO_3^-$) is formed as a product of cysteine catabolism and also arises from the oxidation of cysteamine, which is produced during coenzyme-A degradation. It was given the name taurine because it was first isolated from the bile of the ox, *Bos taurus*.

The high concentration of taurine in cells is an indication that it is an important molecule but not all functions or their importance are known. Cell membranes are impermeable to taurine so that it must be formed in the cell, within which its concentration is very high. The concentration ratio across the membrane is also very high in cells in the retina (400), neurones in the brain (500) and in some tumour cells (7000). Unfortunately, the precise role of taurine in these particular cells is not known. Of the many functions that are known (Huxtable, 1992; Schuller–Levis & Park, 2003) the following is a summary:

- Bile salts consist of taurine linked to bile acids. The salts are essential for digestion and absorption of fat and also of fat-soluble vitamins and cholesterol.
- The maintenance of the structure and function of photoreceptors in the eye depends on taurine. As an antioxidant, it may protect the cell membranes in the retina.

- It is a neuromodulator in the brain.
- It is present at a high concentration in human milk, presumably to ensure that it is available for the developing brain, but its concentration is very low in cow's milk. It is added to the feeding solutions used for premature babies and neonates.
- It is present at very high levels in skeletal and heart muscle (Table 8.2). It is also present at high levels in immune cells.
- Taurine deficiency is rare in adult humans but is common in domestic cats, due to poor absorption from tinned catfood. Consequences of taurine deficiency in cats are cardiomyopathy, retinal degradation, reproductive failure in females, developmental abnormalities and impairment of the immune system. It is possible that a chronic deficiency in humans may have similar effects.
- It is an ideal compound for regulation of osmotic pressure, since it has a low molecular mass, is highly soluble and has no net charge. It serves this function in some of the tissues of elasmobranch fish such as the skate and shark, and in marine invertebrates. Any damage to these tissues releases taurine, which is used as a chemoattractant for predators such as the shrimp, which will attack small fish.

---

with the generation of ATP. The ammonia is converted to urea (Figure 8.4).

Transport of amino acids into the cell

Amino acid metabolism occurs within the cell but, before this can occur, the amino acids must be transported across the plasma membrane. This requires transport proteins, which have three important characteristics:

- Since the intracellular concentration of most amino acids is considerably greater than that in the plasma (Table 8.2), the transport of these amino acids is an energy-requiring process. This is achieved via the Na^+ gradient across the plasma membrane, which is maintained by the ATP-dependent Na^+ pump the Na^+/K^+ ATPase (Figure 8.5). This is similar to that of the transport of glucose across the luminal membranes of epithelial cells in the gut and in the tubules is the kidney cortex.

- There are ten transporters so that some transport more than one amino acid (Table 8.8).

- The properties of some of the transporters are different in different tissues, but this will not be discussed further here.

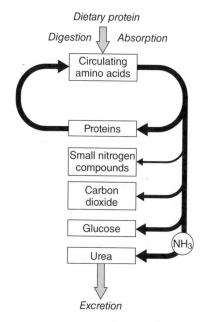

Figure 8.4 *General summary of metabolism of amino acids.* Amino acids in blood can be derived from the diet or hydrolysis of endogenous protein. The nitrogen in the amino acids can be used to synthesise other nitrogen-containing compounds (e.g. glutamine – see Table 8.4) or removed as urea (Chapter 10). The amino acids are also used to synthesise proteins or peptides. The carbon can be converted to CO_2, glucose or triacylglycerol, but, in humans, very little is converted into fat, so triacylglycerol is omitted from the figure.

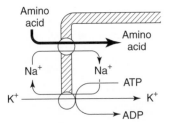

Figure 8.5 *Active amino acid transport into cells.* Amino acids are transported into cells against their concentration gradient coupled to Na$^+$ ion transport down its concentration gradient. The Na$^+$ ion is transported out in exchange for K$^+$-ions, via the Na$^+$/K$^+$ ATPase, Chapter 5).

General principles in amino acid catabolism

Liver, small intestine, muscle and kidney all participate in amino acid catabolism with the liver, under most conditions, playing the major role, but the metabolism of specific amino acids in the other three tissues is of considerable biochemical and physiological importance (see below).

There are three general processes by which amino acids are catabolised:

(i)   by specific catabolic pathways;

(ii) by conversion to other amino acids, which are then catabolised by specific pathways;

(iii) by a combination of transamination plus deamination, known as transdeamination.

The amino acids falling into each category are presented in Table 8.9. The pathways in processes (i) and (ii) are described in Appendix 8.3. The combined process of deamination plus transamination illustrates important principles in amino acid catabolism. These lead to an appreciation of some of the biochemical and physiological functions of amino acids that are important in health and disease.

Deamination

A major 'aim' of amino acid catabolism is removal of the α-NH$_2$ group, which results in the formation of ammonia which is then converted to urea. The removal of the α-NH$_2$ group for most amino acids results in the formation of a carbon-compound, which is usually an oxoacid (e.g. the oxoacid for alanine is pyruvate).

In metabolic reactions, oxidations usually occur by removal of hydrogen: a hypothetical reaction for an amino acid is:

$$\text{amino acid} + X \rightarrow \text{oxoacid} + NH_3 + XH_2$$

where X is the hydrogen acceptor (i.e. oxidising agent). The usual acceptor molecules are either the flavin nucleo-

Table 8.8 Amino acid transport systems in the cell membranes of major tissues

Name of system	Preferred amino acids	Na$^+$-linked
A	Alanine, glycine, proline, serine and others	✓
ASC	Alanine, cysteine, proline, serine, threonine	✗
L	Isoleucine, leucine, methionine, phenylalanine, tryptophan, tyrosine, valine	✗
y$^+$	Arginine, histidine, lysine, ornithine	✗
AG	Glutamate, aspartate	-
Dicarboxylate	Aspartate, glutamate	✓
β	β-Alanine, taurine	✓
N	Asparagine, glutamine, histidine	✓
N^m	Glutamine, asparagine (muscle only)	-
Gly	Glycine, sarcosine	✓

✓ Indicates transport is linked to Na$^+$ ion.

✗ Indicates transport is not linked to Na$^+$ ion.

- Not known.

β-Alanine is a component of enzyme A.

Taurine is discussed in Box 8.1.

tides (FAD or FMN) or the nicotinamide adenine nucleotides (NAD$^+$ or NADP$^+$) (Chapter 9). Despite the fact that there are 21 amino acids, there are only three reactions which use these acceptors. The three reactions are:

• The oxidation of D-amino acids, catalysed by D amino acid oxidase.

• The oxidation of proline, catalysed by proline oxidase.

• The dehydrogenation (i.e. oxidation) of glutamate catalysed by glutamate dehydrogenose (Figure 8.6).

Glutamate dehydrogenation is involved in deamination of most of the amino acids. The first two reactions are not involved in the overall deamination system; they are included here for completeness and because they are of some general interest. The complete biochemical description of these reactions is given in Appendix 8.4. For a few amino acids, e.g. threonine and serine, other specific reactions are responsible for deamination.

D-amino acid oxidase

Although free amino acids and those in proteins in eukaryotes are entirely of the L-form (except glycine, which is not optically active), D-amino acids do occur in nature, for example in bacterial cell walls (D-alanine and D-glutamate). Consequently, they enter the body from bacteria in food and from the digestion of bacteria in the

Table 8.9 Processes involved in catabolism of amino acids

General catabolic process	Amino acid	Nitrogen end product
Amino acids that are converted to other amino acids	Arginine	glutamate
	Asparagine	aspartate
	Glutamine	glutamate
	Histidine	glutamine
	Phenylalanine	tyrosine
	Proline	glutamate
	Serine	glycine
A specific pathway for each amino acid	Glycine	ammonia
	Lysine	glutamate
	Methionine	ammonia
	(Serine)[b]	ammonia
	Threonine	ammonia
	Tryptophan	ammonia
Transamination/deamination	Alanine	glutamate
	Aspartate	glutamate
	Isoleucine[a]	glutamate
	Leucine[a]	glutamate
	Ornithine[a]	glutamate
	Serine	alanine
	Tyrosine[a]	glutamate
	Valine[a]	glutamate

[a] For all these amino acids, the first reaction involves transamination followed by a specific catabolic pathway.

[b] There are several catabolic pathways for serine (Appendix 8.3).

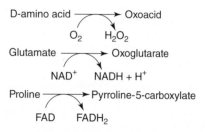

Figure 8.6 *The three dehydrogenase (oxidase) reactions in amino acid degradation.* The enzymes are D-amino acid oxidase, glutamate dehydrogenase and proline oxidase (dehydrogenase). Biochemical details are given in Appendix 8.4.

intestine. It is the role of the D-amino acid oxidase to metabolise them. FAD is the hydrogen acceptor, as follows:

$$\text{D-amino acid} + \text{FAD} \rightarrow \text{oxoacid} + \text{NH}_3 + \text{FADH}_2$$

The FADH$_2$ then reacts with molecular oxygen, a reaction also catalysed by the oxidase:

$$\text{FADH}_2 + \text{O}_2 \rightarrow \text{FAD} + \text{H}_2\text{O}_2$$

The H$_2$O$_2$ is decomposed by the enzyme catalase which occurs in the peroxisomes:

$$2\text{H}_2\text{O}_2 \rightarrow 2\text{H}_2\text{O} + \text{O}_2$$

Hence, the overall reaction is:

$$\text{D-amino acid} + \text{O}_2 \rightarrow \text{oxoacid} + \text{NH}_3$$

The oxoacid is then metabolised in the usual way (see below). High activities of the oxidase are found in liver and kidney where its function is to remove rapidly (i.e. detoxify) the D-amino acids. Failure to metabolise these amino acids could lead to their accumulation in cells with the danger of osmotic effects or interference in the metabolism of L-amino acids.

Proline oxidase (proline dehydrogenase)

Proline is oxidised to form pyrroline-5-carboxylate, and FAD is the hydrogen acceptor. The pyrroline-5-carboxylate is converted, via several reactions, to glutamate:

proline $\longrightarrow$ pyrroline-5-carboxylate $\longrightarrow$ $\longrightarrow$ $\longrightarrow$ glutamate
        FAD   FADH$_2$

The enzyme is present in the mitochondria so that the FADH$_2$ is oxidised by the electron transfer chain. The enzyme is worthy of note since it catalyses the initiating reaction for the major pathway that generates ATP in the flight muscle of two insects that are of some interest to humans: the tsetse fly and the Colorado beetle. Proline is the major, if not the only, fuel for flight in these insects. Both show high activities of proline oxidase in their flight muscles. In addition all other beetles, so far investigated, have high activities of proline oxidase and use proline as a fuel for flight (Box 8.2). Why these insects use proline rather than the more usual fuels, i.e. glucose trehalose or fatty acids, is not known.

Glutamate dehydrogenase

The amino acid glutamate is deaminated in a reaction catalysed by the enzyme glutamate dehydrogenase, using either NAD^+ or $NADP^+$ as the oxidising agent, as follows:

$$glutamate + NAD^+ \rightarrow oxoglutarate + NADH + NH_3$$

The biochemical details of this reaction are given in Appendix 8.4.

Glutamate dehydrogenase, in combination with the process of transamination, plays the major role in deamination of many amino acids.

Transamination

Transamination is a process in which the αNH_2 group of an amino acid is removed and the oxoacid is formed. However, the $\alpha\text{-}NH_2$ group is not lost: it is transferred to the oxoacid of another amino acid:

$$amino\ acid_1 + oxoacid_2 \rightarrow oxoacid_1 + amino\ acid_2$$

For many amino acids, the accepting oxoacid (i.e. $oxoacid_2$) is oxoglutarate, so that transamination results in the formation of glutamate:

$$amino\ acid_1 + oxoglutarate \rightarrow oxoacid_1 + glutamate$$

Three examples are:

(i)   aspartate + oxoglutarate $\rightarrow$ oxaloacetate + glutamate

(ii) alanine + oxoglutarate $\rightarrow$ pyruvate + glutamate

(iii) leucine + oxoglutarate $\rightarrow$ oxoisocaproate + glutamate

The reactions are catalysed by enzymes known as aminotransferases (formerly known as transaminases). For the above reactions, they are (i) aspartate aminotransferase, (ii) alanine aminotransferase and (iii) leucine aminotransferase. Details of these reactions can be found in Appendix 8.4.

Transamination reactions can also occur within, rather than at the beginning of, the pathway of catabolism, so that the substrate for the aminotransferase is a partially degraded amino acid (Appendix 8.3 contains examples of such pathways).

Two questions now arise: (i) what is the fate of the oxoacid, e.g. what is the fate of oxaloacetate, pyruvate and oxoisocaproate, in the above examples, and (ii) what is the fate of the glutamate?

> Aminotransferases have received many names as fashions in nomenclature have changed. Two obsolete names are still used in clinical practice: glutamate-oxaloacetate transaminase (abbreviated to GOT) is now aspartate aminotransferase, and glutamate-pyruvate transaminase (GPT) is now alanine aminotransferase. The new abbreviations are AST and ALT, respectively.

Fate of the oxoacid

The fate of the oxoacid is either (i) formation of a common intermediate of metabolism, i.e. an intermediate within a well-established metabolic pathway (e.g. oxaloacetate or pyruvate, in the above examples), or (ii) conversion to a 'common intermediate', e.g. oxoisocaproate is converted to acetyl-CoA (see Appendix 8.3).

Although amino acid catabolism appears complex, there are two simple but important points (principles) that help in understanding the overall plan:

- The catabolism of all 21 amino acids gives rise to only *six* common intermediates: acetyl-CoA, pyruvate, oxoglutarate, succinyl-CoA, fumarate or oxaloacetate (Table 8.10).

- From these six intermediates, only *three* end-products are produced: carbon dioxide, glucose or fat, and the first two are quantitatively the most important. A simple overview of these processes is given in Figure 8.7. (The details of the reactions that convert the individual amino acids to one of the intermediates are given in Appendix 8.3.)

The processes that give rise to these three end-products are:

- Oxidation of acetyl-CoA through the complete Krebs cycle to produce CO_2 (Chapter 9).

- Gluconeogenesis to form glucose (or glycogen) (Chapter 6).

- Synthesis of fatty acid and thence triacylglycerol, via the fatty acid synthesis and the esterification pathways (Chapter 11).

Oxidation to carbon dioxide Since the only compound for which the carbon-skeleton can be fully oxidised to CO_2, by the Krebs cycle, is acetate (in the form of acetyl-CoA), all of the common intermediates that are fully oxidised to CO_2 must first be converted to acetyl-CoA. Thus, intermediates that enter the cycle at positions other than acetyl-CoA are converted to malate, which then leaves the

Box 8.2 Proline: A fuel for flight in the tsetse fly and beetles

The tsetse fly is a blood-sucking insect which transmits the parasite causing sleeping sickness in humans and cattle. A study of the maximum enzyme activities in the flight muscle showed that the activity of citrate synthase, the first enzyme of the complete Krebs cycle, is lower in the flight muscle of the tsetse than in other insects (Table B2(i)). It could be inferred from this observation that the Krebs cycle is unimportant for ATP generation. However, there are two major points of entry of substrates into the Krebs cycle: at the level of citrate, which is where acetyl-CoA feeds in for complete oxidation, and at the level of oxoglutarate, where other fuels such as glutamine feed into the cycle. In the tsetse fly and beetles, proline feeds into the cycle after conversion to oxoglutarate. As is the case for glutamine in immune and tumour cells (Chapters 17 and 21), the pathway for proline oxidation uses part of the Krebs cycle between oxoglutarate and malate (see below, Figure B2(i)): proline is converted to malate, and then to pyruvate which is transaminated to form alanine. The activity of proline dehydrogenase (also known as proline oxidase) in the flight muscle of the tsetse and the Colorado beetle is high by comparison with that in the flight muscle of other insects (Table B2(i) below).

For each alanine molecule formed from proline in this pathway, 12 molecules of ATP are generated from ADP. What happens to the alanine? It is released from the muscle into the blood (haemolymph) and is then taken up by the fat body, which is equivalent to the liver of higher animals. The alanine is here converted back to proline using ATP provided by fat oxidation. In order to fuel this pathway, the concentration of proline in the haemolymph is very high, about 140 mM! This quantity can generate enough ATP to support flight for about 30 minutes. It is calculated that this would permit flight for a distance of about 1–2 km. After such a flight, the tsetse has to rest whilst the proline is resynthesised from alanine. Hence, clearing of bush by 1–2 km swathes in Africa means that the insect has to rest on the ground, increasing the opportunity for it to enter traps containing insecticide.

So why are tsetse flies and beetles different from other insects? No one has come up with a satisfactory answer. What is surprising is that the pathway for ATP generation in these insects is almost identical to that which generates ATP for tumour cells, except that glutamine rather than proline is the fuel. The interesting question is whether these insects can provide clues as to why tumour cells prefer glutamine and use only part of the Krebs cycle rather than the usual complete cycle. It would not be the first time that the study of insects has provided novel biochemical and physiological information relevant to higher animals (e.g. insects were the first animals in which it was shown that hormones stimulate gene expression).

Table B2(i) Activities of some enzymes in insect flight muscles

Insect	Enzyme activity (μmol of product formed/min per g wet wt. of muscle at 25 °C)		
	Proline dehydrogenase	**Alanine aminotransferase**	**Citrate synthase**
Tsetse fly	40	402	74
Giant waterbug	0.5	4.5	244
Honey bee	1.5	7.2	346
Bumblebee	2.9	8.0	382
Poplar hawk moth	2.4	7.5	430
Fleshfly	3.0	45	345
Blowfly	2.0	40	418

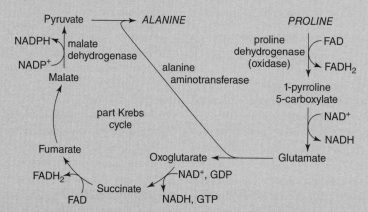

Figure B2(i) *The pathway for conversion of proline and alanine in the flight muscle of the tsetse fly: the major ATP-generating pathway*. Alanine aminotransferase is essential for the proline oxidation pathway in order for glutamate to enter the Krebs cycle as oxoglutarate and pyruvate to be converted to alanine, the end of the pathway. It is assumed that the pathway is the same for the Colorado beetle, but no studies have been reported.

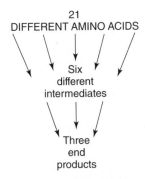

Figure 8.7 *Simplified summary of degradation of 21 amino acids.* The 21 amino acids are degraded to six different common intermediates that are further metabolised to produce only three end-products (CO_2, glucose or fat).

Table 8.10 The six common metabolic intermediates derived from amino acids via the catabolism of their oxoacids

Amino acid source	The six common intermediates
Alanine, glycine, serine, cysteine, tryptophan[a]	pyruvate
Arginine, histidine, proline, glutamine, glutamate	oxoglutarate
Valine, isoleucine,[a] methionine, threonine	succinyl-CoA
Phenylalanine,[a] tyrosine[a]	fumarate
Asparagine, aspartate	oxaloacetate
Leucine, phenylalanine, tyrosine, lysine, tryptophan, isoleucine	acetyl-CoA[b]

[a]Also give rise to acetyl-CoA.

[b]Acetyl-CoA is formed directly from these amino acids, that is, not via pyruvate and pyruvate dehydrogenase.

cycle and is converted to acetyl-CoA, for the complete oxidation. The sequence of the five reactions by which this occurs is as follows:

(i) The conversion of the intermediate to malate via the reactions of the Krebs cycle.

(ii) The transport of malate across the mitochondrial membrane into the cytosol.

(iii) The conversion of cytosolic malate to oxaloacetate, which is then converted to phosphoenolpyruvate or to pyruvate in the reactions catalysed by phosphoenolpyruvate carboxykinase or the malic enzyme, respectively:

$$\text{malate} + NAD^+ \rightarrow \text{oxaloacetate} + NADH$$

$$\text{oxaloacetate} + GTP \rightarrow \text{phosphoenolpyruvate} + GDP + CO_2$$

$$\text{malate} + NADP^+ \rightarrow \text{pyruvate} + NADPH + CO_2$$

(iv) The conversion of the phosphoenolpyruvate (PEP) to pyruvate by the action of pyruvate kinase (as in glycolysis).

$$PEP + ADP \rightarrow \text{pyruvate} + ATP$$

(v) The transport of the pyruvate into the mitochondrion followed by its conversion to acetyl-CoA by pyruvate dehydrogenase.

These processes, except for transport across the mitochondrial membrane, are summarised in Figure 8.8.

Oxidation plus gluconeogenesis Most but not all amino acids are catabolised in the liver. Perhaps somewhat surprisingly, the amount of ATP generated from the oxidation of the amino acids derived from ingested protein on a normal Western diet usually exceeds the amount of ATP

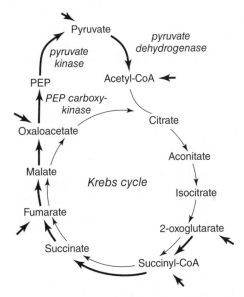

Figure 8.8 *The role of the Krebs cycle in the oxidation of the six common intermediates.* The six short arrows indicate the positions in the cycle where the various intermediates from amino acid catabolism feed into the cycle. Eventually they are all converted to acetyl-CoA for complete oxidation by the cycle. The pathway is indicated by the broader arrows.

that is normally used by the liver, under normal conditions. Consequently, the rate of the catabolic process could be restricted by the rate of utilisation of ATP, which would result in accumulation of oxoacids in the liver. The problem is overcome by increasing the rate of a process that requires ATP hydrolysis and, in addition, also utilises the oxoacids: this process is formation of glucose from the oxoacids (i. e. gluconeogenesis). In fact, the oxidation of approximately

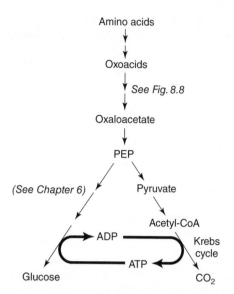

Figure 8.9 *The ATP produced from oxidation of half of the amino acids is used to synthesise glucose from the other half of the amino acids.* This is a general metabolic point, it does not apply in all conditions: the ATP can be used in other processes (e.g. urea cycle). In starvation, ATP is generated from fat oxidation, since oxoacids are not oxidised but are converted to glucose.

> It is important to note that glucose can be synthesised from all the common intermediates *except* acetyl-CoA. Hence, all amino acids except two, leucine and lysine, can give rise to the net synthesis of glucose (Table 8.10).

half of the oxoacids generates sufficient ATP to convert the other half to glucose (Figure 8.9).

The situation is, however, different in starvation. In this condition, it is the degradation of muscle protein that provides the amino acids for gluconeogenesis, so that all the oxo-acids generated (except those for lysine and leucine) are used to synthesise the glucose required for oxidation by the brain. Hence, a process other than amino acid oxidation must generate the ATP required by gluconeogenesis. This process is fatty acid oxidation.

Fat synthesis   The acetyl-CoA produced from amino acid catabolism is also a precursor for fatty acid and triacylglycerol synthesis, both in adipose tissue and liver (details of pathways are given in Chapter 11). Unfortunately, the quantitative significance of this pathway is not known. It is likely to be variable and probably small in humans.

Fate of glutamate

In transamination, the acceptor oxoacid is always oxoglutarate, which is converted to glutamate:

amino acid + oxoglutarate → glutamate + oxoacid

The fate of the glutamate is re-formation of oxoglutarate in the deamination reaction catalysed by glutamate dehydrogenase, in which the NH_2 group in glutamate is removed as ammonia and the oxoglutarate is formed, as follows:

glutamate + NAD^+ → NADH + oxoglutarate + NH_3

Full details of this reaction are given in Appendix 8.4.

Transdeamination

The net result of all these reactions is deamination of the amino acid, as follows:

amino acid + NAD^+ → oxoacid + NADH + NH_4^+

This combination of reactions is known as transdeamination and is the mechanism for deamination of a number of amino acids (Table 8.9). The role of this process in catabolism is shown in Figure 8.10. The ammonia that is produced is converted, almost exclusively, to urea for excretion. Because of the biochemical and clinical significance of ammonia, a whole chapter is devoted to it and to urea formation.

As might be expected, there are a number of exceptions to this general principle:

- For some amino acids, the transamination reaction occurs during the course of, rather than at the beginning of, the catabolic pathway (Appendix 8.3).

- For some reactions, the process of transamination is near-equilibrium. This means that amino acids involved can be synthesised from their oxoacid by transfer of the α-NH_2 group. There are five amino acids in this group:

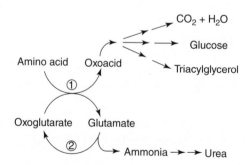

Figure 8.10 *A summary of the processes involved in transdeamination.* (1) transamination (2) glutamate dehydrogenase. The glutamate dehydrogenase reaction results in ammonia production. The oxoacids are further metabolised to CO_2, glucose or fat.

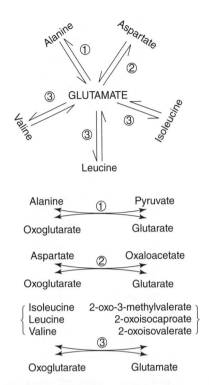

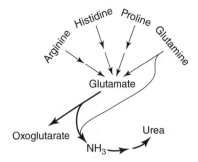

Figure 8.12 *Degradation of arginine, histidine, proline and glutamine.* These amino acids are all converted to glutamate which can then be degraded to produce oxoglutarate and ammonia. For details, see Appendix 8.3.

Figure 8.11 *Five near-equilibrium reactions involved in transamination of five different amino acids.* Three enzymes are involved in these reactions: (1) alanine aminotransferase (2) aspartate aminotransferase (3) branched-chain amino acid aminotransferase, i.e. one enzyme catalyses the three reactions. (The branched-chain amino acids are essential.)

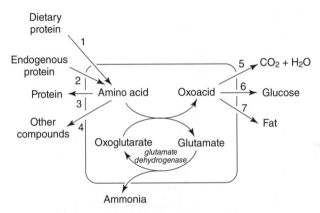

Figure 8.13 *The central role of transdeamination in metabolism of amino acids and further metabolism of the oxoacids in the liver.* The box contains the reactions for conversion of the amino acids to their respective oxoacids. Processes are as follows: (1) digestion of protein in the intestine and absorption of resultant amino acids, (2) degradation of endogenous protein to amino acids (primarily but not exclusively muscle protein), (3) protein synthesis, (4) conversion of amino acid to other nitrogen-containing compounds (see Table 8.4), (5) oxidation to CO_2, (6) conversion to glucose via gluconeogenesis, (7) conversion to fat.

alanine, aspartate and the three branched-chain amino acids, isoleucine, leucine and valine (Figure 8.11). The biochemical importance of this is discussed below.

- Some amino acids are converted to glutamate prior to deamination: these are proline, arginine, histidine and glutamine (Figure 8.12).

Central role of transdeamination

The physiological relevance together with clinical importance of transamination and deamination is wide-ranging. As an aid to understanding the somewhat complex nature of amino acid metabolism, it can be considered (or imagined) as a metabolic 'box' (represented in Figure 8.13). Some pathways feed oxoacids into the box whereas others remove oxoacids; and the ammonia that is released is removed to form urea. The box illustrates the role of transdeamination as central to a considerable amount of the overall metabolism in the liver cell (i.e. protein, carbohydrate and fat metabolism, see below).

There are at least seven pathways that feed into or out of the transdeamination box: these are shown in Figure 8.13. They are:

1. amino acids from dietary protein;

2. amino acids from body protein;

3. amino acids for protein synthesis;

4. amino acids that provide nitrogen and/or carbon for the synthesis of other nitrogen-containing compounds; small compounds, e.g. glutamine and larger compounds (e.g. peptides);

5. oxidation of the oxoacids;

6. conversion of oxoacids to glucose (or glycogen);

7. conversion of oxoacids to fat.

The biochemical details of these processes and their control are given elsewhere in this text.

It is the rates of these various processes that control the magnitude and direction of flux through the transdeamination system and the eventual fate of the amino acids in various conditions. These are discussed for several conditions: the normal fed state; starvation; trauma, surgery and cancer.

Changes in amino acid metabolism in different conditions

The fed state

The protein in the food is digested and the resultant amino acids are absorbed from the intestine (process 1): these will be used for synthesis of protein (process 3) and some nitrogen-containing compounds (process 4). Amino acids not required for these two processes are converted to oxoacids, of which about half are oxidised for ATP generation (process 5) and the other half converted to glucose (or glycogen) (process 6); the ATP generated in process 5 is used in process 6. Triacylglycerol synthesis (process 7) occurs if the formation of oxoacids exceeds those required for oxidation or for glucose synthesis. The quantitative importance of process 7 is probably minimal (Figure 8.14(a)).

Starvation

If starvation lasts for more than 24 hours, the rate of degradation of body protein (process 2) exceeds the rate of protein synthesis (process 3). The resultant amino acids are converted to oxoacids, most of which are converted to glucose (process 6) which is released and used predominantly by the brain (see Chapter 6). In this condition, the ATP required for gluconeogenesis is obtained from the oxidation of fatty acids (Figure 8.14(b)).

Trauma, surgery and cancer

In these conditions, in general, process 2 is accelerated and the resultant oxoacids are converted to glucose or are oxidised. If anorexia is present and the patient is not receiving parenteral nutrition, oxoacids will be converted mainly

to glucose. Process 4 is increased to provide for the nitrogen-containing small compounds and peptides that are required in this condition (e.g. glutamine, arginine, cysteine, cytokines, acute phase proteins) (Figure 8.14(c)).

This brief account of the integration of amino acid metabolism is over-simplified but serves to illustrate how

(a)

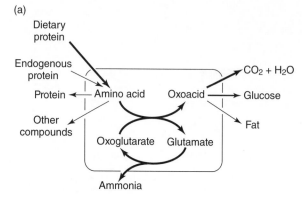

(b)

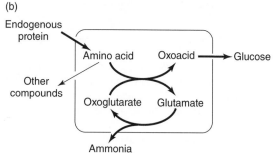

(c)

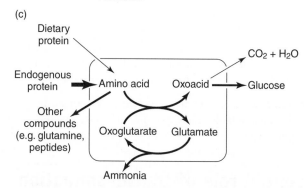

Figure 8.14 (a) *The central role of transdeamination in the fed state.* The emphasis in the fed state is digestion of dietary protein and conversion of oxoacids to glucose or CO_2. Only a small amount of fat, if any, is synthesised. **(b)** *The central role of transdeamination in the starved state.* Endogenous protein is degraded and the oxoacids are converted solely to glucose. **(c)** *The central role of transdeamination during trauma or in the cancer patient.* Endogenous protein is degraded to provide amino acids for (1) other nitrogen-containing compounds, e.g. essential proteins or peptides, amino acids, such as glutamine; (2) glucose, especially if anorexia is present. It is likely that, in starvation and trauma, some protein synthesis occurs, for example, to replace damaged proteins.

changes in amino acid metabolism play a role in the essential physiological processes in these conditions. Evidence for these changes is provided in this and subsequent chapters. Some amino acids are metabolised in tissues other than the liver and description of this metabolism provides a more realistic picture of whole-body protein and amino acid metabolism in the body (see below).

Amino acid metabolism in different tissues

Amino acid metabolism is important in all tissues/organs but especially so in the liver, intestine, skeletal muscle, adipose tissue, kidney, lung, brain, cells in the bone marrow and cells of the immune system.

Liver

The liver is the only organ capable of catabolising all amino acids, with the important exception of the branched-chain amino acids. This makes physiological sense, since it is the only organ in which ammonia can be converted into urea and, moreover, most amino acids that are absorbed by the gut enter the hepatic portal vein for immediate passage through the liver. In fact, of the amino acids that enter this vein, the liver normally removes and catabolises more than 70% although, on a low protein diet, a lower proportion, especially the essential amino acids, is removed, so that they are available for other tissues. The liver plays a major role in the regulation of catabolism of amino acids.

With a minimal protein diet (just sufficient to maintain nitrogen balance), the maximal activities of the enzymes responsible for the degradation of essential amino acids are low. This ensures that the essential amino acids are protected from degradation. As the intake of dietary protein increases, the activities of these enzymes increase, due to acute changes in activity and chronic increases in the amount of enzyme. Acute regulation of essential amino acid oxidation is achieved via changes in amino acid concentrations in tissues. For example, it has been shown that the rate of leucine oxidation in the whole body is proportional to the amount of protein ingested and hence the concentration of leucine in the body (Figure 8.15). The rate of urea production is also directly proportional to protein intake. This is consistent with the view that the intracellular concentrations of amino acids play a role in the control of the rate of catabolism of amino acids, and consequently ammonia and urea formation (Chapter 10).

The overall rate of amino acid metabolism depends on:

- The concentration of amino acids in the liver.

- The activities of the key enzymes that catalyse degradation of the essential amino acids.

- The rate of protein synthesis, in liver and other tissues, which depends on the concentrations of some hormones and the concentration(s) of 'signal' amino acids (e.g. leucine) (Figure 8.16).

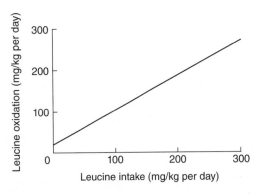

Figure 8.15 *A plot of leucine oxidation against leucine intake.* The plot illustrates a linear relationship between leucine intake and oxidation. The leucine intake is calculated from the protein intake. It is assumed that the same relationship would hold for other amino acids. A similar response is seen for urea production. Data from Young *et al.*, 2000.

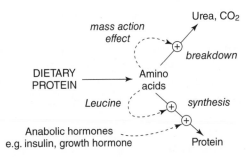

Figure 8.16 *The control of amino acid breakdown and protein synthesis in liver.* The factors in regulation are as follows: (i) the amino acid concentration in the blood regulates the rate of urea production (Chapter 10); (ii) the amino acid leucine, and the anabolic hormones increase the rate of protein synthesis. Mass action is a term used to describe the effect of concentration of substrate on the reaction rate. The control of protein synthesis is discussed in Chapter 20. Control by leucine has been studied primarily in muscle.

Skeletal muscle

The major role of skeletal muscle is movement, which is described and discussed in Chapter 13). Nevertheless, since muscle comprises 40% of the body it is large enough to play a part in control of the blood concentrations of the major fuels: glucose, fatty acids, triacylglycerol and some amino acids. Skeletal muscle contains the largest quantity of protein in the body, which is used as a source of amino acids under various conditions (e.g. starvation, trauma, cancer: see above). It plays an important part in the metabolism, in particular, of branched-chain amino acids, glutamine and alanine, which are important in the overall metabolism of amino acids in the body (discussed below).

Branched-chain amino acid metabolism in muscle and liver

Muscle is quantitatively the most important tissue for uptake of the branched-chain amino acids (BCAAs) but both muscle and liver can oxidise the oxoacids of the BCAAs. Liver cannot utilise the BCAAs, since it lacks the first enzymes in their metabolism (the aminotransferases) but it contains the enzymes for oxidising the oxoacids, which includes the first enzyme, the branched-chain oxoacid dehydrogenase, which oxidises all three of the oxoacids. The lack of the aminotransferases in liver directs these amino acids to muscle for metabolism and protects the nitrogen from conversion to urea. The significance of this is that nitrogen in the BCAAs is transferred to oxoglutarate to form glutamate in muscle and the glutamate is then aminated with ammonia to form glutamine. Muscle is a major tissue for production of glutamine which is also stored in muscle prior to release into the blood for use by many cells and tissues as a fuel and also as a nitrogen donor.

The oxoacids produced as a result of transamination of the BCAAs are either oxidised in the muscle, or, in resting muscle, released into the blood for uptake and oxidation in the liver (Figure 8.17, see also Figure 8.23).

Small intestine

At least five amino acids – glutamine, glutamate, aspartate, asparagine and arginine – are metabolised largely within the enterocytes of the small intestine. The reactions involved in their metabolism are:

- Asparagine and glutamine are converted to aspartate and glutamate, by the enzymes asparaginase and glutaminase, respectively:

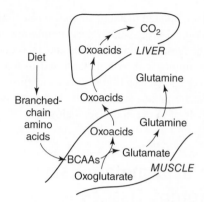

Figure 8.17 *The metabolism of branched-chain amino acids in muscle and the fate of the nitrogen and oxoacids. The α-NH$_2$ group is transferred to form glutamate which is then aminated to form glutamine. The ammonia required for amination arises from glutamate via glutamate dehydrogenase, but originally from the transamination of the branded chain amino acids. Hence, they provide both nitrogen atoms for glutamine formation.*

$$glutamine + H_2O \rightarrow glutamate + NH_3$$

$$asparagine + H_2O \rightarrow aspartate + NH_3$$

- The aspartate and glutamate produced by these reactions, plus those taken up from the lumen, are metabolised to oxaloacetate and oxoglutarate, respectively, as discussed above. The α-NH$_2$ group in these amino acids is transferred to pyruvate to form alanine, which is released and then taken up by the liver, where the NH$_2$ group is converted to ammonia and then to urea.

- The ammonia produced from asparagine and glutamine is released into the hepatic portal vein, for removal by the liver and conversion to urea. The concentration of ammonia in the blood in the hepatic portal vein is about ten times higher than in the hepatic vein, indicating the quantitative importance of the liver in removing this ammonia.

- Arginine is metabolised only as far as citrulline, which is released into the bloodstream; its fate is interesting and is described below.

The biochemical roles of these processes The metabolism of glutamate and aspartate by the enterocytes provides not only ATP, via oxidation of the oxoacids, but can also be considered to be a detoxification process. Both glutamate and aspartate are neurotransmitters in the brain. If their concentrations in blood increase too much, they could interfere with the control of neurotransmitter levels in the brain with possible changes in behaviour or clinical problems (see below). One such phenomenon is 'Chinese Restaurant Syndrome', but there may be other problems, as yet not reparted.

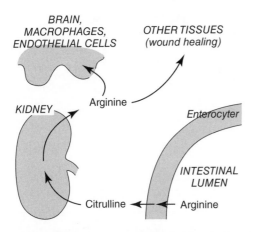

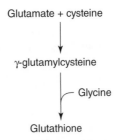

Figure 8.19 *The formation of glutathione.* Glutathione is an important peptide especially in removal of free radicals (i.e. as an antioxidant). It is synthesised in the intestine from glutamate, cysteine and glycine (see Appendix 8.4 for details).

Figure 8.18 *The role of the intestine and kidney in formation of arginine.* The intestine converts arginine to citrulline, to be released into the blood to be taken up by the kidney where it is converted to arginine. An important role of arginine is conversion to nitric oxide, a key messenger molecule, and neuromodulator in the brain (Chapter 14).

crine glands or to the brain, where they interact with neurones that affect appetite.

(iv) Mucus to aid transport and digestion of food in the intestine.

(v) Enzymes and other proteins which are involved in digestion.

The amount of protein synthesised and then released in (iv) and (v) is about 70 g each day. Even under conditions of starvation or malnutrition, proliferation and differentiation of stem cells located in the crypts of the villi are important to provide the cells necessary for replenishment of those lost from the villi. New cells move up the villus to replace those lost at the top. Under these conditions, amino acids are not available from the intestine and have to be taken up from the blood across the basolateral membrane. A low level of amino acids in the blood, due to chronic malnutrition, will prevent or reduce the rate of proliferation of these cells, so that digestion of even the small amount of food ingested during malnutrition, or refeeding after starvation, is difficult. A vicious circle thus results from protein-deficient diets which increase the risk of development of protein-energy-malnutrition. This is especially severe in children but may also contribute to the clinical problems that occur in the elderly whose diets are of low quality.

> Some meals, particularly those served in Chinese restaurants, contain large amounts of glutamate (an additive in cooking, sometimes, which is usually known as monosodium glutamate). In sensitive individuals, the amount of glutamate absorbed overloads the capacity of the enzymes to metabolise the glutamate, possibly due to a low activity of glutamate dehydrogenase in the intestine: symptoms are headache, sweating and nausea. They are short-lived, probably due to metabolism of glutamate in the liver. A transient increase in the blood concentration increases the entry of glutamate into the brain, where it activates neurones in which glutamate is a neurotransmitter.

The citrulline leaves the intestine and is transported to the kidney, where it is converted back to arginine (Figure 8.18). This inter-organ process protects arginine from degradation in the liver. Maintenance of the arginine level in the blood is important since it is a precursor for several compounds that are of particular importance for cells of the immune system and other cells (see Table 8.4) and as the precursor for the formation of the messenger molecule, nitric oxide.

The intestine is the first tissue to 'see' all the amino acids, whereas the liver is the first tissue to catabolise the amino acids that are not removed by intestine. Hence it makes physiological sense for the intestine to synthesise important nitrogen-containing compounds from amino acids, prior to their metabolism in the liver. These include:

(i) Proline, which is formed from glutamate and required for synthesis of collagen.

(ii) Glutathione, which is synthesised from glutamate, glycine and cysteine (Figure 8.19).

(iii) Some peptides that act as local hormones or which are released into the blood, to be transported to endo-

Colon

The two important fuels for colonocytes are glutamine and short-chain fatty acids. The oxidation of both fuels provides ATP for the cells, which is important not only to maintain digestive and absorptive functions but also to maintain membrane structure and hence the physical barrier between the lumen and the blood and peritoneal cavity. This barrier normally prevents significant rates of translocation of bacteria into the peritoneal cavity and thence into the blood. If this barrier is breached, translocation of pathogens and

toxins normally present in the colon can occur which can result in peritonitis and even systemic sepsis (Chapter 18). The specific role of glutamine in these cells is discussed below.

Kidney

The kidney has several important roles in amino acid metabolism:

- *The synthesis of arginine from citrulline.* The latter is produced from other amino acids in the small intestine and then released into the blood. The kidney takes up citrulline and converts it to arginine, which is then released into the blood for use by other tissues (Figure 8.18). Since arginine is a precursor for a number of important compounds, and aids wound healing, this is a significant biochemical role of the kidney.

- *The synthesis of creatine.* In the kidney, guanidinoacetate is produced from arginine and glycine, then released into the blood to be taken up by the liver and methylated to form creatine (Figure 8.20(a)). The creatine is, in turn, taken up by the muscle where it is phosphorylated to produce phosphocreatine, which can maintain the ATP level, especially in 'explosive' exercise. Creatine and phosphocreatine are converted in muscle to creatinine, which is important in clinical practice (Figure 8.20(b)) (Box 8.3).

- *Gluconeogenesis.* The gluconeogenic pathway is present in the kidney, as in the liver. Thus, amino acids (and lactate) can be converted to glucose in the kidney but a major precursor, in acidotic conditions, is glutamine.

- The kidney takes up glutamine and metabolises it to produce ammonia, which is released into the tubules to buffer protons that are secreted into the glomerular filtrate by the kidney, to reduce the hydrogenion concentration in the blood:

> Since these are fundamentally important processes, the transport of protons into the glomerular filtrate and the transport of ammonia can be considered as secretions rather than excretions.

$$NH_3 + H^+ \rightarrow NH_4^+$$

This helps to maintain the normal pH of the blood. The reactions also produce oxoglutarate, which can either be oxidised or be used to synthesise glucose (see below).

Brain

As indicated in other parts of this book, the oxidation of glucose for ATP generation, and the uptake of essential

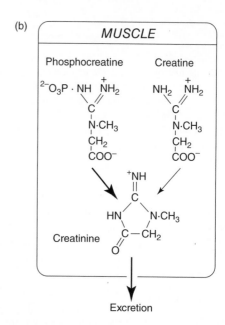

Figure 8.20 (a) *The synthesis of phosphocreatine.* The compound guanidinoacetate is formed from arginine and glycine in the kidney and is then transported to the liver where it is methylated addition of 'CH₃' (see Chapter 15) to form creatine (see Appendix 8.4 for details). Creatine is taken up by tissues/organs/cells and phosphorylated to form phosphocreatine, particularly in muscle. **(b)** *Conversion of phosphocreatine and creatine to creatinine in muscle.* Creatinine is gradually formed and then released into blood and excreted in urine.

fatty acids for phospholipid formation during development of the brain and repair of oxidative damage in the adult brain, are of the utmost importance. However, amino acid metabolism is also crucial. The following points are pertinent to the metabolism of amino acids in the brain:

- Three amino acids, glutamate, aspartate and glycine, are neurotransmitters.

- Some amino acids are precursors for neurotransmitters: glutamate for

> Measurement of arteriovenous differences across the brain shows that it does take up the precursor amino acids for formation of the neurotransmitters including tryptophan and tyrosine. It also takes up the branched-chain amino acids for formation of glutamine.

Box 8.3 Creatine, phosphocreatine and creatinine

Creatine is not an amino acid but is made from glycine, arginine and methionine in a pathway that spans three tissues (Figure 8.20(a)). Arginine and glycine are condensed in the kidney to form guanidinoacetate which is transported to the liver where it acquires a methyl group. The creatine is released and then taken up by tissues, particularly skeletal muscle, which requires a transporter in the plasma membrane for creatine to enter muscle. The activity of this transporter, somewhat surprisingly, is stimulated by insulin. In the muscle, creatine is phosphorylated to form phosphocreatine. This has two roles:

- In cardiac muscle, skeletal muscle and brain it acts as a reservoir of high-energy phosphate which can rapidly re-phosphorylate ADP to maintain ATP levels for short periods, i.e. a buffer for ATP;
- In some cells, it functions as a shuttle to transport ATP from the mitochondria to the sites at which a high rate of ATP hydrolysis occurs; this occurs in cardiac muscle, some skeletal muscles, spermatozoa, in some species, and the retina.

In the muscle, phosphocreatine and creatine undergo cyclisation to form creatinine (Figure 8.20(b)). Since creatinine cannot be metabolised, it is released from muscle and is then excreted in the urine. This biochemical process is useful in clinical practice, since creatinine production is spontaneous and is remarkably constant: 1.7% of the phosphocreatine and creatine in muscle cyclises each day, so that its concentration in blood provides an indication of the glomerular filtration rate, and hence provides an indication of the function (i.e. the health) of the kidney.

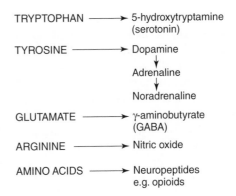

Figure 8.21 *The amino acids that are precursors for neurotransmitters in the brain.* These are discussed in Chapter 14.

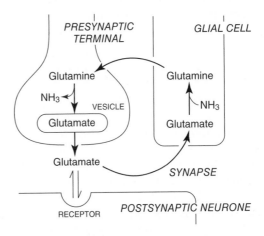

Figure 8.22 *The glutamate/glutamine shuttle for provision of the glutamate neurotransmitter in the brain.* This illustrates another role for glial cells in the brain.

γ-aminobutyrate; tyrosine for adrenaline, noradrenaline and dopamine; tryptophan for 5-hydroxytryptamine; arginine for nitric oxide (Figure 8.21).

- The amino acids, aspartate and glutamate, are not taken up from the blood but are synthesised in the brain. This requires nitrogen (for the $-NH_2$ groups) which is obtained from branched-chain amino acids via transamination, as in other tissues.

- Glutamine also plays an important role in the brain as an amino acid shuttle to maintain the concentration of the neurotransmitter glutamate in the presynaptic terminal. Glutamate is released from the presynaptic neurone and crosses the synapse to bind to the postsynaptic receptor. To terminate the action of glutamate as a neurotransmitter, it is transported out of the synapse into the glial cell, where it is converted back to glutamine. The glutamine is released and taken up again by presynaptic neurone where it is converted back to glutamate and packed into

vesicles in preparation for release into the synapse once more (Figure 8.22).

- The pathways for synthesis of the monoamine neurotransmitters are not, at least in some neurones, saturated with precursor amino acids (tyrosine for formation of noradrenaline plus dopamine; tryptophan for formation of 5-hydroxytryptamine (serotonin)). Marked increases in the blood level of these amino acids can increase their concentrations in neurones which can influence the concentration of the respective neurotransmitters in some neurones in the brain. This may result in changes in behaviour.

- The transport of amino acids into the brain (i.e. across the blood–brain barrier) requires a specific transporter, as for transport of amino acids in other tissues. The transporter that transports tyrosine and tryptophan into the brain also transports the branched-chain amino acids (Table 8.8). Hence, the branched-chain amino acids can compete with tyrosine and tryptophan for entry into brain. Consequently, a marked increase in the blood levels of branched-chain amino acids can lead to decreased levels of tyrosine and tryptophan in neurones and, consequently, their corresponding neurotransmitters. A large increase in the blood levels of branched-chain amino acids, produced by supplementation of the diet with these amino acids, can influence behaviour, probably by influencing the brain levels of amine transmitters. There is some evidence that intake of branched chain amino acids can reduce physical fatigue (Chapter 13).

Insulin-secreting cells in the endocrine pancreas

The rate of insulin secretion by the β-cells in the islets of Langerhans in the pancreas is increased by a number of amino acids: e.g. alanine, aspartate and glutamine.

Glutamine: an amino acid of central importance

Glutamine is an amino acid that can be oxidised to generate ATP in many different cells. Hence, it is considered, along with fatty acids and glucose, as a major fuel in the body, a role which is discussed below and in Chapter 9. To function as a quantitatively significant fuel it must be present in or made available from protein in the diet, and/or synthesised in the body, utilised at a significant rate and its concentrations in the blood maintained reasonably constant, under all conditions. These topics are discussed in five subsections: glutamine in the diet; tissues that synthesise glutamine; tissues that utilise glutamine; liver: a tissue that can both produce and use glutamine; blood glutamine as a fuel. It is estimated that about 100 g of glutamine is metabolised within the body each day, most of that is synthesised in the body. A list of the roles of glutamine in the body is presented in Table 8.11.

Glutamine in the diet

Glutamine is present in peptide bonds in the protein in the food but it is also present in the free form in meat and some root vegetables. It is released by proteolysis in the small intestine and absorbed into the enterocytes. Here some of it is metabolised so that only about 50% or less of the absorbed glutamine enters the blood. However, if the food is supplemented with glutamine or if it is ingested as a bolus, a significant proportion enters the blood. To maintain the physiological blood level under normal conditions, it must be synthesised de novo in the body.

Table 8.11 Roles of glutamine in some cells, tissues and organs

Immune cells
  Precurson for nucleotide synthesis.
  Fuel for immune cells: at rest and during proliferation.
  Enhances T-lymphocyte responses to infection.
  Supports phagocytosis by neutrophils and macrophages.

Small and large intestine
  Precurson for nucleotide synthesis.
  Fuel for colonocytes.
  Fuel for enterocytes and for stem cells in the crypts of villi.
  Maintenance of gut-associated lymphoid tissue (GALT).
  Maintenance of gut barrier, especially that in the colon.

Kidney
  Substrate for gluconeogenesis.
  Formation of ammonia for buffering protons in the filtrate in the tubules.

Skeletal muscle
  Glutamine is synthesised, stored and secreted.

Lung
  Fuel for endothelial cells (largest number of endothelial cells in any organ).
  Fuel for lung alveolar macrophages.

Central nervous system
  Shuttle for maintenance of glutamate as a neurotransmitter.

Data from Biolo *et al.* (2005).

Tissues that synthesise glutamine

The three tissues that fulfil this role are skeletal muscle, adipose tissue and lung. Of these, muscle is quantitatively the most important.

Skeletal muscle

Early studies using indwelling catheters to measure arteriovenous concentration differences across muscle indicated that of those amino acids released, during starvation about 70% is alanine and glutamine (Table 8.12). Yet, the content of these two amino acids in muscle protein is only about 10%. This was the first line of evidence that muscle synthesis glutamine requires both carbon and nitrogen atoms. The carbon atoms are obtained from glucose and the two nitrogen atoms are obtained from the branched-chain amino acids, via transamination and the glutamate dehydrogenase reactions (Figure 8.23). It is estimated that in humans approximately 80 g of glutamine is released from muscle each day. The oxoacids that are produced from the branched-chain acids are either oxidised in muscle or released for use by other tissues (see Figure 8.17).

Table 8.12 Arteriovenous concentration differences in amino acids across the muscles in the leg of humans in the postabsorptive period

Amino acid	Arteriovenous difference[a]	
	μmol/L	Percentage of total amino acids released
Alanine	−70	30
Glutamine	−70	30
Glycine	−24	10
Lysine	−20	9
Proline	−16	7
Threonine	−10	4
Histidine	−10	4
Leucine	−10	4
Valine	−8	3
Arginine	−5	2
Phenylalanine	−5	2
Tyrosine	−4	2
Methionine	−4	2
Isoleucine	−4	2

[a]The concentration difference is that between blood in the femoral artery and that in the femoral vein. The minus sign indicates release from the muscle. Data taken from Felig (1975). It is estimated that about 80 g of glutamine is released each day from skeletal muscle.

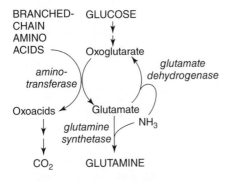

Figure 8.23 *Formation of glutamine from glucose and branched-chain amino acids in muscle and adipose tissue and probably in the lung. Oxoacids may also be released into blood for oxidation in the liver.*

In addition to synthesis, muscle also stores glutamine. It is estimated that the total quantity stored in all the skeletal muscles is about 80 g. The glutamine released by muscle can be utilised by the kidney, enterocytes in the small intestine, colonocytes, all the immune cells and the cells in the bone marrow (Figure 8.24). Details of the pathways of utilisation by these tissues are discussed.

Alanine synthesis in muscle   The synthesis and release of alanine by muscle can be considered as a safety mecha-

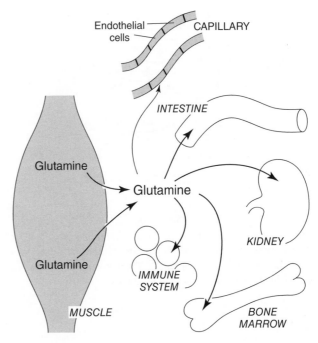

Figure 8.24 *Some fates of glutamine that is released by muscle.* Glutamine is released from the store of glutamine in the muscle but, for immune system and bone marrow, it may also be provided from adipocytes (Chapter 17). It is assumed that glutamine is present as a free amino acid in muscle and that there is a specific transport protein in the plasma membrane that can be regulated.

nism to prevent the accumulation of ammonia. To understand how alanine synthesis can remove ammonia, it must be remembered that the glutamate dehydrogenase reaction removes, as well as forms, ammonia (i.e. it is a near-equilibrium reaction). Hence, the direction of this reaction is dependent on the concentrations of substrates and products so that, at high levels of ammonia, ammonia reacts with oxoglutarate to form glutamate, which can then be transaminated with pyruvate to yield alanine, which is then released into the bloodstream. In rather loose terms, alanine formation in muscle can be considered as a 'nitrogen overflow system'. The fate of the alanine that is released by muscle is removal from the blood by liver and conversion to glucose via gluconeogenesis. The two reactions are as follows:

$$NADH + oxoglutarate + NH_3 \xrightarrow{\text{(i)}} glutamate + NAD^+$$

$$\text{(ii)} \quad glutamate \rightarrow oxoglutarate$$
$$pyruvate \rightarrow alanine$$

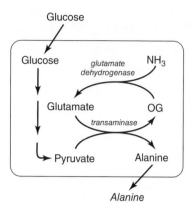

Figure 8.25 *Excess ammonia in the muscle is used to form alanine. Ammonia is released from several reactions and is incorporated into alanine via glutamate dehydrogenase and transamination. OG – oxoglutarate. Alanine is released into the blood from volece it is removed by the liver.*

These reactions are catalysed by:

(i) glutamate dehydrogenase and (ii) alanine-glutamate aminotransferase. The net reaction is, therefore:

$$pyruvate + NH_3 \rightarrow alanine$$

and the alanine is released from the muscle. A description of this nitrogen overflow system is provided in Figure 8.25.

Adipose tissue

The metabolism of the branched-chain amino acids, glutamine and alanine in adipose tissue is similar to that in muscle. Glutamine is synthesised from glucose and branched-chain amino acids and then released from the adipocytes into the blood. Alanine is synthesised from glutamate and pyruvate. Unfortunately, the quantitative importance of adipose tissue for the uptake of branched-chain amino acids and formation of glutamine is not known. It is of interest to note that most lymphocytes and other immune cells in the body are contained within lymph nodes and almost all lymph nodes are present within adipose tissue depots, where adipocytes may provide glutamine, locally, for immune cells. This is a physiological explanation for the ability of adipocytes to form glutamine (Chapter 17). Since adipose tissue is important in the formation and release of glutamine, it would be of interest to know if this changed in patients suffering from obesity. No studies have yet been done. Adipocytes which are numerous in bone marrow, may provide most of the glutamine required by proliferating cells to this tissue.

Lung

The cells in lung are able to synthesise glutamine presumably from glucose and branched-chain amino acids. The quantitative importance of glutamine synthesis and release are not known. It is estimated that the lung, adipose tissue and possibly the liver synthesise and release about 40 g of glutamine each day: about twice this amount is released by muscle. The physiological/biochemical importance of this process in the lung may be to provide glutamine for the large number of resident macrophages present in the lung. Glutamine is essential for phagocytosis by macrophages. In the lung, they remove microorganisms and environmental particles that enter with every breath.

Tissues that utilise glutamine

The major tissues that use glutamine are kidney, small intestine, colon, immune cells and bone marrow cells. Rates of utilisation by these last four tissues cannot be measured *in vivo*. Rates of utilisation by isolated cells have been measured *in vitro* (Chapters 17 and 21).

Kidney

As indicated above, one function of the kidney is to maintain the pH of the blood by excreting (secreting) protons into the glomerular filtrate during acidosis. In order to buffer these protons in the glomerular filtrate, ammonia is also excreted. The buffering occurs as follows:

$$NH_3 + H^+ \rightarrow NH_4^+$$

This buffering action is important to maintain the pH of the urine between 5 and 6. If the system fails, the pH drops to between 1 and 2, which is a very dangerous level. The ammonia is provided from glutamine which is taken up from the blood and metabolised in such a way that both of its nitrogen atoms are released as ammonia. This is achieved by the successive activities of the enzymes, glutaminase and glutamate dehydrogenase, in the epithelial cells of the proximal tubules of the kidney, as follows (Figure 8.26):

$$glutamine + H_2O \rightarrow glutamate + NH_3$$

$$glutamate \rightarrow oxoglutarate + NH_3$$

Full details of the reaction are given in Appendix 8.4 and a detailed discussion of the role of the kidney in regulation of the acid/base balance is given in (Appendix 13.4). The oxoglutarate produced from these successive reactions is either oxidised to yield ATP or, under conditions of starvation, converted to glucose via gluconeogenesis (Figure 8.26).

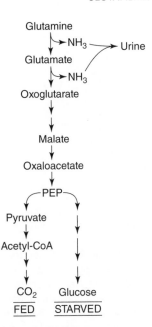

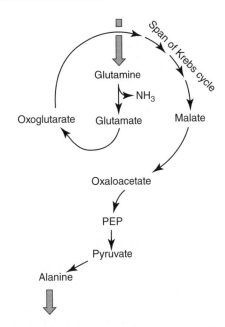

Figure 8.26 *Fate of glutamine in the kidney.* Glutamine is converted to oxoglutarate releasing two molecules of ammonia. The fate of the oxoglutarate is, in the fed condition, oxidation to CO_2 to generate ATP. In starvation, it is converted to glucose via gluconeogenesis. The quantitative importance of the kidney to provide glucose, in comparison with the liver, is not known but it may be significant.

Figure 8.27 *Pathway of glutamine metabolism in the intestinal cells.* Glutamine is metabolised to alanine to generate ATP: alanine is released into the blood to be taken up by the liver.

Small and large intestine

The enterocytes in the small intestine and colonocytes in the large intestine both use glutamine as a fuel (Figure 8.27). In the small intestine, the ATP generated from the conversion of glutamine to alanine is important not only for maintenance of the normal energy-requiring process in the cells but, in addition, for synthesis and secretion of mucus, and enzymes of digestion), and also for absorption of the products of digestion. It is also important as a precursor for the nucleotides required for DNA and RNA synthesis for the proliferation of the stem cells in the crypts of the villi. This process is important for producing cells to replace those lost from the villi by damage incurred during the process of digestion, including desquamation. Thus, a chronic low level of glutamine could impair this process, which would interfere in the normal digestion and absorption of food. Hence, especially on a low protein or inadequate diet, synthesis of glutamine or starvation, is essential to maintain a functioning intestine. This is one reason why enteral feeding of patients who are recovering from trauma or surgery is encouraged in hospitals.

Also of clinical importance, radiation or chemotherapy for the treatment of cancer can cause damage to the villi in the intestine, probably as a result of DNA damage in the proliferating stem cells. Provision of glutamine to the patient undergoing such therapy protects against this damage or encourages repair of DNA during the recovery period between treatments.

Immune cells

Immune cells play major roles in defence against infection and also in recovery from trauma but, even when these cells are resting (i.e. when there is no or little immune activity), they use glutamine at a high rate to generate ATP (Chapters 9 and 17). The pathway for metabolism of glutamine in these cells is shown in Figure 8.28.

The glutamine taken up by these cells is not completely oxidised but converted to aspartate, which is then released. Not only is glutamine a fuel for generating ATP in these cells but it is also a precursor for the nucleotides required for DNA and RNA synthesis when immune cells are proliferating, a major response of immune cells to an infection. Consequently, it is important to maintain the normal level of glutamine in blood at all times. A decrease in the level could result in a poor response to an infection (Chapter 18). This consideration has led to the supplementation of normal diets with glutamine or to the inclusion of glutamine in commercial feeds for patients (Chapter 18). The positions in the pathways for purine and pyrimidine nucleotides synthesis where glutamine provides the nitrogen are shown in Figures 20.9 and 20.10.

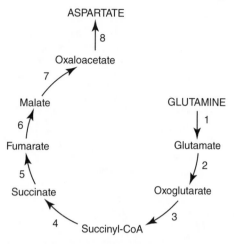

Figure 8.28 *Metabolic fate of glutamine in immune cells.* Glutamine is converted to glutamate and then to oxoglutarate which is converted to oxaloacetate. Enzymes are as follows: (1) glutaminase, (2) glutamate dehydrogenase, (3) oxoglutarate dehydrogenase, (4) succinyl-CoA transferase, (5) succinate dehydrogenase, (6) fumarase, (7) malate dehydrogenase, (8) aspartate aminotransferase. The pathway is presented in such a way as to indicate that conversion of oxoglutarate to oxaloacetate is part of the Krebs cycle. The pathway is known as glutaminolysis.

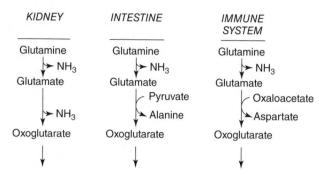

Figure 8.29 *The initial reactions of glutamine metabolism in kidney, intestine and cells of the immune system.* The initial reaction in all these tissues is the same, glutamine conversion to glutamate catalysed by glutaminase: the next reactions are different depending on the function of the tissue or organ. In the kidney, glutamate dehydrogenase produces ammonia to buffer protons. In the intestine, the transamination produces alanine for release and then uptake and formation of glucose in the liver. In the immune cells, transamination produces aspartate which is essential for synthesis of pyrimidine nucleotides required for DNA synthesis: otherwise it is released into the blood to be removed by the enterocytes in the small intestine or by cells in the liver.

Bone marrow cells

Between them, the bone marrow and the small intestine possess the highest number of proliferating cells in the body. The bone marrow contains stem cells which proliferate and differentiate to produce red and white blood cells (Chapter 17). This requires not only this amino acids to support protein synthesis but also glutamine, both as a fuel and as a precursor for nucleotides, as in the other proliferating cells. The pathway for metabolism of glutamine in cells isolated from the bone marrow is similar to that in lymphocytes (Figure 8.28).

Tumour cells

Tumour cells also require glutamine as a fuel for energy generation and as a precursor for the synthesis of purine and pyrimidine nucleotides for DNA and RNA synthesis. The roles and importance of glutamine in tumour cells and possible competition between the cells for glutamine are discussed in Chapter 21. The pathway for the metabolism of glutamine is similar to that in the immune cells.

Comparison of the early stages of glutamine metabolism in kidney, intestine and lymphocytes

Although glutamine metabolism in these various tissues is important in the generation of ATP, the metabolism is different due to other functions of glutamine metabolism in the tissues. The early stages of glutamine metabolism in these tissues are presented in Figure 8.29 for comparison.

Liver: a tissue that can both produce and use glutamine

There are two types of hepatocytes in the liver, the periportal and perivenous cells (Chapter 10) and they metabolise glutamine in opposite directions. The periportal cells possess the enzyme glutaminase that converts glutamine to glutamate and ammonia, which is converted to area: glutamate is converted to glucose. In contrast, perivenous cells possess the enzyme glutamine synthetase, which converts glutamate to glutamine with the uptake of ammonia. This provides a 'safety net' to remove any ammonia from the blood that has 'escaped' the urea cycle (Figure 8.30).

Blood glutamine as a fuel

From the above discussion, it is clear that some tissues synthesise and then release glutamine into the blood and others take up glutamine and use it as a fuel and/or as a nitrogen donor. Hence, glutamine can be considered as a fuel in the blood, alongside glucose and fatty acids. Its oxidation in the Krebs cycle is described in Chapter 9. The

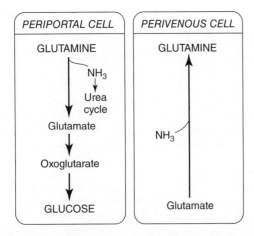

Figure 8.30 *Different roles of periportal and perivenous cells in the liver in respect of glutamine metabolism.* Glutamine is converted to glucose in periportal cells via gluconeogenesis: in perivenous cells, ammonia is taken up, to form glutamine, which is released into the blood. This emphasises the importance of the liver in removing ammonia from the blood, i.e. if possesses two process to ensure that all the ammonia is removed.

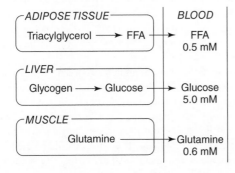

Figure 8.31 *Comparison of glutamine as a fuel in the blood with glucose and fatty acids.* The concentration of glutamine in the blood is similar to that of fatty acid. The amount of glutamine stored in muscle is similar to the amount of glycogen stored in the liver; that is, about 80 g. Mobilisation of each of these stored fuels is discussed in Chapters 6, 7, 17 and 18. It appears that glutamine is stored free in the cytosol. Polyglutamine on vesicles containing glutamine have not been found.

properties of glutamine which enable it to function as a fuel (Figure 8.31) are as follows:

- It is a small molecule.

- It is very soluble in the plasma.

- It has no net charge, so that transport is not limited by any change in electrical potential across a membrane.

- It can be used by many different cells and tissues.

- In these cells, it is metabolised to glutamate, which is converted to aspartate or alanine. The pathways generate NADH and FADH$_2$ which transfer electrons along the respiratory chain to generate ATP.

- The concentration in human blood is relatively high (0.6 mM).

- This concentration is maintained under many different conditions.

- When the blood concentration of glutamine decreases below that which is physiological, functional impairment can result: poor proliferation of immune cells and poor phagocytosis by macrophages.

- Although not all the glutamine is made available to the blood from the digested protein in food, since some of it is metabolised in the enterocytes, it is readily synthesised in glutamine-producing tissues. This is analogous to glucose, which can also be utilised by the enterocytes but is synthesised, via gluconeogenesis, in the liver and in the kidney. Reactions that are involved in the pathway of glutamine synthesis are presented above.

- Glutamine is stored in muscle: approximately 80 g in the total skeletal muscle in the body. This is an amount similar to that of glucose which is stored as glycogen in the liver (Chapters 2 and 6).

- Glutamine, along with glucose, can stimulate insulin secretion by the endocrine pancreas. The significance of this is the regulation of the plasma glutamine concentration is not known.

Urea 'salvage'

The formation and excretion of urea is the primary mechanism by which excess nitrogen, in the form of ammonia, is removed from the body. Surprisingly, it was found that the actual rate of urea synthesis exceeded considerably the rate of excretion of the urea. The interesting question, therefore, is what is the fate of this lost urea? The answer is that urea enters the large intestine, where it is degraded by microorganisms that possess the enzyme urease, which catalyses the reaction:

$$CO(NH_2)_2 \rightarrow CO_2 + 2NH_3$$

The ammonia can then be utilised for amino acid synthesis in some or all of the microorganisms in the intestine, a process requiring the enzyme glutamate dehydrogenase to incorporate the ammonia into glutamate

$$NH_3 + \text{oxoglutarate} \rightarrow \text{glutamate} \rightarrow \text{other amino acids}$$

Thus, nitrogen is provided for the synthesis of all amino acids and hence for synthesis of protein, as these microorganisms proliferate. However, when some microorganisms die, this protein will be degraded and result in formation of amino acids. (Note that a large number of microorganisms are expelled in the faeces.) Some of these amino acids will be absorbed and enter the blood from where they will be taken up by the liver or bypass the liver and be taken up by other tissues. In these tissues, and in the liver, these amino acids can be used for protein synthesis. This can, therefore, be considered as a salvage pathway for amino acids (Figure 8.32). An important question is what is the quantitative significance in humans? The answer is that it may be particularly important when protein intake is minimal (i.e. during malnutrition, especially in children). It is also known to be important in some animals under some conditions (Box 8.4). Hence, it is unfortunate that it is called 'urea salvage' since amino acid salvage is probably its physiological role.

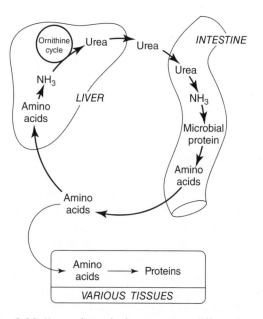

Figure 8.32 *Urea salvage in humans*. Urea diffuses into the colon where it is converted to ammonia in those microorganisms that possess the enzyme urease. The ammonia is used by most microorganisms in the colon to synthesise their protein. Upon death in the colon, these organisms are degraded and the protein is hydrolysed to amino acids, some of which are absorbed by the host to be used for protein synthesis, etc.

Box 8.4 The Long Sleep

Imagine sleeping for between three and five months without getting up to eat, urinate or defecate. The temperature outside is well below zero but you have maintained your temperature between 31 and 35 °C. If you are female, you may well have given birth and suckled your offspring during this time. You wake up and are immediately active – how might you feel? In fact, without external assistance, no human could survive such an ordeal but for the North American black bear or the polar bear it has been a normal winter.

There is no doubt that the bear uses its fat stores for its energy requirement during hibernation. Since it stores so little glycogen, maybe the bear's brain requires little glucose and functions primarily on ketone bodies (Chapter 7). However, one of the most remarkable adaptations to this hibernation must be the bear's nitrogen metabolism. The lack of opportunity to excrete urea might suggest that the bear manages, during this period, to reduce its amino acid metabolism dramatically or tolerate an increased blood urea level. In fact, the rate of amino acid metabolism remains the same, or even increases, while the blood urea level falls. Urea is still produced in these animals and passes, via the urine, into the bear's bladder, but this urine is absorbed back into the bloodstream, from where the urea passes into the large intestine. Here it is hydrolysed to ammonia by microorganisms which possess the enzyme urease (e.g. *ureaplasmsa urealyticum*).

The ammonia is used in the microorganisms to synthesise amino acids, and hence protein. When these die, the protein is digested, releasing the amino acids to be absorbed across the large intestine of the bear to become, once again, available for protein synthesis. Indeed, this may be one reason why the bear does not hibernate deeply; its body temperature remains well above that of the environment to enable bacteria in its colon to remain metabolically active. But how do the microorganisms benefit from their ability to hydrolyse urea? In addition to gaining nitrogen for synthetic processes, it is possible that some microorganisms can generate ATP directly from a proton gradient established across their plasma membrane through the hydrolysis of urea. This underlines the symbiotic nature of relationships between gut microorganisms and their hosts (see Nicholls & Ferguson, 2002).

An interesting possibility, never investigated, is that the bear possesses the ability to synthesise both dispensable and non-dispensable amino acids de novo, from the ammonia released by the microorganisms. As far as is known, no mammal possesses the enzymes necessary to synthesise the indispensable amino acids. It would be interesting if another adaptation to such a long hibernation is development of enzymes necessary to synthesise indispensable amino acids. This possibility is supported by the fact that, despite five months of total starvation, the net loss of protein is negligible; what the mother bear loses, the cubs gain. Polar bears have similar nitrogen metabolism in order to survive winter and produce and feed cubs (see Barboza *et al.* 1997 and Hissa *et al.* 1998).

Photograph kindly provided by Dr. Caroline Pond (Open University, UK)

9
Oxidation of Fuels and ATP Generation: Physiological and Clinical Importance

Like the electrons in a platinum wire ATP can act as an energy currency and link energetically systems of widely different redox potentials . . . The oxidative stages in which oxygen is the final oxidising agent are all complex reactions.

(H.A. Krebs, 1954)

In a resting human, approximately 95% of the total ATP produced in the body is generated aerobically. Only during short-term vigorous exercise (e.g. sprinting, running up a flight of stairs) does this percentage fall below 90%.

It has been considered that glucose and fatty acids (long or short chain) are the major fuels for oxidation in most cells, but the amino acid glutamine is now known to be a major fuel for some cells (Table 9.1).

Overall, the oxidation processes are easy to depict:

$$C_6H_{12}O_6 + 6O_2 \rightarrow 6CO_2 + 6H_2O$$
glucose

$$CH_3(CH_2)_{14}COOH + 23O_2 \rightarrow 16CO_2 + 16H_2O$$
palmitate

$$glutamine + 1\tfrac{1}{2}O_2 \rightarrow aspartate + CO_2 + NH_3$$

These processes are, of course, more complex than this: details are presented in this and other chapters. In brief, there are three phases for the complete oxidation of glucose and fatty acids:

(i) Glucose conversion to acetyl-CoA, via glycolysis, which, apart from conversion of pyruvate to acetyl-CoA, occurs in the cytosol. Fatty acid conversion to acetyl-CoA, via β-oxidation, occurs in the matrix of the mitochondria.

(ii) Oxidation of acetyl-CoA via the Krebs cycle also occurs in the mitochondrial matrix.

(iii) Processes of electron transfer and ATP generation occur within or upon the inner mitochondrial membrane (Figure 9.1; Box 9.1)

The connections between the phases are provided by coenzymes, which become reduced in glycolysis, β-oxidation and the Krebs cycle and, subsequently, transfer hydrogen atoms or electrons into the electron transfer chain. These are ultimately oxidised by oxygen, and ATP is generated.

There are also three phases in the pathway of glutamine oxidation:

(i) Conversion of glutamine to glutamate and thence to oxoglutarate.

(ii) Conversion of oxoglutarate to oxaloacetate via some reactions of the Krebs cycle and electron transfer and ATP generation.

(iii) Transamination of oxaloacetate to aspartate.

Details of all the above processes and their physiological and clinical importance are discussed in this chapter.

The Krebs cycle

The biochemistry of the cycle was elucidated largely by Krebs & Johnson (1937): Krebs was awarded the Nobel Prize in 1953. It is also called the citric acid cycle to acknowledge that others contributed to the biochemical details of the cycle. Acetyl-CoA is one substrate for the cycle but the amount in tissues is very small (2.5 nmol/g wet wt) and the maximum rate of the cycle in, for example,

> Acetyl-CoA is essentially a chemically activated form of acetate (Appendix 9.1). This may be a reason why its concentration is maintained very low, to prevent spontaneous acetylation of proteins.

Functional Biochemistry in Health and Disease by Eric Newsholme and Tony Leech
© 2010 John Wiley & Sons Ltd

Table 9.1 Fuels used by cells, tissues and organs

Cells, tissues, organs	Glucose/glycogen	Fatty acid	Glutamine
Erythrocytes/	✓	✗	✗
immune cells	✓	✓[h]	✓
Brain	✓	✗	✗
Bone marrow cells	✓	✗	✓
Endothelial cells	✓	✗	✓
Small, intestine	✓[a]	✗	✓
Colon	✓	✗[b]	✓
Kidney	✓	✓	✗[c]
Liver	✓[d]	✓	✗[e]
Macrophages	✓	✓[h]	✓
Muscle, skeletal type I	✓[f]	✓	✗
Type IIb	✓	✗	✗
Cardiac	✓	✓	✗
Skin	✓	✗	✓
Stem cells	✓[g]	✗	✓[g]

For the basis and data upon which the information in this table was obtained, see Appendix 9.2. For evidence that glutamine is a fuel in the blood, see Chapter 8.

The sign ✓ indicates utilisation as a fuel; ✗ indicates little or no utilisation as a fuel. For actual rates of utilisation of fuels by some of these cells/tissues see Chapters 13, 17 and 21.

Immune cells include lymphocytes, neutrophils, monocytes.

Studies on macrophages were on those isolated from the peritoneal cavity of the mouse, not on resident macrophages.

[a] and [d] Both the small intestine and liver possess glycolytic enzymes and can convert glucose to lactate but it is not known if glucose oxidation can generate significant amounts of ATP. Liver uses lactate fatty acids and amino acids in the fed condition (Chapter 8).

[b] The colon also uses short-chain fatty acids; it is not known if it uses long-chain fatty acids.

[c] Kidney uses glutamine in acidosis to provide ammonia, but the oxoglutarate that is produced, is oxidised to generate ATP.

[e] Liver can produce or utilise glutamine but probably it is not oxidised to provide ATP (Chapters 6 and 10).

[f] This also includes type IIa fibres.

Type IIb fibres in human muscle are different from the rat and sometime known as type IIx (Chapter 13).

[g] No studies have been carried out on the fuels used by isolated stem cells but, on the basis that they are either proliferating or possess the ability to proliferate rapidly, as with lymphocytes and tumour cells, it is assumed that they utilise glucose and/or glutamine as fuels. If the latter, it could be important in stem cell research and in the clinical use of stem cells.

[h] Some fatty acid oxidation occurs but its quantitative significance is not known.

For some cells or tissues, very little of the glucose or glycogen is fully oxidised so that most of the ATP is generated from the conversion of glucose or glycogen, via glycolysis, to lactate. These include erythrocytes, type IIx muscle fibres and epidermal cells in the skin. Type IIx fibres appear to be specific to humans since they possess almost no mitochondria.

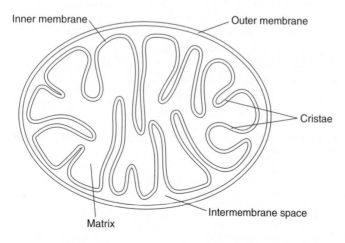

Inner membrane · Outer membrane · Cristae · Intermembrane space · Matrix

Figure 9.1 *Simple diagram of a mitochondrion showing inner and outer membranes, cristae and matrix.*

Box 9.1 Origin of mitochondria

Mitochondria probably arose about 1.5 billion years ago by infection of a proto-eukaryotic cell, which had previously relied solely on an anaerobic process to generate ATP, by a bacterium which possessed the aerobic system for ATP generation. This then developed into a symbiotic relationship: the bacteria were provided with a 'safe haven' and a fuel for oxidation and the eukaryotic cell with the efficient means of ATP generation. The bacteria have changed over many millions of years to adapt to the new environment and the result was the evolution of mitochondria. However, they have retained a number of characteristics that reflect their origin: two membranes; DNA is circular and double stranded; transcription, translation and a protein assembly system are retained within the mitochondria. To facilitate their replication, so as to ensure transmission of mitochondria to the two daughter cells during proliferation of the cell, the genome became reduced in size by transferring most genes from the bacterial genome to that of the nucleus of the host cell. There are at least 70 mitochondrial (i.e. bacterial) genes in the nucleus. The peptides encoded by these genes are synthesised on the ribosomes in the cytosol and are then imported into the mitochondria. Here some of them associate with peptides synthesised within the mitochondria to form the complete mitochondrial proteins. Although the human mitochondrial genome contains 37 genes, only 13 encode for peptides that are components of the proteins that are directly involved in the aerobic generation of ATP. The remainder encode for mitochondrial, ribosomal and transfer RNAs.

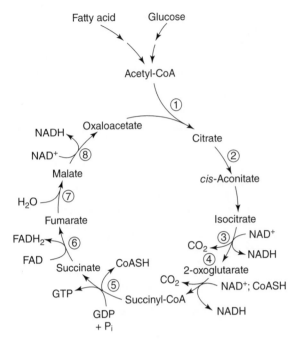

Figure 9.2 *Summary of reactions of the Krebs cycle.* The names of the enzymes are ① citrate synthase, ② aconitase, ③ isocitrate dehydrogenase (there are two enzymes, one utilizes NAD^+ as the cofactor, the other $NADP^+$; it is assumed that the NAD^+-specific enzyme is that involved in the cycle), ④ oxoglutarate dehydrogenase, ⑤ succinyl CoA synthetase, ⑥ succinate dehydrogenase, ⑦ fumarate hydratase, ⑧ malate dehydrogenase.

human skeletal muscle, can be as high as 50 nmol/g wet wt per second. This could very quickly reduce the acetyl-CoA concentration to extremely low levels, which would reduce the rate of the cycle and hence the generation of ATP. The presence of acetyl-carnitine in mitochondria, at a concentration 400-fold greater than that of acetyl-CoA, along with the enzyme carnitine acetyltransferase, acts as a buffer to protect against large changes in the concentration of acetyl-CoA:

> Mutations in the protein that transports carnitine into cells decrease its rate of transport resulting in an intracellular deficiency of both carnitine and acetylcarnitine. The disease is characterised by progressive cardiac myopathy. It usually presents within the first five years of life and is fatal.

$$\text{acetylcarnitine} + \text{CoASH} \rightleftharpoons \text{acetyl-CoA} + \text{carnitine}$$

A summary of the cycle is given in Figure 9.2 and details are presented in Appendix 9.3. However, acetyl-CoA is not only a substrate for the cycle but it is a precursor for a large number of key compounds in the cell so that maintenance of its concentration is important for this role (Appendix 9.4).

The cycle as part of a physiological pathway

The Krebs cycle will only operate when the hydrogen atoms and electrons produced in the cycle enter the electron transfer chain, ultimately to react with oxygen; that is, the two processes must take place simultaneously. A metabolic pathway is defined as a sequence of reactions that is initiated by a flux-generating step. In the cycle, citrate synthase catalyses the flux-generating reaction (Table 9.2) but there is no such reaction in the electron-transfer chain. Consequently, the cycle can be considered to be the first part of a longer pathway, which includes the electron transfer chain (Figure 9.3).

Glutamine oxidation: use of some reactions of the cycle

Glutamine is oxidised by a pathway that utilises some of the reactions of the Krebs cycle, but not all, and the five carbon atoms in glutamine are converted to the four-carbon compound aspartate: that is, glutamine is only partially

Table 9.2 Substrate concentrations and K_m values for some Krebs cycle enzymes

Enzyme	Substrate	Substrate* conc.(μmol/L)	K_m (μmol/L)
Citrate synthase	Acetyl-CoA	100–600	5–10
	Oxaloacetate	1–10	5–10
Isocitrate dehydrogenase (NAD$^+$)	Isocitrate	150–700	60–600
Oxoglutarate dehydrogenase	2-oxoglutarate	600–5900	50–24000

*These values represent the best estimates of concentration within the mitochondrial matrix taken from the available literature. It is difficult to measure concentrations within the mitochondrial matrix, which may explain some of the variation.

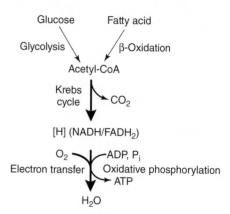

Figure 9.3 *A summary of the Physiological pathway of the Krebs cycle.* The pathway starts with acetyl-CoA, since citrate synthesis is the flux-generating step. The physiological pathway includes the electron transfer chain, since there is no flux-generating step in this chain. The pathway is indicated by the broader lines. The pathway, therefore, starts with acetyl-CoA and finishes with CO_2 and H_2O, which are lost to the environment. Acetyl-CoA is formed from a variety of precursors: glucose and fatty acids are presented in this figure.

oxidised. The pathway is known as glutaminolysis (Figure 9.4, see also Chapter 8) and is present in a large number of cells, most of which use glutamine as a major oxidative fuel (see Table 9.1). For comparison, the positions at which the three major fuels, glucose, fatty acids and glutamine, feed into the cycle are shown in Figure 9.5.

The electron transfer chain

Metabolism of the major fuels, described above, generates hydrogen atoms or electrons. These are oxidised, not directly, but via a series of oxidations and reductions (oxido-reduction reactions or, alternatively, redox reactions) that

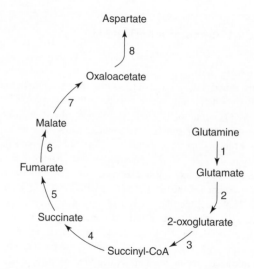

Figure 9.4 *Reactions of glutaminolysis: the pathway for glutamine oxidation.* Reaction 1 is catalysed by glutaminase, reaction 2 by glutamate aminotransferase, and reaction 8 by aspartate aminotransferase: all other enzymes are those of the Krebs cycle (3–7). (See also Chapter 8).

eventually terminate in reduction with molecular oxygen. The actual oxido-reduction reactions are carried out by prosthetic groups or coenzymes of the enzymes that constitute the electron transfer chain, since the side-chain groups of proteins are not well suited for redox reactions. The reactions involve the transfer of electrons or hydrogen atoms (represented as [H]) from a reducing agent (which becomes oxidised) to an oxidising agent (which becomes reduced). For this purpose, a hydrogen atom can be considered to be equivalent to an electron plus a proton, i.e.

$$[H] = H^+ + e^-$$

These enzymes, together with the proteins that generate ATP from ADP and P_i (phosphate), constitute 30–40% of the total protein of the inner mitochondrial membrane.

In summary, the electron transfer chain consists of a series of membrane-bound enzymes that possess different prosthetic groups which become alternately reduced and oxidised as they transfer electrons (or hydrogen atoms), from one carrier to the next in sequence, to oxygen to produce H_2O.

Sequence of carriers

The sequence of the carriers in the chain is shown in Figure 9.6. Each of the components of the chain reduces the next, in sequence, according to the redox potential (Table 9.3). The enzymes and their prosthetic groups are organised into complexes, which can be isolated by gentle disruption of the whole mitochondrion or its inner membrane. Ubiqui-

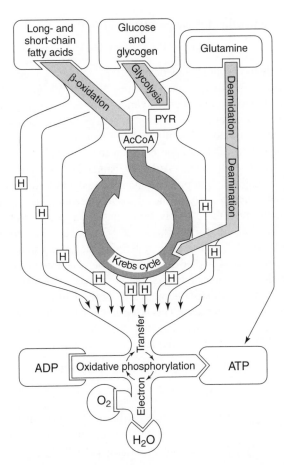

Figure 9.5 *A summary of pathways of the three main fuels and the positions where they enter the cycle.* The figure also shows the release of hydrogen atoms/electrons and their transfer into the electron transfer chain for generation of ATP and formation of water. Glutamine is converted to glutamate by deamidation and glutamate is converted to oxoglutarate by transamination or deamination. The process of glycolysis also generates ATP as shown in the Figure.

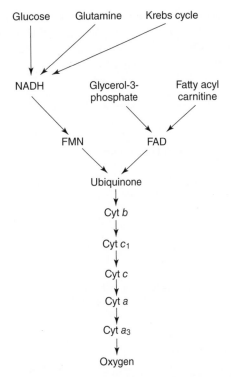

Figure 9.6 *Sequence of electron carriers in the electron transfer chain.* The positions of entry into the chain from metabolism of glucose, glutamine, fatty acyl-CoA, glycerol 3-phosphate and others that are oxidised by the Krebs cycle are shown. The chain is usually considered to start with NADH and finish with cytochrome oxidase. FMN is flavin mononucleotide; FAD is flavin adenine dinucleotide.

Table 9.3 Standard redox potentials of main electron carriers, total change in potential and $\Delta G°$ for oxidation of NADH by oxygen

Carrier	E_o (mvolts)		$\Delta G°$ (kJ/mol)
NAD+/NADH	−320		
Ubiquinone/ubiquinol	+60		
Fe3+/Fe2+ (cyt c)	+220	1140	−224
Oxygen/H_2O	+820		

The value of $\Delta G°$, for the oxidation of NADH by oxygen via the electron transfer chain, is calculated from the equation $\Delta G° = -nF\Delta E$

$\Delta G° = -2 \times 96.5 \times 1.14 = -224$ kJ/mol

none and cytochrome *c* are not part of the complexes; they act as mobile electron-transfer agents linking the complexes within the membrane (Figure 9.7; Table 9.4). The structures of these prosthetic groups and carriers are presented in Appendix 9.5. The existence of a specific sequence of carriers suggests that electrons flow in a linear sequence but this is an oversimplification: for example, some electrons transfer in parallel along identical carriers, others are stored temporarily, to enable one electron rather than two to be transferred. Nonetheless, the complexities do not interfere with the principle underlying electron transfer, so that they are not discussed here (but see the text *Bioenergetics* by Nicholls & Ferguson (2004)). Note that, in the oxidation of these fuels, the use of molecular oxygen is delayed until the terminal reaction, i.e. the cytochrome oxidase, which resides within the inner mitochondrial membrane. Since the catalysis involves a free radical of

oxygen, locating the enzyme within the membrane reduces the risk of release of the free radical, which would cause severe damage to the mitochondrion (see Appendix 9.6 for a discussion of free radicals).

Oxidative phosphorylation

The major role of electron transfer is the generation of ATP from ADP and P_i (oxidative phosphorylation). Since the process occurs within or upon the inner mitochondrial membrane, the latter is an *energy-transducing membrane*, a name also given to membranes in chloroplasts and in bacteria that carry out similar functions.

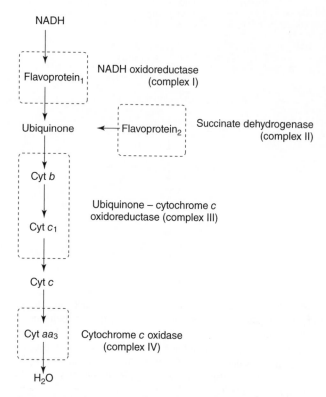

Figure 9.7 *The sequence of electron transfer complexes in the electron transfer chain.* The regions enclosed by broken lines indicate the association of carriers in complexes. Their constituents are listed in Table 9.4. Note that electrons may enter at the level of ubiquinone from sources other than succinate. Complex V is F_oF_1 ATPase (Table 9.4).

The reaction for the generation of ATP is

$$ADP + P_i \rightarrow ATP + H_2O$$

which has a large and positive free energy change: the standard free energy change at pH 7.0 is 30.5 kJ/mol. The actual free energy change, which applies at the physiological concentrations of these reactants within the mitochondria, is considerably higher (about 60 kJ/mol) (for explanation of difference between the standard $\Delta G°$ and the free energy change, ΔG, see Chapter 2). Consequently, this reaction cannot proceed spontaneously; indeed, it proceeds spontaneously in the opposite direction. An important question, therefore, is: how can this large and positive ΔG be converted into a negative ΔG, so that ATP can be synthesised? The principle underlying the answer to this is discussed in Chapter 2. In brief, the reaction must be coupled to another reaction that possesses a negative ΔG, the value of which is numerically larger than the positive ΔG for ATP generation. This is the electron transfer process, which has a large and negative $\Delta G°$ (about −224 kJ/mol) i.e. ATP generation is coupled to electron transfer.

$$NADH + H^+ + \tfrac{1}{2}O_2 \longrightarrow NAD^+ + H_2O$$
$$2.5\,ADP + 2.5P_i \searrow 2.5\,ATP + H_2O$$

A second, related, question is, what is the mechanism of this coupling? The answer, which is central to the process of oxidative phosphorylation, is now presented.

Coupling of electron transfer with oxidative phosphorylation

A major conceptual advance was made when it was appreciated that the coupling between electron transfer and

Table 9.4 Complexes of the mitochondrial electron carriers and ATP generation

Complex no.	Enzyme	Approx.mol.mass (kDa)	Major components	Reduced substrate	Reduced product
I	NADH ubiquinone reductase	8×10^5	FMN, Fe-S protein	NADH	ubiquinol (QH_2)
II	succinate dehydrogenase	1×10^5	FAD, Fe-S protein	succinate	ubiquinol (QH_2)
III	cytochrome bc_1 (ubiquinol-cytochrome c reductase)	3×10^5	cyt b, cyt c_1, Fe-S protein	ubiquinol	cyt c.Fe^{2+}
IV	cytochrome c oxidase (cytochrome aa_3)	2×10^5	cyt a, cyt a_3, copper ion	cyt c.Fe^{2+}	H_2O
V	F_oF_1 ATPase (ATP synthase)	5×10^5	proteins	−	−

Fe-S is the abbreviation for iron–sulphur complex.

QH_2 is sometimes used as an abbreviation for ubiquinol.

oxidative phosphorylation need not be a high energy compound, as in glycolysis, but could be some other form of energy; i.e. a high energy state. In 1961, Peter Mitchell proposed that this high energy state is an electrochemical potential across the inner mitochondrial membrane due to an uneven distribution of protons. The concept became known as the chemiosmotic theory: he was awarded a Nobel Prize for this in 1978.

The energy required to set up the uneven distribution of protons across the inner mitochondrial membrane is transferred from energy made available during transfer of electron along the carriers. In other words, the energy is used to pump the protons from the mitochondrial matrix across the mitochondrial membrane into the inter-membrane space. The electrochemical potential is the gradient of protons plus the electrical potential across the membrane, which is generated by this translocation of protons. The outer surface of the inner mitochondrial membrane is positive, by about 150 to 170 mV; that is, it is the combination of the difference in proton gradient (ΔpH) and the membrane potential ($\Delta\Psi$) which provides the energy for ATP generation. This combination is known as the *proton-motive force* (Δp). Its value is calculated from these two components, according to the equation:

$$\Delta p = \Delta\Psi - 2.303 \, RT/F \cdot \log \frac{[H_i^+]}{[H_o^+]}$$

or

$$\Delta p = \Delta\Psi - 2.303 \frac{RT}{F} \Delta pH$$

where pH is a logarithmic function of [H$^+$] ($-\log[H^+]$), $\Delta\Psi$ is the membrane potential, R is the gas constant, T the absolute temperature (in degrees K) and F the Faraday constant. Since the value of 2.303 RT/F is approximately 60 mV, the equation simplifies to

$$\Delta p = \Delta\Psi - 60 \, \Delta pH$$

The proton motive force is also known as the proton electrochemical potential.

It should be noted that the electrochemical gradient across the mitochondrial membrane is not novel or unusual membrane potential across the plasma membrane of many cells is set up, due to the movement of K$^+$ ions across the membrane. This is particularly important in muscle or nerve (Figure 9.8).

> The proton electrochemical gradient across the mitochondrial membrane is an indication of the extent to which the proton gradient is removed from equilibrium, and it is the electrochemical gradient that is used to generate ATP.

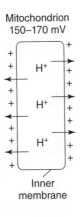

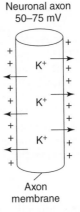

Figure 9.8 *Simple diagram of mitochondrial H$^+$-ion movement and axonal K$^+$-ion movement to establish membrane potentials across membranes.* Note that H$^+$ movement from the mitochondrial matrix to the outer surface of the inner membrane requires a specific proton pump that requires energy, which is transferred from electron transfer, whereas the K$^+$ ion movement occurs via an ion channel with energy provided from the concentration difference of K$^+$ ions on either side of the membrane (approximately 100-fold). The movement of both the protons and K$^+$ ions generates a membrane potential. The potential across the membrane of the nerve axon provides the basis for nervous activity (see Chapter 14).

Mechanism of proton translocation

Protons are translocated across the membrane by what is described as a 'proton pump'. How does the pump operate? The change in redox state experienced by the prosthetic groups of the enzymes in the chain causes conformational changes in the proteins that alter the affinities of some amino acid side-chain groups for protons. In addition, there is a change in the direction in which these groups face in the membrane. Consequently, oxidation results in an association with a proton on the matrix side of the membrane whereas reduction results in reversal of the direction that the side-chain groups face and an increase in

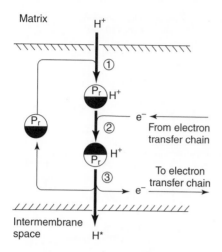

Figure 9.9 *A diagrammatic representation of the mechanism of the proton pump that transfers protons across the inner mitochondrial membrane.* The process can be divided into three parts: (1) the proton carrier (Pr) collects protons from the matrix; (2) acquisition of an electron decreases the affinity of the carrier for the proton and changes its conformation so that the binding site for the proton faces into the inter-membrane space; (3) the carrier discharges the proton into the inter-membrane space and reverts to its former conformation. The electron continues to pass along the transfer chain.

dissociation of the protons into the inter-membrane space (Figure 9.9).

For a proton motive force to develop, the inner mitochondrial membrane must have a very low permeability to protons so that they do not simply flow back down their concentration gradient and dissipate the high-energy state. Indeed, support for the chemiosmotic theory was first provided by the fact that the rate of the leak of protons back across the membrane is very low, although it can occur under special conditions. When it occurs, it is known as uncoupling (Box 9.2).

Coupling of the proton motive force to ATP generation

The inner surface of the inner mitochondrial membrane is covered with many small particles (Figure 9.10) which, when isolated, hydrolyse ATP to ADP and P_i. Under these conditions this isolated protein particle is, therefore, an ATPase, called the *F_1 ATPase*. However, within the mitochondria, the particles (i.e. the enzymes) catalyse the generation, not the hydrolysis, of

> In biochemistry, an enzyme, hormone or process is often named on the basis of the first experimental results that were obtained, even if the name does not describe the actual in vivo function, which is usually discovered after the naming of the compound or process.

Box 9.2 Uncoupling and war

During the First World War, trinitrotoluene (abbreviated to TNT) was the explosive used in shells and bombs. It was noticed that many of the women who packed the explosive into shells suffered from fever and loss of body weight. In some cases the fever was fatal. The cause was not known. Some time later, it was discovered that a very similar chemical, namely dinitrophenol, was an uncoupling agent. The fever and loss of weight were due to uncoupling caused by TNT (i.e. causing a leak of protons back across the mitochondrial membrane).

In 1990, a group of Norwegian soldiers participated in a combat exercise for several days which involved arduous physical activity with limited carbohydrate and sleep. Immediately after completion of the exercise they were tested on a treadmill and their rates of oxygen consumption were much higher than expected for the amount of work performed. These conditions would have increased the fatty acid concentration within the muscle fibres, which could have exceeded the binding capacity of the binding protein within the cell so that the concentrations of *free* fatty acids would increase markedly. It is speculated that this would have damaged the mitochondrial membranes, leading to uncoupling.

If damage had been sufficiently severe and prolonged, as it could be under condition of actual combat, it could result in permanent, or at least chronic, damage to mitochondria with the risk of cell death from either necrosis or apoptosis (Chapter 20). This could cause long-term illness well after the combat had finished.

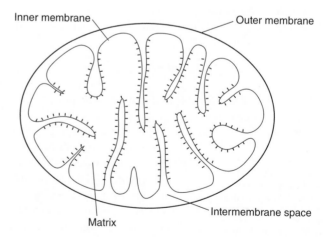

Figure 9.10 *Simple diagram showing the ATPase particles attached to the cristae. The particles are the sites of ATP synthesis.*

ATP. The enzyme should therefore have been termed *ATP synthase* since, in vivo, it catalyses the following reaction:

$$ADP + P_i + H^+ \rightarrow ATP + H_2O$$

The F_1 ATPase consists of a cluster of at least five different proteins which are linked to another complex of proteins within the inner membrane, known as F_o. The combination of the F_1 and F_o protein complexes is Complex V (Table 9.4). Simple diagrams to illustrate their structural relationship and their function in the generation of ATP are presented in Figures 9.11(a) and (b).

It is the movement of protons via the F_o and F_1 complexes, from the intermembrane space to the matrix, that is responsible for ATP generation. The biochemical details of how these complexes respond to the movement of protons to generate the ATP is the field of structural biochemistry, which is not dealt with in detail here (for the details, see Nicholls & Ferguson (2004)). A simplified account of the process is as follows. The substrates ADP and P_i bind to specific sites on one of the proteins of the F_1 ATPase, where they react to form ATP. However, at this stage, no energy has been transferred from the electrochemical gradient, i.e. the formation of ATP from ADP and P_i is a near-equilibrium process. This is converted to a non-equilibrium process, and therefore given direction, by the transfer of energy from the proton motive force. This is achieved as follows. As the protons move, via the F_o complex, en route to the matrix, they result in conformational change in the F_1 protein, which is caused by the transfer of energy from the proton motive force. This change in conformation

results in a marked increase in the dissociation constant of the enzyme for ATP (i.e. the ATP now dissociates much more readily from the enzyme into the matrix of the mitochondria). This change 'pulls' the other reactions to the right, so that ATP is generated from ADP and P_i (i.e. all the reactions now have a direction, which is towards ATP formation). The reactions are as follows:

(i) $\quad ADP_{free} + enzyme \rightleftharpoons enzyme\text{-}ADP$

(ii) $\quad enzyme\text{-}ADP + Pi_{free} \rightleftharpoons P_i\text{-}enzyme\text{-}ADP$

(iii) $\quad P_i\text{-}enzyme\text{-}ADP \rightleftharpoons enzyme\text{-}ATP + H_2O$

$$enzyme\text{-}ATP \xrightarrow{\qquad} enzyme^*\text{-}ATP$$
$$H^+_{in} \qquad\qquad H^+_{out}$$

(iv) $\quad enzyme^*\text{-}ATP \xrightarrow{\qquad} enzyme + ATP$

where enzyme* represents the change in conformation of one of the F_1 proteins that is caused by proton movement. In this conformation, ATP dissociates readily from the enzyme to be released into the matrix of the mitochondria. The overall reaction is, therefore,

$$ADP_{free} + P_{i\,free} \longrightarrow ATP_{free} + H_2O$$

These reactions are shown in Figure 9.12.

The story is, however, not yet complete. There are two major considerations that require discussion of mitochondrial transport. First, almost all the ATP that is generated in the mitochondria is actually used in the cytosol. Consequently, ATP must be transported out of the

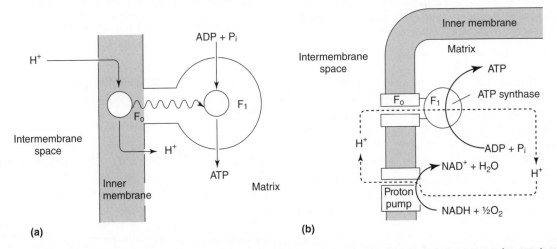

(a)   **(b)**

Figure 9.11 (a) *Simple diagram of the two components of an ATPase particle.* The particle comprises two protein complexes, the F_o and F_1. The latter complex (known as F_1-ATPase) generates ATP from ADP and P_i. The F_o complex is present in the membrane and is responsible for movement of protons across the membrane. The arrow represents some form of communication possibly conformational, between F_o and F_1. **(b)** *Diagrammatic representation of the roles of the F_o and F_1 complexes in generating ATP.* The proton circuit, indicated by the dotted line, provides the energy for ATP generation. However, protons do not pass *through* the F_o protein complex as indicated in this figure. It actually rotates to transfer protons from the intermembrane space to the F_1 complex. Similarly, protons do not pass through the F_1 complex, but they change its conformation, as they move into the matrix.

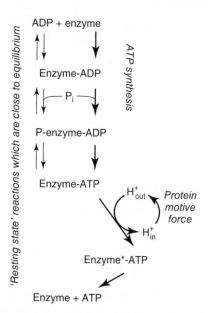

Figure 9.12 *Diagrammatic representation of the reactions by which the F_1 protein complex (F_1-ATPase) generates ATP.* The enzyme (more correctly a protein component of the enzyme complex) binds ADP and P_i and forms ATP in an equilibrium reaction (i.e. no energy is transferred). The movement of protons through the F_0/F_1-complexes changes the conformation of the enzyme so that it readily discharges its ATP into the matrix, i.e. the conformation change caused by the proton motive force increases the dissociation constant for the ATP binding to the enzyme. This 'pulls' all the reactions towards ATP generation, so that net ATP generation results. This is illustrated by (1) the reversible arrows on the left-hand side indicating no net generation of ATP. (2) Irreversible arrows on the right-hand side indicating the effect of the conformation change on the enzyme complex. For explanation of how non-equilibrium reactions provide direction in a process, see Chapter 2.

mitochondria to satisfy the demands in the cytosol and, of course, the ADP and P_i, which are produced in the cytosol, must be transported into the mitochondrial matrix. Second, the major fuels for ATP generation in the mitochondria are pyruvate, fatty acids (and ketone bodies) and glutamine. They are oxidised within the matrix, so they must be transported from the cytosol across the inner membrane into the matrix for oxidation by the Krebs cycle. The processes of transport across the inner membrane are now discussed.

Transport into and out of mitochondria

Mitochondria possess an inner and an outer membrane (see Figure 9.1). The outer membrane contains proteins known as porins that act as pores for entry into the inter-membrane

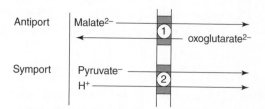

Figure 9.13 *Examples of mitochondrial transport systems for anions.* ① The antiport system transfers malate into but oxoglutarate out of the mitochondrion. ② The symport system transfers both pyruvate and protons into the mitochondrion across the inner membrane. Both transport processes are electroneutral.

space of all metabolites. The outer membrane also acts as a protective sheath for the inner membrane and an attachment site for some enzymes. In contrast, the inner membrane is not only the energy-transducing membrane but is also a barrier. It is freely permeable only to a few small molecules (oxygen, water, carbon dioxide, ammonia, acetate, other small fatty acids and ethanol). The transport of other compounds requires specific transport proteins. Furthermore, some of the compounds that are transported into or out of the mitochondria are anions (e.g. pyruvate⁻, ATP^{4-}, ADP^{3-}), which poses a specific problem. Movement of an ion across the membrane takes with it an electric charge which, in the case of the mitochondria, could disturb the potential that is set up by the proton pumping associated with electron transfer. This is avoided in two ways:

- A similarly charged ion is transported simultaneously but in the opposite direction (known as an antiport process).

- An oppositely charged ion is transported in the same direction (known as a symport process) (Figure 9.13).

The transport is, therefore, electroneutral. If the transport does, in fact, result in the transfer of a net charge, it is termed electrogenic (e.g. the proton pumps in the inner membrane).

Transport of fuels into the mitochondria

The three major fuels for oxidation within the mitochondria are pyruvate, fatty acids in the form of fatty acyl-carnitine, and glutamine – all of which require specific transport processes (Figure 9.14). The hydrogen atoms in the NADH in the cytosol comprise a fourth fuel that must also be transported into the mitochondria for oxidation.

Transport of NADH

The conversion of glucose to pyruvate (glycolysis) generates NADH from NAD^+ in the cytosol so that, to maintain

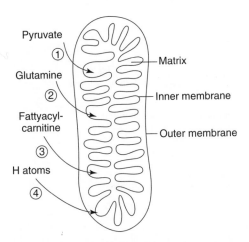

Figure 9.14 *Simple diagram illustrating the transport of the major fuels for the Krebs cycle, and hydrogen atoms for the electron transfer chain, across the inner mitochondrial membrane.*
1. Pyruvate is transported by an anion symport system (Figure 9.13).
2. Glutamine, which has no net charge, is transported by a specific glutamine transporter.
3. Fatty acyl carnitine is transported via a translocase that transports acylcarnitine into and carnitine out of the mitochondrion (Chapter 7).
4. The transport processes for hydrogen atoms are described in Figures 9.17 and 9.18. The biochemical problems in transport of hydrogen atoms are now discussed.

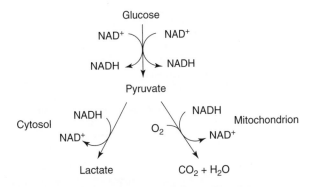

Figure 9.15 *Fate of NADH produced in glycolysis.* In hypoxic or anoxic conditions, pyruvate is converted to lactate with oxidation of NADH. In aerobic conditions, NADH is oxidised as shown in Figure 9.17 or 9.18 and pyruvate is oxidised via the Krebs cycle and the electron transfer chain.

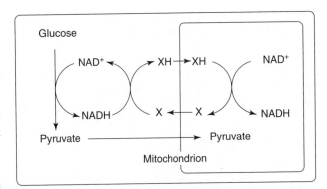

Figure 9.16 *The principle of the transfer shuttle of hydrogen atoms into the mitochondrion.* A dehydrogenase in the cytosol generates XH from NADH. XH is transported into the mitochondrion where a second dehydrogenase catalyses a reaction in which the XH reduces NAD^+ to NADH. X then returns to the cytosol. The nature of XH is considered in Figures 9.17 and 9.18.

the rate of glycolysis, NAD^+ must be regenerated from NADH. Under anaerobic or hypoxic conditions, NAD^+ is regenerated in the reaction catalysed by lactate dehydrogenase in the cytosol:

$$\text{pyruvate} + NADH + H^+ \rightarrow \text{lactate} + NAD^+$$

Under aerobic conditions, the hydrogen atoms of NADH are oxidised within the mitochondrion pyruvate is also oxidised in the mitochondrion (Figure 9.15). However, NADH cannot be transported across the inner mitochondrial membrane, and neither can the hydrogen atoms themselves. This problem is overcome by means of a substrate shuttle. In principle, this involves a reaction between NADH and an oxidised substrate to produce a reduced product in the cytosol, followed by transport of the reduced product into the mitochondrion, where it is oxidised to produce hydrogen atoms or electrons, for entry into the electron transfer chain. Finally, the oxidised compound is transported back into the cytosol. The principle of the shuttle is shown in Figure 9.16.

Two biochemically distinct shuttles, the malate/aspartate shuttle (Figure 9.17) and the glycerol phosphate shuttle (Figure 9.18) are involved in the transfer (Appendix 9.7).

The maximum capacity of these shuttles can be assessed from the maximal in vivo activity of the mitochondrial enzymes involved. This capacity can then be compared with that necessary to re-oxidise NADH in order to maintain the maximal rate of glycolysis. These data demonstrate that there is a remarkable distinction between insect flight muscles in which the glycerol phosphate shuttle is active and vertebrate muscles in which the malate/aspartate shuttle operates. No explanation for this difference has been put forward.

Adenine nucleotide and phosphate transport

The adenine nucleotide transporter is known as a *translocase* – it transports ADP into and ATP out of the mitochondrion in such a way that, when one molecule of ADP is transported in, one molecule of ATP is transported out

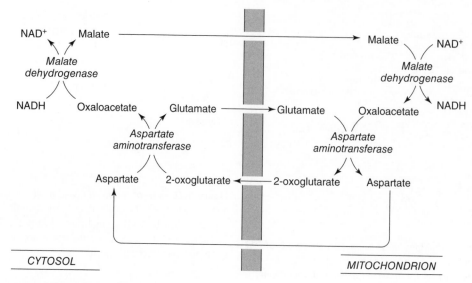

Figure 9.17 *The malate/aspartate shuttle.*
The shuttle involves the following reactions:
(i)   In the cytosol, catalysed by cytosolic malate dehydrogenase, oxaloacetate is converted to malate.
$$\text{oxaloacetate} + \text{NADH} \rightarrow \text{NAD}^+ + \text{malate}$$
(ii) Malate is then transported across the inner membrane, the non-equilibrium step in the shuttle.
(iii) In the mitochondrial matrix, malate is oxidised
$$\text{malate} + \text{NAD}^+ \rightarrow \text{oxaloacetate} + \text{NADH}$$
catalysed by mitochondrial malate dehydrogenase.
(iv) The hydrogen atoms from NADH are then transferred along the electron transfer chain to be oxidised by oxygen.
(v)   The oxaloacetate is then transported from mitochondrion into the cytosol but not directly, since there is no transporter for oxaloacetate in the mitochondrial membrane. This problem is solved by conversion of oxaloacetate to aspartate, by transamination, and it is the aspartate that is transported across the inner mitochondrial membrane to the cytosol, where oxaloacetate is regenerated from aspartate by a cytosolic aminotransferase enzyme.

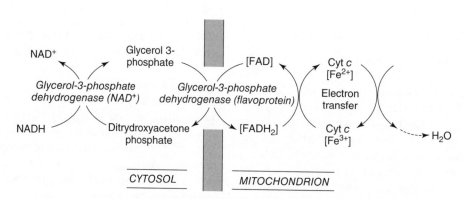

Figure 9.18 *The glycerol phosphate shuttle.* In the cytosol, NADH is oxidised in a reaction in which dihydroxyacetone phosphate is reduced to glycerol 3-phosphate, catalysed by glycerol-3-phosphate dehydrogenase (NAD$^+$ linked):

$$\text{dihydroxyacetone phosphate} + \text{NADH} + \text{H}^+ \rightarrow \text{glycerol 3-phosphate} + \text{NAD}^+$$

The glycerol 3-phosphate is oxidised by a second glycerol-3-phosphate dehydrogenase but this enzyme is located within the inner mitochondrial membrane. It removes electrons from glycerol 3-phosphate and transfers them directly to the electron transfer chain, at the level of ubiquinone, and the dihydroxyacetone phosphate remains in the cytosol (or at least in the intermembrane space from where it diffuses into the cytosol). The non-equilibrium reaction, which gives direction to the overall shuttle, is catalysed by the mitochondrial glycerol phosphate dehydrogenase. There is a suggestion that this shuttle is important is brain, where glutamate and aspartate have specific roles as neurotransmitters (Chapter 14).

of the mitochondrion. The result of this translocation is that three negative charges enter (ADP^{3-}) and four leave the mitochondria (ATP^{4-}) (Figure 9.19). This would result in a net negative charge on the outside of the inner membrane but this is prevented by movement of a proton from the outer side of the membrane into the matrix of the mitochondrion. This results in a reduction in the proton gradient and hence a reduction in the proton motive force (possibly by as much as 25%). This is, however, biochemically beneficial. It results in a higher ATP/ADP concentration ratio in the cytosol than that in the mitochondrial matrix so that the hydrolysis of ATP releases (transfers) sufficient energy to 'drive' all the energy-requiring processes in the cytosol. The ratio is approximately 10 in the cytosol but probably varies from tissue to tissue. The ratio in mitochondria is about one.

The transport of phosphate into the matrix is also important for ATP generation and involves symport transport with hydrogen ions (Figure 9.19).

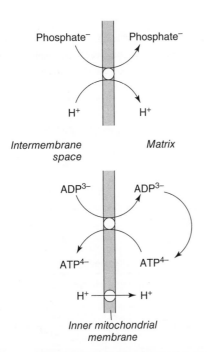

Figure 9.19 *Adenine nucleotide translocase and phosphate transfer into the matrix.* Phosphate is transported into the mitochondria with protons in a symport transport system. The adenine nucleotide translocase transports ADP^{3-} into and ATP^{4-} out of the mitochondria, i.e. it is electrogenic. The charge is neutralised by H^+ movement into the matrix from the proton motive force which utilises about 25% of the energy in the proton motive force.

'Energy' transport in the cytosol: the creatine/phosphocreatine shuttle

In addition to the processes described above, there still remains one further process which, at least in some cells or tissues, is required prior to the utilisation of ATP in the cytosol: that is, the transport of 'energy' within the cytosol, via a shuttle. The transport of ATP out and ADP into the mitochondrion, via the translocase, results in a high ATP/ADP concentration ratio in the cytosol. However, a high ratio means that the actual concentration of ADP in the cytosol is low, which could result in slow diffusion of ADP from a site of ATP utilisation back to the inner mitochondrial membrane. If sufficiently slow, it could limit the rate of ATP generation. To overcome this, a process exists that transports 'energy' within the cytosol, not by diffusion of ATP and ADP, but by the diffusion of phosphocreatine and creatine, a process known as the *phosphocreatine/creatine shuttle*. The reactions involved in the shuttle in muscle help to explain the significance of the process. They are:

- The enzyme creatine kinase is present in two positions in the muscle fibre: within the intermembrane space of the mitochondria and close to the myofibrils.

- The ATP, which is transported out of the mitochondrion, immediately phosphorylates creatine, catalysed by the creatine kinase in the intermembrane space.

  ATP + creatine → phosphocreatine + ADP

- The ADP that is produced is immediately available for transport back into the mitochondrion for generation of ATP.

- The phosphocreatine diffuses to the site of ATP utilisation (i.e. the myofibrils in skeletal muscle).

- At this site, creatine kinase catalyses the phosphorylation of ADP by phosphocreatine with the production of ATP which, in turn, is used by the energy-requiring process; that is, the cross-bridge cycle (Chapter 13).

- The resultant creatine, rather than ADP, diffuses back to the mitochondrion to continue the shuttle.

The concentration of phosphocreatine is usually greater than that of ATP and that of creatine greater than that of ADP so that these metabolites diffuse more rapidly. This is because diffusion depends upon the concentrations of participants and the concentration gradient: the larger the gradient, the greater is the rate of diffusion. Consequently, the shuttle is important in cells where the distance between the sites of ATP utilisation and the mitochondria is large and where the rate of ATP utilisation is high. Two examples are:

- In some muscles, some of the mitochondria are present just beneath the plasma membrane (sub-sarcolemmal mitochondria), e.g. in the leg muscles of marathon and ultramarathon runners. This localisation of some of the mitochondria has the advantage of minimising the distance that oxygen has to diffuse from the capillaries across the plasma membrane to the mitochondria but the disadvantage of increasing the diffusion distance of ADP from the myofibrils to the mitochondria (Figure 9.20).

- In spermatozoa, the phosphocreatine shuttle is present to transfer energy from the mitochondria to the flagellum, which is essential for swimming of the sperm (Figure 9.21) (see also Chapter 19).

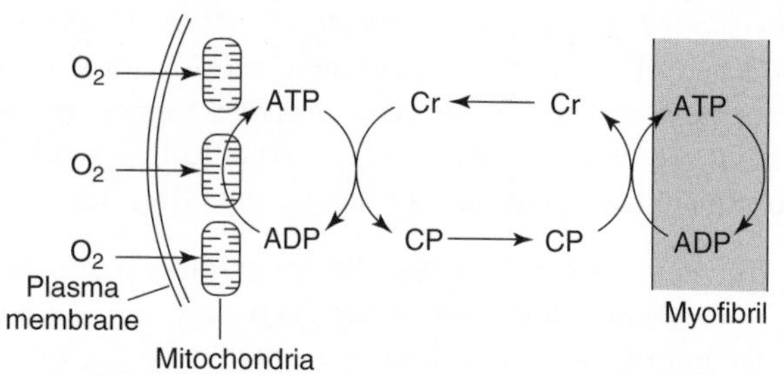

Figure 9.20 *The creatine/phosphocreatine shuttle between subsarcolemmal mitochondria and myosin ATPase in muscle.* The distance between the mitochondria that reside just below the plasma membrane (sarcolemma) and the myofibrils in which the myosin ATPase results in contraction, is long in such muscles. The advantage of the position of these mitochondria is ready access to oxygen and fuel from blood. Such mitochondria are common in endurance athletes.

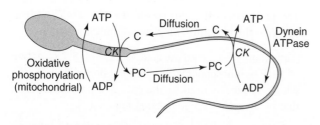

Figure 9.21 *The creatine/phosphocreatine shuttle in spermatozoa.* This shuttle may not be present in all sperm: it will depend upon the distance between the mitochondria and the flagellum. Mitochondria are present in the midpiece just below the head. ATP is required for movement of the flagellum which enables the sperm to swim. Dynein ATPase is the specific 'motor' ATPase, similar to myosin ATPase, that transfers energy from ATP to the flagellum. A deficiency of creatine may explain low sperm motility in some infertile men. CK – creatine kinase. Deficiences of enzymes in the pathway for synthesis of creatine are known to occur (see Appendix 8.3).

Regulation of fluxes

In any cell that depends on aerobic metabolism, if the rate of ATP utilisation increases, the rate of the Krebs cycle, electron transfer and oxidative phosphorylation must also increase. The mechanism of regulation discussed here is for mammalian skeletal muscle since, to provide sufficient ATP to maintain the maximal power output, at least a 50-fold increase in flux through the cycle is required so that the mechanism is easier to study (Figure 9.22).

The approach that is used to identify a regulatory mechanism in any biochemical process is discussed in Chapter 3. In brief, the approach is: (i) identify the regulatory enzymes or processes; (ii) study the properties of the enzymes or processes; (iii) on the basis of the properties, formulate a theory of regulation; (iv) test the theory.

This is now done for the Krebs cycle, electron transfer and ATP generation. Finally, the proposed overall mechanism for regulation of flux from glycogen to CO_2 and H_2O is described. The increased flux also requires an increased supply of oxygen (Chapter 13) (Box 9.3).

The Krebs cycle

Citrate synthase, isocitrate dehydrogenase and oxoglutarate dehydrogenase are key enzymes regulating the flux through the cycle; all three catalyse non-equilibrium reactions (Chapter 3).

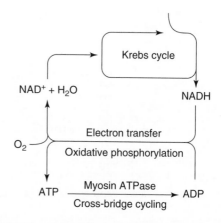

Figure 9.22 *The relationship between ATP utilisation by myosin ATPase and ATP generation by Krebs cycle and electron transfer.* This relationship between the two major energy systems in muscle is critical. The rate of the cycle and electron transfer is controlled, in part, by ATP utilisation by muscle contraction (see below). This is equivalent to a 'market economy' so that the law of supply and demand applies. The greater the demand and hence the use of ATP, the greater is the rate of generation.

Box 9.3 The first experiments on oxygen and life

Joseph Priestley (1733–1804) was a Unitarian minister and schoolteacher but he developed an interest in science and published several books on the history of electricity that won him membership of the Royal Society. For a time he lived next to a brewery, where he saw the large quantities of gas bubbling up from the fermentation vats. This encouraged him to study the gas, which is CO_2, which led to the invention of soda water. He went on to discover ten different gases including nitrogen, ammonia and, most importantly, oxygen in 1774, the discovery of which he published in 1775. (Carl Wilhelm Scheele working in Sweden also discovered oxygen at about the same time.)

The basis of Priestley's discovery was the use of a simple apparatus to collect gas: it consisted of a trough full of mercury over which glass vessels could be inverted to collect the gas. The substance to be heated, mercuric oxide, was placed in another glass vessel on the mercury, and the heating was achieved by focusing the Sun's rays upon it using a lens. He then experimented with the gas given off from mercuric oxide. The following extract is taken from his paper.

EXPERIMENTS IN DIFFERENT FORMS OF AIR

by Joseph Priestley, published in 1775

On the 1st of August, 1774, I endeavoured to extract air from mercurius calcinatus per se; and I presently found that, by means of the lens, air[1] was expelled from it very readily . . . But what surprised me more than I can well express, was, that a candle burned in this air with a remarkably vigorous flame . . .

On the 8th of this month [March, 1775] I procured a mouse, and put it into a glass vessel, containing . . . the air . . . Had it been common air, a full-grown mouse, as this was, would have lived in it about a quarter of an hour. In this air, however, my mouse lived a full half hour; and though it was taken out seemingly dead, it appeared to have been only exceedingly chilled; for, upon being held to the fire, it presently revived, and appeared not to have received any harm from the experiment.

For my further satisfaction I procured another mouse, and putting it into more air . . . it lived three quarters of an hour. However, as it had lived three times as long as it could probably have lived in the same quantity of common air, and I did not expect much accuracy from this kind of test, I did not think it necessary to make any more experiments with mice . . .

. . . Being at Paris in the October following, and knowing that there were several very eminent chemists

in that place . . . I frequently mentioned my surprise at the kind of air which I had got from this preparation to Mr Lavoisier, Mr le Roy, and several other philosophers, who honoured me with their notice in that city; and who, I daresay, cannot fail to recollect the circumstance.

Sadly, Priestley missed the significance of his discovery. From knowledge of Priestley's findings, Antoin-Laurent Lavoisier realised the theoretical importance of them: the gas that he had found that reacted with carbon in food to form a weak acid (carbonic acid) was Priestley's gas. He called the new air 'oxygen', from the Greek words *oxys* (sour, acidic) and *genous* (origin, descent).

With remarkable foresight, Priestley went on in his 1775 paper as follows:

It may hence be inferred, that a quantity of very pure air would agreeably qualify the noxious air of a room in which much company should be confined, and which should be so situated, that it could not be conveniently ventilated; so that from being offensive and unwholesome, it would almost instantly become sweet and wholesome. This air might be brought into the room in casks; or a laboratory might be constructed for generating the air, and throwing it into the room as fast as it should be produced. This pure air would be sufficiently cheap for the purpose of many assemblies, and a very little ingenuity would be sufficient to reduce the scheme into practice . . .

From the greater strength and vivacity of the flame of a candle, in this pure air, it may be conjectured, that it might be peculiarly salutary to the lungs in certain morbid cases, when the common air would not be sufficient to carry off the phlogistic putrid effluvium fast enough. But, perhaps, we may also infer from these experiments, that though pure dephlogisticated air might be very useful as medicine,[2]

Priestley believed that all materials contained an element called 'phlogiston', which was given off when they burned. Air in which things had been burned became less able to support combustion because it was then saturated with 'phlogiston'. Accordingly, Priestley called his gas, in which a candle flame burned brightly, 'dephlogisticated air'.

In fact, this discovery, and the realisation of its significance by Lavoisier, led to the abandonment of the phlogiston theory and, with his other work, established the basis of modern chemistry.

---

[1] Air was the term used for gas at that time.

[2] We might now conclude that his statement about being 'very useful in medicine' is a massive understatement.

Citrate synthase   There are three properties of citrate synthase that are relevant to regulation. The product of the reaction, citrate, is an allosteric inhibitor of the enzyme. The concentration of acetyl-CoA in muscle is well above the K_m of citrate synthase for acetyl-CoA. Consequently, the activity of this enzyme is flux-generating for the cycle plus the transfer of electrons along the electron transfer chain, i.e. the process from acetyl-CoA to molecular oxygen can be considered as a 'transmission sequence', as defined in Chapter 3. In contrast, the concentration of oxaloacetate is well below the K_m, so that variations in its concentration can regulate the enzyme activity and therefore, the flux through the cycle.

Isocitrate dehydrogenase and oxoglutarate dehydrogenase
The activities of both these enzymes are increased by Ca^{2+} ions and the former enzyme is activated by ADP. Oxoglutarate dehydrogenase activity is inhibited by a high mitochondrial concentration ratios of ATP/ADP and succinyl CoA/CoASH. These properties, including the effect of Ca^{2+}, are similar to those of pyruvate dehydrogenase but, for the Krebs cycle enzymes, the changes in the concentration of these regulatory molecules affect the enzyme directly, not via an interconversion cycle (Figure 9.23).

The theory of regulation of the cycle is as follows. First, an increase in oxaloacetate concentration increases the activity of citrate synthase and hence the cycle. The concentration of oxaloacetate is regulated by the activity of the enzyme pyruvate carboxylase, which catalyses the reaction:

$$\text{pyruvate} + CO_2 + ATP + H_2O \xrightarrow[\overset{\text{acetyl-CoA}}{\vdots}]{\oplus} \text{oxaloacetate} + ADP + P_i + 2H^+$$

This enzyme is activated by acetyl-CoA. Hence, an increase in the concentrations of its substrate, pyruvate, and/or acetyl-CoA increase the activity of the enzyme, which increases the concentration of oxaloacetate (Figure 9.24). Thus, in response to a greater demand for ATP (i.e. increased work by the muscle) the activity of citrate synthase is increased by an increased concentration of oxaloacetate and a decreased concentration of citrate. Secondly, the mitochondrial concentration of Ca^{2+} ions is increased, which is due the increased cytosolic concentration which, is controlled by activity of the motor nerves (Chapter 13). This increase in Ca^{2+} ion concentration increases the activi-

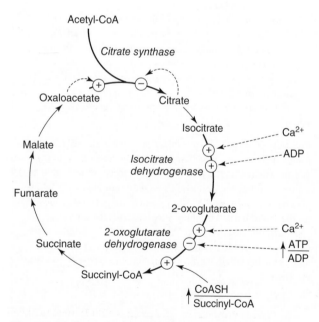

Figure 9.23 *Properties of the three enzymes that control the flux through the Krebs cycle. During physical activity. The CoASH/ succinyl CoA concentration ratio increases whereas that of ATP/ADP ratio decrease. These changes increase the flux through the cycle.*

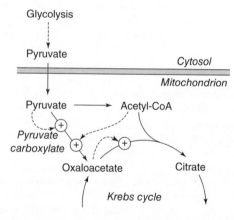

Figure 9.24 *Control of the oxaloacetate concentration and hence the flux through the cycle by pyruvate carboxylase. The activity of pyruvate carboxylase is increased by an increase in its substrate, pyruvate, and its allosteric regulator, acetyl-CoA. Regulation of the activity is important to increase the concentration of oxaloacetate which increases the flux through the cycle. An increase in the rate of glycolysis increases the concentration of pyruvate, and an increase in the rate of fatty acid oxidation increases that of acetyl-CoA. Both result in an increase in the concentration of oxaloacetate and hence in the flux through the cycle, providing coordination between the rates of glycolysis, fatty acid oxidation and the cycle.*

ties of isocitrate dehydrogenase and oxoglutarate dehydrogenase. Thirdly, the fall in the ATP/ADP concentration ratio also increases the activities of these two enzymes (Figure 9.23). The increase in flux through the cycle must occur in concert with increased flux through glycolysis or fatty acid oxidation in order to increase the rate of acetyl-CoA formation. The overall mechanism of control of glycolysis and the Krebs cycle is discussed below.

Regulation of electron transfer and oxidative phosphorylation

The processes of electron transfer and ATP generation are coupled together so that the regulation of the rate of both processes is interdependent. In summary, the rates of both processes are controlled by the concentration ratio ATP/ADP in the mitochondrial matrix: in addition, electron transfer is regulated by the concentration ratio $NAD^+/NADH$.

The major factor initiating these changes in muscle is an increase in the cytosolic concentration of Ca^{2+} due to an increase in nervous stimulation of the muscle. The increase in the cytosolic Ca^{2+} ion concentration changes two factors that are directly relevant to regulation of two processes.

- It simulates the activity of myosin ATPase, which results in contraction of the muscle (i.e. increased cross-bridge cycling). This increases the rate of utilisation of ATP and hence increases the concentration of ADP.

- It increases the concentration of Ca^{2+} ions in the mitochondrial matrix and this stimulates the flux through the Krebs cycle and hence electron transfer. This increases the rate of generation of ATP (Figure 9.25).

These two effects provide the basis for the mechanism of the regulation of electron transfer and ATP generation, as follows.

Control of flow of electrons along the electron transfer chain

An increase in the rate of cross-bridge cycling increases force of contraction of muscle, which decreases the cytosolic concentration of ATP and increases that of ADP. This results, via the adenine nucleotide translocase, in a similar change in direction of ATP and ADP concentrations within the mitochondrial matrix (i.e. a decrease in the ATP/ADP concentration ratio).

The increased flux through the Krebs cycle results in an increase in the matrix concentration of NADH and a decrease in that of NAD^+, i.e. the $NADH/NAD^+$ concentration ratio increases. An important point is that all of these changes depend on the increase in the cytosolic and hence

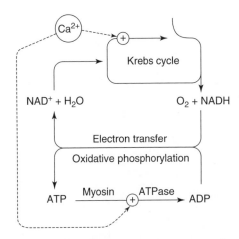

Figure 9.25 *Control of the Krebs cycle and myosin-ATPase by direct effects of Ca^{2+} ions and the resultant effects on electron transfer and oxidative phosphorylation in muscle.* The stimulation of the Krebs cycle by Ca^{2+} ions results in an increase in the $NADH/NAD^+$ concentration ratio, which stimulates electron transfer. The stimulation of myosin-ATPase by Ca^{2+} lowers the ATP/ADP concentration ratio, which also stimulates electron transfer. The Ca^{2+} ions are released from the sarcoplasmic reticulum in muscle in response to nervous stimulation. In addition, generation of ADP by myosin ATPase increases the ADP concentration, which stimulates the cycle. Note that a lack of oxygen will prevent generation of ATP (Chapter 13).

the mitochondrial Ca^{2+} ion concentration, which depends upon release of Ca^{2+} ions from the sarcoplasmic reticulum.

An important point in the regulation of these processes is that all the reactions from mitochondrial NADH, to and including cytochrome c, are near-equilibrium (Figure 9.26(a)) (Appendix 9.8): that is, there is only one reaction in the electron transfer chain that is non-equilibrium – the terminal reaction catalysed by cytochrome oxidase. There is some similarity with the process of glycolysis in which the initial reaction and the terminal reactions are the non-equilibrium reactions (Figure 9.26(b)).

Consequently, the major part of the electron transfer process, including that which generates ATP, can be summarised in one complex equation:

$$NADH + 2cytc\,Fe^{3+} + 2.5ADP + 2.5P_i + 2H^+ \rightleftharpoons$$
$$NAD^+ + 2cytc\,Fe^{2+} + 2.5ATP + 3H_2O$$

The importance of this is that it is not simply the changes in NADH and ADP concentrations that regulate the rate of electron transfer and ATP generation, but also the concentration ratios: $NADH/NAD^+$ and ATP/ADP; that is, the falls in the concentrations of ATP and NAD^+ 'pull' the process to the right whereas the increases in the

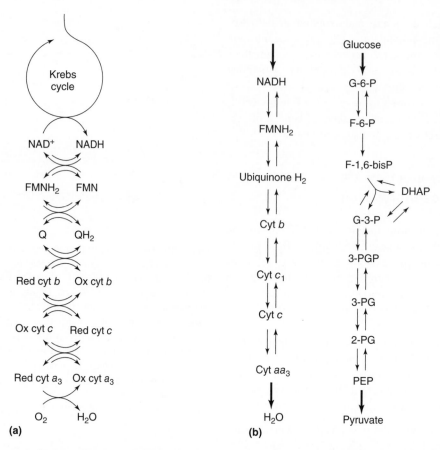

Figure 9.26 (a) *Near-equilibrium and non-equilibrium reactions in the electron transfer chain.* The electron transfer chain is considered to be the latter part of the physiological Krebs cycle (see above). The non-equilibrium processes are the Krebs cycle and the terminal reaction cytochrome oxidase. All other reactions are near-equilibrium, including the ATP-generating reactions. These are not shown in the figure. **(b)** *The similarity of electron transfer chain and glycolysis in the position of near-equilibrium/non-equilibrium reactions, in the two pathways.* In both cases, non-equilibrium reactions are at the beginning and at the end of the processes (see Chapters 2 and 3 for description of these terms and the means by which such reactions can be identified).

concentrations of ADP and NADH 'push' the process to the right as follows:

$$NADH \uparrow + 2\mathrm{cyt}c\mathrm{Fe}^{3+} + 1.5 \uparrow ADP + 3P_i + 2H^+ \rightarrow$$
$$\downarrow NAD^+ + 2\mathrm{cyt}c\mathrm{Fe}^{2+} + \downarrow 1.5 ATP + 3H_2O$$

The arrows indicate the direction of the changes in the concentrations of the factors in the mitochondrial matrix that increase the flux.

Regulation of flux from glycogen to oxygen

A summary of the regulation of the processes of glycogenolysis, glycolysis, Krebs cycle, electron transfer and ATP generation is presented in Figure 9.27. Muscle tissue is used as an example but the basis of the mechanism applies to other tissues.

Regulation is achieved by changes in two factors: the cytosolic Ca^{2+} ion concentration and the ATP/ADP concentration ratio, as follows.

Ca^{2+} ion concentration   As an action potential travels along the muscle fibre and into the interior of the fibre, via the T-tubule system, Ca^{2+} ions are released from the sarcoplasmic reticulum (Chapter 13). This increases the concentration of Ca^{2+} ions in the cytosol which is followed by an increase in concentration within the mitochondria. These increases in Ca^{2+} ion concentration have four effects:

(i)   It stimulates the activity of myosin ATPase, which increases the rate of hydrolysis of ATP, so that its concentration is decreased whereas that of ADP is increased. These changes, via the translocase, increase the concentration of ADP and lower that of ATP in the mitochondria matrix, which stimulates the rate of electron transfer.

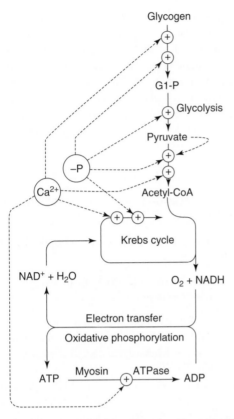

Figure 9.27 *An overview of the mechanisms by which the rates of glycolysis, Krebs cycle, electron transfer and ATP generation are regulated by changes in the Ca²⁺ ion concentration and the ATP/ ADP concentration ratio in muscle.* The symbol ⊖P represents a decrease in the ATP/ADP concentration ratio. An increase in the cytosolic concentration of Ca^{2+} ions and a decrease in the ATP/ ADP concentration ratio increase the activity of phosphorylase and hence glycogen breakdown, pyruvate dehydrogenase and hence pyruvate conversion to acetyl-CoA, and the Krebs cycle. A decrease in the ATP/ADP concentration ratio increases glycolytic flux (via effects on phosphofructokinase and pyruvate kinase). The increase in the NAD⁺/NADH concentration ratio and the decrease in the ATP/ADP concentration ratio, which are consequent upon the increase in Ca^{2+} ion concentration, increase the flux through the electron transfer chain and ATP generation. For further details on control of glycogenolysis and glycolysis, see Chapter 6.

(ii) It stimulates glycogen breakdown and hence the rate of glycolysis is increased. An increase in glycolysis increases the concentration of pyruvate, which, via pyruvate carboxylase, increases the concentration of oxaloacetate which, in turn, increases the flux through the Krebs cycle.

(iii) It increases the activities of isocitrate dehydrogenase and oxoglutarate dehydrogenase; the former lowers the concentration of citrate, which further increases the activity of citrate synthase and hence flux through the cycle.

(iv) Both effects (ii) and (iii) increase the NADH/NAD⁺ concentration ratio in the matrix, which increases the flux of electrons along the transfer chain.

ATP/ADP concentration ratio A decrease in the ATP/ ADP concentration in the cytosol has the following effects:

- It increases the rate of glycolysis from glucose or glycogen (via activation of glycogen phosphorylase, glucose transport and phosphofructokinase (Chapter 6)).

- It increases the activity of isocitrate dehydrogenase and oxoglutarate dehydrogenase and hence the flux through the cycle.

- The changes in the cytosolic ADP and ATP concentrations increase the mitochondrial matrix concentration of ADP and decrease that of ATP, via the adenine nucleotide translocase, which stimulates the flux of electrons along the transfer chain.

The regulation of the overall processes is summarised in Figure 9.27. See also experiments with isolated mitochondria described in Appendix 9.9.

Similarities between control of the human tissue energy economy and national financial economies

Although it is a very large mental jump from the control of ATP generation to the control of financial/national economies, comparison between the two helps provide understanding of the biochemistry of regulation of energy metabolism in muscle. The control of ATP generation involves (i) centrally directed regulation and (ii) regulation by 'market forces', as follows:

(i) '*Central control*' is achieved via changes in the Ca^{2+} ion concentration, which depend on nervous stimulation and therefore eventually the brain.

In processes in other tissues, endocrine glands play the same role as the brain and Ca^{2+} ions on cyclic ATPase the regulatory effectors.

(ii) '*Market economy control*'. In muscle, this is achieved via changes in the ATP/ADP concentration ratio.

To take the anology further disturbance in the market force–control causes problems in national economies and so it does in the biochemical economy in muscle. For example, if the rate of ATP expenditure exceeds the capacity for generation of ATP, fatigue rapidly results. In many people, this leads to ill health: it is a major factor limiting their daily activities and, therefore, their well-being

(Chapter 13). Failure of central control of muscle action can also have severe consequences. Malfunctioning of some parts of the central control mechanism can result in failure of muscles to contract normally and in a coordinated manner, e.g. Parkinson's disease (Chapter 14) or result in excessive fatigue (chronic fatigue syndrome (Chapter 13)).

The physiological importance of mitochondrial ATP generation

Structural biochemistry and enzymology have led to major advances in understanding the mechanism of ATP generation. The emphasis on this aspect of ATP metabolism has, perhaps, led to neglect of the role of ATP, including the processes that lead to its generation, in providing energy to maintain not only the life of cells but their physiological function, and hence the life of each individual human.

This aspect of the subject is now discussed. It is divided into several sections, as follows:

(i)   The amount of ATP that is generated in the various pathways that oxidise the fuels.

(ii)   The amount of glucose required to generate sufficient ATP for particular physiological activities from either aerobic or anaerobic metabolism.

(iii) Importance of ATP generation for the processes that maintain physiological activities in particular cells.

(iv)   Calculation of the maximal capacity of the Krebs cycle in a tissue or organ can be made from oxygen uptake in vivo, or from the catalytic activity of a specific enzyme in vitro. This is of value and is useful in assessing the importance of the cycle in generation of ATP in many cells or tissues.

(v)   Examples of leisure activities that are supported by aerobic ATP generation.

(vi)   The biochemical mechanisms that convert chemical energy into heat for maintenance of body temperature or maintenance of normal body weight, i.e. the amount of adipose tissue.

The amount of ATP generated in various pathways

The number of molecules of ATP that can be generated from the transfer of a given number of hydrogens (or electrons) along the electron transfer chain to oxygen is usually presented as the P/O ratio; that is, the number of molecules of ATP synthesised per atom of oxygen ($\frac{1}{2}$ O_2) consumed (equivalent to the transfer of two electrons along the chain). This can be done experimentally by isolating mitochondria from the tissue or organ, incubating them in a medium that contains various substrates that feed electrons into the electron transfer chain at different positions and then measuring the amount of ATP generated and oxygen consumed by the mitochondria (Appendix 9.9).

The values are as follows:

• For transfer of two hydrogens from NADH to oxygen, the P/O ratio is 2.5.

• For transfer of two electrons from ubiquinol to oxygen, the ratio is 1.5.

• For transfer of two electrons from cytochrome c to oxygen, the ratio is unity.

This means, for example, that ten molecules of ATP are produced from the oxidation of each molecule of acetyl-CoA in the Krebs cycle together with the electron transfer chain. The complete oxidation of one molecule of glucose generates 30 molecules of ATP. The conversion of glucose or glucose-in-glycogen to lactate (i.e. glycolysis) generates two, for each molecule of glucose used or three for each molecule of glucose derived-from-glycogen that is used (Table 9.5).

Table 9.5 Number of ATP molecules generated from the oxidation of one molecule of various fuels

Fuel	Conditions	ATP yield (mol) per mol of fuel utilised
Glucose	complete oxidation	30
Glucose	conversion to lactate[a]	2
Glycogen	complete oxidation (for one glucose molecule-in-glycogen)	31
Glycogen	conversion to lactate (for one glucose molecule-in-glycogen)[a]	3
Glutamine	conversion to aspartate	14
Palmitate	complete oxidation	106
Acetoacetate	complete oxidation	19

Based on the pathways described in Chapters 6, 7 and 9, with the following assumptions:

• Oxidation of two electrons transferred along the electron transfer chain from NADH to oxygen generates 2.5 molecules of ATP.

• Oxidation of two electrons, from $FADH_2$ to oxygen, generates 1.5 molecules of ATP.

[a] No oxidation at all (i.e. *anaerobic* glycolysis, Chapter 6).

Aerobic or anaerobic metabolism?

Approximately ten times more ATP is generated from the complete oxidation of one molecule of glucose, compared with the conversion of one molecule of glucose to lactate. A number of calculations demonstrate the difference that this would make in the use of fuels to provide sufficient ATP for various activities:

- The ATP required by the human brain is obtained from the complete oxidation of glucose: it requires about 4 g of glucose per hour. If the same amount of ATP were to be generated from the conversion of glucose to lactate, approximately 60 g per hour would be required (i.e. almost 1.5 kg of glucose each day).

- Prolonged physical activities require the complete oxidation of a fuel to provide a sufficient rate of ATP generation. A marathon run requires the oxidation of about 700 g of glycogen, provided no other fuel is used. To generate the same amount of ATP from the conversion of glycogen to lactate, approximately 7 kg of glycogen would be required.

Many other physiological processes are supported totally by the complete oxidation of fuels and cannot be supported by glycolysis alone, as calculations similar to the above would testify. Some of these processes are described in the next section.

Mitochondrial ATP generation for the processes that maintain essential physiological activities

Mitochondrial ATP generation (i.e. that from the cycle plus electron transfer) is necessary to support almost all biochemical or physiological processes that are essential to life, as follows:

- Continual pumping of blood by muscles of the atria and ventricles of the heart.

- Breathing requires sustained activity of diaphragm and intercostal muscles.

- Sustained activity of muscles involved in maintaining posture.

- Mastication of foods. Most foods in their natural state require prolonged activity of the masseter muscles involved in closing the jaws.

- Mental activity requires ion transport, the synthesis of neurotransmitters and their release into the synapse.

- Protein synthesis and degradation (i.e. protein turnover).

- Active transport of, for example, glucose and amino acids by the kidney and small intestine.

The common factor in all these processes, which accounts for the need for mitochondrial ATP generation, is that they require ATP utilisation either continuously or for prolonged periods, so that the most efficient process for the use of fuel to generate ATP is necessary.

Calculation of the maximal rate of the Krebs cycle

From the rate of oxygen consumption of a cell, a tissue or an organ, the rate of generation of ATP can be calculated. Oxygen consumption by an organ, in vivo, can be measured from the difference in oxygen concentration in the artery and the vein supplying and removing blood from the organ (the arteriovenous difference) together with the blood flow through the tissue. The results for several organs are presented in Chapter 2. It is also possible from these data to calculate the actual flux through the Krebs cycle, since one turn of the cycle consumes one third of the total oxygen uptake by a cell, tissue or organ, if glucose is the only fuel. It is somewhat less if fat is the fuel.

Unfortunately, it is not always possible to measure oxygen consumption for some cells or tissues in vivo; for example, for cells that circulate in blood or in the lymph (e.g. red blood cells, immune cells), for different cells in one organ (e.g. macrophages in liver, glial cells in the brain, tumour cells in a tumour). However, another method is available that provides data on the maximum capacities of pathways in vivo. This is the measurement of the maximum catalytic activity of a key enzyme of the pathway in an in vitro extract of that cell, tissue or organ. The biochemical principles and assumptions underlying this method are presented in Chapter 3. The enzyme chosen for the Krebs cycle is oxoglutarate dehydrogenase, which catalyses the following reaction

> The oxidation of one molecule of glucose utilises six molecules of oxygen. Of these six, how many are utilized by the conversion of glucose to acetyl-CoA (i.e. aerobic glycolysis) and how many by the Krebs cycle? The oxidation of the two hydrogen atoms from one molecule of NADH or one molecule of $FADH_2$ (or $FMNH_2$) via the electron transfer chain, utilises one atom of oxygen ($\frac{1}{2}O_2$). Conversion of one molecule of glucose to two molecules of acetyl-CoA generates six molecules of NADH, whereas the oxidation of two molecules of acetyl-CoA, via the cycle, generates six molecules of NADH and two molecules of $FADH_2$. Therefore *aerobic* glycolysis is responsible for two of the six molecules of oxygen, and the Krebs cycle is responsible for four of the six, i.e. two-thirds of the total oxygen consumed. Consequently, when glucose is the only fuel, one turn of the cycle utilises one-third of the of the total oxygen consumption. Under maximal conditions of oxygen consumption, the maximal flux through the cycle is indicated by one-third of the total rate of oxygen consumption. A similar calculation shows that when fat oxidation occurs there is a greater contribution from the Krebs cycle.

oxoglutarate + NAD$^+$ + CoASH →

$$\text{succinyl-CoA} + NADH + CO_2$$

The oxygen uptake of a maximally working individual muscle in adult humans has been measured in vivo, enabling the calculation of the flux through the Krebs cycle in that muscle to be made (Appendix 9.10). It is compared with the capacity that is calculated from the maximal in vitro activity of oxoglutarate dehydrogenase, in an extract prepared from a biopsy obtained from that particular muscle in that volunteer. The procedure has also been done for various muscles from seven different animals across the animal kingdom, for comparison with humans (Table 9.6). These results indicate that the maximum in vitro activity of this one enzyme provides a reasonable assessment of the maximum capacity of the Krebs cycle in muscle tissue. This method has been used to study, for example, the

> The physiological usefulness of this method for identifying the fuels that are used and their rates of utilisation in different cells is discussed in other chapters (Chapter 3). For example, measurement of the activity of the enzymes hexokinase and glutaminase in immune cells showed, for the first time, that glucose and glutamine are the major fuels utilised by these cells. This finding has had clinical significance (Chapter 17).

effects of aerobic physical training on the maximum capacities of both glycolysis and the Krebs cycle and hence the maximum capacities of the anaerobic and aerobic pathways for ATP generation, in the quadriceps muscle of male and female volunteers (Table 9.7). The study indicates that aerobic training increases the capacity of muscle to generate ATP from aerobic metabolism. It also shows that the capacity of glycolysis to generate ATP, for a short period, is greater than that from oxidation of glucose. This indicates the importance of glycolysis for generation of ATP at a high rate, if only transiently, under extreme conditions (e.g. anoxia, hypoxia sprinting, climbing).

Examples of leisure activities that are supported by mitochondrial ATP generation

Humans engage in a whole range of leisure activities, many of which depend mainly, if not totally, on mitochondrial generation of ATP. Examples are:

- Prolonged running, jogging, walking, cycling, circuit training, most 'workouts', rowing, sailing, dancing,

Table 9.6 Flux through Krebs cycle as calculated from the maximum catalytic activity of oxoglutarate dehydrogenase, measured in extracts of muscle, and from oxygen consumption by muscles working maximally

Species	Muscle	Maximum Flux through the Krebs cycle[a] estimated from:	
		Oxoglutarate dehydrogenase[b] activity	Oxygen consumption[c] [d]
Locust	flight	24	28
Honey bee	flight	46	52
Silver-Y moth	flight	26	26
Trout	red	2.0	2.4
Pigeon	pectoral	4.2	7.6
Rat	heart	11.0	8.0
Human	quadriceps	5.1	4.1

Data from Crabtree & Newsholme (1975) and Paul (1979) except for human data from Blomstrand *et al.* (1997).

[a] Units are µmol of acetyl-CoA oxidised per min per g fresh wt of muscle at physiological temperatures. These are mean values for a least eight animals.

[b] Enzyme activity measured, in vitro, in extracts of the muscle (for details of assay technique, see Chapter 3).

[c] Uptake measured during prolonged activity: flying for insects and pigeon, working rat heart during perfusion in vitro, leg extension exercise in human muscle working maximally.

[d] The flux through the cycle from oxygen uptake is calculated on the basis that one complete turn of the cycle requires 2 molecules of oxygen for the oxidation of the hydrogen atoms or electrons produced from glucose oxidation (i.e. 1/3 of total oxygen consumption).

For human data, volunteers performed one-legged knee extension during which power output is produced mainly from quadriceps muscle, and maximum work rate was performed for at least one minute (Blomstrand *et al.* (1997))

Table 9.7 Effect of aerobic physical training on the maximum capacity for ATP generation from conversion of glycogen to lactate (glycolysis) and complete oxidation of glucose (the Krebs cycle) in the quadriceps muscle of male and female volunteers

		Calculated maximum rate of ATP generation from glucose or glycogen[a]	
Group	Sex	Anaerobic glycolysis	Oxidation to CO_2
Untrained	male	104	22
	female	87	27
Medium-trained	male	91	35
	female	89	44
Well-trained	male	72	44
	female	61	51

Maximum rates of ATP generation are calculated as follows: for conversion of glycogen to lactate (i.e. anaerobic glycolysis) phosphofructokinase activity is multiplied by three (see Chapters 3 and 6). For complete oxidation of glucose, the activity of oxoglutarate dehydrogenase must be multiplied by two, because two turns of the cycle were required to oxidise one molecule of glucose. The resultant activity of oxoglutarate dehydrogenase is multiplied by 30. Training can double the capacity of the Krebs cycle but reduces glycolysis.

[a] Data are presented as μmol/min per g fresh wt muscle.

Data from Blomstrand *et al.* (1986).

skiing, swimming, ice skating and many ball games. Although mitochondria will generate most of the ATP, the proportion generated from conversion of glycogen to lactate depends on the physical fitness of the participant and the duration of activity. An unfit person will depend more on *anaerobic* glycolysis.

- Several muscles are used in singing; particularly important are diaphragm, intercostal and muscles in the mouth and the vocal cords. If singing is performed for long periods of time, for example during performance or rehearsals, the ATP must be generated from aerobic metabolism.

- In normal conversation, the same muscles are also involved, albeit to a lesser extent. Since, in some conversations, the duration may be considerable, ATP must be generated from aerobic metabolism. Indeed, a study has shown that loud prolonged talking increases oxygen consumption considerably.

Thermogenesis: maintenance of normal body temperature

Body temperature is maintained by the balance between heat production and heat loss. Heat is produced in all biochemical and physiological processes (Chapter 2) so that, in humans, temperature is regulated, under normal conditions, by changes in heat loss from the body. However, if the heat produced is not sufficient to maintain the temperature, despite decreased heat loss, it can be produced by specific processes. These are shivering, substrate cycling or uncoupling of ATP formation from electron transfer in mitochondria.

Shivering The process of shivering produces heat from the ATP hydrolysed during contraction and relaxation of muscles and, in addition, from the increased metabolism necessary to generate the ATP required for the muscles involved in shivering. It is likely that the ATP required to support shivering is generated from aerobic metabolism, since shivering can be maintained, in some conditions, for prolonged periods of time. However, there are no studies reported on shivering and fuels required.

Substrate cycling The primary role of substrate cycling is to improve sensitivity in regulation of biochemical processes (Chapter 3). Although ATP hydrolysis occurs, there is no net metabolic change. Cycling, therefore, transfers the chemical energy in ATP directly into heat. This is the reason that substrate cycles have been called 'futile' cycles. An increase in the rate of cycling generates more heat so that its regulation can help to maintain normal body temperature. Exposure to a cold environment increases the rate of substrate cycling via the stress hormones.

The hormone leptin, which is secreted by adipose tissue, is considered to play a role in the control of the amount of triacylglycerol in adipose tissue by decreasing appetite and by increasing energy expenditure. Leptin increases the rate of the triacylglycerol fatty acid cycle (Chapter 15).

Examples of a substrate cycles are: the glucose/glucose 6-phosphate, fructose 6-phosphate/fructose 1,6-bisphosphate and fatty acid/triacylglycerol cycle. These are described in Chapters 3, 6, 7 and 11. One study has shown a direct role of a substrate cycle in heat generation (Appendix 9.11).

Proton-leak proteins in the mitochondrial membrane

Transport of protons from the intermembrane space into the matrix across the inner membrane of the mitochondria occurs via the ATP synthase complex (F_o–F_1), which generates ATP. Somewhat surprisingly, it was discovered that proteins that transport protons back into the matrix, but without generating ATP, do, in fact, exist and are

> The heat production from strenuous muscle activity can maintain normal body temperature even in the harshest of environmental conditions: Mike Stroud and Ranulph Fiennes, during their Transantarctic expedition, even without excessive amounts of clothing, generated sufficient heat from the muscular effort involved in walking and pulling their sledges, that normal body temperature was maintained (Stroud, 1997).

present in this membrane, at least in some tissues. They are therefore known as 'leak' or uncoupling proteins. Such a leakage of protons dissipates the electrochemical gradient, resulting in transfer of the energy from this high energy state into heat. Whether heat production is the main or the only role of these proteins in mitochondria from 'normal' tissue is not known, since the precise role of these proteins is still debated. Nonetheless, there is one such protein whose only role is generation of heat: this is the specific uncoupling protein that is present in brown adipose tissue.

Brown adipose tissue

The main role of brown adipose tissue is the generation of heat when required. It is also considered that it might play a role in regulation of body weight (see below). It is called adipose tissue because of its high content of triacylglycerol: its brown colour is due to the presence of many mitochondria (Figure 9.28). Its function and hence its metabolism is quite different from that of white adipose tissue. It is present in animals that need, at times, to generate heat, e.g. hibernating animals, in small mammals, par-

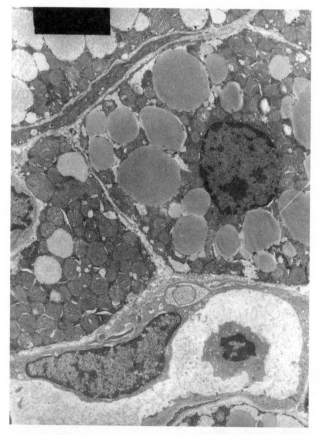

Figure 9.28 *Electron micrograph of brown adipose tissue.* Kindly provided by Dr Caroline Pond of the Open University, UK.

Box 9.4 Maximum capacity of the Krebs cycle in invertebrate muscles versus that in human muscle

The importance of ATP generation from complete oxidation of fuel in insect flight muscle is illustrated by two facts: (a) the microscopic structure of the muscle showing a large number of mitochondria, and (b) the high activity of the enzyme oxoglutarate dehydrogenase. First, an electron micrograph of insect flight muscle shows large numbers of mitochondria, crammed with cristae, and the close association of these mitochondria with the myofibrils and with the oxygen supply system, which emphasises the importance of the functional link between ATP generation in the mitochondria and ATP utilisation by the contractile apparatus (Chapter 13). Second, the data in Table 9.6 indicate that the capacity of the Krebs cycle in insect flight muscle is considerably greater than that in the muscle of even a young well trained adult human!

Even an invertebrate animal that gives no appearance of physical activity possesses a muscle that has a capacity of the Krebs cycle that is similar to that in a muscle of a young adult human. This is the radular retractor muscle of a mollusc, the whelk. Whelks are found on the seashore: they can use their radula continually for very long periods, up to 24 hours in some cases, to rasp flesh off, for example, a fish carcass. A simple dissection of a whelk readily reveals the radular retractor muscle, easily identified by its brilliant red colour. This muscle illustrates the principle that for muscles that are physiologically essential and have to work for long periods of time, the generation of ATP must be from the oxidation of a fuel which requires mitochondria and therefore cytochromes, which is why the radular retractor muscle is red.

This biology/biochemistry of the whelk is of some commercial importance because a related species, the sting winkle, sometimes attacks and devours oysters in oyster farms by boring a hole in the shell in order to digest the protein inside and then absorb the digested peptides and amino acids. It is the radular retractor muscle that is the culprit!

---

ticularly those whose habitat is cold, and in the newborn of some species, including humans, since they do not have the capacity to shiver. It is present in adult humans but the amount is not large and is variable.

It is localised mainly in the dorsal part of the thorax under the shoulder blades, along the spinal cord and around the adrenal glands so that it is close to the vital tissues and organs. These can therefore be rapidly warmed by increased activity in brown adipose tissue.

There is discussion as to whether brown adipose tissue in humans can play a role in body weight control by oxidising fat fuels and converting the chemical energy into heat.

Stimulation of the sympathetic nervous system releases noradrenaline which stimulates mobilisation of fatty acids in adipose tissue. The latter stimulate the activity of the uncoupling protein.

There are several properties of the uncoupling protein that are consistent with this hypothesis.

- The rate of electron transfer in isolated mitochondria from brown adipose tissue, in the absence of ADP and phosphate, is high indicating that the mitochondria are uncoupled (see Appendix 9.9).

- The activity of the ATP synthase is low in comparison with that in mitochondria from other tissues: i.e. mitochondria in brown adipose tissue can generate very little ATP.

- The capacity for the transport of protons across the membrane back into the matrix is about 100-fold greater than that for mitochondria from other tissues (e.g. liver).

Mechanism of control of heat generation

In order to provide heat when it is required to maintain or increase body temperature, a mechanism must exist for the regulation of the activity of this uncoupling protein. The mechanism has been established by following the principles described in Chapter 3. The properties of the uncoupling protein are studied using mitochondria isolated from brown adipose tissue in vitro (as described in Appendix 9.9).

The properties are as follows. (i) The activity of the protein (i.e. the inward transport of protons) is inhibited by ATP. (ii) The activity of the protein is increased by the presence of long-chain fatty acids, since they relieve the ATP inhibition. (iii) When mitochondria, isolated from brown adipose tissue, are incubated in the presence of fatty acids, there is a sharp increase in the rates of electron transfer, substrate utilisation and oxygen consumption, whereas the rate of ATP generation remains low. These studies indicate that the rate of proton transport, by the uncoupling protein, depends on the balance between the concentrations of ATP and fatty acids. (iv) In adipocytes isolated from brown adipose tissue, the rate of oxygen consumption (i.e. electron transfer) is increased in the presence of catecholamines.

The hypothesis that is formulated on the basis of these properties is as follows. The signal for increased thermogenesis, when the body temperature falls, is an increase in the level of catecholamines, probably the local concentration of noradrenaline, which will be increased via stimulation of sympathetic nervous system. The catecholamine increases the activity of triacylglycerol lipase within the brown adipose tissue, by a similar mechanism to that which occurs in white adipose tissue (Chapter 7), i.e. by an

increase in the concentration of cyclic AMP. Hence, the concentration of fatty acids in the cell increases, which has two effects: (i) An increased rate of fatty acid oxidation (β-oxidation) and the Krebs cycle, which increase the rate of electron transfer and hence oxygen consumption. (ii) The proton motive force is immediately dissipated by the increased activity of the uncoupling protein, which is due to the increase in the intracellular concentration of fatty acids.

In summary, the fuel that is oxidised to provide the energy is fatty acids but the energy is converted to heat, via the increased activity of the uncoupling protein. The increased activity of the uncoupling protein is due to the increased concentration of fatty acids (Figure 9.29).

The first in vivo evidence for the involvement of this tissue in heat production was obtained from experiments on hibernating animals. Thermocouples were placed within brown adipose tissue during arousal from hibernation: the

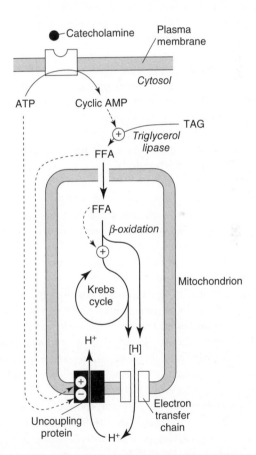

Figure 9.29 *Control of heat production in brown adipose tissue.* Catecholamines increase cyclic AMP concentration which stimulates triacylglycerol lipase which increases the long-chain fatty acid level, which increases the fluxes through β-oxidation and the Krebs cycle, and the activity of the uncoupling protein. Uncoupling decreases the ATP concentration which further increases the activity of the uncoupling.

temperature of brown adipose tissue increased before that of any other tissue.

The effect of ageing on ATP generation

One very obvious effect of ageing is decreased capacity to perform physical activity: the maximal work rate is reduced and fatigue occurs at a lower work rate. This effect is indicated by a progressive decrease in the maximum oxygen uptake of the whole body ($\dot{V}O_{2max}$). The decline in $\dot{V}O_{2max}$ is between 5 and 15% per decade after age 25. This is due, in part, to a decline in the total amount of skeletal muscle, a reduction in the number of mitochondria in a given amount of muscle and a decline in the surface area of inner membranes within the mitochondria. Hence, the capacity of a mitochondrion within the muscle to generate ATP is decreased. The mechanism(s) responsible for these changes in the mitochondria and loss of muscle (sarcopenia) is not known.

Two mechanisms for loss of mitochondrial function have been suggested: (i) damage caused by the chronic production of free radicals within the mitochondria and (ii) somatic mutations in the mitochondrial genome, which progressively accumulate during a lifetime.

Loss of mitochondria will not only decrease the maximal exercise performance but, for any given level of exercise, aerobic generation of ATP will be insufficient, so that a greater proportion must be generated from anaerobic metabolism, i.e. glycogen conversion to lactic acid, so that muscle glycogen and pH are decreased which can result in fatigue (discussed in Chapter 13).

Participation in physical activity is recommended for the elderly, and even for the very elderly, because it helps to reduce the loss of skeletal muscle and probably loss of mitochondria. Loss of mitochondria and their capacity for ATP generation also occurs in other organs and this contributes to the effect of ageing on many essential physiological functions. It is, therefore, an important question as to whether maintenance of some physical fitness not only improves quality of life by affecting muscles directly but also by enabling the body to deal more effectively with conditions such as stress, infection, broken bones or surgery and may delay or even prevent senile dementia. The inability to deal with a minor stress may cause physiological changes that increase the risk of death in the otherwise, healthy elderly (which is called physiological death in Chapter 20).

Ageing is an increasing problem in developed countries. For example, by 2020, there will be 10 million North Americans aged 85 or more. This has enormous implications for health care, hence, maintenance of mitochondrial function in the elderly may be of considerable importance not only for the well-being of the elderly themselves but for the health care cost to the nation.

Overview of ATP generation: from food to essential life processes

To re-emphasise the central role of ATP generation in the biochemical and physiological functions in the body, a summary of the sequence of the biochemical pathways and processes by which the energy in food is used to generate the ATP that supports the essential processes of life is provided in Figure 9.30.

Clinical significance of defects in mitochondrial ATP generation

There are at least three ways in which mitochondrial ATP generation can be impaired: mutations in mitochondrial DNA, mutations in nuclear DNA and effects of toxic compounds. The reactions in mitochondrial metabolism that are affected by some toxic compounds are described in Appendix 9.12.

Cell death

Death of cells can occur in two ways – necrosis or apoptosis – and mitochondria play a role in both processes (Chapter 20).

Necrosis   Acute inhibition of or damage to the electron transfer chain and/or oxidative phosphorylation can be caused by several factors including deficiency of oxygen, excess oxygen (i.e. oxygen free radicals), poisons or pollutants. Any of these can decrease the rate of ATP generation, which results in a fall in ATP/ADP concentration ratio that causes several problems leading to death. For example, a decrease in the rate of ion transport, which can lead to a disturbance in the concentration of intracellular ions resulting in uncontrolled entry of water, osmotic shock and lysis of the cell. Such damage causes the release of factors that can stimulate a local inflammatory response and cause damage to surrounding cells. The process is called necrosis and is discussed in Chapter 20.

Apoptosis   Damage to mitochondria can also result in the release of cytochrome *c* and other factors into the cytosol, which can initiate the process of apoptosis. If a sufficient number of cells in a tissue or organ is lost, the function of the organ or tissue can be impaired, so that it would be less

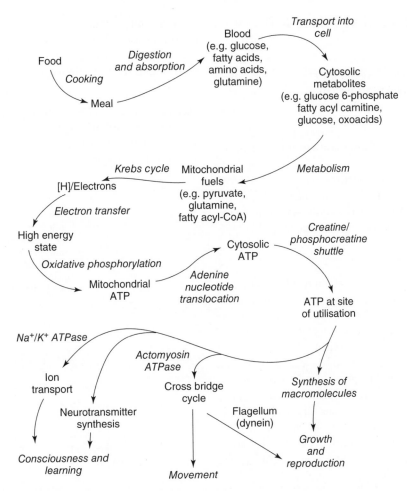

Figure 9.30 *Flow diagram of the 'energy chain' from food to essential processes in human life.* The ATP utilised by the Na⁺/K⁺ ATPase maintains the ion distribution in nerves that is essential for electrical activity and, in addition, maintains neurotransmitter synthesis, both of which provide communication in the brain and hence consciousness, learning and behaviour (Chapter 14). ATP utilisation by myosin ATPase is essential for movement and physical activity. ATP utilisation by the flagellum of sperm is essential for reproduction and ATP utilisation for synthesis of macromolecules is essential for growth.

able to carry out its normal physiological function. For example, death of neurones in the brain can impair mental activity. Since such damage increases with time, this is one factor that is involved in development of senile dementia, which can progress into Alzheimer's disease. Loss of neurones can also result in Parkinson's disease. Such problems that arise from loss of neurones are covered by the term neurodegenerative diseases (Chapter 14), although loss of mitochondrial function is clearly not the only cause of these diseases.

An effect of an excitatory neurotransmitter on the life of neurones

A specific cause of apoptosis of neurons is failure to control the concentration of the neurotransmitter glutamate. It is present in vesicles in the presynaptic terminal of some neurones: the very high glutamate concentration in these vesicles is achieved and maintained by a glutamate pump that is 'driven by' ATP hydrolysis. A fall in the ATP/ADP concentration ratio due to mitochondrial damage can result in failure to maintain the activity of the pump so that uncontrolled release of glutamate from the vesicle into the synapse can occur. So much glutamate can be released into the synapses that it stimulates, massively, the activity of a Ca^{2+}-ion channel, resulting in a marked increase in the cytosolic concentration. The excess Ca^{2+} ions are taken up by the mitochondria, where they react with phosphate to precipitate as calcium phosphate. The precipitate damages the mitochondria, so that cytochrome c is released into the cytosol, and initiates the process of apoptosis in these particular neurones.

Medical conditions that affect ATP generation

A large range of medical conditions can seriously affect aerobic ATP generation. Three examples are given.

Atherosclerosis

Atherosclerosis is not itself a disease but it can lead to a poor supply of blood to tissues/organs, which consequently impairs mitochondrial ATP generation and hence can interfere with the function of that tissue. This leads initially to ill health, but can become sufficiently severe to lead to disease and, eventually, to death (see Chapter 22).

Multiple organ failure

Excessive concentrations of proinflammatory cytokines, produced by macrophages in response to infection or trauma, cause damage to endothelial cells in many organs which leads to local inflammation, thrombosis and poor circulation of blood through that organ. This can occur in any or all the major organs in the body and lead to hypoxia with a decrease in aerobic ATP generation and malfunction of the organs with severe physiological and pathological consequences. When several organs are affected it is known as *multiple organ failure*. It is the cause of death in many patients in intensive care units, even after the patient has recovered from the original infection or trauma (discussed in Chapter 18).

Genetic defects

The energy generation from aerobic metabolism is so important for the vital organs that complete inactivation or loss of any of the proteins involved in this process is not

> The first defect, described in 1962 is, in fact, one of the rarest (Luft's syndrome). It arises from the uncoupling of mitochondria. The resting metabolic rate is markedly raised, there is profuse sweating, fever and generalised muscle weakness. The mitochondria of these patients have an increased permeability, not so much to protons, as in brown adipose tissue mitochondria, but to cations, such as Ca^{2+}, the entry of which similarly dissipates the proton motive force.

compatible with life. If such events occur prenatally, the foetus is aborted. However, partial loss of a protein or partial loss of its activity may still allow generation of ATP, but not sufficient to meet normal demands. Such loss of activity can be due to damage to or mutations in genes, either in the nucleus or mitochondria, that encode for proteins involved in electron transfer or ATP generation.

Defects in mitochondrial proteins were first detected in muscle so that the term mitochondrial myopathy was used to describe these problems but, since they can arise in tissues other than muscle,

the term mitochondrial cytopathy is now used. Patients usually present with signs and symptoms which relate to the organ affected.

Skeletal muscle The usual symptom is muscle weakness and any physical activity can soon result in fatigue.

> There was a considerable increase in interest in adult-onset mitochondrial disorders when the American cyclist, Greg LeMond, who won the Tour de France cycle race three times, announced his retirement from competitive cycling in 1994 because of a 'mitochondrial myopathy'.

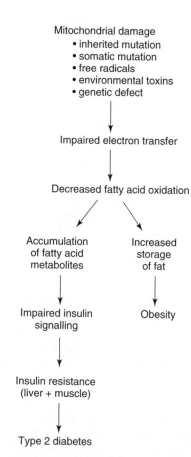

Mitochondrial damage
• inherited mutation
• somatic mutation
• free radicals
• environmental toxins
• genetic defect

↓

Impaired electron transfer

↓

Decreased fatty acid oxidation

↙   ↘

Accumulation of fatty acid metabolites   Increased storage of fat

↓   ↓

Impaired insulin signalling   Obesity

↓

Insulin resistance (liver + muscle)

↓

Type 2 diabetes

Figure 9.31 *Mechanism by which mitochondrial damage can lead to obesity or type 2 diabetes.* Damage to the proteins of the electron transfer chain or the inner mitochondrial membrane can be caused by somatic mutations, free radicals, environmental toxins or inherited genetic defects. The resultant impairment of transfer of electrons along the chain of carriers will decrease the oxidation of fatty acids by muscle and other tissues. Consequently, the fate of the fatty acids derived from fat in the diet will be storage as triacylglycerol in adipose tissue rather than oxidation in muscle and other tissues. This can lead to obesity. Furthermore, failure to oxidise fatty acyl-CoA in muscle can result in accumulation of fatty acid metabolites which can interfere with insulin signalling, resulting in insulin resistance and hence type 2 diabetes.

Brain The usual signs and symptoms include ataxia, dementia, deafness and sometimes seizures. Not surprisingly, there is failure of mental development in children.

Heart The usual sign is cardiomyopathy, leading to heart enlargement and impaired function.

The underlying cause of mitochondrial dysfunction can be investigated in muscle using mitochondria isolated from a biopsy sample. They are incubated with different substrates and the oxygen consumption is measured. Defects in complexes I, III or IV can be identified in this way (Appendix 9.9).

Mutations in mitochondrial DNA

It is well established that mitochondrial DNA suffers more mutations or damage than nuclear DNA. There are at least three reasons for this:

(i)   Proteins (e.g. histones) are present in the nucleus to help maintain the structure of DNA in the nucleus in the form of chromatin, in which form it is protected from damage. Such proteins are not present in mitochondria, so that mitochondrial DNA is not so well protected and more readily damaged by, for example, radiation, free radicals or pollutants. The transfer of electrons along the chain in the mitochondria provides opportunities for production of oxygen free radicals. Furthermore, the mechanism of the catalysis by cytochrome oxidase involves a free radical of oxygen.

(ii) Proteins that halt DNA replication and allow time for repair of damage are not present in mitochondria (e.g. the p53 protein) so that replication of mitochondria is impaired or mutation increase. (Chapter 21).

(iii) The processes that actually repair damaged DNA are less effective in mitochondria than in the nucleus.

In some cases, mutations that result in partial loss of a protein or its activity only cause major problems in adult life. This is explained by the fact that maximum rates of many processes (e.g. ATP formation in muscle) are rarely required by humans in developed countries, so that the rate of the process is sufficient to satisfy most, if not all, of the activities essential to life. However, progressive damage to DNA throughout life may convert a partial deficiency into a total deficiency. This can result in an adult deficiency disease, since symptoms only develop in the adult.

Defects in mitochondria leading to diabetes or obesity

The oxidation of fatty acids occurs within mitochondria, so that any defect in the electron transfer chain will restrict the oxidation of fatty acids. This can result in several clinical problems, especially in developed countries where diets contain a high proportion of fat. For example, one factor that can lead to obesity is a low rate of fat oxidation, since any fat not oxidised is stored as triacylglycerol in adipose tissue. Furthermore, failure to oxidise fatty acid can lead to accumulation of fatty acid derivatives in the muscle or liver cell (e.g. fatty acyl-CoA, diacylglycerol) that can interfere in the insulin signalling process. This leads, therefore, to failure of insulin to stimulate glucose uptake and utilisation in muscle, and glucose conversion to glycogen in the liver. Such changes could lead to chronic elevation in the blood glucose level. This, together with a reduced ability to remove glucose from the blood, are characteristics of glucose intolerance which can, along with other changes, lead to insulin resistance and eventually type 2 diabetes (Figure 9.31). Partial defects in the electron transfer chain could be inherited, acquired by somatic mutations or arise from damage caused by free radicals. The risk of the latter two increases with age, which may explain why the incidence of type 2 diabetes increases with age.

10
Metabolism of Ammonia and Nucleic Acids

The concept of the ornithine cycle arose from the observation that ornithine, citrulline and arginine stimulated urea production in the presence of ammonia without themselves being consumed in the process.

(Krebs, 1964)

It was the quality of his work on the urea cycle, that led the Professor of Biochemistry at Cambridge University F. Gowland Hopkins, to invite Hans Krebs to join his Department in 1933 to escape from the Nazi regime.

Ammonia is generated mainly from the metabolism of amino acids and from the catabolism of purine and pyrimidine bases, which are produced from nucleic acids. Since it is toxic, it must be converted to a non-toxic compound for excretion from the body. This is achieved via the ornithine cycle, more usually known as the urea cycle.

There are six main sources of ammonia in the body:

(i) From the process of transdeamination in the catabolism of some amino acids: it is the reaction catalysed by glutamate dehydrogenase that produces the ammonia.

$$glutamate + NAD^+ \rightarrow oxoglutarate + NADH + NH_4^+;$$

(ii) From the specific pathways for the catabolism of some amino acids.

(iii) From the catabolism of purine and pyrimidine bases, derived from the degradation of nucleic acids.

(iv) Cells in the bone marrow.

(v) From those cells that use glutamine as a fuel (e.g. immune cells, enterocytes, colonocytes).

(vi) From the hydrolysis of urea by some microorganisms in the colon.

A summary is given in Figure 10.1.

Ammonia binds protons to form the ammonium ion (NH_4^+)

$$NH_3 + H^+ \rightleftarrows NH_4^+$$

and, in the cell, about 99% is in the NH_4^+ form but only the NH_3 molecule can diffuse across membranes, so that NH_4^+ requires a carrier protein. It is likely that a carrier is present in most tissues but it has not been studied in detail.

Roles of ammonia

Prior to description of the cycle, it should be noted that ammonia has several important roles in the cell, which include the following:

- It is a substrate for the formation of glutamate from oxoglutarate in a reaction catalysed by the enzyme glutamate dehydrogenase.

$$oxoglutarate + NH_3 + NADPH \rightarrow glutamate + NAD^+$$

Glutamate is required for the synthesis of some non-essential amino acids and key compounds such as glutathione.

- It is also a substrate for formation of glutamine, which is an important fuel in the body; the reaction is catalysed by glutamine synthetase.

$$glutamate + NH_3 + ATP \rightarrow glutamine + ADP + P_i$$

Functional Biochemistry in Health and Disease by Eric Newsholme and Tony Leech
© 2010 John Wiley & Sons Ltd

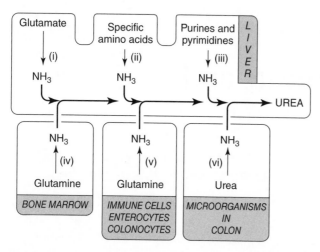

Figure 10.1 *Simple diagram of the sources of ammonia for the urea cycle.* Sources are the liver, bone marrow, immune cells, enterocytes, colonocytes and microorganisms. Numbers refer to list in text.

Table 10.1 Concentration of ammonia in blood and other tissues of the rat

Tissue	Ammonia concentration (mmol/L)
Arterial blood	0.02
Hepatic portal venous blood	0.26
Hepatic venous blood	0.03
Arterial blood in ammonia toxicity	>0.50
Liver	0.71
Skeletal muscle	0.26
Brain	0.34
Heart	0.20
Spleen	0.20

The concentration of ammonia refers to the total ammonia, that is NH_3 plus NH_4^+. For comparison, the concentration of ATP in the cell is 10 mmol/L, that of glucose 6-phosphate is 0.5 mmol/L and that of the amino acid glutamate in plasma is 0.02 mmol/L.

- Ammonia buffers protons in the tubules of the kidney to prevent formation of an acidic urine, when the kidney excretes (secretes) protons to control the pH of the blood.

$$NH_3 + H^+ \rightarrow NH_4^+$$

These topics are discussed in Chapter 8.

Urea synthesis

Despite these important roles of ammonia, the concentration in cells, and particularly in the blood, is maintained low (Table 10.1). Although many cells produce ammonia, liver is the only organ capable of converting it to urea, which is excreted by the kidney. Urea is an excellent molecule for the excretion of ammonia: it is highly soluble, is non-toxic, has a high nitrogen content (47%) and requires relatively little energy for synthesis (0.5 molecules ATP are hydrolysed per atom of nitrogen incorporated into urea). The majority of animals living in fresh water are able to excrete ammonia in a large volume of dilute urine. Terrestrial animals, including humans, do not have access to sufficient quantities of water to make this possible and therefore must convert ammonia to a non-toxic compound.

On an average Western diet, adult humans excrete around 30 g of urea per day but this can easily triple on a protein-rich diet. The reactions and the concept of a cycle were discovered by Krebs & Henseleit (1932). Subsequent work clarified the details of what has become known as the ornithine or the urea cycle.

> In comparison, the vampire bat has a capacity for urea synthesis approximately 1000-fold greater than that of a human. A bat consumes one half of its weight in blood in 10–15 minutes so that the massive amount of protein that is metabolised produces a massive amount of ammonia which must be removed as quickly as possible. Indeed, it also ingests sufficient fluid so that it is too heavy to fly. Hence, it urinates very quickly.

The ornithine cycle

The first important point about the cycle is that ammonia does not enter it directly. Nitrogen atoms enter the cycle via the two compounds carbamoyl phosphate and aspartate. The former is formed in a reaction catalysed by the enzyme carbamoyl phosphate synthetase. This enzyme contributes over 20% to the protein content of the mitochondrial matrix in liver mitochondria, which is more than any other protein and an indication of the importance of this reaction for the removal of ammonia. Carbamoyl phosphate then combines with ornithine in a reaction that is part of the cycle, in a reaction catalysed by ornithine transcarbamoylase, to form citrulline. The latter now acquires the second nitrogen atom, from aspartate, to form arginosuccinate in a reaction catalysed by argininosuccinate synthetase. The argininosuccinate is cleaved, in a reaction catalysed by argininosuccinate lyase, to form arginine, which is hydrolysed to form urea and ornithine, in a reaction catalysed by arginase. The urea diffuses out of the liver and is carried to the kidney for excretion. The ornithine that is produced continues the cycle. A summary of

> The carbamoyl phosphate synthetase (abbreviated to CPS-I) that is involved in the ornithine cycle differs from the enzyme that is involved in pyrimidine synthesis (carbamoyl phosphate synthetase-II). The latter enzyme is cytosolic, requires glutamine for provision of nitrogen, rather than ammonia, and is regulated by different factors (Chapter 20).

the cycle is given in Figure 10.2 and full details of the reactions in Figure 10.3.

Three reactions give rise to these two precursor molecules:

(i) $NH_4^+ + CO_2 + ATP \rightarrow$ *carbamoyl phosphate* $+ ADP$

(ii) $NH_4^+ +$ oxaloacetate $+ NADPH \rightarrow$ glutamate $+ NADP^+$

(iii) glutamate $+$ oxaloacetate $\rightarrow$ *aspartate* $+$ oxoglutarate

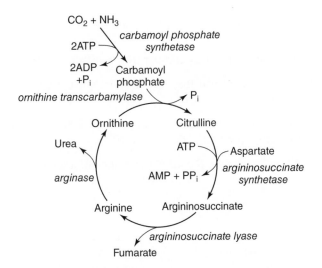

Figure 10.2 *Simple diagram of the urea cycle.*

Reactions (ii) and (iii) are catalysed by glutamate dehydrogenase and aspartate aminotransferase, respectively.

Of the enzymes involved in the cycle, two are present in cells other than the liver, where they are involved in other processes:

• The conversion of citrulline to arginine also occurs in the kidney. Indeed, kidney is the main organ for producing arginine for use by other tissues or organs in the body. Arginine present in the protein in the food is released during digestion in the lumen of the intestine and is then absorbed into the enterocyte. However, in this cell, arginine is degraded to citrulline. The latter enters the blood from where it is taken up by the kidney for conversion to arginine by the same enzymes as in the liver. It is then released into the blood for use in the body (Chapter 8).

• The role of arginase in some cells is not to produce urea but to regulate the concentration of arginine. This is important since arginine is the substrate for the reaction that produces an important messenger molecule, nitric oxide, via the enzyme nitric oxide synthase.

$$\text{arginine} \rightarrow \text{nitric oxide} + \text{citrulline}$$

Changes in the concentration of arginine, via arginase activity, can play a role in regulating the rate of nitric oxide synthesis. Nitric oxide is a messenger molecule that has several roles. One is to increase the activity of the enzyme guanyl cyclase, which increases the concentration of cyclic GMP. The latter causes vasolidation in peripheral

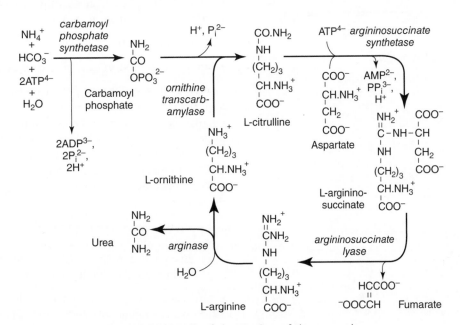

Figure 10.3 *Details of the reactions of the urea cycle.*

arterioles so that nitric oxide can control blood flow through tissues. Nitric oxide synthase is present in the corpora cavernosa in the penis where, via nitric oxide, cyclic GMP plays a role in erection during sexual intercourse. Arginase is also present in the corpora cavernosa, so that it may regulate arginine concentration and therefore influence the formation of nitric oxide and hence sexual activity in the male (Chapter 19). A high activity of arginase or a low level of arginine may be one cause of erectile dysfunction.

Distribution of reactions of the cycle between the cytosol and mitochondria

The reactions of the ornithine cycle are distributed between the two major compartments of the liver cell: mitochondrion and cytosol. Carbamoyl phosphate synthetase and ornithine transcarbamoylase are present within the mitochondrion, whereas argininosuccinate synthetase, argininosuccinate lyase and arginase are present within the cytosol. This compartmentation requires that citrulline is transported out of the mitochondrion, which requires a carrier that also transports ornithine back into the mitochondrion (Figure 10.4). The biochemical significance of the compartmentation is that the enzyme glutamate dehydrogenase is present in the mitochondria, where it catalyses the reaction that generates most of the ammonia in the liver. The ammonia is then immediately available for conversion to carbamoyl phosphate, since the synthetase is also present in mitochondria. The other reactions are present in the cytosol, probably to avoid metabolic congestion in the mitochondria.

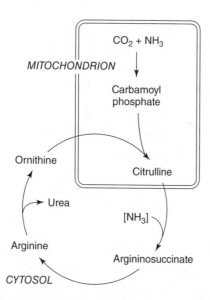

Figure 10.4 *Compartmentation of the reactions of the cycle between the mitochondria and cytosol.*

Table 10.2 Rate of urea production in adult humans on normal, low protein and restored normal diets

Diet	Time on diet (days)	Rate of urea production (g/day)
Normal	0	26
Low protein	2	10
	5	7
	7	3
	10	2
	13	2
Normal	1	17

For sources of information, see Newsholme & Leech, 1983.

Regulation of the flux through the ornithine cycle

It is well established that the flux through the cycle is regulated:

- An increase in protein intake increases the flux whereas a decrease lowers the flux (Table 10.2).

- An increase in the plasma concentration of amino acids, induced artificially by infusion, increases urea production not only in a normal but in a damaged liver. (Table 10.3).

The concentrations of the substrates in the liver of the rat indicates that there is no reaction in the cycle that approaches saturation with the pathway substrate i.e. these is to flux-generating step, so that the cycle cannot be described as a biochemical pathway, at least as defined in Chapter 3. The physiological pathway can be considered to start in either of two processes:

- amino acids entering the liver;

- ammonia released from various cells or tissues.

Consequently, as with some other processes, the physiological ornithine cycle spans more than one tissue. An hypothesis put forward to account for the regulation of the cycle must take into account the supply of ammonia for conversion to urea. The approach used to formulate a theory of regulation is described in Chapter 3.

The properties of the enzymes and the characteristics of the reactions are as follows:

- The K_m of carbamoyl phosphate synthetase for ammonia is assumed to be above that of the ammonia concentration in the mitochondria, so that an increase in the latter should increase the activity of the enzyme.

- An increase in activity of the synthase will increase the concentration of carbamoyl phosphate, which

Table 10.3 Effect of increases in blood amino acid concentration on rate of urea production in normal subjects and in patients suffering from cirrhosis of the liver

Plasma amino acid concn (mmol/L)	Approx. rate of urea synthesis (g/day)	
	Normal subjects	Cirrhotic patients
Control	19	19
5	50	22
10	129	50
15	216	86

The concentrations of amino acids in plasma of normal subjects and in cirrhotic patients were 2.6 and 3.5 mmol/L, respectively. To raise the plasma concentration, amino acids were infused. Note a much lower rate of urea formation in patients with a damaged liver. Cirrhosis is characterised by deposition of collagen in the liver and arises from a variety of causes: a virus infection, deposition of fat in the liver, undernutrition or chronic and excessive consumption of alcohol.

The data take into account, as far as possible, the breakdown of urea by microorganisms in the colon, that is, urea that is synthesised in the liver but which may not appear in the urine.

Data from Vilstrup (1980).

should increase the activity of ornithine transcarbamoylase and hence flux through the cycle.

- The compound, N-acetylglutamate, is an activator of carbamoyl-phosphate synthetase. This compound is synthesised in the liver from glutamate and acetyl-CoA, in a reaction catalysed by the enzyme N-acetyl glutamate synthase (abbreviated to NAGS). N-acetylglutamate is degraded in a reaction catalysed by a de-acetylase. Consequently, the concentration of N-acetylglutamate is maintained by a balance of the activities of the acetyl glutamate synthase and the de-acetylase (i.e. a substrate cycle).

$$\text{glutamate} + \text{acetyl-CoA} \rightarrow \text{N-acetylglutamate}$$

$$\text{N-acetylglutamate} + H_2O \rightarrow \text{glutamate} + \text{acetate}$$

The N-acetylglutamate concentration increases as the flux through the urea cycle increases.

- Ornithine production by the enzyme arginase is essential for maintenance of flux through the cycle, so that an increase in the concentration of ornithine should increase the flux. This mechanism is analogous to regulation of the Krebs cycle by changes in the concentration of oxaloacetate (Chapter 9, Figure 9.23).

Acute regulation of rate of the ornithine cycle

On the basis of these factors, a theory for the acute regulation of the cycle can be put forward. It is based on changes in the hepatic concentrations of the following: ammonia, N-acetylglutamate, ornithine and arginine. The mechanism is put forward in Figure 10.5. The fact that there are four factors involved in the regulation of flux through the pathway, which contains only five reactions, indicates the prime importance of removing ammonia from the blood. Not only is there an acute mechanism, but also a chronic one.

Chronic control of capacity of the cycle

An increase in the protein content of the diet in rats increases the maximum activities of all the enzymes of the cycle in the liver. It is assumed that this represents increased amounts of these enzymes in the liver (Table 10.4). Since a chronic increase in the protein in the diet in humans increases urea production over a long period and also a decrease in protein in the diet decreases urea production, it is assumed that, as in the rat, this is due to changes in the concentrations and therefore activities of urea cycle enzymes.

A 'safety mechanism' for removal of ammonia by the liver

The liver is unusual in that it receives blood from two sources, the hepatic artery and the hepatic portal vein. The two supplies eventually join at the central vein, hence hepatocytes are of two types: those which are exposed to blood derived largely from the hepatic artery, known as the perivenous cells, and those that are exposed to blood largely from the portal vein which are known as periportal cells (see Chapter 8). The periportal hepatocytes contain the enzymes of the ornithine cycle. In contrast, the

Table 10.4 Chronic effects of high and zero protein diets on maximum activities of urea cycle enzymes in the liver of the rat

Condition	Enzyme activities (μmol per min per g)				
	Carbamoyl-phosphate synthetase	Ornithine transcarbamoylase	Argininosuccinate synthetase	Argininosuccinate lyase	Arginase
Normal diet (15% protein)	5.7	223	2.9	4.5	583
High protein diet (30% for two weeks)	9.7	447	4.9	8.1	758
High protein diet (60% for two weeks)	16.0	736	7.8	15.3	991
Zero protein† diet	2.6	60.2	0.8	1.2	350

*Data from Raijman (1976).

Note that total starvation of rats increases the activities of all the urea cycle enzymes.

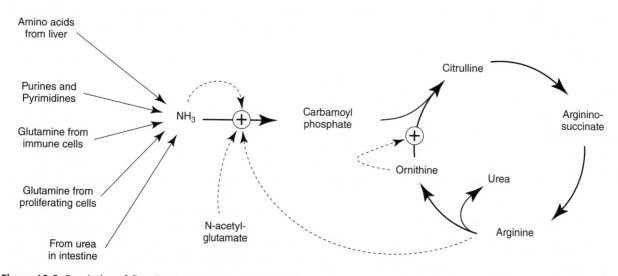

Figure 10.5 *Regulation of flux through the urea cycle.*

• The concentration of ammonia in the liver is not saturating for carbamoyl phosphate synthetase, so that the greater the flux of ammonia into or within the liver, the higher the concentration of ammonia and the higher the activity of the synthetase. The effect of ammonia concentration is, therefore, a mass-action effect.

• N-acetylglutamate is an allosteric activator of the synthetase. The concentration in the liver increases several – fold after a meal.

• An increase in concentration of carbamoyl phosphate and/or an increase in the concentration of ornithine stimulates ornithine transcarbamoylase and increases flux through the cycle.

• Arginine stimulates carbamoyl-phosphate synthetase (an allosteric effect).

This compartmentation of processes in the liver cells occurs with other pathways; glycolysis occurs mainly in the perivenous cells whereas gluconeogenesis occurs mainly in the periportal cells (Chapter 6).

perivenous cells contain low activities of urea cycle enzymes but high activities of glutamine synthetase, which converts ammonia to glutamine. Consequently, any ammonia that 'escapes' the ornithine cycle in the periportal cells will be taken up by the perivenous cells and converted to glutamine:

$$ATP + glutamate + ammonia \rightarrow glutamine + ADP + P_i$$

The glutamine can be used by macrophages in the liver (known as Küpffer cells) or released into the blood (Chapter 8).

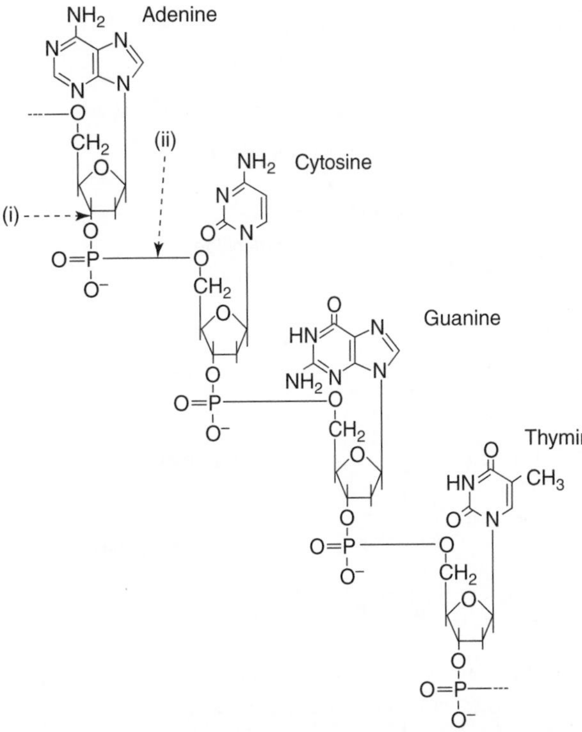

Degradation of nucleic acids, nucleotides, nucleosides and bases: the generation of ammonia

The synthesis of nucleosides, nucleotides and nucleic acids is described and discussed in Chapter 20.

A nucleotide consists of a heterocyclic base linked to a sugar (ribose or deoxyribose) and a phosphate group also linked to the sugar (Figure 10.6). Nucleic acids are polymers of nucleotides linked together by phosphodiester bonds (Figure 10.7). The enzymes that catalyse the breakdown of nucleic acids to nucleotides are nucleases.

Unlike proteins, however, DNA is not broken down until a cell dies, although repair processes do involve exci-

Figure 10.7 *Molecular structure of a section of a single strand of DNA.* Note that interatomic distances are considerably distorted in this representation. Bonds (i) and (ii) are positions of hydrolysis of the phosphodiester bond by different nucleases.

sion of short nucleotide sequences which are then degraded. As an aid to understanding, there is a similar 'numerical reduction' system in nucleic acid and nucleotide degradation as there is in protein degradation. Protein degradation of a peptide chain produces 20 amino acids, which are degraded to six common intermediates, which are finally degraded to three end-products (glucose, triacylglycerol and CO_2) plus, of course, ammonia (Chapter 8). For DNA and RNA, degradation produces eight nucleotides, which are broken down to five bases which finally give rise to three end-products (ribose phosphate, uric acid and two almost identical 3-carbon compounds (malonyl-CoA and methylmalonyl-CoA) plus, or course, ammonia (Figure 10.8).

Nucleases that hydrolyse phosphodiester bonds, within the nucleic acid chain are endonucleases (Figure 10.8). Apart from specificity for DNA or RNA shown by some, nucleases, these enzymes show specificity for:

- Single or double-stranded chains of nucleic acids:

- Phosphodiester links within the chain (endonucleases) or at the end of a chain (exonucleases); for details of nucleases reactions see Appendix 10.1.

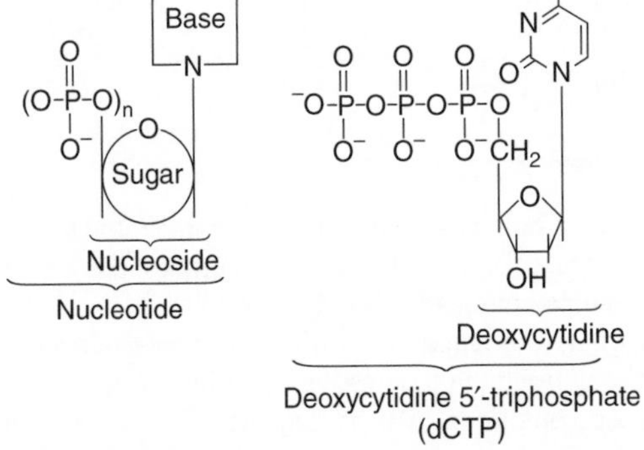

Figure 10.6 *General structure of a nucleoside and nucleotide with a representative example.* Note the sugar is either ribose or deoxy ribose. A base plus a sugar is a nucleoside.

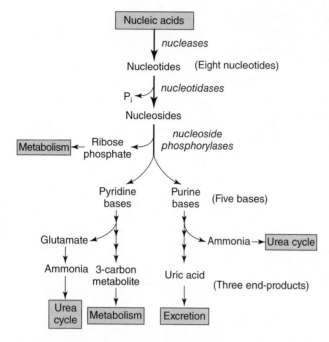

Figure 10.8 *A summary of the reactions involved in the degradation of nucleic acid, nucleotides, nucleosides and purine and pyrimidine bases.* Nucleic acid is hydrolysed by nucleases to form nucleotides, which are hydrolysed to nucleosides. The latter are split into ribose 1-phosphate and a base. The purine bases are converted to uric acid and ammonia. Uric acid is excreted. The pyrimidine bases are converted to 3-carbon intermediates (malonate semialdehyde and methylmalonate semialdehyde). The nitrogen is released as ammonia or used to convert oxoglutarate to glutamate.

• The type of bases on either side of the bond to be cleaved.

Exonucleases hydrolyse the link either at the 3′ end or at the 5′ end of the strand; the former will release nucleotide 5′-phosphates and the latter nucleotide 3′-phosphates.

Nucleotide breakdown

Within a cell, a nucleotidase catalyses the hydrolysis of either a ribonucleotide or deoxyribonucleotide (Figure 10.8). The quantitatively important pathway for degradation of AMP in liver and muscle involves deamination to IMP, catalysed by AMP deaminase, producing ammonia, and subsequent hydrolysis of IMP to inosine. This may be an important source of inosine for synthesis of phosphatidylinositol, a key phospholipid in membranes.

Once produced, nucleosides have three possible fates:

(i)   Hydrolysis to form the free base and ribose 1-phosphate (or deoxyribose 1-phosphate) by the action of

nucleoside phosphorylases. A summary of the processes is given in Figure 10.8.

(ii)   Transport out of the cell into the extracellular fluid, via a nucleoside transporter protein in the plasma membrane;

(iii)   Conversion back to a nucleotide via a salvage pathway (Chapter 20).

The further catabolism of adenine and guanine, which occurs primarily in the liver, involves their progressive oxidation to uric acid (urate) which is excreted through the kidneys. The central intermediate in the pathway is hypoxanthine, which can be formed from adenine, guanine or inosine and is oxidised first to xanthine, and then to uric acid, in reactions catalysed by xanthine oxidase (Figure 10.9). It is during the degradation of the bases that ammonia is produced (Appendix 10.2). Uric acid is responsible for gout (Box 10.1).

The pyrimidine nucleotides (predominantly UMP, CMP, and dTMP) are hydrolysed to their respective bases (uracil, cytosine, and thymine) by reactions similar to those for metabolism of purine nucleotides. The pathways of

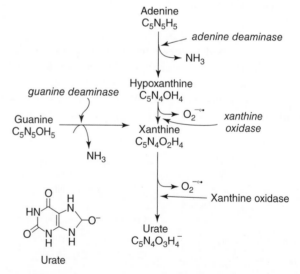

Figure 10.9 *The sequence of reactions in which the purine bases are converted to urate.* The diagram outlines the pathway by which adenine and guanine are converted to uric acid. Note the following: (i) the reactions in which ammonia is released; (ii) the pathway results in progressive oxidations; (iii) during anoxic or hypoxic conditions, AMP is degraded to inosine. When the hypoxic/anoxic tissue is re-oxygenated, inosine is converted to uric acid via hypoxanthine and xanthine, which generate oxygen free radicals that damage the cells, known as reperfusion damage. $O_2^{-\bullet}$ represents the superoxide radical (Appendix 9.6).

Box 10.1 Uric acid and gout

For birds, insects, and reptiles, which have an egg stage during development, so that water availability is severely restricted, the synthesis of a highly soluble excretory product would have serious osmotic consequences; therefore most of the ammonia is converted to the virtually insoluble uric acid (urate). This product can be safely retained in the egg or excreted as a slurry of fine crystals by the adult. In birds that nest colonially this can accumulate in massive amounts; on islands off the coast of Peru cormorants have deposited so much that this guano (hence the name guanine) is collected for use as a fertiliser. Uric acid is less effective as an excretory product, since it has a lower nitrogen content than urea (33%) and is more 'expensive' to synthesise (2.25 molecules ATP per atom of nitrogen). Mammals do produce uric acid but as a product of purine catabolism (see above).

The low solubility of uric acid has unfortunate consequences since at higher than normal concentrations it can crystallise in the body. For example, when the urine is unusually acid, calcium urate stones can form in the kidney and bladder. High levels of uric acid in the blood can result in the formation of urate crystals in the joints, which causes a very painful condition, since it results in inflammation in these joints. Gout is unlikely to develop if the urate concentration remains low (<0.4 mmol/L) but any factor that increases the rate of production or decreases that of elimination by the kidney, so that the concentration rises above 0.6 mmol/L, is very likely to cause gout. The disease is found predominantly in males and it usually affects the elderly (often, in the early stages of the disease, only in the joints of the big toe). It is estimated that there are about one million cases in the USA alone. The Roman emperor Claudius and King Henry VIII were afflicted by gout.

Gout is one of the most ancient diseases: its clinical characteristics have been known for at least 2000 years. It is now very effectively treated with drugs that decrease production of uric acid by inhibition of the enzyme xanthine oxidase in purine degradation (Figure 10.9) (allopurinol), and a drug that increases the excretion of uric acid (probenecid)

The presence of sodium urate in the amorphous deposits in a jug found in 1903, but dating back to the Middle Ages, suggests that it was used to store urine, from which the urate precipitated as the urine evaporated. But why was urine stored in a jug in the Middle Ages? Perhaps because drinking urine was highly recommended at this time for treatment of bubonic plague and other diseases (*Lancet*, July 11, 1942).

A marked increase in the plasma concentration of urate has been linked to a neurological disorder. Lesch-nyhan syndrome which is characterise by mental retardation and a compulsive form of self-mutilation (Chapter 20).

degradation of the bases, summarised in Figure 10.8, generate products, ammonia, malonyl-CoA and methylmalonyl-CoA, which are also produced in amino acid catabolism, so that no distinctive nitrogenous end-products result (for details, see Appendix 10.3).

Ammonia toxicity

The toxicity of ammonia was dramatically demonstrated by experiments carried out as early as 1931: injection of the enzyme urease, which catalyses the conversion of urea to ammonia, into rabbits rapidly caused their death. The normal concentration of ammonia in blood is about 0.02 mmol/L; toxicity becomes apparent at a concentration of about 0.2 mmol/L or above (see Table 10.1). Ammonia toxicity in very young children is usually associated with vomiting and eventually coma. It is almost invariably due to the deficiency of an enzyme of the urea cycle (see below). In adults, ammonia accumulation, and hence toxicity, usually results from damage to the liver caused by poisons, alcohol or viral infection.

Perhaps surprisingly, the reasons why ammonia is toxic remain unknown, although there is no shortage of hypotheses, as follows:

- Several explanations centre on the enzyme glutamate dehydrogenase, which is assumed to catalyse a near-equilibrium reaction in brain mitochondria:

$$glutamate + NAD^+ + H_2O \rightleftharpoons oxoglutarate + NH_3 + NADH + H^+$$

A rise in ammonia concentration would 'push' this equation to the left so that the concentrations of both oxoglutarate and NADH would decrease: the former change could decrease the flux through the Krebs cycle, and the latter change could decrease the rate of electron transfer. Either would lead to a decrease in the ATP/ADP concentration which could impair neural function.

- The excitatory neurotransmitter in brain, glutamate, is synthesised from glutamine in a reaction catalysed by the enzyme glutaminase. This enzyme is inhibited by a high concentration of ammonia, which could lead to a chronic depletion of this excitatory neurotransmitter.

- Ammonia increases the activity of the glycolytic enzyme 6-phosphofructokinase, which could increase the flux through glycolysis. In brain, glycolysis occurs in the glial cells, which provide lactic acid as a fuel for the neurons. An increase in the rate of glycolysis could increase the concentration of lactic acid in the interstitial

Table 10.5 Changes in the concentration of various intermediates of the urea cycle or their metabolites in plasma or urine in various enzyme deficiency diseases in humans

| | Change in the concentration | | |
| | Plasma | | Urine |
Enzyme deficiency	Arginine	Citrulline	Orotic acid
Arginase (AR)	+++[a]	−	−
Ornithine transcarbamylase (OTC)	−	−	++++[c]
Argininosuccinate synthetase (AS)	−	++++++[b]	−
Argininosuccinate lyase (ASL)	−	+++[a]	−
Carbomoyl phosphate synthetase (CPS)	−	−	−

+ Indicates an increase.

− Indicates no change.

[a] Concentration is about 200 μmol/L.

[b] Concentration is about 1000 μmol/L.

[c] Orotic acid is present in urine since carbamoyl phosphate accumulates in OTC deficiency and is broken down to orotic acid.

fluid between the glial cell and the neurone so that the rate exceeds the requirement of the neurone. This could damage the neurone.

Deficiencies of urea cycle enzymes

In view of the toxicity of ammonia, complete absence of any one of the enzymes of the cycle is fatal. Nonetheless, disorders of the cycle do occur, which are caused by a low activity of one of the enzymes or carbamoyl phosphate synthetase. In addition, defects in N-acetylglutamate synthase have been reported, but they are very rare. With the exception of ornithine transcarbamoylase, the deficiencies have an autosomal recessive mode of inheritance. The transcarbamoylase deficiency is inherited as an X-linked dominant trait, usually lethal in male patients. A deficiency of carbamoyl phosphate synthetase, ornithine transcarbamoylase or argininosuccinate synthetase results in accumulation and excretion of citrulline. A deficiency of argininosuccinate lyase results in the accumulation and excretion of argininosuccinate and arginine (Table 10.5). The abbreviations CPSD, OTCD, ASD, ALD and AD stand, respectively, for the deficiencies of these enzymes, where D stands for deficiency.

Symptoms for the first four deficiencies include anorexia, vomiting and lethargy, from which patients may progress to irreversible coma (acute hepatic coma) and death. If infants survive, but remain undiagnosed and untreated for some time, they become mentally handicapped. In arginase

deficiency, there is progressive paralysis with mental retardation, epilepsy and poor growth. The objective of long-term therapy in children with these disorders is to maintain the plasma ammonia concentration as close to normal as possible but provide sufficient protein and essential amino acids and other nutrients to allow for normal growth and intellectual development. Four main approaches are used: reduction in protein intake; correction of any arginine deficiency; increase renal excretion of ammonia; and use of biochemical 'tricks' to remove ammonia.

One trick is the administration of benzoate or phenylacetate. These acids react with glycine or glutamine, respectively, to produce complexes (conjugates) which are excreted in the urine (Figure 10.10). This decreases the plasma level of either of these non-essential amino acids

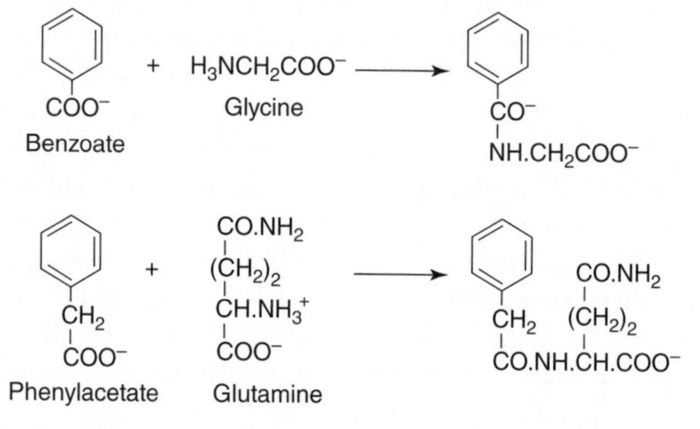

Figure 10.10 The use of benzoate and phenylacetate to lower the concentration of ammonia in patients with a deficiency of a urea cycle enzyme.

so that they are then synthesised, in an attempt to restore their normal blood levels. This usually involves glutamate, so that re-formation of glutamate uses ammonia (see Chapter 8 for a description of the pathways for glycine and glutamine synthesis).

Chronic hepatic coma (encephalopathy)

Since the urea cycle is localised exclusively in the liver, ammonia retention occurs in adults who are suffering from chronic liver disease, including hepatitis, hepatic failure or cirrhosis of the liver. The coma that can result from these conditions is considered to be caused by the high blood ammonia level but other factors may also play a part, e.g. hypoglycaemia or lactic acidosis (Chapter 6) caused by the damage to the liver; accumulation of 'false' neurotransmitters, which are produced by microorganisms in the colon are normally removed by the liver.

The treatment of patients with hepatic coma is designed to reduce the severity of the coma to give a chance for the liver to recover. Hence a primary aim of treatment is to reduce the concentration in the blood of compounds responsible for the coma. Since amino acid metabolism, especially by bacteria in the intestine, can result in the formation of both ammonia and false neurotransmitters, restriction of protein intake and administration of an antibiotic (to reduce the number of colonic bacteria) are recommended treatments. Since high brain levels of the neurotransmitter, gamma-aminobutyrate (GABA), might be responsible for the coma, patients have been treated with antagonists to the GABA receptor, known as the benzodiazepine receptor antagonists (Chapter 14).

Finally, if liver failure is chronic and extensive, the only treatment is liver transplantation.

> False neurotransmitters are amines which are similar enough in structure to normal amine neurotransmitters that they bind to receptors but are much less active or totally inactive (i.e. they are antagonists). One such false neurotransmitter is octopamine, which is formed from tyrosine by decarboxylation followed by side-chain hydroxylation.

11
Synthesis of Fatty Acids, Triacylglycerol, Phospholipids and Fatty Messengers: The Roles of Polyunsaturated Fatty Acids

Sir, scant attention seems to be paid by the medical profession and by food administrators to a very important change in the dietaries of the more civilised countries that has been occurring over recent decades with increasing intensity. I refer to the chronic relative deficiency of the polyethenoid essential fatty acids (E.F.A.). . . . The causes of death that have increased in most recent years are lung cancer, coronary thrombosis and leukaemia: I believe that in all three groups deficiency of E.F.A. may be important. Your readers with stereotyped minds should stop reading at this point . . .

H. Sinclair, (1956)

There is no doubt that cerebral lipids, and EFA-derived LC-PUFAs in particular, have significant direct and indirect actions on cerebral function. Not only does the lipid composition of neural membranes affect the function of their embedded proteins, but also many LC-PUFAs are converted to neurally active substances. There is good evidence that psychiatric illness is associated with depletion of EFAs and, crucially, that supplementation can result in clinical amelioration. As well as challenging traditional views of aetiology and therapeutics in psychiatry, the clinical trial data may herald a simple, safe and effective adjunct to our standard treatments for many disabling conditions.

Hallahan & Garland, 2005

For many years, the biochemistry of fats was a 'poor relation' to that of protein and carbohydrate. Early and recent findings have, however, stimulated an upsurge in the interest in fats in health and disease. For example, it was discovered in 1929 that some fatty acids required by humans were not synthesised in the body and were, therefore, essential components of the diet. In the 1950s, it was discovered that the amount and type of fat consumed by humans could be related to the incidence of coronary heart disease and that, in addition to fuel storage, fats and fatty acids have several key roles in the body, for example as messengers, precursors of messengers, gene regulators, components of phospholipids, and modifiers of the immune response. It has also recently been discovered that deficien- cies of polyunsaturated fatty acids may play a role in devel- opment of neurological and behavioural disorders.

Synthesis of long-chain fatty acids

The synthesis of fatty acids in humans takes place in the liver and adipose tissue. The rates of synthesis are nor- mally relatively low in adults in developed countries, probably because the normal diet contains such a high proportion of fat which reduces the activities of enzymes involved in fatty acid synthesis by decreasing expression

Functional Biochemistry in Health and Disease by Eric Newsholme and Tony Leech
© 2010 John Wiley & Sons Ltd

of genes. It can be increased in the liver by switching to a very high carbohydrate diet but such a diet is not easy to persuade volunteers to consume since some fat is necessary for palatability of meals, so that very few measurements have been carried out on volunteers who have eaten, chronically, a diet rich in carbohydrate. The regulation of fatty acid synthesis is described below. Fatty acid synthesis also occurs during lactation in the mammary gland but only about 100 g each day in humans. In contrast, in ruminants, it can be very large. Dairy cows can readily synthesise 1.5 kg of fat each day, all of which is synthesised from the short-chain fatty acids which are produced by microorganisms in the rumen.

The basic 'building block' for fatty acid synthesis is acetyl-CoA, produced from glucose, fructose or amino acids (Figure 11.1). Acetyl-CoA formation from these precursors occurs within the mitochondrion and so, because fatty acid synthesis occurs in the cytosol, acetyl-CoA must be transported across the mitochondrial membrane. Trans-port is achieved via citrate, formed from acetyl-CoA and oxaloacetate in the first reaction of the Krebs cycle, catalysed by citrate synthase:

$$\text{oxaloacetate} + \text{acetyl-CoA} \rightarrow \text{citrate} + \text{CoASH}$$

Citrate is transported across the mitochondrial membrane by a specific carrier. In the cytosol, acetyl-CoA is re-formed in a reaction catalysed by ATP citrate lyase (Figure 11.3). This reaction involves the hydrolysis of ATP:

> Citrate has a similar role in proliferating (including tumour) cells and in neurones. Acetyl-CoA is a precursor for many compounds in the former and for formation of acetylcholine in neurones (see Appendix 9.4).

$$\text{citrate} + \text{CoASH} + \text{ATP} \rightarrow \text{acetyl-CoA} + \text{oxaloacetate} + \text{ADP} + P_i$$

Oxaloacetate is returned to the mitochondrion, by the shuttle mechanism described in Chapter 9 (Figure 9.17).

Conversion of 2-carbon units to a 16-carbon fatty acid

In principle, fatty acids could be synthesised by the progressive addition of 2-carbon acetyl units to an extending acyl chain and reduction of the carbonyl groups to methylene groups; that is, by reversal of the β-oxidation process, until the 16-carbon, palmitic acid is produced. However, for reasons already considered in this text, synthetic reac-

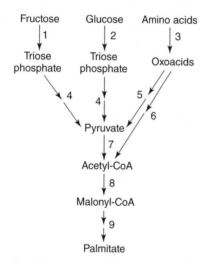

Figure 11.1 *The precursors for fatty acid synthesis.* The immediate precursor is acetyl-CoA, which can be formed from the dietary precursors glucose, fructose or amino acids. The processes are as follows:

1. Fructose is converted to acetyl-CoA as shown in Figure 11.2.
2. Glucose is converted to triose phosphate via the enzyme glucokinase and other glycolytic enzymes in the liver.
3. Amino acids are converted to oxoacids by pathways described in Chapter 8.
4. Triose phosphate is converted to pyruvate via glycolysis.
5&6. Oxoacids are converted to pyruvate and then acetyl-CoA, as shown in Chapter 8 (see Figure 8.8).
7. The pyruvate dehydrogenase reaction. Acetyl-CoA is transported out of the mitochondria, via citrate (Figure 11.3).
8. Acetyl-CoA is converted to malonyl-CoA, catalysed by acetyl-CoA carboxylase.
9. Fatty acid synthase (see text, Figure 11.5).

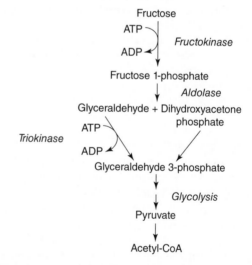

Figure 11.2 *Pathway for conversion of fructose to acetyl-CoA.* The enzyme fructokinase phosphorylates fructose to form fructose 1-phosphate. (The enzyme is present only in the liver.) Fructose 1-phosphate is cleaved by aldolase to form glyceraldehyde and dihydroxyacetone phosphate. Glyceraldehyde is phosphorylated to form glyceraldehyde 3-phosphate, catalysed by the enzyme triokinase. Dihydroxyacetone phosphate is converted to glyceraldehyde 3-phosphate, catalysed by the isomerase. Glyceraldehyde 3-phosphate is converted to pyruvate by the glycolytic reactions (Chapter 6).

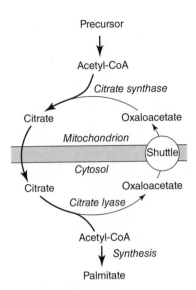

Figure 11.3 *Mechanism of transfer of acetyl-CoA out of the mitochondrion.* In the mitochondrion, acetyl-CoA reacts with oxaloacetate to form citrate, which is transported across the mitochondrial inner membrane. In the cytosol, citrate is split to re-form citrate and oxaloacetate, catalysed by citrate lyase. It has been shown that inhibition of citrate lyase inhibits fatty acid synthesis.

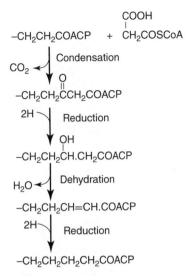

Figure 11.4 *Condensation, dehydration and reduction reactions in fatty acid synthesis.* These reactions constitute the major components of the pathway of fatty acid synthesis and are all catalysed by fatty acid synthase. The reduction reactions, indicated by addition of '2H' in the diagram, involve the conversion of NADPH to $NADP^+$. (The re-conversion of $NADP^+$ back to NADPH occurs in the pentose phosphate pathway.) The condensation reaction results in an increase in size of acyl-ACP by two carbon units in each step. The two carbons for each extension are each provided by malonyl-CoA. ACP – acyl carrier protein.

tions rarely, if ever, proceed by reversal of the corresponding degradative reactions and fatty acid synthesis is no exception.

Specific biochemical points and some differences between synthesis and degradation are as follows:

> Malonyl-CoA is also involved in the regulation of fatty acid oxidation, via inhibition of carnitine palmitoyltransferase. In non-lipogenic tissues, the only role of the carboxylase is provision of malonyl-CoA for regulation of the rate of fatty acid oxidation.

- The chain is progressively extended, not by reaction with acetyl-CoA, but with malonyl-CoA, a three-carbon compound that is formed from acetyl-CoA by a carboxylation reaction catalysed by acetyl-CoA carboxylase:

$$CH_3COSCoA + ATP + HCO_3^- \longrightarrow \underset{\text{malonyl-CoA}}{CH_2.COSCoA} + ADP + P_i$$

- During the subsequent condensation reaction between malonyl-CoA and acetyl-CoA, the carbon dioxide that has been incorporated into malonyl-CoA is eliminated, so that the net result is extension of the chain by two carbon atoms. The process is summarised:

> The advantage of using malonyl-CoA as the condensing agent is that the subsequent release of CO_2 'pulls' the reaction to the right.

$$C\text{-}2 + C\text{-}3 \longrightarrow C\text{-}4 + CO_2$$

- The synthesis of fatty acids involves an ordered sequence of condensations (to build up the chain length), reductions (to convert a carbonyl to a methylene group), dehydrations and a further reduction reaction. (A simplified description of condensation and reduction reactions is presented in Figure 11.4.)

- The reducing agent is NADPH (not NADH).

- Coenzyme A is not involved in the elongation or the reduction reactions: it is replaced by a small protein known as the *acyl carrier protein* (ACP).

- All the reactions in the pathway take place on a multienzyme complex, *fatty acid synthase*, which has seven catalytic activities or functional domains. Details of the individual reactions are presented in Appendix 11.1.

- The end-product of this process is the C-16 saturated fatty acid, palmitate. The elongation of palmitate to longer-chain fatty acids involves another system (see below).

- The overall process can be divided into several stages: activation (or initiation), binding to a macromolecule, elongation, movement of the elongating molecule, termination and post-synthetic modification. Analogous processes occur in the synthesis of polysaccharides and peptides (Chapters 6 and 20. See Table 6.2).

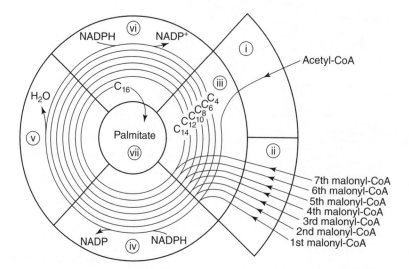

Figure 11.5 *Reactions of the fatty acid synthase complex.* A single multi-subunit enzyme is responsible for the conversion of acetyl-CoA to palmitate. The subunits in the enzyme are: (i) acetyltransferase, (ii) malonyltransferase, (iii) oxoacyl synthase, (iv) oxoacyl reductase, (v) hydroxyacyl dehydratase, (vi) enoyl reductase. Finally, a separate enzyme, thioester hydrolase, hydrolyses palmitoyl-CoA to produce palmitate (vii).

Initiation

The acyl carrier protein is involved in initiation and in all the processes of fatty acid synthesis. The prosthetic group of this protein is 4'-phosphopantetheine, which plays a similar role to that of coenzyme A in other reactions involving carboxylic groups. In the initiation process the acetyl group of acetyl-CoA and the malonyl group of malonyl-CoA are transferred to separate sites on the carrier protein. The acetyl group is then transferred from the carrier to a cysteine residue in another site on the fatty acid synthase, which then catalyses the condensation reaction with the malonyl group. The product remains attached to the carrier protein until all the reactions are completed. A diagram of the sequence of reactions, depicted in a circular fashion, is given in Figure 11.5.

Termination

Once the 16-carbon atom acyl chain is formed, the thioester link between the acyl group and the 4'-phosphopantetheine of the carrier protein is hydrolysed by a thioesterase and palmitate is released. Synthesis of shorter-chain fatty acids, e.g. myristic (a C-14 carbon acid), requires a specific cytosolic thioesterase (thioesterase II) which is present in liver. It hydrolyses the thioester bond when fatty acids reach lengths of less than 16 carbon atoms.

Physiological pathway for fatty acid synthesis

The physiological pathway starts with acetyl-CoA in the cytosol. It is converted to malonyl-CoA, which is the flux-generating step for fatty acid synthesis catalysed by

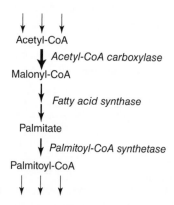

Figure 11.6 *The physiological pathway for fatty acid synthesis: acetyl-CoA to palmitoyl-CoA.* The pathway starts with the conversion of acetyl-CoA to malonyl-CoA in the cytosol, which is the flux-generating step catalysed by acetyl-CoA carboxylase. The pathway can be considered to end with formation of palmitoyl-CoA rather than palmitate, since it has several fates: formation of triacylglycerol and phospholipids or acylation of other compounds.

acetyl-CoA carboxylase (Figure 11.6). Not surprisingly, the activity of this enzyme regulates the rate of fatty acid synthesis (see below).

Source of hydrogen atoms to reduce NADP$^+$

The reduced coenzyme NADPH is required for the reduction reactions shown in Figure 11.5. It is also required for elongation and desaturation of fatty acids. The major source of NADPH for these reactions is the pentose phosphate pathway, which is described in detail in Chapter 6.

The generation of NADPH occurs in two reactions in the pathway. First, in the reaction catalysed by glucose 6-phosphate dehydrogenase:

$$\text{glucose 6-phosphate} + NADP^+ \rightarrow$$
$$\text{6-phosphogluconolactone} + NADPH + H^+$$

The enzyme, 6-phosphogluconolactonase, catalyses the hydrolysis of this product to form phosphogluconate. The oxidation of the phosphogluconate, in a reaction catalysed by phosphogluconate dehydrogenase, generates the second NADPH molecule:

$$\text{6-phosphogluconate} + NADP^+ \rightarrow \text{ribulose 5-P} +$$
$$NADPH + H^+ + CO_2$$

All three enzymes are present at high activities in the liver and in the lactating mammary glands.

Pathway for synthesis of triacylglycerol

The concentration of long-chain fatty acids in the free form (i.e. not esterified in triacylglycerol or phospholipids) is very low in cells. In contrast, the amount in esterified form is very large. Triacylglycerols are synthesised when the three hydroxyl groups in glycerol are esterified with long- or medium-chain fatty acids. The reactants are fatty acyl-CoA and glycerol 3-phosphate. The process is summarised as follows:

$$3 \text{ long-chain acyl-CoA} + \text{glycerol-3-phosphate} + H_2O \rightarrow$$
$$\text{triacylglycerol} + 3 \text{ CoASH} + P_i$$

The individual reactions involved in esterification are shown in Figure 11.7: the enzymes involved and their specificities are as follows:

- The initial acylation at the 1-position of glycerol 3-phosphate is catalysed by glycerol 3-phosphate acyltransferase-1, abbreviated to GPAT-1. This enzyme is specific for a saturated fatty acid (in the acyl form).

- The second acylation at position 2 is catalysed by GPAT-2, which is specific for a fatty acid with one or two double bonds. This produces phosphatidic acid, for which the phosphate must be removed prior to the final acylation.

- The third acylation is catalysed by diacylglycerol acyltransferase (DGAT), which is less specific, so that an unsaturated or saturated fatty acid can be incorporated. The final product is as shown in Figure 11.7.

The resulting triacylglycerol is stored in adipose tissue. In the liver, some is combined with protein and phospholipids to form a complex, known as very low density lipoprotein (VLDL), which is secreted from the liver into the blood. Details of the formation of VLDL are presented in Appendix 11.2. Failure to form VLDL or secrete can cause accu-

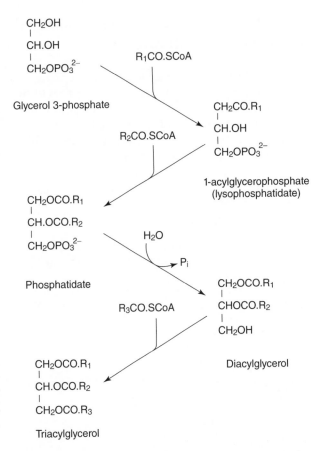

Figure 11.7 *Synthesis of triacylglycerol.* The precursors are glycerol 3-phosphate and long-chain acyl-CoA. R_1 is a saturated fatty acid, R_2 is an unsaturated fatty acid (one or two double bonds) and R_3 is either saturated or unsaturated. The activity of GPAT-1 regulates triacylglycerol synthesis. In all reactions involving RCO.SCoA, the CoASH is released but is not shown in this diagram. P_i – phosphate.

mulation of triacylglycerol: a condition known as 'fatty liver'. This occurs with protein-deficient diets and in some obese patients. The role of VLDL in the mass transport of fat in the blood and the pathology that arises when this falters, and hyperlipoproteinaemia results are described in Appendix 11.3.

Origin of glycerol 3-phosphate

Glycerol 3-phosphate can arise in two ways, either (i) from glycerol, via the enzyme glycerol kinase or (ii) from dihydroxyacetone phosphate, which is produced in glycolysis, by reduction with NADH, catalysed by glycerol-3-phosphate dehydrogenase:

(i) $\text{glycerol} + ATP \rightarrow \text{glycerol 3-phosphate} + ADP$

(ii) $\text{dihydroxyacetone phosphate} + NADH \rightarrow \text{glycerol 3-phosphate} + NAD^+ + H^+$

Regulation of the rates of fatty acid and triacylglycerol synthesis

As indicated above, the flux-generating step for fatty acid synthesis is the conversion of acetyl-CoA to malonyl-CoA, catalysed by acetyl-CoA carboxylase. Consequently, regulation of the rate of synthesis is achieved via changes in the activity of this enzyme. The properties of the carboxylase identify three mechanisms for regulation: allosteric regulation, reversible phosphorylation (an interconversion cycle) and changes in the concentration of the enzyme. (The principles underlying the first two mechanisms are discussed in Chapter 3.)

(1) Citrate is the immediate precursor of acetyl-CoA in the cytosol and is an allosteric activator. Palmitoyl-CoA

is the end-product of the pathway and an allosteric inhibitor. These can be considered as feed-forward and feed-back mechanisms, respectively (Figure 11.8(a)).

(2) There are two protein kinase enzymes that phosphorylate the carboxylase and decrease its activity, cyclic AMP-dependent protein kinase and AMP-dependent protein kinase (Figure 11.8(b)). Since insulin decreases and glucagon increases the cyclic AMP concentration, insulin stimulates and glucagon inhibits the rate of fatty acid synthesis, which are consistent with their physiological effects in controlling the blood glucose levels (Chapters 6 and 12). A change in the concentration of AMP is an indicator of the energy status (i.e. ATP/ADP concentration ratio) in the cell (see Chapter 6). Conditions in which the ATP/ADP concentration

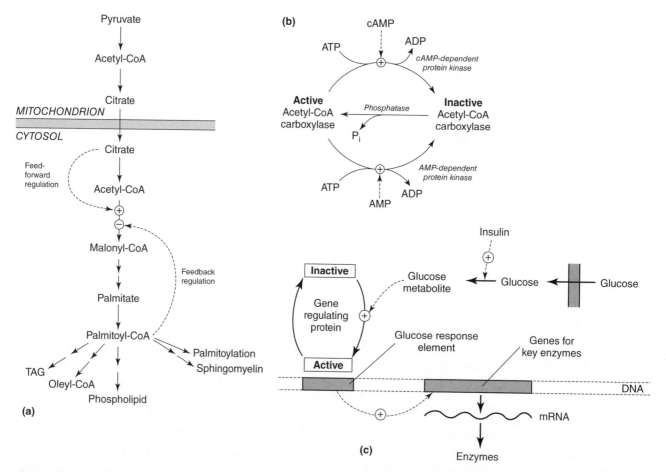

Figure 11.8 **(a)** *Allosteric regulation of acetyl-CoA carboxylase activity by citrate and palmitoyl-CoA.* **(b)** *Regulation of acetyl-CoA carboxylase activity by phosphorylation (inactivation) and dephosphorylation (activation).* There are two phosphorylating reactions catalysed by (i) cyclic AMP-dependent protein kinase, and (ii) AMP-dependent protein kinase: (i) depends on the balance between the two hormones, insulin and glucagon; (ii) depends on ATP/ADP concentration ratio. **(c)** *Regulation of gene expression of enzymes involved in fatty acid synthesis.* An increase in the blood glucose level and insulin concentration increases concentration of a glucose metabolite that activates a gene-regulation protein to enhance expression of genes for key lipogenic enzymes (see also Figure 20.21).

ratio decreases (i.e. energy stress) increase the AMP level, which results in phosphorylation and hence inhibition of acetyl-CoA carboxylase activity and thus decreased rates of fatty acid synthesis. This inhibition helps to maintain the ATP/ADP concentration ratio in the cell.

(3) Finally, a high carbohydrate diet results in activation of a protein that regulates transcription of genes that express enzymes involved in the process of fatty acid synthesis (Figure 11.8(c)).

Triacylglycerol synthesis

The first enzyme involved in triacylglycerol synthesis is glycerol 3-phosphate acyltransferase (GPAT-1). It is saturated with fatty acyl-CoA so that it is the flux-generating step. It is regulated by two mechanisms. First, by the concentration of glycerol-3-phosphate, i.e. its other substrate, that is, GPAT-1 is saturated with glycerol 2-phosphate, so that an increase in its concentration stimulates triacylglycerol synthesis. A high intake of fructose is considered to be one factor that can increase the glycerol 3-phosphate concentration, since utilisation of fructose in the liver bypasses control at the key site for regulating glycolysis, the interconversion cycle catalysed by phosphofructokinase/fructose-1,6-bisphosphatase (see above). This is considered to be of clinical importance in raising the blood triacylglycerol level in children and teenagers who consume excessive amounts of fructose-containing beverages (Chapter 15). Second, GPAT is also regulated by an interconversion cycle (reversible phosphorylation). It is phosphorylated by AMP-dependent protein kinase which decreases its activity. Hence energy stress, as expected, decreases triacylglycerol synthesis as well as fatty acid synthesis. It is also likely that the concentration of GPAT-1 in the liver cell is increased by a high carbohydrate diet as well as the other enzymes involved in fatty acid synthesis (Figure 11.8(c)).

Unsaturated fatty acids

Saturated fatty acids do not contain double bonds in the hydrocarbon chain. Unsaturated fatty acids contain from one to five double bonds. Those with one double bond are known as monounsaturated, those with two as diunsaturated and those with more than two as polyunsaturated fatty acids. A brief summary of the roles of saturated and unsaturated fatty acids is given in Table 11.1. The proportion of these fatty acids in triacylglycerol in human adipose tissue is presented in Table 11.2.

Fatty acid structure and nomenclature

As noted several times in this text, one of the problems for a student entering any biochemical field for the first time is nomenclature: this is very much the case for fatty acids.

Table 11.1 Summary of the roles of long-chain fatty acids

Function	Fatty acid		
	Saturated	MUFA	PUFA
Energy source	++++	++++	++
Membrane constituent	++	+	++++
Messenger[a,b]	–	–	++++
Protein acylation[d]	+	–	+

MUFA – monounsaturated fatty acid; PUFA – polyunsaturated fatty acid.

[a] Arachidonic acid itself can function as a messenger (see below).

[b] The fatty acid is a precursor for fatty acid derived messengers, after it is released from a membrane phospholipid by the action of a phospholipase enzyme (see below).

[c] It can be considered that the saturated long chain fatty acid concentration in blood can act as a messenger for its oxidation. An increase in concentration increases rate of oxidation.

[d] Acylation describes the process by which a fatty acid reacts with another molecule (e.g. glycerol phosphate) or with a protein, to facilitate attachment of the protein to a membrane.

Table 11.2 Summary of fatty acid composition of some adipose tissue depots from human subjects

Adipose tissue depot	Approximate percentage composition		
	Saturated	Monounsaturated	Polyunsaturated
Superficial	40	50	10
Intermuscular	39	51	10
Intra-abdominal	41	49	10

Data from Calder et al. (1992).

Data for superficial adipose tissue are the means for six different depots, those for intermuscular adipose tissue are the mean for three different depots, and those for intra-abdominal adipose tissue are the mean for four different depots. The percentages are remarkably similar.

There are at least sixteen fatty acids of biochemical interest, and to understand how they are named and how their structure relates to function, some basic information is necessary (see also Appendix 7.1).

- Most fatty acids have both systematic and common names. The source of the fatty acid provides the basis for the common name.

- Saturated fatty acids are named, systematically, according to the number of carbon atoms with the addition of the suffix *-anoic*. hence the systematic name for stearic acid, which has 18 carbon atoms, is octadecanoic acid.

- Unsaturated fatty acids are named, systematically, according to the number of carbon atoms, and the position of the double bonds with the addition of the suffix *-enoic*: the positions of the double bonds can be represented in two ways.

  (i) Counting from the carboxyl end, with the carboxyl carbon receiving the number one, the first number presented indicates the total number of carbon atoms. The second number indicates the number of double bonds and the number(s) presented in brackets indicate(s) the position of the double bond(s) in the chain. Thus oleic acid, octadecenoic acid, is 18:1(9) since the double bond is between carbon atoms 9 and 10. It can also be written as Δ^9octadecenoic acid, where Δ^9 indicates a double bond between carbon atoms 9 and 10. This is known as the Δ system of nomenclature. A more complex fatty acid is $\Delta^{5,8,11,14}$ eicosatetraenoic acid or 20:4(5,8,11,14) indicating double bonds between carbon atoms 5 and 6, between 8 and 9, between 11 and 12 and between 14 and 15. The common name for which is arachidonic acid.

  (ii) The position of double bonds can also be identified by counting from the methyl end. This end is known by the Greek letter, omega, which is the last letter of the Greek alphabet, ω. The letter *n* can be used instead of ω. Thus oleic acid, which has the double bond between carbon atoms 9 and 10 from the methyl end, is 18:1ω-9; that is, it belongs to what is known as the omega-9 family. Arachidonic acid has 20 carbons and four double bonds and has its first double bond, counting from the methyl end, between carbon atoms 6 and 7; it is therefore 20:4ω-6 or 20:4n-6, that is, it belongs to the omega-6 family. The unsaturated fatty acid α-linoleic has 18 carbon atoms and 3 double bonds, the first of which is between carbon 3 and 4 and is, therefore 18:3ω-3 and is one of the omega-3 family, sometimes written as 18:3n-3. In an unsaturated fatty acid, with more than one double bond, these bonds are always separated from each other by two single bonds. Hence,

it is unnecessary to specify the position of any double bond except the first. The two families omega-6 and omega-3 have particular medical and lay interest. The terms omega-3 and omega-6 appear on many food items in the supermarkets and in advertisements in the mass media.

Using the omega nomenclature, the major classes of unsaturated fatty acids found in mammalian tissue are ω-3, ω-6 and ω-9. They are not interconvertible (Figure 11.9).

- The unsaturated fatty acids exhibit geometric isomerism, i.e. there are *cis* and *trans* forms (Figure 11.10). The chemical basis for isomerism is discussed in Chapter 3. (Appendix 3.1).

Common names and the source of some fatty acids

Since the systematic chemical names for the fatty acids are complex, common or trivial names are generally used instead. In many cases, these names are derived from the plant from which the acid was first extracted. This section is based on the information provided by Pond (1998) (see Box 11.1).

- **Lauric acid** (12:0) from *Laurus nobilis*, the bay tree.

- **Myristic acid** (14:0) from *Myristica fragrans*, the nutmeg tree.

- **Palmitic acid** (16:0) from *Elaeis guineensis*, the oil-palm.

- **Oleic acid** (18:1) from *Olea europaea*, the olive tree.

- **Arachidonic acid** (20:4) from *Arachis hypogaea*, the peanut.

- **Linoleic** (18:2) and **linolenic acids** (18:3) from *Linum usitatissimum*, the flax plant. This is grown not only for linseed oil but for fibre, to produce linen.

- **Ricinoleic acid** (18:1 with an hydroxyl group on C12) from *Ricinus communis*, the castor oil plant. Castor oil, containing 90% of ricinoleic acid, is poorly digested (because of the hydroxyl group) and acts as a lubricant and a slight irritant to the intestines, hence its use to relieve constipation.

- **Erucic acid** (22:1) from *Eruca sativa*, a plant of the cabbage family. Erucic acid is a significant constituent of the triacylglycerols in some strains of oil-seed rape (*Brassica napus*). Problems can arise in mammalian tissues because of the acid's relatively slow metabolism. In rats fed the oil, triacylglycerols containing erucic acid accumulate in the heart. Strains of oil-seed rape producing virtually no erucic acid are now grown for inclusion

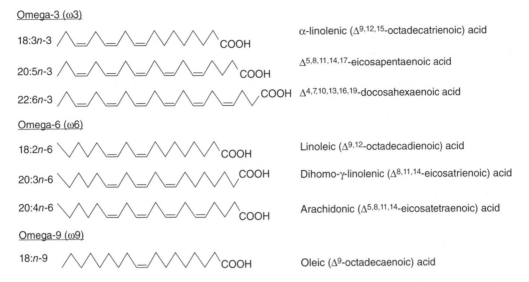

Figure 11.9 *Structures of three omega-3, three omega-6 and one omega-9 polyunsaturated fatty acids.* The structure of a hydrocarbon chain is sometimes presented in this manner, where each peak or trough represents a CH_2 group except at a double bond which links CH groups.

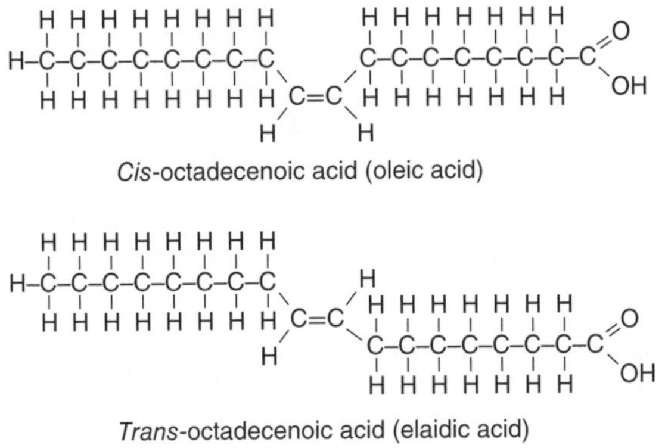

Cis-octadecenoic acid (oleic acid)

Trans-octadecenoic acid (elaidic acid)

Figure 11.10 *Structure of* cis-*octadecenoic and* trans-*octadecenoic acid (an example of* cis-trans *isomerism).* The common name for *trans*-octadecenoic acid is elaidic acid; it is one of the few naturally occurring *trans* fatty acids. The common name for *cis*-octadecenoic acid is oleic acid. Note that the two hydrocarbon chains are separated by the double bond which prevents rotation of the position of the two chains. The structure of the *trans* fatty acid is biochemically unusual and therefore biochemically 'excluded'.

Box 11.1 Significance of plant oils

Plants hold the record for the amount of lipid synthesis that occurs in the world. They are of considerable economic importance, particularly in the production of human and animal food, and currently as a fuel for cars and for generating electricity.

They are used to improve stability and the handling characteristics of food. They are incorporated into cooked or processed foods after the oils are partially hydrogenated. Unfortunately, hydrogenation not only leads to saturation of the double bonds but their displacement to positions in the chain that have no essential fatty acid activity, and/or conversion to the *trans* form (Figure 11.10). The presence of hydrogenated oils in food contributes to the intake of these abnormal fatty acids in developed countries. These fatty acids not only fail to function as essential fatty acids, but they also compete with them and hence increase the requirement for the essential fatty acids.

in animal feeds. However, this strain has a greater susceptibility to insect attack, suggesting that erucic acid in the original oil-seed rape is functioning as a natural insecticide.

Some plants produce a mixture of fatty acids (Table 11.3). The fat in seeds of the cacao tree (*Theobroma cacao*) contains a mixture of stearic and palmitic acids. The fat is known as cocoa butter from its resemblance to the butter produced from cow's milk (see Box 11.2).

The contents of some polyunsaturated fatty acids in meat, fish, and green leaves are given in Table 11.4.

Desaturation of long-chain fatty acids

Desaturation of a saturated bond to produce an unsaturated fatty acid (the conversion of $-CH_2-CH_2-$ to $-CH=CH-$) is catalysed by enzymes known as acyl-CoA desaturases,

Table 11.3 Approximate percentage content of fatty acids in some plant oils

	Percentage of total fatty acids		
Oil	Saturated (16:0)	Monounsaturated (18:1)	Polyunsaturated (18:2n-6)
Soybean	10	25	54
Safflower	7	14	76
Sunflower	7	19	68
Corn	11	24	54
Olive	13	71	10
Canola	4	62	22
Palm	45	40	10
Peanut	11	48	32
Linseed	5	21	16

Only three oils contain some 18:3ω-3; these are soybean (7%), canola (10%) and linseed (54%).

Data from Gurr, *et al.* in *The Lipid Handbook* Chapman and Hall, Chapter 8, 1984.

Box 11.2 Addiction to chocolate, brain cannabinoids and arachidonic acid

The generic name of the cacao tree (*Theobroma*) means 'food of the Gods' and gives its name to a caffeine-like stimulant, theobromine (a methylxanthine). It has been claimed that the theobromine in chocolate is responsible for its addictive characteristics. This is based on the fact that methylxanthines bind to adenosine receptors in the central nervous system and act as antagonists to this neurotransmitter (Chapter 14). However, another group of substances, the amides formed between ethanolamine and unsaturated fatty acids, are also possible candidates for the title of the 'chocolate drug'.

One such amide present in chocolate is anandamide (N-arachidonoyl-ethanolamine). Anandamide, together with N-oleylethanolamine and N-linoleylethanolamine, bind to a receptor. In the brain which also binds tetrahydrocannabinol, the psychoactive compound in cannabis. Could this explain the addiction?

Table 11.4 Content of some unsaturated fatty acids in triacylglycerol in muscle of a fish, a bird and a ruminant and in green leaves

	Amount present in 100 g of total fatty acids			
	Beef	Chicken	Cod	Green leaves
Oleic (18:1)	20	33	11	7
Linoleic (18:2)	26	18	1	16
α-linolenic (18:3)	1	1	Trace	56
Arachidonic (20:4)	13	6	4	0
Long-chain unsaturated (20:5, 22:5, 22:6)	10	0	52	0

The amount is presented in grams. Data from Gurr (1984) (see above).

- Three desaturase enzymes are present in human tissue: they insert double bonds only at positions 9, 6, or 5 (counting from the carboxyl end) so that they are known as the Δ^9-, Δ^6- and Δ^5-desaturases.

- If the substrate is fully saturated, the first double bond is always inserted at position 9 by the Δ^9 desaturase (so that, for example, stearic acid (18:0) is converted to oleic acid (18:ω-9). Thus the Δ^6 desaturase requires the presence of a *cis* double bond at position 9 before it can catalyse desaturation at position 6. Animals do not possess a desaturase that can insert a double bond at a position greater than nine. (Such desaturations are present in plants). It is this fact that determines that such unsaturated fatty acids must be provided in the diet, i.e. they are essential.

several of which exist in human tissues with different specificities for the position of insertion of the double bond. The reaction requires molecular oxygen and NADPH:

$$H^+ + NADPH \longrightarrow NADP^+$$

$$stearoyl\text{-}CoA + O_2 \longrightarrow oleoyl\text{-}CoA + 2H_2O$$

These reactions are of the utmost importance in ensuring the formation of the polyunsaturated fatty acids that are essential for health (see below).

Some characteristics of the enzymes that catalyse desaturation reactions are as follows:

- Desaturation occurs in such a way as to maintain a methylene-interrupted distribution of double bonds (that is, $-CH=CH.CH_2.CH=CH.CH_2-$).

- All double bonds produced by the activity of desaturases, of either plant or animal origin, are *cis* rather than *trans* (see Figure 11.10). (Some *trans*-fatty acids are present in some foods but they are considered to be unhealthy. They have been banned from restaurants in some parts of the USA (see below).

> Ruminant bacteria produce some *trans*-unsaturated fatty acids when long-chain fatty acids are synthesised in the bacteria. These are absorbed by the host so that *trans*-unsaturated fatty acids can be found in adipose tissue and muscle of ruminants.

- The enzymes catalyse desaturation in sequence: for example, the Δ^9 desaturase converts stearate to oleic acid, which is converted, by the Δ^6 desaturase, to the di-unsaturated fatty acid (18:2n-9), which is elongated to form eicosadienoic acid, which can be converted by the Δ^5-desaturase to the tri-unsaturated fatty acid (20:3n-9) which is known as Mead acid (see below).

- Particularly in the metabolism of the polyunsaturated fatty acids, the process of elongation occurs in sequence with desaturation to produce the specific acids required by the tissues in the body.

- The omega-3, omega-6 and omega-9 fatty acid families compete with each other for the desaturase enzymes, especially for the rate-limiting enzyme, the Δ^6 desaturase. The order of preference for the desaturase is α-linolenic acid (omega-3) > linoleic acid (omega-6) > oleic acid (omega-9). Similar competition also occurs with the elongase enzymes. The biochemical, physiological and nutritional significance of these 'rules' is discussed below.

Elongation of long-chain fatty acids

More than 50% of fatty acids in the triacylglycerol in human adipose tissue are longer than 16 carbon atoms (e.g. stearic acid and oleic acid with 18 carbon atoms). An enzyme is present in liver that can increase the chain length of a fatty acid with 16 to 18 carbon atoms, to 20 or even higher. As in the process of fatty acid synthesis (described above), it is malonyl-CoA that extends the length of the chain by two carbon atoms, so that the enzyme acetyl-CoA carboxylase is required for elongation. The reducing agent is also the same, NADPH. An example is the conversion of palmitate to stearate as follows:

$$\text{palmitoyl-CoA} + \text{malonyl-CoA} + 2\text{NADPH} + 2\text{H}^+ \rightarrow$$
$$\text{stearoyl-CoA} + 2\text{NADP}^+ + \text{CoASH} + CO_2$$

Essential fatty acids

There are four facts about essential fatty acids that need to be introduced at this stage.

(i) They contain at least two double bonds and these are in positions greater than Δ^9.

(ii) The essential fatty acids are present in triacylglycerols and phospholipids in foods, particularly in some plants and fish, that are eaten by humans.

(iii) The two essential fatty acids that occur to the largest extent in the diet are linoleic (18:2n-6) and α-linolenic acid (18:3n-3).

(iv) Although these are termed essential fatty acids, they are, in fact, precursors for the major polyunsaturated fatty acids that have essential roles in the body but are present only in small amounts in the diet. Linoleic acid is converted, via elongation and desaturation reactions, to dihomo-γ-linolenic (20:3n-6) and then to arachidonic (20:4n-6) acid. α-Linolenic is converted to eicosapentaenoic (20:5n-3) and then docosahexaenoic (22:6n-3). The pathways for formation of these latter fatty acids, from their dietary precursors, are presented in Figures 11.11 and 11.12. Full details of one pathway are provided, as an example, in Appendix 11.4. For comparison of the two pathways, they are presented side by side in Figure 11.13.

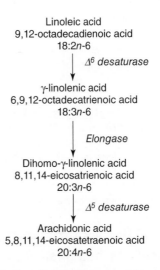

Figure 11.11 *Outline of the pathway consisting of desaturation and elongation reactions that convert linoleic acid into arachidonic acid.*

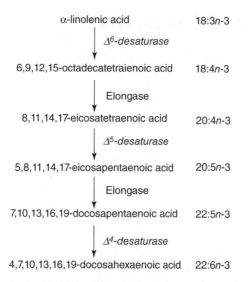

α-linolenic acid   18:3n-3

Δ^6-desaturase

6,9,12,15-octadecatetraienoic acid   18:4n-3

Elongase

8,11,14,17-eicosatetraenoic acid   20:4n-3

Δ^5-desaturase

5,8,11,14,17-eicosapentaenoic acid   20:5n-3

Elongase

7,10,13,16,19-docosapentaenoic acid   22:5n-3

Δ^4-desaturase

4,7,10,13,16,19-docosahexaenoic acid   22:6n-3

Figure 11.12 *Outline of the pathway by which α-linolenic acid is converted to eicosapentaenoic (EPA) and docosahexaenoic acid (DHA).*

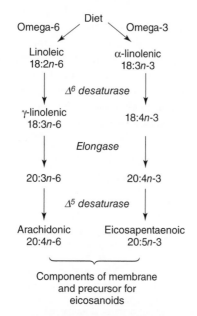

Figure 11.13 *The pathways in Figures 11.11 and 11.12 are presented side by side for comparison. The pathways are known as the omega-6 and the omega-3 pathways. Both linoleic and α-linolenic acids are, in general, present in sufficient amounts in the diet of humans to provide adequate amounts of arachidonic and eicosapentaenoic acids, but the enzymes in the two pathways must be sufficiently active for conversions to occur (see below).*

Discovery of the essential nature of fatty acids in humans

In 1929, Burr & Burr reported that when rats were fed a fat-free diet they ceased to grow and developed scaliness of the tail, dry skin, kidney malfunction and reproductive

failure. The conditions were reversed by feeding small amounts of linseed oil which contains both linoleic and α-linolenic acids. It was some time before essential fatty acid deficiency was observed in humans: it was first seen in children who were tube fed with fat-free diets. Symptoms of a deficiency in an adult were first reported in 1971. After abdominal surgery, a patient was fed intravenously with a feed containing no fat, and a scaly dermatitis developed after 100 days on the diet. The reason for this delay is that adipose tissue in adults contains about 2 kg of essential fatty acids in triacylglycerol molecules and these are released continuously in small amounts. A diet deficient in essential fatty acids would not, therefore, produce symptoms of deficiency until this store was depleted or when lipolysis in adipose tissue was markedly inhibited. The latter condition occurs readily in patients given parenteral feeds containing mainly carbohydrate, which stimulate insulin secretion and hence increase the plasma insulin level. This inhibits lipolysis and the rate of the triacylglycerol/breakdown in adipose tissue is inhibited so that fatty acids including essential fatty acids are not made available within the adipocyte and very little is released into the blood. The symptoms, when they develop, disappear when linoleic and α-linolenic acids are included in the parenteral nutrition.

Signs and symptoms of a deficiency

There are a number of signs and symptoms of essential fatty acid deficiency. They include scaly and thickened skin, alopecia, increased capillary fragility so that bruising readily occurs, poor wound healing, increased susceptibility to infection and growth retardation in infants and children. Some of these symptoms can be explained by deficiency of eicosanoid synthesis and/or failure to complete cell cycles in various tissues (Chapter 20). Consequently, it is important to detect a deficiency before symptoms develop. The principle underlying the method to do this is described:

Detecting a deficiency

A functional method for detection depends upon competition for the activity of the Δ^6 and Δ^5 desaturases between a non-essential fatty acid (e.g. oleic acid) and an essential fatty acid (see above). If the latter is deficient, oleic acid is readily converted, via the desaturases, to Mead acid, since there is little competition (Figure 11.14). Hence the amount of the latter can be used as a marker for deficiency of essential fatty acids, although it is better to use the ratio of double bonds: only three are present in Mead acid (i.e. a triene) but four are present in arachidonic acid (i.e. a tetraene). A ratio in plasma, triene/tetraene >4.0 is an indication of a deficiency of essential fatty acids. This method has shown that a deficiency can occur in a number of conditions which can lead to disease (Table 11.5).

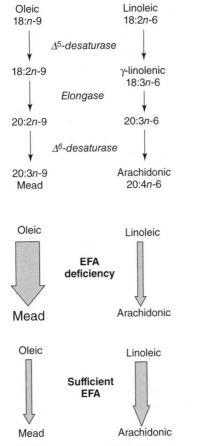

Figure 11.14 *Comparison of flux through the omega-9 pathway with that through the omega-6 pathway in essential fatty acid deficiency and sufficiency when oleic acid is provided. Oleic acid, via the omega-9 pathway, gives rise to Mead acid whereas linoleic acid, via the omega-6 pathway, gives rise to arachidonic acid. The latter contains four double bonds but Mead acid contains only three. The bottom half shows the effect of essential fatty acid deficiency: the conversion of oleic acid to Mead acid is high: the ratio of arachidonic acid to Mead acid (that is, the ratio of tetraene to triene) is high. This is a method used to detect essential fatty acid deficiency. The difference in the size of the arrows is exaggerated in the figure, to illustrate the principle.*

Functions of unsaturated fatty acids

The essential fatty acids have a number of vital functions.

(i)   Esterification of cholesterol that is bound to HDL (e.g. with linoleic acid) which is catalysed by the enzyme, lecithin cholesterol acyltransferase (LCAT), occurs in the plasma (Figure 11.15). It forms a stable HDL complex that transports cholesterol from peripheral tissues to the liver, a process known as reverse cholesterol transport (see Figure 22.10). The process prevents accumulation of cholesterol in cells.

Table 11.5 Conditions in which a deficiency of essential fatty acids can occur

Stress[a]
Trauma[a]
Sepsis[a]
Long-term parenteral nutrition without lipid
Cystic fibrosis[b]
Multiple sclerosis[c]
Anorexia nervosa[d]
AIDS[d]

[a] In these conditions, there can be a sufficient loss of adipose tissue that the store of EFAs in adipose tissue is depleted but there is an increased use of EFAs by some tissues, so that a deficiency can occur.

[b] The deficiency is probably due to poor absorption of EFAs from intestine.

[c] It is not known how a deficiency occurs but undernutrition is likely cause.

[d] The deficiency is probably caused by undernutrition.

(ii)   The presence of polyunsaturated fatty acids in a triacylglycerol lowers its melting point, which prevents the formation of crystals of triacylglycerol within cells, which damage cell membranes.

(iii)   They are essential components of phospholipids, especially those in membranes.

(iv)   The more polyunsaturated fatty acids that are present in phospholipids in a membrane, the greater is the fluidity of the membrane (see below for a discussion of the physiological significance of fluidity).

(v)   They are precursors for the formation of several fatty messengers, including the eicosanoids, docosatrienes and resolvins.

(vi)   Arachidonic acid is a precursor for the formation of a novel neurotransmitter, N-arachidonoylethanolamine, and arachidonic acid itself may be a neurotransmitter.

(vii)   A balance of the essential omega-3 and omega-6 polyunsaturated fatty acids in phospholipids is important in health (see below and Chapters 14 and 15).

(viii)   There is evidence that polyunsaturated fatty acids have an immunosuppressive effect, so that they have been used to treat chronic inflammation disorders (Box 11.3). It is, however, unclear if this is a physiological effect.

These functions are now discussed. The functions that depend upon their presence in phospholipids are discussed in the section on phospholipids below.

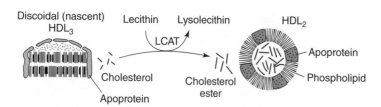

Phosphatidylcholine (lecithin)   Lysophosphatidylcholine (lysolecithin)

Cholesterol   Cholesterol ester (cholesteryl linoleate)

Discoidal (nascent) HDL$_3$ — Lecithin — Lysolecithin — HDL$_2$

LCAT

Cholesterol — Apoprotein — Cholesterol ester — Apoprotein — Phospholipid

Figure 11.15 *The reaction catalysed by lecithin cholesterol acyltransferase (LCAT). Linoleate is transferred from a phospholipid in the blood to cholesterol to form cholesteryl linoleate, catalysed by LCAT. The cholesterol ester forms the core of HDL, which transfers cholesterol to the liver. Discoidal HDL (i.e. HDL$_3$) is secreted by the liver and collects cholesterol from the peripheral tissues, especially endothellial cells (see Figure 22.10). Cholesterol is then esterified with linoleic acid and HDL changes its structure (HDL$_2$) to a more stable form; as shown in the lower part of the figure. R′ is linoleate.*

Box 11.3 Polyunsaturated fatty acids and the immune system

There is considerable evidence that polyunsaturated fatty acids or diets containing fat with a high proportion of these fatty acids can suppress the response of the immune system and hence reduce the severity of chronic inflammation. That polyunsaturated fatty acids could inhibit the response of the immune system was first suggested by Merton *et al.* in 1971. Many other subsequent studies have also shown that addition of polyunsaturated fatty acids to immune cells in culture suppresses proliferation in response to specific or non-specific stimulation. Volunteers were fed a diet high in fat containing polyunsaturated fatty acids, had their lymphocytes removed from the blood and then tested in vitro: their proliferative response to antigen stimulation was markedly decreased. A similar diet reduced the clearance of colloidal carbon from the blood, an index of phagocytosis. Since both proliferation and phagocytosis by immune cells are central to the immune response, it is considered that such experiments support the view that polyunsaturated fatty acids can suppress an immune response (Chapter 17).

Autoimmunity or chronic inflammation are now considered to be the basis of a large number of diseases including multiple sclerosis, polyneuritis, rheumatoid arthritis and type 1 diabetes mellitus. There is strong evidence that supplementation of the diet of patients suffering from rheumatoid arthritis with fish oil is beneficial. Studies on other conditions are in progress (see Calder, 2006).

Prevention of intracellular damage: lowering of the melting point of triacylglycerol reserves

It is important that triacylglycerol, when stored in cells, remains liquid at the temperature of the organism to prevent damage to membranes; for example, tripalmitin (tripalmitoylglycerol) has a melting point of approximately 65 °C and would, therefore, be solid and would crystallise at body temperature. The percentage of oleic acid in triacylglycerol in human adipose tissue is high (Table 11.2). On a diet high in saturated fatty acid, the ability to desaturate stearic acid to produce oleic acid, via the Δ^9-desaturase, is, therefore, important, since it lowers the melting point of the triacylglycerol that is formed from these fatty acids.

> An interesting observation is that the larger the amount of unsaturated fatty acids in the diet of hibernating animals, prior to hibernation, the lower the body temperature falls during hibernation. The lower the temperature, the lower is the metabolic rate, which is important in survival from a prolonged period of hibernation. It is suggested that this is caused by an increase in fluidity of membranes but the mechanism is not known.

Provision of fluidity of membranes

The fluidity of a membrane is difficult to define but it is known to increase the rate of lateral movement of proteins in the membrane and the activity of some membrane proteins, such as ion channels and transporters of fuels (Chapter 5). Fluidity depends, in part, upon the amount and the degree of unsaturation of the fatty acids that are present in the phospholipids of the membrane. In general, the larger the proportion of the polyunsaturated fatty acids in the membrane, the greater is the fluidity of the membrane, and the greater is the rate of the lateral movement of proteins and the greater the activity of ion channels and transporters in the membrane (Box 11.3). Since ion channels are very important in the brain and the retina, it is not surprising that these tissues contain large amounts of the highly unsaturated fatty acids, such as arachidonate, eicosapentaenoic and docosahexaenoic.

Precursors for eicosanoids

Eicosanoids include prostaglandins, leukotrienes, prostacyclins and thromboxanes. They are given their collective name since they contain 20 carbon atoms (Figure 11.16).

They act as local messengers in tissues and they provide communication between one cell type and another within a single tissue or an organ (i.e. they have a paracrine effect). They are synthesised from the polyunsaturated fatty acids that contain 20 carbons (e.g. arachidonate, eicosapentaenoic acids). These fatty acids are generated from the hydrolysis of membrane phospholipids by the activity of a phospholipase which releases the fatty acids from position 2 of the phospholipids. Their roles are described in detail below.

Chronic changes in the type of fat in the diet can change the type of polyunsaturated fatty acids in the phospholipids that are components of membranes and hence change fluidity of the membrane. This might change the activity of the phospholipase and/or the type of eicosanoid produced from

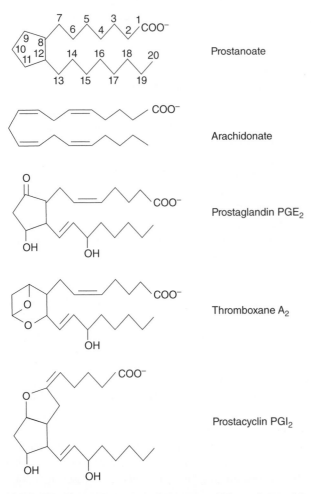

Figure 11.16 *Structure of prostanoate, and three eicosanoids.* Structurally, prostaglandins, thromboxanes and prostacyclins may be regarded as derivates of prostanoate, that is, they contain 20 carbon atoms and a saturated five membered ring (cyclopentane).

Prostanoate

Arachidonate

Prostaglandin PGE$_2$

Thromboxane A$_2$

Prostacyclin PGI$_2$

Box 11.4 Changes in the type of fat in the diet, modification of the structure of phospholipids and hence membrane structure and the activity of membrane proteins

The central dogma in biochemistry is that the information in genes, via messenger RNA, governs the amino acid sequence in proteins which determines their structure, activity and function. Changes in the structure of genes (mutations) can lead to changes in the structure of the proteins and this can change their activity and sometimes their function. One reason for the immense investments in the human genome project was to identify all the genes so that any changes might identify 'abnormal' proteins that might account for particular diseases. However, it is not always appreciated that changes in the structure of phospholipids in membranes can also change the activity of proteins associated with the membrane without any involvement of genes. Modifications in the type of fat in the diet can, over time, change the composition of the phospholipids in membranes and, therefore, their structure which can change fluidity and can result in changes in the activities of proteins, such as transporters and ion channels, or modify the accessibility of receptors to hormones, growth factors or neurotransmitters. (These might be described as 'nutritional' mutations.) In view of the physiological/biochemical importance of such membrane proteins, any significant changes could result in disease or, alternatively, alleviate or even cure a disease. Examples include: changes in membranes of endothelial cells in blood vessels that increase the risk of damage and can lead to atherosclerosis (Figure 22.3); restriction of tumour growth (Chapter 21); reduction in severity of an autoimmune attack (Chapter 17); improvement in neurological disorders (see below).

the membrane phospholipid (see below). Such changes may be of considerable physiological and clinical significance (Box 11.4).

Essential fatty acids in foetal and neonatal development

About 50% of the dry weight of the human brain is lipid, a high proportion of which is polyunsaturated. The uptake of essential fatty acids of both the omega-3 and omega-6 families is, therefore, important in development of the brain of the foetus, which increases in size by about 30% during the third semester. Consequently, the uptake of these fatty acids are particularly important during this period but since the activities of Δ^6 or Δ^5-desaturates are very low or absent in the placenta and in the foetus, the polyunsaturated fatty acids must be provided from mother's blood and transferred to the foetus via the placenta. The essential fatty acids are transported across the placental membrane by fatty acid transport proteins, which have a high affinity for arachidonic, eicosatetraenoic and docosahexaenoic acids. These are present in chylomicrons on VLDL in mother's blood. They are made available by the activity of lipoprotein lipase which resides in capillaries on the maternal side of the placenta. It has been suggested that a deficiency of polyunsaturated fatty acids in the diet of the mother prior to and/or during pregnancy could account for some neurological and behavioural disorders in the subsequent child or adult. This is discussed below and in Chapter 15.

Human milk is rich in essential fatty acids of both the omega-3 and omega-6 families. This suggests that the activity of the Δ^6-desaturase may be too low in the infant to provide a sufficient amount of these fatty acids for development of tissues, particularly the brain and retina. It has been shown that development of visual acuity in infants is dependent upon the presence of docosahexaenoic acid in mother's milk. Hence, it is recommended that breast-feeding should be carried out for as long as 12 months after birth. These fatty acids are now added to commercial infant feeds (Chapter 15, Table 15.8).

Importance of Δ^6 desaturase in health and disease

The pathways of conversion of linoleic and α-linolenic fatty acids to their respective fatty acid end-products (arachidonic and eicosapentaenoic acid) are described in Figures 11.11 and 11.12. The initial reaction in these pathways is catalysed by the same enzyme, the Δ^6 desaturase which is the rate-limiting step for the pathways. This is important for good health, but there are a number of factors or conditions that can decrease the activity of this desaturase. These are: a low intake of zinc, high intakes of alcohol, saturated fat or *trans*-fatty acid, and high blood levels of cholesterol or stress hormones (catecholamines, cortisol, thyroxine). Diabetes mellitus and ageing also decrease the activity of Δ^6 desaturase. Since this enzyme has a preference for the omega-3 fatty acids, it is the synthesis of arachidonic acid that can be particularly affected by a low activity of the enzyme; that is, the rates of formation of dihomo-γ-linolenic acid and arachidonic acids are decreased (see Figure 11.11).

Consequently any of the above factors or conditions could result in failure to produce sufficient amounts of these polyunsaturated fatty acids, which could result in modification of the type of fatty acids present in phospholipids in membranes, and hence the structure of the membranes.

γ-Linolenic acid: a bypass mechanism

The question arises whether inhibition of the desaturase for this particular pathway can be overcome. The answer is yes. The product of the Δ^6 desaturase when desaturating linoleic acid is γ-linolenic acid. Supplementation of the diet with γ-linolenic acid, which bypasses the Δ^6 desaturase reaction, has been used to increase the formation of dihomo-γ-linolenic acid and arachidonic acid (Figure 11.17). This procedure has some similarity to that used to overcome a deficiency of the enzyme that synthesise the neurotransmitter dopamine in the brain. The deficiency gives rise to Parkinsons disease (Chapter 14). (Figure 11.17).

There are several sources of γ-linolenic acid including evening primrose, borage and blackcurrant oils. Indeed, the evening primrose is grown commercially for the production of γ-linolenic acid. It is available in most health food shops and pharmacies.

It is claimed that chronic deficiency of arachidonic acid can lead to a number of medical problems that can be overcome by supplementation of a normal diet with evening primrose oil. Supplementation is claimed to lead to an alleviation of eczema; reduction in premenstrual tension and breast pain during menstruation; improvement in some chronic inflammatory and autoimmune diseases; reduction in blood pressure in hypertensive patients; and reduction in blood cholesterol levels. As might be expected, these claims are controversial, but they serve to illustrate how basic biochemical information can lead to a considerable lay interest in a subject.

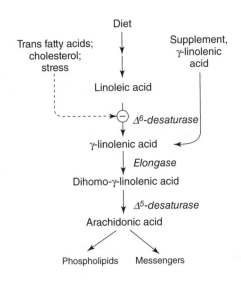

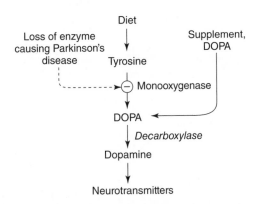

Figure 11.17 *Supplementation of diet with γ-linolenic acid to overcome a deficiency of Δ⁶ desaturase: Supplementation of a diet with DOPA to overcome a deficiency of monooxygenase in Parkinson's disease. Δ⁶ desaturase is a rate-limiting enzyme in the synthesis of arachidonic acid. Supplementation of diet with γ-linolenic acid bypasses this enzyme. Damage to neurones in the brain that use dopamine as a neurotransmitter causes a deficiency of rate-limiting a supplement – enzyme, tyrosine monooxygenase, which is bypassed by a supplement, DOPA (dihydroxyphenylalanine). DOPA (usually, described as L-DOPA) is considered by the medical profession as a drug but, in reality, it is a dietary supplement.*

Phospholipids

Phospholipids are phosphorus-containing lipids: they are compound-lipids in which both fatty acids and phosphoric acid are esterified to an alcohol. They are divided into two classes: phosphoglycerides (also known as glycerophospholipids) in which the alcohol is glycerol, and sphingolipids, in which the alcohol is sphingosine. The structures of these two alcohols are very different: glycerol possesses

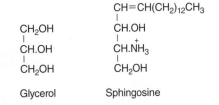

Figure 11.18 *Structure of glycerol and sphingosine.*

three hydroxyl groups; sphingosine is an amino alcohol, which possesses an amino group as well as a hydroxyl group. (Figure 11.18). The main phosphoglyceride in the plasma membrane is phosphatidylcholine; the main sphingolipid is sphingomyelin. Most membranes contain mixtures of glycerophospholipids and sphingolipids. The exact composition depends on the source of the membrane. Although wide variations occur, the physical properties of the membrane remain much the same.

In phosphoglycerides, two long-chain fatty acids are esterified through their -COOH groups to the hydroxyl groups at the 1- and 2- positions of glycerol 3-phosphate. A base is attached to the phosphate of the glycerol phosphate: the bases are ethanolamine, choline, inositol or serine. The base provides a positive charge and the phosphate a negative charge. These charges form the hydrophilic part of the molecule whereas the fatty acids form the hydrophobic part. The structure of a phosphoglyceride, including the common bases, is presented in Chapter 5 but also in Figure 11.19 where it can be compared with sphingomyelin.

Synthesis of phosphoglycerides

The pathways for the synthesis of phosphoglycerides and triacylglycerol are identical up to the formation of diacylglycerol, after which they diverge. Hence there is an important branch-point at diacylglycerol (Figure 11.20).

Due to the specificities of the acyltransferases in the pathways, the fatty acid at position one of glycerol is saturated whereas that at position 2 is monounsaturated (e.g. oleic acid), although in most glycerophospholipids, the fatty acid at position 2 is polyunsaturated (e.g. arachidonic or eicosapentaenoic acids). This is important for the

> The phospholipid molecules are such that in aqueous media they spontaneously form extended bilayers with a hydrophobic core. Although membrane proteins vary enormously they all form compact structures. This minimises the surface of interaction with the lipid, so that, although protein may account for 30–80% of the weight of the membrane, it does not affect the basic physical properties of the lipid bilayer.

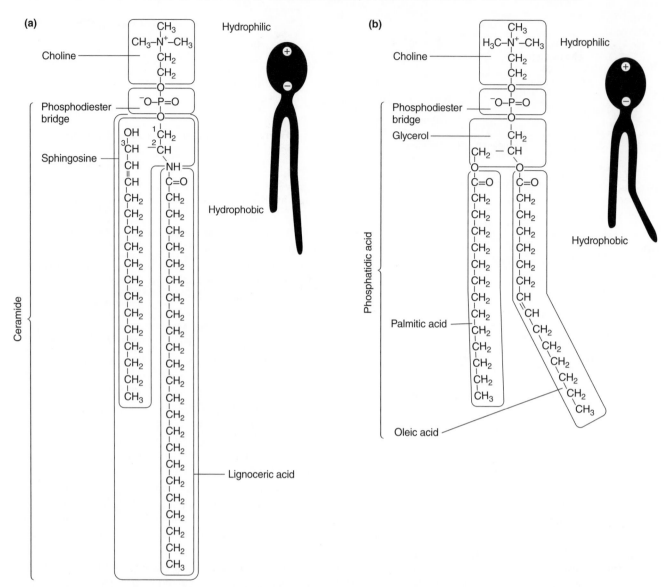

Figure 11.19 (a) *The structure of sphingomyelin.* The hydrophobic unit comprises the hydrocarbon chain of sphingosine and a fatty acid esterified to the amino group of sphingosine. Other fatty acids can replace lignoceric (e.g. behenic acid) and ethanolamine can replace choline. A variety of sphingoglycolipids are present in membranes. For example, a sugar molecule can replace phosphocholine: in cerebrosides, the sugar is a monosaccharide; in gangliosides, the sugar is an oligosaccharide incorporating N-acetylneuraminic acid (Appendix 11.5). **(b)** The structure of a phosphoglyceride for comparison.

various roles of phospholipids. The means by which a change from mono- to polyunsaturated acid is achieved is described below.

The three quantitatively important phosphoglycerides are phosphatidylcholine, phosphatidylethanolamine and phosphatidylinositol.

Phosphatidylcholine   There are three sequential reactions in this synthetic pathway:

(i)   Choline is phosphorylated, by the enzyme choline kinase:

$$\text{choline} + \text{ATP} \rightarrow \text{phosphocholine} + \text{ADP}$$

(ii)   Phosphocholine reacts with a nucleoside triphosphate (cytidine triphosphate) in a reaction, catalysed by cholinephosphate cytidylyltransferase, which produces cytidine diphosphocholine:

$$\text{phosphocholine} + \text{CTP} \rightarrow \text{CDPcholine} + \text{PP}_i$$

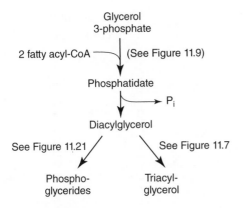

Figure 11.20 *Diacylglycerol as a branch point in the synthesis of phosphoglycerides and triacylglycerol.*

(iii) The phosphocholine group is transferred from cytidine diphosphocholine to the diacylglycerol, in a reaction catalysed by choline phosphotransferase.

$$\text{CDP-choline} + \text{diacylglycerol} \rightarrow$$
$$\text{phosphatidylcholine} + \text{CMP}$$

Phosphatidylethanolamine is synthesised in a similar manner but phosphatidylinositol and phosphatidylserine are synthesised differently (Figure 11.21).

Phosphatidylcholine can also be produced from phosphatidylethanolamine by methylation which involves the methylating agent S-adenosylmethionine as follows:

$$\text{Phosphatidylethanolamine} + 3\text{'CH}_3\text{'} \rightarrow$$
$$\text{phosphatidylcholine}$$

Where 'CH$_3$' is an abbreviation for S-adenosylmethionine which donates methyl groups (Chapter 15, see Figure 15.3).

The importance of this reaction is suggested from the observation that, if the level of the methylating agent is low (which can be caused by deficiency of vitamin B$_{12}$ and/or folic acid), this can restrict formation of phosphatidyl

choline. It may be, in part, the cause of the neurological disorder that occurs in vitamin B$_{12}$ deficiency (Chapter 15).

Phosphatidylinositol This phospholipid is synthesised in a totally different manner from the above two. The compound phosphatidate, rather than diacylglycerol, is the reactant in the reaction:

(i)   phosphatidate + CTP $\rightarrow$ CDP-diacylglycerol + PP$_i$

The CDP diacylglycerol reacts directly with inositol as follows.

(ii) CDP-diacylglycerol + inositol $\rightarrow$ phosphatidylinositol + CMP

Phosphatidylserine This phospholipid is synthesised from phosphatidylethanolamine by exchange as follows:

$$\text{Phosphatidylethanolamine} + \text{serine} \rightarrow$$
$$\text{phosphatidylserine} + \text{ethanolamine}$$

Figure 11.21 summarises the synthesis of all four glycerophospholipids. Details of the reactions are presented in Appendix 11.6.

Mechanism for changing the type of fatty acid at position 2 of phosphoglycerides

Activities of two enzymes lead to the insertion of a different fatty acid at position 2 of the glycerol component of the phosphoglyceride:

First, the acyl group at position 2 is removed by the action of a phospholipase.

$$\text{phosphatidylcholine} + \text{H}_2\text{O} \rightarrow \text{lysophosphatidylcholine} +$$
$$\text{fatty acid}$$

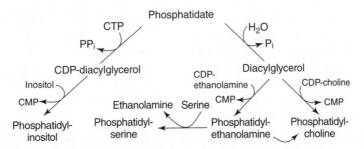

Figure 11.21 *Outline of synthesis of phosphatidylinositol, phosphatidylserine, phosphatidylethanolamine and phosphatidylcholine.* Note in the synthesis of phosphatidylinositol, the free base, inositol, is used directly. Inositol is produced in the phosphatase reactions that hydrolyse and inactivate the messenger molecule, inositol trisphosphate (IP$_3$). This pathway recycles inositol, so that it is unlikely to be limiting for the formation of phosphatidylinositol bisphosphate (PIP$_2$). This is important since inhibition of recycling is used to treat bipolar disease (mania) (Chapter 12, Figure 12.9). Full details of the pathway are presented in Appendix 11.5. Inositol, along with choline, is classified as a possible vitamin (Table 15.3).

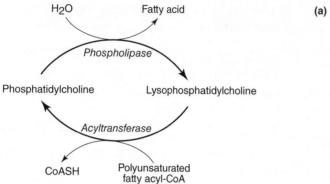

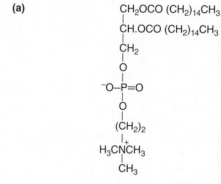

(a)

(b)

Figure 11.22 *The phosphatidylcholine/lysophosphatidyl cycle.* It is unlikely that the two enzymes will be simultaneously catalytically active unless they are spatially. Separated but how the activities are controlled is not known.

Secondly, the hydroxyl group now available at position 2 is re-esterified with another fatty acid in a reaction catalysed by an acyltransferase; that is, the enzyme catalyses a re-esterification (i.e. acylation) with a different acyl-CoA. In many cases, this is a polyunsaturated fatty acid (such as arachidonoyl-CoA) and forms the more unsaturated phosphoglyceride:

$$\text{lysophosphatidylcholine} + \text{arachidonyl-CoA} \rightarrow$$
$$\text{phosphatidylcholine} + \text{CoASH}$$

This emphasises the importance of the cell containing sufficient free arachidonic acid for such an exchange to take place. The combination of these two reactions, if they occur simultaneously, is a cycle – the phospholipid/lysophospholipid cycle (Figure 11.22).

There are at least three roles for the exchange of fatty acids (i.e. the cycle).

(i) *Formation of a 'typical' membrane phospholipid* Phospholipids are produced from diacylglycerol. The enzymes involved in synthesis of diacylglycerol insert a fatty acid at position 2 of the glycerol phosphate that contains only one or two double bonds. However, most phospholipids in the membranes of a cell contain fatty acids with four or five double bonds (e.g. arachidonic) which play significant roles in the membrane, e.g. to increase the fluidity (see above) to supply polyunsaturated fatty acids for formation of messenger molecules (see below).

(ii) *Formation of palmitoyl phosphatidylcholine* This is an unusual phospholipid since both hydroxyl positions of the glycerol component are esterified with the same saturated fatty acid, palmitate, so that a completely 'saturated' phospholipid is produced (Figure 11.23 (a)). This phospholipid forms a complex

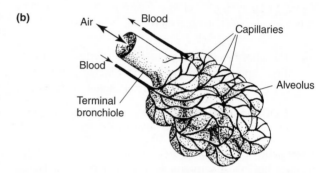

Figure 11.23 (a) *Structure of dipalmitoyl phosphatidylcholine.* **(b)** *Aggregates of alveoli.* The advantage of dipalmitoyl phosphatidylcholine is that it spreads at physiological temperatures to form a thin fat film over the whole inner surface of the alveolar sacs, which reduces surface tension. There are 300–400 million alveoli in each human lung.

with a protein and sphingomyelin (see below), to produce pulmonary surfactant, which coats the surface of the epithelial cells of the alveoli in the lung (Figure 11.23 (b)). Its role is to lower the surface tension of the aqueous surface of the cells in the alveoli so that, during inspiration, they open sufficiently for uptake of oxygen from that in the air in the lung and the loss of carbon dioxide. Of particular clinical importance, the two enzymes that catalyse these reactions are only produced in the late stages of foetal development so that, in premature infants, lack of this particular phospholipid results in respiratory distress (Box 11.5).

(iii) *Replacement of a damaged unsaturated fatty acid in a phospholipid* The polyunsaturated fatty acids in membrane phospholipids are readily oxidised or peroxidised by the action of free radicals (e.g. the superoxide radical). This disturbs the structure of the membrane which can interfere with the activities of proteins in the membrane. Damaged fatty acids are removed and then replaced by an undamaged fatty

Box 11.5 Pulmonary surfactant and respiratory distress syndromes

Pulmonary surfactant is a complex mixture of lipids and proteins. Surfactant contains 90% phospholipid and 10% protein. These proteins and lipids are synthesised in and secreted from type II cells of the lung. This complex forms a monolayer on the aqueous surface of the alveoli which lowers the surface tension so allowing the alveoli to open up during inspiration and preventing their collapse after each expiration. This is essential for sufficient rates of loss of CO_2 from the blood to the air and uptake of oxygen in the opposite direction to satisfy the metabolic needs of the body. Most of the phospholipid is dipalmitoylphospholipid which is synthesised via the phospholipid cycle. Although pure dipalmitoylphospholipid lowers the surface tension, additional phospholipids are required to confer 'spreadability' so that the monolayer covers the whole of the surface of the alveoli.

Failure to synthesise sufficient surfactant or the synthesis of abnormal surfactant, so that surface tension cannot be lowered, may play a role in several conditions: respiratory distress syndrome of the newborn; sudden infant death ('cot death'); and adult respiratory distress syndrome. The enzymes involved in the synthesis of surfactant only appear during the third trimester of pregnancy, so that surfactant is not produced in premature babies and they have difficulty breathing.

The metabolic disturbances in these conditions are as expected from a shortage of oxygen: dependence of tissues on glycogen conversion to lactate (which results in lactic acido-sis) and a decrease in the ATP/ADP concentration ratio leading to failure of ion pumps, so that K^+ ions are not taken up by cells (a raised K^+ ion concentration in the blood can result in cardiac arrest). This is speculated to be one cause of death in sudden infant death syndrome. The condition of respiratory distress syndrome in premature babies is characterised by a rapid, grunting respiration with indrawing of both ribs and sternum. The concentration of palmitoyl phosphatidylcholine in amniotic fluid samples is measured in an attempt to predict the risk of this syndrome of the newborn in abnormal pregnancies.

Artificial surfactant, which contains dipalmitoylphosphatidyl choline and some palmitic acid to provide for spreadability, is now commercially available for instillation into the lung. Administration of steroids to the mother prior to birth of the premature infant is also carried out.

An adult human synthesises about 5 g of surfactant each day. Some (perhaps 50%) is recycled by the type II cells of the lung. This recycling process regulates the amount of surfactant in the lung which is very important since either too much or too little can restrict exchange of the gases. In addition, once inside the cell, any damage to the protein or phospholipids in the surfactant, e.g. caused by free radicals in the environment, can be repaired. Another role of the surfactant is to coat inhaled dust particles or microorganisms, which facilitates their phagocytosis by pulmonary macrophages.

acid via the action of the cycle. This is similar to the removal of a damaged nucleotide in DNA and its replacement by an undamaged nucleotide, a process known as nucleotide excision repair. The damage to DNA may also be caused by the same factor, free radicals (Chapter 20, page 463).

Synthesis of sphingolipids

Sphingolipids contain the long-chain amino alcohol sphingosine in place of glycerol. However, sphingosine is not involved in the synthesis of sphingolipids in the way that glycerol is involved in the synthesis of phosphoglycerides.

The precursors for the synthesis of sphingomyelin, are serine and palmitoyl-CoA.

In summary, the synthetic pathway is as follows:

serine + palmitoyl-CoA → sphinganine →
 dihydroceramide → ceramide → sphingomyelin

Details are presented in Figure 11.24.

Myelin is the material that surrounds some neurons and their axons but it is not a single compound. It is a complex of sphingolipids and protein. It is synthesised by Schwann cells in peripheral nerves and by oligodendrocytes in the brain. It is secreted by these cells in such a way that it surrounds the axons of the nerves where it provides for the high rates of conduction of electrical activity along nerves, so that damage to myelin disturbs this conductivity leading to disease (e.g. multiple sclerosis).

> Ceramide is now included in some skin creams which are claimed to be beneficial to the skin, preventing drying and reducing the effects of ageing. Linoleoyl-ceramide is present in the skin where it restricts water permeability, preventing excessive water loss from the skin.

Fatty messenger molecules

Fatty messenger molecules include: diacylglycerol, ceramide, platelet activating factor, the eicosanoids and resolvins. A summary of the reactions that produce some of these messengers is given in Appendix 11.7.

It is re-emphasised that the fatty acids that are precursors for many of these messengers are derived from the phospholipids that are components of the plasma membrane. This is possible since the amount of phospholipid in the

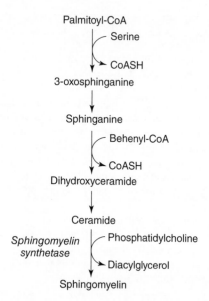

Figure 11.24 *A summary of the reactions involved in synthesis of sphingomyelin.* Reaction between serine and palmitoyl-CoA produces 3-oxosphinganine, which is converted to sphingamine. Attachment of a long-chain fatty acid to the amino group of sphinganine produces dihydroxyceramide. Ceramide reacts with phosphatidylcholine: the phosphocholine component forms an ester bond with the hydroxyl group at position one of ceramide.

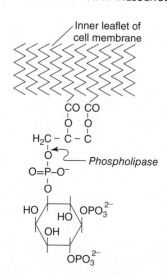

Figure 11.25 *Structure of phosphatidylinositol bisphosphate and position of hydrolysis by phospholipase.* Phosphatidylinositol 4,5-bisphosphate (PIP$_2$) is a component of the inner leaflet of a cell membrane from which inositol 1,4,5-trisphosphate (IP$_3$) is released by hydrolysis catalysed by phospholipase where indicated. The diacylglycerol is retained in the membrane.

plasma membrane is enormous in comparison with the amount of fatty acid that is required as a precursor. It is assumed that the fatty acid, which is released and converted to the messenger, is replaced as soon as possible to prevent damage to the membrane.

Three messengers are discussed here: diacylglycerol, inositol trisphosphate and the eicosanoids.

Diacylglycerol and inositol trisphosphate

The phosphorylated phospholipid, phosphatidylinositol bisphosphate, is present in cell membranes. On hydrolysis by a phospholipase, it produces two products, inositol trisphosphate and diacylglycerol (Figure 11.25), as follows:

phosphatidylinositol bisphosphate →
inositol trisphosphate + diacylglycerol

using abbreviations, this is summarised as

$$PIP_2 \rightarrow IP_3 + DAG$$

Note that the DAG remains in the membrane: it is not released. Inositol trisphosphate binds to receptors on the endoplasmic reticulum which opens Ca^{2+} ion channels in the reticulum and Ca^{2+} ions are released. This increases the cytosolic Ca^{2+} ion concentration and leads to activation of a number of processes in different cells. After removal

of inositol trisphosphate from the membrane, diacylglycerol remains within the membrane. The enzyme protein kinase-C attaches to cell membranes where it is activated by the diacylglycerol. The enzyme is usually part of signalling pathways which, in some cases, means that the enzyme must dissociate from the membrane to perform its signalling role. How it retains its activation is not known.

Eicosanoids

Since eicosanoids are a whole family of local mediators or hormones that have a wide range of effects, a full section is devoted to them. They include prostaglandins, thromboxanes, prostacyclins, hydroperoxy fatty acids, leucotrienes and lipoxins. They are termed eicosanoids because they contain 20 carbon atoms (*eikosi* is the Greek word for twenty) (see Figure 11.16 and Appendix 11.8 for structures). The precursors for their synthesis are homo-γ-linolenic acid (20:3), arachidonic acid (20:4) and eicosapentaenoic acid (20:5). The number and positions of the double bonds in eicosanoids depend on these precursors (Figure 11.21). This is important in differentiating the different roles of the eicosanoids. The enzyme cyclooxygenase acts on these three different precursors to produce different prostaglandins (Figure 11.26).

Arachidonate is used as an example of the processes involved in the synthesis of the eicosanoids. The con-

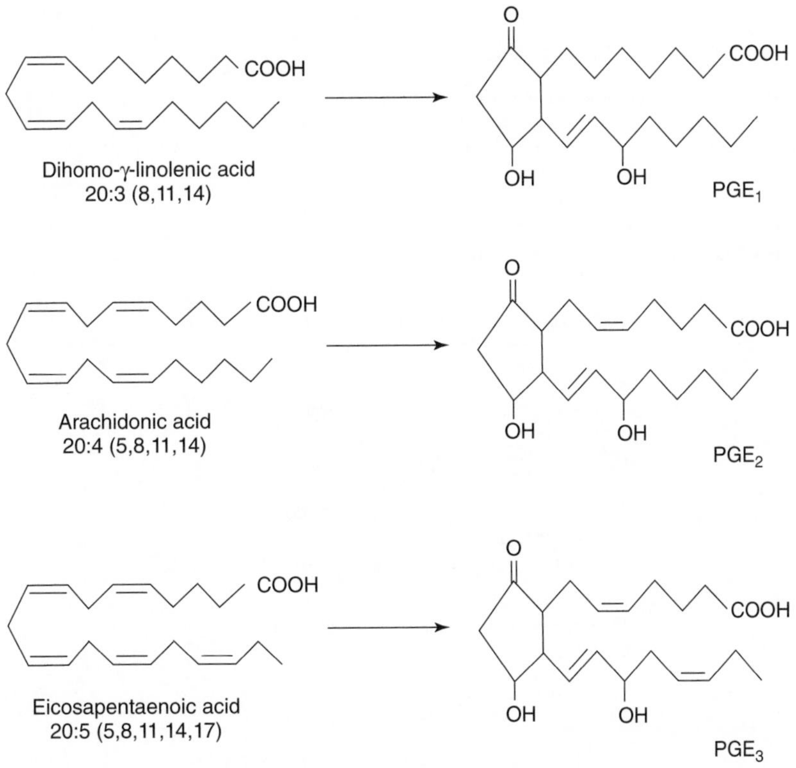

Figure 11.26 *The structures of the prostaglandin E series produced from three polyunsaturated fatty acids containing 20 carbon atoms but a different number of double bonds.* The number of double bonds in the three different acids produces prostaglandins of the E series with a different number of double bonds outside the cyclopentane ring. It is this number which influences the function of the prostaglandin and similarly the function of prostacyclins and thromboxanes (see text). Note, PGE_1 has one double bond, PGE_2 has two double bonds and PGE_3 has three double bonds outside the cyclopentane ring.

centration of free arachidonate in a cell is very low ($<10\ \mu mol/L$) so that virtually no eicosanoid synthesis could take place unless the acid is made available from the fatty acids held in esterified form in the phospholipids present in membranes. This is carried out by the action of a phospholipase which catalyses the flux-generating step for formation of eicosanoids. It is the phospholipid in the membrane that provides a 'reservoir' of the polyunsaturated fatty acid for eicosanoid formation. This is analogous to the large amount of ATP in the cell that provides substrate for cAMP formation

$$ATP \rightarrow cyclic\ AMP + PP_i$$

A summary of the sequence is as follows:

phospholipid ⟶ arachidonic acid → eicosanoids
→ metabolites

(The principle underlying such a messenger system is discussed in Chapters 3 and 12.) The flux-generating step is the reaction catalysed by the phospholipase indicated by the broader arrow in the above sequence.

Synthesis of eicosanoids

A summary of the processes for producing the eicosanoids from the polyunsaturated fatty acid, arachidonic acid, is presented in Figure 11.27. The two enzymes separate for synthesising the prostanoids or the leucotrienes are cyclooxygenase and lipoxygenase, respectively. Whether prostanoids or leucotrienes are produced in any given tissue will depend on the relative activities of these two enzymes in that tissue.

Prostaglandins, thromboxanes and prostacyclins

Cyclooxygenase converts arachidonic acid first to prostaglandin G (PGG) and then to PGH prior to formation of prostaglandins, thromboxanes and prostacyclins. The structures of the intermediates and some of the end-products of these conversions are provided in Figure 11.28.

Leucotrienes, hydroperoxy fatty acids and lipoxins

The leucotrienes, lipoxins and hydroperoxy fatty acids are also synthesised from arachidonic acid. The initial enzyme

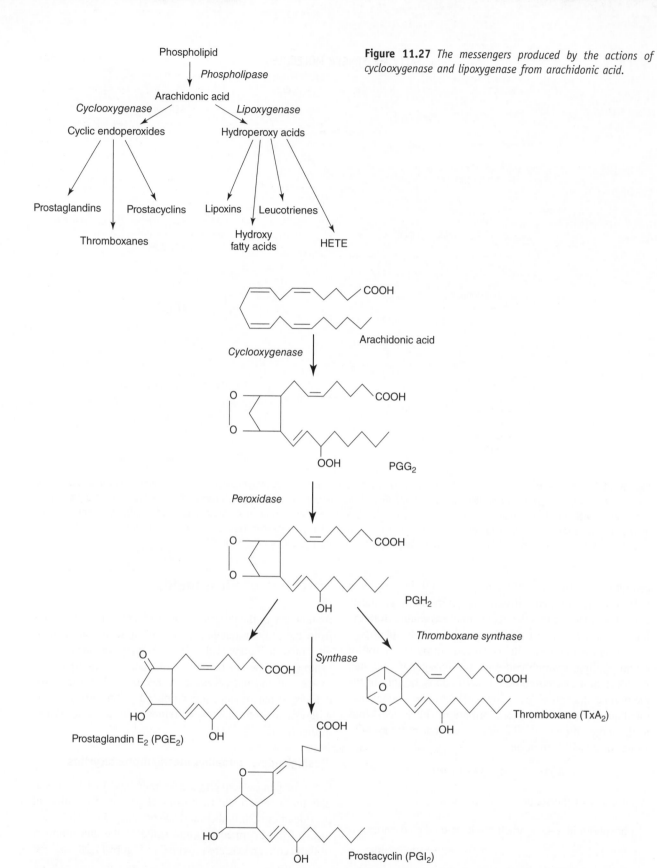

Figure 11.27 *The messengers produced by the actions of cyclooxygenase and lipoxygenase from arachidonic acid.*

Figure 11.28 *Pathway for conversion of arachidonic acid to prostaglandins (PGE_2), prostacyclin (PGI_2) and thromboxane (TxA_2). The prostaglandins PGG_2 and PGH_2 are intermediates in the pathway.* Cyclooxygenase catalyses conversion of arachidonic acid to a cyclic peroxide, PGG_2 which is then reduced to form the prostaglandin alcohol, PGH_2. This is substrate for an isomerase and two synthases to produce another prostaglandin, a thromboxane and a prostacyclin, respectively. There are two cyclooxygenases, that is, isoenzymes, cyclooxygenase-1 (abbreviated to COX-1) and cyclooxygenase-2 (COX-2). The isoenzymes are distributed differently in different tissues (see text). COX-1 is present in most cells and is constitutive (i.e. present in the cell at all times). COX-2 is present in immune cells, particularly macrophages, but is inducible, i.e. its synthesis is stimulated by inflammation. This has considerable clinical significance – see below.

in the pathway is lipoxygenase (LO). The structures of the intermediates in one pathway and some of the products are presented in Figure 11.29.

Regulation of the activity of phospholipase A₂

It is the phospholipase A₂ that catalyses the flux-generating step and consequentially is regulated by a variety of factors, including cytosolic Ca²⁺ ions, shear stress in some cells (e.g. endothelial cells in arteries), and cytokines, some of which increase the activity whereas others decrease it (Figure 11.30). Glucocorticoids decrease the activity (Figure 11.30) which is clinically very important. (Many

inflammatory conditions are treated with steroid hormones.)

Once released, the arachidonic acid will be further metabolised by the enzymes, cyclooxygenases or lipoxygenase, according to the activity of each enzyme, which will vary from tissue to tissue. It is likely that the mechanisms for regulating both cyclooxygenase and lipoxygenase are changes in the enzyme concentration by activation of genes. Cyclooxygenase concentration is increased by various stimuli, such as shear stress; due to flow of blood in post endothelial cells, and by cytokines in immune cells.

Whether prostacyclin or thromboxanes are produced depends upon the cell type: for example, endothelial cells contain enzymes for prostacyclin formation whereas

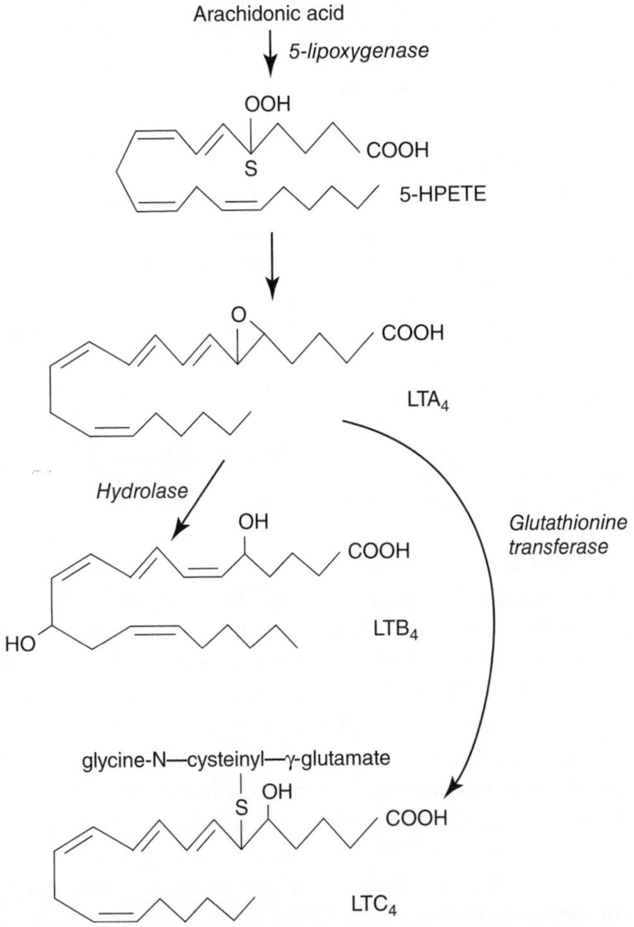

Figure 11.29 *Details of the pathway for conversion of arachidonic acid to leucotrienes A₄, B₄ and C₄. There are three lipoxygenases, 5-, 12- and 15-lipoxygenase. The number indicates the position in the polyunsaturate fatty acid to which lipoxygenase adds a hydroperoxy group (OOH). The 5-lipoxygenase adds the group to the carbon atom at position 5 of arachidonic acid (see Fig 11.16). Lipoxygenase acts upon arachidonic acid to form hydroperoxyeicosatetraenoic acid (HPETE) which is converted to leucotrienes, lipoxins and hydroxyeicosatetraenoic acid (HETE). Leucotrienes differ from prostanoids in that they have no cyclopentane ring, but they have a conjugated triene structure, that is, three double bonds separated from each other by single bonds.*

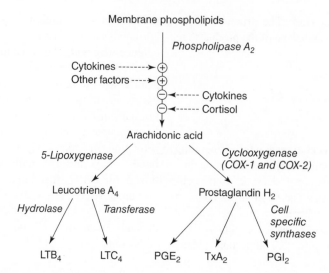

Figure 11.30 *Mechanisms of regulation of phospholipase A₂.* In all these processes described above, it is phospholipase A₂ that carries out the hydrolysis of membrane phospholipid. Cytokines are local hormones produced by immune cells, T-lymphocytes and macrophages (Chapter 17). Other factors relate to shear stress in endothelial cells and those that stimulate release of granules from mast cells. Eicosanoids are present in the granules and they must be re-synthesised after degranulation in the mast cells. Here the enzymes described above must be present in mast cells.

platelets contain enzymes for the synthesis of thromboxanes (see above). As with other hormones eicosanoids bind to receptors, stimulate an effector mechanism, are then internalised and metabolised within the cell.

Roles of prostaglandins, thromboxanes and prostacyclins

The interest in eicosanoids is due to their physiological effects which are the basis of this functions. The term prostaglandin arose since it was thought that they were produced in the prostate gland but, in fact, they are produced and secreted by seminal vesicles, a discovery that came too late to influence the name. Some indications of the roles of these messengers were suggested by early observations on effects of semen and control of pain by various agents. For example, in the early 1930s, semen was found to contain lipids that cause uterine muscle to contract and increase blood pressure in vivo due to contraction of vascular smooth muscle. Some therapeutic effects of semen have been known for centuries. In Ancient China it was used to treat chronic pain in the stomach (due, probably, to a peptic ulcer); some North African tribes used it to initiate parturition.

In the 1970s the synthetic pathway for prostaglandins was elucidated and some of the biological effects identified. On the basis of this work, S. Bergström, B. Samuelsson and J. Vane were awarded the Nobel Prize in 1982.

Eicosanoids are involved in control of a number of physiological processes that are essential to life. Some information on each role is provided below but further detail can be found in other chapters: clotting of blood (Chapters 17 and 22); menstruation and parturition (Chapter 19); secretion of protons in the stomach (Chapter 4); pain and fever (Chapter 18).

Clotting of blood

Platelets convert arachidonic acid into thromboxane (TxA), which stimulates aggregation of platelets. In contrast, endothelial cells convert arachidonic acid into prostacyclin, P61, which inhibits aggregation of platelets. The extent of platelet aggregation, therefore, depends upon the balance between thromboxane and prostacyclin formation by these two cell types. These different sources of the two messengers allow some pharmacological discrimination. Aspirin is an irreversible inhibitor of both cyclooxygenase isoenzymes of so that very low concentrations of aspirin inhibit the enzymes. Consequently, the inhibition can only be overcome by synthesis of new enzyme, in the absence of the inhibitor. Since platelets have no 'machinery' for the synthesis of protein, recovery of activity requires formation of new platelets, so that the inhibition is chronic. In contrast, the inhibition of the enzyme in endothelial cells can be overcome relatively rapidly by synthesis of new protein. Platelets are formed from megakaryocytes and the process takes some time, so the inhibition is more effective in platelets than in endothelial cells. Hence, low but regular intake of aspirin is sufficient to inhibit the enzyme in platelets but not in endothelial cells. Chronic medication with low dose aspirin is recommended to minimise the risk of thrombosis and hence cardiac infarction or stroke in patients who have an increased risk of thrombosis (e.g. those suffering from atrial fibrillation). It can be used as an alternative to the drug warfarin but it is usually considered that warfarin is the more effective drug (Box 17.2).

Menstruation

Menstruation is defined as the periodic (about 28 days) flow of blood, mucus and some tissue fragments from the uterus caused by necrosis of the endometrium (Chapter 19). The endometrium and myometrium synthesise and secrete prostaglandins, which initiate the process. Prostaglandin ($PGF_{2\alpha}$) stimulates contraction of smooth muscle of the myometrium resulting in severe vasoconstriction of the arterioles and hence hypoxia and, eventually, ischaemia. This is followed by necrosis of the endometrium which initiates menstruation. Disorders of menstruation, dysmenorrhoea (painful menstruation) and menorrhagia (excessive or prolonged menstruation) are, in some cases, caused by an imbalance or an excess of prostaglandins.

It is through the local generation of eicosanoids that intrauterine contraceptive devices are effective, since they stimulate production of prostaglandins by the myometrium which increases its contractility, preventing implantation.

Parturition

Prostaglandins are involved, along with oxytoxin and steroids, in control of parturition. They result in contraction of the uterine muscles and dilation of the cervix (the neck of the womb). This effect is very powerful since the maximum dilation is an increase in diameter of about 10 cm. Prostaglandins were first used for induction of abortion in the UK in 1972.

Secretions of protons into the stomach

Prostaglandins inhibit the secretion of protons by the parietal cells in the stomach, which is normally increased in response to food and the hormone gastrin. Consequently, inhibition of prostaglandin synthesis by aspirin or other similar drugs results in increased secretion of protons by the stomach, which can result in considerable gastric discomfort and can, if chronic, lead to the development of a peptic ulcer. Consequently, there is some conflict between the use of such inhibitors to relieve chronic pain (see below), in diseases such as arthritis, and the risk of development of ulcers.

The two isoenzymes of cyclooxygenase are distributed differently: Macrophages contain COX-2 whereas cells in the stomach contain COX-1. Consequently, a pharmacological means of selective inhibition of the effect on proton secretion but maintenance of the analgesic effect has been an aim of the pharmaceutical industry for some time. Unfortunately, although a successful drug was developed, it had serious side-effects, the cause of which is of biochemical interest (Box 11.6).

Pain

Prostaglandins do not cause pain directly but sensitise pain receptors to the algesic effect of other stimuli. One value of pain is that, in the part of the body suffering from trauma or infection, it restricts mobility reducing the risk of further damage and thus helps the repair process. Although prostaglandin synthesis inhibitors are very effective in reducing pain, the control of pain can be controversial since, if it leads to activities that should have been restricted by the pain, it can result in further damage. This is a particular problem that confronts the sports medicine physicians and physiotherapists. Control of pain may allow a sportsperson to participate in the sport with minimal restriction, but may result in further damage which delays full recovery. A similar controversy exists in relation to control of fever in patients suffering from infection (see below).

Fever

Fever is the abnormal rise in body temperature in response to infection or trauma. It is accompanied by increase in heart and breathing rates, intense thirst, loss of appetite and nausea. The role of fever is to reduce growth of bacteria. The cells in the temperature regulatory centre in the hypothalamus are influenced by several cytokines including interleukin-1 (Figure 11.21). The cytokine stimulates the activity of a phospholipase in the cell membrane of these cells, which results in the release of arachidonic acid, which is converted to PGE_2. This results in increased heat production and decreased heat loss, leading to fever. Drugs that inhibit brain cyclooxygenase reduce fever (e.g. aspirin). An important question, not yet resolved, is whether such drugs, although reducing some of the symptoms of the infection, actually delay some of the responses of the body that play a role in combating infection, e.g. restriction of iron available to the pathogen (Chapter 18, page 425).

A balanced intake of omega-3 and omega-6 fatty acids and health

The two essential fatty acids of most biochemical and clinical interest are arachidonic acid, which gives rise to prostanoids of series 2 and leucotrienes of series 4, and eicosapentaenoic acid, which gives rise to prostanoids of

Box 11.6 Ancient and modern cyclooxygenase inhibitors: chronic pain and gastric injury, and myocardial infarction

Inhibitors of cyclooxygenase inhibit the formation of prostaglandins and therefore decrease the level of prostaglandins which reduces the intensity of pain. Acetylsalicylic acid (aspirin) is a potent inhibitor of cyclooxygenase and has been used over many years to control pain. In fact, it was produced by the Bayer Company as early as 1898 to control pain. Prostaglandins also reduce proton secretion by the parietal cells in the stomach, so that aspirin increases proton secretion causing discomfort but, if secretion is prolonged and excessive, it can lead to a peptic ulcer. Aspirin inhibits the activities of both isoenzymes (COX-1 and COX-2) but COX-1 is present in the parietal cells whereas COX-2 is present in macrophages. It is the prostaglandin secretion by macrophages that causes chronic inflammation and pain (e.g. pain in the joints in rheumatoid arthritis). Consequently, a drug was developed that selectively inhibits COX-2 activity, so that prostaglandins would still be produced in the parietal cells to control acid secretion but would not be produced by the macrophages, which would result in selective control of the pain. The

expected result was achieved, that is, pain was reduced and gastric discomfort avoided. Unfortunately, the drug had adverse side-effects which, in some patients, included myocardial infarction so that the drug was withdrawn from the market in 2005. The side-effects are explained by the distribution of the different isoenzymes of cyclooxygenase and the fact that cyclooxygenase produces not only prostaglandins but also thromboxanes and prostacyclins. Both of these messengers control the aggregation of platelets and, therefore, the clotting of blood. The problems arose from the fact that both isoenzymes are present in platelets, which produce thromboxane, but only COX-2 is present in endothelial cells, which produce prostacyclin. Consequently, the drug inhibited prostacyclin formation in endothelial cells but thromboxane was still produced in platelets. Since thromboxanes stimulate aggregation of platelets but prostacyclin inhibits it, the drug resulted in thrombosis with the increased risk of a myocardial infarction (Figure 11.31).

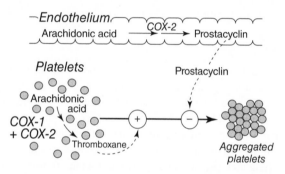

Figure 11.31 *Effects of endothelial cells, via prostacyclins, and platelets, via thromboxanes, on aggregation of platelets.* Aggregation of platelets plays an important role in thrombosis. The drug inhibited COX-2 and hence prostacyclin formation. However, both COX-1 and COX-2 are present in platelets so COX-1 continues to produce thromboxane.

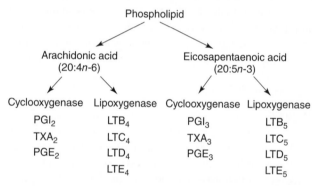

Figure 11.32 *Formation of prostaglandins of the 2 and 3 series and leucotrienes of the 4 and 5 series from arachidonic acid and eicosapentaenoic acid.*

series 3 and leucotrienes of series 5 (Figure 11.32). The significance of this is discussed below.

The sources of these fatty acids in the cells are those that are present at position 2 of the membrane phospholipids. The proportion of these two in the phospholipid depends to a large extent on the type of fatty acids in the triacylglycerol in the diet, that is, the amount of the omega-6 (linoleic acid) and that of the omega-3 (α-linolenic acid).

A chronic change in the diet can influence the composition of membrane phospholipids and such a change can be important in health. The ratio of the amount of these two fatty acids consumed (i.e. omega-6/omega-3) in a typical

modern Western diet is approximately 15:1 in favour of omega-6. This high ratio is due to the consumption of meat from commercially produced livestock, margarines and fat spreads produced from seed oils and the use of such oils in preparation of snack and fast foods. (see Tables 11.3 and 11.4). This value is considered to have been much lower in the diet of early humans (hunter-gatherers), possibly as low as 4:1, which was due to the high intake of vegetables and green leaves.

The prostanoids generated from an omega-6 fatty acid diet are series 2 whereas those generated from an omega-3 fatty acid-rich diet are of the series 3. The benefits of such a change are discussed in Chapter 22 but summarised here.

Cardiovascular disease Thromboxanes of the series 2 have strong positive effects on aggregation of platelets whereas those of the 3-series have much less effect. Hence, a low chronic intake of omega-3 fatty acid (linolenic) and a high intake of omega-6 fatty acid (linoleic) increases the formation of thromboxanes of the series 2 and increases the formation of prostaglandin of series 2 which has a lower vasodilatory effect, which increases the risk of thrombosis. Both changes, therefore, increase the risk of thrombosis. Obviously, to reduce the risk, the intake of omega-3 fatty acids should be high (Figure 11.32). This is discussed in Chapter 22 (see Figure 22.9).

Leucotrienes and inflammation

The biochemical and physiological effects of leucotrienes are less well defined than those of the prostanoids. Leucotrienes are powerful chemotactic compounds that are released, for example by damaged cells, and by platelets and mast cells to attract immune cells to sites of damage or immune activity. They are involved in inflammation, probably attracting neutrophils and macrophages to sites of inflammation (Chapter 17, Figure 17.37). Leucotrienes of series 5 have less effect on inflammation and prostaglandins of series 3 have greater inhibitory effect on inflammation compared with those of series 2. These differences in the extent of the effects of different prostanoids and leucotrienes may be part of the explanation for the beneficial effect of diets containing fish oils on chronic inflammatory disorders such as arthritis (Figure 17.32, Box 11.3).

Fatty acids in neurological and behavioural disorders

It has been known for some time that omega-3 fatty acids are quantitatively important components of the membranes of neurones in the brain. The remarkable number of axons and dendrites that link the neurones in the brain means that the number of membranes in this organ is enormous. The requirement for the omega-3 fatty acids in the brain will be high in at least two conditions:

- The developing brain in the foetus and the young child (see above).

- The repair of damaged membranes in the adult brain. Since the brain requires a substantial amount of oxygen and ATP is generated primarily from mitochondrial oxidation, the generation of oxygen free radicals will be

large, and will cause some damage to the omega-3 fatty acids. These damaged fatty acids will be replaced by undamaged omega-3 fatty acids via the phospholipid/lysophospholipid cycle described above (Figure 11.22).

In view of these factors, it has been suggested that a deficiency of omega-3 fatty acid in the diet will decrease the concentration of these fatty acids available for synthesis of the required phospholipids in the body, including the brain. If this was a chronic deficiency it could increase the risk of development of some disorders, including depression, schizophrenia and attention deficit syndrome. There is some evidence that this is the case.

Neurological disorders

The three disorders depression, schizophrenia and dementia are discussed in relation to the possible involvement of omega-3 fatty acids in these diseases, that is a deficiency may be involved in the aetiology of these diseases.

Depression

Three different types of study have provided evidence that omega-3 fatty acids are involved in depression.

- Epidemiological studies have been carried out to investigate if there is a relationship between the amount of fish consumed and the incidence of depression. This was done for nine different countries. An inverse relationship was observed between the amount of fish consumed and the incidence of depression.

- The amount of omega-3 fatty acids in the membrane of erythrocytes is less in patients with depression compared with that in controls.

- In some studies, supplementation of the diet with omega-3 fatty acids has produced an improvement in this condition, especially in women.

Schizophrenia

The amounts of both omega-3 and omega-6 fatty acids are low in erythrocytes from patients suffering from schizophrenia. As with depression, supplementation of the diet of such patients with omega-3 fatty acids has been claimed to improve the condition.

Dementia

Post mortem studies have shown that the level of polyunsaturated fatty acids in some areas of the brain is decreased (particularly the hippocampus, striatum and cortex). In addition, it is known that the fluidity of membranes in isolated cells is decreased in patients with Alzheimer's disease.

In serum, cholesterol ester is transferred from high density lipoproteins to other lipoprotein particles and is taken up by the liver. The cholesterol ester is composed of cholesterol and a polyunsaturated fatty acid which is formed in the blood by a reaction catalysed by the enzyme lecithin-acyltransferase (Figure 11.15). The fatty acid in the phospholipid lecithin and, therefore in the ester, is an indication of the fatty acid that is present and available in the liver when the lecithin is synthesised. The amount of eicosapentaenoic acid in the serum cholesterol ester is decreased in patients with Alzheimer's disease, suggesting a decreased level of the omega-3 fatty acid in the liver. Epidemiological studies showed that fish consumption is inversely correlated with cognitive impairment. Finally supplementation of the diet with the omega-3 fatty acid, docosahexaenoic fatty acid, for as little as one year, produced a cognitive benefit in patients.

Behavioural disorders

Aggression

There is evidence that the level of an omega-3 fatty acid (docosahexaenoic acid) is low in the tissues of subjects who exhibit violent behaviour. Supplementation with this fatty acid has been shown to reduce aggression in a normal population of university students.

Antisocial behaviour

The antisocial behaviour of young adults in prison was studied. Supplementation of the diet with capsules containing vitamins, minerals and essential fatty acids reduced antisocial behaviour in comparison with control subjects. A much larger study is currently being planned.

Attention deficit disorder

Patients, usually children, suffering from this disorder exhibit both impaired attention to an activity, in which they are involved, and frequent changes from one activity to another without completing the first. Associated problems include excessive restlessness, running and jumping, noisiness and talkativeness. In the UK, the prevalence is up to 2% of the school-age population, while in the USA it is 3–5%.

There is some evidence that, in these patients, the interconversion between the polyunsaturated fatty acids is disturbed, which restricts the formation of eicosapentaenoic and docosahexaenoic acids. Such children are less likely to have been breastfed (breast milk contains these omega-3 fatty acids); they are more likely to suffer from allergies associated with essential fatty acid deficiency and also dry skin and hair and the membranes of the erythrocytes contain less omega-3 fatty acids compared with normal children. So far, the results of supplementation of the diet of these children with this disorder have not been conclusive.

12
Hormones: From Action in the Cell to Function in the Body

'One evening in January or February, 1922, while I was working alone in the Medical Building, Dr. J.B. Collip came into the small room where Banting and I had a dog cage and some chemical apparatus. He announced to me that he was leaving our group and that he intended to take out a patent in his own name on the improvement of our pancreatic extract. This seemed an extraordinary move to me, so I requested him to wait until Fred Banting appeared, and to make quite sure that he did I closed the door and sat in a chair which I placed against it. Before very long Banting returned to the Medical Building and came along the corridor to this little room. I explained to him what Collip had told me and Banting appeared to take it very quietly. I could, however, feel his temper rising and I will pass over the subsequent events. Banting was thoroughly angry and Collip was fortunate not to be seriously hurt. I was disturbed for fear Banting would do something which we would both tremendously regret later and I can remember restraining Banting with all the force at my command.'

Except for a veiled but important reference in Banting's 1922 account, there are no other useful written records of this incident. Clark Noble once drew a cartoon, unfortunately now lost, of Banting sitting on Collip, choking him; he captioned it 'The Discovery of Insulin'.

(Bliss, 1983)

Hormones play a fundamental role in the regulation and integration of biochemical and physiological processes. By common usage, they fall into two classes, endocrine and local. Endocrine hormones are chemical messengers that are synthesised in cells, usually grouped together in glands, and are secreted into the blood to elicit responses within cells of one or more target tissues. Local hormones are chemicals released by one cell to influence another cell in the same tissue or organ. This action is known as paracrine but some endocrine glands may fall into this category (e.g. pituitary gland) and it includes messengers not normally considered as local hormones, e.g. neurotransmitters. The problem is further compounded by the fact that some cells release factors that enter the interstitial space but then affect biochemical processes within the same cell: this is known as an autocrine effect. As most of these factors control growth, they are usually known as growth factors. At the other extreme, chemicals which communicate between individuals of the same species, travelling between them via water for aquatic animals or via air for terrestrial animals are also hormones, known as pheromones. Consequently, hormones are difficult to define.

Endocrine hormones: traditional and novel

The study of hormones is known as endocrinology, and it arose from both clinical and experimental observations. The physical and emotional effects of castration in the male had long been recognised and in the nineteenth century the signs and symptoms of a number of diseases, which are now known to be caused by endocrine dysfunction, were described. In 1871, Charles Fagge attributed

> Endocrinology is the study of the regulation and function of the endocrine glands and the action of hormones secreted by the glands. The word 'endocrine' derives from the Greek and was first used in 1913 to describe glands which secrete substances into the blood, unlike the exocrine glands, which release their products through a duct. The concept of 'internal secretion' had been developed earlier, and the term 'hormone' for 'chemical messengers' or effectors was introduced in 1905.

Functional Biochemistry in Health and Disease by Eric Newsholme and Tony Leech
© 2010 John Wiley & Sons Ltd

cretinism to congenital absence of the thyroid gland. In 1873, William Gull described a similar but acquired disease in adults and, in 1883, Emil Kocher, a Swiss physician, described the same clinical picture after surgical removal of the gland. In 1891, George Murray reported successful treatment with thyroid extract. Many studies on the effects of removal of apparently functionless glands led to identification of the endocrine system. These studies were followed by isolation and purification of the active compounds, determination of their chemical structure and investigations into their mode of action.

For many years it was considered that there were seven major endocrine glands: the hypothalamus, pituitary, thyroid, adrenals, gonads, parathyroids and the Islets of Langerhans in the pancreas (Figure 12.1). However, since then, other endocrine glands have been discovered. The original seven are termed *traditional* and the more recent, *novel*.

> William Gull was a British physician who also described anorexia nervosa in 1866.

Traditional endocrine glands and hormones

The hypothalamus

The hypothalamus is part of the brain. It is an important regulatory centre and produces a range of peptide-releasing and release-inhibiting hormones, which control the secretory activity of the pituitary.

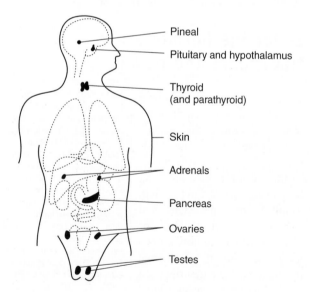

Figure 12.1 *Location in the body of some glands.* The pineal secretes melatonin, which is discussed in Chapter 14. Skin should be included since it secretes a precursor of vitamin D, now considered to be a steroid hormone.

The pituitary gland

The gland, which is anatomically and functionally connected to the hypothalamus, is divided into two parts, anterior and posterior.

Posterior pituitary Two hormones, vasopressin and oxytocin, are synthesised in the hypothalamus and then transported through nerve axons to the posterior pituitary, where they are stored until released. Vasopressin acts on the kidney to conserve water. Its secretion is stimulated by thirst and a decrease in blood pressure. Secretion of oxytocin initiates uterine contraction for parturition. It also stimulates milk ejection from the mammary glands.

The anterior pituitary Six peptide hormones are secreted by this gland: follicle-stimulating hormone (FSH), luteinising hormone (LH), thyroid-stimulating hormone (TSH), adrenocortitrophic hormone (ACTH), growth hormone and prolactin. The last promotes lactation. ACTH regulates secretion of glucocorticoids by the adrenal cortex. Excess production of ACTH, for example by pituitary tumours, over-stimulates the adrenal cortex, leading to excess cortisol secretion, a cause of Cushing's syndrome. Growth hormone stimulates fatty acid mobilisation by adipose tissue and fat oxidation in muscle. Although growth hormone was named for its ability to enhance linear growth, particularly during puberty, this effect is due to stimulation of the secretion of a peptide, insulin-like growth factor, by the liver but it only responds to growth hormone if there is sufficient intake of carbohydrate and proteins (i.e. if there is adequate nutrition).

The thyroid gland

The gland is situated in the neck across the front of the trachea. It secretes thyroxine (T_4), which is converted to the active form of the hormone, triiodothyronine (T_3), in peripheral tissues. It stimulates metabolic activity in tissues so that it increases heat production (for example, by stimulating protein turnover and substrate cycles).

A deficiency of iodine results in insufficient synthesis of thyroxine and a low plasma level, so that secretion of TSH is increased in an attempt to stimulate synthesis but this results in enlargement of the thyroid gland (goitre). Iodine deficiency in pregnancy impairs brain development in the foetus, causing mental retardation (known as cretinism). Indeed iodine deficiency is one of the major public health issues worldwide: an estimated 200 million people are affected.

The adrenal glands

The glands are situated above each kidney and consist of two parts, an outer cortex and inner medulla. The cortex produces cortisol and aldosterone. The rate of cortisol secretion is controlled by ACTH, whereas aldosterone secretion is regulated by the kidney, via the renin–angiotensin system (Chapter 22). The adrenal medulla is a specialised part of the autonomic nervous system and secretes catecholamines (adrenaline and noradrenaline) in response to stress.

The gonads

Cells in the ovaries and testes produce the sex hormones: oestrogen and progesterone in the female, and testosterone and androstenedione in the male (Chapter 19).

The parathyroid glands

There are four parathyroid glands, which are situated behind the thyroid. They produce parathyroid hormone (PTH), a peptide which interacts with vitamin D to control the level of calcium in the blood. PTH stimulates release of calcium from bone and increases the uptake of calcium by the kidney tubules from the glomerular filtrate.

The Islets of Langerhans

The endocrine pancreas comprises clumps of cells (Islets of Langerhans) distributed throughout the exocrine gland. The islets contain three different specialised cells. The alpha cells produce glucagon, the beta cells produce insulin and the gamma cells produce somatostatin, which inhibits secretion of the other hormones. Paul Langerhans (1847–88) was a German physician and pathologist who first described the islets in 1869.

Novel endocrine glands

The novel endocrine glands are the skin, gastrointestinal tract, adipose tissue, kidney (juxtaglomerular apparatus in the cortex which secretes renin that indirectly controls aldosterone secretion, via angiotensin-II), pineal gland, which secretes melatonin, and the heart (cardiac myocytes in the atria, which secrete atrial natriuretic peptide).

The skin

Ultraviolet light causes a chemical change in dihydrocholesterol to produce cholecalciferol, a precursor of vitamin D. The latter conforms better to the definition of a steroid hormone than a vitamin. Indeed, the classification of vitamin D as a vitamin is an historical accident. The precursor is released from the skin and is further modified in the liver and kidney to form dihydroxycholecalciferol, which is the active form of the hormone (see Chapter 15 for the reactions). It increases calcium absorption from the gut and is essential for normal bone formation. Lack of the steroid results in rickets in children and osteomalacia in adults. The steroid or its precursor may also be involved in the development of tumours (Chapter 21).

In addition, the apocrine glands in the skin secrete pheromones (see below).

Gastrointestinal tract

The hormones GIP (glucose-dependent insulinotrophic polypeptide) and GLP (glucagon-like peptide) are secreted by the intestine in response to ingestion of glucose and/or protein. They stimulate insulin secretion, particularly the initial phase of secretion that occurs within the first 20 minutes. They also slow gastric emptying, allowing a more gradual absorption of the products of digestion. GLP also suppresses secretion of glucagon so that it reduces or prevents stimulation of glycogenolysis in the liver, which further helps to minimise changes in blood glucose level after a meal.

The stomach secretes a peptide known as ghrelin that increases appetite, stimulating energy intake.

Adipose tissue

Adipose tissue was at one time considered to be an inert storage depot for triacylglycerol. It is now known to be an extremely active endocrine gland. Adipocytes secrete several hormones in response to the amount of triacylglycerol in the cell, and in response to cytokines and growth factors. The hormones include adiponectin, leptin, resistin and also cytokines and growth factors. The concentration of adiponectin in blood is very high, approximately 1000-fold greater than any other hormone, and it increases as the fat mass decreases and vice versa. One effect is to increase the sensitivity of cells to insulin; another is to inhibit non-essential ('luxury') processes that utilise ATP in the cell. Consequently, it decreases energy expenditure. Leptin is secreted by adipose tissue also in response to an increase in the amount of triacylglycerol in the adipocyte. The effect of leptin is on the hypothalamus which responds by restricting food intake (an anorexic effect) and increasing energy expenditure. Since it may play a role in development of obesity, it is discussed in Appendix 15.4. Leptin also initiates pubertal development and the menstrual cycle in young females. It is considered that leptin acts as a signal from adipose tissue to the brain to indicate that fuel stores are sufficient so that the changes necessary for reproduction in females can take place. Resistin antagonises the effect of insulin on glucose uptake in muscle and adipose tissue. It has been suggested that it is responsible for insulin resistance in some conditions but its precise role in the body is unknown. Some metabolic effects of traditional and novel hormones are summarised in Table 12.1.

Table 12.1 Effects of hormones on rates of synthesis and breakdown of carbohydrate, fat and protein in the whole body

	Glycogen synthesis in liver and muscle	Glycogen breakdown in liver	Fat synthesis in liver	Fat breakdown in adipose tissue	Protein synthesis in muscle	Protein breakdown in muscle
Traditional hormones						
Insulin	+	– –	+	– –	+	– –
Glucagon	–	++	–	+	NK	NK
Thyroxine	NK	+	NK	+	+	+
Cortisol	+	NK	+	NK	–	+
Growth hormone	+	NK	–	++	+	NK
Catecholamines	–	++	NK	+++	+	NK
Insulin-like growth factor 1	NK	NK	NK	NK	+	NK
Novel hormones						
Adiponectin	NK	NK	–	+	NK	NK
Leptin	NK	NK	–	+	NK	NK
Resistin	–	+	NK	NK	NK	NK
Testosterone	NK	NK	NK	+	++	NK
Oestrogen	NK	NK	+	+	NK	NK

+indicates stimulation; – indicates inhibition; NK indicates not known.
Some of the data from Brodsky (2006).

The action, effects and functions of a hormone

Despite the diversity of hormones – endocrine, paracrine, autocrine or pheromones – they all influence intracellular events by intracellular signalling mechanisms, and the number of such mechanisms is remarkably small. In this chapter, the signalling mechanisms, the effects and the functions of hormones are described, discussed and illustrated by reference to selected hormones. In addition, the principles underlying the mechanisms by which hormones affect the biochemistry in cells are discussed.

It is first necessary to clarify terminology in this field. The terms *action*, *effect* and *function* of a hormone are often used indiscriminately. Since there are no clear-cut definitions, this can be confusing. Consequently, definitions, even if only operational, are put forward here and are used throughout this chapter.

The *action* of a hormone is defined as the primary effect on a cell, usually the binding of the hormone to a specific receptor and the resultant interaction between the hormone–receptor complex and an effector system within the cell. The *effect* of a hormone is an experimental observation that is made either in vitro or in vivo; it can be molecular, biochemical or physiological but, when a sufficient number of effects are established, a relationship between the action and effects can be drawn. This can best be described as a 'pyramid' (Figure 12.2). The *function* of a hormone is an

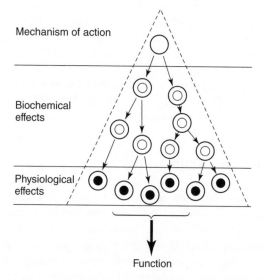

Figure 12.2 *A hormone pyramid*. This diagram illustrates how a single point of action leads to several biochemical effects which, in turn, lead to physiological effects that combine to bring about the function of the hormone.

inference that is made from its biochemical and/or physiological effect(s) in the whole body or from the pathology that results from removal of the hormone. Therefore, interpretation of the function, or the name given to the hormone, is limited by the system under study. These are usually ascribed to the first observation that is made but this can dominate, even when later observations may be more physiologically or clinically relevant. Fox example:

- When a factor isolated from the anterior pituitary was observed to promote growth, especially in animals from which the pituitary had been removed, it was given the name growth hormone, implying a role in growth, but other hormones and factors also play a role. However, if the changes in fat metabolism had been measured first, the hormone might have been termed 'fat-mobilising hormone' since it stimulates release of fatty acids from adipose tissue.

- The demonstration in 1921 that insulin lowered the blood glucose level in diabetic dogs and humans led to the inference that its main function was the control of the blood glucose level. However, once the diversity of its effects was discovered, it was realised that its major function is stimulation of anabolic processes (e.g. synthesis of RNA, protein, fat and carbohydrate). Hence, lack of insulin prevents anabolism so that catabolism dominates. Indeed, this was known well before 1921. Regulation of the blood glucose level is just one component of its function, albeit a very important one (see below).

> Prior to isolation of insulin, patients were treated by diet, consisting largely of boiled vegetables to control the glycosuria and presumably the blood glucose level (Allen, 1919). However, patients simply 'faded away' and probably died from effects of loss of muscle protein (Chapter 16). About 2000 years ago diabetes was aptly described: 'A portion of flesh is excreted in the urine' (Aretaeus of Cappadocia (30–90 AD).

The relationship between all three definitions can also be interpreted as a linear sequence of events from binding to the receptor through to the whole-body response (Figure 12.3). The sequence has some similarities to that which is initiated by a growth factor for example control of proliferation end tumour development (Chapter 21 (see Figure 21.3)).

Action of hormones

The action of a hormone is, first, binding to a receptor to form a hormone–receptor complex. The binding is reversible (i.e. equilibrium binding) as with other receptors. Second, the complex stimulates an effector system which then activates a signal transducing mechanism that results in a change(s) in a biochemical process(es) within the cell.

Receptors

The receptors can either be located on the cell surface or within the cell. Not surprisingly, the mechanisms of the effector systems are, in principle, quite distinct for the two

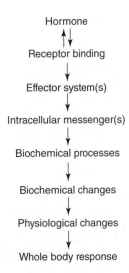

Figure 12.3 *The hormone sequence, from hormone receptor to the response in the whole body.* This diagram provides an outline of the processes on which discussion of hormone action in this chapter is based.

types of receptor. Since the *first* change produced by the hormone is a change in activity of the effector system, the messenger produced by the effection system is sometimes known as the *second* messenger, but this can be confusing since there might be several messengers in sequence.

Cell surface

For a cell surface receptor, the binding of the hormone must change the activity of an effector system that results in a change in concentration of an intracellular messenger. This is achieved by location of the effector system on the cytosolic side of the membrane, although the receptor and effector system may be combined in one complex.

Intracellular

Only if the hormone is lipid soluble, and can therefore transfer rapidly across the cell membrane, can the receptor be within the cell. Examples of such hormones include the steroid hormones (e.g. sex hormones, adrenal steroids and dihydroxycholecalciferol) and thyroxine (i.e. triiodothyronine).

Effector systems

Despite the existence of a large number of hormones, only four effector systems for linking hormone binding to an intracellular effect have been described:

- The hormone–receptor complex directly increases the activity of an enzyme, usually a protein kinase (Figure 12.4) or a phospholipase (Figure 12.5).

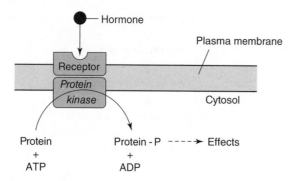

Figure 12.4 *Effector mechanism: direct activation of a protein kinase*. The protein kinase is the cytosolic component of the hormone receptor. Activation of the kinase results in phosphorylation of a protein which initiates the effects of the hormone. An example of this is the effector mechanism of insulin, which is a tyrosine kinase.

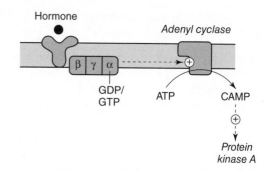

Figure 12.6 *Effector mechanism: activation of adenyl cyclase via a G-protein*. The activation of adenyl cyclase, which resides on the cytosolic side of the cell membrane, is mediated through the membrane-bound G-protein system (see below for biochemistry and role of the G-protein). An example is adrenaline binding to the β-receptor.

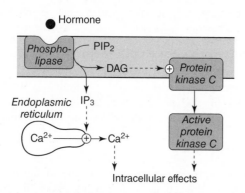

Figure 12.5 *Effector mechanism: activation of a membrane-bound phospholipase*. An example is activation of a membrane-bound phospholipase which hydrolyses phosphatidylinositol bisphosphate (PIP_2) and results in the formation of the two messengers, inositol trisphosphate (IP_3) and diacylglycerol (DAG). Messenger IP_3 binds to a receptor on the endoplasmic reticulum that results in release of Ca^{2+} ions into the cytosol. DAG, which remains within the membrane, activates protein kinase-C at the membrane surface. When the kinase leaves the membrane, it is unclear how it remains active or loss of activity is prevented, so that it can phosphorylate proteins in the cytosol or even the nucleus. An example is adrenaline binding to the α-receptor in the liver, in which Ca^{2+} ions stimulate glycogenolysis.

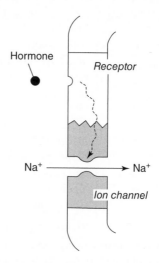

Figure 12.7 *Effector mechanism: opening or closing of an ion channel*. The binding of the hormone opens an ion channel. The link between the receptor and the ion channel may be via a messenger molecule that causes a conformational change in one of the proteins in the channel or may be via phosphorylation of one of the proteins in the channel. An example of opening an ion channel is that of a pheromone on a sensory cell in the olfaction epithelium (see below).

- The complex activates a gene (i.e. it is a transcription factor) (Figure 12.8).

The biochemical and physiological effects of a hormone

The effector systems result in changes in messenger systems that then cause the effects of the hormone. Not surprisingly, the effects of the hormone depend on which hormone is being considered. To illustrate this four hormones – insulin, cortisol, adrenaline and glucagon – are discussed.

- The complex activates a G-protein, which then activates a membrane-bound enzyme (e.g. adenyl cyclase, which converts ATP to cyclic AMP; Figure 12.6).

- The complex activates (opens) an ion channel which results in an increase in the intracellular concentration of an ion (e.g. Na^+ ion). The ion then interacts with a protein that results in an increase in the rate of a biochemical process (Figure 12.7).

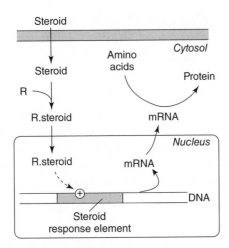

Figure 12.8 *Effector mechanism: activation of a specific gene by hormone–receptor complex binding to DNA.* A steroid is used to illustrate the mechanism. The hormone enters the cell and binds to its receptor (R) in the cytosol, the hormone–receptor complex enters the nucleus and binds to a specific sequence in the DNA that stimulates transcription of a gene or genes: the resultant increase in mRNA increases the synthesis of specific proteins. The binding site on the DNA is specific and is usually termed a response element. Thyroxine (i.e. triiodothyronine) also uses this effector mechanism. Activation of genes, RNA processing to produce mRNA and translation are described in Chapter 20 (see Figures 20.20, 20.21 and 20.22).

Insulin

In the 1950s, it was discovered that adrenaline stimulated glycogen breakdown in muscle by activation of the enzyme glycogen phosphorylase and this was achieved by an effect of a simple messenger molecule, cyclic AMP (identified by E. Sutherland for which he was awarded the Nobel Prize in 1971). In the 1960s, it was discovered that insulin stimulated glycogen synthesis in muscle by activating the enzyme glycogen synthase. Consequently, by analogy with adrenaline, it was considered that insulin would activate the synthase via a 'simple' messenger molecule. Despite much work and many possible candidates for messengers, this proved *not* to be the case. The signalling pathway, even to stimulate glycogen synthesis, is biochemically complex. This complexity is not surprising since the major function of insulin is to stimulate anabolic and inhibit catabolic processes involved in carbohydrate, fat and protein metabolism.

Mechanism of action

The action is the binding of insulin to its receptor, part of which is located on the external surface of the plasma membrane; the other part is on the cytosolic side. This is a tyrosine kinase. Binding of insulin to its receptor activates the tyrosine kinase, which leads to phosphorylation of tyrosine residues on a protein, known as the insulin-receptor substrate (IRS) which can lead to activation of the enzyme phosphatidylinositol 3-kinase. Consequent upon this, other proteins are phosphorylated, particularly protein kinase B, which leads to many of the metabolic effects. The tyrosine kinase activity also leads to activation of the MAP-kinase cascade, which increases the transcription of some genes which stimulates many of the anabolic effects of insulin. (MAP-kinase is not involved in any of the metabolic effects of insulin.) This cascade is discussed in Chapter 21 (see Figure 21.9).

Biochemical effects of insulin

Five of the major biochemical effects of the hormone illustrate the processes that are modified by insulin.

- Activation of a factor(s) that controls initiation of protein synthesis.

- Activation of transport of some amino acids into muscle.

- Activation of a protein kinase that is responsible for many of the metabolic effects, for example increased activity of glycogen synthase, increased translocation of the glucose transporter molecules to the plasma membrane and increased activity of acetyl-CoA carboxylase.

- Activation of cAMP phosphodiesterase, which decreases levels of cAMP and thus the activity of protein kinase A. This results in decreased rates of lipolysis in adipose tissue and glycogenolysis and gluconeogenesis in liver.

- Activation of a ribosomal protein kinase, which increases the rate of translation of mRNA.

Physiological effects

The rates of five major pathways are stimulated by insulin are.

- Glycolysis in muscle and liver (Chapter 6).

- Glycogen synthesis in muscle, an leading to increased store of glycogen.

- Lipogenesis in liver and possibly adipose tissue, leading to increased synthesis of phospholipids and triacylglycerol.

- Protein synthesis in muscle leading to an increase in size of muscle.

- VLDL synthesis and secretion by liver leading to increased deposition of fat in some tissues (adipose tissues and muscle).

The rates of five major pathways are inhibited by insulin:

• gluconeogenesis in the liver;

• glycogen breakdown in liver;

• lipolysis in adipose tissue and hence fatty acid release;

• fatty acid oxidation in muscle and liver;

• protein degradation in muscle.

An attempt has been made to incorporate the mechanism of action, the biochemical and physiological effects, together with the proposed function of insulin, into a 'biochemical pyramid' which is presented in Figure 12.9.

Cortisol

The hormone cortisol is 11β, 17α, 21-trihydroxy-4-pregnene-3,20 dione. It is the major glucocorticoid synthesised and secreted by the human adrenal cortex. Cortisol enters the cell then binds to its receptor. The cortisol–receptor complex enters the nucleus and binds to DNA.

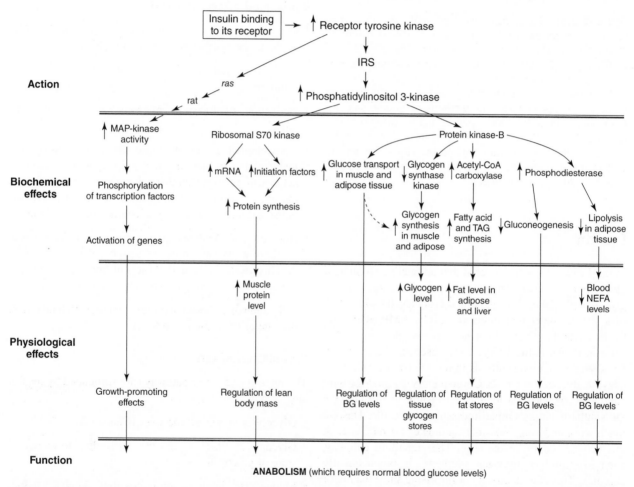

Figure 12.9 *The insulin pyramid: a diagram illustrating the action, biochemical and physiological effects and the function of insulin.* The tip of the pyramid is the binding of insulin to its receptor which activates the cytosolic component of the receptor, a tyrosine kinase, which activates the insulin-receptor substrate (IRS), which activates the enzyme, phosphatidylinositol 3-kinase, which results in the metabolic effects, and *ras*, which stimulates the mitogen-activated kinase cascade (see Figure 21.9). These lead to biochemical effects which lead to physiological effects, almost all of which lead to anabolic changes. The diagram is not comprehensive: the information is only a guide to the action and biochemical effects of insulin as an example of the principle of a hormone pyramid. BG is an abbreviation for blood glucose. It should be clear from this figure that regulation of the blood glucose level is only one factor of the many that is regulated by insulin. For details of current understanding of insulin action and effects, readers should consult recent reviews.

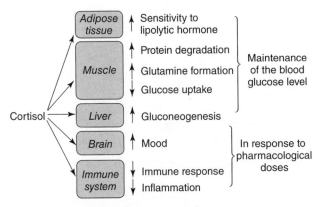

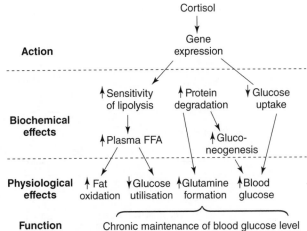

Figure 12.10 *The effects of cortisol.* These can be divided into two classes. (i) Physiological levels of cortisol result in changes that maintain the blood glucose level. (ii) Higher levels (pharmacological doses) have an anti-inflammatory effect and a central effect on wellbeing. It is the breakdown of protein that provides amino acids, which along with glucose are used to synthesise glutamine (Chapter 8).

Figure 12.11 *The cortisol pyramid.* On the basis of the biochemical and physiological effects, the function of the hormone is the chronic maintenance of the blood glucose and glutamine levels. FFA – long-chain fatty acids.

This leads to activation of several genes, the transcription of which produces mRNA for the synthesis of the proteins to carry out the effects of the hormone. A scheme describing this type of effect of a hormone is provided in Figure 12.8.

Effects of cortisol

The biochemical and physiological effects of cortisol are summarised in Figure 12.10. Higher than normal concentrations of cortisol improve mood and reduce the activity of the immune system. Hence, they are routinely used to reduce chronic inflammation. There are, however, side-effects of these high levels: increased levels of blood glucose, obesity and retention of water giving rise to 'moon face'.

Functions of cortisol

The function of cortisol is to prepare the body for, and respond to, chronic stress. It is particularly important in maintenance of the blood glucose level, by reducing glucose utilisation via effects on the glucose transporter, by increasing the sensitivity of lipolysis in adipose tissue to lipolytic hormones (e.g. adrenaline, growth hormone) and stimulation of protein degradation in muscle (e.g. to provide amino acids for gluconeogenesis). Thus the function of cortisol is important for survival in prolonged starvation and after trauma. On the basis of this discussion, a 'biochemical pyramid' for cortisol is presented in Figure 12.11.

Adrenaline

The action of adrenaline (or noradrenaline) involves binding to an extracellular receptor, of which there are two classes, the α- and β-receptor. When the hormone binds to the β-receptor, the hormone–receptor complex activates adenyl cyclase, which catalyses the formation of cyclic AMP from ATP.

$$ATP \rightarrow cyclic\ AMP + PP_i$$

This is achieved via the G-protein system, which is discussed below.

Biochemical effects

The increase in the concentration of cyclic AMP activates protein kinase A. In skeletal muscle, protein kinase A phosphorylates and activates phosphorylase kinase which, in turn, phosphorylates and activates the enzyme phosphorylase. This catalyses the degradation of glycogen so that the glucose residues are converted to glucose 1-phosphate which then forms glucose-6-phosphate to enter glycolysis. These biochemical effects form the enzymatic cascade for the activation of glycogen breakdown in muscle. The activation steps, catalysed by the kinase reactions, are reversed by inactivation steps (dephosphorylation reactions) catalysed by a protein phosphatase. The details of these reactions are described and discussed in Chapter 3 (Figure 3.12). Although adrenaline activates phosphorylase, via the cascade, under resting conditions,

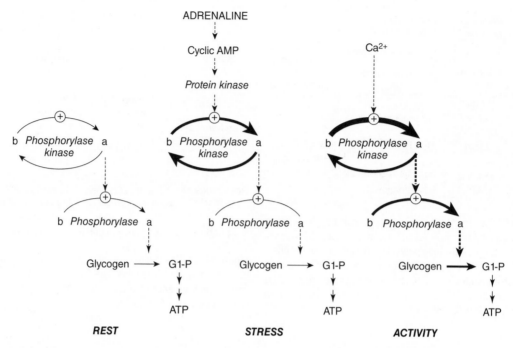

Figure 12.12 *Adrenaline increases cycling between inactive phosphorylase b kinase and active phosphorylase b kinase.* The three conditions considered are muscle at rest, muscle responding to stress and muscle responding to nervous stimulation (i.e. physical activity). The effect of an increase in adrenaline on glycogen breakdown in the absence of physical activity is small. However, the cycling rate between phosphorylase kinase *b* and *a* is high during stress due to cycling rate and the increased level of cyclic AMP. This increases the sensitivity to an increase in Ca^{2+} ion concentration, so that the rate of glycogen breakdown is now very large. The small effect of adrenaline alone on the rate of glycogen breakdown but the marked increase when adrenaline is combined with physical activity was first reported by the group of Carl Cori in 1965. This effect of stress (i.e. catecholamines and sympathetic nervous stimulation) is similar to that proposed for increased substrate cycling between fructose 6-phosphate and fructose bisphosphate in control of glycolytic flux in the muscles of a sprinter in a competitive event (Figure 3.31).

the effect on the rate of glycogenolysis is quite small. The largest activation and the highest rate of glycogen breakdown occur in muscle when it contracts in the presence of adrenaline (see below).

In adipose tissue, the increased concentration of cyclic AMP activates the hormone-sensitive lipase to increase the rate of lipolysis and hence fatty acid release from adipose tissue. This increases the plasma level of fatty acids and hence their oxidation by muscle (see Chapter 7).

In liver, adrenaline binds to the α-receptor, and the hormone–receptor complex activates a membrane-bound phospholipase enzyme which hydrolyses the phospholipid phosphatidylinositol 4,5-bisphosphate. This produces two messengers, inositol trisphosphate (IP_3) and diacylglycerol (DAG) (Figure 12.5). The increase in IP_3 stimulates release of Ca^{2+} ions from the endoplasmic reticulum into the cytosol, the effect of which is glycogen breakdown and release into the blood (see Figure 12.5 and Chapter 6).

Physiological effects

The physiological effects of adrenaline are as follows.

- It increases the amount of phosphorylase in the *a* form in muscle, which stimulates glycogen breakdown, but its major role is to 'prime' the phosphorylase cascade for a marked increase in activity, if and when the muscle becomes physically active. This is achieved by increasing the rate of cycling between the active and inactive forms of phosphorylase *b* kinase via the increase in concentration of cyclic AMP. This is a mechanism for increasing the sensitivity of phosphorylase kinase to the increase in Ca^{2+} ion concentration (Figure 12.12). This is equivalent to the effect of increasing the flux through a substrate cycle in improving sensitivity in control of, for example, glycolytic flux (see Chapter 3: Figure 3.31).

- It increases the release of fatty acids from adipose tissue which raises the plasma level of long-chain fatty acids, to provide a fuel for muscle, if it becomes physically active (Chapter 13). It also increases the cycling between triacylglycerol and fatty acids in adipose tissue.

- It increases glycogen breakdown in liver, resulting in an increased rate of release of glucose by the liver which helps to maintain or increase the blood glucose concentration to provide fuel for muscle.

- It also increases substate cycling in glycolysis in muscle (Chapter 3: Figure 3.31).

- The activation of protein kinase A in heart leads to phosphorylation of a number of proteins which together result in changes that increase cardiac output (Chapter 22). An important clinical effect of adrenaline.

Function

The function of adrenaline is to mobilise all fuels that can be used by muscle to provide ATP to support physical activity in response to stress (i.e. 'fight or flight' response). And to increase sensitivity of regulation of enzymes involved in control of the rate of processes that generate ATP. The biochemical effects in the heart increase cardiac output, in preparation for 'fight or flight'.

Glucagon

Glucagon is secreted by the α-cells in the Islets of Langerhans in response to a decrease in the concentration of blood glucose. It binds to a receptor in liver and adipose tissue which activates adenyl cyclase and raises the intracellular level of cAMP, which activates protein kinase A (Figure 12.13).

> Although many effector systems utilise ATP or the equivalent, directly or indirectly, the rate at which ATP is used by such processes is trivial in comparison with many other processes. For example, in rat heart, the rate of cyclic AMP formation and degradation represents only 0.05 per cent of the normal rate of ATP turnover.

Effects

The result of increasing the activity of protein kinase A in liver is activation of glycogen phosphorylase which increases glycogen breakdown and leads to an increase in the rate of glucose release. The increased level of cyclic AMP stimulates gluconeogenesis via inhibition of phosphofructokinase and activation of fructose bisphosphatase (Chapter 6). It also activates hormone-sensitive lipase in adipose tissue which stimulates lipolysis and the release of fatty acids by the tissue, which raises the plasma level of fatty acids. This leads to an increase in their rate of oxidation in muscle and liver. In muscle, fatty acid oxidation inhibits glucose utilisation. It also leads to an increase in the rate of ketone body formation (Figure 12.13). (See Figure 16.6).

Function

The function of glucagon is to respond rapidly to an acute fall in the blood glucose level by stimulating glucose release by the liver and fatty acid release by adipose tissue,

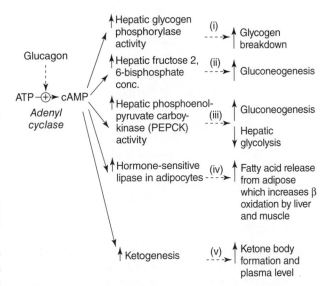

Figure 12.13 *Action and effects of glucagon.* Glucagon binds to its receptor on the plasma membrane of the liver which activates adenyl cyclase. The resultant cyclic AMP activates protein kinase which results in phosphorylation and activation of:
(i)   Glycogen phosphorylase, which increases hepatic glycogen breakdown and glucose formation and release by the liver.
(ii) Phosphofructokinase-2, which increases formation of fructose 2,6 bisphosphate that activates fructose bisphosphatase which stimulates gluconeogenesis (see Figure 16.6).
(iii) Cyclic AMP also binds to the cyclic AMP response element on DNA in the nucleus of hepatocytes and activates the gene for phosphoenolpyruvate carboxykinase, which increases the concentration of this enzyme, which increases the rate of gluconeogenesis (Chapter 6).
(iv) Hormone-sensitive lipase in adipose tissue which increases fatty acid release and increases the plasma fatty acid concentration and hence oxidation in muscle and liver.
(v)   Fatty acid oxidation in the liver which increases rate of ketogenesis, the plasma ketone body level and oxidation.

and ketone body formation all of which will help to restore and maintain blood glucose concentration (Chapters 6 and 16: see Figure 16.6). Hence, injection of glucagon is used to overcome, rapidly, hypoglycaemia caused by an overdose of insulin in diabetic patients.

Integration of the effects of these various hormones on the regulation of the blood glucose level

Changes in the blood levels of these hormones all contribute to regulation of blood glucose level in several conditions. After a meal glucose utilisation is increased, since insulin stimulates glucose uptake by muscle and inhibits release of fatty acids from adipose tissue. Physical activity

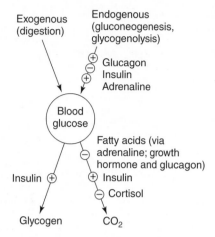

Figure 12.14 *The integration of the effects of insulin, adrenaline, glucagon and cortisol on maintenance of the blood glucose level.* Glucose is formed via glycogenolysis in the liver which is controlled by adrenaline, insulin and glucagon. Glucose is also formed via gluconeogenesis which is regulated by glucagon and insulin. Cortisol reduces glucose utilisation by decreasing the number of glucose carriers in the plasma membrane of muscle, whereas insulin has the opposite effect. An increase in the plasma level of fatty acids (due to growth hormone, adrenaline and glucagon) decreases glucose utilisation by muscle, via the glucose/fatty acid cycle (Chapter 16, see Figures 16.4, 16.5 and 16.6).

decreases the plasma concentration of insulin and increases plasma levels of adrenaline and growth hormone, which results in an increase in the release of glucose by the liver and mobilisation of fatty acids from adipose tissue. Finally, starvation results in decreased levels of insulin and increased levels of growth hormone and cortisol which stimulate fatty acid release from adipose tissue and increase the plasma level, which increases their oxidation, which inhibits glucose utilisation (Figure 12.14). From this figure, it can be seen that insulin is central to the regulation of the blood glucose level so that it is not surprising that the discovery of insulin depended, particularly, on measurement of the changes in blood glucose level in diabetic animals (Box 12.1).

Pheromones

Pheromones are volatile chemicals that allow communication between individuals via air or water, over a distance which can be quite long. That is, one individual animal produces and emits a chemical that changes the biochemistry and physiology of another member of the same species. This can result in changes in social or sexual behaviour. The pheromones are known more generally as smells or scents. Communication via pheromones is well

developed in many animals but appears less so in primates, although it is possible that their importance is underestimated, since olfactory receptors and pheromones have been studied in less detail in humans than for example, sound or sound receptors.

Humans possess glands that produce potential pheromones, known as scent or apocrine glands. They are associated with hair follicles primarily in the underarm or the anogenital regions (pubic hair) and on the skin of the abdomen. The role of the hair in these locations may be to facilitate release of the pheromone into the atmosphere.

Detection of pheromones

Although pheromones can be considered as a special form of odorants (scents), their actions, effects and functions have similarities to those of hormones. They bind to a specific receptor which then activates an effector system, which initiates an action potential. They bind to specific sensory cells, the neurones, in the olfactory epithelium, which is located on the roof of the nasal cavities. The epithelium consists of three types of cells, basal, supporting and sensory cells (neurones). The neurones are bipolar, that is they possess a single dendrite, which extends from the cell body to the surface of the olfactory epithelium, and an axon that forms a synapse with a nerve that transfers information to the olfactory centre in the brain. The epithelium is covered with a thick layer of mucus, in which the pheromones dissolve. The mucus contains proteins that bind the pheromone(s) for delivery to the olfactory receptors and then to remove them once they have been detected.

Receptors are located on non-motile cilia that project from the dendrite of the neurone into the mucus layer. It is the cilia that possess the receptor for the pheromone and respond to it via an effector system that results in opening of a Na^+ ion channel. This depolarises the membrane across the sensory cell which, if of sufficient magnitude, leads to generation of action potential along the axon with which it forms a synapse. The effector system is adenyl cyclase and the generation of cyclic AMP, a process that involves the G-protein (Figure 12.15).

The cyclase produces cAMP which results in opening of a Na^+ ion channel in the membrane of the sensory cell. If a sufficient number of Na^+ ions enter, this depolarises the membrane and initiates an action potential along the axon to the olfactory nerve. Further effects depend upon interaction between the nerves and synapses within the olfactory centre in the brain. This can result in physiological effects in other parts of the body which define the function of the pheromone. The effects of pheromones on the sexual responses of men and women are discussed in Chapter 19 (see Figure 19.17).

Box 12.1 History of the events leading to the discovery of insulin

In 1889 von Mering & Minkowski showed that removal of the pancreas of a dog rapidly resulted in diabetes. Soon after, Edward Schäfer proposed that pathological changes in the part of the pancreas known as the Islets of Langerhans were responsible for the diabetes. In 1908, George Zuelzer, a Berlin physician, found that if pancreatic extract was injected intravenously into a dog that had had its pancreas removed, the sugar level in the urine could be reduced provided the injections were given daily. Zuelzer then administered the extract to eight diabetic patients which improved their condition but the extract ran out and impurities resulted in fever. Ernest Scott, in the Department of Physiology, University of Chicago, used alcohol to inhibit the digestive enzymes in a pancreatic extract from which he obtained material that also decreased sugar excretion in diabetic dogs. Lack of a suitable assay prevented him carrying out measurements on blood glucose. Scott submitted his research for a higher degree but did not have the opportunity to continue the work.

In the late 1890s and early 1900s, Nicolas Paulesco in Bucharest also prepared extracts of the pancreas which he injected into diabetic dogs. Paulesco reported the successful isolation of a hormone, which he claimed was anti-diabetic, in a paper in *Comptes Rendus de la Société de Biologie*. In this and a subsequent paper, he described how this new hormone, 'pancreine', dramatically lowered the blood glucose level in normal and diabetic dogs when given intravenously. Further purification of the extract was, however, required before injections could be given subcutaneously in patients; before this was achieved, the Canadians had won the race!

After returning from service in the armed forces in the First World War, Frederick Grant Banting practised, in London, Ontario, as an orthopaedic surgeon and as a demonstrator at the University. In October 1920 he read a review article which suggested that a substance secreted by the pancreas might be capable of alleviating diabetes. He approached Professor John Macleod of the University of Toronto who provided not only facilities in his Department and advice but the services of a final year biochemistry student, Charles Best. Banting ligated the pancreatic duct in order to cause atrophy of the exocrine pancreas and then made an aqueous extract of the gland. Best had access to a new sensitive assay for glucose, with which he showed that the injection of this extract into diabetic dogs dramatically lowered their blood glucose levels, reduced their glycosuria and, also, stimulated glycogen synthesis in the liver of a rabbit. Macleod suggested that an ethanol extraction procedure, used earlier by Scott, might provide better material. By Christmas of 1921 an extract was available and Banting and Best injected each other with no untoward effects. In January 1922, Leonard Thompson, a 12-year-old, who was dangerously ill with diabetes and not expected to live much longer, received the first injection but it caused so much local irritation that treatment was stopped. It was clear that the extract had to be improved. Working in Macleod's department, at the time, as a Rockefeller Fellow, was Professor J.B. Collip of the University of Edmonton. By refining the ethanol extraction procedure, he was able to obtain a much purer preparation of the protein. When this was given to Leonard Thompson, only a few days after the original injections had been withdrawn, there was a dramatic effect; he was restored to excellent health in a very short time. Mass production of insulin was put in the hands of the Eli Lilly Company and within a year they were producing enough insulin to satisfy the requirements of all the diabetics in North America.

In 1925 Banting and Macleod were awarded the Nobel Prize for Medicine. Banting shared his award with Best and Macleod shared his with Collip. (See also page 253.)

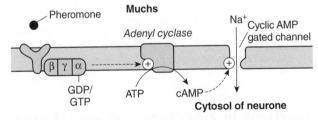

Figure 12.15 *Effector mechanism for a pheromone.* The mechanism is similar to that shown in Figure 12.6 except cyclic AMP leads to opening of a Na^+ ion channel which will result in depolarisation and initiation of an action potential. It is of interest that the physiology/biochemistry is opposite to that involved in the detection of light, in which the signalling system results in hyperpolarisation of a nerve ratter that hypopolarisation (Chapter 15: see Figure 15.10).

The presence of pheromones in humans is indicated by the following:

- Blind individuals can often recognise people by their scent.

- It is claimed that some individuals can recognise whether a garment has been worn by a male or female.

- Couples can discriminate between their own and their partner's natural scent and also that of a stranger.

- Mothers can recognise the natural scent of their own infant and vice versa.

- Women living in close contact (e.g. sleeping in the same dormitory) often synchronise their menstrual cycles.

- The enormous size of the perfume industry suggests a relationship between olfaction and, hence pheromones, and social activity and sexual stimulation.

Kinetic principles that apply to hormone action

Although many hormones are involved in control of many biochemical and physiological processes, the kinetic principles underlying the mechanisms by which this is achieved are similar if not identical for each hormone. These principles are outlined under four separate headings.

- Factors that affect the formation of the hormone–receptor complex.

- The reactions in the pathway or process that are affected by the hormones.

- The kinetic regulation of the change in the concentration of the messenger molecule in relation to the change in the activity of the effector system.

- The role that G-proteins play in the action of some hormones.

Formation of the hormone–receptor complex

The hormone interacts with its receptor by equilibrium binding. The dissociation constant is k_2/k_1.

$$H + R \underset{k_2}{\overset{k_1}{\rightleftarrows}} HR \overset{k_3}{\longrightarrow} Effector \rightarrow\rightarrow cellular\ response$$

where H is the hormone and R is the receptor, and k_3 governs the response of the effector to the hormone – receptor complex (HR).

An increase in the plasma concentration of the hormone increases its binding to the receptor in a hyperbolic manner so that, at a high concentration of hormone, the receptor approaches saturation with the hormone. However, in vivo,

This, at one time, was described as 'spare receptors' but this is a misinterpretation of the role of the relationship. The response of the effector system is most sensitive to the change in hormone concentration when it is below the value of the dissociation complex. This principle is the same as that for the binding of a regulatory molecule to an enzyme (Chapter 3).

the number of receptors is much greater than required to bind all the hormone molecules so that, under physiological conditions, the binding sites never approach saturation with hormone. The concentration of the hormone–receptor complex, which governs the magnitude of the cellular response, can be increased by four factors.

(i)   An increase in receptor number. If the receptor is a cell surface receptor, the number is known as receptor density.

(ii) A decrease in the dissociation constant of the receptor for the hormone.

(iii) An increase in the concentration of the hormone. A rapid effect of the hormone on the biochemistry in the cell requires that the hormone has a short half-life, so that variations in the plasma concentration can occur quickly. The half-lives for insulin, adrenaline, triiodothyronine and cortisol, for example, are minutes rather than hours.

(iv) For the same concentration of hormone, a change in the response of the effector system (the rate constant k_3, which can include the G-protein coupling system – see below) can change the magnitude of the effect of the hormone on the target system (known as a post-receptor effect).

These principles are similar to those that govern the relationship between an enzyme and its catalytic activity. For the hormone, R is equivalent to the enzyme, H to the substrate, and hormone–receptor complex to the enzyme–substrate complex. The activity of the substrate effector system is similar to the transition state. The cellular response to the hormone is similar to the catalytic role of the enzyme in the cell (Chapter 3),

$$E + S \underset{k_2}{\overset{k_1}{\rightleftarrows}} ES \overset{k_3}{\longrightarrow} P \longrightarrow cellular\ response$$

For an enzyme, the factors that can increase the catalytic activity and hence the cellular response are similar to above:

(i)   The concentration (number) of enzyme molecules.

(ii) A decrease in the K_m.

(iii) An increase in the concentration of substrate.

(iv) An increase in the rate constant k_3 (governed by the number of molecules in the transition state).

Reactions in a pathway that are affected by hormones

The role of many hormones, and particularly those discussed in this chapter, is to change the flux through a biochemical pathway or process. In order to identify the mechanism by which the hormone affects the biochemistry in the cell, it is necessary to know or to predict which reaction(s) in the biochemical process is changed by the hormone (or its intracellular messenger). This question is answered by reference to the kinetic structure of the process and how the transmission of a steady-state flux can be modified (as described in Chapter 3, in discussion of how a pathway or process is regulated). In a simple process, the activity of the enzyme or process that catalyses the flux-generating step is likely to be a site at which hormones will have an effect. Three examples are given (see Chapter 3, for discussion of flux-generating steps):

(i) The three hormones – adrenaline, glucagon or vasopressin – increase the rate of glycogen breakdown in the liver by increasing the activity of hepatic phosphorylase, the flux-generating step for this process.

(ii) Insulin decreases the rate of fatty acid mobilisation from adipose tissue by inhibiting the activity of triacylglycerol lipase, the flux-generating step for this process. Adrenaline and growth hormone increase the rate by stimulating the activity of the lipase (Chapter 7).

(iii) Cortisol increases protein degradation in muscle. The flux-generating step is protein degradation by one of the proteolytic systems in muscle, probably the proteasome system (Chapter 8).

Maintenance of the steady-state flux in a pathway or process that is affected by a hormone

As discussed above, to change the flux through a pathway or process, hormones must change the activity of the enzyme that catalyses the flux-generating step but, in order to maintain a steady-state flux and maintain the concentrations of intermediates in the process as constant as possible, a hormone may also change the activity of an enzyme that catalyses another reaction further along the process (i.e. downstream of the flux-generating step). This will then ensure a 'smooth' and coordinated change of flux (Figure 12.16). Two examples are given.

(i) Insulin increases the rate of peptide (protein) synthesis in muscle by stimulation of three reactions: (a) transport of amino acids into the cell; (b) the process of initiation; and (c) elongation of the peptide chain (Figure 12.17). (For details of peptide synthesis and its regulation, see Chapter 20).

(ii) Insulin increases the acute rate of glycogen synthesis in muscle by stimulation of two processes: (i) glucose transport into the cell; and (ii) the activity of glycogen synthase (Figure 12.18).

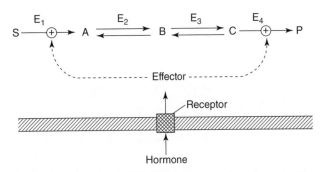

Figure 12.16 *Multiple sites of hormone effects on the same biochemical process.* The hormone binds to its receptor activating the effector system which increases the activity of two separate reactions in the same biochemical pathway (process) to increase flux through the pathway. This means that the flux can change without large changes in the concentrations of intermediates in the pathway, i.e. activation of E_1 and E_4 ensures increased flux from S to the product P with little change in the concentrations of A, B or C.

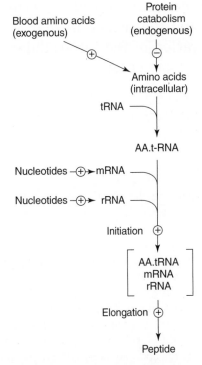

Figure 12.17 *Sites at which insulin stimulates protein synthesis in a muscle.* The sites are indicated by $\oplus$. Insulin has its anabolic effect on protein synthesis in muscle by affecting six processes or reactions: (i) it inhibits protein degradation in the muscle; (ii) it stimulates amino acid transport from the blood into the muscle; (iii) it stimulates the initiation-reaction of the pathway for protein synthesis, i.e. formation of the complex (tRNA-amino acid-mRNA-ribosomal RNA); (iv) it increases the rate of mRNA synthesis, and therefore the number of mRNA molecules; (v) it stimulates ribosomal RNA synthesis; (vi) it stimulates elongation of the peptide (see Chapter 20).

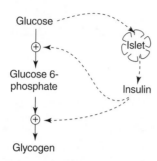

Figure 12.18 *Sites at which insulin stimulates glycogen synthesis in muscle.* An increase in the blood glucose level, after a meal, increases secretion of insulin from the β-cells in the Islets of Langerhans. Insulin increases the transport of glucose into the muscle fibre and the activity of glycogen synthase (Chapter 6). The result is that insulin increases the rate of glycogen synthesis without marked changes in concentrations of glucose 6-phosphate, glucose 1-phosphate or UDP-glucose in the liver.

The quantitative relationship between the change in activity of the effector system and the change in the concentration of the intracellular messenger molecule

The principle of this relationship is discussed in Chapter 3. It is repeated here since it is relevant to control by hormones and several examples are given. The relation is formalised as follows:

$$S \xrightarrow{E_1} X \xrightarrow{E_2} B$$

where E_1 represents the effector system, X represents the intracellular messenger, E_2 is the step that metabolises (inactivates) or removes X, and compound B is further metabolised.

The kinetics of this relationship are straightforward: the effector system (enzyme activity) is a zero order process (i.e. the substrate S saturates the enzyme E_1 whereas the inactivation reaction E_2 is a first order process. The consequences of the kinetics of this system is that the magnitude of the change in E_1 results in precisely the same quantitative change in the concentration of X. For example a fivefold increase in E_1 produces a fivefold increase in the concentration of X. This relationship is shown in Box 12.2. Three examples are given.

1. Cyclic AMP

ATP is converted to cyclic AMP catalysed by adenyl cyclase (E_1) and cyclic AMP is degraded to AMP by a phosphodiesterase (E_2)

$$ATP \xrightarrow{E_1} \text{cyclic AMP} \xrightarrow{E_2} AMP \longrightarrow$$

Box 12.2 Theoretical analysis of the quantitative changes in the steady-state concentration of a messenger molecule that are the result of changes in the activity of a zero order process

The process is described as follows

$$S \xrightarrow{E_1} X \xrightarrow{E_2} B \longrightarrow$$

E_1 is a zero order process whose activity can be changed. X is the messenger molecule. Reaction (E_2) is catalysed by an enzyme that obeys Michaelis Menten kinetics, the K_m of which for X is assumed to be 0.1 arbitrary units.

Activity of E_1	Concentration of X^a	Concentration of X^b
1	1.0	0.1
2	2.0	0.2
5	5.0	0.5
10	11.0	1.0
50	100.0	5.3

^{a}The maximum activity of E_2 is 100-fold greater than E_1.
^{b}The maximum activity of E_2 is 1,000-fold greater than E_1.

The equation governing this situation is

$$[X] = \frac{V_1 \times K_m}{V_2 - V_1}$$

where V_1 and V_2 are the maximal activities of enzymes catalysing reactions E_1 and E_2. The reaction system is considered to be similar to that controlling the intracellular concentration of the messenger molecules, cyclic AMP and cyclic GMP, cytosolic Ca^{2+} ion, neurotransmitter in synapses. In the cyclic ATP system, reaction 1 represents the adenyl cyclase reaction, reaction 2 represents the phosphodiesterase reaction and X represents cyclic AMP. This principle is the basis for many messenger systems, see, for example, Chapter 14: Figures 14.13 and 14.14).

The adenyl cyclase reaction is zero order, that is, it is saturated with ATP. The phosphodiesterase reaction is first order, at least at the concentrations of cyclic AMP normally found in the cell, and the maximum activity is greater than that of adenyl cyclase.

2. Cyclic GMP

Similarly, cyclic GMP is generated from GTP by the enzyme guanyl cyclase and is converted (inactivated) by a phosphodiesterase to GMP and the same principle applies, since the cyclase catalyses a zero order process and phosphodiesterase a first order process.

$$GTP \xrightarrow{\quad E_1 \quad} cyclic\ GMP \xrightarrow{\quad E_2 \quad} GMP$$

A further characteristic of this principle is that, if the activity of phosphodiesterase is decreased, the concentration of cyclic GMP will increase to an extent dependent upon the extent of the decrease in activity. This characteristic has been made use of by the pharmaceutical industry. Cyclic GMP has a vasodilatory effect and this is the case for the arterioles that supply blood to the corpus cavernosum in the penis, which controls the erection of the penis. Drugs were developed (e.g. sildenafil) that inhibits cyclic GMP phosphodiesterase and hence increases the cyclic GMP level which results in vasodilation of the arterioles and an increase in the supply of blood to the spongy tissue of the corpus cavernosum, which expands resulting in erection. This drug has been found to be effective in some patients suffering from erectile dysfunction. This can be a particular problem in diabetic patients and more elderly men (Chapter 19).

3. Inositol trisphosphate (IP$_3$)

Phosphatidylinositol bisphosphate (PIP$_2$) is a component of the cell membrane which is hydrolysed by a phospholipase to produce inositol trisphosphate (IP$_3$).

$$PIP_2 \longrightarrow IP_3 \longrightarrow IP_2 \longrightarrow IP \longrightarrow inositol$$

Enzyme E$_1$ is the phospholipase A$_2$, for which there is an excess of substrate in the plasma membrane i.e. a zero order process. (For details of this process, see Chapter 11). E$_2$ is a phosphatase, which catalyses a first order process. In fact, IP$_2$ can be hydrolysed to produce IP$_1$ which is further hydrolysed to produce free inositol. The latter is 'salvaged' by using it to re-form phosphatidylinositol in the phospholipid synthetic pathway and then phosphorylated to produce PIP$_2$ (Chapter 11, Figure 11.21). These reactions are not just of biochemical interest but are involved in the treatment of bipolar disease a mental disorder.

The disorder is caused by an excessive concentration of a monoamine neurotransmitter in the synaptic cleft between neurones in some parts of the brain. It can be treated by decreasing the rate of neurotransmitter release from the presynaptic terminal into the synaptic cleft. This is achieved by exocytosis which is stimulated by an increase in Ca^{2+}-ion concentration, which is achieved by electrical activity in the presynatic neurone that opens Ca^{2+}-ion channels (Chapter 14, page 315). Another means of increasing the Ca^{2+}-ion concentration is via the effect of IP$_3$ on Ca^{2+}-ion channels is the entoplasmic reticulum (see Figure 12.5). This provides a means of treating the disorder. Lithium ions inhibits hydrolysis of IP$_2$ and IP$_1$, so inositol is not produced. This reduces the amount of PIP$_2$. (Figure 12.19).

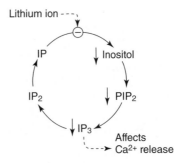

Figure 12.19 *Phosphatidylinositol bisphosphate cycle and treatment of bipolar disease.* The metal ion lithium inhibits inositol monophosphate phosphatases and, therefore, inhibits the flux from IP$_3$ to inositol, so that the concentration of the latter decreases. This can restrict formation of phosphatidylinositol the bisphosphate (PIP$_2$) so that the amount in the membrane decreases and the phospholipase no longer catalyses a zero order reaction. The extent of the decrease in the IP$_3$ concentration will depend on how far the process is removed from zero order. This may explain the well-known variability in the response of patients to lithium which is probably dependent on the patient taking the precise dose of the drug (Chapter 14).

If the amount of phosphatidylinositol bisphosphate in the membrane is reduced, so that it is no longer saturating for the phospholipase (i.e. the process is no longer zero order), the change in concentration of IP$_3$ is no longer quantitatively related to the change in the activity of the phospholipase so that the concentration is decreased to an extent dependent on the reduction of the amount of PIP$_2$ in the membrane.

In summary, the release of neurotransmitter from a presynaptic neurone into a synaptic cleft occurs via the process of exocytosis, which is regulated by the increase in Ca^{2+} ion concentration in the presynaptic terminal. The increase in Ca^{2+} ion concentration is achieved by release of Ca^{2+} ions by opening of the Ca^{2+} ion channel in the endoplasmic reticulum, which is controlled by the concentration of IP$_3$. Failure to release inositol from the inositol phosphates reduces the free inositol concentration, which interferes in the synthesis of PIP$_2$. The phospholipase no longer catalyses a zero order reaction. Consequently, sufficient IP$_3$ to activate the ion channel is released in the presynaptic neurone, so that less neurotransmitter is released into the synaptic cleft (Figure 12.19).

G-proteins and hormone receptor–effector systems

The G-proteins are a family of guanine nucleotide (GTP or GDP) binding proteins that are a component of several hundred hormone-effector systems or ligand-binding

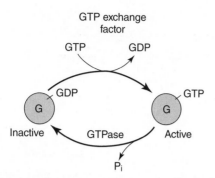

Figure 12.20 *The G-protein activation/inactivation cycle.* When the G-protein is associated with GTP, it is active: when it is associated with GDP it is inactive. The hormone–receptor complex is the GTP exchange factor, which exchanges GDP for GTP to convert the inactive form to the active form. A GTPase activity inactivates the G-protein by hydrolysing GTP.

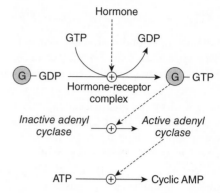

Figure 12.21 *Effect of the hormone–receptor complex on activation of the G-protein and the resultant effect on the activation of adenyl cyclase.* The hormone bind to the receptor to produce the hormone–receptor complex that activates to G-protein.

systems. When GTP is bound, the G-protein is active but, when GDP is bound, it is inactive (Figure 12.20). The activation step is the exchange of bound GDP by GTP. The hormone–receptor complex is, in fact, a GTP–GDP exchange system. Inactivation is achieved by GTP hydrolysis, which is catalysed by a GTPase activity. In fact, the GTPase activity is an intrinsic activity of the G-protein. For the action of some hormones the G-protein links the hormone receptor complex to adenyl cyclase, as shown in Figure 12.6. The relationship is that the hormone–receptor complex acts as the GTP exchange factor and converts inactive G-protein to active G-protein. The latter then activates adenyl cyclase and cyclic AMP is formed from ATP (Figure 12.21). This activation will last until the intrinsic GTPase activity is greater than that of the GDP–GTP exchange, i.e. the hormone–receptor complex. As the concentration of hormone is reduced (caused by a lower rate of secretion from the gland), the hormone dissociates from the receptor, and exchange of GDP/GTP fails to take place, so that the GTPase activity dominates and hence the G-protein activity and, therefore, adenyl cyclase activity is decreased and the cyclic AMP concentration falls (Figure 12.21) (Box 12.3).

The 'story' of the G-proteins is extended further. The G-protein in such processes is a trimer, consisting of three subunits, α-, β- and γ-units. In the inactive state, the G-protein is in the form of a trimer but, when it is activated, by exchange of GDP with GTP, the trimer dissociates and it is the free form of the α-subunit that is active. Once the GTP is converted to GDP, via the GTPase activity, the subunits re-associate (Figure 12.22). Presumably, the formation of the trimer plays a role in the inactivation of the α-subunit. By reference to the information provided in Chapter 3, it should be clear that the active–inactive inter-

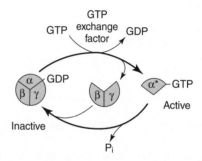

Figure 12.22 *Dissociation/association of G-proteins associated with GTP/GDP exchange.* The G-protein is trimeric when GDP is bound. Conversion of GDP to GTP allows the α-subunit to dissociate from trimer. The 'free' α-subunit is now biochemically active and it migrates along the inner surface of the cell membrane to activate adenyl cyclase. Hydrolysis of GTP to GDP allows the subunits to re-associate and the α-subunit is now inactive (see Figures 12.6 and 12.15).

change of the G-protein is equivalent to a reversible phosphorylation system, i.e. an interconversion cycle.

The G-protein system as an interconvertible enzyme cycle

The principle underlying the changes in activity of a G-protein is similar to that of an interconversion cycle (Chapter 3). The 'classic' example of an interconversion cycle is that between the two forms of the enzyme phosphorylase phosphorylase *a* and *b*. The interconversions between *b* and *a* are catalysed by a protein kinase and a protein phosphatase. The similarities are as follows.

- The enzyme phosphorylase is activated by phosphorylation. The exchange of GDP by GTP is equivalent to a phosphorylation.

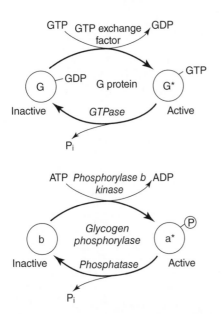

Figure 12.23 *Comparison of the G-protein cycle and the phosphorylase interconversion cycle.* The comparison should help to explain the principle underlying the concept of the interconversion cycle and to understand more readily the G-protein system.

- The active forms of both proteins are inactivated by dephosphorylation reactions: the active form of phosphorylase is inactivated by a protein phosphatase; the active G-protein is inactivated by the GTPase, which is a phosphatase. A comparison between the two is made in Figure 12.23.

Box 12.3 Bacterial toxins and G-proteins

Some bacterial toxins have their effects on humans by interfering with the activity of the G-protein, that is G-proteins are involved in the action of some toxins, e.g. cholera toxin and pertussis toxin. Cholera toxin is a protein (mol. mass 87 kDa) consisting of A and B subunits, produced by the bacterium *Vibrio cholerae*. Subunit B binds to the plasma membrane of an intestinal cell and then the A subunit enters the cell. This subunit possesses ADP-ribosyltransferase activity, an activity which transfers the ADP-ribose from NAD^+ to the α-subunit of G-protein,

$$\alpha\text{-subunit} + NAD^+ \rightarrow \alpha\text{-ribose-ADP} + nicotinamide$$

ADP ribosylation results in inhibition of GTPase activity and hence maintains the α-subunit in the active form. The constant activity of the G-protein results in an increase in adenyl cyclase activity and therefore a chronic increase in the cyclic AMP level. This stimulates an ion channel in the enterocyte which results in a loss of Na^+ ions and hence water from the cells into the intestine. This leads to diarrhoea and a massive loss of fluid from the body which can be sufficiently severe to result in death. Since 2000 there have been epidemics in South America and parts of central Africa. Infection is usually caused by drinking water contaminated with faecal matter. Treatment consists of hydration with rehydration fluids (Chapter 5).

Pertussis toxin is a protein secreted by the bacterium *Bordetella pertussis* which causes whooping cough. The toxin enters a cell where it also catalyses ADP-ribosylation of the α-subunit of G-protein.

IV ESSENTIAL PROCESSES OF LIFE

13
Physical Activity: In Non-Athletes, Athletes and Patients

In my opinion exercise physiology is from these viewpoints particularly important because an exercise situation in various environments provides a unique opportunity to study how different functions are regulated and integrated. In fact, most functions and structures are in one way or another affected by acute and chronic (i.e., in a training program) exercise. Therefore, exercise physiology is to a high degree an integrated science that has as its goal the identification of the mechanisms of overall bodily function and its regulation. It is regrettable that so few pages in standard textbooks of physiology for medical students are devoted to discussions of the effects of exercise on different functions and structures. That may explain why a majority of physicians do not recommend regular physical activity!

(Ästrand, 1997)

Football is the most popular sport in the world with approximately 40 million organised players (and c.100–150 million participants in total) ... In football maximum potential needs to be activated in aerobic capacity, anaerobic alactacid power, aerobic regenerative capacity ... repair mechanisms, creativeness, intelligence and personality.

(Ekblom, 1994)

The term post-operative fatigue describes the state of overall drowsiness felt by patients who have undergone surgery: it can last for many days after the operation.

(McGuire *et al.,* 2003)

Movement is an essential component of life. If something moves, it is assumed to be alive. External movement depends upon skeletal muscle whereas internal movement depends upon smooth and cardiac muscles. The latter muscles are discussed in other chapters. The term 'physical activity' is another term for external movement. Hence, it depends on skeletal muscle. This chapter contains a discussion of the biochemistry of skeletal muscle, including its involvement in everyday activities. This leads into the biochemistry underlying both the benefits and hazards of physical activity. In addition, the biochemical bases of the common problem of fatigue and the debilitating condition of chronic fatigue are discussed.

The mechanical basis of movement by skeletal muscle

When a force acts on a rigid bar on one side of a fulcrum, a load can be moved on the other side of the fulcrum; this arrangement is known as a lever. Muscle and bone form a lever: the muscle generates the force, the fulcrum is the joint and the rigid bar is the bone (Figure 13.1). It is the process of shortening of a muscle, also termed contraction, that generates the force. However, for movement to occur, the process of contraction must be repeated in a cyclic fashion to provide for movement so that a muscle must lengthen as well as shorten. When a muscle is relaxed (i.e. not contracting) it can be extended (lengthened) by some force, usually due to the operation of an opposing muscle. In the upper arm, for example, the biceps extend the triceps and the triceps extend the biceps (Figure 13.2). Coordination between these muscles is essential to ensure smooth movement. Failure can result in one muscle tearing the other.

Contraction does not necessarily imply shortening but refers only to activation of the force-generating process. In order for shortening to occur, the force generated by a muscle must be greater than the force opposing the shortening. If the two forces are equal, there will be an increase

Functional Biochemistry in Health and Disease by Eric Newsholme and Tony Leech
© 2010 John Wiley & Sons Ltd

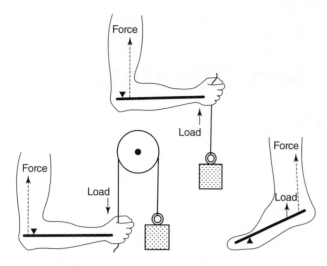

Figure 13.1 *Muscle and bone as a lever.* The figure shows the arm and leg as levers. The triangle (▼) indicates the position of the fulcrum.

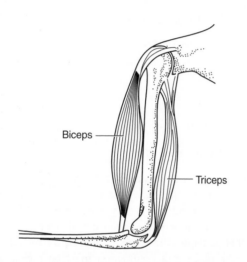

Figure 13.2 *Antagonistic muscles: their role in movement.* The action of antagonist muscles is exemplified by the actions involved in delivering a punch. The arm is first flexed by contraction of the biceps muscle and simultaneous relaxation of the triceps muscle. The arm is then rapidly extended by contraction of the triceps muscle and relaxation of the biceps muscle. Such movements involve considerable nervous coordination.

in tension without shortening and the contraction will be isometric. It is also possible to increase the length despite development of tension, a situation known as 'eccentric contraction' (Table 13.1). The body contains approximately 600 skeletal muscles that comprise about 40% of normal body weight. Nonetheless it must be appreciated that muscle plays roles other than movement. These include the supply of amino acids by breakdown of muscle protein in some conditions to provide fuels for other tissues and

Table 13.1 Different muscle actions

Type of exercise	Description of action or 'contraction'	Change in length
Dynamic	Concentric	Decrease
	Eccentric	Increase
Static	Isometric	None

Eccentric contraction can most readily be appreciated by holding a weight in one hand close to the shoulder with the arm bent and lowering the arm from the elbow; the biceps are developing tension to hold the weight but increasing in length as the arm is lowered.

synthetic processes (including glucose formation during starvation). Its ability to take up considerable amounts of glucose plays a significant role in the regulation of the blood glucose concentration after a meal. It also synthesises and stores glutamine, an important fuel for the immune and other cells.

The versatility of muscle and the power output that can be achieved is demonstrated by athletic feats performed by many animals. The migratory bar-tailed godwit flies almost 6500 miles non-stop in 7 days, the monarch butterfly can fly 2000 miles at an average speed of 75 miles per day and mature eels make the epic journey from the Sargasso Sea to the rivers of Europe to spawn – and back again to die. Sprinters are also found in the animal world. A cheetah can accelerate to 45 miles per hour in 2 seconds and reach 70 miles per hour during the chase. Although all these athletic performances far outstrip those of humans, it must not be forgotten that humans are all-purpose animals. Individuals of this species have sprinted 100 m in under 10 seconds, run 5110 miles in less than 107 days (Max Telford ran from Anchorage, Alaska, to Halifax, Nova Scotia, in 1977), flown for 1 mile (with the aid of artificial wings but no additional power) and swum 21 miles non-stop across the English Channel. In addition, many human activities, such as playing a musical instrument, typing a manuscript or striking a moving ball, demand great precision of movement and an extraordinary coordination between muscles and the brain, which is achieved by no other single species.

Structure of muscle

A muscle consists of groups of muscle bundles that join into a tendon at each end. The muscle bundles in the quadriceps are the vastus medialis, rectus femoris, vastus intermedialis and vastus lateralis. Each bundle is separately wrapped in a sheath of connective tissue. Each muscle is composed of many fibres, packaged into bundles of about

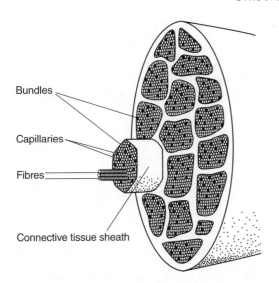

Bundles

Capillaries

Fibres

Connective tissue sheath

Figure 13.3 *The arrangement of capillaries, fibres and bundles in a muscle.*

Muscles differ widely in both their physiological properties (e.g. speed of contraction) and their metabolic properties (e.g. oxidative or glycolytic capacities). Originally, muscles were distinguished by the depth of their red colour, caused by the contents of myoglobin and cytochromes. Red muscles contain more myoglobin, which binds oxygen in the cell, and more cytochromes indicating that ATP is generated largely from oxidation. White muscle contains much less myoglobin and cytochromes and relies on glycolysis for ATP generation. Redness, however, is a poor indicator of the biochemical and physiological characteristics, although the terms red and white are still occasionally used. Since muscles contain fibres that differ physiologically and biochemically, it is more useful to classify fibres rather than whole muscles.

Fibre types

Fibres in human skeletal muscle are classified into two types: I and II. The Type I fibres are oxidative and generate force more slowly than type II fibres. Type II fibres have a lower oxidative capacity (relying more on glycolysis to generate ATP), a higher myosin ATPase activity and a greater power output. Type II fibres can be further subdivided into type IIA and type IIB. Type IIA are somewhat similar to type I and generate most of their ATP from oxidative processes but glycolytic capacity is still moderate. In contrast, type IIB fibres generate ATP almost exclusively from glycolysis and have the highest power output of any fibre although this can be maintained only for short periods (i.e. explosive muscles) (Table 13.2). Fibre types can be identified by histochemical methods performed on a biopsy sample (Appendix 13.1). Microtechniques allow individual fibres to be isolated and studied so that, for example, fuel contents and enzyme activities can be measured. The different fibres have different properties, and different proportions are present in different individuals. The biochemical and physiological characteristics of the different fibres explain why some individuals are good sprinters whereas others are good endurance athletes. These differences are most marked in

> In a strap muscle there is little connective tissue within the muscle since most is concentrated in tendons at the end. Such muscles produce the tenderest meat: fillet steak is the psoas muscle of a cow. Pennate muscles contain more connective tissue within the muscle itself and hence form the cheaper cuts of meat.

1000 fibres (Figure 13.3). In some muscles (e.g. the psoas) the bundles run parallel to the long axis of the muscle. These are known as strap muscles and they contract quickly. In the muscles of the lower limbs and trunk, the bundles are arranged in a chevron fashion, oriented at an angle to the length of the muscle. These muscles are known as pennate. They develop greater tension but they contract more slowly than strap muscles.

Long, cylindrical fibres within the bundles correspond to the cells of the other tissues. These muscle fibres vary enormously in length from about 0.4 to 10–15 cm and in diameter from about 0.01 to 0.1 mm. Each fibre contains mitochondria and many nuclei (since it develops from a number of cells). It is surrounded by a plasma membrane (the sarcolemma).

A fibre is packed with longitudinally arranged myofibrils that contract. This is the contractile unit of the fibre; it is 1–3 μm thick in diameter. Each myofibril consists of about 1000–2000 filaments, which are known as myofilaments.

Muscles are attached to bone by tendons: tough, strong strands of connective tissue which transmit the force from muscle to bone. In addition, because of their elastic properties, they convert some kinetic energy into potential energy which is then used during the next stroke. This reduces the amount of ATP and hence the amount of fuel required to power running (Box 13.1).

> The maximum rate of glycogenolysis in type II fibres, as measured by difference in glycogen levels, is identical to the maximum activity of the enzyme phosphorylase measured in extracts of the same individual type II fibres in vitro. The maximum in vitro activity of this enzyme provides an excellent indication of the maximal capacity for glycogen breakdown even in individual fibres.

> If your muscles have an average fibre composition, there is no chance of you winning an Olympic gold medal in the marathon or the 100 metre sprint. Although the percentage of type I and type II fibres is largely inherited, training can result in type IIc fibres, which are uncommitted fibres, being converted to type IIA or type IIB according to the type of training.

Box 13.1 The Achilles tendon: springs in the leg

Much of the energetic cost of running on the flat is expended against the pull of gravity since, at each step, the centre of gravity of the body is raised. Not all energy is lost, however. The transfer of potential into kinetic energy is the basis of the children's toy, the pogo-stick. Similarly, the elastic components of the leg allow some of the energy used to raise the body's centre of gravity to be stored as potential energy and transferred back to kinetic energy for use in the next stride. The Achilles tendon is the main spring, stretching by about 1.5 cm (6% of its length), as the foot hits the ground. Some of the kinetic energy in the stride is transferred to potential energy in the tendon. More than 90% of this potential energy is also transferred back to kinetic energy in the next stride. The arch of the foot is another spring. As it flattens, it transfers some 15% of the kinetic energy into potential energy which is transferred back to kinetic energy in the next stride.

Kangaroos put the pogo-stick principle to particularly effective use and elite athletes are much more 'kangaroo-like' than non-elite (see Figure 13.B). It has been calculated that is to these springs the effect of raise the efficiency of conversion of the chemical energy into mechanical energy from about 25% to more than 40%, a remarkable increase. It is unclear if tendon elasticity can be improved by training. Unfortunately, with ageing, this energy-recycling property of the tendons is progressively lost.

It is this principle that explains why a rubber ball travels further by bouncing. It is well known, however, that a ball does not bounce well in a field of mud. Nor do elite track athletes. Steve Ovett,* a world record holder for the mile and 1500 m races, illustrated this point when he described his performance in a cross-country championship early in his career (1970):

The race was a disaster. It was pouring with rain all day and by the time my event started the course was churned into a quagmire and at one point looked like a river. I am a fluid runner and this was a course for sloggers ... It was desperately tough ... I think I finished twenty-seventh!

(Ovett & Rodda 1984)

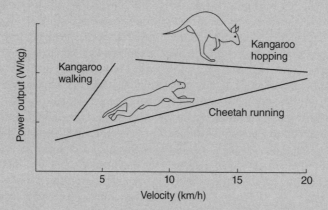

Figure 13.B In the hopping kangaroo, the effect of the long Achilles tendon is that faster locomotion does not use more ATP! Indeed, the faster the kangaroo hops the more efficient it becomes in relation to use of chemical energy. In contrast, the kangaroos when walking follows the usual response to a faster pace, more ATP is used. As the kangaroo lands, the tendon is stretched, converting kinetic energy into potential energy. The cheetah is also economical, possibly because of its flexible back and length of stride. (Goldspink, 1977)

---

*Steve Ovett was Olympic 800 metre champion in the Moscow Games in 1980 and held the world record for the 1500 m event from September 1983 until July 1985, and also that for the one mile in 1980 and 1981.

Table 13.2 Properties of human muscle fibre types and their capacities for fuel utilisation

Property	Type I	Type IIA	Type IIB
Myosin ATPase activity, after pre-incubation at pH 10.6	−	+++	++++
Myosin ATPase activity, after pre-incubation at pH 4.3	+++	−	−
V_{max} of myosin ATPase	low	high	high
Speed of contraction	slow	fast	fast
Capacity of pathway for glycogen conversion to lactic acid	moderate	moderate	high
Capacity of pathway for glycogen conversion to CO_2	high	moderate	low
Glycogen store	moderate–high	moderate–high	moderate–high
Triacylglycerol store	high	moderate	very low
Capacity to oxidise fatty acids	high	moderate	very low
Capillary density	high	moderate	low
Content of mitochondria	high	high–moderate	low

Table 13.3 Fibre composition of muscle in different types of athletes (see also Figure 13.4)

Data from Helge et al. (2000):	Percentage fibre composition		
	Type I	Type IIA	Type IIB
Sedentary controls	52	36	12
Track athletes			
800–1500 m	58	37	5
3000–10 000 m	64	33	3
Road/cross-country runners	73	27	0
Data from Noakes (1986):			
Sprinters	26	–	–
Weight lifters	44–49	–	–
Middle distance runners	45–52	–	–
Elite half-marathon runners	54	–	–
Elite oarsmen	60–90	–	–
Elite distance runners	79–88	–	–
Cross-country skiers	72–79	–	–

Type I is also known as slow twitch fibre. Type IIA is also known as *fast twitch*, and fibre type IIB is also known as *fast twitch*. Data are from world-class athletes and from fibres in the quadriceps muscle.

Noakes → did not separate type II fibres into A and B. Thus the muscles of sprinter comprise 74% type II fibres, probably primarily type IIB.

top-class (elite) athletes. Top-class sprinters have over 70% of type II fibres, whereas top-class long distance runners have a high proportion of type I fibres (Table 13.3 and Figure 13.4). A similar trend is seen in domestic animals: horses bred for racing have a very high proportion of type II fibres; the heavy hunters, which are bred for their stamina, have a high percentage of type I fibres.

Structure of the myofibril

Each fibre contains an array of parallel myofibrils; each consisting of overlapping thick and thin filaments that form repeating units (sarcomeres) along the length of the fibre (Figure 13.5). The thick filaments are composed almost entirely of the protein myosin, whereas the thin filaments contain actin as well as troponin and tropomyosin.

Cross-bridges are the heads of myosin molecules that extend from the surface of the thick filaments towards the thin filaments. During contraction, the cross-bridges connect with the thin filaments and, by changing their angle of contact, exert a force on the filaments. A cycle of these changes, driven by ATP hydrolysis, results in shortening of the myofibrils and hence contraction of the muscle fibres (Figure 13.6).

Proteins involved in muscle action

Myofibrils contain four major proteins: myosin, actin, tropomyosin and troponin.

Myosin

Myosin proteins are motor proteins and generate the force for movement. Myosin II is present in skeletal muscle where it forms filaments, binds to actin and hydrolyses ATP.

It is a large protein that consists of six polypeptide chains, two identical heavy chains (each 220 kDa) and two pairs of light chains (each ~20 kDa). The two heavy chains form a long coiled unit known as the tail, which associates laterally with other myosin molecules to form the thick filament. At one end of the chain are the two globular heads which form the cross-bridges. Each head has two domains, one on either side. One possesses the ATPase activity whereas the other possesses the actin-binding site.

> The long tail of myosin contains a high proportion of the amino acids leucine, isoleucine, aspartate and glutamate. These are released upon the degradation of myosin by intracellular proteases and peptidases and they provide nitrogen for the synthesis of glutamine. It is then stored in muscle and is a very important fuel for immune cells (Chapter 17).

In contrast, myosin I, which is not present in muscle, possesses only one head region and a short tail. Its role in cells may be involved in movement associated with membranes (endocytosis, phagocytosis).

Actin

In skeletal muscle, actin is highly organised into filaments. The actin molecules, known as G-actins (each 42 kDa) form a pair, known as F-actin, which polymerise to form the thin filaments in the myofibrils (Figure 13.6).

Tropomyosin

Tropomyosin is a long helical molecule (70 kDa) which extends along the long axis of the actin filament (Figure 13.7). Each tropomyosin molecule covers seven actin monomers and plays a central role in the regulation of muscle contraction.

Troponin

Troponin is present only in skeletal muscle; it is a complex of three monomeric proteins: troponin C, troponin I and

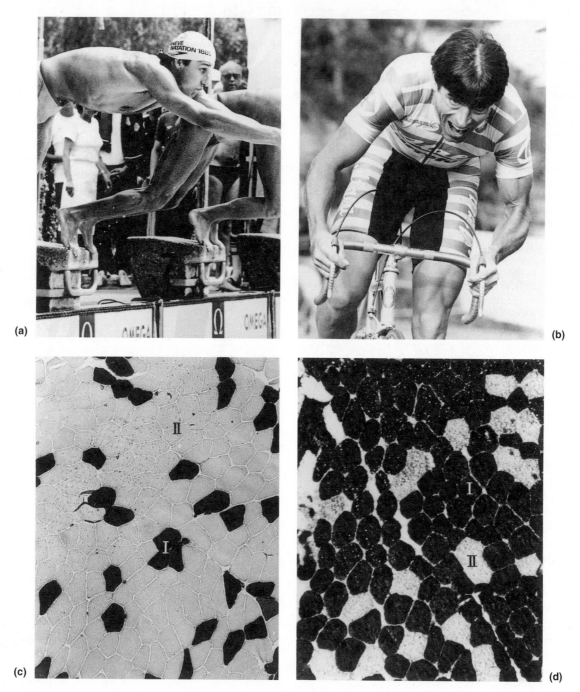

Figure 13.4 *Electron micrographs of the different fibres in different athletes.* The fibre composition (type I and type II) of two selected top athletes. **(a)** A swimmer, whose speciality is the 50 metre crawl sprint. **(b)** A professional world-class cyclist of the 'roller' type. **(c) and (d)** Cryostat sections of the swimmer's and cyclist's vastus lateralis stained for myosin ATPase, after preincubation at pH 4.3. Type I fibres stain dark, type II fibres remain unstained. **(c)** Almost all of the swimmer's fibres are type II. **(d)** Almost all of the cyclist's fibres are type I. Photographs kindly provided by Professor Hans Hoppeler, Department of Anatomy, University of Bern, Switzerland. Published in *Strength and Power in Sport*, ed. P.V. Komi, Blackwell Science (1992), pp.39–63.

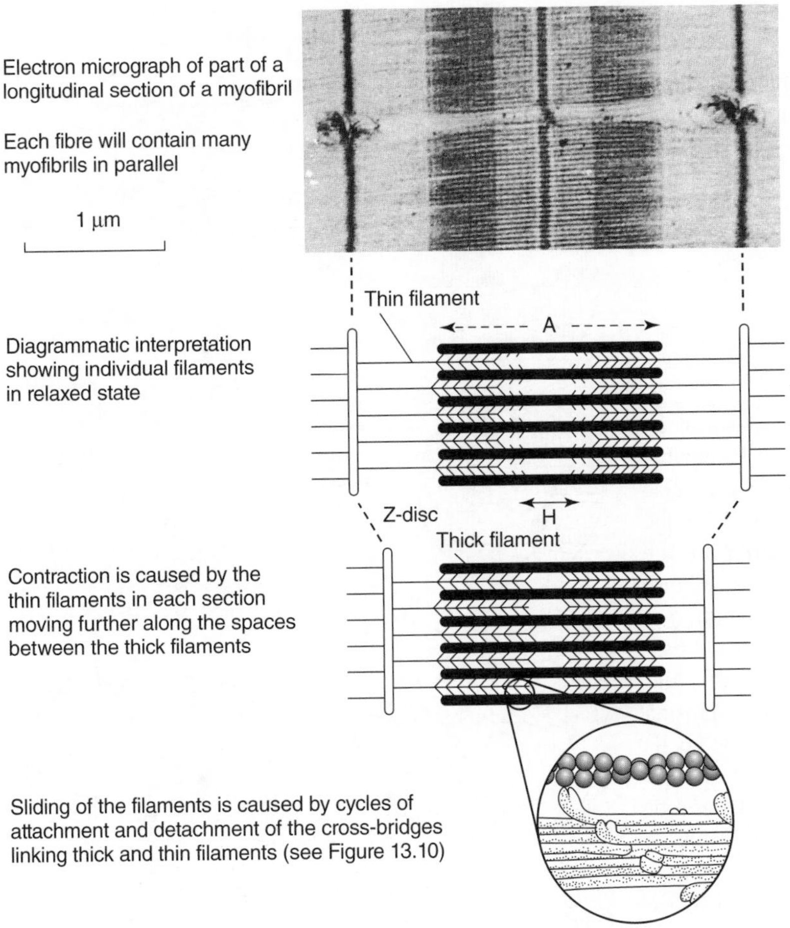

Electron micrograph of part of a
longitudinal section of a myofibril

Each fibre will contain many
myofibrils in parallel

1 µm

Diagrammatic interpretation
showing individual filaments
in relaxed state

Thin filament

A

Z-disc

H

Thick filament

Contraction is caused by the
thin filaments in each section
moving further along the spaces
between the thick filaments

Sliding of the filaments is caused by cycles of
attachment and detachment of the cross-bridges
linking thick and thin filaments (see Figure 13.10)

Figure 13.5 *Electron micrograph of part of a longitudinal section of a myofibril. Identification of components and the mechanism of contraction.* When a muscle fibre is stimulated to contract, the actin and myosin filaments react by sliding past each other but with no change in length of either myofilament. The thick myosin strands in the A band are relatively stationary, whereas the thin actin filaments, which are attached to the Z discs, extend further into the A band and may eventually obliterate the H band. Because the thin filaments are attached to Z discs, the discs are drawn toward each other, so that the sarcomeres, the distance between the adjacent Z-discs, are compressed, the myofibril is shortened, and contraction of the muscle occurs. Contraction, therefore, is not due to a shortening of either the actin or the myosin filaments but is due to an increase in the overlap between the filaments. The force is generated by millions of cross-bridges interacting with actin filaments (Fig. 13.6). The electron micrograph was kindly provided by Professor D.S. Smith.

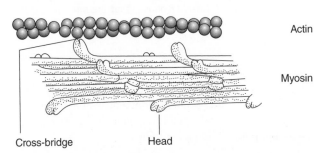

Cross-bridge   Head

Figure 13.6 *Myosin and actin molecules and myosin cross-bridges.* Each kind of filament is composed of a different protein: myosin in the thick filaments and actin in the thin filaments. In the case of actin, the individual F-actins are more or less spherical but a large number of these combine to produce a long chain, two of which wind around each other, rather like a rope, to produce the thin filament. The myosin molecule is more complex and shaped somewhat like a golf club. To form the thick filament, the shafts aggregate to leave the heads protruding on all sides. These heads form the cross-bridges and are responsible for 'pulling' the thin filaments into the spaces between the thick filaments (see Figure 13.5).

Actin

Myosin

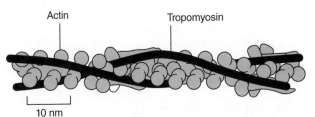

Actin   Tropomyosin

10 nm

Figure 13.7 *A diagram of the actin helix showing position of the tropomyosin.* Both actin chains are flanked by tropomyosin molecules, which are long string-like molecules that span seven actin monomers. The troponin complex is attached to the tropomyosin but is not shown. From this diagram, it should be clear how the tropomyosin molecule can conceal the actin-binding sites for the myosin cross-bridges in the relaxed condition. A small conformational change in tropomyosin exposes the sites for attachment of the cross-bridges.

troponin T (Tn-C, Tn-I and Tn-T). The proteins have distinct but coordinated roles in the regulation of contraction: Tn-C binds Ca^{2+} ions, Tn-I binds actin and Tn-T binds tropomyosin. When troponin C binds Ca^{2+} ions, it causes a conformational change in the troponin complex that is transmitted to tropomyosin. The latter modifies its position on actin so that the binding sites of the actin helix are available to bind the myosin cross-bridges.

Cardiac and smooth muscle

Although the actin–myosin interaction is the basis of all muscle contraction, the proteins are organised somewhat differently in cardiac and smooth muscles. In cardiac muscle, which is the endurance muscle *par excellence*, the contractile proteins are organised into myofibrils but the fibres containing them are short and branched (Figure 13.8). In smooth muscle, myofibrils are absent and so is striation. Nevertheless, power is still generated by cross-bridge cycling between actin and myosin, with the filaments of the former being anchored to the plasma membrane. The tissue is composed of spindle-shaped cells rather than fibres, each with a single nucleus. The power output of smooth muscle is lower than that of skeletal or cardiac muscle and shortening is about tenfold slower. This is appropriate for gentle movements, which include squeezing food through the intestine, changing pupil size and altering the diameter of arterioles.

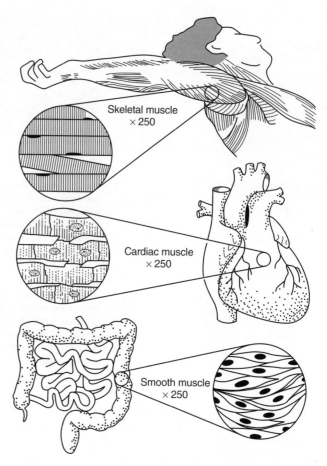

Figure 13.8 *Cardiac, skeletal and smooth muscle fibres.* Cardiac and skeletal muscle fibres appear striated because each fibre is packed with longitudinally arranged myofibrils which lie side by side in almost perfect register (Figure 13.5). In contrast, smooth muscle fibres contain fewer myofibrils that are uneven in diameter and length and are not in register.

Mechanism of contraction: the cross-bridge cycle

The interaction between actin and myosin leads to contraction and the generation of force, in some cases very great force (Figure 13.9). The process is known as the cross-bridge cycle, the components of which are:

- One head of the myosin cross-bridge attaches to the actin filament.

- The other head hydrolyses ATP.

- The angle of the cross-bridge changes, producing movement of the actin filament.

- The cross-bridge detaches from the actin filament.

- The cross-bridge moves into a position where it can again attach to the actin filament and repeat the cycle.

Each cross-bridge undergoes its own cycle of movements, independently of the other cross-bridges (Figure 13.10). At any one instant during contraction, only about 50% of the cross-bridges in the filament-overlap region are attached to actin and, therefore, only about 50% of the maximum possible force is produced.

Regulation of contraction

In the relaxed state, myosin heads are prevented from interacting with actin by tropomyosin, which conceals the actin-binding sites. When the muscle receives a nervous impulse, the sarcolemma depolarises and an action potential is generated. The process, known as excitation–contraction coupling, then leads to exposure of the actin-binding sites, the binding of the cross-bridges and generation of force.

Membrane excitation: the neuromuscular junction

On reaching the muscle, the axon of the motor nerve divides into many branches. Each branch of the axon forms a single junction, known as the neuromuscular junction, with a single muscle fibre, so that a single motor nerve controls the activity of many muscle fibres within one muscle. The branch of the nerve plus the dependent muscle fibre constitute a motor unit (Figure 13.11). The greater the number of fibres that contract at the same time, the greater the force that is developed by the muscle. However, all the fibres in a given muscle are very unlikely to be active at the same time since the force might be too great and damage the muscle.

At the neuromuscular junction, the terminus of the axon is separated from the sarcolemma by a cleft about 4 nm wide. When an action potential arrives at the terminus, it activates a voltage-sensitive Ca^{2+} ion channel. This results in Ca^{2+} ions diffusing into the terminus increasing the intracellular Ca^{2+} ion concentration, which stimulates exocytosis of acetylcholine from the terminus into the cleft. The acetylcholine diffuses across the cleft and binds to receptors on the motor end-plate (Figure 13.12) on the muscle side of the cleft. The binding of acetylcholine to

Figure 13.9 *A soldier wounded at the Battle of Corunna (a battle in the Peninsular War) suffering simultaneous contraction of all muscles after infection with the bacterium,* Clostridium tetani. Both agonist and antagonist muscles are active in this condition. The bacterium is found in the earth, especially in places where animal faeces have been present. Bacteria invade the body through a wound, especially in soldiers in battle. The bacterium secretes a toxin that is absorbed into the motor nerves which then become acutely responsive to mild stimuli. It can lead to death unless treated (from Bell 1824). The toxin is now used in cosmetic manipulation to stimulate contraction of muscles in the face to tighten the skin which removes or conceals wrinkles (Botox).

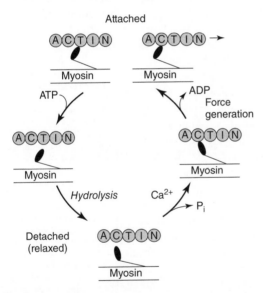

Figure 13.10 *Model of the cross-bridge cycle.* The myosin head binds ATP upon detachment of the cross-bridge from the actin. The myosin ATPase then hydrolyses the bound ATP to ADP and phosphate (P_i), both of which remain attached to the ATPase. When phosphate is released, the myosin cross-bridge (i.e. myosin head) binds tightly to the actin. The myosin head then undergoes a conformational change which is the power stroke. This moves the actin filament and hence generates force. The head is detached from the actin when it releases the ADP and then binds a new ATP molecule, which initiates another cycle. If the fibre contains little or no ATP, the cross-bridge does not detach from the actin, a condition known as rigor. Complete loss of ATP occurs on death, when the rigor condition is known as *rigor mortis*.

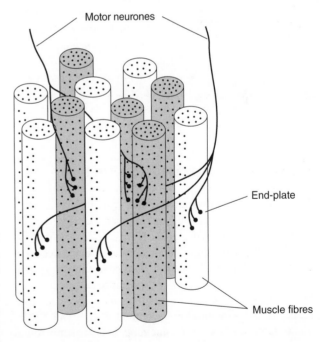

Figure 13.11 *Motor units.* The diagram shows two motor units. In reality, the fibres would be much more tightly packed. In large fast nerves (e.g. supplying limb muscles), a single motor unit will innervate many fibres (up to 1000) via a synapse on each fibre.

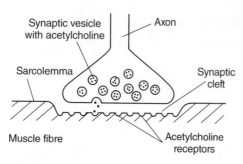

Figure 13.12 *A motor end-plate.* The axon terminates very close to the muscle. They are separated by a small gap (the synaptic cleft). When the nerve is stimulated, acetylcholine is released into the cleft where it diffuses across the cleft, and then binds to receptors located on the muscle side of the cleft and initiates an action potential along the sarcolemma.

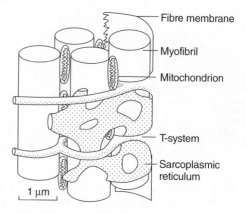

Figure 13.13 *A diagrammatic three-dimensional view of part of a single muscle fibre showing the T-tubule system and the sarcoplasmic reticulum.* The T-tubules are located within the fibre and are attached to the reticulum. This is a sheet of anastomosing flattened vesicles that surround each myofibril like a net stocking.

the receptor opens a Na⁺ ion channel, and Na⁺ ions move into the fibre, causing a depolarisation of the motor end-plate. This initiates an action potential that spreads along the sarcolemma (see pp. 310–311 and Figure 14.5).

The action of acetylcholine is brought to an end by the enzyme acetylcholinesterase, which catalyses the hydrolysis of acetylcholine (Chapter 3). As the concentration of acetylcholine falls, it dissociates from the receptor sites and the Na⁺ ion channels close, returning it to its resting potential. After a short interval, the end-plate can respond to the arrival of a new pulse of acetylcholine, released by another action potential along the motor nerve.

> Myasthenia gravis is an autoimmune disease in which antibodies are present in blood and bind to the acetylcholine receptor on the motor end-plate. This prevents the muscles from contracting so that, due to lack of use, they become weak and fatigue easily. In particular, there is difficulty in speaking, swallowing and chewing food.

Excitation–contraction coupling

The action potential in muscle fibre lasts only 1–2 milliseconds and is completed before there is any sign of mechanical activity. This indicates that the electrical activity along the sarcolemma does not act directly upon the contractile proteins. Instead, a transverse tubular system (T-system) invaginates the fibre and carries the depolarisation of the sarcolemma into the fibre, ending in close proximity to the sarcoplasmic reticulum (SR). The SR is homologous to the endoplasmic reticulum, which is found in most other cells. It forms a series of sleeve-like structures around each myofibril and is the major store of Ca^{2+} ions in the muscle (Figure 13.13). Enlarged regions of the reticulum, known as terminal cisternae, straddle a T-tubule to form a complex known as a triad. Depolarisation of the

T-system membrane results in an interaction between the T-tubules and the SR at specialised junctions, known as excitation–contraction coupling units. Here the two membranes are linked by a 'foot process' consisting of two Ca^{2+} ion channels that function in concert.

Depolarisation of the sarcolemma leads to depolarisation of the T-tubule membrane, which increases the activity of a voltage-sensitive Ca^{2+} ion channel and hence increases the local concentration of Ca^{2+} ions. This increase causes activation of a Ca^{2+} ion channel in the SR membrane, so that cytosolic Ca^{2+} ion concentration increases further. This process, therefore, acts as a positive feedback mechanism, so that the Ca^{2+} ion concentration in the cytosol increases very rapidly from about 0.1 µmol/L to 2–10 µmol/L (i.e. about 100-fold). This increase in Ca^{2+} concentration initiates the cross-bridge cycle by binding to troponin C as described above. It also activates myosin ATPase. Hence contraction takes place (Figure 13.14) (Appendix 13.2).

After a short period, Ca^{2+} ions are pumped back into the SR and the cytosolic Ca^{2+} con-

> The channel within the SR membrane is inhibited by a compound known as ryanodine, so that it is known as the *ryanodine-sensitive Ca^{2+}-channel*. A mutation in the ryanodine receptor is responsible for the sensitivity, in some individuals, to the anaesthetic halothane. This sensitivity results in severe hyperthermia, a condition known as malignant hyperthermia. The explanation is that the modified receptor allows a massive Ca^{2+} ion release from the SR, which is then pumped back into the SR. Thus, the rates of both release and uptake are increased, i.e. the rate of the cycle is increased so that the rate of ATP hydrolysis is very high, resulting in heat generation, and hence hyperthermia. This is analogous to an increase in rate of substrate cycles which also leads to transfer of chemical energy from ATP to heat (Chapter 2). How halothane effects the receptor cause to this release of Ca^{2+} ions is not known.

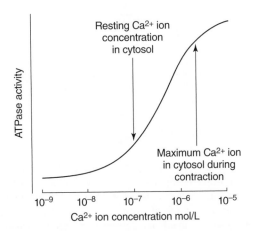

Figure 13.14 *The effect of a change in the Ca^{2+} ion concentration on in vitro activity of myosin ATPase activity*. At rest, the cytosolic Ca^{2+} ion concentration is about 0.1 μmol/L, at which concentration myosin ATPase activity is low. Nervous stimulation of the fibre increases the cytosolic Ca^{2+} ion concentration to about 2–10 μmol/L, with the half maximum change at about 0.5 μmol/L. Hence, on the basis of this property of myosin ATPase in the test tube, the in vivo change in Ca^{2+} ion concentration should result in almost total activation of the enzyme. Approximately 50% activation of the myosin ATPase, when measured in vitro, occurs at about 0.5 μmol/L Ca^{2+} ion concentration. A similar Ca^{2+} ion concentration in the cytosol of the fibre results in 50% of the maximal force of contraction (Appendix 13.2).

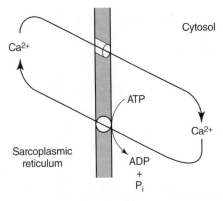

Figure 13.15 *A diagram representing the Ca^{2+} ion cycle between the sarcoplasmic reticulum (SR) and the cytosol*. The Ca^{2+} ion release from the SR is down a concentration gradient of about 1000-fold. The re-uptake of Ca^{2+} ions back into the SR, therefore, requires ATP hydrolysis to provide the energy to overcome this gradient (i.e. 'active' transport). This is a transport cycle, equivalent to a substrate cycle (Chapter 3). The release of Ca^{2+} ions is via a Ca^{2+} ion channel. See text for details of activation of the release of Ca^{2+} ions.

centration is returned to its resting level. Since the concentration of Ca^{2+} in the SR is about 1 mmol/L, which is about 1000 times higher than that in the cytosol, the flow of Ca^{2+} ions out of the SR is down a large concentration gradient, so that the cytosolic Ca^{2+} ion concentration can change markedly and quickly. Consequently, energy is required to pump Ca^{2+} ions back into the SR, ATP hydrolysis. Hence, the release and uptake of Ca^{2+} ions constitutes a transport cycle which utilises ATP (Figure 13.15) and as much as 30% of the ATP consumed in the muscle during physical activity may be used in this cycle.

> It would be 'economic' in the use of ATP, if a means could be developed to reduce the necessity for this Ca^{2+} ion cycle. Some insects have developed a mechanism to activate flight muscles without Ca^{2+} ions, so that the amount of the SR is dramatically reduced (see Chapter 14). It is the more evolutionarily advanced insects that have developed this biochemical 'trick'; it has not been developed by humans or any other vertebrate, as far as is known.

The accumulation of Ca^{2+} ions in the SR is facilitated by binding with a protein, calsequestrin, the major protein in the lumen of the terminal cisternae but absent from the rest of the SR. It associates with the membrane close to the Ca^{2+}-ATPase pump in order to bind Ca^{2+} ions as they enter the SR and facilitate their uptake.

The contraction–relaxation cycle

It is troponin that responds to the increased Ca^{2+} ion concentration. The Ca^{2+} ions bind to Tn-C, which then binds to Tn-I and causes a conformational change in Tn-T. This results in a conformational change in tropomyosin that exposes sites on the actin filament for binding with the myosin head of the cross-bridge. The process can be summarised as follows (see also Figure 13.16):

$$SR \longrightarrow Ca^{2+} \longrightarrow troponin \longrightarrow tropomyosin \longrightarrow$$
$$actin \longrightarrow myosin \longrightarrow myosin\ ATPase \longrightarrow cross\ bridge\ cycle$$

The decrease in the cytosolic Ca^{2+} ion concentration reverses these changes, so that the actin-binding sites are concealed and the relaxed state of the muscle is restored.

Varying the force of contraction

The force generated by skeletal muscle can be varied by increasing the number of fibres that are activated by increasing the number of motor units that are stimulated – a process known as *recruitment of fibres*. The force generated by the muscle also depends upon the extent of fatigue. Signals for fatigue can originate in the brain and inhibit nervous stimulation of the muscle, and/or can originate in the muscle and inhibit the myosin ATPase (see below).

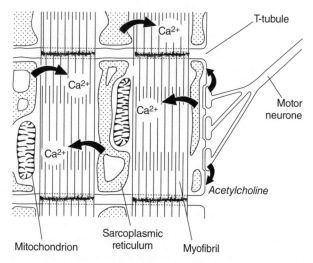

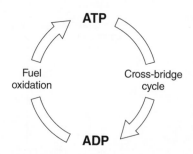

Figure 13.16 *A summary of the control of muscle contraction by the motor neurone.* When an electrical impulse arrives at the junction between a nerve axon and a muscle fibre, a small amount of acetylcholine is released. This initiates an action potential which is transmitted throughout the fibre via the T-tubules. This causes the sarcoplasmic reticulum to release Ca^{2+} ions which initiate contraction of the myofibrils via changes in troponin and tropomyosin. Thus sites on the actin for binding of the myosin cross-bridges are exposed.

Figure 13.17 *The ATP/ADP cycle.* The hydrolysis of ATP provides energy for other energy-requiring activities in the fibre. ATP generation must occur at the same time as ATP is being used, since the actual amount of ATP in muscle is such that it would last for less than about one second in a sprint. Indeed, it is the increase in ADP and P_i concentrations in the fibre, as the cross-bridge cycling takes place, that is one factor that increases the rate of ATP generation (see Chapter 9). In the muscle fibre, it is the conversion of glycogen to lactic acid or the break down of phosphocreatine, the complete oxidation of glucose, glycogen or fatty acids, in the mitochondria, that generate ATP from ADP and P_i (see Chapters 6, 7 and 9, and below).

- the availability of oxygen for the muscle;
- the intensity of the exercise;
- the duration of the exercise.

Type of fibre and fuel used

All fibre types can use phosphocreatine, particularly type IIB fibres which also convert glycogen to lactic acid. In type I and IIA fibres, most of the glycogen is completely oxidised but there is still a large capacity to convert glycogen to lactic acid to generate extra ATP, if it is required (see Table 9.7). The fate of the glycogen will depend, however, on the intensity of the exercise and the supply of oxygen (Figure 13.18).

Although glucose can be converted to lactic acid to generate ATP, the capacity of this process is limited by the activity of hexokinase. The activity of hexokinase in muscle is so low that glycolysis from glucose cannot generate sufficient ATP to power the muscle during sprinting. In contrast, the capacity to convert glycogen to lactic acid is considerably greater (about tenfold) and sufficient to provide all the ATP required to satisfy sprinting (Chapters 3 and 6).

The amount of fuel stored

Fuels are stored within the muscle, liver and adipose tissue (Table 13.4) and amounts vary according to the nutritional status of the subject and the previous physical activity (see also Chapter 2). Triacylglycerol that is stored within the muscle can be used but the most significant fat fuel is long-chain fatty acids, derived from triacylglycerol in adipose tissue.

Fuels for muscle

A fuel is a compound that leads to generation of ATP from ADP and P_i (Figure 13.17). The fuels used by muscle differ according to fibre, the type of activity and the conditions under which the activity takes place. For example, in athletics, the fuel used varies according to the particular event, from the 100 m sprint to the marathon.

Fuel choice

A variety of fuels are available to generate ATP for muscle activity: phosphocreatine; glycogen (which can be converted to lactic acid or completely oxidised to CO_2); glucose (from liver glycogen, transported to the muscle via the blood and completely oxidised to CO_2); triacylglycerol within the muscle (completely oxidised to CO_2); and fatty acids from triacylglycerol in adipose tissue (completely oxidised to CO_2).

At least six factors interplay to determine what fuel is selected:

- the type of muscle fibre involved;
- the amount of fuel stored (e.g. glycogen in muscle or liver);
- the transport of fuel to the muscle;

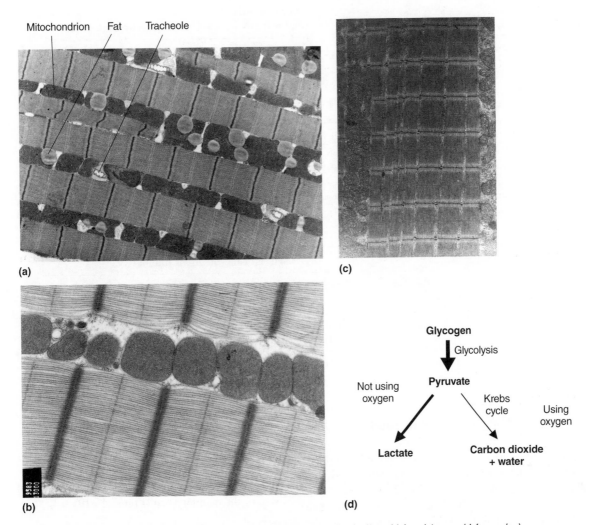

Figure 13.18 *Electron micrographs of different muscles from different animals ('aerobic' and 'anaerobic' muscles).*

(a) A longitudinal section of the flight muscle of an insect (a locust) showing the abundance of mitochondria interspaced between the myofibrils. Droplets of fat are shown which provide fuel for the mitochondria, when glucose in the haemolymph (blood) is depleted. Also shown are the tracheoles, which are in direct contact with the exterior and so provide oxygen for the mitochondria. Note the very close association of both the supply of oxygen and fuel to the mitochondria, which generates the ATP, and also the close association of mitochondria to the myofibrils, which utilise the ATP. This provides the structural basis for the ATP/ADP cycle (Figure 13.17).

(b) A longitudinal section of pectoral muscle of the pigeon ('aerobic') showing the large number of mitochondria between the myofibrils. The pigeon can fly very long distances.

(c) A longitudinal section of the pectoral muscle of a pheasant ('anaerobic') showing no or almost no mitochondria. The pheasant flies only short distances usually to escape from a predator.

Electron micrographs kindly provided by Dr Belinda Bullard.

(d) *'Aerobic' and 'anaerobic' metabolism in muscle.* The figure provides an indication of the glycolytic process for ATP generation, which is the major process for generation of ATP in pheasant pectoral muscle, i.e. glycogen is converted to pyruvate which is converted to lactate. For the other two muscles glycogen (or glucose) provides pyruvate, for oxidation via the Krebs cycle, which generates most, if not all, of the ATP in insect and pigeon flight muscles.

Table 13.4 Approximate amounts of various fuels and energy content in muscle, liver and adipose tissue of a standard male human (body weight 75 kg containing 20 kg muscle)

Fuel	Tissue	Approx. content in total body tissue (g)	Energy content (kJ)
Phosphocreatine	Muscle	–[a]	15
Glycogen	Muscle	500	8 000
	Liver	80	1 280
Triacylglycerol	Muscle[b]	–	17 500
	Adipose	9 000	31 500

[a] The weight is not relevant to its role as a fuel. See also Table 2.2.

[b] The amount of triacylglycerol in muscle varies according to fibre composition.

[c] An approximate value.

See also Table 2.2. The percentage of the body weight as fat is 2–7%, elite male marathon runners. The average in female marathon runners is 15% with the lowest at 6% (for reference, see Noakes 1986).

Table 13.5 Arterial concentrations of some fuels during prolonged physical activity in human volunteers[a]

Duration of exercise (min)	Arterial concentrations (mmol/L)				
	Glucose	Lactate	Fatty acids	3-Hydroxybutyrate	Glycerol[b]
Rest	4.4	0.40	0.40	0.04	0.07
10	4.2	0.66	0.41	0.05	0.08
60	4.2	0.49	0.63	0.09	0.12

[a] Physical activity was running on a treadmill at 65% $\dot{V}O_{2max}$. Blood was taken via an indwelling catheter. Data from Kiens *et al.* (1993).

[b] The concentration of glycerol is an indication of the rate of release of fatty acids from adipose tissue.

Transport of the fuel to the muscle

Blood-borne fuels are glucose, which is derived from liver glycogen, and fatty acids derived from adipose tissue. Uptake depends on the flow of blood through the muscle, the concentration of the fuel in the blood and the demand for ATP within the muscle. During sustained exercise the flow of blood to the muscle can increase up to 50-fold and the rate of utilisation of the fuel can increase to a similar extent, yet the concentration of the fuels in blood remains remarkably constant (Table 13.5).

Oxygen supply to the muscle

For a given amount of glucose metabolised, the number of ATP molecules generated from mitochondrial oxidation is about tenfold greater than that from the conversion of glycogen to lactic acid (glycolysis). The former process requires oxygen, so that the availability of oxygen is crucially important for generation of ATP. It is the major factor limiting performance in athletic events longer than about 400 metres and also limits physical activity in many non-athletes and in patients.

Oxygen supply to the muscles depends on the following factors:

- The flow of blood through the muscle, which depends upon cardiac output and dilatation of the arterioles that supply blood to the muscle.

- The number of capillaries that supply each fibre and the size of the capillary bed (Figure 13.19).

- The concentration (partial pressure) of oxygen in the blood in the capillaries.

- The uptake of oxygen from the air in the alveoli of the lung; this can be reduced by altitude, pollution and in patients suffering from injury to the lung (e.g. emphysema).

- The number of erythrocytes in the blood.

- The volume of the blood.

- The concentration in the erythrocytes of 2,3-bisphosphoglycerate that influences the dissociation curve of oxyhaemoglobin (Chapter 6).

Physical activity increases blood flow through the muscle by stimulation of cardiac output, by increasing the diameter of arterioles (vasodilatation) that supply the muscle and by decreasing the flow through abdominal and pelvic viscera and the liver (vasoconstriction) so that more is

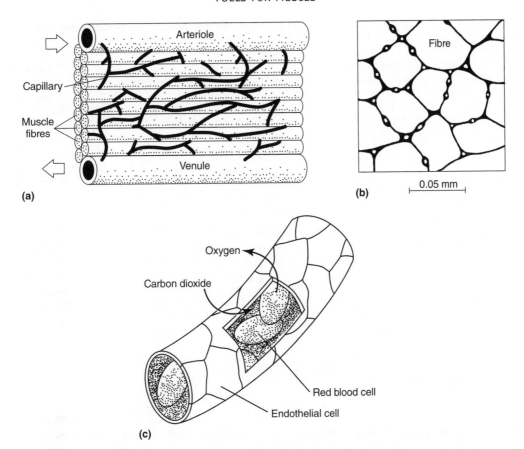

Figure 13.19 *(a) The capillary supply to muscle, (b) the capillary density in a muscle and (c) gas exchange in a capillary.*
(a) Capillaries meander between the muscle fibres from the arteriole to the venule to provide oxygen and fuel to the fibres.
(b) The distribution of capillaries between fibres (based on a human muscle). The more capillaries that surround each fibre, the greater the capillary density and the greater the capacity for oxygen and fuel supply to individual fibres.
(c) The oxygen and carbon dioxide enter and leave the capillaries either through or between the endothelial cells.

Box 13.2 Blood doping

One means of improving endurance in athletic events is by increasing the number of erythrocytes in the blood, known as 'blood doping'. The procedure involves the removal of about 1 litre of blood from the athlete. The blood is stored at a low temperature in the presence of glycerol to protect the erythrocytes from damage due to the freezing. The total number of red cells in the body is thus reduced but it takes about 5 weeks for the number to be restored to normal. The frozen blood can then be warmed and reinfused, increasing the total number of erythrocytes by about 10%. Experimental studies have shown that, soon after reinfusion, the maximum oxygen uptake by the athlete ($\dot{V}O_{2max}$) is increased by about 5%. In competition, a difference of 5% easily covers the margin between winning and losing, hence this procedure is banned in all sports. More recently, this means of increasing the number of erythrocytes has been replaced by use of the hormone erythropoietin (EPO). This hormone stimulates the production of erythrocytes in the bone marrow. Both manipulations are banned by the authorities controlling competitive athletics.

---

available for muscle. The limitation on physical activity imposed by the number of erythrocytes is demonstrated by the beneficial effect of 'blood doping' on athletic performance, i.e. the process of artificially increasing the number of erythrocytes in the blood (Box 13.2). It is also demonstrated by the negative effect on physical activity due to loss of blood or during anaemia.

Some of the athletic events in which oxygen supply to muscle is limiting and can affect ATP generation are given in Table 13.6.

There are several conditions in non-athletes during which most if not all ATP is generated from phosphocreatine and conversion of glycogen to lactic acid. These include:

Table 13.6 Indications of the oxygen supply to muscle, its limitations for performance and the pathways for ATP generation in different athletic events

Athletic event	The contribution of oxygen supply to ATP generation	Major pathways for ATP generation	Extent of oxygen limitation for performance
100–200 m	very low or zero	• phosphocreatine + ADP → creatine + ATP • glycogen → lactic acid	not limiting
400–1500 m	moderate to high	• glycogen + O_2 → CO_2 + H_2O • glucose + O_2 → CO_2 + H_2O • glycogen → lactic acid	low to moderate
5 km 10 km	high very high	• glycogen + O_2 → CO_2 + H_2O • glycogen + O_2 → CO_2 + H_2O • glycogen → lactic acid[a]	severely to very seriously limiting
Marathon, ultramarathon	almost total	• glucose + O_2 → CO_2 + H_2O • fatty acid + O_2 → CO_2 + H_2O • glycogen + O_2 → CO_2 + H_2O	limiting (together with content of muscle glycogen)[b]

[a]This pathway makes a minor contribution to ATP generation, especially during a 10 km race, but may be significant in the sprint to the tape (wire) at the finish of the race.

[b]Oxygen is probably seriously limiting in elite marathon runners but possibly not in non-elite runners.

Table 13.7 Estimates of percentage contribution of glucose, glycogen and fatty acids to ATP generation over four hours of physical activity

Time of exercise (hour)	Percentage contributions to total ATP generation		
	Blood glucose	Blood fatty acids	Muscle glycogen
0.5	27	37	36
1.0	35	45	20
2.0	35	50	15
4.0	30	62	8

Data from Noakes (1986).

- Very unfit individuals, during any form of physical activity.

- During very brief but very high-intensity exercise (sprinting, running upstairs), when there is not sufficient time for the arterioles to dilate to permit increased blood flow to the muscle.

- In elderly people, where ageing has resulted in loss of mitochondria and oxidative capacity in the muscles.

Oxygen supply to muscle in patients suffering from atherosclerosis

Well-known medical conditions in which oxygen supply to muscles is reduced are an attack of asthma, emphysema or heart failure. However, probably the most common condition is atherosclerosis. If the femoral arteries are affected by atherosclerosis, ATP generation from fuel oxidation may not be sufficient to satisfy the energy requirements of even mild physical activity (e.g. walking) and rapidly results in fatigue and pain. This is known as intermittent claudication. If the coronary arteries are affected, the rate of ATP generation may not be sufficient to support the increased demand for ATP necessary for the heart to pump the extra blood required by skeletal muscle during physical activity. The activity must be curtailed and the extra work of the heart causes pain in the chest, angina pectoris (see Chapter 22).

Duration of exercise

As the duration of low-intensity exercise increases, the contribution of fat oxidation to ATP generation rises and parallels the reduced contribution of glycogen (Table 13.7). For very prolonged activities, when all the muscle and liver glycogen has been used, fatty acids become the only fuel available. However, fatty acid oxidation can generate ATP that no more than the rate provides about 60% of which required for an activity that approaches the maximum power output.

Intensity of exercise

As intensity increases, the contributions of blood glucose and glycogen to ATP generation increase. Above about 50% of maximum power output, the rate of glycogen utilisation increases almost exponentially since some glycogen is converted to lactic acid at an increasing rate.

Fuels for various athletic events and games

Experiments cannot be performed on athletes during competitions, but athletes, non-athletes and patients can be studied in the laboratory. Estimates of the proportion of ATP generated from which fuels in various athletic events are based on such laboratory experiments and knowledge of exercise biochemistry and physiology. Some estimates are provided in Table 13.8. Quantitative information on the maximal capacity of some pathways for generating

ATP in muscle can be assessed by the measurement of the maximal activities of key enzymes in extracts of human muscle (see Appendix 13.3).

The sprints

Only two fuels are used during the 100 m sprint – glycogen and phosphocreatine. The glycogen is converted exclusively to lactic acid. Both fuels contribute to the generation of ATP. The evidence for this observation is that the maximum activity of phosphorylase, can generate about 3–4 units of ATP. However, the maximum rate at which ATP can be utilised is about 10 units (Hultman *et al.* 1990; Cheetham *et al.* 1986).

> The maximum activity of phosphorylase in human quadriceps muscle is 50 μmol/min per g or 0.83 μmol/s/g. This is equivalent to an ATP production of about 3 μmol per s per g (one molecule of glucose-in-glycogen when converted to lactic acid generates three molecules of ATP) so the rate of ATP generation is (0.83×3.0) approximately 3 μmol per s per g fresh weight (Appendix 13.3).

As the distance extends to 400 m, oxidation of some of the glycogen occurs, producing about 25% of the ATP generated in this event. Some of the oxygen required comes from oxymyoglobin that is present in the muscle (see below).

Table 13.8 Estimates of percentage contribution of different fuels to the generation of ATP in various athletic track events and in a marathon, an ultramarathon, a 24-hour running race and three ball games

| Event (metres) | Phosphocreatine | Glycogen conversion to | | Blood glucose (from liver glycogen) | Blood fatty acids (from adipose tissue triacylglycerol) |
		Lactic acid	CO$_2$		
100	50	50	0	–	–
200	25	65	10	–	–
400	12	63	25	–	–
800	5	65	30	–	–
1500	–	65	35[a]	–	–
5000	–	60	40	–	–
10000	–	24	76	–	–
Marathon (42.2 km)	–	–	70	10	20
Ultramarathon (e.g. 100 km)	–	–	35	5	60
24 h race	–	–	10	2	88
Soccer game	10	40	20	–	30
Tennis game	5	85	10	–	–
Basketball game	5	80	15	–	–

Some blood glucose will be used in 5, 10 km events but the amount is likely to be small in comparison with glycogen. The amount used in the marathon will be greater than this but primarily in the non-elite runner.

In the 5 and 10 km events, phosphocreatine will be used perhaps at the beginning of the event, but especially during the final sprint to the tape when ATP/ADP concentration decreases.

Estimates based on knowledge of exercise biochemistry, functional biochemistry and data presented by Astrand & Rodahl (1986) and Bangsbo *et al.* (1990).

The estimates for tennis and basketball are very approximate. In a very long tennis game, fatty acids will contribute.

Middle distances

As the distance and time taken for the event extends, oxidation becomes increasingly important in the generation of ATP, demonstrated by the decline in the rate of glycogen breakdown in muscle (Table 13.9). As oxidation of glycogen takes over from glycogen conversion to lactate, less glycogen breakdown is required to provide the same amount of ATP.

In some events rates of oxidation of glycogen are limited by the supply of oxygen from the blood to the muscle, so that conversion of glycogen to lactic acid can generate additional ATP, which will allow a faster pace. How much additional ATP can be generated from this process is difficult to calculate.

It should be noted that the capacity of glycolysis is very large in muscle, even in type I fibres, and much greater than the oxidative process. Consequently, glycolysis must be restrained in events such as 10 km or longer, since utilisation of even a small proportion of the capacity of glycolysis would consume all of the glycogen before the finish and result in severe fatigue. Precisely how this is achieved is not known.

Table 13.9 Rates of glycogen breakdown in muscle at specific times after beginning of physical activity

Time after beginning of activity (min)	Rate of glycogen breakdown (μmol/min per g wet wt)	Estimation of extent of oxidation of glycogen to CO_2
0.5	40	++
1.0	26	+++
2.0	14	+++++
3.0	12	+++++
5.0	9	++++++

Data from Hultman (1986).

At a constant pace (i.e. constant power output) and therefore constant rate of ATP utilisation, the rate of glycogen breakdown decreases with time. This is due to the increase in blood supply to the muscle (vasodilation) which increases oxygen supply so that ATP generation from complete oxidation of glycogen, rather than conversion of glycogen to lactic acid, becomes increasingly more important.

The marathon

The popularity of the marathon as a physical challenge, rather than a competition, has raised the question of the origin of this event (Box 13.3).

A full marathon (42.2 km), for a normal person, run at whatever speed, requires the oxidation of about 700 g of glycogen – about 5 g/min for elite runners. However, since only about 500 g can be stored in liver plus muscle, another fuel must contribute. This is fatty acid, derived from tria-

---

Box 13.3 History of the marathon

Pheidippides was one of the *hemerodromoi*, intercity messengers, of Ancient Greece. Humans can travel faster than horses over rough ground, so runners were an important part of the communication system in Ancient Greece. In 490 BC, when invasion of Athens by the Persians was imminent, Pheidippides was dispatched from Marathon to Sparta with a message requesting help. It is recorded that he arrived in Sparta (about 140 miles away) the day after setting out. His mission was in vain but the Athenians managed to defeat the Persians at Marathon without help from the Spartans. Pheidippides carried a second message to Athens, some 26 miles distant, in order to prevent the Athenians surrendering to the Persian fleet. It was to commemorate this feat that Baron de Coubertin included a race from Marathon to the Olympic stadium in Athens in the first modern Olympic Games in 1896.

A disquieting aspect of the report of Pheidippides' achievement is that, after giving his message to the Athenians, he dropped dead from his exertions. In fact, the account of the events given by the Greek historian Herodotus makes no mention of Pheidippides' death. Herodotus was writing only a generation after the run and it seems unlikely that he would have failed to record such a dramatic incident. It was left to Lucian, writing well over six centuries later, to invoke artistic licence and embellish the story with Pheidippides' death.

The first marathon race, at the summer Olympics in 1896, was won by a Greek postal messenger, Spiridon Louis. Of the 25 (male) starters, only nine finished, eight of them Greek, but the marathon's place in the modern Olympics was assured. Races took place in the USA soon after the 1896 Games; in New York, on 20 September 1896, and in Boston on 19 April 1897. Surprisingly, the distance was not precisely set and varied around 25 miles. In the London Olympics of 1908 it was decided to set the distance at 26 miles from the start at Windsor Castle to the Royal Box in the White City Stadium but, at the request of Queen Alexandra, the start was moved back closer to the castle so that the Royal Family could get a better view. The total distance was then 26 miles 365 yards, which was only fixed as the internationally accepted marathon distance in 1924. The first marathon for women at the Olympic Games was run in September 1984 in Los Angeles (see Box 13.6).

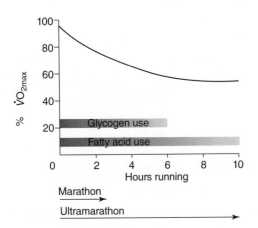

Figure 13.20 *The use of glycogen and/or fatty acids during a prolonged running event (an ultramarathon).* The distance of an ultramarathon is usually >50 miles. In the early part of the run, both glycogen and fatty acids are the fuels oxidised by the muscle. After several hours, glycogen is exhausted and fatty acids are the only fuel used. As fatty acid oxidation cannot provide more than about 60% of the ATP required for maximum power output, if the athlete is running at about 70 or 80% of the maximum, the output (i.e. the pace) must slow. Hence the rate of oxygen consumption ($\dot{V}O_2$) falls to about 60% of maximum ($\dot{V}O_{2max}$), as shown in the Figure. The data on which the plot is based are from Davies & Thompson (1979).

Table 13.10 The predicted effect of short repetitive bursts of activity followed by short rest periods, repeated for 30 minutes, on the extent of fatigue

Duration of the total exercise (min)	Extent of fatigue		
	(a)	**(b)**	**(c)**
5	–	+	+++
10	–	++	++++
20	–	++	+++++
30	–	++	+++++

(a) Represents 10 s exercise bursts, followed by 20 s rests, repetitively over 30 minutes; (b) 30 s exercise bursts, followed by 60 s rests, repetitively over 30 minutes; (c) 60 s exercise bursts, followed by 120 s rests, over 30 minutes.

The effects are interpretations based on the results reported by Åstrand & Rodahl, 1986.

cylglycerol in adipose tissue, although that in muscle also contributes. However, the maximum rate of fatty acid oxidation cannot generate ATP at a sufficient rate to sustain a power output above about 60% of the maximum aerobic power output. Most non-elite athletes run a marathon at a pace achieving about 60% of this power but elite athletes at 80% or even higher. For both elite and non-elite athletes, depletion of their glycogen stores before the end of the event would force a slowing of the pace. For an elite athlete, the pace would fall from about 19 km/h to about 11 km/h and the athlete would undoubtedly quit the race at that point, disgusted with his or her performance (Figure 13.20). Hence the importance of storing as much glycogen as possible before the event.

lactate (which indicates production of lactic acid in the muscle) was shown to depend on both the duration of the burst of activity and that of the rest period. These findings are as follows: repetitive bursts of 10 seconds followed by rests of 20 seconds resulted in minimal glycogen breakdown in the muscle over the 30-minute period, whereas bursts of 60 seconds followed by rests of 120 seconds results in depletion of most of the glycogen in the muscle over the 30-minute period (Table 13.10). Since the glycogen level in the muscle is one factor that influences fatigue, the duration of the rest periods is very important to reduce the fatigue that results from the bursts over the course of the event.

Consequently, the relationship between duration of the bursts and that of the rest periods is vitally important in minimising fatigue during most or all ball games. The metabolic changes that are considered to occur in the burst and rest periods are presented in Figure 13.21.

> It is tempting to speculate that the popularity of the 4, 4, 2 formation in football (soccer) is that it allows players to have longer 'rest' periods between bursts. It encourages passing the ball between players rather than running between different positions (i.e. the ball 'does the work').

Multiple sprint sports

Most team games (e.g. soccer, rugby, hockey, football) and some individual sports (e.g. squash, tennis) involve intermittent high-intensity bursts of exercise interspersed with rest periods (i.e. mostly less intense periods), although the whole period of activity can be very long. These are known as multiple sprint sports. In experiments that involved a study of short bursts of activity followed by short rest periods, repeated for 30 minutes, the increase in blood

Very prolonged activity

Quantitative studies of energy metabolism and fuel utilisation in very prolonged physical activity were performed by Mike Stroud during his Antarctic expedition with Ranulph Fiennes in 1992. One part of the study was measurement of energy expenditure by the dual isotope techniques (Chapter 2). Calculated over the whole expedition, the average daily energy expenditure of Stroud and Fiennes was 29 MJ but on particularly arduous days it increased to

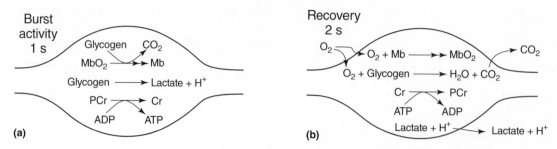

Figure 13.21 (a) *Metabolic processes used to generate ATP during a short burst of physical activity.* A short burst is, for example, a short sprint in a soccer game, a short game in a tennis match.

ATP is generated in three metabolic processes:
(i) Oxidation of glycogen, for which the oxygen is provided mainly from the oxymyoglobin (MbO_2) present in muscle.
(ii) Glycogen conversion to lactic acid.
(iii) Transfer of phosphate from phosphocreatine (PCr) to ADP.

(b) *Metabolic processes that occur in the recovery after a burst of activity.* Four processes are important:
(i) Myoglobin (Mb) is re-oxygenated with oxygen taken up from blood.
(ii) Some glycogen is oxidised and ATP generated, with oxygen taken up from the blood.
(iii) From the ATP generated in (ii), phosphocreatine (PCr) is reformed from creatine (Cr).
(iv) Lactic acid escapes from the muscle into the blood and the pH in the muscle is restored from about 6.5 to the resting value (pH 7.1). The lactic acid in the blood will either be taken up and oxidised by the heart or taken up by the liver and converted to glucose (gluconeogenesis) (see Figure 6.8).

49 MJ – the highest level that has ever been reported. It is about seven times the resting energy expenditure of normal adult males (see Table 2.4).

The proportions of different fuels used to provide energy for this event must be identical to those carried on the sledges, which were: 57% fat, 35% carbohydrate and 8% protein. The proportions taken in seven other previous expeditions in Antarctica were similar. A greater proportion of energy is obtained from fat compared with other prolonged activities (e.g. ultramarathon runs), since fuel has to be transported by the athlete and therefore, the energy/weight ratio of the fuel is important. Migratory birds and insects practise similar energy economy during their very prolonged flights (Chapter 2) (see also, Tables 15.5 and 15.6).

Fatigue

Although fatigue is common and everyone 'knows' what it is, it is difficult to define. One definition is 'the inability to maintain the required power output'. Fatigue can be divided into two classes, peripheral and central. Peripheral fatigue arises within the muscle whereas central fatigue arises within the brain or the motor nerves. Although fatigue affects most people at some time in their life, there is no acceptable biochemical mechanism(s) to explain it. There are, however, several hypotheses.

Peripheral fatigue

The sequence of events within muscle that leads to activation of the cross-bridge cycle, as described above, is

$$SR \longrightarrow Ca^{2+} \longrightarrow troponin \longrightarrow tropomyosin \longrightarrow$$
$$actin \longrightarrow myosin \longrightarrow myosin\ ATPase \longrightarrow contraction$$

Inhibition or failure to activate any one of these factors could result in fatigue. The primary change within a muscle fibre that results in fatigue is a decrease in the ATP/ADP concentration ratio. This arises when the demand for ATP by physical activity exceeds the ability of the biochemical processes within the fibre to generate ATP at a sufficient rate to satisfy this demand. The raison d'être for fatigue is to restrict the extent of the physical activity so that the ATP/ADP ratio does not fall to such low values that sufficient energy cannot be transferred to power processes that are essential to the life of the cell (e.g. maintenance of the ion balance within the cell). Two key questions arise:

1 What factors limit the generation of ATP so that it does not meet the demand of the cross bridge cycle for ATP hydrolysis. An imbalance between demand and generation soon decreases the ATP/ADP concentration ratio.

2 What is the mechanism(s) by which a fall in the ATP/ADP concentration ratio leads to fatigue?

Table 13.11 Content of glycogen and concentration of phosphocreatine, ATP and lactate in muscle before and after sprinting/strength training exercises

	Glycogen content (% of initial content)	Concentration in muscle (μmol/g wet weight of muscle)		
		Phosphocreatine	**ATP**	**Lactate**
6–10 s sprint				
Pre-sprint	100	21	6.8	1.0
Post-sprint	87	5.0	5.0	21
Strength exercises				
Pre-exercise	100	20	6.0	5.0
Post-exercise	70	12	5.0	20

A sample of quadriceps muscle was obtained by biopsy in each case. The data presented are means for two volunteers for sprinting and six volunteers performing strength exercises.

Sprinting was for 6 seconds on a treadmill. (Cheetham *et al.* 1986).

Strength exercises comprised four sets of weight lifting (6–12 repetitions per set). Subjects were bodybuilders (data from Essen-Gustavsson & Tesch, 1990). Note the very small charges in content of glycogen and concentration of ATP but the large charges in lactate (and hence proton concentration) and phosphocreatine concentrations.

Answers are as follows.

1 At least four factors that can limit the generation of ATP have been proposed:

   (i) A marked decrease in the phosphocreatine concentration that occurs in muscle, especially during explosive activities, such as sprinting.

   (ii) A decrease in the glycogen content that occurs in muscle during prolonged activity.

   (iii) The limitation in fatty acid oxidation; i.e. the maximum rate of ATP generation from fatty acid oxidation can provide only about 60% of maximum. The reason for this is not known. Hence, if a high power output is maintained, when fatty acid is the only fuel, the ATP/ADP concentration ratio will soon fall and fatigue will result.

   (iv) Excessive rates of conversion of glycogen to lactic acid result in accumulation of protons (H^+ ions), resulting in a marked fall in intracellular pH. The protons cause inhibition of glycolysis, and oxidative phosphorylation and hence decrease ATP generation (Appendix 13.4).

2 To investigate what factors might provide a mechanism to explain fatigue, measurements have been carried out on biopsy samples of muscle taken from volunteers performing five different fatiguing activities.

   • Sprinting for a short period of time on a treadmill (Table 13.11)

   • Strength training exercises, i.e. weight training by body builders (Table 13.11).

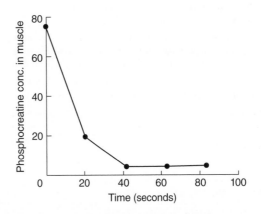

Figure 13.22 *The decrease in phosphocreatine concentration in the muscle during stimulation.* Electrical stimulation of muscle in the laboratory is used to mimic sprinting activity in the muscle. Data from Hultman & Sjöholm (1983). The units for the concentration of phosphocreatine are μmole per gram dry weight of muscle taken by biopsy. Note apparent differences in concentration when data are presented as wet or dry weight.

   • Electrical stimulation of muscle to mimic sprinting (Figure 13.22).

   • Endurance activities (Table 13.12 and Figure 13.23).

   • The effect of the initial content of glycogen in the leg muscles of soccer players on the intensity of activities during a game (Table 13.13). The conclusions from these studies are that both explosive and endurance activities decrease markedly the content of fuels in muscle:

Table 13.12 Effect of different types of physical activity on contents of glycogen and concentrations of phosphocreatine, ATP, ADP, phosphate and lactate, and on the pH in muscle at exhaustion

'Interpretation' of the type of activity[a]	Glycogen content (% of initial content)	Concentration (µmol/g wet weight)				Muscle pH	Estimated ATP/ADP concentration ratio
		Phosphocreatine	ATP	Phosphate	Lactate		
Rest	100	22.0	6.2	1.0	0.75	7.1	60
(a) Sprint (100–400 m)	85	1.9	5.2	27	23	6.6	5
(b) Short/middle distance (1–2 km)	81	4.2	5.0	22	28	6.6	5
(c) Endurance (21 km)	11	7.2	5.7	25	3.5	7.1	5

Data from Sahlin (1992).

[a]The types of exercise have been 'interpreted' from those originally used and published by Sahlin in 1992 as follows:

(a) Sahlin studied isometric contraction in which all of the energy is generated from conversion of glycogen to lactate, since this type of contraction prevents bloodflow.

(b) Sahlin studied intense short-term (6 min) cycling to fatigue at intensity close to 100% maximum.

(c) Sahlin studied endurance cycling which lasted 75 min at 75% $\dot{V}O_{2max}$ to exhaustion.

In (a) the sprint, and middle-distance running there is little change in the glycogen content but a marked decrease in that of phosphocreatine and a marked increase in that of phosphate and a decrease in pH. The precise and relevant concentration of ADP in muscle is not easy to measure, since most of it is bound. Hence the data are not presented but the concentration can be calculated from the change in phosphocreatine concentration. This indicates that the decrease in the ATP/ADP ratio is tenfold.

In the endurance event, the pH does not change but the content of phosphate increases whereas the contents of glycogen and phosphocreatine decrease markedly. The change in phosphate is considered to be responsible for fatigue. Studies using ^{31}P NMR spectroscopy indicate that the resting content of phosphate is about 1 mmol/L and increases more than 20 fold at exhaustion in these physical activities.

Note that a fall in PH from 7.1 to 6.6 represents more than a 10-fold increase in proton concentration.

Note that the estimated ATP/ADP concentration in muscle cytosol is considerably higher than that reported for rat liver (Chapter 9, which is about 10).

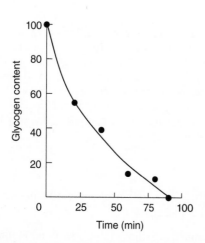

Figure 13.23 *Depletion of muscle glycogen content during prolonged physical activity.* The units of glycogen are µmols glucose-equivalent per gram fresh muscle. Glycogen content was measured in biopsy samples taken from the vastus lateralis muscle. Exhaustive physical activity was performed on a bicycle ergometer. Exhaustion coincided with the glycogen content when it was close to zero. (Data from Hermansen *et al.,* 1967).

- Sprinting lowers the phosphocreatine concentration (Figure 13.22) (Tables 13.11 and 13.12).

- Endurance activities lower the phosphocreatine concentration and the glycogen content.

These decreases in contents of fuel could account for failure to generate sufficient ATP in the muscle, so that the ATP/ADP concentration ratio decreases markedly (Table 13.12) (Figure 13.23).

From these studies, there are several factors that change sufficiently to provide a mechanism for fatigue and they are not mutually exclusive:

1 An increase in the concentrations of phosphate, since it inhibits myosin ATPase.

2 The decrease in ATP/ADP concentration ratio, since it decreases the energy released on hydrolysis of ATP which could decrease two processes that would result directly in fatigue: the cross bridge cycle and the Na^+/K^+ ion ATPase (Figure 13.24).

3 An increase in the proton concentration since it inhibits three processes that are part of the sequence by which Ca^{2+} ion release from the SR activates the cross bridge cycle (Figure 13.25).

Table 13.13 Changes in glycogen content, distance covered and intensity of activity during two soccer games in which the average content of glycogen in the quadriceps muscles of players before the game is different

Content of glycogen μmol/g			Total distance covered (km)	Intensity of activities (% during game)	
Prior to game	Half-time	End of game		Low ('resting' activity)	High ('maximal burst's)
50	6	0	10.0	50	15
100	32	9	12.0	27	24

Data from Saltin (1973).

Note that the players who had a higher content of muscle glycogen before the game performed almost twice the amount of high activity bursts (i.e. sprints) and had a lower percentage of 'resting' activities (i.e. low activity) compared with the players who started with a low content of muscle glycogen.

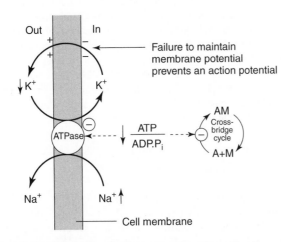

Figure 13.24 *Two key processes that are affected by decrease in ATP/ADP concentration ratio. A low ATP/ADP concentration ratio reduces the energy that is made available when ATP is hydrolysed to ADP. This can have at least two effects that would result in fatigue:*

(i) The energy available would not be sufficient to maintain a particular rate of cross-bridge cycling so that the power output would decrease, i.e. could not be maintained.

(ii) The energy available would not be sufficient to maintain a high activity of the Na^+/K^+ ATPase in the sarcolemma, so that the extracellular K^+ ion concentration increases, and the intracellular Na^+ ion concentration increases which increase the resting membrane potential along the sarcolemma, so that it is difficult to initiate and maintain an action potential. Hence, electrical activity in the motor unit is restricted and cross-bridge cycling must decrease.

A – actin; M – myosin.

Central fatigue

Events within the brain can limit power output, a phenomenon known as central fatigue. It is demonstrated experimentally when the maximal effort that can be achieved voluntarily is less than that which

Overcoming or reducing central fatigue may be part of the basis of what sports psychologists recognise as 'the will to win'.

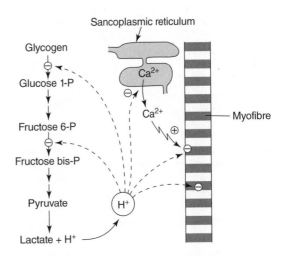

Figure 13.25 *Processes in the muscle fibre that are inhibited by protons (H^+) that could directly or indirectly reduce the rate of cross-bridge cycling. The increase in proton concentration inhibits (i) Ca^{2+} ion release from the reticulum (ii) Ca^{2+} ion binding to troponin-C (iii) Ca^{2+} ion activation of the cross bridge cycle (Donaldson et al. 1978).*

can be achieved when the nerves supplying the muscle are stimulated directly by electrical impulses. A stimulation for contraction of a muscle is initiated in the *motor cortex* of the brain, then travels as an action potential along the nerves in the spinal cord to the muscle. This is the *motor control pathway* which consists of a central and peripheral sequence of events (Figure 13.26).

Information transfer between two nerves in the brain occurs at synaptic junctions, across which chemical messengers (neurotransmitters) diffuse. The neurotransmitter binds to a receptor on the postsynaptic neurone, changing its membrane potential. If the membrane potential decreases, this either initiates an action potential or increases the likelihood that a further depolarisation, from stimulation by another nerve, will initiate an action potential. Such a neurotransmitter is described as excitatory. In contrast, if it increases the membrane potential, it reduces the likelihood of initiation of an action potential, such a

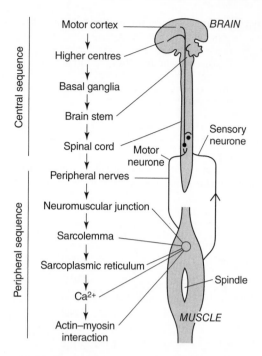

Figure 13.26 *The motor control pathway for stimulation of voluntary contractions.* The motor control pathway begins in the motor cortex in the brain and the sequence continues through the spinal cord to the muscle and the myofibrils. It has two components, central and peripheral. One or both components can be involved in biochemical mechanisms of fatigue (see below).

neurotransmitter is described as inhibitory (see Figure 14.11). One mechanism for influencing the likelihood of developing an action potential is by changing the neurotransmitter concentration in the nerves. If the concentration of the neurotransmitter increased, more is released into the synapse and, if it is an inhibitory neurotransmitter, this would inhibit the development of an action potential.

Three hypotheses have been put forward to account for central fatigue: the hypoglycaemic hypothesis; the dopamine hypothesis; and the 5-hydroxytryptamine hypothesis.

The hypoglycaemic hypothesis

Glucose is an essential fuel for the brain and, if the blood concentration falls, uptake by the brain decreases and less fuel is available for ATP generation in the neurones. This results in a decrease in the ATP/ADP concentration ratio. Consequently, less energy is released on ATP hydrolysis, so that less is available for synthesis, transport of neurotransmitter within the nerve and release into the synapse. Hypoglycaemia could, therefore, reduce the effectiveness of neurotransmitters which would reduce stimulation of the motor control pathway. The result would be inhibition of muscle contraction (Figure 13.27).

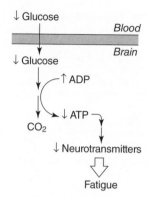

Figure 13.27 *A possible mechanism by which a low blood glucose level could give rise to central fatigue.* A low blood glucose level reduces the rate of glucose utilisation in the brain which decreases the ATP/ADP concentration ratio in the presunaptic neurone. This reduces the energy available for synthesis of neurotransmitters, packaging of neurotransmitter molecules into vesicles and exocytosis of neurotransmitter into synaptic cleft. This decreases electrical activity in postsynaptic neurones and hence in the motor pathway.

The dopamine hypothesis

This hypothesis proposes that physical activity increases tyrosine transport into the presynaptic neurone, where it is used to synthesise dopamine, increasing its concentration. Hence, when the nerve is stimulated, more dopamine is secreted into the synapse. If it is an inhibitory neurotransmitter, within the motor control pathway, this will reduce electrical activity in the pathway and hence reduce contraction (see Figure 13.28).

The 5-hydroxytryptamine hypothesis

This hypothesis, similarly, proposes that physical activity increases tryptophan transport into the presynaptic neurone, where it is used to synthesise 5-hydroxytryptamine. Hence, when the nerve is stimulated, more 5-hydroxytryptamine is released into the synapse and, if this is another inhibitory transmitter in the motor control pathway, it will inhibit contraction (Figure 13.28). This is one of several effects of 5-hydroxytryptamine in the brain which are probably achieved via different receptors on different neurones. All three hypotheses are summarised in Figure 13.29.

Combination of central and peripheral fatigue

On the basis of the proposed mechanisms for peripheral fatigue and for central fatigue, it is possible to combine the mechanisms to provide an overall summary of all the

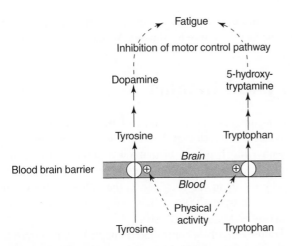

Figure 13.28 *A possible mechanism by which increased levels of tryptophan and/or tyrosine can occur in neurones and lead to fatigue.* The mechanism proposes that physical activity increases the entry of tryptophan or tyrosine into the neurones which increases the concentration of the neurotransmitters, 5-hydroxy-tryptamine or dopamine, respectively. The neurotransmitters are present in vesicles in the presynaptic terminal (Chapter 14). (The pathways for the formation of 5-hydroxytryptamine and dopamine are described in Chapter 14.) This enhances the amount release into the synapses which decreases the excitation of 5-hydroxytryptamine or dopamine neurones in the motor control pathway. It is assumed that they are inhibitory neurotransmitters, they will reduce electrical activity in the motor control pathway and hence nervous stimulation of muscle fibres. This results in fatigue. Mechanisms by which physical activity might result in increased entry of these amino acids into the brain are presented in Appendix 13.5.

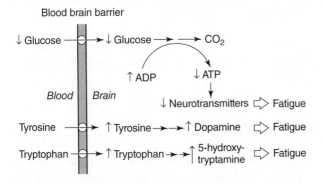

Figure 13.29 *A diagram summarising three possible mechanisms for central fatigue.* It is possible that all three mechanisms could occur simultaneously, giving rise to severe fatigue.

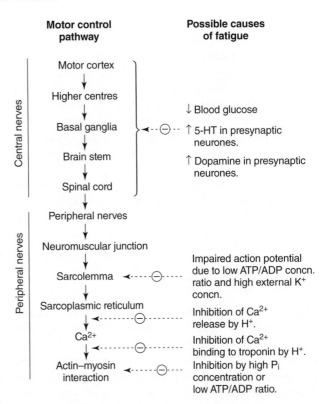

Figure 13.30 *A summary diagram illustrating the mechanisms by which central and peripheral fatigue can cause fatigue.* Central fatigue could be caused by changes in concentration of blood glucose, 5-HT or lumaine concn. in presynaptic neurones.

possible mechanisms that could result in physical fatigue (Figure 13.30). These mechanisms are not mutually exclusive, so one or many of these proposed mechanisms could operate and result in fatigue.

Fatigue in patients

Fatigue accompanies viral infection, sepsis, trauma or major surgery. The cause of this fatigue is not known: it may be peripheral, central or both. Studies on biopsy samples of patients with trauma show a reduction in the muscle ATP concentration, which could be responsible for peripheral fatigue, as explained above (Chapter 18 : Table 13.3). Central fatigue has been identified in three different clinical conditions, post-polio syndrome, multiple sclerosis and after spinal cord injury, but has not been investigated in other conditions.

Fatigue in elderly people

A common complaint to the physician from elderly people is that they become fatigued easily even during mild physical activity such as washing, dressing or cooking. If severe they may require help with these daily tasks. The increasing proportion of elderly people in the population will be as increasing financial burden to social and health care services. Minimising such fatigue could be important on both a personal and an economic level.

Peripheral fatigue in elderly people is likely to be caused by several factors, such as age-related loss of muscle (sarcopenia), reduced numbers of mitochondria, poor nutrition and decreased elasticity of tendons and ligaments (so that the energy conservation achieved via the tendons is lost (Box 13.1). Whether central fatigue could be a further cause of fatigue in elderly people has not been explored.

These effects of ageing can be reduced by regular exercise. There is now considerable encouragement for elderly people to participate in regular mild exercise (Chapter 9).

Chronic fatigue syndrome

Chronic fatigue syndrome has only recently been recognised as a distinct clinical condition. The name replaces the original term, myalgic encephalomyelitis (ME). It is characterised by a combination of severe fatigue and exhaustion after minimal physical activity, together with increased mental effort to carry out any activity. Unlike the patients described above, those with this syndrome experience fatigue not only when they exercise, but even at rest. Diagnosis is difficult, since the symptoms overlap with other illnesses, especially depression.

The fatigue in these patients is probably central, since muscle responds normally to local electrical stimulation although not to voluntary contraction. However, the cause is unknown and there is no satisfactory treatment. If central fatigue is the cause, it may be due to an increased sensitivity to dopamine or 5-hydroxytryptamine or some other inhibitory neurotransmitter not so far identified.

One factor that has been ignored so far is the role of the glial cells in the brain to 'predigest' the glucose taken up from the capillaries and release of the 'digested' product for uptake by the neurones, i.e., the glial cells take up glucose, convert it into lactic acid and release it for uptake and oxidation by the neurones (see Chapters 6, 9 and 14). Impairment of glycolysis in these glial cells would result in impairment of the function of the neurones. If the glial cells that supply the neurones in the motor control pathway were especially susceptible to a low blood glucose level, this could result in failure of nerves in the motor control pathway (e.g. a progressive decline in the number of glucose transporters in the plasma membrane of the glial cells).

Chronic fatigue syndrome was first identified among nursing and medical staff at the Royal Free Hospital in London and was originally termed *Royal Free Hospital syndrome or illness*. In fact, it had been diagnosed two cen-

turies before, as recently discovered from a search through the medical literature by Jones & Wessely (Box 13.4).

Physical training

The techniques and principles of training were once the province of physiology, psychology and physiotherapy but now biochemistry is making a significant contribution. Discussions of training regimens and the principles underlying them can be found in the following books and reviews: Whyte (2006), Elliot (1999), Dick (1997), and Spurway & MacLaren (2006).

Training involves controlled damage, to which muscles and nerves respond, bringing about functional improvement. Many systems in the body limit athletic performance and training regimens need to be sufficiently variable to stress all of them. They include motor nerves, cardiac output, the central mechanism in fatigue, processes involved in ATP generation and regulation of these processes.

Training for strength and power

Strength training involves movement of muscles against an increasing load which is performed in sets and then repeated with fixed rest intervals. In response, the cross-sectional area of muscle increases due to an increase in the number of myofibrils within a fibre.

There is also an increase in the contents of phosphocreatine and glycogen. These are depleted during each training session and replenished during the recovery period. This illustrates the importance of rest during the training period and also of nutrition between sessions.

Moreover, training enhances the maximum activities of key glycolytic enzymes so that glycolytic capacity is increased.

Training for endurance

There are probably as many different endurance training regimens as there are endurance athletes. Training is designed to increase: the number of capillaries surrounding each muscle fibre (capillary density); the number of mitochondria; and the activities of the Krebs cycle and activities of β-oxidation pathway enzymes (Figure 13.31). Training is likely to affect most of the organs of the body but especially the cardiovascular system, the nervous system and the respiratory system. Despite the remarkable increase in knowledge and understanding of the biochemistry of physical activity, these is no satisfactory answer to

Box 13.4 Early reports of chronic fatigue syndrome

Although chronic fatigue syndrome is generally considered to be a recent illness, compounded by contemporary stresses particularly after a viral attack, this appears not to be the case. Edgar Jones and Simon Wesseley researched the medical literature and discovered a report of a condition resembling chronic fatigue syndrome. An extract from their paper follows.

Attention has recently been drawn to the often striking similarities between chronic fatigue syndrome and myalgic encephalomyelitis and the condition formerly known as neurasthenia, a term first coined independently in 1869 by the neurologist George Beard and psychiatrist E. Van Deusen ... . In this article, we suggest that conditions resembling either chronic fatigue syndrome or neurasthenia have an even older provenance, and draw attention to the cases of two British soldiers who served in the Crimean War* and in India at the time of the mutiny.** We surveyed the first 4,000 pension files of the Royal Hospital, Chelsea to identify soldiers who had been medically discharged from the army with unexplained symptoms ... .

Charles Dawes at the age of 18 was posted to the Crimea ... . And served there for six months ... . He then went with his regiment to India and remained there for six-and-a-half years. Having turned to the United Kingdom in May 1864, Dawes suffered from increasing fatigue. He reported the following symptoms: exhaustion, weakness, tremor, pains in his legs when walking,

and pain in his joints, particularly knees, elbows, and shoulders ... . The established nature of his functional impairment, together with the presence of multiple somatic symptoms, would today have qualified Dawes for the diagnosis of chronic fatigue syndrome.

The second case concerns Farrier Major John Dyer who in June 1862 was medically discharged from the Eighth Hussars with 'general debility, the result of long and arduous service'. Dyer had enlisted in 1833 at the age of 16 and served throughout the Crimean campaign ... In the absence of detailed medical records, it is difficult to reach a firm conclusion, although the presentation suggests chronic fatigue syndrome. Dyer and the 10 other cases of debility in the Chelsea pension files imply that Dawes may not have been an isolated example ... .

The two clinical histories suggest that chronic fatigue syndrome and its earlier classifications can be traced through two lineages: one military and one civilian. Had Sergeant Dawes presented in 1915, he might have been diagnosed as experiencing disordered action of the heart, and had he been a Vietnam veteran, he might have been classified as having post traumatic stress disorder or suffering from the effects of exposure to Agent Orange.

Edgar Jones & Simon Wessely, 'Case of chronic fatigue syndrome after Crimean war and Indian mutiny', *British Medical Journal* (1999) **319**, 1645–1647.

*The Crimean War was fought between Britain and Russia on the Crimean Peninsula, 1853–1856.
**The Indian Mutiny, 1856–1858, took place when a local population in India rose up against the British rule.

---

the common question, 'Why do the Kenyans perform so well in endurance events?' (Box 13.5).

Development of muscle

Myoblasts are embryonic stem cells destined to develop into skeletal muscle fibres. They proliferate and fuse with one another to form myotubes which possess several nuclei. They then produce the proteins that are characteristic of skeletal muscle. Although a mature skeletal muscle fibre is normally retained for a lifetime, the muscle proteins, particularly myosin, do change in limited ways. Fibres may lengthen through addition of sarcomeres (and shorten through their loss), and cross-sectional area can increase by the splitting of myofibrils and subsequent addition of contractile proteins to the daughter myofibrils. However, the actual number of fibres is constant from birth and is unaffected by training.

Differentiated fibres are not capable of proliferation, but a small population of myoblasts (satellite cells) persists in mature muscle. They can be stimulated to proliferate and fuse with existing fibres to increase the number of nuclei present and restore the critical ratio of nuclei to cytoplasm that has been reduced by fibre enlargement.

Response to injury

Provided that blood supply and innervation remain intact, skeletal muscle heals well after injury or disease. Damage to fibres causes endothelial cells to secrete general growth factors (e.g. fibroblast growth factor, insulin-like growth factors) and growth factors specific to muscle development which stimulate proliferation of satellite cells. These then migrate to the site of injury to form myotubes, as in foetal development. If, however, the number of satellite cells is low, perhaps due to severe muscle damage or denervation, fibre regeneration is overtaken by the formation of scar tissue (predominantly extracellular collagen secreted by fibroblasts) and this prevents fibre regeneration.

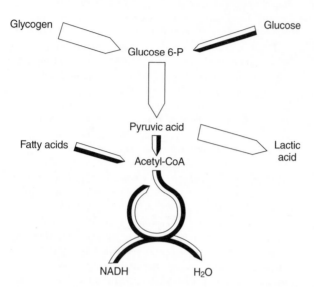

Figure 13.31 *A diagram illustrating the effect of aerobic training on key metabolic process.* The activities of hexokinase, pyruvate dehydrogenase, β-oxidation of fatty acids and the Krebs cycle plus electron transfer chain in the mitochondria are all increased. The effects are indicated by the dark components of the arrows. These wide-ranging effects on metabolism could explain how aerobic training has a profound effect on wellbeing and explain some of the health benefits of such physical activity such as reduction in fatigue, fatty acid oxidation reducing obesity and increased glucose utilisation reducing the risk of type II diabetes (see below).

The number of satellite cells in skeletal muscle declines with age so that recovery after injury is slower in the elderly. Moreover, no satellite cells are present in cardiac muscle, and damage (e.g. after a heart attack) is not repaired but is replaced by scar tissue. These are no reports of severe muscle damage in top sports personalities berg treated with embryonic stem cells. Rapid recovery from such damage, may be financially rewarding to both the athlete and the club, so that such treatment may be considered in the future.

Underperformance syndrome: overtraining

It has been known in sports medicine that too much training can bring on a condition known as overtraining or *underperformance syndrome* (UPS). Symptoms include sleep disturbance, increased perception of effort, decreased effectiveness of the immune system and unexpected poor performance in training or in competition. At present, there is no satisfactory biochemical or physiological explanation for the syndrome. Identification of a biochemical marker to indicate the risk of development of UPS would be of great advantage to a coach, as the only current cure for UPS is complete rest for several weeks or even months. This is difficult for an athlete to accept and despite suffering from UPS the athlete sometimes continues to

> The unexplained underperformance syndrome occurs in cycling, rowing, swimming and middle- and long-distance runners. A high proportion of elite athletes have experienced it. In contrast, it appears much less frequently in athletes who participate in 'explosive' sports.

Box 13.5 Why are the Kenyans so good?

Kenyan runners have fascinated the athletic world in recent decades. Kenya has produced some of the world's top class middle- and long-distance runners over this period. An interesting question is whether features in their physiology or biochemistry could explain this phenomenon. To answer this question, a detailed study of Kenyan and Scandinavian runners was carried out by Professor Bengt Saltin and his colleagues in north-west Kenya (Eldoret), which is 2000 metres above sea level. The Scandinavians were acclimatised to the altitude before measurements were made but all the Kenyans were resident. A summary of the result is as follows: (Saltin *et al.*, 1995a,b; Saltin, 1997):

- The $\dot{V}O_{2max}$ of Kenyan runners was slightly larger than that of the Scandinavian runners (at sea level, mean values for $\dot{V}O_{2max}$ for Kenyan and Scandinavian runners were 79.9 and 75.5 cm^3/kg per min, respectively).

- Maximum values of heart rate varied between 180 and 203 – the mean value was slightly higher for the Kenyans.
- Running economy, an index of the contribution to performance made by the springs, the Achilles tendon and the arch of the foot, was better in the Kenyans.
- There were no major differences in the fibre composition or fibre diameter of the muscle or the capillary density between Scandinavian and Kenyan runners.

The conclusion from the study was that there is no dramatic or sensational difference. A number of small differences, particularly in the elasticity of their tendons, may explain their dominance of distance events. More 'efficiency' from the tendon reduces the amount of muscle and thus the size of the calf and hence there is less wind resistance and less inertia during running.

train, allowing the condition to worsen. No such marker has yet been identified.

Health benefits of physical activity

Many disorders benefit from exercise (Pederson & Saltin, 2005). These include asthma, cancer, chronic heart failure, coronary artery disease, chronic obstructive pulmonary disease (COPD), depression, type 1 diabetes mellitus, type 2 diabetes mellitus, hypertension, intermittent claudication, osteoarthritis, osteoporosis, rheumatoid arthritis and obesity.

Exercise also promotes physical, mental and social wellbeing and delays the infirmities and disabilities that develop with age but these are effects that are difficult to study and to define (see Haskell, 1996; Pederson & Saltin, 2005). Recommendations for beneficial exercise are given in Pederson & Saltin (2005).

> A recent change in working practice in an office may be one factor contributing to an increase in the problem of obesity. It is calculated that if an office worker (60–75 kg in weight), working 8 hours a day for 5 days a week, exchanges 2 minutes of walking from one office to another each hour for the same time sitting at a computer sending e-mails to the offices, annual energy expenditure is decreased by an amount equivalent to 0.5 kg of fat. The result over several years, could be 'creeping obesity'.

Health hazards of physical activity

Exercise is not without risks: these include sudden cardiac death, hyperthermia, hypothermia, hypoglycaemia, hyponatraemia, a reduction in the effectiveness of the immune system, overuse injury and interference in the reproductive system in females. Whether severe physical activity affects the reproductive system in males is sometimes discussed but these are no reports in the scientific literature.

Hyperthermia

During a marathon, the rate of heat production can be more than tenfold greater than at rest and sufficient to raise the core body temperature by 1°C every eight minutes, if no cooling occurs. The core temperature is normally regulated so precisely that it does not rise more than about 1°C. The main mechanism for cooling is evaporation from the skin. Endurance runners can produce one litre of sweat per hour which removes about 2.4 MJ of heat. The energy used and therefore converted into heat in a marathon is about 12 MJ

and to dissipate this amount of heat a runner requires the loss of about 3 litres of sweat. It is important therefore to take sufficient fluid and electrolytes (Na^+ and especially K^+) soon after the event to replace that lost in sweat.

Hyperthermia can lead to heatstroke, which occurs when the core temperature increases to about 42°C (as measured by a rectal thermometer). It is characterised by irritability, aggressive behaviour, disorientation, unsteady gait and a glassy stare and will eventually result in collapse and coma. It can be fatal unless treated within minutes. There are several well-known examples of heatstroke in athletes. In the Olympic Games in London in 1908, Dorundo Pietri collapsed in the final stages of the marathon and was in coma for 2 days. Jim Peters collapsed in the stadium in Vancouver before reaching the finishing line of the marathon in the Empire Games in 1954, and Tommy Simpson collapsed and died while cycling in a mountainous region of the Tour de France cycle race in 1959.

One aspect of exercise that increases the risk of heatstroke is that the loss of heat by sweating requires a proportion of the cardiac output to be directed away from muscle to the skin. In this way, heat is lost from the blood as it circulates close to the skin. However, even in severe hyperthermia, the blood supply delivering fuel and oxygen to the muscles takes preference over that to the skin. In this way, symptoms of hyperthermia are overriden and physical activity is maintained so that the core temperature can rapidly increase to dangerous levels. (Neilsen et al. 1990).

Hypothermia

Hypothermia occurs when the body temperature falls below 35°C (rectal measurement). It causes death by reduced cardiac output so that sufficient blood does not flow to the brain. In some countries (including the UK) death from hypothermia is not uncommon in elderly people, usually due to low environmental temperature coupled to low physical activity even within their own home. Malnutrition, undernutrition and poor hypothalamic control of both heat production and also loss probably contribute. Nonetheless, it can occur also in young and physically fit individuals when exercise is taken during cold weather with inadequate clothing, especially when fatigue causes a reduction in activity and hence a fall in heat generation. Activities in which hypothermia can occur are walking, hiking, climbing in mountainous regions and marathon running, especially in countries where the environmental temperatures in winter and spring are low. Hypothermia can, of course, be prevented with adequate insulation. Even in the coldest of conditions, if sufficient physical exercise is taken, enough heat can be produced to balance that lost provided that clothing provides sufficient insulation. This was illustrated by Mike Stroud and Ranulph Fiennes during their trans-Antarctic expedition in which walking and pulling a sledge were sufficient to maintain body temperature despite very low environmental temperatures (Stroud, 1997). However, prolonged hypothermia can cause frostbite in the extremities (toes and fingers). The low temperature reduces blood flow and the rate of cooling can be very rapid. The lack of blood results in decreased oxygen supply and failure to maintain ATP generation, and the ATP/ADP concentration ratio falls, leading to necrosis. See Chapter 20 for discussion of neurosis and frostbite.

Hypoglycaemia

Hypoglycaemia is defined as a blood glucose level <3 mmol/L. However, there are reports of some endurance athletes suffering levels below this with no obvious adverse effects. Most athletes are aware of the importance of a high-carbohydrate diet prior to exercise to ensure adequate stores of glycogen in muscle and liver. It is less frequently appreciated, however, that low stores of glycogen in these tissues prior to physical activity can increase the risk of hypothermia. Hence, adequate clothing, a previous diet high in carbohydrates and a source of carbohydrate available during the exercise are essential for prolonged exercise in cold conditions (e.g. climbing mountains, prolonged walking or running in cold regions or cross-country skiing). The neurones involved in thermoregulation appear to be particularly dependent on glucose and hence very sensitive to a low blood glucose level. Ingestion of ethyl alcohol inhibits glucose formation in the liver, which can be particularly dangerous when the environmental temperature is low (Chapter 14).

Decreased effectiveness of the immune system

Much evidence substantiates the connection between prolonged intense physical activity and increased incidence of infections. For example, a sixfold increase in infections was observed in participants in the Los Angeles marathon during the week after the race compared with a control group who had undergone a similar level of training but had not competed in the marathon. In addition, many studies have shown a high incidence of infections in military personnel undergoing prolonged intensive training, especially in recruits who are not physically fit.

Several explanations have been suggested, including a decrease in the number of immune cells, an increase in the level of stress hormones (especially cortisol, which depresses the immune system), and a decreased level of glucose and glutamine in the blood (Chapter 17).

In contrast, regular mild physical activity can increase the effectiveness of the immune system, although the mechanism underlying this effect is also unknown.

Soccer, contact sports and motor neurone disease

Motor neurone disease (or amyotrophic lateral sclerosis) is an adult-onset degenerative disorder of the nervous system, with a progressive course that eventually results in death. The cause is unknown but several factors have been implicated, including trauma and participation in sport. A retrospective study on 24 000 former Italian professional soccer (football) players indicated a high incidence of motor neurone disease as a cause of death (Chio *et al.*, 2005). Brain damage caused by repeated heading of the ball, in association with a genetic predisposition, may play a part in the increased incidence in soccer

Motor neurone disease is often called amyotrophic lateral sclerosis in the USA, where it is also known as Lou Gehrig's disease after the famous New York Yankees baseball player who developed the disease in 1939.

players. It is possible that a similar increased risk is also present in other contact sports including rugby, American football, boxing and ice hockey, but these are no reports of any studies.

Boxing and post-traumatic encephalopathy (the punch-drunk syndrome)

Post-traumatic encephalopathy (punch-drunk syndrome) and even death due to brain damage are well known in boxing. The damage causes cerebral atrophy and enlargement of the cerebral ventricles, which are associated with progressive deterioration of the intellect (dementia), a slow, shuffling gait and other features resembling those of Parkinson's disease. Boxing causes concussion due to jarring of the brain within the rigid skull cage, which kills brain cells. Not surprisingly, professional boxing has been banned by law in some countries. Post-traumatic encephalopathy also occurs in steeplechase jockeys after repeated head injuries due to falls.

The female triad

The triad is the combination of disordered eating, amenorrhoea and osteoporosis that occurs in female endurance athletes, ballet dancers, gymnasts and lightweight rowers.

These are activities in which a low body weight is an advantage. A particular sequence of events may be responsible (Figure 13.32). A poor diet and/or disordered meals, together with prolonged physical activity, leads to negative energy balance which changes the blood levels of hormones. Decreased levels of insulin and increased levels of glucocorticoids, catecholamines and growth hormone cause the mobilisation of fatty acids and hence loss of weight. The imbalance also changes the normal pattern of secretion of gonadotrophin-releasing hormone (GnRH). This hormone controls the synthesis and secretion of the hormones that control the menstrual cycle: luteinising hormone (LH) and follicle-stimulating hormone (FSH) secreted by the pituitary. These, in turn, stimulate the synthesis and secretion of the steroids, oestrogen and progesterone, by the ovary. These steroid hormones control the changes in the endometrium of the uterus during the menstrual cycle, and failure to maintain their levels disturbs the cycle to such an extent that it eventually fails (amenorrhoea). The decrease in the oestrogen level increases the activity of the cells (osteoclasts) that degrade the matrix of bone and release calcium phosphate. This eventually results in loss of the structure of bone and bone density (osteoporosis) which can increase the risk of fractures.

Once the triad has developed, it is difficult to treat. The initial treatment is restoration of the normal energy balance, by increasing energy intake and decreasing energy expenditure (i.e. less physical activity). However, both of these changes present difficulties to the dedicated athlete, whose daily life may centre on physical training and irregular meals. The late identification of the problem in female athletes this was possibly caused by the delay in allowing females to participate in athletic running events including the marathon so that the relatively high incidence of this problem in female athletes was not apparent until fairly recent (Box 13.6).

Skeletal muscle diseases

Skeletal muscle comprises about 40% of body mass and plays an important part in life, but specific pathological problems relating directly to the contractile process are rare. Perhaps the foetus does not survive with defective skeletal muscles.

Pain, muscular weakness, cramps and ease of fatigue are the most usual symptoms of muscular disease. In most cases, it is diseases of the vascular or nervous system or problems with the processes providing energy within the muscle that are responsible for clinical problems with muscles. Other clinical problems include the muscular dystrophies, myotonic disorders, inflammatory myopathies and disorders of neuromuscular transmission (see Walton, 1996). The best known is Duchenne muscular dystrophy.

Muscular dystrophy

Muscular dystrophies are characterised by variable degrees of muscle weakness and degeneration. The most common forms are Duchenne (severe) and Becker (benign), both caused by mutation in the same gene in the X chromosome. In Duchenne muscular dystrophy, the first symptom is muscle weakness in early years of life which gradually worsens so that patients are unable to walk by the age of 10. Death from cardiac or respiratory insufficiency usually occurs before the age of 25. In Becker muscular dystrophy, weakness and wasting become apparent between 5 and 25 years but, although severely disabled, patients can have a normal lifespan.

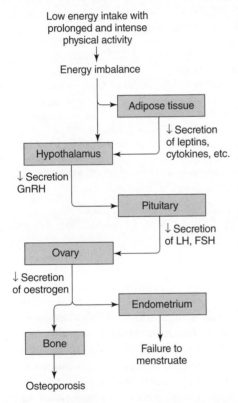

Figure 13.32 *The female triad: disordered eating, osteoporosis and menstrual cycle disturbance.* A low energy intake and high energy expenditure can lead to endocrine changes that result in osteoporosis and disturbance in the menstrual cycle. A low energy intake in female runners, rowers or gymnasts in order to maintain a low body weight can lead to other disturbances in eating such as bulimia or anorexia nervosa. Stress may contribute to energy imbalance and may have additional effects on the hypothalamus.

FSH – follicle-stimulating hormone; GnRH – gonadotrophin-releasing hormone; LH – luteinising hormone.

Box 13.6 History of women's participation in athletic competition

In recent years, the performances of women in long-distance running events have engendered more interest from spectators and the mass media than those of men. It is interesting to note, however, that it is only relatively recently that women have been 'allowed' to participate in running events, as Nina Kuscik relates.

The modern Olympics were devised by men for men; in 1896 anything else would have been quite shocking. Nevertheless a woman did, it is claimed, run unofficially with the 25 male competitors in that first marathon. It was not until the Amsterdam Olympics of 1928 that three track events – the 100 m, 800 m and 4 × 100 m relay – were opened to women, together with the discus and high jump. As it turned out, this first women's Olympic 800 m event was to set back the course of women's athletics by many years since some of the participants in that race collapsed. Although the failures probably resulted from inadequate preparation (as they would have done for male athletes) they inevitably received more publicity than the winner, Lina Radke, from Germany. This provided support for the view that women should not be allowed to participate in such events. Indeed the 800 m event for women was not reinstated until the 1960 Olympic Games.

The 1960s brought . . . the realisation that running long distances was no more harmful to women than to men. Since this realisation did not extend to governing bodies of athletics, the decade was one of 'unofficial' female participation, particularly in the marathon.

In December 1963, two women from California, Lyn Carman and Merry Lepper, hid at the start of the Western Hemisphere Marathon in Culver City, California, and 'jumped into' the race when it started. Irate

officials failed to prevent them completing the race and Lepper finished in 3 h 37 min 7 s. In 1964, a Scottish woman, Dale Greig, set an unofficial world best of 3 h 27 min 45 s, and was the first woman to break the 3 h 30 min barrier. During the race, an ambulance followed her all the way! Roberta Gibb Bingay 'jumped in' at the start of the 1966 Boston Marathon and completed the race in 3 h 21 min, ahead of two-thirds of the otherwise all-male field, and this became news across the USA. Male prejudice was apparent in the race director's comment: 'Roberta Gibb Bingay did not run in the Boston Marathon, she merely covered the same route as the official race while it was in progress.' In 1967 Kathrine Switzer applied to run in the Boston Marathon without indicating her sex. She was assumed to be male and turned up for the race wearing a hood. When the hood was removed to reveal her sex at the beginning of the race, the race director ran after her but was unable to catch her!

Finally, in October 1970, the Road Runners Club of America (RRCA) arranged the first marathon for women in the USA.

In April 1972 women were allowed to participate in the Boston Marathon. West Germany hosted the first international marathon for women, in Waldniel, on 22nd September 1974. In 1979, Joan Benoit from the USA won the first Olympic marathon race for women in Los Angeles in 1984 in 2 h 24 min 52 s. The current world's best time is held by Paula Radcliffe, 2 h 13 min 0 s achieved in the London Marathon.

Nina Kuscik, (1977) The history of women's participation in the marathon. *Ann. New York Acad. Sci.*, **301**, 862–876.

In the early stages of the disease (around 2–4 years), affected boys are not disabled and indeed some of their muscles, notably gastrocnemius, are larger than normal. However, biochemical examination of blood samples reveals exceptionally high activities of muscle enzymes, such as creatine kinase. By the age of 5–7, the muscles are considerably wasted and strength is diminished. Electron micrographs of samples from such boys suggest that the cause of these symptoms is incomplete sarcomeres, disordered myofibrils and necrosis.

In 1987 the gene responsible for muscular dystrophy was identified, leading to the isolation of a protein (dys-

> Duchenne muscular dystrophy is named after the eighteenth century French neurologist and physician, Guillaume Duchenne, who made the first thorough study of the disease.

trophin) that is either totally absent (Duchenne), or partially absent (Becker). Dystrophin is a large protein (427 kDa) that occurs on the inside of the plasma membrane of all normal striated muscles and in some other tissues, including non-striated muscle and certain neurones. Although its precise function is not known, the mutant form is associated with structural abnormalities of the sarcolemma and the degradation of the myofibrils. The link between the abnormalities of the membrane and degradation is not known. One theory is that the activity of a Ca^{2+} ion channel in the membrane is increased and excessive entry of Ca^{2+} ions into the fibre elevates the concentration in the cytosol. A Ca^{2+}-dependent proteolytic enzyme, calpain, is stimulated and leads to the degeneration within the fibre.

14
Mental Activity and Mental Illness

Intuitions sometimes occur during a sleep and a remarkable example is quoted by Cannon. Otto Loewi (1873–1961), professor of pharmacology at the University of Graz, awoke one night with a brilliant idea. The next day he went to his laboratory and in one of the most definitive experiments in the history of biology, brought proof of the chemical mediation of nerve impulses.

(Beveridge, 1950)

Dementia has reached epidemic proportions with an estimated 4.6 million new cases worldwide each year . . . it is a true disease, cause by exposure to several genetic and non-genetic risk factors.

(Smith, 2008)

There is no doubt that between them mental illness and the mental effects of recreational drugs are of major medical and lay interest and of considerable public concern. The first section of this chapter provides an account of the biochemistry and physiology of normal mental activity, to enable a better understanding of how far changes in the biochemistry can explain mental illnesses and the effects of recreational drugs.

Mental activity

The term 'central nervous system' is sometimes used as a synonym for the brain, but it also includes the spinal cord. Indeed, the word 'system' implies the entirety of the tissues working together to achieve a single function. Usage, however, has validated its division into the central and peripheral nervous systems, and even the subdivision of the latter into the autonomic nervous system and the voluntary nervous system.

All mental activity is achieved by two types of cells, neurones and glial cells, and approximately 50 different chemicals that provide communication between the neurones. The neurones are specialised cells that generate differences in electrical potential that are transmitted along their surface to carry information. Within the brain, integration and coordination is brought about by around 10^{11} neurones, comprising 99% of all the neurones in the body. These central nervous system neurones are characterised by the very large number of connections (known as synapses) that they form with other neurones (on average around one thousand). If each communicates with one thousand others, a human brain contains approximately 10^{14} synapses. Their organisation within the brain results in considerable localisation of function. These locations contain clusters of neurones with similar properties and functions, known as nuclei. The brain and spinal cord are divided into grey and white matter. Neurones and some non-myelinated nerve fibres are present predominantly in the grey matter. In this part of the brain, the neurone cell bodies are separated by dendrites, axons and glial cells and a considerable capillary bed is present. It is the dull grey colour in fresh tissue that explains the name. In contrast, white matter contains bundles of myelinated axons, some non-myelinated axons, a few capillaries and only a little extracellular space. It is mainly this part of the brain in which electrical communication transmits information between neurones, between sensory organs and neurones, and between neurones and other tissues (Figure 14.1).

At each synapse, chemical messenger molecules (known as neurotransmitters) are released from one neurone to influence the generation of electrical activity in adjacent neurones. In the brain, the action of these neurotransmitters

Functional Biochemistry in Health and Disease by Eric Newsholme and Tony Leech
© 2010 John Wiley & Sons Ltd

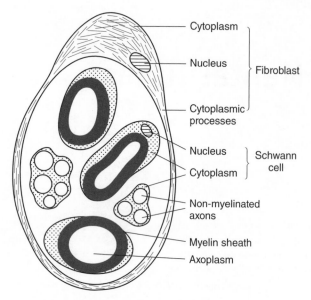

Figure 14.1 *A diagram, based on an electron micrograph, of a cross section of a nerve bundle.* It shows three axons enclosed in myelin sheaths. The sheaths completely surround the axoplasm. Between the myelinated axons are clusters of non-myelinated axons. Except for smaller diameter and absence of myelination they are structurally similar to myelinated axons. (Drawn from an electron micrograph of Telford & Bridgman, 1995.)

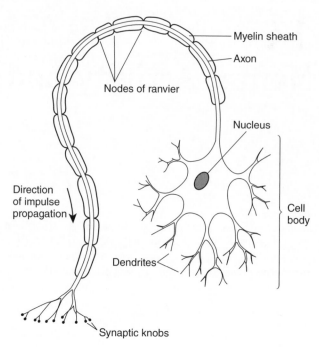

Figure 14.2 *A myelinated axon.* Some neurones lack the myelin sheath. (See below for description of the myelin sheath.) The cell body may be elsewhere other than at the end of the axon.

is modulated by a wide range of other molecules (neuro-modulators) creating numerous opportunities for modifying communication. Synapses are therefore the basis for mental activity, so that disturbance of their action gives rise to pathology. They also present opportunities for pharmacological interventions.

The way in which impulses are generated and propagated is now largely understood, as are the principles by which synapses operate and the means by which impulses, that are controlled by synapses, influence some processes within the brain. Even functions as complex as learning and memory can now be explained, at least in part, from biochemical knowledge of the activities within the brain. What is not understood are the mechanisms by which the biochemistry of the billions of connections in the brain provides for consciousness and determines personality.

Cells in the brain

To provide for all the functions of the brain, there are three classes of cells: neurones, glia and endothelial. Most of the emphasis in this chapter is on neurones and glia. These latter cells can be considered to act as 'nanny' and 'nurse' to neurones. The endothelial cells provide part of the barrier between the blood and the brain. The microglia,

specific glial cells, are phagocytes, i.e. they are the resident macrophages in the brain.

Neurones

Despite the diversity of neuronal structure, it is possible to describe a generalised neurone (Figure 14.2). The cell body (also known as the perikaryon) contains a nucleus, cytosol and the usual array of organelles but its surface is extended into one or, more usually, many processes. One of these, the axon, which arises at an axon hillock, a conical projection on the cell body, transmits impulses away from the cell body. In some neurones, motor neurones (i.e. those controlling movement) in particular, the axon can be enormously elongated (one to two metres). The shorter processes, along which impulses are transmitted to the cell body from other neurones or receptors, are known as dendrites. These are usually much-branched and may, in addition, be covered with short projections, known as 'spines', further increasing the area over which chemical communication between neurones can occur. Neurones are unipolar, bipolar or multipolar (Figure 14.3).

Most, but not all, of the axons of both peripheral and central neurones are wrapped in a sheath of material known as myelin (which is a complex of protein and phospho-

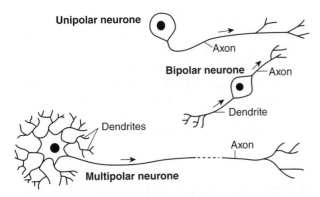

Figure 14.3 *Unipolar, bipolar and multipolar neurones.* Unipolar describes a neurone with only one emerging process (an axon); bipolar describes a neurone with an axon and one dendrite (found in sense organs); and multipolar describes a neurone with one axon but many dendrites. Multipolar neurones are the most abundant in the CNS. In motor neurones the axon may be very long.

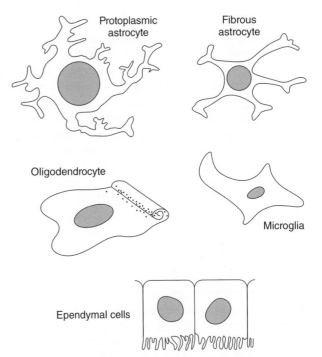

Figure 14.4 *Different types of glial cells.* Astrocytes 'connect' capillaries and neurones. Fibrous astrocytes, with less branching and more filamentous processes, occur mainly in white matter while protoplasmic astrocytes are located principally in the grey matter. Oligodendrocytes form the myelin sheath by 'wrapping' themselves around axons. The connection between the myelin sheath and the oligodendrocyte is permanent and provides material for the myelin sheath. Microgliocytes (microglia) are the phagocytes of the nervous system. The ciliated ependymal cells line the cavities of the central nervous system.

lipids; see Chapter 11). The function of this sheath is protection and acceleration of transmission of electrical activity along the axon. Myelin is derived from the membranes of oligodendrocytes, glial cells in the brain. Schwann cells perform a similar role in peripheral nerves. These cells become wrapped around the axon and provide the myelin which then completely surrounds the axon.

In myelinated peripheral nerves, the sheath is interrupted by small gaps, the nodes of Ranvier. It is these nodes that increase markedly the velocity of the action potential as it travels along the axon (see below). Myelinated and non-myelinated axons, together with their Schwann cells, are shown in Figure 14.1.

Glial cells

For every neurone in the brain there are about ten glial cells, which are diverse in both structure and function (Figure 14.4). 'Glia' is derived from the Greek word for glue since it appears to hold the neurones together. The three major types are astrocytes, oligodendrocytes and microglia. Astrocytes have a large cell body from which a number of processes radiate, sometimes terminating in thickenings or end-feet which surround capillaries and the neurones (see below). During development, monocytes invade and differentiate into microglia. Although glial cells do not transmit impulses they are involved in most other functions: guiding the migration of neurones during development; transfer of nutrients to neurones (see below); structural support; myelination (oligodendrocytes); removal of invading microorganisms and necrotic tissue from sites of injury (microglia); removal of neurotransmitters that are

outside the synapse and also the breakdown products of neurotransmitters; and maintenance of the ionic environment surrounding the neurones.

Endothelial cells

Since fuel oxidation is essential to provide energy for the brain and since glucose is the only fuel (under most conditions), the supply of glucose and oxygen via blood is vitally important.

Endothelial cells of the cerebral capillaries are linked by tight junctions, sufficiently effective that there is no space between them. A well-developed basal lamina surrounds these capillaries and the external surface of the capillaries is almost entirely covered by processes of the astrocytes

If microorganisms cross this barrier, the membranes enveloping the brain – the meninges – can become infected, resulting in meningitis.

known as end-feet. The plasma membranes of the endo-thelial cells and the astrocytes plus the basal lamella provide a barrier known as the blood–brain barrier. This prevents access to the neurones of molecules other than oxygen, nutrients and amino acids. Nonetheless, these compounds have to pass through the endothelial cells and astrocyte to gain access to the neurones. This requires specific transport proteins in each of the membranes. Only the hypothalamus escapes the protection afforded by the blood–brain barrier, since one of its roles is to 'assess' the concentration of some compounds in blood to allow it to regulate functions such as water balance, appetite, satiety, resting energy expenditure, reproductive function and sleep.

Immune cells

The resident macrophages are the microglial cells. Lymphocytes and neutrophils are present in cerebrospinal fluid. They assist the microglia to detect and kill invading organisms. Samples of this fluid can be obtained by a technique known as lumbar puncture, analysis of which can be used to aid the diagnosis of pathology in the brain (e.g. from presence of enzymes, other proteins, immune cells, antibodies or microorganisms).

Blood supply

Most of the blood reaching the brain does so through the left and right carotid arteries but about 15% enters via a pair of vertebral arteries that branch from the subclavian arteries. These arteries are interconnected via the circle of Willis from which arise the three cerebral arteries. The cerebellum is supplied by branches from the vertebral arteries. Being unusually thin-walled, cerebral arteries are particularly vulnerable to damage from hypertension, which can lead to a stroke. For the most part, however, blood pressure in the cerebral vessels is stabilised, despite fluctuations in general arterial pressure, by regulatory changes occurring within the brain. An increase in electrical activity in a specific part of the brain leads to localised increases in blood flow. For example, voluntary movements of the mouth cause an increase in blood flow in the corresponding parts of the cerebral cortex.

The brain lacks connection with the lymphatic system. The interstitial fluid drains into the perivascular space, which surrounds arteries and veins, and from there into the sub-arachnoid space where it mixes with the cerebrospinal fluid. This is secreted by the choroid plexuses, which are capillary-rich outgrowths into cavities within the brain, known as ventricles. From the ventricles, cerebrospinal fluid flows through channels to the surface of the brain and

to the spinal cord. There is continuous absorption of cerebrospinal fluid into veins through arachnoid villi. About one litre is produced each day.

An important function of cerebrospinal fluid is mechanical, supporting the brain and cushioning it from blows to the cranium. It also provides another means of removing unwanted substances from the brain, including any neurotransmitters or metabolites that have leaked from synapses.

Electrical communication

The resting potential

Most cells possess an electrical potential across their plasma membrane, which is positive on the external surface. This is known as the resting potential. The neurone is no exception: it has a potential of between 50 and 75 millivolts (mV). The resting potential arises from the following:

- The intracellular concentration of K^+ ions within the neurone is considerably greater than that outside.

- The K^+ ion channels in the plasma membrane are open (see Chapter 5 for discussion of ion channels: Figure 5.3).

- The K^+ ions diffuse out of the cell down their concentration gradient which results in a positive charge on the outer surface of the neurone and a negative charge on the inside.

- The concentration of Na^+ ions is greater on the outside of the neurone than on the inside but the Na^+ ion channel is closed in the resting state.

- Under these conditions, the concentration difference of K^+ ions is largely responsible for the potential difference. On the basis of the quantitative difference in concentration of ions, the value of the potential can be calculated from the Nernst equation (Appendix 14.1).

The action potential

Information is transmitted along neurones in the form of electrical impulses, i.e. transient changes in potential across the cell membrane which travel from the cell body along the axon to the axon terminus (Figure 14.5). Impulses are of a fixed size but vary in frequency to convey information.

An impulse is initiated by a sudden depolarisation (i.e. marked reversal of the resting potential). The depolarisation is achieved by the sudden opening of a number of Na^+

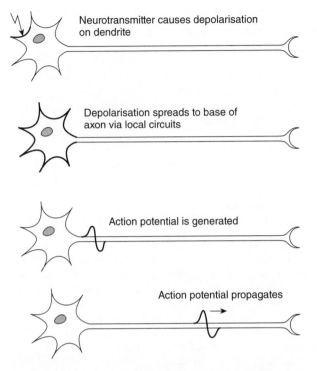

Figure 14.5 *Generation of an action potential.* Binding of the neurotransmitter (an excitatory transmitter – see below) to the postsynaptic membrane causes depolarisation. This spreads to the base of the axon where, if the depolarisation is of sufficient magnitude, an action potential is generated and then travels along the axon.

ion channels. The Na^+ ions diffuse into the cell down their concentration gradient. This neutralises the negative charge on the inside of the membrane so that the potential difference is lost. After depolarisation, the following sequence of changes occurs:

- Depolarisation results in the opening of further Na^+ ion channels in the membrane.

- More Na^+ ions diffuse into the cell, causing a further decrease in potential (i.e. greater depolarisation).

- This depolarisation opens more Na^+ ion channels in the membrane, so that more Na^+ ions enter.

- This process, therefore, provides positive feedback and results in a threshold response to the initial depolarisation.

- This then initiates sequential depolarisation along the whole axon, the action potential. The membrane potential therefore swings from one determined by the K^+ ion concentration difference across the cell membrane to one dominated by the Na^+ ion concentration difference. Since the Na^+ ion concentration is higher outside the neurone, the membrane potential becomes transiently reversed.

Propagation of the action potential along the axon occurs because the voltage change opens the Na^+ ion channels ahead, i.e. the channels are voltage gated (Figure 14.5; see Box 14.1).

A result of the action potential is that some K^+ ions leave and some Na^+ ions enter the axon. The process would, were it not reversed, eventually abolish the ion concentration differences and hence the resting and the action potentials. The activity of an enzyme in the plasma membrane, the Na^+/K^+ ATPase, prevents this happening. It transports Na^+ ions out and K^+ ions into the axon with hydrolysis of ATP (Chapter 5: Figure 5.10).

Chemical communication

The basis of the transmission of information by electrical communication is described above. A limitation of electrical communication is that it lacks flexibility; once an action potential has been generated it is 'conducted' along the neurone without change. Chemical communication provides an opportunity for modifying this transmission. It occurs at synapses, the junction between two nerve cells across which signals are transmitted via chemical messengers. The two neurones on either side of the synapse are known as the presynaptic and postsynaptic neurones, respectively. When an impulse arrives at the terminal of the former, it causes release of a chemical that diffuses across the cleft then binds to a receptor on the membrane of the postsynaptic neurone, which results in a change in the membrane potential. According to the type of chemical, it can either enhance the ability of the nerve to initiate an action potential in the postsynaptic neurone or decrease the likelihood of initiation of an action potential (see below).

Neurotransmitters

The gap across the synapse is so small that the chemical messenger (neurotransmitter) crosses the cleft in less than a millisecond. Within the brain there are more than 50 neurotransmitters, which include amino acids, amines, purines, peptides and some gases. In contrast, in the peripheral nervous system there are only two, acetylcholine and noradrenaline. One of several questions concerning the concept of neurotransmitters is whether they differ, in principle, from local hormones (See below and Chapter 12).

It is important, therefore, to have criteria to identify a chemical in the brain that acts as a neurotransmitter. These are as follows:

Box 14.1 Biochemical explanations for some of the physiological terms associated with electrical activity

Physiologists use terms that are not used by biochemists but are explained by biochemistry, which helps in the understanding of electrical activity. There are three physiological terms that may be unfamiliar biochemical students: (i) excitability; (ii) gated; and (iii) refractory.

Excitability

A characteristic of cells is that the intracellular concentration of K^+ ions is high, whereas the extracellular concentration is low, so that K^+ ions move down the concentration gradient through the open K^+ ion channels and set up a positive charge on the outside of the cell and a negative charge on the inside (i.e. an electrical potential). However, nerve and muscle cells are different from other cells since a decrease in this membrane potential can be transmitted along the membrane as an action potential. Thus, the difference between the potential across the membrane of the erythrocyte and that of the nerve is that the latter can change; in fact, the potential charge reverses, i.e. it is depolarised and this depolarisation can be transmitted along the membrane of the nerve or muscle cell. This is what is meant by excitability, i.e. the ability to transmit an action potential.

Gated

In order to set up this potential and change it, the opening or closure of the ion channel must be regulated. However, the channel can only be open or closed (i.e. either active or inactive in comparison with an enzyme). This is analogous to a gate, so when biochemists talk about enzymes being regulated, physiologists talk about ion channels being '*gated*'. For example, a 'voltage-gated' ion channel is one in which changes in potential (i.e. voltage) regulate the state of the channel, i.e. the term voltage-gated is the same as 'voltage-regulated'.

Refractory

Immediately after the passage of an action potential, the Na^+ ion channels close spontaneously and cannot be re-opened for a period of time. This is the refractory period. An action potential therefore cannot proceed in the opposite direction, i.e. it is unidirectional, which imposes a direction on propagation of the whole action potential. This is analogous to the means by which directionality is achieved in a metabolic pathway or a signalling sequence of reactions: within either process there is at least one irreversible (non-equilibrium) reaction which provides directionality (Chapter 2).

- It should be present within a presynaptic neurone and it should be present in vesicles in the terminus of the neurone.

- It should be released from this neurone into the synapse by exocytosis and this should be stimulated by an increase in Ca^{2+} ion concentration (Chapter 5).

- It should bind to a specific receptor on the surface of the postsynaptic neurone and elicit a response.

- There should be a mechanism for rapid termination of its action, usually enzymatic degradation or removal of the neurotransmitter from the cleft.

- Its action should be mimicked by electrical stimulation of the presynaptic neurone.

This is sometimes described as the 'classical' approach, which appears to be straightforward and sufficiently comprehensive. Unfortunately, however, it does not take into account recent developments in chemical communication within the brain, which are:

(i)   In addition to the 'classical' neurotransmitter, other chemicals are stored in the vesicles, and are released along with the neurotransmitter into the synaptic cleft. It is not clear if they act as an additional neurotransmitter, change the response to the neurotransmitter, i.e. they act as a modulator or act as 'chemical ballast' in the vesicle.

(ii) The gases nitric oxide and carbon monoxide provide chemical communication within the brain. They differ from 'classical' neurotransmitters in that they are not stored in vesicles, their secretion depends on synthesis, and their diffusion is not restricted to a synapse, i.e. they can diffuse in any direction across the extracellular space.

(iii) Some neurotransmitters (e.g. noradrenaline, 5-hydroxytryptamine) are also present in extra-synaptic vesicles; that is, in the extracellular space between neurones.

(iv) Some neurones secrete enzymes that bind to other cells and may change or extend behaviour of neurones or glial cells. Can these be considered as neurotransmitters or neuromodulators?

Such additions to the chemical communication systems within the brain markedly increase the flexibility of chemical communication. Consequently, it is suggested that the term 'neurotransmitter' should be restricted to an agent that transfers a chemical message between neurones within the bounds of a specific structure, the synapse. For all other chemicals, the term neuromodulator can be used. Unfortunately, the distinction between the two is far from clear-cut so that confusion can easily arise. It is suggested by some physiologists that they all should be termed 'neuroactive compounds', although the terms neurotransmitter and neuromodulator are still commonly used and they are used in this chapter and in this book. The structures of some well-known neurotransmitters are provided in Figure 14.6.

Classification of neurotransmitters

Despite the difficulties in terminology, outlined above, it is possible to divide neurotransmitters in the brain into classes, termed types I, II and III:

Type I neurotransmitters These are the fast-acting neurotransmitters, which include glutamate, aspartate, 4-aminobutyrate and glycine. Around 90% of synapses in the brain use this class of neurotransmitter.

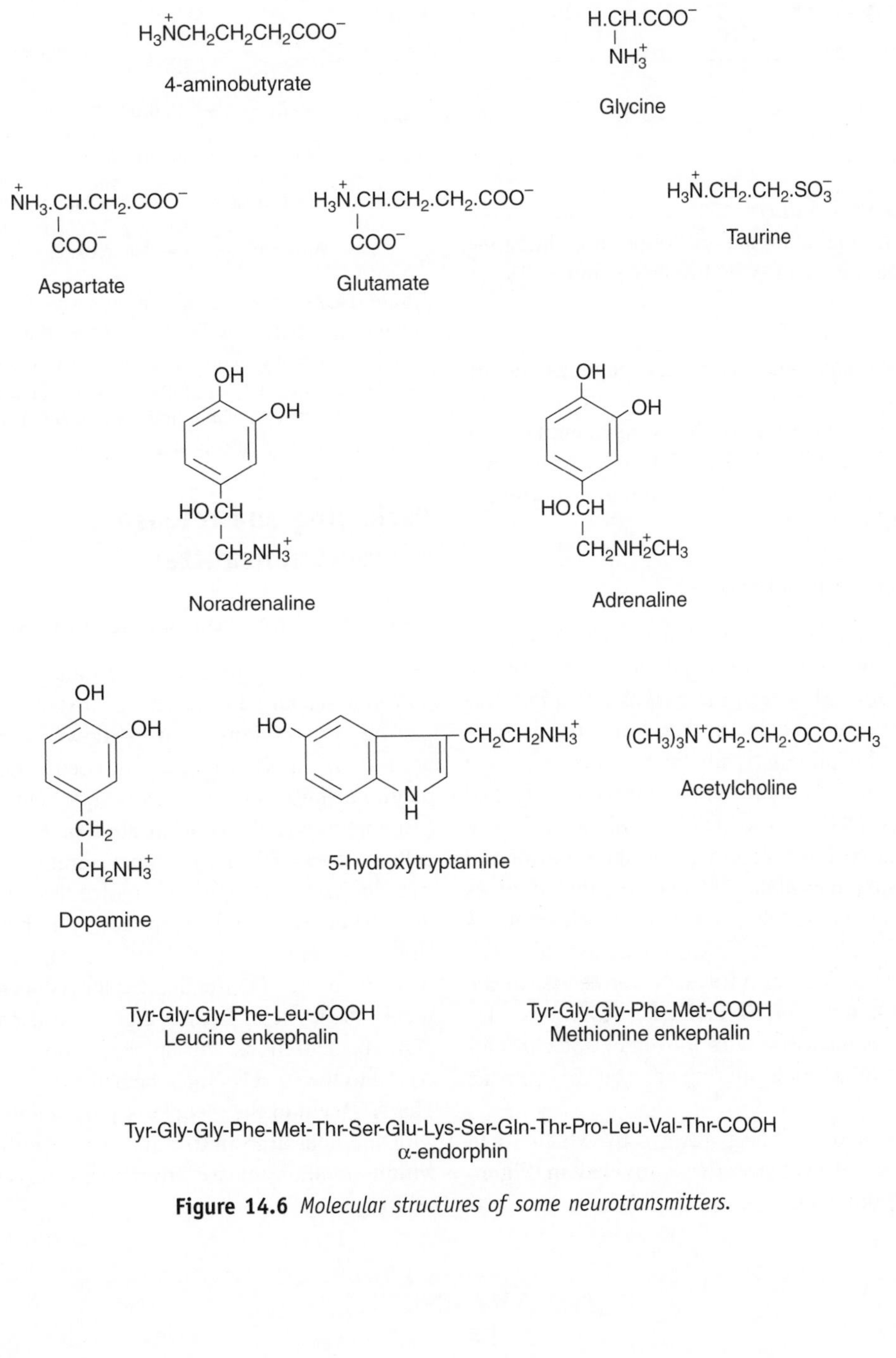

$\overset{+}{H_3}NCH_2CH_2CH_2COO^-$

4-aminobutyrate

$H.CH.COO^-$
$|$
$\overset{+}{NH_3}$

Glycine

$\overset{+}{NH_3}.CH.CH_2.COO^-$
$|$
COO^-

Aspartate

$H_3\overset{+}{N}.CH.CH_2.CH_2.COO^-$
$|$
COO^-

Glutamate

$H_3\overset{+}{N}.CH_2.CH_2.SO_3^-$

Taurine

Noradrenaline

Adrenaline

Dopamine

5-hydroxytryptamine

$(CH_3)_3N^+CH_2.CH_2.OCO.CH_3$

Acetylcholine

Tyr-Gly-Gly-Phe-Leu-COOH
Leucine enkephalin

Tyr-Gly-Gly-Phe-Met-COOH
Methionine enkephalin

Tyr-Gly-Gly-Phe-Met-Thr-Ser-Glu-Lys-Ser-Gln-Thr-Pro-Leu-Val-Thr-COOH
α-endorphin

Figure 14.6 *Molecular structures of some neurotransmitters.*

Table 14.1 Some neuropeptide transmitters in the brain

Name	Abbreviation	Number of amino acids
Angiotensin	AT	9
Bombesin	BB	14
Bradykinin	BK	9
Cholecystokinins	CCK	4 or 8
β-Endorphin	–	26
Enkephalins (leu- & met-)	–	5
Neurotensin	NT	13
Neuromedin-K	NK	10
Oxytocin	OT	9
Somatostatin	ST	14
Substance-K	SK	10
Substance-P	SP	11
Vasoactive intestinal peptide	VIP	28
Vasopressin	VP	9

Type II neurotransmitters These are slow-acting neurotransmitters, including acetylcholine, noradrenaline, adrenaline, dopamine and 5-hydroxytryptamine (Figure 14.6).

Type III neurotransmitters These are peptides (neuropeptides), and most are considered to be neuromodulators rather than neurotransmitters. As a further complication, some of these are also found in the intestine, where they act as local hormones or even endocrine hormones (Chapter 4 Table 14.1).

Synthesis of neurotransmitters

The neurotransmitters are synthesised from their precursors in the place where they are required, i.e. the presynaptic nerve terminal. However, the enzymes that catalyse the reactions involved in the synthesis are, themselves, synthesised via the protein synthetic machinery on the rough endoplasmic reticulum, within the cell body and are then incorporated into vesicles in the Golgi. The vesicles are released from the Golgi and then transported along the axon via microtubules, in a process that involves the molecular motor protein, kinesin, which requires hydrolysis of ATP. The process is known as 'axonal flow'. The speed can be as high as 5 μm/s. Nevertheless, in the longest neurones, the journey can take several weeks. (This is slow in comparison with movement of chromosomes along a similar molecular spindle during mitosis: Chapter 22.)

A summary of some of the pathways by which some neurotransmitters are synthesised is provided in Figure 14.7. Details are provided in Appendix 14.2.

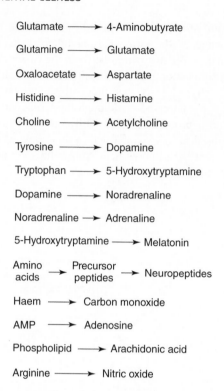

Figure 14.7 *Summaries of the pathways for synthesis of some neurotransmitters.* This figure indicates that to provide for the synthesis of many neurotransmitters the brain must take up a considerable quantity of amino acids. In fact, the brain contributes significantly to whole body amino acid turnover, and cannot be ignored in any calculations of turnover.

Packaging and release of neurotransmitters

Packaging of neurotransmitter in vesicles

Within the presynaptic nerve terminal, the neurotransmitters are transported in vesicles. Most of the information on packaging neurotransmitters is known for the amines. There are several important points. The concentration of the amine neurotransmitters in vesicles is very high, about 0.5 mol/L (for comparison the concentration of ATP in the cytosol of cells is about 0.01 mol/L). Since amines all bear a net positive charge, to package these molecules together the charge must be neutralised. It is, in fact, ATP that neutralises this charge. Several amine molecules can be neutralised by one ATP molecule. Consequently, the concentration of ATP within the vesicles is also high. In order to achieve such high concentrations, the uptake of both monoamines and ATP into the vesicles must be active (i.e. ATP hydrolysis). The ATP within the vesicle is presumably released along with the neurotransmitter. Its fate is probably hydrolysis within or after release from the synapse (e.g. by glial cells).

Release of neurotransmitters into synaptic cleft

In an average synapse, some 20–30 vesicles are docked at sites on the presynaptic plasma membrane ready for fusion. Approximately 10 times more vesicles are held in reserve in the cytosol. When the action potential arrives at the terminal, the vesicles that are attached to the membrane discharge their contents into the synaptic cleft by exocytosis. The whole process involves several stages:

- Docking of the vesicles not already docked on the membrane;

- Fusion with the membrane;

- Formation of a small pore between the two membranes that have fused;

- Discharge of the contents into the cleft through the fusion pore.

The arrival of the action potential at the presynaptic terminal opens voltage-dependent Ca^{2+} ion channels in the plasma membrane so that the Ca^{2+} ions enter the cytosol down their concentration gradient. This results in activation of a Ca^{2+}-binding cytosolic or a membrane protein. This facilitates movement of the vesicles to the membrane and formation of a fusion pore through which the neurotransmitter is discharged into the synaptic cleft (i.e. exocytosis). This occurs within about 0.1 ms of the arrival of the depolarisation (Figure 14.8). The process of exocytosis lasts for only a short time, since the Ca^{2+} ion concentration in the cytosol is rapidly lowered due to the Ca^{2+} ion extrusion from the cell (Appendix 14.3).

Summary of processes involved in formation, release and inactivation of a neurotransmitter

The first step is the synthesis of the enzymes that are involved in the formation of the neurotransmitters, which occurs on the rough endoplasmic reticulum in the cell body. They are then transferred to the terminus of the presynaptic neurone by axonal transport. Here the neurotransmitters are synthesized prior to packaging into vesicles. The contents of the vesicles are, upon stimulation of the presynaptic neurone, released into the synaptic cleft. After binding to the postsynaptic receptor they are inactivated either by uptake from the cleft back into the presynaptic neurone or by enzymic degradation (Figure 14.9). After exocytosis, the membrane of the vesicle can be recycled back into the presynaptic neurone for re-filling and further exocytosis (see Figure 14.8).

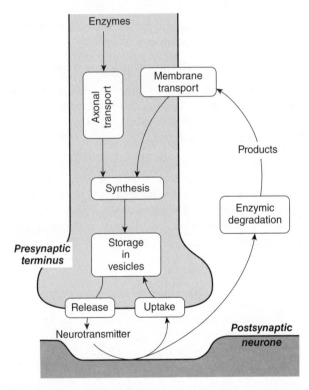

Figure 14.9 *Axonal transport of enzymes, neurotransmitter synthesis, storage in vesicles, release and uptake by presynaptic neurone or enzymic degradation.* The neurotransmitter in the synaptic cleft may be removed by the presynaptic neurone (i.e. recycling), by the postsynaptic neurone or by glial cells (not shown). Alternatively, the neurotransmitter may be degraded, and therefore inactivated, by enzyme action. For example, acetylcholine is degraded by acetylcholinesterase in the synaptic cleft (Chapter 3). One of the products, choline, is transported back into the neurone to be reacted with acetyl-CoA to re-form acetylcholine. The vesicle, once empty, may also be recycled for re-packaging (Figure 14.8).

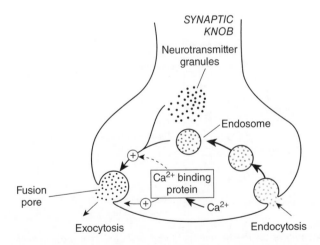

Figure 14.8 *Simple diagram of release of neurotransmitter and recycling of the vesicles in presynaptic neurone.* After exocytosis, the membrane recycles to form a new vesicle which is re-filled with neurotransmitter. The Ca^{2+} ion binding protein may control packaging, formation of fusion pore and release of neurotransmitter.

Effect of neurotransmitter on the postsynaptic membrane

After release by exocytosis, the neurotransmitter diffuses across the cleft and binds to a receptor on the membrane of the postsynaptic neurone. The binding affects an ion channel, which changes the polarisation of the postsynaptic membrane (see below).

Factors affecting the extent of depolarisation

It is the extent of binding of the neurotransmitter to the receptor which determines the magnitude of the change in polarisation. This is determined by several factors:

- The concentration of neurotransmitter in the synaptic cleft, which depends upon the rate of discharge of vesicles into the cleft and/or the rate of removal from the cleft or degradation within the cleft.

- The affinity of the receptor for the neurotransmitter (i.e. the value of the dissociation constant).

- The number of receptors in the postsynaptic membrane.

- The effectiveness of the coupling between the receptor and the effector system, i.e. the ion channel, e.g. the activity or capacity of a protein kinase that modifies an ion channel (see below).

These are similar, in principle, to the factors that determine the binding of a hormone to a receptor and the resultant change in the rate of a hormone-dependent biochemical process (Chapter 12). The significance of these factors in the neurological disorders and effects of recreational drugs are discussed below.

The eventual result of binding of the neurotransmitter is a change in activity (i.e. opening or closing) of an ion channel. This can occur directly or indirectly. The direct mechanism is via a ligand-gated ion channel, abbreviated to LGIC (also known as an ionotropic receptor). The indirect mechanism is a G-protein coupled receptor, abbreviated to GPCR (also known as a metabotropic receptor).

The ligand-gated ion-channel (LGIC) receptor

This rather cumbersome name indicates that the neurotransmitter binds a receptor that directly affects an ion channel, i.e. the neurotransmitter binds to one of the proteins that constitute the channel (Chapter 5, see Figure 5.4).

The G-protein coupled receptor (GPCR)

The binding of a neurotransmitter to this receptor changes the activity of an ion channel, indirectly. Binding to the receptor increases the activity of a G-protein (via exchange of GDP for GTP – see Chapter 12). This active G-protein activates adenyl cyclase which increases the concentration of cyclic AMP which leads to a change in the activity of protein kinase A. This phosphorylates an ion channel resulting in a change in activity, i.e. opening or closing the channel (Figure 14.10). The effect is reversed by a protein phosphatase, i.e. the activity is controlled by reversible phosphorylation (an interconversion cycle). Such a sequence of processes means that the action is much slower than that of the LGIC receptor.

It is likely that the advantage of regulation of ion channel activity via the interconversion cycle is an improvement in sensitivity of the change in activity of the ion channel to the change in concentration of the neurotransmitter in the

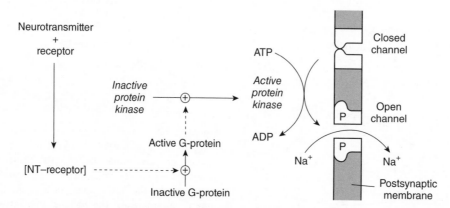

Figure 14.10 *Diagrammatic representation of regulation of the opening of an ion channel by phosphorylation of a protein in the channel.* The neurotransmitter–receptor complex functions as a nucleotide exchange factor to activate a G-protein which then activates a protein kinase. This is identical to control of G-proteins in the action of hormones (Chapter 12, see Figure 12.21). Phosphorylation of a protein in the ion channel opens it to allow movement of Na$^+$ ions. The formation of the complex, activation of the G-protein and the kinase takes place on the postsynaptic membrane. An example of the structural organisation and the involvement of a G-protein is shown in Chapter 12 (Figure 12.6).

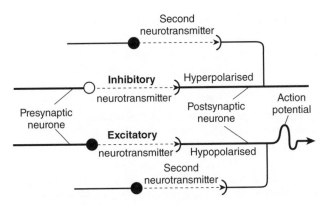

Figure 14.11 *Effects of excitatory and inhibitory neurotransmitters on initiation of an action potential in response to a second neurotransmitter.* If the neurotransmitter released from the presynaptic membrane is inhibitory, it will reduce the likelihood that the second neurotransmitter will initiate an action potential. If the neurotransmitter is excitatory, it will increase the likelihood that the second neurotransmitter will initiate an action potential in the postsynaptic neurone. The second neurotransmitter arises from synapses of other axons.

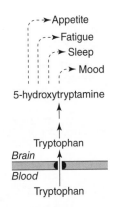

Figure 14.12 *Various physiological conditions that are modified by 5-hydroxytryptamine, probably due to different receptors acting on different neurones for the neurotransmitter.* The physiological significance of 5-hydroxytryptamine on mood is discussed below, its role in fatigue is discussed in Chapter 13 (See Figures 13.29 and 13.30). For its role in appetite, see Appendix 15.4.

cleft. The biochemical basis of interconversion cycles in improving sensitivity is discussed in Chapter 3.

Excitatory and inhibitory neurotransmitters

Binding of the neurotransmitter to the receptor on the postsynaptic membrane can have one of two opposing effects. (i) It can decrease the membrane potential by opening Na^+ ion channels. If the resultant hypopolarisation is of sufficient magnitude, it will either initiate an action potential or make it more likely that the binding of a different neurotransmitter will initiate an action potential. This is described as an excitatory neurotransmitter (e.g. glutamate). (ii) It can close an ion channel to result in hyperpolarisation which will decrease the likelihood that the binding of a different neurotransmitter will initiate an action potential. In this case, it is known as an inhibitory neurotransmitter (e.g. gamma-aminobutyrate) (Figure 14.11 and Appendix 14.4).

Different receptors for the same neurotransmitter

Henry Dale first established that there were two different receptors for acetylcholine, the muscarinic and nicotinic, and it is now known that there are five different classes of muscarinic and an ill-defined number of nicotinic receptors. Different receptors were originally identified by pharmacological methods but molecular biological techniques now provide very sensitive and precise means of identify-

ing different receptors for the same neurotransmitter, so that very many have now been discovered. The role of different receptors is to vary the response to the same neurotransmitter according to which neurone is involved in the pathway at a particular location in the brain. Although the physiological roles of many or most of these different subclasses of receptor are not known, there is no doubt that they provide even more flexibility and variation in information transfer. Thus it is now known that 5-hydroxytryptamine affects several physiological processes, presumably via different classes of receptor (Figure 14.12).

The kinetics of neurotransmitter communication

In the brain there are over 50 neurotransmitters, the role of which is to convey information either quickly or slowly and either acutely or chronically across a synapse. The changes in the concentration of neurotransmitters in the synaptic cleft are central to the process of information transfer in the brain, as indicated by the pathology that arises when the concentrations are disturbed. The concentration depend upon the rate of release into the cleft and the rate of removal or inactivation. The kinetics of such a sequence is discussed in chapter 12 for control of the concentration of second messenger (Box 12.2). These kinetics are, therefore, applied to the neurotransmitter system and, in addition, it is compared with two 'classical' second messenger systems, cyclic AMP and Ca^{2+} ions.

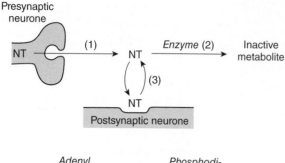

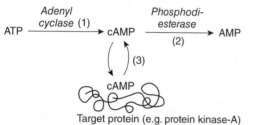

Figure 14.13 *The kinetic sequence of reactions that control the cyclic AMP concentration, and its binding to the effector system, and the kinetic sequence that controls the concentration of a neurotransmitter and its binding to the receptor on the postsynaptic membrane.* Processes (1) are reactions catalysed by adenyl cyclase, and exocytosis. Reactions (2) are catalysed by phosphodiesterase and, for example, acetylcholinesterase. Reactions (3) are the interactions between the messenger and the effector system: both the latter are equilibrium binding processes. (See Chapter 12 (p. 266) for discussions of equilibrium binding.)

The events in the neurotransmitter sequence can be represented by the same general scheme as for second messenger:

$$S \xrightarrow{(1)} X \xrightarrow{(2)} P \xrightarrow{(3)}$$

In this process, X represents the neurotransmitter. Reaction (1) is the zero order process, i.e. the release of the neurotransmitter, which is the process of endocytosis from the presynaptic neurone. The process (2) is the reaction that results in decreasing the concentration of the neurotransmitter, which terminates this effect: it is a first order process (see Box 12.2). Two reactions can carry out process (2), enzymatic conversion to an inactive product in the cleft or removal of neurotransmitter from the cleft by uptake into the presynaptic terminal, a glial cell or the postsynaptic neurone. It is unclear whether these two processes for termination can exist together in one synapse or whether only one process occurs in one synapse. To demonstrate that the kinetic sequence that controls the neurotransmitter concentration in the cleft is similar, in principle, to other well-established sequences, those for cyclic AMP and Ca^{2+} ions are presented alongside that for a neurotransmitter in Figures 14.13 and 14.14. A major difference is that the neurotransmitter sequence is physically bounded by the synapse whereas the cyclic AMP sequence, in the case of protein kinase, is restricted to the cytosol as is the messenger Ca^{2+} ions. There is remarkable similarity in the concen-

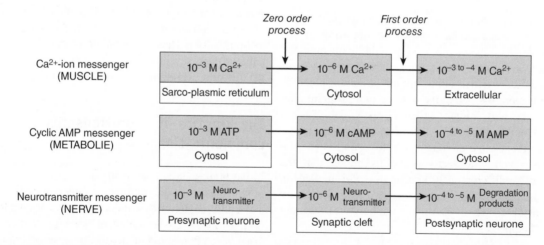

Figure 14.14 *Quantitative biochemical similarities between the concentrations of precursors for messenger formation, the concentrations of messengers and the concentration of degradation products.* These are represented by Ca^{2+} ions, cyclic AMP and a neurotransmitter. The precise concentrations of neurotransmitter in the presynaptic neurone terminal are not known; if it were not localised in vesicles, it would be approximately 10^{-6} M/L so that within the vesicle it is estimated to be about 10^{-3} M/L. Similarly the precise concentration in the synaptic cleft is also a guesstimate. On the basis of the activity of exocytosis and the K_m of the degradation process, it would be possible to calculate the concentration within the synaptic cleft (see Chapter 12 considerable for an example). There is Box 12.2 similarity between the substrate concentration for the zero order processes, the concentrations of messenger and the 'breakdown' products for each process.

The increase in the concentrations of Ca^{2+} ions and cyclic AMP upon stimulation is usually, between 10 and 100-fold. It is assumed that a similar increase in neurotransmitter concentration will occur in the cleft but this has not been reported.

trations of the substrates for the zero order processes, in the concentrations of the messengers and in those of the degradation or inactivated products (Figure 14.14).

Fuels and energy metabolism in the brain

Despite comprising only 2% of body weight, the brain takes 20% of the resting cardiac output to supply glucose and oxygen. The blood supply is required to generate sufficient ATP for various energy-requiring processes in the brain. These include not only maintenance of ion gradients for generation of electrical impulses, on which neural communication is based, but also to synthesise the many different neurotransmitter molecules (some of which are peptides) to package them into vesicles, to move them within the cell, release them into the synaptic cleft and then remove them from the synapse.

In the fed state, the only fuel used by the brain is glucose, derived from the blood. In prolonged starvation or chronic hypoglycaemia, ketone bodies are used which reduce the rate of utilisation of glucose by the brain but, even so, glucose still provides about 50% of the energy. Consequently, under all conditions, maintenance of the blood glucose concentration is essential for the function of the brain: the mechanisms that are responsible for this are discussed in Chapters 6, 12 and 16.

The brain takes up glucose and oxygen in stoichiometric amounts which indicates that glucose is completely oxidised to CO_2. However, this does not mean that lactate is *not* produced or that neurones actually oxidise glucose (see below). Of the two major groups of cells in the brain, neurones and glial cells, most of the energy-consuming processes occur in the neurone but the glial cells play a vital role in the provision of nutrients for the neurones.

From arteriovenous difference, the utilisation of glucose by human brain is approximately 0.32 mmol/min which, if completely oxidised would consume (6 × 0.32) or 1.92 mmol/min of oxygen. The rate of oxygen uptake by the human brain is also measured from arteriovenous difference: it is 2.1 mmol/min (i.e. very close to that calculated, 1.92).

Provision of fuel for the neurone

The traditional model for the processes by which glucose is made available for the neurones is that glucose is transported across the blood–brain barrier into the extracellular fluid, from where it is taken up by neurones (and glial cells). However, there are several problems with this model:

- The capillaries are lined by endothelial cells with extremely effective tight junctions, which constitute part

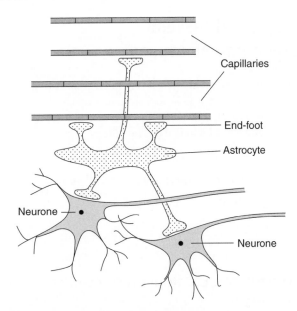

Figure 14.15 *A diagram to illustrate how the end-feet of an astrocyte are in direct contact with capillaries and with neurones. It is likely that the endfeet of the astrocytes will extent over most of the surface of the capillaries, since they provide one component of the blood-brain barrier.*

of the blood–brain barrier, hence glucose must pass through the endothelial cell prior to entering the brain.

- The end-feet of astrocytes are in very close contact with the endothelial cells, so that they are interposed between the capillaries and the extracellular space; in fact 90% of the capillary surface is covered with astrocyte feet (Figure 14.15).

- Neurones are also surrounded by astrocyte end-feet.

- The extracellular glucose concentration in the brain of the rat is about 0.5 mmol/L whereas the intracellular concentration is estimated to be 2 mmol/L. Since there is no *active* transport into the brain, some form of compartmentation of glucose must exist.

A more recent model is that glucose is transferred from the blood through the endothelial cells and then directly into the astrocytes, so that the concentration of glucose in the astrocyte is similar to that in the blood (>2 mmol/L) and that astrocytes then feed the neurones. How is this done? In astrocytes, glucose is converted into lactate, which is then released for uptake by the neurone. After uptake, lactate is converted to pynuvate which is completely oxidised via the Krebs cycle to generate the ATP required. Hence, the astrocyte obtains its energy from glycolysis but most is saved for the neurone (see also Chapter 9). Any interference in glycolysis in astrocytes could have serious consequences for the neurone.

Mental illnesses: biochemical causes

Disorders of the brain affect a very large number of people. It is estimated that about one in five women and one in seven men in the UK suffer from mental illness. The biochemical causes include the following:

• disturbances in neurotransmitter systems;

• impairment of electrical activity;

• energy stress;

• disturbance of membrane structure.

Primary defects in neurotransmitter systems

Schizophrenia

Schizophrenia is a psychotic illness and is one of the most common psychotic disorders (a mental illness in which the sufferer loses contact with reality). About half a million people in the UK suffer from schizophrenia. It affects mainly adolescents and young adults, and there is a genetic component to the disease. Lay terms that have been used for the disorder are insanity, lunacy and madness. Hospitals that catered for such patients were formerly known as lunatic asylums.

Patients suffering from schizophrenia have difficulty with thought processes rather than with mood. The patient loses touch with reality, emotional responses are either blunted or inappropriate and behaviour can be abnormal. Patients feel that control of their thoughts has been lost to some outside agency. They may also suffer from delusions (completely unrealistic ideas), hallucinations (the patient vividly experiences sensations that are not real) or hears voices when no one is present. Sometimes the voices convey derogatory and critical messages. Whereas disorders of mood are usually episodic, schizophrenia is usually a chronic illness. Recreational drugs such as lysergic acid diethylamide (LSD) also cause hallucinations (see below) and the effects of taking such drugs may bring about a psychosis that is similar to schizophrenia.

The term 'schizophrenia' was introduced by the Swiss physician Eugen Bleuter to replace the earlier term *dementia praecox*. It derives from the two Greek words *schism*, 'a split', and *phren*, 'the mind', to indicate the apparent splitting of the mind. One part remains in touch with reality whereas the other part is out of touch. It is, however, characterised by its symptoms rather than by biological markers. The current hypothesis to account for the most

prominent symptoms which, to a student entering the field for the first time, can appear to be surprisingly straightforward for such a severe disorder, is an overactivity of the dopamine neurotransmitter system in one part of the forebrain. Overactivity could be due to one of three changes from the normal:

• An abnormally high concentration of dopamine in the synapses, which is due either to an excessive rate of release from the presynaptic terminal or an abnormally low rate of removal of dopamine from the synaptic cleft.

• An increase in the number or the affinity of receptors for dopamine in the postsynaptic neurones in parts of the forebrain.

• An increase in the activity of the effector system for the dopamine neurotransmitter (e.g. a protein kinase – see Figure 14.10).

Consistent with this proposal, some post-mortem analyses of the brains of patients have shown either an increased concentration of dopamine or an increase in the number of dopamine receptors in the forebrain. These analyses have provided no indication of degeneration of neurones in the brain, but gross structural abnormalities are present: the ventricles are enlarged and the cortex is thinner compared with 'normal' brains. This suggests a disorder of development of the brain but how this could give rise to overactivity of the dopamine system in the forebrain is not known.

Drugs that are successful in treating the disease act as dopamine receptor blockers and are known as antipsychotics or neuroleptics (e.g. chlorpromazine, haloperidol). Antipsychotic drugs reduce some of the symptoms, especially the delusions and hallucinations. A side-effect of the drugs is that they can result in symptoms similar to those seen in patients with Parkinson's disease. This is not surprising, since the hypothesis to explain Parkinson's disease is too low a concentration of dopamine in a specific area of the brain (see below).

Affective disorders

Disorders that are characterised by changes in mood are known as affective disorders, which are depression and mania, now known as unipolar and bipolar affective disorders, respectively. Mood is considered to depend upon the concentration of an amine neurotransmitter in some parts of the brain.

Depression Depression is characterised by feelings of profound sadness, insomnia, and diminished appetite and sexual desire. It is not an uncommon disorder; it is estimated that 10–20% of the population in developed countries will suffer from depression at some time in their life. It is caused by an abnormally low concentration of a catecholamine and/or 5-hydroxytryptamine in some parts of the brain. This was first proposed in the 1950s when hypertensive patients were treated with the drug reserpine and it was noted that they became severely depressed: reserpine was known to decrease the concentration of monoamines in the brain of experimental animals. The low concentration could be caused by a low rate of release into the synaptic cleft or too high an activity of the process that removes the neurotransmitter, or both. A low effectiveness of the neurotransmitter could also be due to a low number of receptors in the postsynaptic membrane. Consistent with this hypothesis, some recreational drugs that improve mood (i.e. result in feelings of euphoria) bind to amine receptors (e.g. amphetamines, ecstasy – see below).

It should be possible to treat the disease by increasing the concentration of the neurotransmitter in the synaptic cleft. There are, in principle, three ways in which this could be achieved. (i) Neurotransmitter synthesis could be increased. (ii) The rate of exocytosis could be increased. (iii) Removal of neurotransmitter from the synapse could be inhibited. Drugs that affect process (iii) have been developed. The tricyclic antidepressants and the specific serotonin (5-hydroxytryptamine) reuptake inhibitors (abbreviated to SSRIs) inhibit uptake of the neurotransmitter into the presynaptic on postsynaptic neurone. The most prescribed SSRI is fluoxetine (Prozac).

Bipolar disorder (mania) This disorder is characterised by marked swings in mood from depression to elation. It is considered to be caused by too high a concentration of a catecholamine, and/or 5-hydroxytryptamine in the same parts of the brain that are affected in depression. The effectiveness of the neurotransmitter could be reduced by inhibiting exocytosis or by blocking the binding of the neurotransmitter to its receptor. This has been achieved by use of the antipsychotic drug which blocks dopamine receptors. However, the most frequently used drug is a lithium salt. Its mode of action is interesting since it interferes with regulation of Ca^{2+} ion concentration in the cytosol of the presynaptic terminal and fence in exocytosis. This is achieved indirectly by inhibiting a phosphatase that degrades inositol phosphates so that the inositol concentration is markedly decreased and phosphatidylinositol bisphosphate cannot be synthesised. This lowers the concentration of the bisphosphate in the plasma membrane so that insufficient inositol trisphorphate is formed to stimulate Ca^{2+} ion release from the reticulum (see Chapter 12: Figure 12.19).

Secondary defects in neurotransmitter systems

Neurodegenerative disorders are due to chronic damage to neurones from which they fail to recover and eventually die. This results in failure of the neurotransmitter system in which these neurones are involved. Since the damage progressively accumulates throughout life, neurodegenerative disorders increase with age. Consequently, as the lifespan of the population progressively increases in developed countries, the proportion of the population suffering from these disorders is increasing.

Alzheimer's disease

The disease is named after Alois Alzheimer (1864–1915), a German neurologist who first described it in 1906: it was previously known as presenile dementia. It is now the commonest form of dementia in the elderly, although it can affect people earlier, in the middle part of life. It is estimated to affect more than 15 million people worldwide. Signs and symptoms of the disease are as follows: progressive impairment of cognitive function; behavioural change, such as aggression and depression, and loss of short-term memory (Box 14.2).

There are two main hypotheses to explain the damage to neurones in Alzheimer's disease:

- The presence of neurofibrillary tangles within the neurones (both in the cell body and the axons). Tangles are

---

Box 14.2 A snapshot of one problem of Alzheimer's disease: loss of short-term memory

The novelist, Iris Murdoch, developed Alzheimer's disease in the mid 1990s. Her husband, John Bayley, in his book *A Memoir of Iris Murdoch*, recounts an incident which neatly encapsulates the problem of memory loss. Despite some misgivings, he takes Iris to an Oxford academic party, which usually consists of alcoholic drinks and conversation. The latter is usually a superficial description of occupations between two guests who are not known to one another. Bayley describes how, from a suitable but discreet distance, he keeps an eye on Iris's 'performance'. She is in conversation with an Insurance Adjustment officer who is describing what his job entails. Bayley writes, 'During a break in conversation, I hear her say: "What do you do?" From the face opposite her it is evident the question has been repeated several times in the last few minutes. Undiscouraged he begins all over again.' Bayley continues, 'Some people might actually find it restful at a party to talk to someone more or less with Iris's condition. I think I should myself.'

filaments of proteins twisted around each other. The major protein in the tangle is known as tau, which is phosphorylated to an abnormal extent and contains attached ubiquitin molecules. This suggests that the neurone is 'attempting' to degrade the protein by 'tagging' it with ubiquitin (see Figure 8.2). The protein tau is normally a component of microtubules and hence it is involved in axonal transport. This suggests that degradation of tubules is one of the first changes that give rise to neuronal damage but there is no explanation as to why the tubules are being degraded.

- A protein known as the amyloid precursor protein (APP) spans the plasma membrane of the neurone. It possesses an extracellular domain but its function is unknown. The extracellular domain is partially hydrolised by proteolytic enzymes, known as secretases. One of the products is the amyloid peptide, of which there are two forms. The larger form, contains 42 amino acids and readily polymerises to form plaques in the extracellular space, damaging the neurones. Some sufferers possess a mutated form of the APP protein which more readily produces the larger peptide upon proteolysis, so that more toxic plaques are produced. It is the progressive accumulation of these plaques that is considered to be one cause of Alzheimer's disease.

The only known change in neurotransmitter metabolism so far detected is a deficiency of acetylcholine in the brain. This has been shown in post-mortem studies on the brains of patients with Alzheimer's disease. Some success in reducing the symptoms of the disease has been obtained with drugs that inhibit the activity of acetylcholinesterase leading to an increase in the acetylcholine concentration, but the improvement is minimal so that its use is controversial.

Of considerable biochemical interest, several factors are associated with Alzheimer's disease but their significance is not known. An elevated plasma level of homocysteine is associated with the disease and it is considered as a risk factor for development of the disease. The damage may be caused by a degradation product of homocysteine a thiolactone which causes oxidative damage to cells (Chapter 22; see Figure 22.7).

Epidemiological studies have shown an association between high levels of plasma cholesterol and increased incidence of Alzheimer's disease. A possible explanation is that if cholesterol enters the plasma membranes of neurones it might reduce membrane fluidity, which could enhance cleavage of APP and hence increased formation of the larger peptide and hence the plaques. Cholesterol can reduce fluidity by filling the gaps in the membrane between the polyunsaturated fatty acids (Chapter 11). Consistent with this idea is that the use of the statin drugs, which lower the plasma cholesterol level, appears to decrease the risk of developing Alzheimer's disease. A high plasma level of the lipoprotein, apoE, is also associated with Alzheimer's disease. ApoE is necessary for transport of cholesterol into cells, there, cholesterol can form cholesteral ester which, in excess can damage cells (see Chapter 22).

Parkinson's disease

James Parkinson, a British physician, published the first description of this disorder in an article entitled 'An Essay on Shaking Palsy' in 1817. Since then, it has been known as Parkinson's disease. Patients with this disease suffer from muscular rigidity, rhythmic tremor, slowness of movement and stiffness of the trunk. Post-mortem studies show that the degeneration occurs in neurones in the substantia nigra region of the brain stem. Dopamine is the neurotransmitter in neurones in this region and, as expected, the level of dopamine in this area is low (For treatment, see Figure 11.17). It is suspected that degeneration results from loss of activity of a protein(s) in the electron transfer chain, so that ATP generation is not sufficient to sustain the normal functioning of the cell and it dies from necrosis. Since there is no evidence of a genetic basis, viruses, environmental pollutants or free radicals have been suggested as agents causing damage to the protein in the electron transfer chain.

Evidence for a direct effect of a toxic agent on the electron transfer chain in mitochondria in this region of the brain has come from an unexpected source. In 1976, a number of drug addicts in California were admitted to hospital with symptoms characteristic of Parkinson's disease. When a medical history was taken, it was noted that all had injected the same batch of a designer drug, which was an analogue of the drug pethidine. It was discovered that this batch contained an impurity, 1-methyl-4-phenyl-1,2,3,6-tetrahydropyridine (MTPT). The oxidation of this compound is catalysed by the enzyme amine oxidase in glial cells to produce a compound (MPP^+) (Figure 14.16) which has a structure similar to dopamine. Hence, upon release from glial cells, it is transported into these neurones by the dopamine transporter. Once within the neurone, it enters the mitochondria where it inhibits the activity of complex 1 in the electron transfer chain, probably in a similar manner to the compound amytal, so that electron transfer and hence ATP generation is inhibited. The resultant decrease in the ATP/ADP concentration ratio results in death of these neurones by necrosis. Since the neurones in the substantia nigra contain dopamine, these are particularly sensitive to MPP^+ and their loss results in Parkinson's disease.

The herbicide paraquat is structurally similar to MPP^+. Paraquat is toxic since it can generate superoxide free

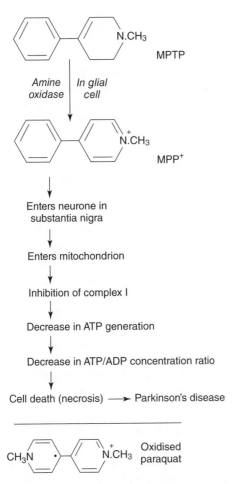

Figure 14.16 *A proposed mechanism by which MPTP can result in Parkinson's disease.* MPTP is 1-methyl-4-phenyltetrahydropyridine and is oxidised to MPP$^+$ (1-methyl-4-phenylpyridinium). This is the agent that destroys dopaminergic neurones. Note the similarity between the structure of MPP$^+$ and paraquat.

radicals that damage the electron transfer chain, but it might also inhibit complex 1 in the electron transfer chain (Figure 14.16).

Impairment of electrical activity

The biochemical basis underlying the normal functioning of the brain is discussed above, where it is emphasised that of the two forms of communication, electrical and chemical, it is the latter that provides most flexibility. Not surprisingly, therefore, there are more disorders due to disturbances in neurotransmitter metabolism than there are due to disturbances in electrical communication. Nonetheless, damage to the structure of the myelin sheath around the axon which interferes with electrical communication can lead to the severe disorder multiple sclerosis.

Multiple sclerosis

Multiple sclerosis is due to damage to myelin in some axons of the brain and spinal cord. This demyelination blocks or slows the transmission of electrical activity in the affected pathway. The underlying cause is in multiple sclerosis is not known, but one suggestion is that it is an autoimmune disease. A virus, or exposure to pollutants, damages the axons, which exposes proteins in the myelin sheath to immune cells that recognise them as foreign (i.e. non-self compound, see Chapter 17), precipitating an immune attack on the myelin sheath. For unknown reasons, certain sites are particularly vulnerable: the optic nerves, causing loss of vision; the spinal cord, causing limb weakness; nerves controlling the bladder sphincter, causing impaired urinary control; and the brain stem, causing loss of balance and coordination. The severity of multiple sclerosis is variable and may depend the ability of the Schwann cells to synthesise new myelin. No treatment is known to reduce the frequency of attacks or enhance the recovery from each attack.

There is some evidence suggesting an association between multiple sclerosis and low exposure to sunlight and hence poor vitamin D status. Thus there is a greater incidence of multiple sclerosis in Scandinavia and a lower incidence in countries nearer the equator. Exposure to sunlight is a first step in formation of vitamin D in the skin (Figure 15.12). There is also some evidence that exposure to sunlight protects against colon and breast cancer but a mechanism for such an effect is not known (Chapter 21). It is tempting to speculate that there is some common factor that protects against the development of multiple sclerosis and tumours, and that vitamin D, an analogue or a metabolite is responsible. It is of interest, therefore, that a deficiency of vitamin O can also result in myopathy, apparently unrelated to its effect on calcium mobilisation from bone.

Polyneuritis

Polyneuritis is a disorder of the peripheral nerves. It involves damage to the myelin sheath. The condition is due to inflammation of the axonal membrane caused by viral or bacterial attack, i.e. an autoimmune disease. Guillain–Barré syndrome is one form of polyneuritis and is an example of an autoimmune disease caused by immune mimicry in response to a bacterial or viral antigen. It is discussed in Chapter 17.

Energy stress

Failure of neurones to generate sufficient ATP to maintain the ATP/ADP concentration ratio can lead to sufficient abnormalities of metabolism, particularly maintenance of the intracellular ionic balance, that necrosis and cell death result (Chapter 20). This could account for some of the loss of neurones in both Alzheimer's and Parkinson's diseases. The loss of the capacity to generate ATP in the brain could be due to hypoglycaemia, oxidative stress or a pathological disorder.

Another condition in which energy stress plays a role in disturbance of neural activity is stroke. In stroke, either haemorrhagic or ischaemic, the oxygen supply to parts of the brain is reduced. Since ATP generation is totally aerobic in neurones, this results in complete failure of ATP generation.

Oxidative stress

It is considered that the brain and nervous tissue are more susceptible to oxidative damage than other tissues or organs. Chronic oxidative stress may lead to degeneration of neurones and hence the degenerative diseases. Susceptibility may be due to the exclusive dependence on mitochondrial ATP generation, with the opportunity of formation of oxygen free radicals from the continuous flux of electrons along the electron transfer chain and the presence of the high concentration of oxygen in the neurones (hence they are totally dependent on aerobic ATP generation.). In addition, several areas of the brain have a high content of iron which, if released due to damage to neurones, can increase the risk of formation of the dangerous hydroxyl radicals. Frequent damage to the brain as a result of trauma (e.g. boxing, soccer) may result in iron release, neuronal damage and illness such as dementia pugilistica and motor neurone disease (See discussion in Chapter 13).

Hypoglycaemia

A low blood glucose level can cause a low rate of glucose uptake by the glial cells and a low rate of lactate formation for the neurones, which can then result in failure of some higher functions of the brain. Behavioural changes, loss of consciousness and eventually coma can result (Chapter 6). It can also contribute to central fatigue (Chapter 13: Figure 13.27). It has been suggested that a chronic low blood glucose level could be an explanation for chronic fatigue syndrome. This could be a particular problem if hypoglycemia is associated with a partial deficiency of the glucose transporters in the astrocytes in the brain.

Migraine

The main feature of migraine (from the Greek *hemicrania*) is a severe one-sided headache. It can last for up to 24 hours and is frequently accompanied by photophobia and nausea. The initial cause is a decrease in blood flow to specific parts of the brain, due to release of 5-hydroxytryptamine by aggregated blood platelets. This results in local vasoconstriction of the intracerebral arteries (since this neurotransmitter causes contraction of the arterial smooth muscle) and the resulting cerebral hypoxia causes disturbances which are responsible for the aura or prodromal phase. The vasoconstriction results in the production of local vasodilators (e.g. adenosine) which, although normalising the blood flow in the intracerebral arteries, causes marked vasodilation in the extracerebral vessels. These changes, in concert with release of prostaglandins from the endothelial cells, increase the sensitivity of pain receptors in the brain, which results in the headache and other symptoms.

Certain foods can trigger a migraine attack by effects on neurotransmitter release or metabolism in the brain. For example, a number of foods contain tryptamine which is known to cause release of other amines (dopamine, noradrenaline and 5-hydroxytryptamine) from both nerve terminals and platelets. This release could initiate the sequence of events that results in the migraine attack. Elimination of such foods from the diet can decrease the number of headaches. Compounds that discourage platelet aggregation (e.g. aspirin) may prevent such attacks.

Disturbances in membrane structure

Approximately 60% of the dry weight of the brain is fat, a considerable proportion of which is polyunsaturated fatty acids that are present in plasma membranes. It would not be surprising if replacement of the unsaturated acids by the saturated fatly acids in membrane structure due to a dietary deficiency of polyunsaturated fatty acids played some part in development of mental illness. Indeed, it has been found that supplementation of a normal diet with polyunsaturated fatly acids can improve some mental disorder (see chapter 11).

Attention deficit disorder

This disorder is characterised by inappropriate levels of activity, a high frequency of periods of frustration and distraction and hence inability to sustain attention and to concentrate on one activity for a prolonged period of time. A surprising finding is that amphetamine administration, which normally increases or facilitates activity, rapidly and markedly improves behaviour. Patients become calm and their alertness is enhanced. A drug that has been used is methylphenidate (Ritalin). One interesting and recent development is the improvement in the condition by supplementation of the diet with polyunsaturated fatty acids, particularly the omega-3 acids in fish oils (See Chapter 11).

Recreational drugs

Herodotus (c. 484–424 BC), a Greek historian, wrote *Histories* which contained an account of the life in most of the nations of the known world. He was probably the first to describe the use of the recreational drug, hashish, in Book IV as follows.

And now for the vapour-bath. On a framework made up of three sticks, meeting at the top, they stretch pieces of woollen cloth, taking care to get the jams as perfect as they can, and inside this little tent, they place a dish with red-hot stones on it. They then take some hemp seed, enter the tent and throw the seed onto the hot stones. It immediately begins to smoke, giving off a vapour unsurpassed by any vapour bath one could find in Greece. The Scythians enjoy the experience so much that they howl with pleasure.

It has long been known that many plants produce alkaloids, which are nitrogen-containing substances that have potent effects when consumed by animals. Many are valuable drugs (e.g. morphine quinine, atropine). There appears to be no biochemical reason why plants produce these compounds.

In this section, a brief description of some recreational drugs is provided in order to give some background and to link the information in the first part of this chapter with the effects of recreational drugs. Those described here are opium (heroin), cannabis, the stimulants, amphetamines and cocaine, ethanol and the hallucinogens (LSD, mescaline, psilocin, also known as 'magic mushrooms'). The molecular structures of some these drugs are presented in Figure 14.17.

Opiates

Opiates include morphine, heroin, methadone and pethidine.

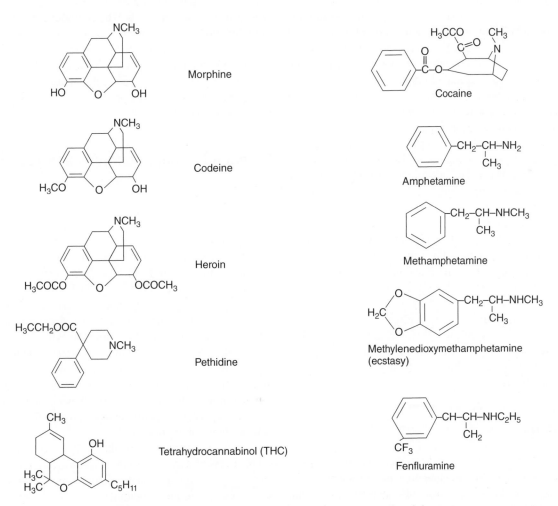

Figure 14.17 *Molecular structures of some common recreational drugs.*

Opium is the dried latex obtained from the seed capsule of the poppy *Papaver somniferum*. The major chemical in opium is morphine, long known as an extremely powerful analgesic, (which is also found in the poppy) albeit one that can lead to addiction. A related molecule, differing only in the addition of a methyl group, is codeine – a much milder and less addictive analgesic. Chemical modification, acetylation, of morphine produces heroin, a very widely abused agent which, among other physiological effects, induces a state of euphoria. The rather specific effects of these alkaloids suggest that they are interfering with one particular process in one particular part of the brain, the thalamus. Specific binding sites for morphine have been found in the thalamus. This in turn suggested that there would be a natural ligand for these receptors and, in due course, a family of peptides known as endorphins were identified as the first peptides affecting the brain.

Signals from pain receptors are transmitted to the brain where they are processed in the thalamus and then passed on to the sensory cortex where the sensation of pain originates. Receptors for endorphins (and for the opiates) are abundant in these regions and it is likely that a natural role of endorphins is to interfere with the transmission of impulses through these regions. The opiates appear to do the same, acting not so much to alter the pain threshold but altering the patient's attitude to pain.

Only one antagonist is known, naloxone, which is used clinically to treat opiate overdoses and, experimentally, to investigate whether physiological or biochemical actions are opiate-mediated. One example of its use is to support the hypothesis that β-endorphin is responsible for the analgesic effects of acupuncture. Not only does low frequency electroacupuncture increase β-endorphin levels in cerebrospinal fluid but naloxone nullifies the analgesic effect of this treatment.

Opiate receptors are linked, via the G-protein system, to K^+ ion channels and to voltage-gated Ca^{2+} ion channels. Binding of the opiates results in opening of K^+ ion channels and hyperpolarisation, so that it is more difficult to initiate an action potential, i.e. they behave like inhibitory neurotransmitters (see above). The binding also results in inhibition of the opening of the Ca^{2+} ion channels in response to depolarisation. That is, both effects are inhibitory on the nervous activity in the brain, which may explain their analgesic effects.

Cannabis

The psychoactive compounds in the hemp plant, *Cannabis sativa*, are known as cannabinoids: the principal one is tetrahydrocannabinol. It is used as a recreational drug either in the form of dried leaves and flowers (cannabis,

Table 14.2 Common (slang or street) names for some drugs discussed in this chapter

Cannabis	marijuana, puff, blow, pot, draw, weed, shit, hashish, spliff, tackle, wacky, ganja
LSD and magic mushrooms	mushies, 'shrooms, acid
Amphetamines	speed, wiz, uppers, billy, amph
Ecstasy	E, doves, disco biscuits, echoes, hug drug, burgers, fantasy
Cocaine	coke, Charlie, snow, C
Heroin	smack, gear, brown, horse, junk, scag, jack

LSD is the abbreviation for lysergic acid diethylamide.

marijuana) or as a resin (hashish) extracted from the plant (Table 14.2). These contain up to 10% tetrahydrocannabinol. It heightens sensory awareness and induces feelings of relaxation and wellbeing similar to those produced by ethanol. Similar to the opiates, a specific cannabinoid receptor is present in the brain, which is coupled, also via a G-protein, to adenyl cyclase and to Ca^{2+} ion channels. Tetrahydrocannabinol decreases the enzyme activity and closes the Ca^{2+} ion channels. The compound present in the brain that binds to the cannabinoid receptor is anandamide, which is an amide of arachidonic acid and ethanolamine. Of interest, compounds similar to anandamide are present in chocolate which may explain the pleasure gained from eating this product and support the claim that it is addictive (see Chapter 11 Box 11.2).

Stimulants: amphetamines and cocaine

Stimulants increase mental alertness and may decrease physical fatigue. The latter effect is likely to be central rather than peripheral since stimulant drugs interfere with brain neurotransmission. Among those most widely used are amphetamines, which are chemically related to the catecholamines (Figure 14.17). Amphetamines increase the amounts of adrenaline, noradrenaline or dopamine in specific parts of the brain, either by releasing them from local extracellular vesicles, by stimulating exocytosis or by interfering with their removal from the synaptic cleft. The immediate effect of amphetamines is an increase in breathing and heart rate but prolonged use can lead to psychological disturbance, including depression. Stimulant use can also impair thermoregulation mechanisms and lead to a risk of hyperthermia, particularly when vigorous exercise is taken in a hot environment.

Cocaine is obtained from the coca tree (*Erythroxylon coca*). It has had a legitimate medical use as a local anaesthetic but has now been superseded by more effective and safer agents. Ecstasy, methylenedioxymethamphetamine (MDMA), is a synthetic amphetamine with no medical use.

It is currently in vogue as a drug that staves off fatigue, induces euphoria and reduces inhibitions. Although sudden death has been attributed to it, this is probably due to dehydration and hyperthermia. Long-term use may result in damage to the neurones in the brain that use 5-hydroxytryptamine as a neurotransmitter.

Ethanol

Ethanol is the major intoxicant in drinks that are made by the fermentation of sugar solutions by yeasts and is the most widely used non-therapeutic drug. A small sherry or whisky can raise the concentration of ethanol in the blood to 2–4 mmol/L. Ethanol is taken up and metabolised by the liver but, if large quantities are consumed, the liver cannot increase the rate of utilisation and the plasma concentration can reach 0.1 M/L, which can damage the liver (see below).

Most of the alcohol that is taken up by the liver is oxidised to ethanal (acetaldehyde). Three separate enzymes, or enzyme systems, can catalyse the formation of acetaldehyde:

(1) $CH_3CH_2OH + NAD^+ \rightarrow CH_3CHO + NADH + H^+$

(2) $CH_3CH_2OH + NADPH + H^+ + 2O_2 \rightarrow CH_3CHO + 2H_2O_2 + NADP^+$

(3) $CH_3CH_2OH + H_2O_2 \rightarrow CH_3CHO + 2H_2O$

The enzymes responsible are (1) alcohol dehydrogenase, (2) microsomal ethanol-oxidising system (MEOS), (3) catalase. At low concentrations of ethanol, the dehydrogenase is most important since its K_m is much lower than that of the other two systems.

Even if no alcohol is ingested, some actually enters the body from the fermentation of sugars by microorganisms in the colon. The amount will depend on the type of microorganisms present in the colon and the amount of complex carbohydrate and fibre that is eaten (Chapter 4). Both vary from one individual to another. The ethanol produced in the colon enters the blood and is taken up from the hepatic portal vein by the liver. It is estimated that the concentration of ethanol derived from the colon in the hepatic portal blood in some individuals may be as high as 0.5 mM/L but is insignificant when compared with that which can result from alcoholic beverages. Nonetheless, it may have important physiological effects. It is likely that this alcohol will be metabolised largely by alcohol dehydrogenase in the liver. A chronic production of alcohol from the colon may maintain a significant concentration (i.e. activity) of this enzyme in the liver. This may explain why some individuals are able to drink alcoholic beverages without obvious neurological effects until large quantities are ingested.

It is noteworthy, here, that one commercial source of purified alcohol dehydrogenase is horse liver. There is a considerable quantity of this enzyme in horse liver. It is likely that alcohol arises from fermentation in the caecum of the high content of fibre present in the food of the horse. Hence its liver will be exposed chronically to high concentrations of alcohol, and the liver adapts by synthesising large quantities of the enzyme.

The further oxidation of ethanal to acetate is catalysed by aldehyde dehydrogenase:

$$CH_3CHO + NAD^+ + H_2O \rightarrow CH_3COO^- + NADH + 2H^+$$

A very high activity of mitochondrial aldehyde dehydrogenases (together with its low K_m) ensures very efficient oxidation in the liver so that the concentration of acetaldehyde in blood remains very low. Nonetheless, it is possible that some of the pathological effects of ethanol are due to acetaldehyde (ethanal). In contrast, a large proportion of the acetate escapes from the liver and is converted to acetyl-CoA by acetyl-CoA synthetase in other tissues:

$$CH_3COO + ATP + CoASH \rightarrow CH_3COSCoA + AMP + PP_i$$

In most tissues, this acetyl-CoA is oxidised via the Krebs cycle to provide ATP. One litre of table wine has an energy content of around 3000 kJ, so that in some individuals as much as 10% of the daily energy requirement can be provided by ethanol (see Table 2.3).

The final reactions to be considered in the metabolism of ethanol in the liver are those involved in reoxidation of cytosolic NADH and in the reduction of NADP$^+$. The latter is achieved by the pentose phosphate pathway which has a high capacity in the liver (Chapter 6). The cytosolic NADH is reoxidised mainly by the mitochondrial electron transfer system, which means that substrate shuttles must be used to transport the hydrogen atoms into the mitochondria. The malate/aspartate is the main shuttle involved. Under some conditions, the rate of transfer of hydrogen atoms by the shuttle is less than the rate of NADH generation so that the redox state in the cytosolic compartment of the liver becomes highly reduced and the concentration of NAD$^+$ severely decreased. This limits the rate of ethanol oxidation by alcohol dehydrogenase.

The NAD$^+$/NADH concentration ratio in the cytosol of the liver is maintained at a value of about 1000 but oxidation of ethanol can lower this ratio by at least tenfold. Many dehydrogenase reactions are close to equilibrium so that, for those that react with NAD$^+$/NADH, the concentrations of all the other substrates and products will be affected by a change in the NAD$^+$/NADH concentration ratio. Hence, a decrease in the NAD$^+$/NADH concentration ratio will lower the concentration of the oxidised reactant and raise that of the reduced reactant of all the dehydrogenation reactions in the cytosol:

$$NAD^+ + \text{reduced reactant} \rightleftharpoons NADH + \text{oxidised reactant} + H^+$$

If either of these reactants has an important metabolic role in the tissues, marked changes in their concentration could produce abnormal effects (see below).

Pathological effects of ethanol

There are many pathological effects of excessive alcohol consumption. The most important are: neuronal and gonadal dysfunction, hypoglycaemia, 'fatty liver', and hepatitis, which can eventually lead to cirrhosis of the liver.

Neuronal dysfunction The transient effects of ethanol are reduction in activity in the brain, similar to effects of volatile anaesthetic agents (e.g. ether). Indeed, ethanol has been used in the past as an anaesthetic. An effect of chronic ethanol consumption is neurological degeneration resulting in dementia, loss of memory and peripheral neuropathies. The dementia is called the Wernicke–Korsakoff syndrome type (or Wernicke's encephalopathy). One explanation for this is a deficiency of thiamine (vitamin B$_1$), which is necessary in the reactions catalysed by pyruvate dehydrogenase and oxoglutarate dehydrogenase and hence the oxidation of glucose and ATP generation are decreased. The deficiency may be the result of a poor diet, since ethanol oxidation can provide a considerable proportion of the daily energy requirement, leading to malnutrition in alcoholics.

Acute neurological effects are probably caused by alcohol 'dissolving' in the plasma membrane and increasing fluidity of membranes (similar to the action of the volatile anaesthetics) which affects ion channels and receptors, particularly neurotransmitter receptors. Three proposed biochemical explanations for effects of ethanol are:

(i) Enhancement of the action of γ-aminobutyric acid which is an inhibitory neurotransmitter. This is similar to the action of benzodiazepines (Appendix 14.4).

(ii) Inhibition of Ca^{2+} ion entry into neurones by inhibition (blocking) of voltage-gated Ca^{2+} ion channels, particularly those involved in control of exocytosis of the neurotransmitters from presynaptic terminals (see above).

(iii) Ethanol directly modifies the activity of receptors for the neurotransmitter, glutamate, which is an excitatory neurotransmitter. Excessive stimulation of glutamate receptors can that use glutamate as a neurotransmitter cause death of neurones (Chapter 9).

Gonadal dysfunction There is evidence of reduced gonadal function in men who suffer from alcoholism. This manifests itself as impotence, testicular atrophy, gynaecomastia, sterility and changes in distribution of body hair. Indeed, it has been claimed that alcoholism is the most common cause of non-functional impotence and sterility. There are two suggestions for the cause of this problem.

- A deficiency of liver function reduces metabolism of steroid hormones so that the blood concentration of oestrogens in men is increased.

- Ethanol oxidation, via alcohol dehydrogenase, reduces testosterone secretion, due to a high NADH/NAD$^+$ ratio is the Leydig cells in the testes.

Fatty liver, hepatitis and cirrhosis Chronic consumption of ethanol can lead to the development of 'fatty liver', which is the deposition of excess triacylglycerol in the liver. It is still unclear whether this deposition actually causes damage to the liver or if it is a consequence of the damage. In either case, the damage may result in hepatitis and, eventually, cirrhosis. As the 'fatty liver' develops there is a decrease in the amount of endoplasmic reticulum (especially the rough endoplasmic reticulum which is involved in protein synthesis) and, in addition, damage to the mitochondria. This cellular damage may be caused by a chronically elevated concentration of ethanal within the liver cell that is eventually severe enough to cause death of hepatocytes. Cell damage and death will result in an inflammatory response, that is, infiltration and activation of lymphocytes and macrophages. This condition is then known as hepatitis. If this is not treated it is likely to result in the formation of fibrous tissue with the development of cirrhosis, when the amount of functional liver becomes markedly reduced. The rates of the urea cycle, and hence removal of ammonia, and of detoxification by the liver become insufficient resulting in nitrogen retention and hepatic coma (see Table 10.3 and Chapter 8).

Hypoglycaemia After 24 hours of starvation, hepatic gluconeogenesis is vital for the production of glucose and maintenance of the blood sugar level. Ethanol is a potent

> Shakespeare (*Macbeth*, Act II, scene III) was clearly aware of the effects of ethanol on gonadal function.
>
> Macduff: What three things does drink especially provoke?
> Porter: Marry, sir, nose painting, sleep, and urine. Lechery, sir, it provokes and unprovokes. It provokes the desire but it takes away the performance.
>
> Biochemistry provides the explanation for Porter's observations: a decrease in the plasma concentration of testosterone could account for the inability to perform whilst the reduced extent of feedback inhibition on the pituitary would lead to an increased rate of secretion and hence an increase in the blood concentration of luteinising hormone, which has been claimed to increase the degree of sexual arousal in men.
> Ethanol can also explain sleep, due to the neurological changes, reddened nose, since it results in cutaneous vasodilation, and urine production, since it has a diuretic effect.

inhibitor of gluconeogenesis and this effect could contribute to the hypoglycaemia that can occur on consumption of ethanol, especially the 'executive lunch syndrome' after the overnight fast and a missed breakfast (Chapter 6). Hypoglycaemia is also common in many chronic alcoholics, particularly after a period of excessive drinking. Under these conditions, the alcoholic usually does not eat, so that gluconeogenesis would be essential to maintain the blood glucose concentration. Severe hypoglycaemia and concomitant lactic acidosis, due to failure to oxidise lactic acid by the liver, may be the two most important factors that precipitate the coma and collapse of the alcoholic who is 'on the bottle'.

Tolerance

Tolerance is an adaptation to a drug, leading to a need to increase the dose required to produce a given physiological effect. It can develop rapidly and often precedes development of physical dependence. *Dependence* is a different phenomenon, much more difficult to define and measure, which involves two separate components, namely *physical* and *psychological* dependence.

- Psychological dependence is an intense craving for the drug, especially for a drug that induces pleasurable effects on the body.

- Physical or physiological dependence is associated with a clear-cut *physical withdrawal syndrome* (or *abstinence syndrome*), which can be reproduced in experimental animals and appears to be closely related to tolerance in humans. In experimental animals, abrupt withdrawal of morphine, after chronic administration for a few days, causes increased irritability, loss of weight and a variety of abnormal behaviour patterns such as body shakes, writhing, jumping and signs of aggression. These reactions decrease after a few days, but abnormal irritability and aggression persist for many weeks. Human addicts show a similar abstinence syndrome, somewhat resembling severe influenza, with yawning, pupillary dilation, fever, sweating and piloerection (causing goose pimples – hence the origin of the term 'cold turkey' used to describe the effects of withdrawal).

Hypothesis to account for tolerance and dependence

One characteristic of the action and effects of hormones provides the basis for a biochemical explanation for tolerance. A chronic effect of some hormones is a reduction in the number of the hormone receptors in the plasma membrane of the tissue upon which the hormone is interacting (e.g. incubation of adipocytes with insulin for about one or two hours decreases the number of insulin receptors in the plasma membrane of the adipocyte). This phenomenon is sometimes known as 'down-regulation'. It is explained by internalisation (endocytosis) of the hormone–receptor complex, which is then degraded, although in some cases only the hormone is degraded and the receptor is recycled. It has been proposed that one mechanism for preventing down-regulation of hormone receptors is pulsatile secretion of hormones (e.g. growth hormone, gonadotrophins). This minimises the period during which the hormone interacts with the receptor, but the biochemical effect of the hormone is still maintained, that is pulsatile secretion restricts or prevents the reduction in the number of receptors. If this is the case, it suggests that down-regulation of receptors is a spontaneous biochemical process that will normally occur unless local biochemistry is modified to prevent the effect.

This phenomenon may be the basis for tolerance. The effect of a neurotransmitter (and therefore a drug) is achieved by binding to a receptor which then activates an effector system that produces, directly or indirectly, a messenger that stimulates intracellular biochemistry. The effector system is usually an ion channel or a protein kinase, which is identical to the mechanism of action of many hormones (Chapter 12). A chronic presence of neurotransmitter in the synapse and hence interaction with the postsynaptic receptors could result in a reduction in the number of these receptors. This phenomenon also applies to drugs that mimic the action of neurotransmitters, either by binding to the receptor or by releasing neurotransmitters into the synapse. Consequently, in order to maintain the same effect of the drug, as the receptor number decreases, the dose of the drug must be increased to match the decrease in receptor number. That is, the phenomenon of 'down-regulation' provides a biochemical explanation for tolerance.

15
Nutrition: biochemistry, physiology and pathology

Unfortunately, even in the face of accumulated knowledge, standard hospital care knowingly produces energy and nutrient deficits while otherwise attempting to provide acceptable care ... Disregard for the effect of malnutrition on care seems primarily due to both undergraduate and postgraduate medical training programmes that present only limited aspects of nutrition and in a fashion that makes unclear any relevance to medical practice.

(Tucker, 1997)

As we enter the 21st century and the new millennium, malnutrition acting directly or indirectly remains the single most important factor impairing health and productivity of large human populations.

(Gopalan, 2000)

There are many excellent texts on nutrition. This chapter, therefore, focuses not on nutrition per se but on how biochemistry helps us understand well established and less well established aspects of nutrition and how such knowledge fits in with other subjects discussed in this text. There is now considerable medical and lay interest in what is meant by healthy and unhealthy diets. Nutrition has become a major issue in the medical sciences and in clinical practice. It is also of concern to politicians, particularly in the link between nutrition and 'Western' diseases such as cardiovascular disease, obesity, cancer and neurological problems. In this chapter an attempt is made to provide a biochemical basis for discussion of nutrition and development of these conditions. To this end, biochemical explanations for nutritional advice and the recommendations from national bodies are provided. Similarly, explanations for the recommendations designed for different populations, different conditions and activities (physical and mental activity, the elderly, the young, during pregnancy and space flight) are discussed. Finally, the biochemistry of malnutrition, undernutrition and overnutrition is discussed.

Although it is usually considered that mastication and digestion are the first processes in the conversion of food into suitable components to be absorbed by the body, cooking of food is also very important: it makes the macronutrients and micronutrients more readily available for digestion and absorption. The biochemistry involved in cooking is described in Chapter 4.

Basic information required for discussion of some biochemical aspects of nutrition

Of the 92 naturally occurring chemical elements, about 21 are needed to build the human body and to allow it to function. Six of them – carbon, hydrogen, oxygen, nitrogen, sulphur and phosphorus – come from carbohydrate, fat and protein and from some vitamins. The remaining 15 elements, known as minerals, are required in widely differing amounts, from grams of sodium and potassium to milligrams, or less, of trace elements. It is not clear, however, if any of the so called ultratrace elements are actually needed, so that the total number needed is not precise. Minerals and vitamins are known as micronutrients. To satisfy all of these requirements, the average person consumes about 60 tons of food in a lifetime.

Macronutrients

Macronutrients are carbohydrate, fat and protein. The approximate amounts ingested each day are given in Table 15.1.

Functional Biochemistry in Health and Disease by Eric Newsholme and Tony Leech
© 2010 John Wiley & Sons Ltd

Table 15.1 Average daily intake of macronutrients and energy by adults in the UK

	Intake g/day	
	Male	**Female**
Carbohydrate		
Total	272	193
Sugars	115	86
Starch	156	106
Non-starch polysaccharide	11	12
Fat		
Total	102	73
Omega-3 fatty acids	2	1
Omega-6 fatty acids	14	10
Protein		
Total	58	50
Energy (kcal/day)	2450	1600
(MJ/day)	10.3	7.0

For source of data see Table 15.2.

Table 15.2 Amounts of protein recommended for subjects, United Kingdom 1991

	Amount (g/day)	
Age	**Male**	**Female**
0–12 months	14	14
1–3 years	15	15
4–10 years	24	24
11–14 years	42	41
15–18 years	55	45
19–50 years	55	45
50+ years	53	46

For precise amounts see *Dietary Reference Values for Food Energy and Nutrients for the United Kingdom, 1991.*

These recommendations are based on the complete digestibility of milk or egg protein. Protein from plant sources may be slightly less digestible, and the UK Department of Health recommends that vegetarians and vegans multiply the above figures by a factor of 1.1.

Similar recommendations apply in the USA.

Carbohydrate

In developed countries, carbohydrate provides about 45% of the energy ingested. It comprises starch, sucrose, lactose, fructose and glucose with traces of maltose and trehalose. The structure of these carbohydrates and how they are digested, absorbed into the bloodstream and further metabolised is described in Chapters 4 and 6.

Fat

In developed countries, fat provides about 40% of the total energy consumed but in some individuals this percentage may be much higher. Almost all of this is in the form of triacylglycerol, containing mainly long-chain but also some short-chain fatty acids. The structure, digestion, absorption and eventual fate of the products of absorption are described in Chapters 4 and 5 and the metabolism of fat is discussed in Chapter 7.

Protein

Protein intake is very variable. It represents about 15% of the energy ingested in developed countries. The recommended protein intakes for adults and children are given in Table 15.2. Digestion to peptides and amino acids and absorption of these in the intestine is described in Chapter 4. The metabolism is discussed in Chapter 8.

Vitamins

The early work on micronutrients in human nutrition was directed towards identifying those components of the diet essential for healthy life. These can now be divided into two groups: vitamins and minerals.

Vitamins are complex organic molecules required in relatively small quantities. They either cannot be synthesised in the body at all, or they are synthesised in amounts too small so that they must be present in the diet to provide optimal levels. For some vitamins, a proportion of the requirement can be synthesised by the intestinal microorganisms. If adequate amounts of a vitamin are not present in the diet, or if they are not adequately absorbed, a deficiency disease develops. Since many of the vitamins provide cofactors or prosthetic groups, in some cases the symptoms of the disease can be traced to a low activity of a specific enzyme due to the low level of the vitamin.

About 12 chemicals are widely recognised as vitamins (Table 15.3), although some are actually families of inter-convertible chemical forms. Choline, inositol and bioflavonoids are considered to be possible vitamins. The essential fatty acids and essential amino acids are excluded

> Before 1900, many experiments suggested that dietary components other than protein, carbohydrate, fat and minerals were needed for survival. However, it was Frederick Gowland Hopkins who provided the evidence that minute amounts of unknown substances, present in normal foods, were essential for normal healthy life. His eminence in the scientific community ensured that the work and ideas were accepted: he was awarded the Nobel Prize for his work in 1929.

Table 15.3 Vitamins: their sources and roles

Vitamin	Sources of vitamin	Roles of vitamin
Water-soluble vitamins		
B_1 (thiamine)	Fresh vegetables, husk of cereal grains, meats, especially liver	Energy metabolism, synthesis of myelin
B_2 (riboflavin)	Milk and meat products	Energy metabolism
Nicotinic acid (niacin)	Liver, lean meats, cereals, legumes	Energy metabolism
B_6 (pyridoxine)	Meats, cereals, lentils, nuts, some fruits and vegetables	Amino acid metabolism
Pantothenic acid	Liver, meats, cereals, milk, egg yolk, vegetables	Energy metabolism
Folates	Yeast, liver, fresh green vegetables	Nucleotide metabolism
B_{12} (cobalamin)	Meat, fish, poultry, milk and milk products; produced also by bacteria in the large intestine	Nucleotide metabolism
Vitamin C	Fresh fruits and vegetables	Synthesis of connective tissue and repair of tissue; antioxidant
Fat-soluble vitamins		
Vitamin A (retinol)	Animal tissue, liver, green plants	Component of the pigment involved in vision
Vitamin D (calciferol)	Produced in skin by action of ultraviolet light during summer	Controls calcium uptake into the body and mobilisation of calcium from bone
Vitamin E (tocopherol)	Vegetable seed oils, milk, eggs, meat	Antioxidant
Vitamin K (naphthoquinone)	Green leafy vegetables, meats, dairy produce	Activation of blood-clotting factors
Possible vitamins		
Biotin	Yeast, organ meats, muscle meats, dairy products, grains, fruits; produced by microorganisms	Amino acid metabolism
Choline	Meat, egg yolk, legumes, cereals	Component of a neurotransmitter Phospholipid in cell membranes
Inositol		Phosphatidylinositade bisphosphate is present in plasma membranes, where it is hydrolysed to produce inositol trisphosphate, an important intracellular messenger

for historical reasons and because they are needed in rather larger amounts than 'true' vitamins. The recommended daily allowances (RDA) of vitamins per day for adults range from 3 µg for vitamin B_{12} to 60 mg for vitamin C. Four vitamins – A, D, E and K – are fat soluble and the rest are water soluble. Fat-soluble vitamins must accompany a meal that contains some fat for their effective absorption.

The advice on vitamins for the general population is that supplementation is unnecessary for a normal diet. However, vitamin deficiency can occur in the elderly (due to poor nutrition, and lack of sunlight) the very young, the malnourished, when food absorption problems exist or when there is an exceptional demand as in pregnancy. Some of the general points about vitamins are:

- The water-soluble vitamins (B and C) are not stored to any large extent in the body, so any effects caused by their deficiency will become apparent quickly.

- In contrast, the fat-soluble vitamins (A, D, E and K) are stored in the body.

- There is no firm evidence that very high doses of any vitamin produce beneficial effects unrelated to their normal vitamin function.

- Very large doses of vitamins can be harmful: an excess of vitamin A can cause foetal malformations and can be fatal (a risk which is reduced if the vitamin is taken in the form of β-carotene); massive amounts of vitamin C can cause gastrointestinal problems.

- Excessive consumption of alcohol may impair the uptake of some water-soluble vitamins.

- Some drugs, including oral contraceptives, and smoking tobacco, increase the body's demand for vitamins B, C and E.

- A recent concern is that a moderate but chronic vitamin deficiency, in what is considered to be an adequate diet, may increase the risk of development of some diseases.

The structures and specific roles of the vitamins, together with the symptoms associated with their deficiency, are described in Appendix 15.1. Seven vitamins are selected for discussion in the text, as they are of particular biochemical, physiological and pathological significance.

> Compared with other vitamins, the chemical structures of both folic acid and B$_{12}$ are complex. They are prosthetic groups for the enzymes that catalyse the transfer of the methyl group (–CH$_3$) between compounds (one-carbon metabolism). The –CH$_3$ group is chemically unreactive, so that the chemistry for the transfers is difficult, requiring complex structures for catalysis.

Folic acid and vitamin B$_{12}$

The functions of folic acid and vitamin B$_{12}$ are very closely linked, especially in what is known as 'one carbon' metabolism or methyl group transfer.

Folic acid

Folates are present in nearly all natural foods. Those with the highest folate content include yeast, liver and other organ meats, fresh green vegetables and some fresh fruits. The term folic acid arose from its first identification in leafy vegetables (e.g. spinach).

The folates are a family of compounds, all containing a pteridine group (see Figure 15.1). The pteridine group is the prosthetic group in several of the many enzymes that transfer a single carbon atom including the methyl group –CH$_3$ from compound to compound. The structural formulae of several of the group are presented below (Figure 15.2). The major source of the one-carbon for these reactions is serine (see below). It is important, for example, in two key reactions in nucleotide synthesis: de novo purine

nucleotide synthesis and the synthesis of deoxythymidine from deoxyuridine (in the monophosphate form) (see below and Chapter 20: Figure 20.8 and 20.12 (a), respectively).

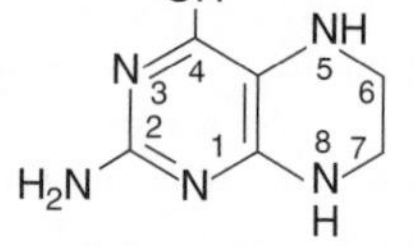

Folic acid deficiency Folic acid deficiency can be due to poor diet, chronic alcoholism, inadequate absorption (e.g. damage to the upper third of small intestine), use of drugs such as anticonvulsants, or extra demand during pregnancy or lactation. Unfortunately, folates are susceptible to oxidation and 50% to 90% of the folate content of foods can be destroyed by protracted cooking or other processing such as canning. An intracellular deficiency of folic acid can also arise if methionine synthesis is impaired. Folic acid deficiency results in anaemia. This is known as megaloblastic anaemia, since it results in the formation of large, abnormal forms of the cells that are precursors for erythrocytes. This is because erythrocyte development arrests at this stage due to lack of purine and pyrimidine nucleotides for DNA synthesis. Anaemia is characterised by weakness, tiredness, irritability, headache and palpitations.

Vitamin B$_{12}$ (cobalamin)

Vitamin B$_{12}$ is not found in plants. The primary source is microorganisms and the usual dietary sources for humans are meat and meat products (including shellfish, fish and poultry) and to a lesser extent milk and milk products. However, normally the microorganisms in the colon provide most of the requirement.

The structure of cobalamin is more complex than that of folic acid (Figure 15.2 and 15.3). At its heart is a porphyrin ring containing the metal ion cobalt at its centre. In catalytic reactions the cobalt ion forms a bond with the one-carbon group, which is then transferred from one compound to another. Vitamin B$_{12}$ is the prosthetic group of only two enzymes, methylmalonyl-CoA mutase and methionine synthase. The latter enzyme is particularly important, as it is essential for the synthesis of nucleotides which indicates the importance of vitamin B$_{12}$ in maintenance of good health.

There are several steps in the absorption of vitamin B$_{12}$. In the stomach and lumen of the small intestine it is hydrolysed from its (peptide) links with the proteins of which it is a component. It then attaches to gastric intrinsic factor, which is a glycoprotein of molecular mass about 50 000 kDa, to form a complex. This protects the vitamin from being damaged by acid in the stomach. The complex is carried into the ileum, where it binds to a receptor on the surface of the absorptive cells and is released from the intrinsic factor within the absorptive cell. In the portal venous blood, it is transported to the liver bound to the vitamin B$_{12}$-binding protein, which also protects the vitamin.

Figure 15.1 *Structure of pterin.* In the folates, pterin is linked to 4-aminobenzoic acid through at methylene bridge at position 6.

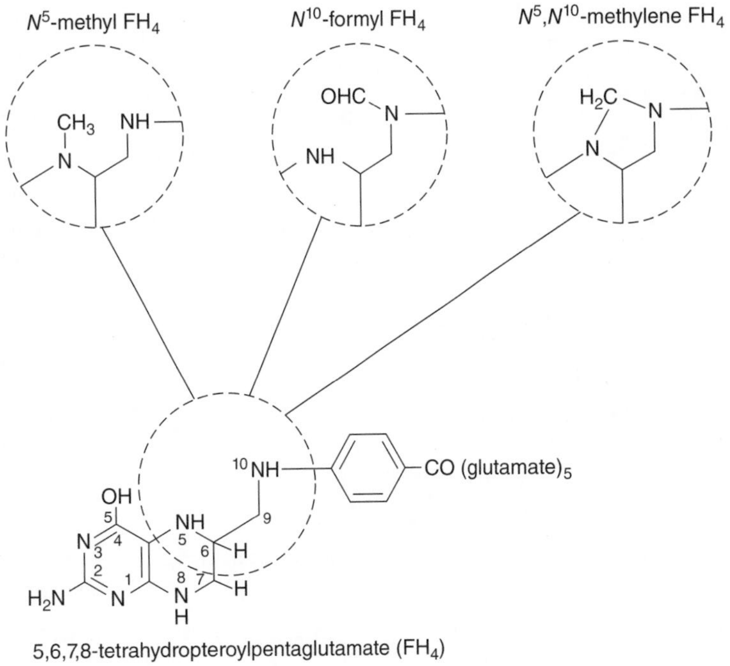

Figure 15.2 *Structural formula of tetrahydrofolate and representation of derivatives involved in single carbon transfer.* The tetrahydrofolate is always part of a complex with several glutamate residues. The parent compound, pteroylglutamate (folate) lacks four hydrogen atoms, one each from carbon atoms 5, 6, 7 and 8. Tetrahydrofolate can exist in any one of three oxidation states, as shown; they are inter-convertible through oxidereduction reactions. Each plays a individual and different role is synthesis of key compounds (See below).

5,6,7,8-tetrahydropteroylpentaglutamate (FH$_4$)

Vitamin B$_{12}$ deficiency Deficiency, although rare, results in two serious problems: megaloblastic anaemia (which is identical to that caused by folate deficiency) and a specific neuropathy called B$_{12}$-associated neuropathy or cobalamin-deficiency-associated neuropathy (previously called, subacute combined degeneration of the cord). A normal healthy adult can survive more than a decade without dietary vitamin B$_{12}$ without any signs of deficiency since it is synthesised by microorganisms in the colon and then absorbed. However, pernicious anaemia develops fairly rapidly in patients who have a defective vitamin B$_{12}$ absorption system due to a lack of intrinsic factor. It results in death in 3–4 days. Minot and Murphy discovered that giving patients liver, which contains the intrinsic factor, and which is lightly cooked to avoid denaturation, cured the anaemia. For this discovery they were awarded the Nobel Prize in Medicine in 1934.

The anaemia in B$_{12}$ deficiency is caused by an inability to produce sufficient of the methylating agent *S*-adenosylmethionine. This is required by proliferating cells for methyl group transfer, needed for synthesis of the deoxythymidine nucleotide for DNA synthesis (see below and Chapter 20). This leads to failure of the development of the nucleus in the precursor cells for erythrocytes. The neuropathy, which affects peripheral nerves as well as those in the brain, is probably due to lack of methionine for methyl transfer to form choline from ethanolamine, which is required for synthesis of phosphoglycerides and sphingomyelin which are required for formation of myelin and cell membranes. Hence, the neuropathy results from a

deficiency of myelin or the presence of damaged myelin in the nerves in the brain.

Methyl group transfer

Small methyl groups are important in the stracture of some small compounds, nucleotides, some bases in DNA molecules and in postranslational modification of amino acids in proteins. The transfer of a single carbon atom is important in the synthesis of purine nucleotides. The compounds involved in the whole process of methyl group transfer, and are carbon metolism, are methionine, homocysteine, serine and the vitamins, folic acid and B$_{12}$.

The compounds that are the immediate methyl group donors are methyltetra hydrofate (CH$_3$–FH$_4$) and *S*-adenosyl methionine (SAM) (see Figure 15.2). These are involved in, at least, five key reactions or processes which are summarised in Figure 15.4. Complexity arises in the topic of methyl group transfer in formation and reformation of the methylating compounds *S*-adenosylmethione and methyl tetrahydrofolate. There are four important reactions in the formation utilisation and then the reformation of *S*-adenosylmethionine as follows:

(i)   The amino acid methionine is the compound that transfers the –CH$_3$– group but only in the 'active' form, which is *S*-adenosyl methionine (SAM), a reaction formed by interaction with ATP.

(ii) The methylation reaction is abbreviated as follows:

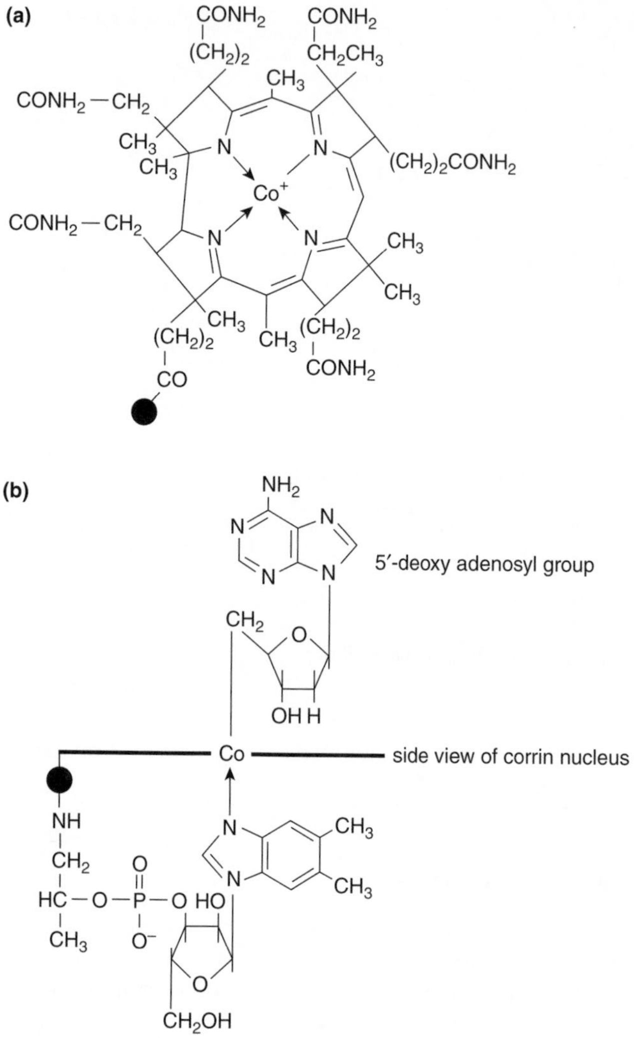

(a)

(b)

Figure 15.3 *Structural formula of deoxyadenosylcobalamin (coenzyme B₁₂).* (**a**) A 'plan' view of the corrin nucleus with substituents. (**b**) The position of the remaining two ligands of the cobalt atom. No attempt is made to show correct stereochemical relationships. Related compounds have different groups in place of the 5'-deoxyadenosyl group: cyanocobalamin, (vitamin B₁₂)–CN; hydroxycobalamin, (vitamin B₁₂)–OH; methylcobalamin, (vitamin B₁₂)–CH₃.

$$SAM + X \rightarrow X\text{-}CH_3 + SAH$$

Where SAH is *S*-adenosyl homocysteine (see Figure 15.4)

(iii) SAH is converted to homocysteine, by loss of the adenosine, which is converted back to ATP (via AMP and adenosine kinase).

(iv) A methyl group is transferred from methyl folic acid to homocysteine to produce methionine. The latter is

once again activated to continue the methyl group transfer cycle. The reactions are presented in Figure 15.5.

The methyl group for all these reactions is obtained primarily from the degradation of the amino acid serine as it is converted to glycine. This sequence of reactions is sometimes known as the folic acid or tetrahydrofolate cycle (Figure 15.6).

Methylation and chronic diseases

A high level of homocysteine is an indication of a low rate of conversion of homocysteine to methionine and hence a low level of the methylating agent *S*-adenosyl methionine. The latter, plus the 'toxic' effects of homocysteine, provides a link between the reactions listed above and three diseases: cardiovascular disease (Chapter 22), senile dementia (Chapter 14) and cancer (Chapter 21).

Cardiovascular disease A chronic partial deficiency of folic acid and/or vitamin B₁₂ could decreases the activity of methionine synthase (see Figure 15.5) resulting in a low level of methionine and a high level of homocysteine in cells and in the blood. Homocysteine is converted spontaneously to the 'toxic' thiolactone (see Figure 22.7) which damages endothelial cells. This initiates atherosclerosis by allowing the entry of LDL-cholesterol into the subendothelial space where it forms fatty streaks (see Figure 22.3). Damage to endothelial cells also increases the risk of thrombosis (see Figure 22.5).

Neurodegeneration High plasma levels of homocysteine are a risk factor for Alzheimer's disease. There are two mechanisms by which homocysteine could damage neurones.

(i) Thiolactone could damage the plasma membrane.

(ii) Homocysteine can increase the affinity of N-methyl-D-aspartate receptors for glutamate. This could overexcite synaptic receptors which results in opening of Ca²⁺ ion channels, causing the entry of excess Ca²⁺ ions (see Chapter 14).

Furthermore, the two pathways that normally degrade homocysteine are absent from the neurone and glial cells and so homocysteine can accumulate in the brain (Chapter 8, Appendix 8.2). Consequently, the maintenance of adequate intake of folic acid and vitamin B₁₂ over many years, to ensure low levels of homocysteine, may help to protect neurones and reduce the risk of development Alzheimer's disease.

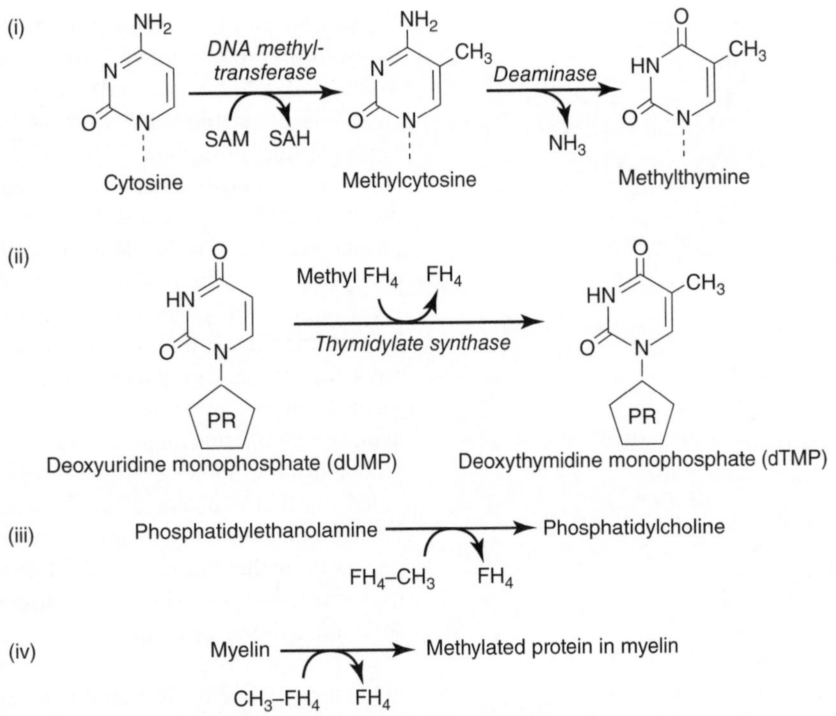

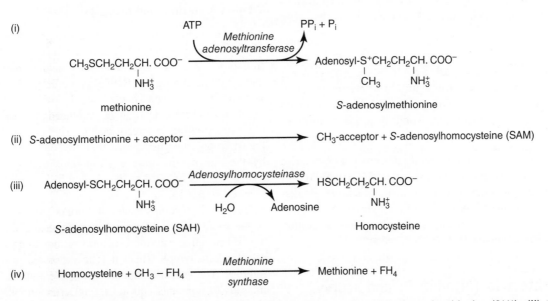

Figure 15.4 *Four compounds that are methylated either by SAM or methyl FH₄ (CH₃ FH₄).* The processes are: (i) cytidine bases in DNA; (ii) methylation of deoxyUMP to produce deoxythymidine monophosphate; (iii) formation of phosphatidylcholine from phosphatidylethanolamine (Chapter 17); (iv) methylation of a protein in myelin. The base cytidine is methylated in DNA to produce methylcytidine which, if deaminated, produces methylthymidine in DNA. Methylation of the bases can modify gene transcription (see text). (PR: 5′-phosphoribose).

Figure 15.5 *Four reactions involved in methylation.* The reactions are: (i) formation of *S*-adenosylmethionine (SAM); (ii) transfer of methyl group to an acceptor; (iii) conversion of *S*-adenosylmethionine to homocysteine; (iv) conversion of homocysteine to methionine using methyl tetrahydrofolate as the methyl donor with the formation of FH₄.

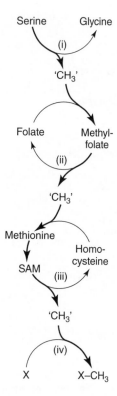

Figure 15.6 *Formation of methyl tetrahydrofolate and SAM from serine.* Reaction (i) is described in Appendix 8.3. Reaction (ii) is Figure 15.5 and the several reactions represent in reaction (iv) are discribed in Figure 15.4.

Cancer

A deficiency of methylation is a risk factor for development of a tumour cell. Two explanations for this have been put forward:

(i) Cytosine bases in DNA are methylated by DNA methyl-tranferase enzymes, using SAM as the methyl donor (see above). Since methylation restricts transcription of genes, a low activity of methyltransferase could lead to less methylation and hence activation of oncogenes.

(ii) Methylcytosine can be readily deaminated to give rise to methylthymidine. It is estimated that a high proportion of mutations that give rise to human genetic diseases are due to conversion of methylcytidine bases to methylated thymidine bases, possibly due to activation of the deaminase (Figure 15.4, reaction (i)).

Ascorbic acid (vitamin C) and scurvy

Vitamin C has been selected for discussion because of its historical interest (Box 15.1). Its structure is shown in Figure 15.7 and it is obtained primarily from plants, especially fresh, rapidly growing fruits and vegetables. It acts as an antioxidant, particularly in association with vitamin E, and as a chelating agent (a metal binding agent) which facilitates the absorption of iron from the intestine. It also functions as a cosubstrate in a number of oxido-reduction reactions, including the post-translational hydroxylation of proline necessary for formation of collagen, lack of which causes scurvy. This disease caused much misery and a great number of deaths among sailors on long voyages. The severity of the disease is not always appreciated nowadays because it is relatively rare. The clinical features include fatigue, lassitude, bleeding gums, depression, dry skin, xerophthalmia, impaired iron absorption and impaired wound healing. James Lind, a Scottish naval surgeon, provided the first evidence of the benefit of drinking lemon juice to combat scurvy. Indeed, he may have been the first person in medical history to conduct a controlled clinical trial when, in 1747, he tested a variety of remedies on 12 scorbutic sailors. One pair remained untreated while five other pairs were given various treatments, only the two given lemon juice made a speedy recovery. When Captain James Cook supplied lemon juice and fresh fruit and vegetables to his crew on the second voyage round the world, not a single member died from scurvy, although the voyage lasted for three years.

Vitamin A and vision

Vitamin A (retinol and retinal)

The term 'vitamin A' now includes several biologically active compounds, including retinol and retinal. The term 'retinoids' refers to retinol and its natural metabolites and synthetic analogues. Retinal is the active form of the vitamin in the visual cycle (see below). Liver is an excellent source of vitamin A. Another is the carotenoid pigments present in green plants. The carotenoids are known as provitamins A, since they are converted to retinol by the enzyme β-carotene 15,15′-dioxygenase to form retinal. This is reduced to the alcohol by retinol dehydrogenase (Figure 15.8).

Hydrolysis of retinyl ester to retinol occurs in the lumen of the small intestine from where it is absorbed with the aid of bile salts, esterified to form retinyl ester and then released into lymph where it is incorporated into chylomicrons. The action of lipoprotein lipase converts chylomicrons to remnants and the retinyl ester remains in the remnants to be taken up by the liver, where it is stored as the ester until required. On release from the liver, it is transported in blood bound to retinal binding-protein.

Box 15.1 From the treatise on scurvy by Lind

In 1753 Lind published his famous *Treatise on Scurvy*. Since this is such a remarkable publication, some parts of it are presented here to illustrate the care he took in describing the terrible symptoms of scurvy and the significance of the discussion at that time.

PREFACE
The subject of the following sheets is of great importance to this nation; the most powerful in her fleets, and the most flourishing in her commerce, of any in the world. Armies have been supposed to lose more of their men by sickness, than by the sword. But this observation has been much more verified in our fleets and squadrons; where the scurvy alone, during the last war, proved a more destructive enemy, and cut off more valuable lives, than the united efforts of the French and Spanish arms. It has not only occasionally committed surprising ravages in ships and fleets, but almost always affects the constitution of sailors; and where it does not rise to any visible calamity, yet it often makes a powerful addition to the malignity of other diseases. It is now above 150 years since that great sea-officer, Sir Peter Hawkins, in his observations made in a voyage to the South Sea, remarked it to be the pestilence of that element. He was able, in the course of twenty years, in which he had been employed at sea, to give an account of 10,000 mariners destroyed by it.

CHAPTER 11: The diagnostics or signs
In the second stage of this disease, they most commonly lose the use of their limbs; having a contraction of the flexor tendons in the ham with a swelling and pain in the joint of the knee. Indeed a stiffness in these tendons, and a weakness of the knees, appear pretty early in this disease, generally terminating in a contracted and

swelled joint. They are subject to frequent languors; and when long confined from exercise, to a proneness to faint upon the least motion of the body; which are the most peculiar, constant, and essential symptoms of this stage.

Some have their legs monstrously swelled, and covered with one or more large livid spots, or ecchymoses; others have hard swellings there in different places, extremely painful; and others I have seen, without any swelling, have the calf of the leg quite indurated.

They are apt, upon being moved, or exposed to the fresh air, suddenly to expire. This happened to one of our people, when in the boat, going to be landed at Plymouth hospital. It was remarkable he had made shift to get there without any assistance, while many others were obliged to be carried out upon their beds. He had a deep scorbutical colour in his face, with complaints in his breast. He panted for about half a minute, then expired.

Scorbutic people are at all times, but more especially in this stage, subject to profuse haemorrhages from different parts of the body; as from the nose, gums, intestines, lungs, &c., and from their ulcers, which generally bleed very plentifully. Many at this time are afflicted with violent dysenteries, accompanied with exquisite pain; by which they are reduced to the lowest and most weakly condition: while others I have seen, without a diarrhoea or gripes, discharge great quantities of pure blood by the anus.

The gums are for the most part excessively fungous, with an intolerable degree of stench, putrefaction, and pain; sometimes deeply ulcerated, with a gangrenous aspect.

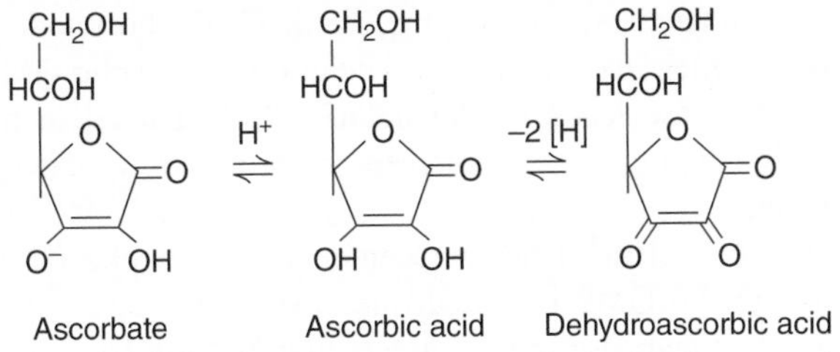

Figure 15.7 *Structure of ascorbic acid.*

Retinoids play a role in the regulation of differentiation and development. This they achieve by binding to retinoid receptor-proteins in the nucleus and inhibiting the expression of transcription factors that regulate proliferation. Hence, deficiency of vitamin A can result in impaired differentiation and hence foetal malformation and spon-

taneous abortions. Any treatment involving retinoids is, therefore, contraindicated during pregnancy. Retinoids have been used in the treatment of cancer but their toxicity precludes long-term use.

A deficiency of vitamin A results in night blindness. A chronic deficiency results in a thickening of membranes in the cornea which, if untreated, can lead to blindness through perforation of the cornea and loss of the lens. It is estimated that half a million children develop blindness due to vitamin A deficiency every year. Refeeding malnourished children can produce a deficiency of vitamin A (see below).

Biochemistry and physiology of vision

The retina contains two distinct types of photoreceptor cell to detect light: cones and rods. The rods are specialised

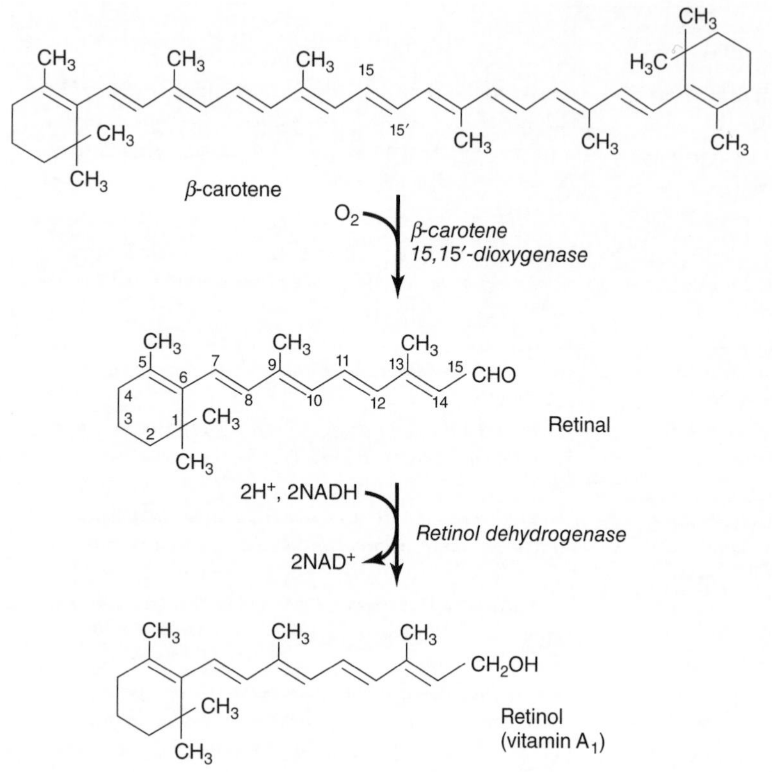

Figure 15.8 *Formation of vitamin A from β-carotene.*

Rods absorb maximally at 509 nm, blue-sensitive cones at 420 nm, green-sensitive cones at 530 nm and red-sensitive cones at 565 nm. Colour blindness results from mutations that occur in the genes for the three different colour-sensitive opsin molecules.

for vision in dim light whereas the cones are specialised for colour vision in bright light. There are three different types of cones which detect blue, green or red light respectively. The molecule that absorbs the light is retinal, *cis*-retinal, which is linked covalently (via a lysine residue) to a protein, opsin. The complex of retinal and opsin is known as rhodopsin, which is also referred to as the photopigment. There are four different opsins: rhodopsin in the rods, and red-, green- and blue-sensitive opsins in the three different cones. Each cell contains about 10^9 molecules of rhodopsin. These different opsins modify the absorption spectrum of the *cis*-retinal resulting in responses to different wavelengths of light.

The absorption of light by *cis*-retinal converts it to the *trans*-form which induces a conformational change in opsin. One photon catalyses the isomerisation of one molecule of

cis-retinal to *trans*-retinal (Figure 15.9). On this change, the retinal dissociates from the opsin. The latter is now 'free' to initiate the sequence of events that leads to the detection of light (Figure 15.9). The opsin interacts with a membrane trimeric G-protein, known as transducin, which results in an exchange GDP for GTP on the transducin. This activates the trimeric G-protein in the usual way, i.e. by dissociation of the α-β-γ complex, which releases the α-subunit (see Figure 12.22). It is now active and activates an enzyme, cyclic GMP phosphodiesterase, which decreases the concentration of cyclic GMP. This decrease leads to closing of a Na^+ channel in the membrane of the photoreceptor cell (by restricting opening).

The Na^+ channel has a receptor site for cyclic GMP; when cyclic GMP is bound, the channel is closed. This leads to a decrease in the intracellular Na^+ ion concentration, resulting in hyperpolarisation of the cell membrane. This decreases the release of the neurotransmitter glutamate into the synapse that connects the photoreceptor cell to the bipolar neurones. In this specific case, a decrease in the neurotransmitter concentration in the synapse is a signal that results in depolarisation of the bipolar cell. The action potential in the bipolar cells communicate with ganglion cells, the axons of which form the optic nerve. Thus action potentials are generated in the axons which are

An increase in the activity of cyclic GMP phosphodiesterase produces a decrease in the concentration of cyclic GMP *only* if the activity of the enzyme guanyl cyclase remains constant. The principle underlying this requirement is discussed in Chapter 12,

  GTP → cyclic GMP → GMP

The system also depends upon the eventual conversion of GMP back to GTP. (Box 12.2)

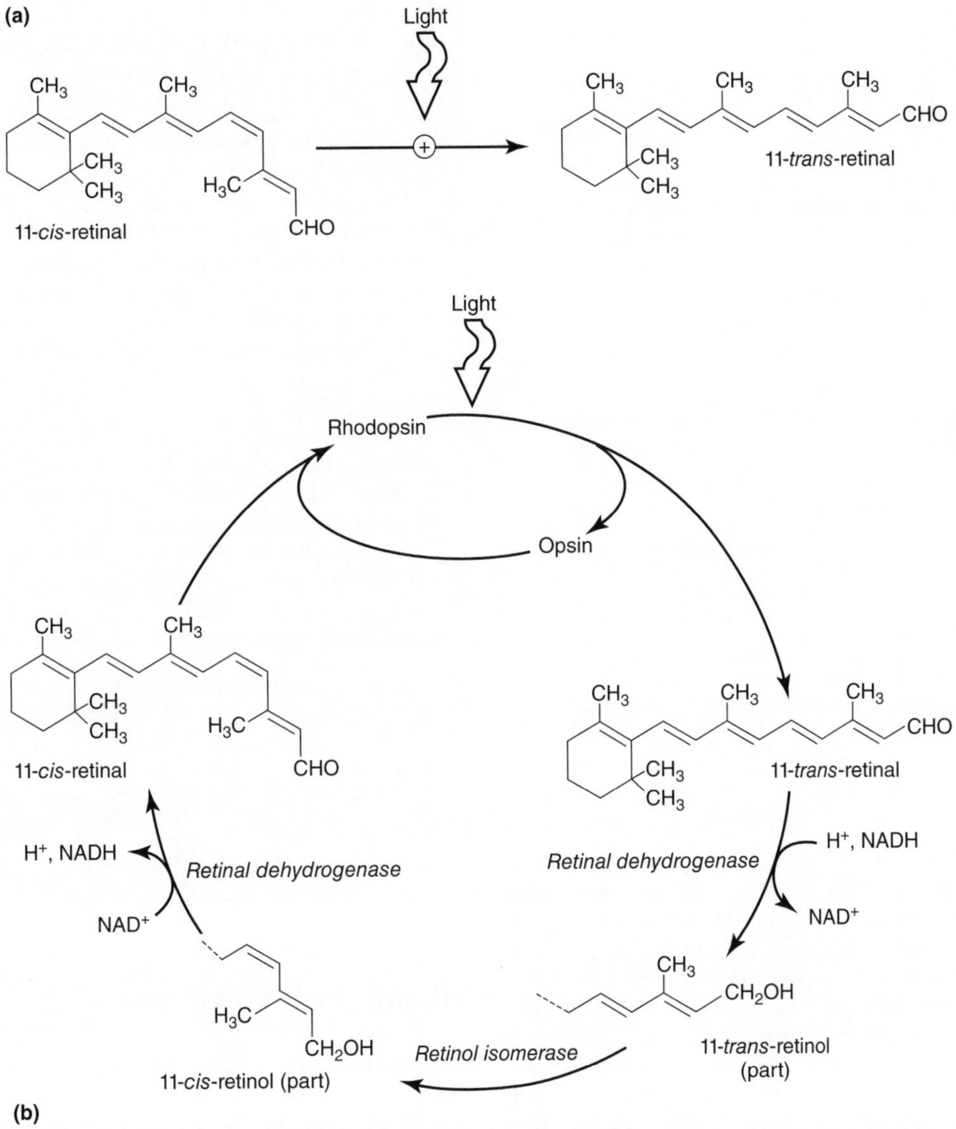

Figure 15.9 **(a)** *The* cis- *and* trans-*retinal interconversions in the detection of light*. Within the photoreceptor cell, light is detected by the conversion of *cis*-retinal to *trans*-retinal, components of the light-sensitive pigment rhodopsin. This apparently small chemical change is sufficient for *trans*-retinal to dissociate from rhodopsin. **(b)** *The* cis/trans *cycle*. To continue the process, *trans*-retinal must be converted back to *cis*-retinal.

transmitted along the optic nerve to the visual cortex in the brain (Figure 15.10).

For vision to continue, *cis*-retinal must be regenerated. The *trans*-retinal is reduced to *trans*-retinol (i.e. the aldehyde is converted to alcohol) and is isomerised to *cis*-retinal: this is oxidised to *cis*-retinol (see Figure 15.9(b)). Two different cells are involved: the oxidation of retinal to retinol occurs in the photoreceptor cell. The retinol is then released and is taken up by the adjacent epithelial cell where it is isomerised to *cis*-retinol and then reduced to

cis-retinal. This returns to the photoreceptor cell to reassociate with the opsin to form the rhodopsin which is now ready to respond to another photon (Figure 15.11). The retinal and retinol are transported across the interstitial space on a binding protein, presumably to protect such key molecules. Although the process has been described for a single retinol molecule, in reality signals from thousands of rods, each containing millions of molecules of rhodopsin, are triggered simultaneously and their signals integrated in the brain.

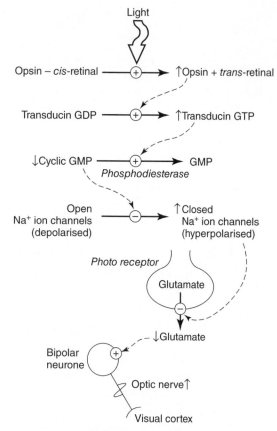

Figure 15.10 *The biochemical / physiological mechanism by which light initiates an action potential in the optic nerve.* Light is responsible for the conversion of opsin–*cis*-retinal to opsin (Box 15.2). Transducin is a G protein. Opsin results in the conversion of inactive transducin (i.e. GDP bound) to active transducin (as GDP is ex exchanged for GTP). Via transducin and cyclic GMP diesterase, light results in closure of Na⁺ ion channels by decreasing the concentration of cyclic GMP. This results in hyperpolarisation of the membrane of the photoreceptor. This results in a decrease in glutamate release and in this particular case it is considered that a decrease in glutamate in the synapse actually results in hypopolarisation of the bipolar neurone which stimulates the ganglionic neurone and therefore initiation of an action potential in the optic nerve. A physiologist, Baylor (1987), has commented that some physiologists were reluctant to accept the idea that the hyperpolarisation modulates synaptic transmitter release. Evidence was then obtained, however, that the hyperpolarisation which is the effect of light actually lowers the rate of transmitter release whereas the rate in darkness is high. A reduction in the release rate is a perfectly acceptable kind of signal, and, in this case, causes postsynaptic depolarisation and an action potential. Arrows adjacent to participants in the process indicates the change in activity or concentration caused by light.

Box 15.2 Similarities between the visual and the glycogenolytic cascades

It is instructive to note that the biochemistry of the reactions that initiate the visual cascade and the glycogenolytic cascade is similar. The cyclic AMP-dependent protein kinase complex comprises the regulatory and catalytic components (R and C) for which the regulatory signal is the concentration of cyclic AMP. This binds to the regulatory component of the kinase (the R subunit) which then dissociates from the R–C complex. The C is now catalytically active and catalyses the initial reaction in a cascade sequence which leads to activation of the target protein (phosphorylase).

The photoreceptor complex comprises *cis*-retinal and opsin (R–O) for which the regulatory signal is light. This interacts with the regulatory component of the complex R, which then dissociates (in the form *cis*-R) from the O component. The free O is now catalytically active and catalyses the initial reaction in the visual cascade sequence which leads to activation of the target system (i.e. produces an action potential in the optic nerve).

$$(R–C) + cAMP \rightarrow R\text{-}cAMP + C$$

catalytic subunit which initiates cascade

$$(R–O) + light \rightarrow R^* + O$$

catalytic subunit which initiates cascade

R* is the *trans* form of retinal.

Vitamin D: the hormone

The classification of vitamin D as a vitamin is largely an historical accident. It conforms much more to the definition of a hormone: it is produced in one part of the body and released into the blood to affect other parts of the body in this case, a novel endocrine hormone (Chapter 12).

In the skin, cholesterol is converted to 7-dehydrocholesterol by desaturation of the 9,10-carbon bond and ultraviolet light breaks this bond to produce cholecalciferol (Figure 15.12). The cholecalciferol is transported via the bloodstream to the liver where the first step in the activation of the hormone occurs, namely hydroxylation by a monooxygenase to produce 25-hydroxycholecalciferol, which is transported to the kidney where a further hydroxylation takes place at the 1-position to produce 1α,25-dihydroxycholecalciferol, which is the active form of the hormone (Figure 15.13).

Cholecalciferol can be produced in sufficient quantities in humans provided that the subject is adequately exposed to sunlight. In primitive humans, living with very few clothes and exposed to sunlight for considerable periods of

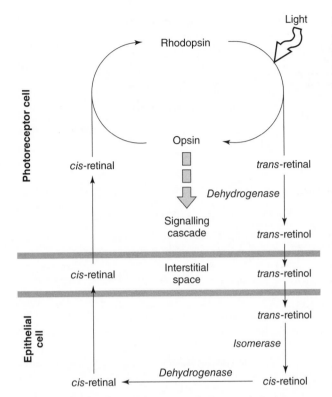

Figure 15.11 *The biochemical reactions that result in the conversion of* trans-*retinal to* cis-*retinal, to continue the detection of light.* To continue the process, *trans*-retinal must be converted back to *cis*-retinal. This is achieved in three reactions: a dehydrogenase converts *trans*-retinal to *trans*-retinol; an isomerase converts the *trans*-retinol to *cis*-retinol; and another dehydrogenase converts *cis*-retinol to *cis*-retinal. To ensure the process proceeds in a clockwise direction (i.e. the process does not reverse) the two dehydrogenases are separated. The *trans*-retinal dehydrogenase is present in the photoreceptor cell where it catalyses the conversion of *trans*-retinal to *trans*-retinol which is released into the interstitial space, from where it is taken up by an epithelial cell. Here it is isomerised to *cis*-retinol and the same dehydrogenase catalyses its conversion back to *cis*-retinal. This is released by the epithelial cell into the interstitial space from where it is taken up by the photoreceptor cell. This *cis*-retinal then associates with the protein opsin to produce the light-sensitive rhodopsin to initiate another cycle. The 'division of labour' between the two cells may be necessary to provide different NADH/NAD$^+$ concentration ratios in the two cells. A high ratio is necessary in the photoreceptor cell to favour reduction of retinal and a low ration in the epithelial cell for the oxidation reaction (Appendix 9.7).

time, sufficient cholecalciferol would have been synthesised in the skin. Only when they began to live in colder climes and covered most of the body with clothes would dependence upon a dietary source of cholecalciferol develop. Hence, it was then discovered as a vitamin. The vitamin/hormone has several physiological roles: it increases the concentration of a specific Ca^{2+} ion transport protein in the enterocytes of small intestine to increase Ca^{2+} ion uptake. This increases the Ca^{2+} ion concentration in the plasma which is required for mineralisation of bone. A deficiency of the vitamin/hormone results in osteomalacia (i.e. a deficiency of mineral in the bone), also known as rickets in children. The growth of the skeleton and the role of vitamin D is discussed in Appendix 15.2.

Vitamin E (tocopherol)

The most important source of vitamin E is vegetable seed oils (e.g. corn oil, sunflower seed oil, wheat germ oil). Other sources include milk, eggs, meat and leafy vegetables. Vitamin E is a group of eight naturally occurring compounds called tocopherols (Figure 15.14).

For a long time vitamin E was a vitamin in search of a disease because its deficiency led to a host of symptoms. Once it was discovered that vitamin E is a scavenger for oxygen free radicals (an antioxidant), a biochemical explanation for its function was possible. Damage by oxygen free radicals (reactive oxygen species or ROS) can occur in any tissue that uses or is associated with oxygen – and the extent of damage depends on what is termed 'oxidative' stress and vitamin E protects against such damage. It is particularly important in protection of polyunsaturated fatty acids in phospholipids from oxidation, which can damage the structure of membranes, particularly in endothelial cells, and the structure of lipoproteins in the blood. A detailed discussion of the biochemistry of vitamin E as an antioxidant is given in Appendix 9.6.

Deficiency of vitamin E is rare; it can occur from abnormalities in lipid absorption as well as dietary deficiency. Its deficiency affects the muscular system, causing dystrophy and paralysis and, if the heart is affected, death by myocardial failure. This is probably caused by demyelination of axons due to oxidative damage. Vitamin E is incorporated into chylomicrons within the enterocyte, so that its uptake into cells requires the activity of lipoprotein lipase.

Naphthoquinones (vitamin K)

The vitamin K requirement is met from the diet (vitamin K_1) and microorganisms in the intestine (vitamin K_2). The richest dietary source of vitamin K is green leafy vegetables but it is also present in meat and dairy produce. The structural formulae of vitamin K_1 (phylloquinone) and vitamin K_2 (menaquinone) are given in Figure 15.15. It was discovered in 1929 by Henrik Dam in Copenhagen, who discovered that it was necessary for the clotting

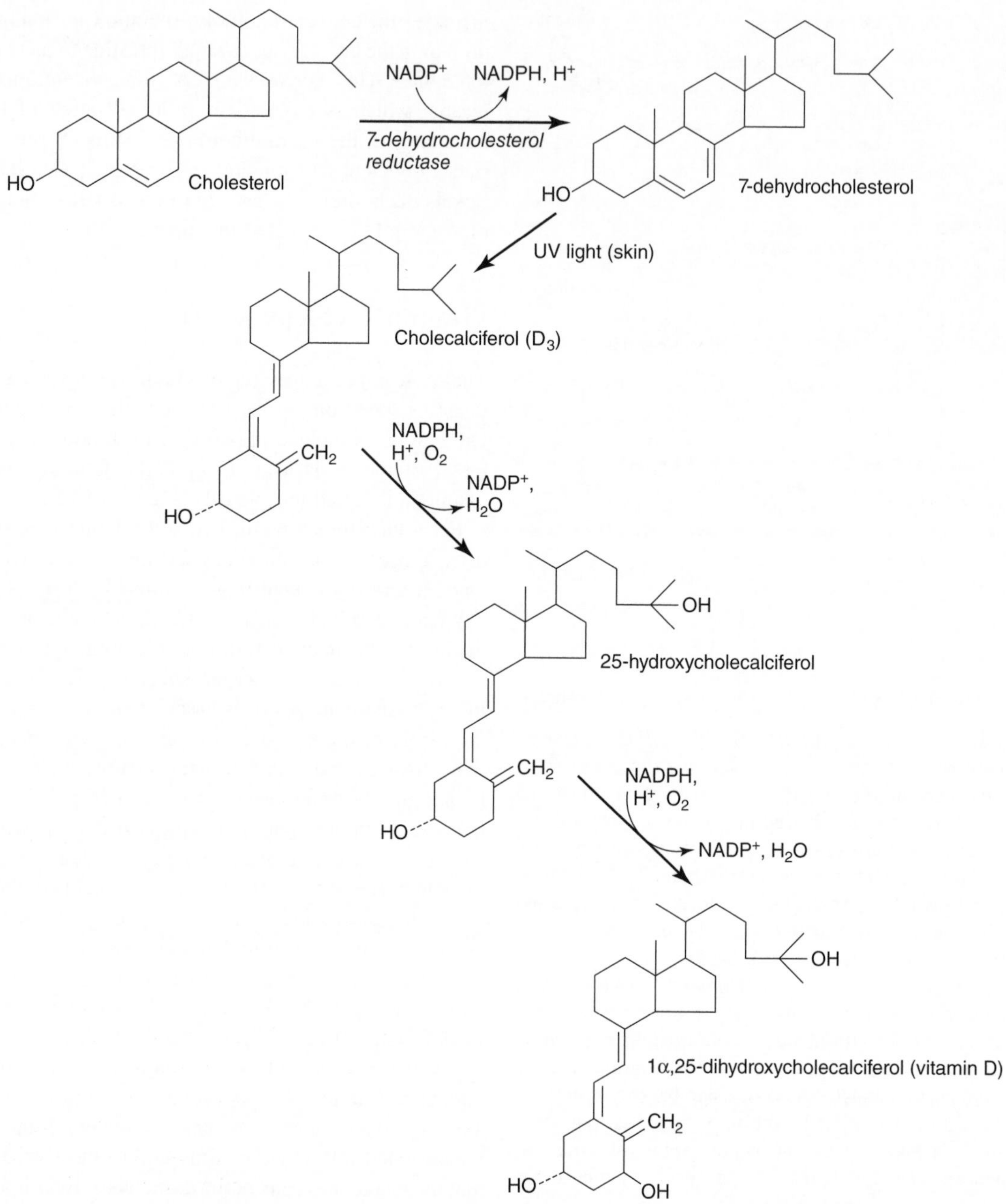

Figure 15.12 *The conversion of cholesterol to vitamin D.*

of blood (hence its name vitamin K, for Koagulation factor).

Primary vitamin K deficiency is uncommon, not only because of the presence of vitamin K in many plant and animal tissues but also due to its production by micro-organisms in the intestine. Broad spectrum antibiotics, however, can lead to low levels of vitamin K due to loss of microorganisms in the intestine. Newborn infants can become deficient since it does not cross the placenta, hence

is not stored in the foetus, and the intestine is sterile during the first few days of life.

Vitamin K is a component of the carboxylase enzyme that carboxylates the amino acid glutamate in proteins to form γ-carboxyglutamate, which binds calcium ions; i.e. it catalyses a post-transcriptional modification. Proteins so carboxylated include clotting factors (Factors II, VII, IX, and X) and two proteins in bone: oesteocalcin (known as matrix-gln-protein) and bone gln protein (BGP). The

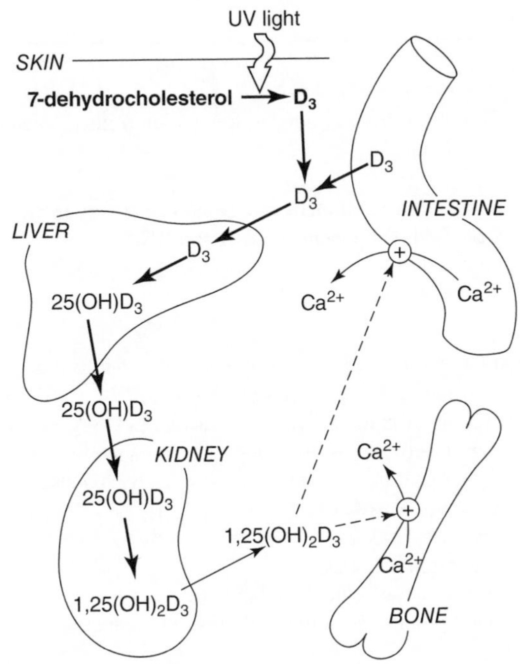

Figure 15.13 *The roles of different tissues in production of the active form of vitamin D.* Two major effects of vitamin D are presented: release of calcium from bone and uptake of calcium from the intestine.

role of the vitamin in blood clotting are discussed in Chapter 17 (Box 17.2).

Minerals

A list of minerals, sources and a summary of some roles are given in Table 15.4. Shortage of some of these minerals in the diet causes well-known and well-defined deficiency diseases such as anaemia, due to lack of iron, and goitre, due to lack of iodine. As with vitamins, *partial* deficiencies

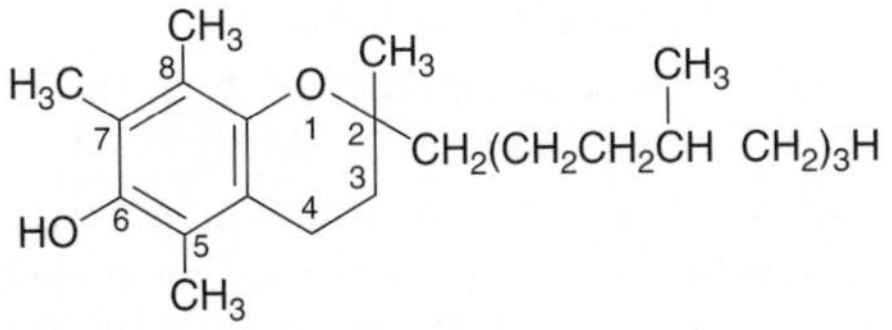

Figure 15.14 *Structural formula of α-tocopherol.* In β-tocopherol, the 7-methyl group is absent, in γ-tocopherol the 5-methyl group is absent, and in δ-tocopherol, both methyl groups are absent. In the tocotrienols, the side-chain at position 2 is replaced by:

$$CH_2(CH_2C=CCH_2)_3H$$
$$|$$
$$CH_3$$

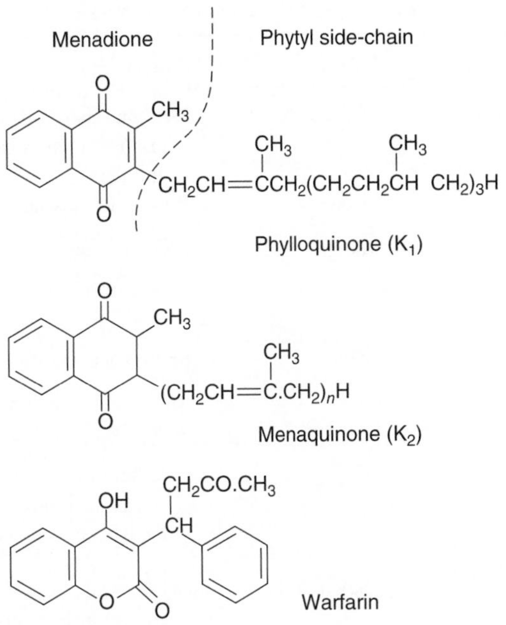

Figure 15.15 *Different forms of vitamin K and the structure of an antagonist, warfarin.*

of some minerals may impair biochemical processes leading to possible clinical problems. For example a partial deficiency of magnesium may impair function of the heart. Metal ions exert their biochemical effects largely through binding to proteins. Zinc is an important metal for protein and DNA structure and also in the storage of insulin.

Electrolytes (Na^+, K^+ and Cl^-)

The concentrations of sodium, potassium (and chloride) ions in the body are high and make the largest contribution to the electrical charge of cells; hence they are known as electrolytes. They have two important roles: maintenance of the total solute concentration in the cell which prevents excessive movement of water into or out of cells through osmosis; and the controlled movement of these ions across cell membranes acts as a signalling mechanism (e.g. the action potential in neurones and muscle, Chapter 14). Severe disruption of sodium or potassium levels in the body interferes with this signalling mechanism and with osmotic balance in cells.

A normal diet contains a sufficient amount of both sodium and potassium. The body has specific mechanisms for regulating the Na^+ ion concentration, which is important in the control of blood pressure (Chapter 22). However, there is less potassium in the normal diet and it may not always be sufficient (e.g. if the sweating rate is high). Fruit and fruit juices are good natural sources of this electrolyte.

Table 15.4 Some mineral: their source and roles

	Metal	Source	Role
Major metals	Calcium	Milk and dairy products; grains; green vegetables and fruit	Rigidity of bone and teeth. Regulation of metabolism Blood clotting
	Iron	Present in most animal and plant foods	Involved in >300 metabolic reactions; energy metabolism Structural component of DNA and RNA
	Magnesium	Vegetable, fruit and grains; meat products; dairy products	Involved in >300 essential metabolic reactions: metabolism ATP/ADP
Trace metals	Zinc	Red meats, shellfish, wholegrain cereals	Involved in many metabolic reactions; stabilisation of structure RNA, DNA and ribosomes Binding of some transcription factors to DNA Stabilisation of insulin complex in storage granules
	Copper	Liver and organ meats; grains, legumes, nuts, seeds (esp. cocoa powder)	Component of many enzymes, especially oxidases Amine oxidases Lysyl oxidases Cytochrome oxidase
	Selenium	Organ meats, seafood, muscle meats, cereals and grains	Component of glutathione peroxidase Iodothyronine deiodinase, thioredoxin reductase
	Manganese	Unrefined cereals; nuts; leafy vegetables; tea	Component of some enzymes: glycosyl transferase, arginase, pyruvate carboxylase
	Molybdenum	Milk, milk products; dried legumes or pulses; liver and kidney; grains	Prosthetic group of enzymes; aldehyde oxidase Xanthine oxidase Electron transfer chain enzymes
	Phosphorus	Meat, grains	As phosphate in all energy reactions in cells

Calcium

A 60 kg adult contains 1000–1200 g of calcium: more than 99% is in the bones and teeth. About 1 g is in the plasma and extracellular fluid and 6–8 g in the tissues, sequestered in calcium storage vesicles. The calcium concentration in the blood is about 2.5 mmol/L, about 50% as the free ion and the rest bound to plasma proteins.

A major function of Ca^{2+} ions is the regulation of certain chemical reactions within cells. It is an excellent messenger molecule since its intracellular concentration is very low (10^{-7} mol/L) whereas its extracellular concentration is high (10^{-3} mol/L). Hence, any change in the activity of a transporter can change the cytosolic Ca^{2+} ion concentration rapidly and markedly. There are several processes involved in control of the intracellular Ca^{2+} ion concentration (Chapters 13 and 22).

Salts of calcium, especially calcium phosphate, are responsible for the rigidity of bone and teeth but this rigidity does not mean that bone is biochemically inert. Far from it, for cells within the bone are continuously taking up and releasing calcium and phosphate, and the small amounts that are lost from the body must be replaced.

Copper

Copper is a component of many enzymes including amine oxidase, lysyl oxidase, ferroxidase, cytochrome oxidase, dopamine β-hydroxylase, superoxide dismutase and tyrosinase. This latter enzyme is present in melanocytes and is important in formation of melanin controlling the colour of skin, hair and eyes. Deficiency of tyrosinase in skin leads to albinism. Cu^{2+} ion plays an important role in collagen formation.

Since Cu^{2+} ions can catalyse formation of the dangerous hydroxyl radical, its concentration in both the intra- and extracellular compartments is maintained at very low levels by binding to the protein metallothionine. Deficiency of copper results in defects in formation of connective tissue, which may cause cardiovascular problems and poor bone formation.

Iron

Iron is important in so many metabolic reactions that it has been described as the 'metal of life'. Since it is so important, it is discussed in detail in a separate section below.

Main sources of iron are green vegetables (e.g. spinach), avocado, lentils, potatoes, liver and meats. A balanced diet containing these sources is normally sufficient for maintenance of the iron content in the body, unless losses of iron are large. There is, however, some concern about the iron content in the diet of some teenagers in developed countries. Since much of the energy consumed by many youngsters is now from snack foods, pizzas and soft drinks, in which the content of iron is very low, anaemia is increasing in teenagers and in young adults.

The free iron in the food enters the intestinal mucus from which it is absorbed by the epithelial cells via a transporter protein. This absorption is decreased by tea, coffee and phytate (inositol hexaphosphate) present in cereal fibre. Iron combined in haem is absorbed directly by epithelial cells after being released from haem-containing protein. The free iron is released in these cells by the enzyme haem oxidase. The free Fe^{2+} is then bound to paraferritin and released into the blood where it is bound by ferritin. The three reactions are as follows.

$$Haem + O_2 \xrightarrow{\text{haem oxygenase}} Fe^{3+} + CO + biliverdin$$

$$(Fe^{2+}\text{-paraferritin}) \rightarrow paraferritin + Fe^{2+}$$

$$Fe^{2+} + ferritin \rightarrow (Fe^{2+}\text{-ferritin})$$

The quantity of iron stored in the body varies considerably: in a 50 kg person it is about 1.4 g. Much of this is present in the erythrocytes but is released upon death of the cells and is re-utilised so it is conserved very effectively.

Magnesium

Magnesium is present in most foods but green plants are an important source since the chlorophyll molecule contains magnesium.

Magnesium as the Mg^{2+} ion is directly associated with energy metabolism since ATP must complex with the Mg^{2+} ion before it can undergo any enzyme-catalysed reaction. In normal circumstances, magnesium deficiency rarely occurs but there is interest in the possibility that a partial deficiency may result in problems with the function of the heart.

Selenium

Selenium is present in meat, seafood and cereals. The former two contain the highest levels. It is present in soil as inorganic selenium that enters the food chain via plants. In plant protein, it is present as selenomethionine and in animals as selenocysteine: this difference is due to the metabolism of selenomethionine in the liver as part of the normal catabolic pathway for methionine (Chapter 8). Somewhat surprisingly, selenocysteine is incorporated into protein via a specific tRNA which possesses a UCA anticodon for this amino acid.

Selenium in this form is present in three enzymes: glutathione peroxidase, iodothyronine deiodinase and thioredoxin reductase. Deficiency of selenium therefore decreases the activity of these three enzymes and results, at least in experimental animals, in liver necrosis and muscular dystrophy. In humans, it is known to be a cause of a particular form of cardiomyopathy known as Keshan disease which affects children and women. This cardiomyopathy was first described in China in 1979. It is also considered that a deficiency of selenium is a risk factor for cancer.

Zinc

Red meat, shellfish and wholegrain cereals are good sources of zinc (most of the zinc is in the bran and germ of the cereal). Zinc is a component of more than 100 enzymes which carry out a wide range of cellular functions and most of the zinc is present in muscle. A deficiency of zinc is associated with slow wound healing, decreased appetite, loss of taste and smell and decreased immune function.

Phosphorus

Phosphorus is present in large amounts in protein-rich foods and in cereals.

Phosphorus is a critically important element in every cell of the body and also in the form of hydroxyapatite in bone and in all other functions as phosphate. The concentration of phosphate in blood is 1.0 to 1.5 mmol/L existing as $H_2PO_4^-$ and HPO_4^{2-}; the equilibrium between the two acts as a proton buffer

$$HPO_4^{2-} + H^+ \rightleftharpoons H_2PO_4^-$$

Ultratrace minerals

These are classified as those with a requirement below one μg (microgram) per day. Elements in this class include: boron, chromium, fluoride, iodine, molybdenum, nickel, selenium, cobalt and manganese. Cobalt is part of vitamin B_{12} (see above). However, there appear to be no recommended dietary intakes for any of these except molybdenum.

The biochemistry, physiology and pathology of iron

The synthesis of haemoglobin in the bone marrow requires about 20–25 mg of iron each day. This is obtained either from the diet or from the iron that is recycled from the degradation of senescent erythrocytes (they survive for only about 120 days) which are phagocytosed by macrophages in the liver and spleen. The iron released from the

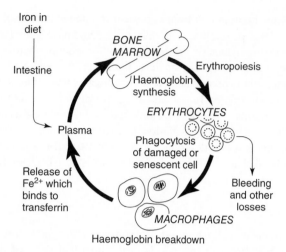

Figure 15.16 *The iron cycle in humans.* Iron in the diet is absorbed via the enterocyte into the bloodstream where it binds to the protein transferrin in the plasma. The iron in transferrin is taken up into the cells in the bone marrow for synthesis of haem, which complexes with globin to form haemoglobin during development of cells to produce erythrocytes (the process is known as erythropoiesis). The life of red cells in the blood is about 120 days. During this period the red cells are damaged (i.e. they become senescent red cells). These senescent cells are phagocytosed by macrophages, where the haemoglobin is broken down. The resultant iron is released and enters the bloodstream to be bound and transported around the body in a complex with transferrin. Some of the iron is, however, stored in the macrophages and other cells in a complex with the protein ferritin. The iron cycle is very effective. Approximately 25 mg of iron circulates each day, but only 1 mg is lost. This is made up from that taken up from the dietary iron in the intestine. Loss is somewhat higher in premenopausal women so that intake must be correspondingly higher.

haemoglobin is initially stored in the macrophages, prior to release(Figure 15.16). Ferritin is the protein that stores the iron: one molecule of ferritin can accommodate about 4000 iron atoms. Apoferritin binds the iron (Fe^{2+}), then transports it into the cavity.

The protein that transports iron around the body in blood and lymph, and indeed within the cell, is transferrin (500 kDa). It has two binding sites for iron (Fe^{2+}): when no iron is bound, it is known as apotransferrin. Transferrin picks up not only the iron absorbed from the intestine but also that released from the macrophages and then transports it to the cells that require it, which is primarily the cells in the bone marrow but also other cells that are proliferating. For uptake into the cells, the transferrin binds to a receptor on the plasma membrane and then the complex enters the cell by endocytosis. The iron is released from the complex in the cytosol where it is bound by the intra-

cellular ferritin before the iron is transported to the mitochondria, for the synthesis of haem, which is required for synthesis of cytochromes. Iron is also required for the synthesis of the non-haem iron that also plays a role in electron transfer in the mitochondria (Chapter 9).

Loss of iron from the body

The normal loss of iron is about 1 mg per day, which is normally matched by the intake. It is, of course, greater in premenopausal women due to menstruation. It may also be increased by trauma or major surgery or by donation of blood. The amount lost can be calculated on the basis that 1 mL of packed erythrocytes contains about 1 mg of iron.

Menstruation An average loss of blood in menstruation is about 100–200 mL which represents a loss of about 200 mg of iron over the month.

Blood donation Donors usually give about 500 mL of blood every two months. The loss is a little over 200 mg.

Pregnancy Pregnancy results in 'loss' of iron for the mother in four ways:

- to the foetus (about 300 mg);

- in bleeding during birth (150 mg);

- in the placenta during birth (100 mg);

- in milk during lactation (0.5–1.0 mg daily).

These represent a total loss to the mother of 1.25 mg per day, spread over the 15 months of pregnancy plus lactation.

Regulation of iron metabolism in cells

The magnitude of these losses emphasises the importance of regulating the amount of iron in cells, especially in those cells in which it plays a key role: erythrocyte precursor cells and proliferating cells. Three proteins are involved in the regulation of the iron concentration in a cell:

- transferrin receptors in the plasma membrane;

- apoferritin in the cytosol;

- iron-regulating protein (IRP).

It is the first two proteins that are regulated by the protein, IRP. This is achieved, somewhat surprisingly, at the level of translation rather than that of transcription. The regulation is discussed in three sections:

(a) the effect on changes in the flux-generating concentration of apoferritin in the cell;

(b) the actual mechanism of regulation of these concentrations;

(c) the mechanism by which the concentration of iron is sensed.

(a) The two changes that occur when the iron level in the cell is *low* are:

- an *increase* in the number of transferritin receptors (i.e. the receptor density in the membrane) in the plasma membrane of the cell;

- a *decrease* in the concentration of apoferritin in the cytosol of the cell.

Both changes should lead to restoration of the free iron to normal levels. This iron is then available for use.

(b) Regulation is achieved via changes in the concentration of mRNA. For the transferrin receptor, the level of mRNA is increased, when the iron level is low, by an increase in stability. For the apoferritin, the rate of translation of the mRNA is decreased so that the apoferritin concentration is decreased, which increases the concentration of free iron in the cell. These changes are achieved by the binding of the iron-regulating protein to receptors (the iron-response element), present on the mRNA molecules (Figures 15.17 and 15.18).

(c) A surprising fact is that the iron-regulating protein is a Krebs cycle enzyme, namely aconitase. This protein, therefore, has two roles: a catalytic one in the Krebs cycle and a regulatory one for the iron concentration in the cell: it has been termed a 'two-faced enzyme or protein'.

As with some other enzymes involved in the generation of ATP in the mitochondria (e.g. oxoglutarate dehydrogenase), aconitase possesses an iron–sulphur complex in its

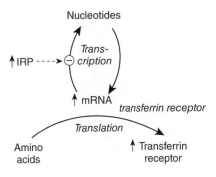

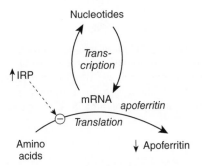

Figure 15.18 *The role of the iron-regulating protein on translation of mRNA for the transferrin receptor number in the membrane and the concentration of apoferritin.* IRP decreases the rate of degradation of mRNA for translation of the transferrin receptor which increases the number of the receptor molecules. IRP decreases directly the rate of translation of mRNA for apoferritin.

active site (Chapter 9). For aconitase, the structure of this complex changes according to the concentration of iron. When it is high, the complex contains four atoms of iron and the enzyme is catalytically active; when it is low, one iron atom is lost and the enzyme activity is also lost. This transforms the enzyme into the IRP (Figure 15.19). Aconitase is present in both the cytosolic and mitochondrial compartments and plays the same roles in both, i.e. as a catalyst and as an iron-regulating protein.

Iron and the regulation of haem synthesis

An extremely important role of iron is the synthesis of haem for formation of erythrocytes and also for proliferating cells for synthesis of the mitochondrial enzymes that contain haem (e.g. cytochromes). The flux-generating enzyme in the synthesis of haem is aminolevulinic acid synthase (ALS) (Figure 15.20). If the cellular iron concentration is low, the concentration of this enzyme is increased in an attempt to maintain the rate of synthesis. As with the other two proteins, the concentration of ALS is controlled at the level of translation in a similar manner to that for transferrin, i.e. by increased stability of the mRNA, which is achieved by the binding of the IRP to the mRNA.

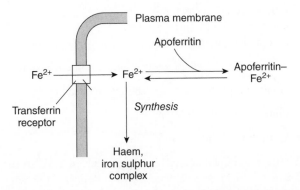

Figure 15.17 *The factors that influence the concentration of free iron in the cell.* The two factors are the number of transferrin receptors in the plasma membrane and the concentration of apoferritin.

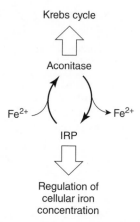

Figure 15.19 *Regulation of the aconitase/iron-regulating protein balance*. A low iron concentration increases the amount of the iron-regulating protein, the role of which is to regulate the iron concentration in the cell. A high iron concentration lowers the amount of the iron-regulating protein and increases that of aconitase.

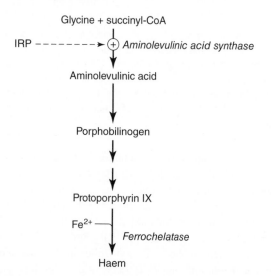

Figure 15.20 *Control of the rate of haem synthesis*. The concentration of the enzyme aminolevulinic acid synthase, the first enzyme in the synthesis of haem, and the flux-generating enzyme, is increased by IRP. This ensures an adequate rate of synthesis of haem, even though the iron level in the cell may be low. This is achieved by stimulation of translation. Full details of the pathway are presented in Appendix 15.3.

Iron and pathogens

The iron ion is highly reactive and readily catalyses oxidative/peroxidative processes and interacts with oxygen to form the oxygen free radical (superoxide) that damages cell membranes, proteins and DNA. To prevent such destructive events and still safely deliver oxygen, virtually all iron is maintained tightly bound to the proteins involved in oxygen transport and to the enzymes involved in energy metabolism. Excess free iron is precluded by the presence of the high iron-binding capacity of the proteins transferrin, lactoferrin and ferritin.

Pathogens, when they invade, therefore, encounter a low free iron environment in the host. Consequently, they must compete for iron to survive and multiply, primarily by making iron-chelating proteins (siderophores) with extremely high affinities for iron that can strip bound iron from host proteins, as well as by making iron-siderophore receptors and transport proteins to transport the iron into the pathogen. Some pathogens synthesise receptors for transferrin that resemble the natural receptors of the host, enabling them to acquire iron directly from the transferrin in blood fluids.

It is suggested that one reason why the body temperature increases during an infection is to restrict the iron available to the pathogens. The increase in temperature favours uptake by the immune cells and by the liver (Chapter 18).

A healthy diet

It is not possible to define a healthy diet. Consequently the recommendations made by national nutrition committees are general; they are designed to reduce what is considered to be unhealthy and increase what are considered to be the healthy components.

The report by the National Advisory Committee on Nutrition Education (NACNE), published in 1983 in the UK, recommended the following:

| Although some 'modern' diets are often considered to be unhealthy, the benefits of improved nutritional value of such diets over the past 100 years cannot be overlooked. The increases in energy, protein and micronutrients in food are one of the factors responsible for improved defence mechanism in the body which decreased the incidence of infectious diseases. These factors have also helped to reduce the parasite load in many countries. The diet has also led to an increase in the size of children, teenagers and adults. |

- Carbohydrate, preferably complex, should form 50–60% of the energy intake.

- No more than 30% of energy should come from fats.

- A high proportion of these should be unsaturated.

- Fibre intake should be higher.

- Salt intake should be moderate.

Similar recommendations have been made in the USA (see *Dietary Goals for the United States*, a Report of the Select Committee on Nutrition and Human Needs, US Senate 1977). In addition to these, the UK Department of Health has issued guidelines for healthy eating. These are: enjoy your food; eat a variety of different foods; eat the right amount for a healthy weight; eat plenty of foods rich in starch and fibre; eat plenty of fruit and vegetables; do not eat too many foods that contain a lot of fat; do not eat, too often, sugary foods or sugar-containing drinks; alcohol should be drunk sensibly.

The benefits of these diet guidelines can be accounted for from our knowledge of metabolism.

- Enjoyment of food improves mood and this can improve appetite in individuals who need to eat more (e.g. patients recovering from illness and surgery, elderly people and young children).

- A variety of food should ensure provision of all the essential components of the diet, including micronutrients (see above), essential fatty acids and amino acids.

- When the energy content of food is calculated, the intake can be adjusted to provide sufficient but not excess energy.

- Starch should provide a high proportion of the energy content of the diet since most foods containing starch contain some non-digestible starch, one form of 'fibre'; there are several reasons why fibre is important (Chapter 6).

- The digestion of fat produces chylomicrons and VLDL – metabolism of the latter produces LDL, high concentrations of which increase the risk of development of atherosclerosis (Appendix 11.3) (Chapter 22).

- Fatty foods also usually contain relatively high levels of cholesterol.

- Sugary foods cause tooth decay due to metabolism of sugar by bacteria in the mouth and can result in increased levels of triacylglycerol leading to resistance to insulin and hence increasing the risk of developing type 2 diabetes and obesity, see above. They can also, if they increase the blood glucose level above normal, be a risk factor for development of atherosclerosis (by causing glycosylation of various proteins including the protein in LDL.

- Alcohol impairs some metabolic processes, including gluconeogenesis which, in some conditions, is an essential process for wellbeing and normal behaviour (Chapters 6 and 14).

Nutrition for specific activities or conditions

Physical activity

It has been known for centuries that nutrition is important for physical activity. The importance of ingestion of carbohydrate is now appreciated by most if not all participants in sporting and other physical activities. This is because the level of glycogen in muscle is so very important to provide fuel for prolonged activities. It is important, also, to fuel the activities in hard training sessions and competitive ball games. To replenish the glycogen stores, the diet must contain enough carbohydrate to provide 60–70% of energy. There are, however, at least two nutritional problems with this:

(i) The diet must be designed specifically with this value in mind, since most Western diets contain only 30–40% carbohydrate. Special menus for meals are therefore required by people engaging in prolonged arduous physical activity.

(ii) Even in highly motivated athletes, this essential intake of carbohydrate is not always appreciated (Newsholme et al., 1984).

In contrast, when food has to be carried for very prolonged arduous endurance activities, a higher intake of fat is essential (see below). These points are discussed in Chapter 13.

The ultimate physical endurance study

It might be considered that running an ultramarathon (various distances from 50 to >100 miles), completing a 24-hour race, racing in the Tour de France or climbing Mount Everest are the ultimate in endurance activities. However, these activities may seem physically insignificant when compared with transantarctic expeditions, such as that carried out by R. Fiennes and M. Stroud in 1992–93. Not only was it a remarkable endurance feat but they carried out studies on energy expenditure, body composition and nutrition. To illustrate this feat, we can do no better than quote from Mike Stroud's paper.

During the southern summer of 1992/93, Sir Ranulph Fiennes and I set out from the Atlantic coast of the Antarctic, aiming to perform the first crossing of the continent unaided by other men, animals or machines. Behind us we each towed sledges weighing 222 kg which contained 100 days of food, fuel and essential survival equipment.

In the first three weeks we travelled slowly, for 10 to 12 hours a day, across the 350 km Filchner ice-shelf – a region of glacial ice covering the sea bed. We reached the coast of mainland Antarctica on Day 20 when we began a demoralising ascent over rough, wind-carved ice. Over the next 15 days we travelled a further 340 km, slowly gaining height to reach the flatter Polar plateau at 3000 m altitude. It was then only 550 km to our first target, the South Pole.

We eventually reached the Pole on Day 68 of the expedition, arriving in what were becoming familiar temperatures of below –40 °C combined with winds gusting above 40 knots. The resultant windchill factor was around

−90 °C. Inevitably, the work and the terrible conditions had taken their toll. We weighed ourselves at the Pole on small load-cell scales, and found that we had both lost over 20% of our starting body weights despite having eaten 23 MJ per day.

After leaving the Pole we travelled a further 480 km over the plateau to arrive at the Transantarctic Mountains, but during that part of the journey the debilitation from our continued energy deficit became so bad that we were forced to increase our intake to around 28 MJ per day ... It was not to be. By Day 95, even through increasingly clouded consciousness, we could see that the chances of success were virtually negligible. With increasing vulnerability to the effects of our undernutrition we decided to call it a day – disappointed despite completing the longest self-contained journey ever made.

The nature of our journey also dictated that the food should weigh a minimum and hence should contain a high fat content.

This requirement, however, conflicted with that of achieving a maximum exercise performance, for to perform many hours of work each day, the diet should have been high in carbohydrate. This is because manhauling, despite being of fairly low intensity in exercise physiological terms, would still lead to marked muscle glycogen depletion in 10 to 12 hours of exertion daily.

In the end, a compromise diet containing 23 MJ as 57% fat, 35% carbohydrate and 8% protein was selected. The diet consisted of porridge fortified with butter in the morning, soup with added butter during two brief stops in the day, a flapjack biscuit with butter after stopping in the tent, and a freeze dried meal with butter in the evening. All in all, a lot of butter! We also ate four small chocolate bars daily and drank coffee in the mornings and tea and chocolate in the evenings.

Stroud (1994)

Unlike other endurance activities, those like that of Stroud and Fiennes require food to be carried (or towed on sledges), so that energy-dense food is required; i.e., food which contains a high proportion of fat. The remarkable daily energy expenditure achieved by Stroud and Fiennes is given in Table 15.5. Especially on days 21–30, it was in excess of that recorded in any other form of endurance activity (Table 15.6). Not surprisingly, during this trek they lost body mass. The loss of fat and lean body mass is given in Table 15.7, where it is compared with losses that have occurred during prolonged starvation.

Mental activity

The fuel used by the brain is glucose obtained from the blood. The adult brain uses about 5 g of glucose every hour. Thus, although only comprising only 2% of the body weight, the brain uses as much glucose as skeletal muscle (at rest) which is 40% of the body weight (Chapter 14). However, muscle can increase its uptake of glucose more than 20-fold during physical activity whereas, in contrast, increased mental activity has very little effect on total glucose uptake by the brain. This is explained by the fact that most neural activity in the brain is involved in control of the basic functions of the body. Therefore, although increased activity of higher order functions of the brain (thinking, learning, speaking) might require more energy and consequentially more glucose, the effect on total brain glucose uptake is negligible.

Not surprisingly, the intake of sufficient carbohydrate to satisfy this demand is vital. If carbohydrate intake is low, glucose is provided in the short term (<24 hours in the adult) by breakdown of liver glycogen. After this period, glucose must be synthesised via the process of gluconeogenesis primarily from amino acids released from the breakdown of muscle protein and from glycerol released

Table 15.5 Percentage of energy derived from carbohydrate, fat and protein and total energy of food taken in several Antarctic expeditions

| Expedition leader | Year of expedition | Percentage energy from | | | Approx. total energy (MJ) supplied for the expedition per man per day |
		Carbohydrate	Fat	Protein	
Scott[a]	1910	42	39	19	18
Amundsen[b]	1910	NK	NK	NK	18
Mear/Swan	1985	34	57	9	18
Pentland	1992–93	35	57	8	21
Stroud	1992–93	35	57	8	23

[a] The second expedition to reach the South Pole which sadly ended in disaster on the return journey (see Cherry-Garrard, 2003).

[b] The first expedition to reach the South Pole.

NK – not known.

Information kindly provided by Ranulph Fiennes (personal communication).

Table 15.6 Daily energy intake and expenditure during very prolonged exercise

Event	Energy intake (MJ/day)	Composition of diet (% total energy)			Energy expenditure (MJ/day) as measured by	
		CHO	Fat	Protein	Energy intake[a]	Isotopes[b]
1005 km run (9 days)	25.0	62	27	11	25	–
Military mountain training (11 days)	13.0	49	38	13	20	21
Tour de France (20 days)	25.0	62	23	15		
Transantarctic expedition						
Day 0–5	20	37	55	8	29–38	29–35
Day 51–95	22	34	57	9	24–27	19–23
Day 21–30	22	34	57	9	c.46	45–49

[a] Assuming no change in body weight, or if it occurs adding the energy calculated from the amount of body fat consumed.

[b] See Chapter 2.

CHO – carbohydrate.

Data from Stroud (1998).

Table 15.7 Loss of body mass, fat and fat-free mass during a prolonged Antarctic expedition and in one individual fasting to death

Condition	Subject	Initial body wt.	Wt. loss (kg)	Fat loss (kg)	FFM loss (kg)[b]
Antarctic	RF	95.6	24.6	16.7	7.9
crossing	MS	74.8	21.8	12.2	9.6
Fasting to death	1 lean	61.0	25.0	36.0[a]	25.0

[a] At death no fat was left and 40% of body protein was lost (assuming no loss of water or bone). RF and MS lost about 20% of body protein.

FFM – fat free mass (see Chapter 2).

[b] Data from Stroud (1998) and from those presented in Chapter 16.

from triacylglycerol in adipose tissue. (During prolonged starvation in adults or short-term starvation in children, the brain uses ketone bodies, which replace some of the requirement for glucose as a fuel, Chapter 16) Glycogenolysis and gluconeogenesis are essential under these conditions to maintain the blood glucose level. If it falls below about 3 mmol/L, the brain fails to function normally; resulting in behavioural changes. Mental concentration may be lost and consciousness can become clouded. Conditions that give rise to hypoglycaemia are discussed in Chapter 6. What is not always appreciated is the high rate at which muscle and liver glycogen can be degraded in physical activities that involve repetitive short bursts of activity over a prolonged period. For example, it has been shown that all the glycogen in leg muscle of a soccer player is utilised in one match (Chapter 13). Modern dancing in clubs, discos and at parties may fall into this category. Low content of carbohydrate in fast foods, combined with glycogen breakdown in physical activity, might readily result in hypoglycaemia. For example, if most of the muscle and liver glycogen is depleted in the normal adult, it can be calculated that approximately 10 large beefburgers need to be consumed to refill these stores!

Pregnancy

Nutrition during pregnancy provides macro- and mitronutrients for both the mother and the foetus. This is important not only for the wellbeing of the foetus but also the health and wellbeing of the subsequent child. It has been shown that poor maternal nutrition can impair cognitive function as the child grows. More surprisingly, poor maternal nutrition can adversely affect the health of the subsequent adult (Box 15.3). Despite this, the increase in energy and protein requirements during pregnancy is quite small: the WHO recommends an extra energy intake of about 10% (1.3 MJ) and an extra protein intake of about 60 g per day. The intake of most vitamins in a normal diet is adequate. However, if the mother has a low activity of enzymes involved in methylation as well as a deficiency of folic acid, the synthesis of purine and pyrimidine nucleotides can be impaired. This can lead to several problems including neural tube defects, anaemia, retarded intrauterine growth, premature birth and even abortion. To avoid this, mothers are recommended to supplement their diet with folic acid for three months after conception and, if possible, four weeks before.

Box 15.3 Maternal nutrition and adult degenerative disease

In many parts of Britain in the 1930s, excellent records of the birthweight of babies were made. When the causes of death in the subsequent adults were studied and medical records investigated, a strong correlation between low birthweight, stroke incidence of mortality from cardiovascular disease and diabetes was observed (Barker, 1998). It may be that low birthweight is due to undernutrition of the mother.

Research is now in progress to determine what aspect(s) of maternal nutrition is responsible, e.g. is it low intake of energy, protein, micronutrient or poor-quality nutrition or any combination? Another intriguing question is what is the metabolic mechanism(s) underlying these findings? Several suggestions have been made: one example is as follows. If the foetus is not provided with sufficient glucose from the mother, the deficit could be made up by gluconeogenesis in the foetal liver. In response to this the capacity for gluconeogenesis in the foetal liver is increased by increasing the proportion of those hepatocytes with the capacity for gluconeogenesis (perivenous hepatocytes) at the expense of those with the glycolytic capacity (periportal). At a critical stage in the development of the foetus, the change would be permanent and hence would persist in the infant and the adult. The capacity for gluconeogenesis would be permanently increased, and this could lead to inappropriate rates of gluconeogenesis in fasting and after a meal, and consequently hepatic insulin resistance, especially on a Western diet. In other words, undernutrition of the mother induces changes in the metabolism of the foetus during a time of critical metabolic development which are carried over into the baby and are maintained into adulthood. These are called 'programmed' changes (or 'foetal' programming). This is a concept that might apply to many aspects of metabolism in adults that may lead to disease.

An adequate supply of essential fatty acids (EFA) is vital for brain development. The brain develops very rapidly during the last three months of pregnancy (third trimester) and its weight increases more than fourfold. This increase requires a considerable amount of essential fatty acids for formation of the phospholipid required for the myelin sheaths of developing axons. Indeed, it is estimated that the requirement for these fatty acids during the whole of pregnancy is more than half a kilogram. This is required to satisfy the demands for the growth of the foetus, placenta, mammary gland and increased maternal blood volume.

Infants

Nutrition is also of vital importance in infancy; the synthesis of structural components containing carbohydrates, proteins, nucleic acids and phospholipids takes place at high rates as organs develop and mature. Carbohydrate and fat are required for provision of energy and energy expenditure in the newborn is three to four times higher than in the child.

Milk

Milk is a remarkably complex biological fluid. It contains several hundred different molecules including enzymes and also different cells, e.g. immune cells. In particular, it contains proteins, fat, lactose, both the indispensable and dispensable amino acids, essential fatty acids, micronutrients, cholesterol and phospholipids. Milk is essential for the first four months of life. The major protein in milk is

Table 15.8 Approximate contents of selected constituents in mature human milk

Constituent (per litre)[a]	Content
Energy (MJ)	3.0
Total protein (g)	9.0
Casein (g)	5.7
Total casein[b] (g)	11
Immunoglobulin A (g)	1.0
Protein nitrogen (g)	1.45
Non-protein nitrogen (g)	0.45
Lactose (g)	67
Glucose (g)	0.3
Oligosaccharide (g)	14
Triglyceride (% total fat)	98
Fatty acids (%)	88
Saturated fatty acids (%wt)	45
Medium chain	6.0
Polyunsaturated (%)	15
omega-3 (%)	1.5
omega-6 (%)	13

[a]All values are expressed as per litre of milk with the exception of fats which are presented as percentage of either milk volume or weight of total fats.
[b]Total casein includes β- and κ-casein.

casein. An important property of milk is the ease with which the proteins and fats can be digested. Some of the contents of human milk are presented in Table 15.8.

Milk production in the mother is approximately 750 mL per day for the first 4–6 months of lactation and the energy content is about 2.8 kJ/mL, so that the infant is provided with about 2 MJ per day. On a daily basis, this is greater than that required by the mother during pregnancy so that lactation represents a considerable drain on the mother's fuel reserves if nutrition is poor. During the first five days after birth, the milk produced is known as colostrum. This contains protein, immunoglobulins (especially IgA), cytokines and memory B and T lymphocytes, together with lysozyme and lactoferrin to inhibit the proliferation of pathogens. In these first few days, the permeability of the intestine is high, providing the opportunity for the proteins and lymphocytes to pass from mother to infant across the intestinal barrier.

The elderly

Experiments on nutrition for the elderly are difficult to carry out and to interpret. Hence, it is difficult to distinguish whether any frailty in the elderly is due to malnutrition or underlying disease. Undoubtedly, some of the normal effects of ageing affect physiological function and hence nutritional requirements. These include:

- A decrease in muscle mass occurs normally during ageing (known as sarcopenia) and decreases resting energy expenditure so that less food is required.

- A decrease in fitness, due to less physical activity, results in ease of fatigue, which further reduces physical activity.

- The lower intake of food can result in a deficiency of micronutrients and essential fatty acids (especially vitamins B_{12} and folic acid – see below).

- Less activity in the open air and poor nutrition can result in a deficiency of vitamin D and be responsible, in part, for osteoporosis.

- A decline in immune function occurs which increases the risk of infection. It is not clear if this is caused by lack of a specific nutrient or a decline caused by ageing.

- Senile dementia can result in an even lower intake of food with further diminution of vitamin levels, so that a vicious circle can develop.

These facts lead to obvious nutritional advice for the elderly: vitamin and mineral supplements, increased intake of fruit and vegetables and provision of appetising food in an attempt to overcome reduced appetite.

There is a possibility that a partial deficiency of some vitamins (especially vitamin B_{12} and folic acid) impairs the activity of key enzymes in the brain which might be one factor in the development of senile dementia and even Alzheimer's disease (see above for discussion of homocysteine in this context).

Space flight

Humans have been flying in space for only about 50 years, so that knowledge of the long-term nutritional requirements in space is minimal. Such information, however, will be essential for the very long flights now being planned, e.g. to Mars. Opportunities for studying physiology, biochemistry and nutrition during long flights have only been possible on astronauts in the US Skylab Mission and in the Russian Space Station, Mir. Unfortunately, ground-based models for simulation of space flight, such as prolonged bed rest, have very limited value.

The biochemical/physiological findings that may influence nutritional requirements are:

- Loss of body mass of 1–5% occurs due to loss of fluid, muscle, bone and adipose tissue.

- Energy expenditure during flight, measured by the dual isotope technique, indicates that there is very little difference to that before flight. Although resting energy expenditure is higher during flight, this is compensated for by less physical activity. There was no additional energy requirement during space walks.

The diet provided is high in protein (about 114 g per day) and 60% of the energy is in the form of carbohydrate with 15% in fat. Some foods are fresh (e.g. fruit) but most are dehydrated.

These findings indicate that, at least for the energy requirements of astronauts, no special nutrition is required. Whether it will be possible to prevent loss of muscle and bone during prolonged flights by nutritional supplements is being considered. This approach is supported by the fact that supplementation or changes in diet can have beneficial effects in some diseases (Chapter 18 and see below).

Overnutrition

Overnutrition is difficult to define but is synonymous with excessive intake in the following discussions. Recent emphasis has been directed towards diets that contain excess energy in relation to expenditure, which result in obesity. This is a major problem in developed and increasingly in underdeveloped countries. There is particular concern about the marked increase in obesity in children, which can lead to major health problems in later life and will result in a massive increase in financial expenditure on health provision in the future: some obese children are now developing type 2 diabetes as young as 12. The topics obesity and type 2 diabetes are discussed in Appendix 15.4.

Overnutrition of some specific nutrients increases the risk of various diseases:

- Overnutrition of saturated fats and cholesterol is a risk factor for development of atherosclerosis.

- Saturated fat is a risk factor for breast cancer.

- Meat is a risk factor for cancer of the colon.

- Refined carbohydrate is a risk factor for several diseases.

Experiments on small experimental animals suggest that overnutrition per se may reduce lifespan: experimental animals fed on energy-deficient diets live longer than those on control diets. An explanation for this is awaited with interest.

Malnutrition

Malnutrition is also difficult to define since it can cover many different diets. The term is normally reserved for diets low in energy and/or protein but it is often used as a synonym for general undernutrition. According to this definition, some individuals in developed countries can be considered to be malnourished: for example, the elderly, the homeless and children from poor families. The term has also been used to describe diets that, although sufficient in energy, protein and micronutrients, are of poor quality and unhealthy. Some current Western diets may fall into this category and there is increasing concern about the effects of 'junk' or 'fast' foods that are now eaten in large quantities in developed, and in some underdeveloped, countries (Box 15.4).

Undernutrition

At the World Food Summit Conference in 1996, it was reported that 500 million people worldwide were estimated to be suffering from undernutrition.

Box 15.4 The diets of children and teenagers in developed countries and chronic diseases

There are at least three concerns about the diet of children in developed countries since they may lead to disease in adulthood or even earlier. The concerns are related to:

- the fat content;
- the *trans* fatty acid content;
- the sugar (sucrose) content.

Fat content   The large amount of fat (as a high percentage of energy intake) is one factor that can lead to obesity, which increases the risk of developing type 2 diabetes. The increased availability of 'fast food' and snacks that contain a high percentage of fat are temptations to children, in whom appetite is large, to accommodate sufficient intake of food to support growth. If the energy intake is higher than expenditure, obesity can result.

Trans *fatty acids*   The phospholipids in the plasma and in membranes of all cells contain long-chain polyunsaturated fatty acids (PUFA). During periods of growth and development of organs, PUFAs are required for phospholipid synthesis. The PUFAs are, of course, obtained from dietary triacylglycerol and phospholipids. The double bonds in most natural fatty acids are *cis* not *trans* Nonetheless *trans* fatty acids do occur in dietary fats. If the diet contains *trans* fatty acids, they might be incorporated into the phospholipids along with the *cis* fatty acids and hence into membranes. The presence of these abnormal fatty acids will modify the structure of the phospholipids which could impair the function of the membrane. There are two main sources of *trans* fatty acids in the diet: foods produced from ruminants contain *trans* fatty acids due to the activity of bacteria in the rumen; commercial hydrogenation of oils results in conversion of some *cis* into *trans* bonds. These artificially hydrogenated fats are used in the preparation of children's favourite foods such as potato crisps (chips), biscuits and pastries, and fast food such as beefburgers. Cells particularly exposed to such fatty acids are the endothelial cells. Damage to membranes of endothelial cells can lead to local inflammation and predispose to atherosclerosis.

Sugar   The hydrolysis of sucrose in the intestine produces both glucose and fructose, which are transported across the epithelial cells by specific carrier proteins. The fructose is taken up solely by the liver. Fructose is metabolised in the liver to the triose phosphates, dihydroxy-acetone and glyceraldehyde phosphates. These can be converted either to glucose or to acetyl-CoA for lipid synthesis. In addition, they can be converted to glycerol 3-phosphate which is required for, and stimulates, esterification of fatty acids. The resulting triacylglycerol is incorporated into the VLDL which is then secreted. In this way, fructose increases the blood level of VLDL (Chapter 11).

The metabolism of VLDL by lipoprotein lipase in the capillaries in many tissues results in the formation of low density lipoprotein (LDL), which is atherogenic, so that diets high in sucrose are a risk factor for development of atherosclerosis. Many children in developed countries now consume large quantities of soft drinks containing sucrose or fructose. According to the above discussion, this could lead to atherosclerosis in later life.

The most obvious cause of undernutrition is prolonged starvation, which is most dramatic in parts of the world where war, drought or other disasters have brought about failed harvests or mass migration.

Starvation

Starvation covers a number of different situations and a definition is therefore not possible. Voluntary starvation is known as fasting, a term usually restricted to short periods of starvation (e.g. overnight) or one or two days. In general, the term 'starvation' is taken to mean prolonged inadequate intake of energy and protein; other terms have also been used including inanition, wasting, marasmus or cachexia, usually without any clear definition. In addition, the weight loss that occurs during metabolic stress has been termed 'cytokine-induced malnutrition': this is a special type of 'starvation' discussed in Chapter 16.

The obvious symptom of starvation is a loss of weight, which is most easily assessed by a decrease in the body mass index (BMI). Indeed, it is an excellent predictor of death from starvation. A value of BMI below about 13 in men and about 12 in women is not compatible with life. These values coincide with a loss of about 50% of lean body mass. The major causes of death from malnutrition in developed countries are pneumonia, other infections or heart failure.

A detailed account of the metabolic changes that occur in starvation and the role they play in ensuring survival, from even a prolonged period of starvation, is presented in Chapter 16.

Feeding the severely malnourished

It is extremely important to select the correct type of nutrition when feeding severely malnourished individuals or patients. The provision of food can be dangerous unless carefully controlled as it can lead to what is known as the 'refeeding syndrome'. This is characterised by a rapid increase in extracellular volume, due to increased sodium intake, and decreased blood levels of phosphate and potassium due to increased levels of insulin which stimulate the entry of these into muscle. (The latter changes are also seen when type 1 diabetic patients in a severe hyperglycaemic state are treated with insulin.) A recommended refeeding schedule is as follows:

(i)   Establish normal fluid and electrolyte levels.

(ii) Provide a mixed diet to maintain energy expenditure.

(iii) Gradually increase the energy content.

(iv) Increase protein intake together with an increase in energy to encourage protein synthesis and increase lean body mass (i.e. positive nitrogen balance).

Undernutrition in children

The prevalence of undernutrition in children is staggering. It is estimated that globally more than 200 million children younger than 5 years are undernourished. Undernutrition is most obvious in the underdeveloped or developing countries, where the condition often takes severe forms. Images of emaciated bodies in famine-struck or war-torn regions are tragically familiar. Milder forms are present, even in developed nations. In 1992 an estimated 12 million children in North America consumed diets that were significantly below the recommended allowances of nutrients established by the US National Academy of Science (see Chapter 16).

A special form of undernutrition that particularly affects children is known as protein–energy malnutrition (PEM) which is a spectrum of syndromes from marasmus (lack of energy intake) to kwashiorkor (deficient protein intake).

Marasmus and kwashiorkor

Malnourished children can have two very different appearances. In marasmus, the limbs are wasted and the whole of the body assumes a shrunken 'skin and bone' appearance, as muscle is sacrificed to support more vital tissues. The shrunken cheeks are caused by a loss of the Biclet fat depots. In kwashiorkor, children have 'pot-bellies' partially explained by generalised oedema and hepatomegaly. Both have brittle, bleached hair, skin lesions and a deeply apathetic demeanour. Although the two conditions intergrade they show some geographical separation with kwashiorkor being restricted to tropical and subtropical regions while marasmus can occur anywhere.

Marasmus is considered to be due to inadequate food intake. It is not usually the quantity but the quality of the food that is deficient, e.g. low nutritional value of bulky vegetables. Kwashiorkor is considered to be caused, more specifically, by a low-protein diet. This condition frequently develops at the time of weaning when protein-rich milk is replaced by protein-deficient solid food. It did not appear in the medical literature until 1934 when it was reported by Cicely Williams who studied the condition while she was working among tribes of Western Africa. She gave it the name *kwashiorkor*, which was used by the Ga tribe to describe the condition that develops when the baby is taken away from mother's breast, usually because another baby has been born. It has generally been held that the oedema is a consequence of a low plasma albumin concentration and a reduction in the colloid osmotic pressure which reduces the movement of water from tissue fluid back into capillaries. The low albumin level results from a decreased rate of synthesis of albumin by the liver. However, if marasmus is due entirely to lack of energy intake, the biochemical changes should be similar to those produced by starvation, but this is not the case. In marasmus, by comparison with kwashiorkor, major metabolic disturbances are few.

Doubts have, however, been expressed about protein deficiency being the sole cause of kwashiorkor. There is only a poor connection between its occurrence and the protein content of the diet and recovery from kwashiorkor is not simply related to protein consumption. Other factors may be involved, such as:

- low intake of micronutrients, leading to impaired ability to deal with free radicals and the damage they cause (Appendix 9.6);

- chronic damage due to toxins present in food, for example aflatoxins, which are produced by fungi that grow on foods stored in warm humid conditions (Chapter 21);

- decreased effectiveness of the immune system due to failure of proliferation of lymphocytes caused by poor nutrition (lack of energy, amino acids and essential fatty acids) (Chapters 17 and 20). The decreased effectiveness of the immune system leads to increased incidence of infections, especially of the gastrointestinal tract, which causes severe diarrhoea, a major cause of death in these children.

Metabolic characterisition of kwashiorkor

The most common characteristics of kwashiorkor are:

- Very low reserves of vitamin A. This leads to an important effect of refeeding. The requirement for vitamin A increases when the children are supplied with protein so that they can quickly become vitamin A deficient, resulting in blindness (see above). Hence, the vitamin must be administered with the protein.

- The serum protein concentrations, especially of albumin, are low in kwashiorkor but normal in marasmus.

- The plasma fatty acid levels are increased in both but especially in kwashiorkor.

- The blood level of glucose is normal but decreases rapidly during short-term starvation (within 6 hours), probably due to low stores of liver glycogen.

- Excretion of creatinine, 3-methylhistidine and urea are low, indicating a low rate of protein degradation.

- Fatty liver occurs in kwashiorkor, probably due to lack of protein in the diet, which reduces the synthesis of the structural protein for VLDL (apolipoprotein B). The increased triacylglycerol produced in the liver from fatty acids removed from the blood (i.e. the inter-tissue triac-

ylglycerol/fatty acid cycle, Chapter 7) cannot be secreted and accumulates in the liver.

- The ratio of essential to non-essential amino acids is high in kwashiorkor but normal in marasmus. The cause of this may be low activities of the enzymes for metabolising the essential amino acids. These are required for any protein synthesis that must take place even in kwashiorkor.

- The plasma concentration of growth hormone is increased, whereas that of insulin is decreased.

- The level of triiodothyronine (T_3) is decreased whereas that of reverse T_3 (rT_3) is increased. This is similar to the situation in starvation (Chapter 16).

Functional foods and nutraceuticals

Over the past few years it has become apparent that some cells require specific nutrients to maintain their function. Administration of these nutrients or their precursors can help a patient fight a disease and also aid recovery if these cells are directly or indirectly involved in the disease. An important question is whether these compounds can be regarded as nutrients or drugs. A nutrient can be defined as a substance which maintains and supports any physiological or metabolic process. A drug is a substance that prevents or influences the progression of a disease. Combining these definitions leads to the terms 'nutraceutical' or 'functional food'. These terms are synonymous for foods that can prevent or treat diseases. However, there are differences.

Functional foods

These are defined as any food that has a positive impact on the health of an individual, their physical performance or state of mind, in addition to its nutritional value.

Three conditions define a functional food:

- It is a food (not a capsule, tablet or powder) derived from naturally occurring ingredients.

- It can and should be consumed as part of the daily diet.

- It has a particular function when ingested, e.g. enhancement of a defence mechanism, prevention or amelioration of a specific disease, recovery from a specific disease, or slowing of the ageing process.

There are several diseases or physiological/pathological conditions for which functional foods are claimed to have a beneficial effect. Mood or physical endurance can be improved, in some conditions, by high carbohydrate intake; the incidence of cardiovascular disease and cancer can be reduced by ingestion of fruit and vegetables which provide free radical scavengers; immune function can be improved by intake of sufficient protein (see Table 15.9).

Nutraceuticals

Nutraceuticals are used by the medical profession as part of the treatment of specific conditions. (Many are available in commercial preparations.) They include glutamine, arginine, nucleosides and polyunsaturated fatty acids. The explanation for their benefit is described in Chapter 18.

Nutrition for patients with genetic disorders

Patients suffering from metabolic disorders such as phenylketonuria (PKU), branched-chain ketoaciduria (maple syrup urine disease, MSUD), urea and ammonia disorders or glycogen storage disease require formulations manufactured specifically for each disease (Elsas & Acosta, 2006). (Appendix 15.1).

Vegetarian diets

Vegetarian diets consist largely of plant foods – grains, legumes, fruits, vegetables, nuts and seeds. Several different diets are covered under this term: two main ones are a *vegan* diet which contains no animal products whatsoever, and a *lacto-ova-vegetarian* diet, which is a vegan diet that contains dairy products and eggs.

A problem with vegan diets is whether they can provide sufficient protein and some micronutrients. Foods recommended are: whole grains and grain products e.g. wheat, millet, barley, rice, rye, oats, maize, wholemeal breads and pastas); pulses and products made from them (e.g. peas, beans, lentils, tofu); fresh vegetables (including green leafy vegetables and salads); fresh and dried fruits; nuts; seeds (e.g. sunflower, sesame and pumpkin).

Vegetables should be cooked as lightly as possible, and the cooking water (which contains minerals and vitamins) used in soups and stocks. Salads and raw vegetables are particularly good sources of vitamins and minerals. Seeds are more nutritious if they are ground or milled. Dried fruits are good sources of some minerals.

Two micronutrients of particular concern in vegan diets are vitamin B_{12} and iron which are normally obtained from meat. A dietary supply of vitamin B_{12} can be ensured from sources such as yeast extracts or soya protein.

Table 15.9 Suggested mechanisms by which ingestion of some plant foods decreases risk for development of some degenerative diseases and cancer

Plant foods	Diseases in which risk is decreased	Mechanism by which risk is decreased
Whole grains, beans, vegetables: all containing high content of fibre	Cardiovascular disease, cancer	Fibre has two effects: (i) lowered cholesterol levels (ii) increased production of volatile fatty acids by microorganisms in colon
Legumes	Cardiovascular disease	Presence of saponin which decreases cholesterol absorption from the gut
Soy	Cardiovascular disease	Presence of some proteins that increase number of LDL receptors
	Cardiovascular disease, cancer	Antioxidants
Nuts	Atherosclerosis	Increase in ratio of monosaturated/saturated fatty acids which causes a decrease in the plasma level of LDL-cholesterol
Wheat grain, legumes	Colon cancer	Contains digestion-resistant starch and other non-digestible carbohydrates which increase fermentation in colon and hence production of volatile fatty acids
Soy protein	(i) Osteoporosis (ii) Breast and ovarian cancer	(i) Presence of phyto-oestrogens which inhibit damage to DNA (ii) Phyto-oestrogens compete with oestrogen for receptors on breast and ovarian cells

There is evidence that adherence to a vegetarian diet reduces the incidence of some types of cancer and cardiovascular diseases. The possible mechanisms by which some of the components of these and other diets may have a beneficial effect are given in Table 15.9.

Eating disorders

The term 'eating disorder' is defined as a persistent disturbance of eating which impairs health or psychosocial functioning or both and which is not secondary to any general medical disorder or any other psychiatric disorder. The most common disorder is obesity, which is discussed in Appendix 15.4. Anorexia nervosa and bulimia nervosa share many characteristics. Anorexia nervosa has been recognised for at least 100 years, whereas bulimia nervosa was first described in 1979 despite the frequency of its occurrence. Bulimia can affect up to 20% of the female population in some groups, notably students. The biochemical and physiological changes in the patients, as far as they are known, are what are expected for starvation or semi-starvation. Adaptations are expected to include reduced energy expenditure and increased rate of fatty acid oxidation to provide energy and also to maintain the blood glucose level via the glucose/fatty acid cycle. Discussion of these disorders is unfortunately limited due to the paucity of information relating to these patients. Useful parameters, for example, to measure would be plasma levels of fatty acids, glycerol and amino acids, especially glutamine and branched-chain amino acids.

Anorexia nervosa

Anorexia nervosa is a disease that mainly affects women, usually aged 10–30 years, and is characterised by three main features:

- An overriding pursuit of thinness, achieved through a reluctance to eat and a disturbed eating pattern. Even when a weight goal has been achieved, the individual still feels overweight so that a new lower weight goal is then set.

- A body mass of less than 85% of the ideal. In extreme cases, anorexics can lose 60% of their body weight and achieve weights of below 35 kg (77 pounds).

- Menstrual cycle disturbances, including amenorrhoea, are a response of young women to prolonged negative energy balance.

Anorexia nervosa was first described as a distinct clinical entity by Sir William Gull in 1873. He described 'a peculiar form of disease' occurring mostly in young women, usually around the onset of puberty, who refused to eat and suffered extreme emaciation. Gull's description remains largely accurate today: 90–95% of patients suffering from anorexia nervosa are females and the peak incidence is about 18 years of age.

Self-inflicted starvation was not uncommon in early religious groups, so that anorexia nervosa may have occurred in the medieval periods but Richard Martin is usually credited with the first medical description in 1689: 'I do not remember that I ever did in all my practice see one, that was conversant with living so much wasted with the greatest degree of consumption (like a skeleton only clad with skin)'.

Many sufferers of anorexia nervosa show an active interest in food (for example, preparing it for others), but not in eating, so it is unclear whether they are lacking in appetite or whether they repress the sensation of hunger. Patients frequently deny that they are suffering from an illness and only seek medical attention, when weight loss is profound and after pressure from family or friends. Long-term follow-up studies indicate that anorexia nervosa is associated with significant morbidity and a mortality ranging from 4% to 18%. The life-threatening complications in anorexia nervosa are largely those seen in starvation, with loss of cardiac muscle (Chapter 16) and electrolyte imbalance. Indeed, cardiac arrhythmias are not uncommon. Loss of mineral density in bone (osteoporosis) is common in anorexic patients and has usually been attributed to decreased levels of oestrogen but hormone replacement therapy does not always restore bone density. Another probability is a deficiency of vitamin D.

The personality and psychological traits of those with anorexia nervosa – potential high achievers, with high self-expectation and a propensity for self-denial – help to explain why anorexia is more prevalent in the higher social groups, and in women of above average intelligence. Of interest, the personality traits that characterise elite athletes – high self-expectation, perfectionism, persistence and independence – are similar to those of many anorexic patients. If the athlete is a young woman distance runner in whom a light body confers competitive advantage, then the body image is important. It is no surprise therefore that of a group of 93 elite women runners, 13% reported a history of anorexia nervosa, compared with 0.2–1.1% in the general female population. Furthermore, the prevalence of eating disorders in female athletes correlates strongly with the importance of body leanness in the particular sport pursued. Many anorexic patients are hyperactive and some are known to maintain their abnormally low body weight by excessive exercise.

During the Second World War when food was scarce, anorexia nervosa was relatively rare. It has been reported that in Italy between 1939 and 1945 there were no patients admitted to the clinic for treatment of anorexia nervosa. The psychological fear of overeating appears to be removed when food is scarce or when social circumstances clearly demonstrate a need and a positive role for women, as occurred in the Second World War.

Studies have been carried out on levels of hormones involved in the response to starvation and in control of the menstrual cycle in anorexic patients:

- T_3 levels in the blood are low with a decreased rate of formation from thyroxine but increased conversion of the latter to reverse triiodothyronine (rT_3).

- Plasma cortisol levels are increased and the normal diurnal variation is lost.

- The plasma growth hormone level is increased.

- The plasma levels of luteinising and follicle-stimulating hormones and oestradiol are low. There can be a complete loss of pulsatile secretions of these hormones, or a loss of sleep-related pulsatile secretion, as occurs in puberty.

These changes are similar to those that occur in starvation (see Chapter 16) and can be considered as physiological adaptations to this condition. Some are also similar to the triad syndrome which is prevalent in some female endurance athletes (Chapter 13).

Treatment

In principle, treatment is straightforward; in practice, it is often very hard to carry out. The aims are to:

- Establish a healthy eating pattern so that a gradual but progressive increase in body mass occurs.

- Remove the psychological factors that may have precipitated and maintained the disease. This may be as simple as establishing better relationships with family members or may require prolonged cognitive behavioural therapy.

Bulimia nervosa

The important difference between bulimia and anorexia nervosa is that bulimia involves frequent binges with the consumption of very large amounts of food and a sense of loss of control of food consumption. Other characteristic features are extremes of behaviour to control body weight, including self-induced vomiting and misuse of laxatives, and, as with anorexic patients, extreme attitudes to body shape and weight. Many bulimic patients have cravings for food and difficulty controlling eating after a meal which suggests an impairment of the satiety mechanism. Patients are almost invariably female and are usually older than those with anorexia. Bulimia is considered to have become much more frequent over the last 25 years and is now the most common eating disorder encountered in psychiatric practice. The self-induced vomiting may be sufficiently frequent to cause erosion of tooth enamel due to gastric acid and it can also result in alkalosis (the blood pH is above 7.4) and hypokalaemia (a low plasma level of potassium). Treatment is the same as that for anorexia although antidepressants are sometimes prescribed.

16
Starvation: Metabolic Changes, Survival and Death

When the prisoners-of-war in the Far East were freed, a picture of malnutrition was revealed which was the result of inadequate food, debilitating disease and enforced labour during three-and-a-half years of captivity. The story is here told by medical officers who were themselves prisoners, and the material they collected with little or no conventional apparatus makes no mean contribution to science. Valuable nutritional studies were made also on groups of prisoners after liberation, by medical relief teams, and at bases overseas during evacuation.

(Bennet, 1951)

Even in the United States, the richest nation in the world, thirty five million people are considered 'food insecure', that is, they are not sure where there next meal is coming from . . .

(Vernon, 2007)

The use of prolonged starvation for treatment of obesity has posed a fascinating problem: that man is capable of fasting for periods of time beyond which he would have used all his carbohydrate resources and all his protein for gluconeogenesis in order to provide adequate calories as glucose for the central nervous system.

(Owen *et al.*, 1967)

One difficulty in discussing starvation is the different terms used to describe it: for example, starvation, fast, undernutrition, inanition, wasting, cachexia or marasmus. Each can mean different things in various contexts. The term 'fast' is sometimes used to describe the overnight period of starvation prior to breakfast. The terms 'marasmus', 'wasting' and 'cachexia' are usually used to describe pathological conditions. For the purposes of this chapter the term 'starvation' is defined as the deprivation of food and the metabolic changes that occur in relation to macronutrient deficiency, whether starvation is voluntary or not. It is not always possible to distinguish between starvation and undernutrition: consequences of specific nutrient deficiencies are dealt with in other chapters.

Prolonged periods of starvation, or undernutrition, are all too frequent in the less developed countries of the world, especially when they are affected by natural disaster or war. The effect on children is particularly severe and the diseases that result are known collectively as protein-energy malnutrition, which is a spectrum of diseases from marasmus to kwashiorkor. They are discussed in Chapter 15. Undernutrition can also occur in the less privileged citizens of developed countries, such as the poor, elderly or homeless. Consequently, the biochemical response to starvation is as important today as it would have been in our evolutionary past. Survival for prolonged periods of starvation involves behavioural changes that result in decreases in energy expenditure and biochemical changes that result in changes in rates and types of fuels that are used.

Very few scientific studies have been carried out on humans during prolonged starvation. One of the first concerned a single individual and later studies confirmed that the responses observed in this 'classical' experiment were not atypical. Tragically, the Second World War provided the opportunity to study the effects of starvation on much larger populations, for example in the Western Netherlands, the Warsaw Ghetto and German and Japanese

Functional Biochemistry in Health and Disease by Eric Newsholme and Tony Leech
© 2010 John Wiley & Sons Ltd

concentration camps (Box 16.1). For a review of metabolism in starvation see Hoffer (2005).

Between 1944 and 1946, Ancel Keys and his co-workers conducted experiments on 32 male volunteers at the University of Minnesota who ate meals containing only two-thirds of their normal intake. After 4 weeks their resting energy expenditure was decreased by 40%, while physical activity and the thermic response to the meals were also reduced. In a similar study in Cambridge, England, normal subjects were underfed by half for 42 days, resulting in a mass loss per subject of about 6 kg. Each participant used about 1.6 MJ less energy each day (measured by the dual isotope technique) compared with the control period. Again, reduction in energy expenditure was achieved through falls in both resting energy expenditure and physical activity. The remainder had to be met from the endogenous fat and protein. In some other animals there are even greater energy economies (the Emperor penguin is a fascinating example, Box 16.2).

Box 16.2 Starvation and the Emperor penguin

Photograph Kindly provided by Dr. Mike Stroud, University of Southampton. It is Dr. Stroud who is surrounded by the Emperor Penguins in Antartica.

At 40 kg, the Emperor penguin is the largest member of its family. Despite its diet of fish and squid, it breeds (nesting is hardly the appropriate term) many miles from the sea on a high ice plateau in order to escape from predators such as leopard seals. After courtship and mating, the male is left to incubate the egg while the female returns to feed to replenish her fat stores, which she has used to produce eggs, and to bring back food for the chick. Starvation may extend to 100 days for the male and, if the return of his mate is delayed, he then feeds the chick with a curd secreted from his oesophagus, a further drain on his fuel reserves.

As with humans, the secret of surviving long-term starvation is the fat store, which in the case of the male emperor penguin constitutes 30% of its body mass, or 82% of the total fuel reserves of the body, established by a period of frenzied feeding (hyperphagia). During the period of starvation, oxidation of fat provides 96% of the penguin's energy requirements, leaving protein to supply the remaining 4% – a proportion similar to that seen during the starvation of an *obese* human. Although the levels of plasma fatty acids and 3-hydroxybutyrate are considerably lower than in humans, the blood glucose level is maintained at about 13 mmol/L (a high level but one which is characteristic of birds).

The other aspect of the Emperor penguin's survival strategy is the reduction in energy expenditure – an adaptation not easy to achieve in a blizzard at −30 °C! At the breeding site, the birds huddle together with each taking its turn to do duty on the outside. This group-insulation reduces the energy that must be expended to maintain body temperature. By keeping relatively still in such flocks, the penguins may save as much as 30–40% of their normal energy expenditure.

Box 16.1 Starvation in the Western Netherlands, 1944–45

The Netherlands were occupied by the Nazis from the spring of 1940, but until September 1944 the rationing system provided the population with near-adequate nutrition. However, from September 1944 until after the liberation in May 1945, the population of the Western Netherlands (more than 4 million) suffered severe malnutrition. This was because on 17 September 1944 the Dutch government in London declared a general railway strike in Holland and, in retaliation, the occupying forces prohibited all transport of food from areas of food production (the north and east) to the west of the country. Although this embargo was removed in November 1944, so that the food could be sent to the west by water transport, this was soon prevented by the cold weather because the waterways froze over and the limited stocks of food were rapidly depleted. It is estimated that well over 10 000 people died from starvation alone during this period and probably an equal number from diseases related to starvation. Even in the midst of the war, the situation was so bad that the senior war leaders could not fail to note it. In April 1945, the British Prime Minister Winston Churchill sent a telegram to US President Roosevelt warning that 'The plight of the civil population in occupied Holland is desperate. Between two and three million people are facing starvation . . .'

A detailed study of the problems caused by starvation, the response to refeeding and the best methods for refeeding was carried out by an international team of medical and nutritional experts whose findings were published in 1948 by The Netherlands Government (Burger *et al.*, 1948).

Mechanisms for the regulation of the blood glucose concentration

Despite the importance of decreasing energy expenditure in starvation, the need to maintain the blood glucose concentration at a near normal level is the basis for many of the metabolic changes that occur during starvation or carbohydrate deprivation. This is apparent from the observation that provision of exogenous energy by infusion of fat during several days of starvation did not affect the changes in metabolism (Klein & Wolfe 1992). Indeed, in the absence of disease it is the need to maintain the blood glucose concentration that results in death from starvation. Consequently, near-death metabolism is discussed in this chapter. Prior to a discussion of carbohydrate, fat and protein metabolism during starvation, details of the mechanisms for maintenance of the blood glucose concentration are given. The metabolic mechanisms and the concepts derived from them were developed over a period of 20–30 years from about 1960. They are divided into four sections, which are summarised in Figures 16.1 to 16.6 and briefly described here:

(1) The basic facts are as follows: glucose is taken up by muscle, in which it is converted to pyruvate and then oxidised by the Krebs cycle. The store of triacylglycerol in adipose tissue is mobilised as fatty acid by the process of lipolysis (Chapter 7). The fatty acids are released into the blood, from where they are taken up by muscle and oxidised. This oxidation results in inhibition of glucose uptake and metabolism. Glucose is also taken up by adipose tissue and its metabolism results in inhibition of fatty acid release which lowers the level of fatty acids in the blood. Hence, fatty acid metabolism in muscle inhibits glucose uptake by muscle, and glucose metabolism in adipose tissue inhibits fatty acid mobilisation. This is the basis of the concept of the glucose/fatty acid cycle (Figure 16.1), the effect of which is conservation and maintenance of the blood glucose concentration. A consequence of the cycle is that, if fatty acid oxidation is inhibited, the blood glucose level falls in starvation (Figure 16.2).

(2) Changes in hormone levels in starvation can extend the control provided by the cycle. The levels of glucagon and growth hormone are increased, which stimulates lipolysis, and the level of insulin is decreased, which decreases rates of glucose uptake but increases lipolysis and hence fatty acid mobilisation (Figures 16.3 and 16.4).

(3) Malonyl-CoA inhibits fatty acid oxidation in muscle. Insulin increases the concentration of malonyl-CoA in muscle and so inhibits fatty acid oxidation. A fall in

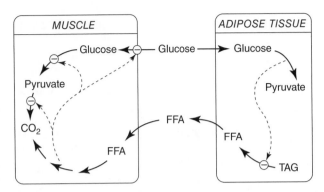

Figure 16.1 *The glucose/fatty acid cycle.* The dotted lines represent regulation. Glucose in adipose tissue produces glycerol 3-phosphate which enhances esterification of fatty acids, so that less are available for release. The effect is, therefore, tantamount to inhibition of lipolysis. Fatty acid oxidation inhibits pyruvate dehydrogenase, phosphofructokinase and glucose transport in muscle (Chapters 6 and 7) (Randle *et al.* 1963).

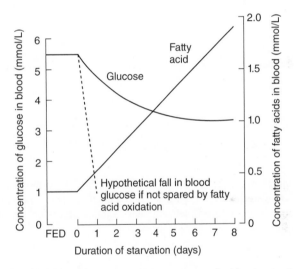

Figure 16.2 *Reciprocal relationship between the changes in the concentrations of glucose and fatty acids in blood during starvation in adult humans.* As the glucose concentration decreases, fatty acids are released from adipose tissue (for mechanisms see Figure 16.4). The dotted line is an estimate of what would occur if fatty acid oxidation did not inhibit glucose utilisation. Such a decrease occurs if fatty acid oxidation in muscle is decreased by specific inhibitors.

the insulin level, therefore, decreases that of malonyl-CoA which stimulates fatty acid oxidation and restricts further glucose uptake by muscle (Figure 16.5).

(4) The effects of the glucose/fatty acid cycle and those of the hormones on the cycle, on the regulation of the blood glucose level, can be extended by two further changes in metabolism. (i) Fatty acids are taken up by liver and converted to ketone bodies, which are released

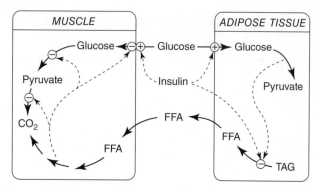

Figure 16.3 *Effects of insulin on the glucose/fatty acid cycle.* Insulin enhances glucose metabolism by stimulating glucose uptake by muscle and adipose tissue and by inhibiting lipolysis in adipose tissue (see Chapter 12 for the mechanism of these effects). The effect of glucose metabolism on lipolysis is via stimulation of fatty acid esterification via glycerol 3-phosphate.

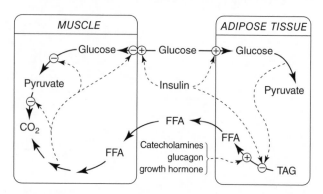

Figure 16.4 *Effect of several hormones on the glucose/fatty acid cycle.* Catecholamines, glucagon and growth hormone stimulate lipolysis in adipose tissue and hence antagonise the effects of insulin.

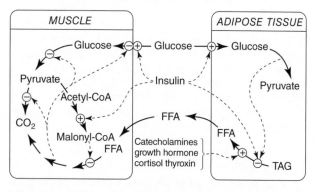

Figure 16.5 *Effect of malonyl-CoA on the glucose/fatty acid cycle.* Malonyl-CoA is an inhibitor of fatty acid oxidation, so that it decreases fatty acid oxidation in muscle and thus facilitates glucose utilisation (See Figure 7.14). Malonyl-CoA is formed from acetyl-CoA via the enzyme acetyl-CoA carboxylase, which is activated by insulin. Insulin therefore has three separate effects to stimulate glucose utilisation in muscle.

by the liver and taken up by muscle and oxidised: the effect of this is to supplement the effect of fatty acids on glucose utilisation. (ii) In the liver, amino acids, lactate and glycerol can be converted to glucose, via gluconeogenesis, which further helps to maintain the blood glucose concentration. The changes in the levels of glucagon and insulin during starvation result in increased rates of formation ketone bodies and glucose in the liver (Figure 16.6).

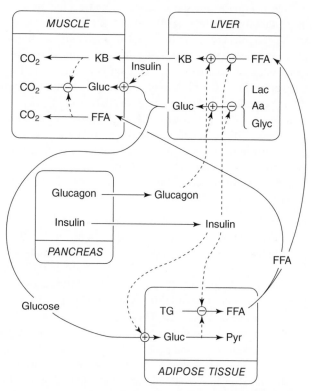

Figure 16.6 *Extension of the glucose/fatty acid cycle by inclusion of ketone body formation and gluconeogenesis.* The liver has three indirect effects on the glucose/fatty acid cycle which help to conserve the blood glucose and maintain its normal level.

(i)   Fatty acids are precursors for ketone bodies. They are released from liver (See Figure 7.19), taken up by muscle and their oxidation inhibits glucose (Gluc) utilisation. The rate of ketone body formation depends upon the plasma concentration of fatty acids: the greater the concentration, the greater is the rate of ketogenesis. In addition, insulin inhibits ketogenesis whereas glucagon enhances it (Chapter 7).

(ii) The precursors for glucose formation in the liver are lactate (lac) and amino acids (aa), which form glucose via gluconeogenesis. Glycerol (Glyc) also forms glucose via glycogenolysis (See Figures 6.22 and 23). Glucagon increases the rate of gluconeogenesis (Figure 12.13).

(iii) Insulin inhibits glycogenolysis and gluconeogenesis. Glucagon opposes the effects of insulin and therefore helps to maintain the blood glucose level so that it has the same end result as that of fatty acid oxidation (See Figure 12.14).

Table 16.1 Energy provided from the oxidation of carbohydrate, fat and protein after 12 and 60 hours starvation in normal lean subjects

| | Energy provided (kJ/hour) | | | |
| | 12 hr starvation | | 60 hr starvation | |
Fuels oxidised	Energy	% of total expenditure	Energy	% of total expenditure
Glucose/glycogen	125	41	20[a]	6
Fat	146	46	234	75
Protein	42	13	59	19

Data taken from Carlson *et al.* (1994). Subjects were normal lean adult males.

The resting energy expenditure is almost the same at both 12 and 60 hour starvation (about 310 kJ/hour).

[a] Only glucose (which will have been produced from gluconeogenesis).

Metabolic responses to starvation

The first detailed metabolic studies on prolonged starvation (38 days) were carried out by George Cahill and colleagues in the 1960s. They studied obese females who starved for 38 days in order to decrease body mass. Measurements were made of arteriovenous differences for various fuels and blood flow across the brain, skeletal muscle and kidney. Similar studies over such a prolonged period of starvation have not been done on lean subjects, so it must be emphasised that the interpretation of these results applies only to obese humans. Metabolic studies on starvation in lean normal volunteers have been carried out for a maximum of 8 days (Tables 16.1 and 16.2). Changes in metabolism during prolonged starvation in the lean humans will be qualitatively similar to those in the obese, but may be quantitatively different. The metabolic changes presented in Figures 16.7, 16.8, 16.9 and 16.11 are likely to be the same in both. Some differences between lean and obese are discussed below.

On the basis of the results with the obese, starvation can be divided arbitrarily into five phases: the postabsorptive period, early starvation, intermediate starvation, prolonged starvation and, finally, the premortal period. Although these are characterised by different metabolic patterns, the transition from one period to another is gradual. Some of the changes in the postabsorptive period and early starvation are described elsewhere in this book but they are brought together in this chapter for completeness.

Postabsorptive period

The postabsorptive period begins several hours after the last meal, when the contents of the small intestine have been absorbed. If the meal contained a high proportion of digestible carbohydrate, the postabsorptive period will last three to four hours and will terminate either with another meal or with the onset of early starvation. Towards the end of the absorptive period, the concentration of glucose in the hepatic portal blood will fall, leading to an increase in the proportion of phosphorylase in the active form in the liver. This will cause glycogen breakdown and the release of glucose by the liver, which will be used by all tissues for mechanism of control of the changes, see Figures 6.30 and 6.31 (Figure 16.7).

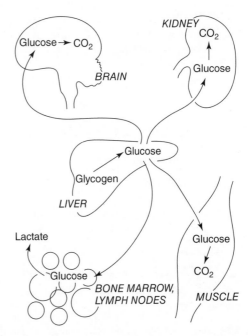

Figure 16.7 *Pattern of fuel utilisation during a short fast.* This is the metabolic condition after, for example, the overnight fast.

Early starvation

Early starvation commences after the postabsorptive period and persists until about 24 hours after the last meal, so that it normally includes the overnight fast. It is characterised by glucose release from the liver (from both gluconeogenesis and glycogenolysis) and mobilisation of fatty acid from adipose tissue. The rates of glucose utilisation by tissues change throughout the period. In the early stages most tissues utilise glucose but during later stages mainly the brain and *anaerobic* tissues use glucose. This is because fatty acid oxidation by muscle increases, which reduces the rate of glucose uptake, through the operation of the glucose/fatty acid cycle (Figure 16.8). Studies in normal lean subjects show that after 12 hours of starvation, about 45% of the resting energy expenditure is obtained from fat oxidation and 41% from glucose (Table 16.1).

> The importance of liver glycogen in the maintenance of the blood glucose concentration during the overnight fast is emphasised by the severe morning hypoglycaemia in a child who failed to store glycogen in the liver due to a deficiency of the enzyme glycogen synthase (See Box 7.2).

Arteriovenous differences show that after the overnight fast the liver of a lean adult releases glucose at a rate of about 8 g/h and the brain utilises more than half of this (5 g/h). Direct measurement of liver glycogen levels in human biopsy samples shows that the mean rate of hepatic glycogen breakdown between 4 and 24 hours of starvation is about 3 g/h. The shortfall between glycogen breakdown and glucose release must be made up by gluconeogenesis, which accounts for about 60% of the glucose released.

Intermediate starvation

This phase is entered with the depletion of liver glycogen stores and is characterised by two important metabolic features:

- High rates of lipolysis and ketogenesis to provide the two fat fuels, fatty acids and ketone bodies (See Figures 7.6 and 7.7).

- An enhanced rate of hepatic gluconeogenesis to maintain the blood glucose level and provide the fuel for the brain and other tissues (Figure 16.9).

After 60 hours of starvation in lean subjects, fat utilisation (i.e. ketone bodies plus fatty acids) accounts for three-quarters of the energy expenditure (Table 16.1); a value which will rise even higher as starvation continues. Much of this increase is accounted for by hydroxybutyrate oxidation (the major ketone body) since, by 60 hours of starvation, the plasma concentration of hydroxybutyrate has increased 26-fold compared with a threefold increase in the concentration of fatty acid (the glucose concentration falls by less than 30%). By eight days of starvation there has been a sixfold increase in fatty acid concentration, whereas the concentration of hydroxybutyrate has increased about 50-fold (Table 16.2). The changes in these three major fuels in obese subjects during starvation for 38 days are shown in Figure 16.10.

Once the glycogen has been depleted, gluconeogenesis is the only source of glucose. In order to avoid hypoglycaemia (or hyperglycaemia due to over-compensation), rates of gluconeogenesis must be precisely matched to the demand for glucose. In fact, blood glucose concentration remains remarkably constant during intermediate and prolonged starvation. Changes in glucose supply are complex as hepatic glycogenolysis is replaced by gluconeogenesis with an increasing contribution from renal gluconeogenesis. The demand changes, too, as the brain uses other fuels (i.e. ketone bodies). Little is known about the mechanisms of control of these metabolic changes and adaptations.

> Studies using C^{13}-NMR spectroscopy show that, after 54 hours of starvation in normal lean subjects, gluconeogenesis accounts for 96% of glucose released from the liver (Rothman *et al.*, 1991).

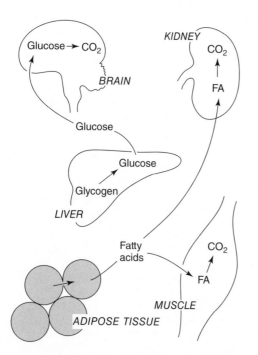

Figure 16.8 *Pattern of fuel mobilisation and utilisation during early starvation*. This is starvation over a period of about 24 hours; liver glycogen stores are nearly depleted and fatty acid mobilisation is taking place.

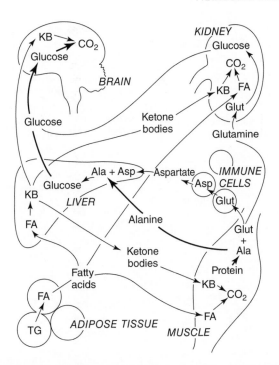

Figure 16.9 *Pattern of fuel mobilisation and utilisation after several days' starvation.* During this period, marked changes occur. (i) The liver produces ketone bodies that will be used by aerobic tissues including the brain (see Chapter 7). Liver glycogen is exhausted, so that gluconeogenesis takes over the formation of glucose. (ii) Body protein (mainly muscle protein) breaks down to provide amino acids that are metabolised in the liver where they are converted to glucose (and urea – see Chapter 8). The rates of breakdown of protein degradation, gluconeogenesis and oxidation of glucose by the brain are high and hence lines and arrows are thicker in this figure. Not all the amino acids are metabolised; some are converted to glutamine (Glut) and used by cells of the immune system and kidney. In immune cells, glutamine produces aspartate which is a precursor for gluconeogenesis. Glutamine also produces glucose in kidney (Chapter 8). Glut is the abbreviation for glutamine. Emphasis is made in this figure on release of alanine and glutamine but most amino acids that are released contribute to glucose formation (See Figure 8.12).

Prolonged starvation

There is no clear-cut division between intermediate and prolonged starvation but after about three weeks the following will apply:

- The rise in blood ketone bodies continues during intermediate and prolonged starvation. After about 20 days the concentration reaches a plateau at about 8 mmol/L, mostly hydroxybutyrate (Figure 16.10).

- The amino acid precursors for gluconeogenesis are provided from the degradation of muscle protein.

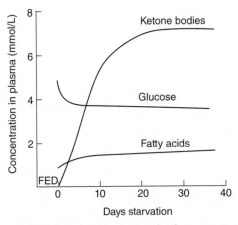

Figure 16.10 *The progressive changes in the concentrations of glucose, fatty acids and ketone bodies in the plasma during prolonged starvation in obese females.* The largest relative change is in the concentration of ketone bodies, particularly in contrast to that of glucose; the concentration changes in glucose are very small due to the glucose/fatty acid cycle, together with the effects of ketone bodies, and gluconeogenesis.

Table 16.2 Plasma concentrations of some compounds during starvation in normal lean adult males

		Plasma concentration (mmol/L)			
Study	Time of starvation (hours or days)	Glucose	Glycerol	FFA	Hydroxybutyrate
(a)	12 hr	5.2	0.05	0.49	0.09
	60 hr	3.8	0.13	1.05	2.40
(b)	Fed (control)	5.5	–	0.30	0.10
	2 d	4.1	–	0.82	0.55
	4 d	3.6	–	1.15	2.89
	6 d	3.5	–	1.18	3.98
	8 d	3.5	–	1.88	5.34

Data are from two separate studies: (a) Carlson *et al.* (1994) and (b) Cahill (1970). FFA – long chain fatty acids.

Since much less glucose is required by the brain, the rate of gluconeogenesis falls and hence the rate of protein degradation falls. In the obese, the energy provided from the oxidation of glucose that has been provided by amino acids from protein degradation is as little as 5% of the total (it is much higher in the lean; see below).

The rise in the plasma level of hydroxybutyrate leads to an increase in its rate of oxidation by the brain which reduces the rate of glucose utilisation by this organ from about 100 g to about 40 g per day (see below). Of this 40 g, about half is produced from glycerol.

Glucose precursors during starvation

The major gluconeogenic precursors are lactate, glycerol and amino acids. At rest, lactate arises from glucose metabolism in *anaerobic* tissues and muscle and is converted back to glucose by the liver, using energy from fat oxidation, for further use by these tissues (the Cori cycle). Indeed, gluconeogenesis from lactate accounts for one-third of the glucose released by the liver, so that the Cori cycle is important during starvation. The cycling of glucose means that the Cori cycle is 'self-contained', i.e. the glucose is not siphoned off for complete oxidation. Instead, glycerol and some amino acids give rise to glucose for oxidation. During prolonged starvation, the concentration of glycerol in plasma increases almost threefold (Table 16.2) due to increased lipolysis in adipose tissue. For a normal lean human, after 60 hours of starvation, about 25% (20 g) of the glucose used by the brain is generated from glycerol. It is quantitatively an important glucose precursor in starvation, at least in the obese.

It might be assumed that the glucose required by the brain during the *early days* of starvation arises from amino acids that are released by the hydrolysis of muscle protein. This is, however, not the case: only half is obtained from protein. The other half is obtained from the *free* amino acids in muscle, especially glutamine. Muscle contains a large store of free glutamine, which is released and taken up by the liver, converted to glutamate and oxoglutarate and then to glucose. Of the store of glutamine in muscle, about half is released in the first few days of starvation. After this period, breakdown of muscle protein provides all the amino acids to continue the formation of glucose which is used primarily by the brain. Although muscle protein breakdown continues through starvation, the rate decreases progressively due to the use of ketone bodies by the brain, which replaces some of the requirement for glucose (Figure 16.11).

It seems likely that the changes in the rates of fatty acid mobilisation, ketogenesis and gluconeogenesis are coordinated by increases in the plasma levels of glucagon and

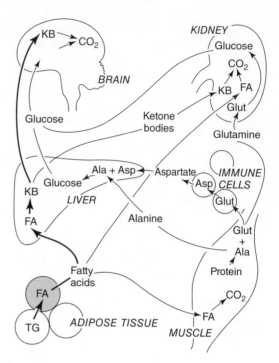

Figure 16.11 *Pattern of fuel utilisation during prolonged starvation.* The major metabolic change during this period is that the rates of ketone body formation and their utilisation by the brain increases, indicated by the increased thickness of lines and arrows. Since less glucose is required by the brain, gluconeogenesis from amino acids is reduced so that protein degradation in muscle is decreased. Note thin line compared to that in Figure 16.9.

growth hormone, together with the decreased level of insulin (Table 16.3).

Responses to starvation in lean adults

There are some important differences between lean and obese subjects in their responses to starvation.

- In lean subjects, amino acid oxidation, via glucose formation and glucose oxidation, provides almost four times more energy than in the obese subjects (Table 16.4). That is, the obese lose protein much more slowly, which may be an important factor favouring survival of the obese in starvation. This is consistent with the fact that, from the data available, obese subjects have survived starvation approximately four times longer than the lean (about 300 days versus 60–70 days; Table 16.5).

- The plasma ketone body level increases twice as fast over five days' starvation in lean subjects compared with obese: this may be an attempt by the body to restrict protein degradation in the lean by providing an alternative fuel to glucose for the brain as quickly as possible.

Table 16.3 Plasma concentrations of some hormones during starvation in normal lean adult males

Time of starvation (h)	Plasma concentration					
	Insulin pmol/L	Glucagon µg/L	Growth hormone µg/L	Adrenaline pmol/L	Noradrenaline pmol/L	Cortisol pmol/L
12	53	160	2.0	140	600	340
60	26	300	11.0	300	1,220	410
8 days	8.3	–	–	–	–	–

Data from Carlson *et al.* (1994). Subjects were normal lean adult males.

Table16.4 The contribution of amino acid oxidation to total energy requirement during starvation in lean and obese subjects

Condition	Approximate fasting urinary N loss (g/day)	Approximate metabolic rate (MJ/day)	Mean value of energy provided by protein oxidation (% of total)
Lean	8.4	5.0	18
Obese	3.8	8.0	5

The data are means from three different lean individuals and three different obese individuals (Henry 1992).

1 g of nitrogen is equivalent to 6.25 g of protein and the oxidation of 1 g of protein provides 16.7 kJ.

The energy provided by protein oxidation, as a percentage of the total, is calculated as follows:

$$\frac{\text{urinary nitrogen excretion (g)} \times 6.25 \times 16.7}{\text{total energy expenditure}} \times 100$$

Table 16.5 Duration of starvation prior to death in lean and obese male adults

Condition	Number of subjects	Length of starvation (days)	Reference
Lean	9[a]	57–73	Elia (1992)
Lean	1	63	Meyers (1917)
Obese	1	382	Stewart & Fleming (1973)
	several	100–249	Elia (1992)

[a] Northern Ireland hunger-strikers in the Maze Prison in Belfast in the 1980s. For one hunger-striker the body weight before the hunger strike was 61.4 kg (BMI was 20.75); at death, body weight was 36.4 kg (BMI was 12.3) (see Peel, 1997).

Table 16.6 Percentage weight loss of organs during starvation

Organ	% loss
Brain	7
Liver	29
Lungs	29
Heart	40
Muscle	41
Kidney	49
Pancreas	49

Data from Elia (1992). It is unclear which organ fails first to result in death.

- The fact that lean people lose more protein than obese during starvation may pose a health problem when normal weight people try to lose weight rapidly by short-term starvation or severe energy restriction. It appears to pose a problem if the starvation is repetitive. The reasons for such behaviour include improvement of performance (for jockeys, boxers, lightweight rowers, etc.), change of appearance (models) and eating disorders. Loss of protein may contribute to morbidity in such professions and mortality in patients suffering anorexia nervosa since protein loss from essential organs will impair their function (Table 16.6).

Response to starvation in children

There are two important differences to these findings in young children:

- In normal young children, the contribution of amino acid oxidation to energy requirement in starvation is about 7%, similar to that in the obese. In malnourished children, who have a protein-energy deficiency, it is even lower (4%). This suggests that a mechanism exists to protect muscle protein from degradation in children. Such a mechanism may involve a faster and greater increase in ketone body formation in children compared with adults (Chapter 7).

- Ketogenesis is particularly important even in a short period of starvation, since the requirement of glucose by the brain of a child is similar to that of an adult, yet the store of glycogen in the liver is much less. Consequently, the store would last for a much shorter time, requiring an alternative fuel for the brain.

Sequence of metabolic changes from intermediate starvation to death

The changes in metabolism during a prolonged period of starvation provide a basis for discussion of the changes likely to occur as death approaches.

(i)   As soon as liver glycogen is depleted, the free amino acids in muscle are released to be deaminated and converted to glucose to be used primarily by the brain.

(ii) Once the free amino acids have been depleted, muscle protein is broken down to provide amino acids that are released and converted to glucose in the liver. The rate of protein degradation is high in order to provide sufficient glucose for the brain.

(iii) During prolonged starvation, the rate of protein degradation is progressively decreased as some of the energy requirement by the brain is provided by oxidation of ketone bodies, so that much less glucose is oxidised.

The ketone bodies are provided from fatty acids mobilised from adipose tissue.

Autopsies carried out on men who have died from starvation reveal virtually no visible fat. Post-mortem measurements on one lean subject who died of starvation indicated that most of the fat, but only about 40% of protein, had been used and the BMI had fallen from 20 to 10. Oxidation of all of the fat and 40% of the protein would have generated 300 and 80 MJ respectively, a total of 380 MJ which, assuming an energy requirement of 6 MJ per day, would last approximately 63 days, in good agreement with the data in Table 16.5. An obese subject weighing 140 kg (about twice the ideal body weight) would possess about 60 kg of fat which, on the above assumption, could provide energy for about 350 days.

(iv) When the fat stores are depleted, ketone body formation can no longer take place, so that the brain now utilises only glucose.

(v)   To provide this glucose for the brain, the breakdown of protein in muscle and perhaps other tissues (Table 16.6) increases. The rate of degradation increases to a rate that is similar to that in the earlier stages of starvation. This results in a relatively large loss of body protein. Loss of about 50% of body protein is not compatible with life. That is, when fat stores are depleted, life cannot be maintained for more than

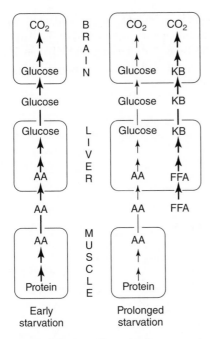

Figure 16.12 *Role of ketone bodies leading to a reduction in the high rate of protein degradation in prolonged starvation.*

about 14 days, the time during which about 50% of the body protein will be broken down.

Causes of death

Once all the fat has been used, the rate of degradation of protein, now the only fuel, must increase considerably. The exact cause of death in such circumstances is not always clear: a study of starvation in the Western Netherlands in 1944–45 (see Box 16.1) raised a number of possibilities:

• Loss of myofibrillar protein from the diaphragm and intercostal muscles limits the ability to cough, reducing the efficiency of fluid removal from the lungs and bronchioles and so increasing risk of infection. Pneumonia is a likely cause of death.

• During starvation, the response of the immune system to infection is slow and the peak level of antibodies is lower than normal, increasing susceptibility to disease. This is probably caused by a decrease in the plasma levels of all amino acids which are required not only for protein synthesis, when cells proliferate, but also for production of cytokines and acute phase proteins. Synthesis of glutamine from branched-chain amino acids is also necessary to maintain a supply of this important

fuel and nucleotide precursors for the cells of the immune system.

• A deficiency of amino acids results in decreased production of albumin in the liver, lowering its concentration in the plasma and hence the colloid osmotic pressure. Consequently, fluid is lost from the blood and its viscosity increased, so that the heart has to work harder to pump the same quantity of blood and eventually it may not be able to cope, especially as cardiac muscle is lost in prolonged starvation (Table 16.6). Death will then result from cardiac failure.

Progressive decrease in protein degradation in starvation

The amino acids that are made available as a result of protein degradation in starvation are used as precursors of glucose, which is required for the brain. The decline in starvation-induced protein degradation is a result of the decreased requirement for glucose by the brain due to the increase in utilisation of ketone bodies. The question arises, therefore, as to the mechanism by which the protein breakdown in muscle is reduced. Two answers, which are interdependent, have been put forward: (i) decreased metabolic activity in tissues, and (ii) a decrease in the plasma level of thyroxine and hence triiodothyronine.

Decreased metabolic rate

During starvation there is a decrease in the resting energy expenditure which is due to reduced metabolic activity in most tissues including, perhaps surprisingly, the brain. The decrease in muscle protein breakdown may, therefore, be a simple consequence of a decrease in the rate of glucose oxidation, so that gluconeogenesis from the amino acids is decreased.

The hormone triiodothyronine (T_3) accelerates both total energy expenditure and protein degradation. The hormone secreted by the thyroid gland is thyroxine, which is converted to the active hormone T_3 in a process that removes an iodine atom from the 5′ position of the thyronine ring. If, however, an iodine atom is removed from the 3′ position, the result is the formation of reverse-T_3

This does not mean that the thyroid hormones are normally detrimental to survival in starvation. Laboratory animals are protected from factors such as marked fluctuations in ambient temperature, the need to find food, and from predators: such problems in the wild require the action of triiodothyronine, to increase the sensitivity of regulatory mechanisms to aid the response to such problems. High rates of energy expenditure are therefore, essential for survival in the wild!

(rT$_3$) (Figure 16.13), which has no biological activity. In prolonged starvation, conversion of thyroxine to T$_3$ is reduced but conversion of thyroxine to reverse-T$_3$ is enhanced. This change may be responsible for the coordinated fall in the energy expenditure and in the rate of protein degradation in starvation. Studies in experimental animals are in support of this role of T$_3$. Thyroidectomy doubles the time that starving animals can survive and decreases the rate of nitrogen excretion during starvation (Figure 16.14).

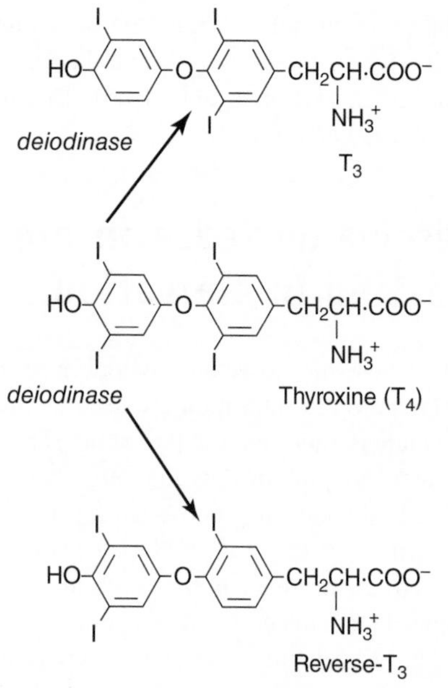

Figure 16.13 *Deiodination of thyroxine in peripheral tissues: formation of T$_3$ and reverse-T$_3$.* Thyroxine is 3,5,3',5' tetraiodothyroxine (T$_4$) which is converted to 3,5,3' triiodothyronine (T$_3$) in normal condition but is converted to 3,3',5' triiodothyronine (reverse T$_3$) during starvation. The prime (') indicates the carbon number to which the iodine is attached in the second ring.

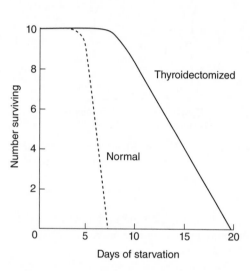

Figure 16.14 *Effect of thyroidectomy on survival of rats during starvation.* Thyroidectomy protects rats from starvation; that is, the number or rats surviving prolonged starvation is much larger when the thyroid gland is removed (Goldberg *et al.* 1978).

17
Defence Against Pathogens: Barriers, Enzymes and the Immune System

This combination of an apparently infinite range of antibody specificity associated with what appeared to be a nearly homogenous group of proteins astonished me and still does.

(Porter, 1972)

To us [Pasteur's] views are standard, but at the time medical tradition thought otherwise. We have to imagine the scientific world before the germ theory of disease. When bacteria were found in wounds or sick people it was really thought of as an unimportant byproduct of the true disease, which came somehow mysteriously from within and had to run its course. This is why doctors were so upset when Pasteur and others suggested that by not washing their hands between touching diseased corpses and touching healthy or somewhat healthy patients, they might be spreading disease. To the doctors this was preposterous. How could minute organisms cause disease in creatures so much larger?

(Bodanis, 1988)

Louis Pasteur (1822–1895) established that some diseases are caused by microorganisms and not, as was thought to be the case at the time, by miasmas, evil spirits or divine intervention. Despite the scepticism of physicians, as illustrated above, Pasteur's work inspired a British surgeon, Joseph Lister, to introduce antiseptic methods into surgery in order to prevent wound infection. In March 1865, he operated for the first time using an antiseptic, carbolic acid (phenol). There were no infections after the surgery – a rare occurrence at that time.

The human body is an excellent habitat for numerous pathogens and parasites. There are two elaborate defence

> A pathogen is defined as a living organism or virus which causes disease. It is broader than the term microorganism, which excludes multicellular parasites and, in some definitions, also viruses which, strictly, are not living.

mechanisms: barriers, both physical and chemical, and the immune system. The barriers prevent the entry of pathogens and parasites into the body and the immune system eliminates them if they do penetrate the barriers. The obvious physical barriers are the epidermis in the skin, the epithelial cells that line the mucosal membranes in the gut, lung, reproductive and urinary systems and the blood–brain barrier.

The tight junctions between epithelial cells provide a major physical barrier.

The chemical barriers are also important but are sometimes neglected in discussions of defence. To some extent they are specific for each tissue or organ and examples are as follows. Enzymes, such as lysozyme, which digests part of the bacterial coat, are present in many secretions (e.g. tears, milk); acid in the stomach, in the vagina, on the surface of the skin; alkali in the small intestine; mucus in the intestine, respiratory tract and vagina.

Vomiting and diarrhoea remove pathogens from the stomach and intestine and urination eliminates them from the bladder. Mucus in the lumen of the intestine, in the vagina and in the bronchioles of the lung not only restricts access of pathogens to the epithelial cells but also localises them for attack by immunoglobulin-A.

When the physical barrier is breached

Damage to the physical barriers generates a point of entry for pathogens and, in addition, risks the loss of blood. The

---

Functional Biochemistry in Health and Disease by Eric Newsholme and Tony Leech
© 2010 John Wiley & Sons Ltd

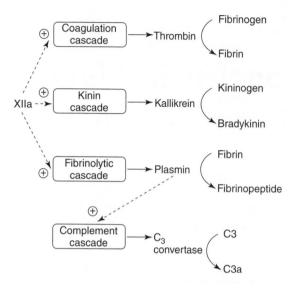

Figure 17.1 *Summary of the four cascades that result from trauma or bleeding and the reactions they catalyse.* These are all activated by the blood clotting factor, XIIa (also known as the Hageman factor). Details of each cascade are presented in Figures 17.2, 17.4 and 17.6. Factor XII is activated by collagen and negatively charged surfaces to form the active form, XIIa.

initial response is activation of four cascades: clotting of blood, fibrinolysis and the kinin and complement systems (Figure 17.1). Each step in these cascades involves conversion of an inactive proteolytic enzyme into an active enzyme, which then catalyses the next reaction in the sequence. All these enzymes are serine proteases. Such cascades bring about large increases in the activity of the final target enzyme in a short time. Indeed, the increase

can be so large that, if uncontrolled, it could lead to unwanted or even dangerous side reactions. To minimise this risk, mechanisms exist to restrict or diminish the amplification, e.g. proteins that inhibit these proteolytic enzymes are present in the blood: they are named serine protease inhibitors (abbreviated to serpins). They bind very tightly to the enzyme and the complex is removed and destroyed by the liver. Of these cascades, blood clotting and fibrinolysis are discussed in this section; the kinins and complement systems are discussed in the immune section.

Clotting of blood

The ability of blood to clot is so important that there are two separate pathways that converge to ensure sufficient activation of the final clotting enzyme, thrombin. It converts fibrinogen to fibrin which, along with the platelets, forms a clot that prevents further loss of blood and restricts entry of pathogens (Box 17.1) (Figure 17.2).

The history of blood clotting

Blood clotting first received serious scientific attention in the seventeenth century, when Malpighi and Borelli proposed that a blood clot was formed from the fluid part of the blood. In the eighteenth century, John Hunter established that clots were formed from the 'coagulable lymph'. In 1845, Buchanan showed that a series of enzymes was involved and by 1900 it was accepted that the soluble substance in blood, fibrinogen, was converted into the solid fibrin clot by an enzyme, thrombin. From that time many

---

Box 17.1 Haemophilia, Queen Victoria and Grigori Rasputin

Haemophilia is characterised by repeated, spontaneous and traumatic haemorrhages. It is divided into two classes, haemophilia A and haemophilia B.

Haemophilia A is also known as *classical* and is caused by lack of factor VIII. It is now treated by replacement with factor VIII which, when prepared from human blood, was not without its problems. It is now prepared by genetic engineering.

Haemophilia B is also known as Christmas disease. It is caused by absence of factor IX. It was called Christmas disease because this was the surname of the first patient diagnosed with type B and the case was first published in the Christmas edition of the *British Medical Journal*.

Both are due to mutations in genes in the X-chromosome. Haemophilia is inherited as an X-linked recessive trait and females do not suffer from the disease (since the other X-

chromosome is normal). There is, therefore, a 50% chance of the male offspring inheriting the abnormal gene and developing the disease whereas a female offspring is a carrier.

Queen Victoria was a carrier of the mutated gene for factor IX: her son Leopold suffered from haemophilia B. Queen Victoria's daughter Alexandra (Alix) was a carrier. She married Nicholas II, Tsar of Russia, and so became Empress of Russia. Her son Alexis suffered from haemophilia. The Russian mystic and faith-healer, Grigori Rasputin, gained great influence with the Empress when he claimed to have saved the life of her son on many occasions. Consequently, he had considerable influence with both the Tsar and, hence, the politicians in Russia. All members of the Russian Royal family, including Alexis, were murdered in 1918 during the Russian Revolution. As a consequence, the mutated gene was not carried on in this family.

Intrinsic pathway   **Extrinsic pathway**

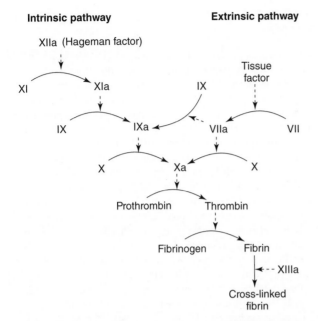

Figure 17.2 *The intrinsic and extrinsic pathways involved in blood clotting.* Both pathways converge to activate thrombin. Solid arrows represent biochemical conversions whereas dotted arrows represent either catalytic or activating actions. Fibrin is formed as monomers which polymerise to form fibrils. Within the fibrils, the fibrin monomers associate laterally which is facillitated by active XIII (ie XIIIa). Thrombin activates XIII.

'factors' involved in blood clotting were discovered but a systematic relationship between them was unclear until 1964, when R.F. Macfarlane organised the factors into a series of enzyme-catalysed reactions that led to the sequential activation of the final enzyme thrombin. He realised that this provided amplification and called the process a 'cascade'. This was the first use of this term to describe a sequence of enzyme catalysed reactions that lead to amplification. Sadly, the term cascade is sometimes used to describe a sequence of reactions or processes in which no amplification occurs: an unfortunate devaluation of this very useful term.

The following is taken from Macfarlane's Nature paper (1964).

After years of confusion, it seems that a relatively simple pattern is emerging from present theories of blood coagulation ... of the often-asked question: 'Why is coagulation so complicated?' resolves itself into the question: 'Why are there so many stages?' Haemostatic efficiency seems to depend on the sudden conversion of fibrinogen to fibrin, and this is brought about by an explosive generation of thrombin; a more gradual generation of thrombin is haemostatically ineffective ... Thus from one stage to the next a greater amount of proenzyme is activated, and it may be suggested that the physiological need to link the minute physical stimulus of surface contact with the final enzymatic explosion has resulted in the evolution of a biochemical amplifier in which enzymes are analogous to photomultiplier or transistor states ... which stops the whole process and prevents over-running. A safeguard of this sort is a biological necessity, since clotting must be limited to areas of injury and not invade other parts of the vascular system, if disastrous thrombosis is to be avoided.

Note that Macfarlane foresaw the need for a safeguard, which we now know is provided by the serpins. Furthermore, the sequence of reactions proposed by Macfarlane in 1964 is almost identical to the current one.

Fibrinolysis

The role of the fibrinolytic system is to dissolve any clots that are formed within the intact vascular system and so restrict clot formation to the site of injury. The digestion of the fibrin and hence its lysis is catalysed by the proteolytic enzyme, plasmin, another serine proteinase. Plasmin is formed from the inactive precursor, plasminogen, by the activity of yet other proteolytic enzymes, urokinase, streptokinase and tissue plasminogen activator (tPA) which are also serine proteinases. These enzymes only hydrolyse plasminogen that is bound to the fibrin. Any plasmin that escapes into the general circulation is inactivated by binding to a serpin (Box 17.2).

> Human recombinant tPA and streptokinase are used as thrombolytic agents: e.g. they are infused into the bloodstream as soon as possible after a heart attack.
> Lipoprotein (a), which is a risk factor for coronary heart disease, interferes in the action of tPA (Chapter 22: Table 22.1).

$$\text{Plasminogen} \xrightarrow[\text{Urokinase}]{tPA} \text{plasmin}$$

The immune system

The role of the immune system is to detect and destroy compounds or organisms that are not part of the host (non-self) but have no effect on compounds that are part of the host (self compounds). In addition, it can identify and destroy dead or dying cells and tumour cells.

The system consists of a variety of cells and various soluble components. A list and summary of these components is provided in Appendices 17.1 and 17.2. The reader should refer to these Appendices when information about the cells or soluble components is necessary to understand the function under discussion.

The immune system operates at two overlapping levels: the *innate* (or non-adaptive) response, which is rapid but

---

Box 17.2 Anticoagulants: cattle, rats, leeches and humans

A strange cattle disease emerged in North Dakota and Alberta in the 1920s. The cattle bled to death; the cause was found to be the hay that the animals had eaten. The hay contained a haemorrhagic factor, which was isolated and identified as 3,3-methylenebis (4-hydroxycoumarin, known as dicoumarol) by Karl Link of the University of Wisconsin. A 3-substituted 4-hydroxycoumarin, which was 5–10 times more active as an anticoagulant than dicoumarol, was synthesised. It was then used as an anticoagulant for killing rodents, which are very sensitive to it. Its value as an anticoagulant in clinical use was only recognised after a US army recruit attempted to commit suicide by taking rat poison. The royalties from its use as an anticoagulant and rat poison were used to support the work of Wisconsin Alumni Research Foundation (WARF), hence the origin of the name, warfarin. It is the most widely used anticoagulant, after aspirin. Patients on anticoagulant therapy have their thrombin activity measured regularly to check the clotting activity of the blood. Even so, haemorrhage is still a problem for patients whose medication includes warfarin.

The mechanism of action of warfarin becomes apparent from the molecular details of the blood clotting process. The clotting factors VII, IX, and X require Ca^{2+} ions for maximum activity. These proteins possess a novel amino acid which binds Ca^{2+} ions. It is a modified form of the amino acid glutamate, as γ-carboxyglutamate (Gla) which possesses two carboxyl groups rather than one. The insertion of a second carboxyl group into the glutamate occurs after translation and is catalysed by a carboxylase enzyme for which vitamin K is an essential cofactor. The coumarins interfere with the action of vitamin K in this process. This results in failure to modify the glutamate molecule so that the clotting factors cannot bind Ca^{2+} ions which reduces their activity markedly, so that the rate of clotting is slow which can lead to haemorrhage.

Leeches live off blood: they attach themselves to the skin and inject an anticoagulant (hirudin) which ensures a good flow of blood. In areas infected with leeches, repeated attacks can result in anaemia. For about 1000 years physicians applied leeches (*Hirudo medicinalis*) to the skin of patients since it was believed that removal of blood by the leech was beneficial. In the West, they were used to treat inflammation but this practice declined in the early part of the nineteenth century, although millions of leeches are still used in some parts of the world. Recently, surgeons have found a novel use for leeches: that is, to remove blood from an engorged part of the body (one leech can easily remove 15–30 mL of blood). For example when replacing an amputated finger, blood can readily enter the wound but, until the veins open up, it is not easy for the blood to escape. A bite from the leech will bleed for up to 24 hours and during this time the excess blood is readily removed by the leech which gives time for the veins to open.

---

non-specific, and the *adaptive*, which is a slower but more specific response to an invading pathogen.

Innate immunity

The innate immune response is the first line of defence, reacting immediately to damage or an invasion by a pathogen but it is limited in scope, since it responds in an identical manner to whatever organism invades. It acts as a delaying mechanism to prevent a huge proliferation of the organism until the more specific mechanisms of the adaptive response come into action. Neutrophils are the first cells to arrive at the site of damage, attracted by chemotactic factors that are released by damaged cells. Endothelial cells possess receptors on their cell surface that recognise certain structures on the surface of pathogens. On binding to these structures, endothelial cells express adhesion molecules on their surface. The neutrophils attach to these adhesion molecules, then squeeze between the endothelial cells into the interstitial fluid, where they destroy the invading organisms (Figures 17.3(a)&(b)). This is achieved by release of free radicals or by phagocytosis (see below for details of killing). Monocytes are also attracted to this site

and enter the interstitial fluid, where they are activated to mature into another phagocyte, the macrophage. Both the neutrophils and macrophages identify the pathogens by the presence of compounds that have specific structural patterns on their surface (e.g. lipopolysaccharide, glycosylated peptides, bacterial lipoproteins). The phagocytes possess specific receptors that bind to these compounds. They are known as pattern recognition receptors (or toll-like receptors, TLRs). Binding of the pathogen stimulates the process of phagocytosis.

Inflammation

Inflammation is a local and early response of a tissue to a noxious stimulus, such as physical injury or infection. It results in an increase in the number of immune cells in the area of damage or infection which kill pathogens, remove damaged or dead cells and initiate the healing process. The well-known characteristics of inflammation are redness, heat, swelling and pain. *Redness* is due to increased blood flow to the damaged area caused by vasodilation of small arterioles, which facilitates an increase in the number of immune cells in the damaged area and facilitates provision

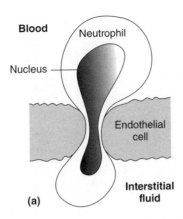

Blood

Neutrophil

Nucleus

Endothelial cell

Interstitial fluid

(a)

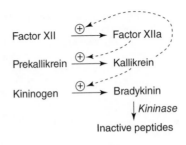

Factor XII $\xrightarrow{\oplus}$ Factor XIIa

Prekallikrein $\xrightarrow{\oplus}$ Kallikrein

Kininogen $\xrightarrow{\oplus}$ Bradykinin

$\downarrow$ *Kininase*

Inactive peptides

Figure 17.4 *The kinin cascade.* Solid arrows represent biochemical conversions whereas dotted lines represent catalytic activity. Note that kallikrein provides amplification in the process (positive feedback).

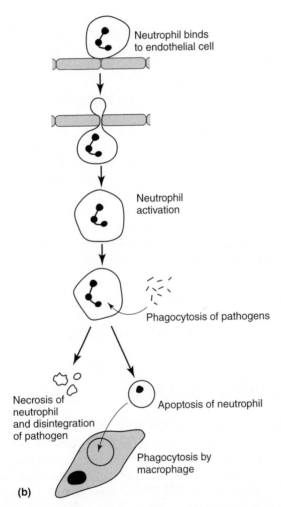

Neutrophil binds to endothelial cell

Neutrophil activation

Phagocytosis of pathogens

Necrosis of neutrophil and disintegration of pathogen

Apoptosis of neutrophil

Phagocytosis by macrophage

(b)

Figure 17.3 (a) *A neutrophil squeezing between endothelial cells to enter the interstitial fluid.* This demonstrates the extreme plasticity of neutrophils (and other immune cells) which enables them to pass through spaces much smaller than their normal diameter. Such entry requires chemotactic signals. **(b)** *Diapedesis.* A neutrophil enters the interstitial space, where it is activated and then phagocytoses bacteria. Finally the neutrophil suffers either necrosis or apoptosis and then phagocytosis.

of fuel and oxygen to support these cells. *Heat* is due to the increase in local metabolism caused, in part, by immune activity. *Swelling* is due to vasodilation and increased permeability of capillaries, which leads to leakage of serum. *Pain* helps to minimise use of the area and so protect it from further damage. The messenger bradykinin plays an important role in these changes.

The kinin system

Bradykinin is a small peptide that is released from a precursor, kininogen, by the action of the proteolytic enzyme kallikrein, which itself is formed from a precursor, prekallikrein, by the action of the blood clotting factor, XIIa (Figure 17.4). Bradykinin is responsible for the pain, vasodilation and increased permeability of the blood vessels by stimulating formation and release of prostaglandins and prostacyclins from the endothelial cells (see Chapter 11).

Bradykinin is inactivated by removal of two amino acids from the C-terminal end of the peptide, by the enzyme kininase (Figure 17.4). Of clinical interest, kininase II is identical to angiotensin-converting enzyme (ACE) which converts angiotensin to angiotensin-II, the messenger that plays a key role in increasing blood pressure, by causing vasoconstriction of arterioles. Inhibitors of ACE are used to treat hypertension by lowering the concentration of angiotensin-II but, in addition, they increase the concentration of bradykinin. Since the latter results in vasodilation, ACE inhibitors have a dual beneficial effect in treating hypertensive patients (Figure 17.5).

Complement

Complement is the collective name for a group of about 20 plasma proteins that participate in the immune response, and kill pathogens. Most of these proteins are proteolytic

John Hunter noted in 1792 that blood resisted putrefaction for a longer period if serum was present, that is another bactericidal factor complemented the effect of other factors. It was later identified and called *complement*.

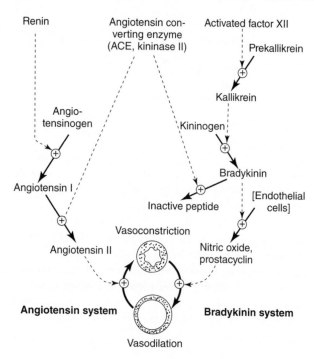

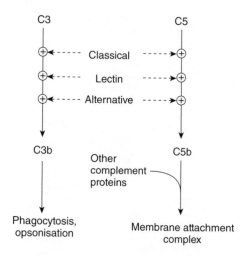

Figure 17.7 *Some complement proteins and a summary of their activation and their effects*. See text and Figure 17.24 for details.

Figure 17.5 *The dual vasodilatory effect of inhibitors of the angiotensin-converting enzyme (ACE)*. The ACE inhibitors not only inhibit ACE but also the kininase which degrades bradykinin. (See also Chapter 22).

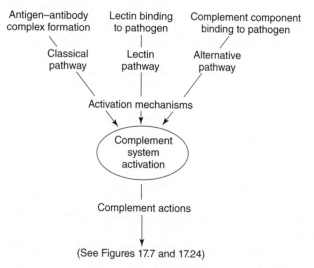

Figure 17.6 *A summary of the three pathways for activation of complement*. See text for details.

enzymes. They are organised into three separate cascades, activated by three different mechanisms (Figure 17.6):

- *Classical pathway*, activated by binding to an antibody–antigen complex.

- *Alternative pathway*, stimulated by binding to a bacterial cell surface.

- *Lectin pathway*, stimulated by binding to a lectin. Lectins belong to a family of proteins called collectins, which are present in blood and bind to bacteria. One lectin, known as the mannose binding lectin (MBL), binds to a sequence of mannose sugars that are part of the carbohydrate on the cell surface of some bacteria. It is the lectin–bacteria complex that activates one of the complement proteins. The components of the pathway are prefixed with a 'C' and a number.

Any one of these pathways, or all three, result in proteolytic cleavage of a protein known as C3 convertase to produce the active form, C3b. The latter is involved in different mechanisms that kill bacteria (Figure 17.7).

Adaptive immunity

The two major components of the immune system are soluble factors and cells. The former include complement, cytokines and antibodies. The cells include lymphocytes, neutrophils macrophages and eosinophils (see Appendices 17.1 and 17.2). The most numerous are the lymphocytes.

Two major classes of lymphocytes are T- and B-cells. Although both develop from the pluripotent stem cell in the bone marrow, B-cells remain in the marrow to mature, prior to release into the blood (hence their name, B-cells). In contrast, immature T-lymphocytes are released from the bone marrow and travel to the thymus, where they develop and mature prior to release into the blood (hence their

Table 17.1 Full name, abbreviation, CD classification and functions of T-lymphocytes

Full name	Abbreviation	CD nomenclature	Functions
Helper T-cell 1	T_H1	CD4+	Coordination of immune response, particularly activation of phagocytosis by macrophages.
Helper T-cell 2	T_H2	CD4+	Coordination of immune response, particularly activation of antibody production by B-cells and responsible for switching class from IgM to IgE.
Cytotoxic T-cells	T_c	CD8+	Killing of host cells infected with viruses.
Natural killer cells	NK	CD56	Killing of host cells infected with virus: killing of 'abnormal' cells displaying abnormal proteins on their cell surface, e.g. tumour cells

There is no agreed convention on the terminology to be used.

The names helper cells, cytotoxic cells and natural killer cells indicate the function of the cells and are, therefore, useful. The more 'formal', nomenclature is the CD classification (e.g. CD4+ve cells and CD8+ve cells). CD stands for the *cluster of differentiation*, which is a technique for distinguishing between different protein molecules attached to different cells: its value is that it can identify very precisely the type of immune cell (Appendix 17.3). Thus the helper cells express CD4 molecules on their surface, whereas cytotoxic cells express CD8 molecules on their surface.

The T-helper cells are further divided into two types, T-helper 1 (T_H1) and T-helper 2 (T_H2) (see below for discussion). Of the total lymphocyte population in the body (approx. 10^{12} cells), the proportion of T-cells is 50–60%, that of B-cells is 10–15% and that of natural killer cells is <10%.

Emil Adolf von Behring (1854–1917) was a German microbiologist who discovered that diseases such as tetanus and diphtheria are produced not by the bacteria but by toxins produced by the bacteria, which circulate in the blood. He showed that immunisation with the toxin was sufficient to protect against the disease. He treated children with diphtheria by injecting serum from a horse immunised with diphtheria toxin. He recognised that the protective effect was due to substances circulating in the blood, from which he coined the term *antibody*. He was awarded the Nobel Prize in 1901 for the development of the diphtheria antiserum.

name T-cells). There are three types of T-cells, known as T-helper, T-cytotoxic and natural killer cells, also known as CD4+, CD8+ and NK cells, respectively. A summary of the different terms that are used to identify them is given in Table 17.1.

B-cells

The function of B-lymphocytes is to synthesise and secrete antibodies, proteins that bind to foreign peptides, proteins or carbohydrates (i.e. to non-self compounds), known as antigens. The binding of the antibody to the complementary antigen is very specific and very tight. The site on the antibody that binds the antigen is small, about 10 amino acids, and is three-dimensional in nature. The part of the antigen to which the antibody binds is the antigenic site or antigenic determinant. They are small, so that a single bacterium possesses several hundred antigenic determinants on its surface, to which the several hundred complementary antibodies will bind. Two important points about antibodies are:

- There are two forms – one is free in the circulation and the other is part of the plasma membrane of the B-lymphocyte. This constitutes a receptor on the B-cell that binds the antigen to elicit a response from the cell.

- Each type or class of B-cell produces antibodies of unique specificity. The number of B-cells in this class can rapidly expand: when antigen binds (i.e. upon recognition of antigen) it results in marked expansion of the original number of B-cells (known as clonal expansion). For each type of cell, prior to stimulation by its complementary antigen, there are about 1000 identical cells (what is known as a clone of B-cells) which will increase many fold upon antigen binding. Furthermore, there are at least 10^{10} clones.

From this very brief description of B-cells, and their antibodies, four key questions arise. The answers are such that they provide more information to help explain the role of B-cells in the immune response. (1) What are the properties of the B-cell receptor? (2) In an infection by a pathogen, there are likely to be a greater number of pathogens, and therefore antigens, than there are B-cells that are able to identify the pathogen. Since the B-cells are therefore, outnumbered, how do they produce an effective defence? (3) What is the structure of an antibody? (4) Since the B-cells in one clone produce only one antibody with unique specificity, yet all the B-cells arise from a single ancestor cell in the bone marrow (Figure 17.8) since there are at least 10^{10} B-cell clones, how is it possible that the genes that encode antibodies in cells in one clone can differ from those in cells in other clones: i.e. what is the mechanism(s) for producing the infinite range of antibody specificities that astonished the Nobel laureate R. Porter.

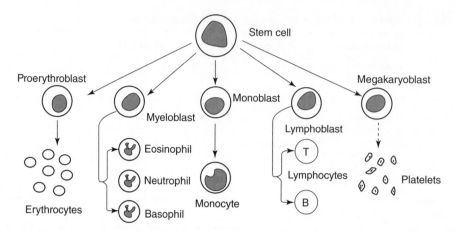

Figure 17.8 *Development of immune cells and erythrocytes in bone marrow from a pluripotent haemopoietic stem cell.*

The B-cell receptor

The properties of the B-cell receptor are similar to those of many other receptors (e.g. hormone receptors, Chapter 12). There are many antigen receptors of the same specificity present on a single B-cell. They recognise a three-dimensional conformation, not a linear sequence, of amino acids; the binding of an antigen leads to activation, which results in proliferation of the B-cell: the more receptors and the greater the affinity for the antigen, the more likely is activation to occur. If a B-cell has already met its complementary antigen, it will have developed a greater affinity for the antigen and there will be more of these cells in the clone, so that the response to a subsequent invasion by the same pathogen is quicker and more effective (see below).

The cellular cascade

In any battle, when the defence is outnumbered by the enemy, more troops are brought into the battle from the reserve. However, in the immune system, there are initially no reserve troops. When an antigen binds to its complementary antibody-receptor on B-cells, these are strongly stimulated to proliferate (clonal expansion). In addition, not only does the number of daughter cells increase but each quickly develops into what is known as an effector (or plasma) cell, in which the protein synthetic machinery increases through the development of the rough endoplasmic reticulum, so that there is a large increase in the number of antibodies synthesised and secreted. A simple description of the sequence of events is as follows:

B-cell → activated B-cell → many daughter cells → effector cells → antibodies
    | | | |
antigen   *proliferation*   *expansion*   *synthesis*
                        of ER   *and secretion*

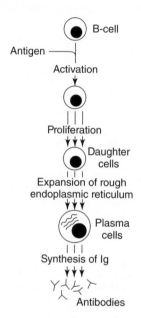

Figure 17.9 *A summary of the B-lymphocyte cascade.* See text for details. The increase in the number of arrows represents the effect of activation and subsequent responses.

This is described as a 'cellular cascade' since it has some similarity to an enzyme cascade (see Chapter 3): that is, a single B-cell produces a very large number of effector cells which produce a very large number of antibodies (Figure 17.9).

Several additional factors facilitate the process. A single species of pathogen possesses many different antigens and a single antigen is likely to possess several antigenic determinants so that many different B-cell clones are involved in the defence against one type of pathogen. Moreover, local signals released by damaged endothelial cells further stimulate the cascade. Proliferation occurs within the lymph nodes, where the local concentrations of cytokines,

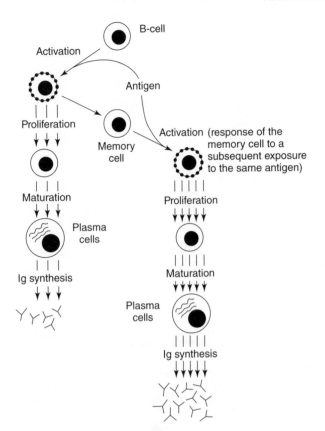

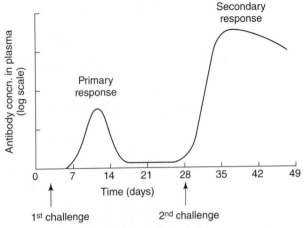

Figure 17.11 *Secondary response to an antigen challenge.* Note the antibody concentration is a logarithm scale, so that the increase in concentration of antibody is considerably greater than would appear in the diagram.

Figure 17.10 *B-lymphocyte cascade including the response of the memory cell.* The memory cell is produced in the first response to the antigen. The response of the memory cell, shown here, is to a subsequent exposure to the antigen. It provides a faster and greater response to the same antigen. This is represented by the greater number of arrows and greater number of antibodies that are produced. Plasma cells are sometimes termed, effector cells.

the hormones that stimulate the cascade, are very high (see below).

A limitation is that the effector cells only have a short half-life (2–4 days) so that a subsequent invasion by the same pathogen would require the establishment of another cellular cascade by a few remaining naïve B-cells, i.e. it would be a slow response. To overcome this, during the first invasion, the cascade also results in the production of B-cells that have a memory of the antigens of the invading pathogen (Figure 17.10). Consequently, the cascade presented above can now be extended to take into account the memory cell:

The first invasion The response is: B-cell → activated B-cell → many daughter cells → plasma cells + memory cells → antibodies

A subsequent invasion The modified response is Memory cells → activated memory cells → daughter cells → effector cells → antibodies

Memory cells have two advantages in fighting a pathogen. (i) There are a larger number of memory cells than naïve cells. (ii) The antibody-receptors on the memory cells have a greater affinity for the antigen. Consequently, a subsequent invasion by the same pathogen, known as the secondary response, is quicker and greater (Figure 17.11). The response is usually so prompt and efficient that, after the first exposure, the same organism can never again survive in the host. Immunisation makes use of this phenomenon (see below).

Antibody structure

Before details of the structure are given, two important general points about the structure must be emphasised. (a) The B-cell receptor for the antigen shares the same structure as the antibody. (b) Although there are a massive number of different antibodies, the basic structure is the same: the variation between the antibodies lies in a small portion of the molecule, the variable region.

Antibodies are proteins known as immunoglobulins (Ig) because they fold to assume a globular structure. They are Y-shaped proteins, comprising a stem and a fork. The molecule is made up of four peptide chains: two identical long chains (known as heavy chains or H) and two short (known as light chains or L). The stem of the fork contains only the heavy chains whereas the fork contains parts of the H chain and the whole of the light chains (Figure 17.12). The chains are held together by disulphide bridges and noncovalent

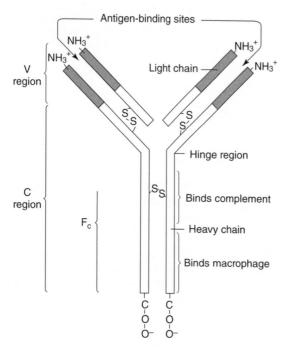

Figure 17.12 *Structure of an antibody.* V-region is the variable region and C-region is the constant region. Fc is the portion of the antibody that contains the effector domains. S-S is the sulphydryl link between the chains. NH₃⁺ represents the amino terminus of each chain.

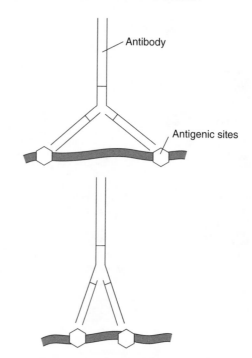

Figure 17.13 *Role of the hinge region in binding of the antibody to sites at different distances apart on the antigen.* The hinge provides flexibility in binding to different sites on the antigen.

> The molecular mass of the H chain is about 50 kDa and that of the L chain about 25 kDa: thus the molecular mass of whole molecule is about 150 kDa.

forces. At the N-terminal, each chain possesses a variable region (V-region). In the other region of each chain, the amino acid sequence is the same, so that it is known as the constant or C-region.

Within the V-region, there are small regions in which the variation is very high – the *hypervariable region*. It is the three-dimensional structure provided by both the hypervariable and the variable regions of both chains that produces the antigen-binding site. This results in a remarkably large number of different antibody molecules.

Antibodies possess three functional regions. First, there are two identical sites, one at the end of each fork, for binding the antigen. Second, a hinge region below the fork provides flexibility so that one antibody can bind to two identical sites on an antigen but which are not immediately adjacent, which is achieved by opening up the hinge (Figure 17.13). Third, within the constant part of the molecule there is the Fc region, which carries out two effector functions: (i) one region binds to and activates complement; (ii) the other region binds to monocytes/macrophages to activate phagocytosis.

For the antibody that acts as a receptor of the B-cell, the constant region links the antigenic sites to the membrane

and initiates the intracellular changes that result in activation, i.e. proliferation of the B-cells so that more plasma cells are produced. The mechanism is discussed in the section on 'Activation' below (p. 404).

Classes of antibodies   There are five different classes of antibodies, IgA, IgD, IgE, IgG and IgM, in which the amino acid sequences of the constant region of the antibodies are different (Figure 17.14). The antibodies in these classes have different functions, increasing further the diversity of the molecules (Table 17.2). Furthermore, the class can change during an infection to improve the response to a particular pathogen. When B-cells are stimulated, they first secrete IgM but then change to IgG, IgA or IgE, a process known as class-switching.

In summary, for all the immunoglobulin molecules within one class, the amino acid sequence in the constant region of the chains is identical; the massive number of different molecules within each class is accounted for by the amino acid sequences in the variable and hypervariable regions of both the light and heavy chains.

Mechanisms for producing variations in the structure of the antigen-binding sites

There are four mechanisms for producing the huge variability in antibodies:

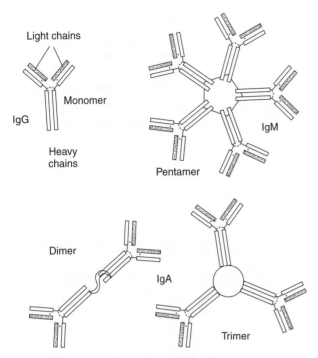

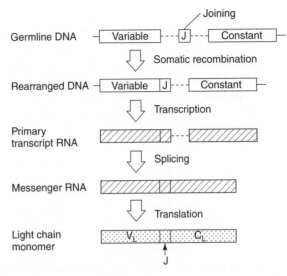

Figure 17.15 *Combinational diversity to produce the gene for the variable region in the light chain.* During somatic recombination, different segments of genes are brought together randomly to produce the gene for the V-region which is joined to that of the constant region by the J segment, which produces even more variation. For description of transcription, splicing and translation see Chpater 20.

Figure 17.14 *Structure of IgG, IgM and IgA classes of antibodies.* Note that IgA can exist as a dimer or trimer.

(i) *Somatic recombination (combination diversity)* (Figure 17.15) The genes for variable regions of both the H and L chains are present in the genome as separate segments that are not contiguous. The complete gene for the variable region is constructed from random recombination of these gene segments. This is then joined to the gene for the constant region to produce the complete gene for each chain of the antibody. The V-region of the light chain is made up from two gene segments. One is known as the variable segment (V), for which there are many segments in germline DNA. The other is the *joining segment* (J),

Table 17.2 Functions of the different classes of antibody

Class	Function
IgA	The major antibody in external secretions such as milk, tears, saliva, mucus of the bronchial, intestinal and urogenital tracts. It is secreted along with a transport protein to aid transport across membranes, which also protects the antibody against digestion or denaturation in extracellular environments
IgD	Present on immature B-lymphocytes and is involved in their maturation (not discussed in this text)
IgE	Provides protection against parasites via activation of eosinophils, and by binding to mast cells, where it binds to parasites as they pass by the mast cells: when sufficient IgE molecules have bound the parasite, the mast cells discharge their granules which contain toxic biochemicals, which attack the parasites, and histamine, chemotactic factors and cytokines to promote local inflammation. Unfortunately, IgE can cause allergic responses (see below)
IgG	The major antibody in the serum. Since it is a small molecule a high proportion passes out of the blood into lymph and then enters back again into the bloodstream to provide a constant surveillance of body fluids
IgM	One of the two predominant classes that form the receptor on the surface of the B-cell. It is the first antibody to be produced in blood in response to an invasion by a pathogen. It has a large number of binding sites: ten compared with two for IgG, so that it binds very effectively, particularly to polyvalent antigens such as those on the surface of bacteria and other microorganisms. This results in agglutination of the complex with the bacteria which is ideal for phagocytosis by macrophages. The IgM–antigen complex also activates the complement cascade, for which it is much more effective than IgG

which joins the variable region to the constant region to produce the whole gene for the light chain. Random recombinations of the V and J segments (known as V–J recombinations) occur prior to joining the constant region. The J segments are separated from the constant region gene by non-coding DNA, which is removed, and the genes for the variable and constant regions are joined by RNA splicing, after transcription. A similar process occurs in formation of the gene for the variable component of the heavy chains but is more complex since there are three gene segments: variable (V), joining (J) and an additional one, the diversity segment (D). Random recombinations of the three segments, V–(D)–J, result in considerable variation.

(ii) *Junctional diversity* Insertion of a random number of nucleotides at the end of each separate section occurs during recombination of the various segments. An enzyme, terminal deoxynucleotide transferase, catalyses these insertions which can be as large as 10 nucleotides on either side of the junction.

(iii) *Endogenously produced point mutations* These mutations are produced within the variable region gene to generate hypervariability. The mutations are point mutations caused by specific enzymes: for example, a deaminase enzyme can convert a cytidine nucleotide into a thymidine nucleotide (Chapter 20). This occurs during proliferation of the B-cells in the lymph node after stimulation by binding of the antigen. If the mutation results in an increase in the affinity of the B-cell receptor for the antigen, these cells will be activated at a much lower level of antigen, so that more high affinity cells will proliferate to continue the fight against invading bacteria. Since this process continues during a prolonged infection, it results in even more high affinity cells that are better able to deal with the infection. In addition, a very small quantity of antigen may remain in the body so that this process of increasing affinity will continue for some time after the initial infection. This is analogous to the process of natural selection. The fittest of the B-cells, i.e. those that produce a successful attack, are selected during the infection. The mutation, which is described as a *hypermutation*, improves the immune response and the progressive improvement is known as *maturation of the immune response*.

Finally, association of the different V-regions of the light and heavy chains, in the forks of the antibody, produces further variation during formation of the complete antigen-binding site. A summary is presented in Figure 17.16.

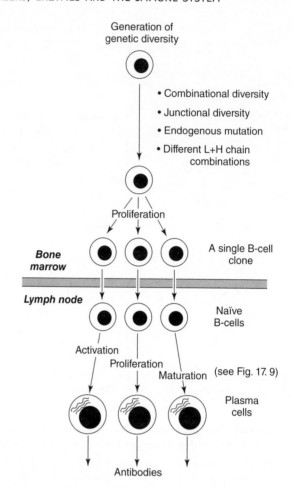

Figure 17.16 *Summary of the mechanisms for producing variable regions of the antibody and the response to antigenic stimulation.* The four mechanisms for producing genetic variation occur in the bone marrow and proliferation of the genetically unique B-cell gives rise to the large number of identical cells in a single B-cell clone. In the lymph node, the naïve cells in this clone are activated by binding to the complementary antigen which is followed by proliferation and maturation to produce the plasma cells. Natural selection to produce the 'fittest' lymphocyte to 'deal with' the antigen and the formation of memory cells occurs in the lymph node.

T-cells

The T-cells are divided into three classes: helper T-cells (T$_H$-cells), cytotoxic T-cells (T$_c$ cells) and natural killer cells (NK). The latter two are involved in killing pathogens, which is discussed below. The helper T-cells coordinate the whole immune response. These cells are discussed first.

As with B-cells, the first response of the T-cells is to bind to an antigen and it does this via a receptor, the T-cell receptor, which is of prime importance in the immune

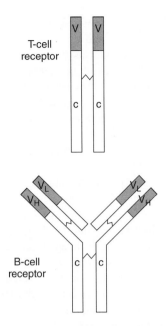

Figure 17.17 *T-cell and B-cell receptors.*

response to an invasion by a pathogen. The best way to describe the properties of the T-cell receptor is to compare them with the B-cell receptor.

As with the B-cell receptor, that for the T-cell is present in the plasma membrane but the two receptors differ in several ways:

- T-cells do not secrete antibodies.

- The T-cell receptors are dimers and possess only one antigen-binding site (Figure 17.17).

- They are only cell surface receptors, they are not free in body fluids.

- They recognise small peptide antigens that must be bound to a specific protein, the major histocompatibility complex protein (MHC) (see below).

- They respond to a linear sequence, not a three-dimensional structure, of amino acids in the peptide antigen.

- Variations in the structure of the variable region of the molecule are achieved by the same mechanisms as for the B-cell receptor, with the exception of endogenous mutations, which do not occur, i.e. variation is produced by processes (i), (ii) and (iii) mechanisms described above.

Some of the problems faced by these cells in response to an invasion by pathogens are as follows:

(a) The T-cell receptor responds to an antigen only on direct contact between the two.

(b) It is likely that, for any given infection, there will be a greater number of individual pathogens invading the body than there are naïve T-cells that possess complementary receptors to bind the antigens derived from the pathogens (see below).

(c) Pathogens can invade at any site in the body so that, in order to detect an invasion, T-cells would need to 'patrol' many tissues in the body.

(d) Once detected, the pathogen must be destroyed. That is, the pathogens must be detected, captured and killed.

All these activities cannot be achieved by T-cells alone, so that other cells must be recruited. It is the role of the T-helper cells to coordinate the activities and localise the response to specific sites within the body (i.e. the lymph nodes).

To understand how this is achieved, two new subjects must be introduced: the antigen-presenting cell and the major histocompatibility complex (MHC). In brief, the antigen-presenting cells 'patrol' the areas most likely to be invaded and, when the invading pathogens are detected, they are phagocytosed, killed, digested and products of digestion (the antigens) are transferred to the surface of the cell antigen-presenting all where they can interact with the T_H cells. This initiates the processes that constitute the whole adaptive immune system, which can then be brought to bear on the invading pathogens. Details of this process are provided below, after discussion of the major histocompatibility proteins (MHC proteins).

Natural killer cells (large granular lymphocytes)

The cells are called natural killer cells since they kill without a requirement for MHC class I or class II proteins. This characteristic is sometimes termed the *MHC-independent immune response*. Their principal role is in defence against viruses and other intracellular microorganisms: that is, they kill host cells infected with a virus or other pathogen. They also kill tumour cells. Thus they have a role in the process of immune surveillance for tumour cells.

Major histocompatibility complex

The major histocompatibility complex (MHC) is, in fact, a set of genes which code for three classes of proteins – class I, class II and class III. The class I proteins are involved in identifying cells that are infected with a virus. Class II proteins are involved in the interactions between T_H cells and antigens. Class III proteins are the complement proteins. Although the name major histocompatibility complex actually refers to the genes, it is the proteins that are produced by the genes that are relevant to discussion here.

The term *histocompatibility complex* (from the Greek word *histo*, meaning 'tissue') is somewhat misleading. It arose from the fact that they were discovered as proteins that caused the rejection of tissues or organs transplanted from a donor to a recipient. There is considerable polymorphism within the MHC genes and hence within the MHC proteins, so that, in most cases, the MHC proteins of a donor are seen as non-self by the immune cells of the recipient. This results in rejection of the transplanted tissue or organ. Hence the alternative, more appropriate, name for the MHC proteins is human leucocyte antigen (HLA). The normal function of these proteins is, obviously, not histocompatibility, it is to direct the response of the immune system to remove the invading pathogens. A better term, perhaps, would be an *immune directing protein*. It is another example of how an inappropriate name is given to a protein or process prior to its physiological function being discovered.

MHC Class II proteins and antigen-presenting cells

The class II proteins are produced mainly within specific cells, known as antigen-presenting cells (APC). Their role is to prepare antigens derived from the pathogens in such a way that they can interact with the T-helper cell receptor. Since this is their only role, they are sometimes called 'professional antigen-presenting cells', i.e. other immune cells can also present peptides to T-helper cells (e.g. macrophages, B-lymphocytes) but this is not their prime role. The professional APCs are also known as dendritic cells.

The professional antigen-presenting cells patrol tissues throughout the body, especially in areas where pathogens are most likely to gain access (e.g. skin, lung, intestine and urinary-genital system) and are also located in strategically important sites, such as lymph nodes. These cells detect and engulf an invading pathogen (phagocytosis) but, since it is far too big to be presented on the surface of the cell, antigens must be prepared from the pathogen. Within the cell, the pathogen proteins are partially degraded. The degradation proceeds only as far as small peptides containing 8–12 amino acids. The peptides then associates with the MHC-II proteins and the complexes are transported to the plasma membrane, where the peptide, still attached to the MHC protein, is positioned on the outer surface of the antigen-presenting cell. This localisation increases the chances that it will meet a T-helper cell that possesses the complementary T-cell receptor for the antigen. Once the complementary T-cell receptor is found, the peptide binds and the T-cell responds as described below. This process is described by the term, an *MHC-class II directed immune response*.

In summary, the APCs patrol the tissues where pathogens are likely to invade, they 'capture', phagocytose and digest the proteins of the pathogen. At this stage, the antigen-presenting cells leave the tissue and travel, via the lymphatic system, to the nearest lymph node where they meet the T-helper cells. This localisation increases the chance that T_H cells will meet the antigen on the presenting cell for interaction between the two. The binding of the antigen to the receptor activates the T-cell, which responds by synthesising and secreting cytokines and by proliferating (Figures 17.18 and 17.19). This binding is aided by the MHC protein and other proteins on the surface of the T-cell.

> Antigen-presenting cells are also called dendritic cells because of their long branched processes on dendrites. They were discovered as early as 1868 by Paul Langerhans, after whom they are sometimes named. It was 100 years later that it was appreciated that they had a role in the immune system.

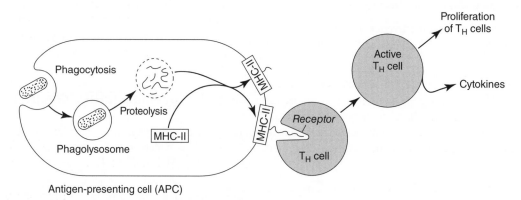

Figure 17.18 *Roles of APC, MHC-II and T-helper cell.* The APCs phagocytose bacteria, digest them, and transfer the resultant peptides plus MHC-II proteins to the surface of APC. The role of this is to present the peptide, as an antigen, to the T_H cells. The binding activates T_H cells which then secrete cytokines. The activated T_H cells now proliferate to produce many more identical T_H cells to bind more of the antigens on the APCs. The roles of the cytokines are discussed below.

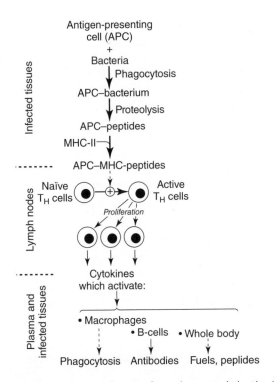

Figure 17.19 *Sequence of events from phagocytosis by the APC, activation of the T$_H$ cell and subsequent immune response.* Events in the infected tissues, in the lymph node, in the plasma and in the whole body are summarised Peplides include acute phase proteins. For effects on the tissues in the whole body, see Chapter 18: Table 18.2.

The binding of the antigen to the T-cell receptor activates the T-cell which synthesises and secretes cytokines. The cytokines secreted by the T$_H$ cells include IL-2 and TNFα which do the following:

- They stimulate proliferation of the T$_H$ cells in the same clone (clonal expansion).

- They stimulate the proliferation of B-cells that are already responding to the antigen to increase the number of effector cells and hence the production of antibodies: that is, they stimulate the cellular cascade process.

- They activate phagocytosis by macrophages.

- They stimulate response of the whole body (e.g. increase the rates of glycogenolysis in the liver and proteolysis in muscle) to provide fuels for the immune cells (see below for details).

There are a number of possible advantages to such a process:

- Digestion of the proteins of the pathogen by the APC provides more than one peptide (antigen) which are then presented on the surface of the presenting cells. This increases the chance that the cell will meet the T-helper

cell that possesses the complementary receptor for one or more of the peptides.

- A small peptide is easier to transport within the cell and be presented with the MHC protein on the cell surface compared with a large protein.

- When a small antigen is complexed with the MHC protein it is protected from proteolysis or other damage (e.g. effects of free radicals) within the presenting cell.

- The MHC protein helps to stabilise the binding between the T-cell receptor and the peptide.

T-helper-1 and T-helper-2 cells

The activation of T$_H$ cells by the APC peptide is even more complex than described above. The activation of the naïve T$_H$ cell can result in production of one of two types of cell: T$_H$1 or T$_H$2. What determines which subset is produced is not known. As expected, the two have different functions. This division provides some further specificity in response of the immune system to a pathogen. This is achieved via the type of cytokines secreted by each type of cell.

T$_H$1 cells The cytokines secreted by the T$_H$1 cells stimulate phagocytosis by macrophages. In particular, they stimulate the activity of macrophages that are infected with pathogens. Some pathogens are able to evade the normal killing mechanisms by preventing fusion of the phagosome with the lysosome to produce the phagolysosome and hence the production of free radicals by the macrophage. Hence the pathogens survive and proliferate in the phagosome. (They include the mycobacteria, e.g. those responsible for tuberculosis, leprosy and listeria.) The cytokines secreted by the T$_H$1 cells stimulate fusion of these two vesicles so that the pathogens can now be killed within the phagolysosome.

T$_H$2 cells The T$_H$2 cells secrete cytokines (IL-4, -6 and -10) that stimulate B-cells. They can cause a class switch from IgG to IgE antibodies. The latter activate eosinophils, which are important in killing parasites that are too large to be phagocytosed. Unfortunately, they are also involved in allergic reactions (see below).

MHC Class I proteins: identification of cells infected with a virus

Class I proteins are present in all nucleated cells. These cells, at some time during their lifetime, are likely to be infected by a virus. The role of class I proteins is to identify host cells that are infected so that these cells, and hence the viruses within them, can be killed. The primary role of the MHC class I protein is to identify infected cells for killing.

The processes prior to killing are as follows. A virus binds to the surface of a host cell, inserts its genome (DNA or RNA) and then uses the cell's machinery for the synthesis of viral DNA (or RNA), viral messenger RNA and hence viral proteins. These associate to produce new viruses that are then released to infect other cells. As viral proteins are produced within a cell, some of them are degraded by the proteolytic enzyme complex, the proteasome, in the cytosol (Chapter 8). However, proteolysis is not complete; it proceeds only as far as short peptides containing about 8–9 amino acids. These are then transported by a specific protein (TAP protein) into the endoplasmic reticulum where they associate with the MHC protein to produce a complex, which is transported into the Golgi. Here the complex is incorporated into a vesicle, which is transferred to the plasma membrane. The complex inserts into the membrane in such a way that the peptide is orientated towards the outer face. It is the peptide that is the antigen. As indicated above and re-emphasised here, the major role of the MHC class I protein is to identify the cell as one infected with virus, so that when this infected cell meets a cytotoxic T-cell, which possesses the complementary receptor to the peptide, it binds, via the receptor, to the infected cell. This activates the killing process. The binding of the antigen to the T-cell receptor is aided by the MHC protein and other proteins on the surface of the T-cell. This ensures that it is only the infected cell to which the cytotoxic cell binds and then kills. This then is the preliminary phase prior to killing, which is described below.

Cytokines

Cytokines are proteins that are secreted by a variety of cells to regulate processes in other cells, both near to and far from the cells in which they are synthesised and released. Consequently, they should be considered as intercellular messengers performing the same roles as endocrine and local hormones (Chapter 12). However, they were first discovered as messengers within the immune system, so that they are most clearly associated with immune cells. They do, however, have significant effects on other cells. There are at least 50 cytokines, and a table that describes their sources, actions and effects is only of value for reference purposes. They can, however, be separated into several different groups: interleukins, chemokines, interferons (IL), tumour necrosis factors (TNF), colony stimulating factors, transforming growth factors and growth factors. (See also Chapter 21 for roles of some of these in tumour development and cancer.)

In this book, discussion is limited to their significance in coordination of the responses to an infection or in trauma. Although it is an oversimplification from the biochemical and clinical points of view, the cytokines can be separated into two classes: those that are involved in the innate response (secreted by the macrophages, particularly during inflammation), and those involved in the responses of the adaptive system (secreted by the T-helper cells). Some of those secreted by T-lymphocytes and their effects are shown in Figure 17.20. Those secreted by macrophages and their effects are described in Figure 17.21.

There are three general points that must be made about cytokines:

(i)   There is considerable redundancy in effects of cytokines and, in addition, the effect of any one cytokine can be increased in the presence of others. Consequently, demonstrating the role of each individual cytokine is difficult.

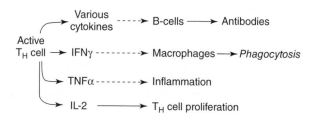

Figure 17.20 *Cytokines produced by activated T$_H$ cells and some of their effects.* The cytokines produced have several functions: activation of B-cells, macrophages and cytotoxic T-cells. The cytokines, along with endocrine hormones, stimulate responses in the whole body (e.g. lipolysis in adipose tissue, protein degradation in muscle, acute phase protein production in liver, Chapter 18).

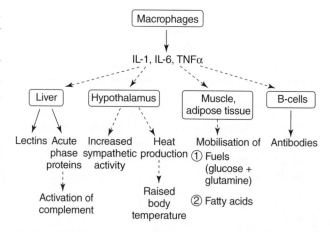

Figure 17.21 *Cytokines secreted by macrophages and their effects.* Solid arrows represent secretions; broken arrows indicate stimulation of a process.

(ii) It is also possible to divide cytokines into two classes: proinflammatory and anti-inflammatory, so that the response of the innate immune system is a balance between the effects of these two classes of cytokines. The proinflammatory include IL-1, IL-6 and TNFα, whereas the anti-inflammatory include IL-1 receptor antagonists (i.e. they bind and block IL-1 receptors) and IL-10.

(iii) An excess of the proinflammatory cytokines can stimulate an inappropriate immune response which, if severe, can cause illness or even death.

Mechanisms for killing pathogens

As with killing in modern warfare, that in the defence of the body uses different systems and different weapons, and collaboration and communication between the different systems is essential. In addition, the weapons must kill without incurring substantial damage to the host tissues (i.e. minimal collateral damage). However, death of some host cells is essential in defence of the body as a whole.

The systems used are divided into two classes:

• killing by soluble agents;

• killing by cells.

There are only two soluble agents involved in killing, antibodies and complement (see above) but there are six types of cells: (i) B-lymphocytes, discussed above; (ii) cytotoxic T-cells; (iii) natural killer cells; (iv) eosinophils plus mast cells; (v) neutrophils; and (vi) macrophages.

One of the reasons there are more types of cells than soluble factors is that they must have the capacity to distinguish between those pathogens that are present in the extracellular fluids (e.g. blood and lymph), those that are present within host cells (e.g. viruses, some bacteria) and those that are too large to be phagocytosed (some parasites e.g. helminths, worms). The mechanisms of killing, therefore, must be sufficiently flexible to deal with all of these pathogens.

Killing by soluble agents

Antibodies

Toxins (Table 17.3) have to bind to a cell surface, prior to entering it, but this is prevented by interaction with the antibody, a process known as *neutralisation*.

Antibodies do not kill pathogens directly but have at least three roles which can lead to killing.

(i) Bacteria possess a large number of antigenic sites enabling many antibodies to bind to the surface of a bacterium, so that it becomes 'coated' with antibodies, a process known as opsonisation (Figure 17.22). Antibodies can also link separate bacteria to form one large complex (Figure 17.22). This antibody–bacteria complex stimulates several killing processes (Figure 17.23). When a bacterium is opsonised, the Fc region of the antibody binds to a receptor (the Fc receptor) on a macrophage, which stimulates phagocytosis of the antibody–bacterium complex (Figure 17.23). The bacterium is then killed by several weapons within the

Table 17.3 Examples of bacterial toxins and effects *in vivo*

Disease	Organism	Toxin	Effects *in vivo*
Cholera	*Vibrio cholerae*	cholera	Results in activation of adenylate cyclase which elevates cAMP, leading to changes in intestinal epithelial cells that cause loss of water and Na⁺ ions from the cells to the lumen of the intestine
Anthrax	*Bacillus anthracis*	anthrax	Increases vascular permeability leading to oedema, haemorrhage and circulatory collapse
Botulism	*Clostridium botulinum*	botulinum	The bacterium is the cause of a fatal form of food poisoning due to release of its toxin. The toxin also blocks release of acetylcholine leading to paralysis of some muscles
Food poisoning	*Staphylococcus aureus*	staphylococcal	Acts on intestinal neurones to induce vomiting
Tetanus	*Clostridium tetani*	tetanus toxin	Blocks neurotransmitter release in the brain leading to chronic contraction of muscles (See Figure 13.9)
Gas gangrene	*Clostridium perfringens*	clostridia toxin	Activates phospholipase to break down cell membranes. (Major cause of death in First World War and American Civil War)

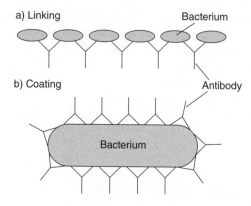

Figure 17.22 *Antibodies form a complex with bacteria either by linking antigenic sites on several bacteria or by binding to different antigenic sites on a single bacterium.*

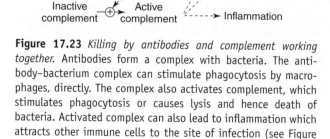

Figure 17.23 *Killing by antibodies and complement working together.* Antibodies form a complex with bacteria. The antibody–bacterium complex can stimulate phagocytosis by macrophages, directly. The complex also activates complement, which stimulates phagocytosis or causes lysis and hence death of bacteria. Activated complex can also lead to inflammation which attracts other immune cells to the site of infection (see Figure 17.24).

macrophage (Table 17.4). This is one of the effector roles of the Fc region of the antibody (see Figure 17.12).

(ii) The antibody–bacteria complex activates a complement cascade.

Complement

Complement kills or facilitates killing as follows.

(i) Complement proteins also bind to the surface of a pathogen in such a way that most of the surface is coated with the proteins, enabling phagocytosis by macrophages. This depends on the interaction between a complement protein bound to the surface of the pathogen and the receptor that is present on the macrophage.

(ii) When the cascade is activated, several of the complement proteins form a complex, known as the membrane attack complex (MAC) that inserts into the bacterial cell membrane to produce a pore, allowing unrestricted exchange of ions between the bacterium and the environment and resulting in lysis of the bacterium (see Figures 17.7 and 17.24) (Table 17.4).

Table 17.4 Summary of processes involved in inactivation of toxins, killing of bacteria and viruses

Progressive response	Bacterial toxins	Bacteria in extracellular fluid		Viruses in cells
First	antibody–toxin complex formed	antibody–antigen complex formed		Formation of viral peptides and MHC-peptide complex
Second	restriction to extracellular fluid	opsonisation with antibody	complement activation	MHC-I–viral peptide complex presentation on cell surface
Third	uptake by resident macrophages	uptake by macrophage	insertion of MAC into bacterial membrane	activation of cytotoxic lymphocytes by MHC-viral peptide
Fourth	digestion of toxin	three mechanisms of killing (see below)	lysis of organism	killing of host cell, and hence viruses, by granzymes (apoptosis)

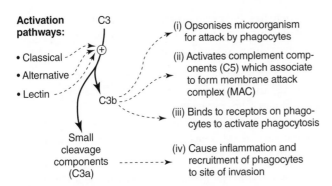

Activation pathways:

• Classical
• Alternative
• Lectin

C3

C3b

Small cleavage components (C3a)

(i) Opsonises microorganism for attack by phagocytes

(ii) Activates complement components (C5) which associate to form membrane attack complex (MAC)

(iii) Binds to receptors on phagocytes to activate phagocytosis

(iv) Cause inflammation and recruitment of phagocytes to site of invasion

Figure 17.24 *Activation of C3 convertase and effects of activation.* There are three pathways that activate complement and four mechanisms that facilitate the killing of pathogens (see text). C3 is the convertase enzyme.

Killing by cells

The killing by cells can be divided into three classes.

(i) Bacteria in the extracellular fluid, when opsonised, are phagocytosed by macrophages and then killed.

(ii) Host cells infected with viruses are killed by cytotoxic T-cells or natural killer cells and released viruses are neutralised by antibodies. (Table 17.4).

(iii) Large parasites present in the blood are killed by eosinophils and mast cells.

In addition, some bacteria survive and proliferate within phagocytes but these can be killed, when phagocytes are specifically activated, by the usual 'weapons' within the phagocyte (see below).

Phagocytosis

Macrophages and neutrophils are the major phagocytes: they possess receptors that recognise the presence of specific molecules or groups of molecules on the surface of pathogens (for example, lipopolysaccharide) or when they are in the form of a complex with an antibody or in a complex with a complement component. Once bound to the receptors, actin filaments of the phagocyte polymerise and part of the plasma membrane surrounds the pathogen or complex which then draws

More than a century ago Elie Metchnikoff discovered that during the inflammatory response, some leucocytes could engulf microorganisms: he termed the process *phagocytosis*, from the Greek *phagein* meaning 'to eat', and *kutos* meaning 'a cell'. Metchnikoff studied the cellular mechanisms of immunity, whereas Paul Ehrlich studied the humoral side of immunity: they shared the Nobel Prize for Physiology and Medicine in 1908.

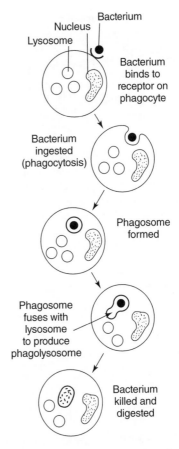

Bacterium

Nucleus

Lysosome

Bacterium binds to receptor on phagocyte

Bacterium ingested (phagocytosis)

Phagosome formed

Phagosome fuses with lysosome to produce phagolysosome

Bacterium killed and digested

Figure 17.25 *Diagram of the events occurring when a bacterium is ingested by a phagocyte.* See text for details.

the pathogen into the cell, to produce a cytoplasmic vesicle, termed a phagosome. Fusion of the phagosome with a lysosome results in the formation of a phagolysosome, in which the pathogen is killed (Figure 17.25). Four killing agents are used by these cells.

(i) The superoxide radical ($O_2^{-\bullet}$) is produced from oxygen by the enzyme complex, NADPH oxidase, which comprises a small number of respiratory chain components that form a sequence in the membrane of the phagolysosome (Chapter 9). The chain transfers electrons from NADPH to oxygen, which is reduced to form the superoxide radical, $O_2^{-\bullet}$ (Figure 17.26). The process is sufficiently quantitative that phagocytosis can be measured by the uptake of oxygen that is required to produce the free radical. This is known as the oxidative burst. The NADPH is produced from the metabolism of glucose through the pentose phosphate pathway; the key enzyme is glucose-6-phosphate dehydrogenase, which is very active in these phagocytes. Free radicals are discussed in Appendix 9.6.

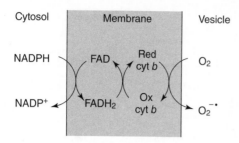

Figure 17.26 *Generation of the superoxide radical by NADPH oxidase.* This limited electron transfer chain is described in Chapter 9, where roles of FAD and cytochrome *b* are discussed. Red – reduced; Ox – oxidised.

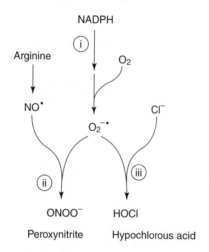

Figure 17.27 *Summary of the mechanisms used by phagocytes to generate agents that kill phagocytosed pathogens.* These are (i) formation of the superoxide radicals, (ii) formation of peroxynitrite by interaction between nitric oxide and the superoxide radical, (iii) formation of hypochlorous acid by interaction between the superoxide radical and the chloride ion.

(ii) Hypochlorous acid (HOCl) is produced in a reaction between $O_2^{-\bullet}$ and a chloride ion, which is catalysed by the enzyme myeloperoxidase.

$$O_2^{-\bullet} + Cl^- \rightarrow HOCl$$

Macrophages can produce so many chemicals/biochemicals that they can be considered as chemical 'factories'. This synthetic activity may explain, in part, the high metabolic rate of activated macrophages, to generate the ATP required for synthesis of these compounds. The fuels are glucose and glutamine.

(iii) Nitric oxide, which is produced from arginine in the reaction catalysed by nitric oxide synthase, reacts with $O_2^{-\bullet}$ to produce the very toxic peroxynitrite free radical (Figure 17.27).

(iv) Proteolytic enzymes within the lysosome also help to kill the pathogens by degrading proteins in the membrane. (See Chapter 8 for discussion of proteolytic digestion in lysosomes.)

The free radicals produced by the macrophage are restricted to the phagolysosome to prevent damage to the cytosol and other organelles.

In addition to these directly toxic agents, macrophages secrete several substances which increase the potency of their antipathogenic role:

(a) The enzyme lysozyme, which breaks down a peptide–polysaccharide complex in the bacterial cell wall.

(b) A colony-stimulating factor that stimulates the production of monocytes in the bone marrow which are then activated to produce more macrophages at the site of infection (see Figure 17.8).

(c) Complement proteins, which maintains at a high level in the area of invasion.

(d) Proinflammatory cytokines (e.g. interleukin-1, interleukin-6, tumour necrosis factor), responsible for inflammation, which provides a local environment that is beneficial for killing pathogens (see Figure 17.21).

Tumour necrosis factor is so named because it can result in necrosis of tumour cells *in vivo* and *in vitro*. However, this is not a physiological action: attempts to use TNF for cancer treatment were abandoned since it is toxic. It was originally termed cachectin, since it causes cachexia, i.e. wasting of body protein, which is a better term since it indicates its physiological role, but tumour necrosis factor is the name that persists.

Killing of pathogens within cells

Some pathogens invade, survive and proliferate within host cells. These include viruses, bacteria and parasites. Viruses are unique since to proliferate they require the nucleic acid and protein synthetic machinery of the infected cells. Hence, once the virus has infected the host cell, the most effective means of killing the virus is to kill the infected cell. Viruses that escape from the dead host cell are neutralised by binding to the antibodies, and the antibody–virus complex is phagocytosed by macrophages.

Host cells are killed by cytotoxic T-cells or natural killer cells by the processes of necrosis or apoptosis. Necrosis leads to release of cell contents that can sufficiently disturb the tissue to initiate a local inflammatory response. However, the cell killed by apoptosis is then phagocytosed, which does not cause local disturbance, so that inflammation does not occur (Chapter 20). Apoptosis is achieved by two mechanisms: release of toxic granules by the cytotoxic cells or by the binding of the cytotoxic cell to the host cell, via its death receptor protein (see below).

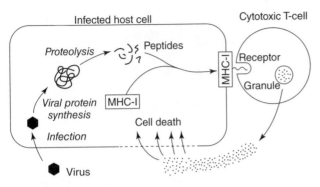

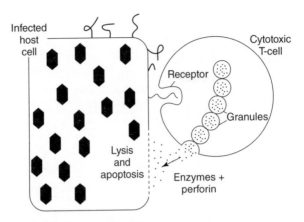

Figure 17.28 *Killing of a virus-infected host cell by a cytotoxic T-cell.* The protein synthetic machinery of the host cell is used to synthesise viral proteins. However, the protein of the virus is digested to produce peptides that form a complex with MHC class I protein, which is presented on the cell surface of the host cell, to which the cytotoxic T-cell binds, releasing the contents of the granules, which kill the host cell (see text).

Figure 17.29 *Organisation of granules within the cytotoxic T-cell prior to release.* The organisation ensures that the released enzymes plus perforin are very close to the infected cell so that killing is restricted to the infected cell. This minimises damage to neighbouring non-infected cells. Death is due to either lysis or apoptosis.

Cytotoxic T-cells   Cytotoxic T-cells and NK cells contain granules which contain enzymes and biochemical compounds that destroy host cells. It is essential that the infected host cells are identified prior to destruction. (The mechanism involves the MHC-class I proteins, as described above.) Once identified, the T_c cells bind to the MHC-class I protein via a receptor that responds to both the MHC-I and a peptide derived from the virus. Upon binding, the T_c cell releases the contents of its granules that result in death of the host cell (Figure 17.28). An additional feature is that the T_c cell binds to the host cell in such a way as to ensure close proximity between them. After this, the granules are organised within the T_c cell so that, when released, they are directed to the smallest space between the two cells. This provides for the most effective killing of the infected cell without killing or damaging the surrounding host cells that are not infected (Figure 17.29). (This is generally known as minimising 'collateral damage'.)

The granules contain two types of proteins that result in death. First, compounds that produce holes (pores) in the membrane of the cells: these are the proteins, perforin and granulysin. Both insert into the membrane to produce the pores. These were once considered to result in death by lysis (i.e. exchange of ions with extracellular space and entry of water into the cell). However, it is now considered that the role of the pores is to enable enzymes in the granules, known as granzymes, to enter the cell. These granzymes contain proteolytic enzymes. They result in death of the cell by proteolysis but, more importantly, activation of specific proteolytic enzymes, known as caspases. These enzymes initiate reactions that result in 'programmed cell death', i.e. apoptosis (Chapter 20).

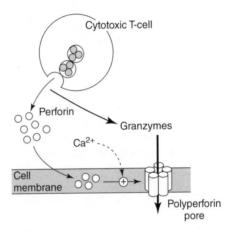

Figure 17.30 *Polyperforin formation which produces a pore in the plasma membrane of the host cell that allows entry of the cytotoxic proteins.* Perforin is released from the granules which enter the plasma membrane of the host cell and Ca^{2+} ions stimulate polymerization to produce polyperforin, which assembles to produce the pore, through which granzymes enter the cell.

Thus, the perforin (and granulysin) and the granzymes function as a 'biochemical unit' to kill the host cell (Figure 17.30).

The death ligand and the death receptor   There is, on the surface of the cytotoxic T_c cell, a ligand that binds to a specific receptor on the infected cell known as a death receptor. The ligand is known as the FAS ligand (or death ligand). This binding results, via activation of intracellular proteases, in stimulation of the caspase system, which initiates apoptosis (Figure 17.31). (See Chapter 20 for description of apoptosis.)

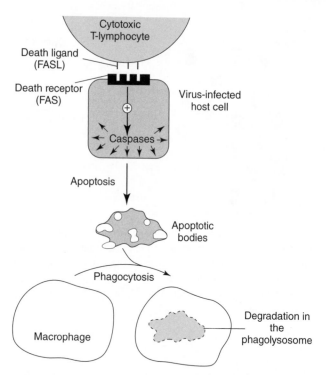

Figure 17.31 *Death of virally infected host cells via the death receptor on the host cell to which is bound the death ligand on the surface of the cytotoxic T-cell.* The interaction between the ligand and the receptor results in activation of caspases that induce apoptosis. The latter process results in disintegration of the cell, the results of which are apoptotic vesicles that are phagocytosed and destroyed by the macrophages.

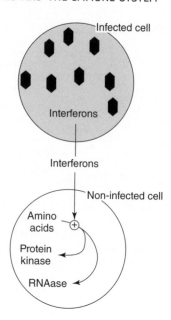

Figure 17.32 *An early warning system.* Virally infected cells release interferons to warn neighbouring cells to prepare for a viral attack. Discovery of these antiviral agents in the 1950s led to great expectations for cures for many diseases. Sadly, it never came to pass.

Killing of intracellular bacteria and large parasites in the extracellular fluid

The immune system not only deals with bacteria in the extracellular fluid and viruses within cells in both the tissues and in blood, but with two further types of pathogens which escape killing by the systems described above. Consequently, two additional mechanisms have evolved to tackle them. These pathogens are (i) the parasites that are too large to be phagocytosed and (ii) the bacteria that, having been phagocytosed by macrophages, avoid being killed, then live and thrive within the macrophages. The large parasites include worms and flukes, for example blood flukes (of the genus *Schistosoma*, e.g. *S. mansoni*, which is widespread in Africa and causes the disease bilharzia), and parasitic worms, such as nematodes (e.g. filariae).

Killing of bacteria infecting the macrophage

In order to understand this activity, it is first necessary to appreciate the different roles of T-cells. The ability of T-cells to activate B-cells was discovered well before it was

An early warning system   If an infected cell is not killed immediately, it becomes filled with viruses, which kill the host cell, and are then released to infect other cells. Consequently, to minimise the spread of infection an *early warning system* exists to inform neighbouring cells to prepare for a possible invasion by viruses. This system involves the cytokines, α and β interferons, which are released by infected cells. The interferons stimulate two defence responses in the neighbouring cells. First, they stimulate synthesis' of a protein kinase which is activated by the viral genome. This kinase phosphorylates and inactivates the initiation factor (IF) of the host's protein synthetic 'machinery', which results in inhibition of the synthesis of all proteins in the cell, including viral proteins (see Chapter 20). Secondly, the cells synthesise an enzyme that activates an RNAse that cleaves both viral and host cell RNA but particularly viral mRNA, to prevent protein synthesis and stop viral proliferation (Figure 17.32).

Box 17.3 Tuberculosis

Tuberculosis has been a major cause of death in humans (and other animals) since the earliest recorded history and certainly before that. Tuberculosis caused the death of about two million people in 2004. It is the leading cause of death in those suffering from AIDS.

Several well-known figures died of tuberculosis, including Frederyk Chopin, who died in 1849 after suffering many years with the disease; John Keats, who was both a poet and a physician, died of the disease in 1821; and George Orwell, who, fortunately, managed to finish his book, *1984*, before dying. Edward VI died from the disease at the age of six; and the three Brontë sisters died within seven years of each other.

It is well established that overcrowding, poor nutrition, stress and general poor health increase the risk of contracting tuberculosis. The conditions of the poor in the large cities during the industrial revolution in Britain were ideal for development of tuberculosis, as recorded by Friedrich Engels in his book *Conditions of the British Working Classes*.

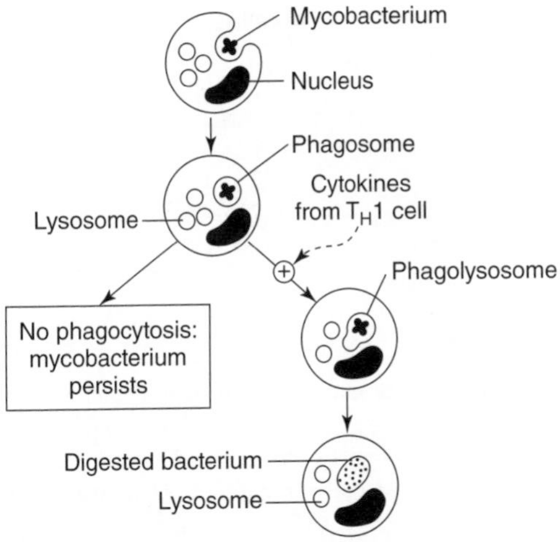

Figure 17.33 *Killing of myobacteria within a macrophage.* Cytokines from T_H1 cells stimulate fusion of the phagosome, in which myobacteria thrive, with the lysosome, so that the mycobacteria are killed within the phagolysosome by the agents described above.

recognised that there is a class of T-cells that activate macrophages, and can stimulate them to kill bacteria that have infected and live within the phagocytes. This class consists of the T-helper-1 (T_H1) cells. The T-helper-2 cells activate B-cells that are involved in killing the large parasites (see below).

Mycobacteria are bacteria that can form filamentous branching structures. Those that are pathogenic to humans include *Mycobacterium leprae*, which causes leprosy, and *M. tuberculosis*, which causes tuberculosis (Box 17.3). Like other bacteria, they are phagocytosed by macrophages to be incorporated within the phagosome. With 'normal' bacteria, the phagosome fuses with the lysosome to form the phagolysosome, in which the bacteria are killed (see above). However the mycobacteria prevent fusion of the phagosome with the lysosome and hence avoid death. They thrive within the phagosome. Somehow, naïve T-cells detect the infection by a mycobacterium and are activated to form a T_H1 subclass of helper-lymphocyte. This subclass secretes a totally different set of cytokines that stimulate fusion of the phagosome and lysosome in the macrophage to form the phagolysosome, in which the mycobacteria are then killed (Figure 17.33). How these cytokines result in fusion of the two organelles is not known.

Killing of large parasites

The significance of large parasites as pathogens in humans is indicated by the evolution of two 'new' cells and one 'new' antibody. These are the eosinophil, the mast cell and IgE (Appendices 17.1 and 17.2). It is the collaboration between these cells and the IgE antibodies that enables the body to kill these parasites.

The antigens on these parasites stimulate B-cells to produce IgE antibodies which do two things: (i) they coat (opsonise) the parasite and (ii) they bind to the mast cell via a specific receptor. The parasite is sufficiently large that many eosinophils bind to it, via the IgE antibodies. Many of the eosinophils are so close that they are in direct contact with the plasma membrane of the parasite. The lysosome of the eosinophil can fuse with the plasma membrane and the contents of the lysosome are released in close proximity to the parasite, which is then damaged or killed by proteolytic and other enzymes (Figure 17.34).

In the meantime, the IgE molecules bind to the mast cells and then become attached to the parasite as it passes by or approaches the mast cell. When a sufficient number of IgE molecules are attached to the parasite, the mast cell is stimulated to discharge the contents of its granules, which contain chemicals that can kill or damage the parasite and biochemicals that attract other immune cells to the area to further attack the parasite (Figure 17.35).

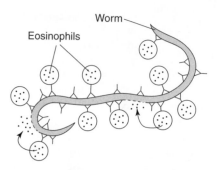

Figure 17.34 *Killing of large parasites by eosinophils.* A parasite worm is surrounded by eosinophils which bind to the IgE molecules that are bound to the worm. The eosinophils kill the parasite worm by release of the contents of their lysosomes.

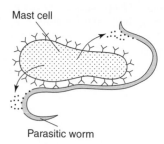

Figure 17.35 *Role of IgE and mast cells in killing large parasites.* The parasitic worm binds to IgE antibodies that are attached to a mast cell, which then releases its granules, the contents of which kill the parasite.

Allergy

It is unclear why certain foreign proteins can also stimulate the B-cells to secrete IgE antibodies, to result in allergy or hypersensitivity. The terms are used interchangeably, although the latter is usually restricted to milder forms of the response. The term anaphylaxis is used to describe the severe response (Box 17.4). Both reactions arise in genetically susceptible individuals and they are precipitated by exposure to environmental antigens such as pollen, some organic compounds, tobacco smoke, animal hairs or even components of some common foods such as milk and cereals.

To help explain an allergic response, discussion is divided into four stages.

1 An antigenic protein enters either the blood or the interstitial fluid, where it is taken up by an antigen-presenting cell (APC), digested and the peptide fragments are presented, along with MHC-II molecules, on the surface of the APC.

2 The peptide binds to the T_H2 cell which, in response, secretes the specific cytokines that activate the B-cell

Box 17.4 The Mediterranean and the discovery of anaphylaxis

The discovery that the immune system could produce acute problems, such as allergy and anaphylaxis, was made on a boat in the Mediterranean in 1901. A French physiologist, Charles Ricket, accompanied the Prince of Monaco on his yacht during a cruise in the Mediterranean. At that time, Ricket was working on blood transfusion and sensitisation of the immune system to foreign proteins. During the cruise he found that the venom of a jellyfish known as the Portuguese Man of War (*Physalia*) was toxic to rabbits and chickens, which were present on the yacht (presumably to provide meat during the cruise). On returning to France, he began to study the response of dogs to injections of small doses of this venom. He had found previously that a second or third injection of an agent provided protection against the effects of the agent. That is, they provide prophylaxis. However, the effect of the venom did not protect animals from subsequent larger doses of the venom; in fact, it rendered the animals much more sensitive so that a second small dose would result in collapse and death of his dogs. In 1903, he termed the response *anaphylaxis*: that is, a very severe hypersensitivity which is to some extent the opposite of prophylaxis. He showed that it could be produced in most animals, even in response to harmless proteins such as egg albumin. The mechanism underlying anaphylaxis is now understood. Ricket was awarded the Nobel Prize in 1913 for its discovery.

cascade and the resultant plasma cell secretes IgE antibodies that are complementary to the antigen that has entered the body on the allergen.

3 The IgE antibodies bind to mast cells, so that mast cells are now primed to respond to the next exposure to the antigen, i.e. the mast cells are sensitised.

4 Upon the next exposure, the antigen binds to the IgE antibodies on the mast cell. If a sufficient number of the antigens bind, the mast cells respond and release their granules into the local extracellular fluid. The contents of the granules are prostaglandins, cytokines, leucotrienes, histamine, thromboxanes and enzymes including proteases. This immediately sets up a local inflammatory reaction – the allergic response (Figure 17.36).

If on subsequent exposure to the antigen, so that most if not all mast cells in the body are exposed to the antigen, the total number of the granules released from the mast cells may be so large that it can result in severe vasoconstriction in the whole body, inducing a massive acute respiratory and circulatory failure. This may be sufficiently severe to result in death. Thus a sting from a large insect such as a bee or wasp can inject enough antigen into the

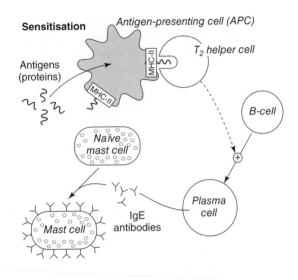

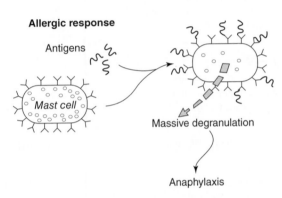

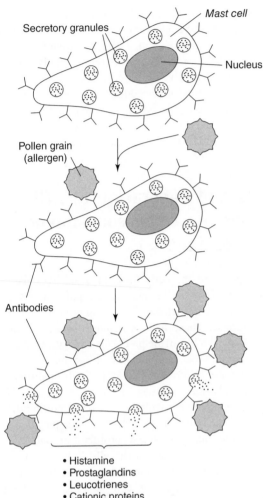

Figure 17.36 *The allergic response.* A protein that enters the body is taken up by an APC, digested and the peptides presented along with MHC-II protein on cell surface. This peptide binds to its complementary receptor on the T_H2 cell, which produces cytokines that stimulate B-cells to proliferate and produce plasma cells that secrete IgE antibodies. The latter bind to mast cells. This is the process of sensitisation. Upon subsequent exposure to the antigenic protein, the antigens bind to the IgE antibodies on the mast cells to produce degranulation. This results in release of the factors that cause the allergic response. If degranulation is massive the response will be severe resulting in anaphylactic shock.

Figure 17.37 *Hay fever.* Pollen grains bind to sensitised mast cells which degranulate, releasing the active biochemicals that result in the allergic response. Serotonin is 5-hydroxytryptamine.

blood to produce anaphylactic shock. The first recorded case is that of King Menes of Egypt, who died of a wasp sting in the 26th century BC. In very rare cases, it has also occurred after immunisation: an additional protein present in the vaccine is usually the cause.

Hay fever is the inflammatory reaction to inhaling pollen, especially grass pollen. Symptoms include copious secretions of mucus in the nose, and itchy and watery eyes. The pollen grains enter interstitial fluid through the mucous membranes and initiate secretion of IgE antibodies. On the next exposure to pollen, the grains bind to the IgE antibodies on the mast cell, which results in degranulation, giving rise to the allergic response (Figure 17.37).

Asthma Many mast cells are located below the skin and the mucous membranes. An antigen that enters through these barriers has ready access to the mast cells to give rise to the allergic response. Allergens in each breath will bind to mast cells in the bronchi and bronchioles. Discharge of the contents of the granules results in vasoconstriction of the smooth muscle in the bronchioles which leads to asthma. One hypothesis to account for the increase in asthma is excessive cleanliness of the environment in which children live and play which underexposes mast cells to different IgE antibodies. Consequently, exposure to a new antigen over sensitise mast cells to the new IgE antibodies, so that subsequent exposures result in severe degranulation and an allergy i.e. asthma.

Dust has been recognised as a trigger for asthma for many years and the allergens are present in the faeces of the house dust mite, of genus *Dermatophagoides*. They are members of the order Arachnida (close relatives of ticks and spiders). They live off human skin that has been shed and hence are present in bedding and carpets. Since they are not visible to the naked eye and are difficult to control, they can be overlooked as the cause of this distressing disease. They may also be a cause of perennial rhinitis and atopic dermatitis.

Fuels and generation of ATP in immune cells: consequences for a patient

The response of the body to an invasion by a pathogen involves not only the immune system but other tissues in which changes in metabolism are essential for a satisfactory response. These include adipose tissue, muscle and liver, all of which provide precursor molecules (i.e. building materials) for the synthesis of the macromolecules required for proliferation (Table 17.5). In addition, they provide fuels, the oxidation of which generates ATP for the immune cells, particularly to support proliferation of the cells in the lymph nodes but also for the precursor cells in the bone marrow where proliferation increases to produce new immune and other cells (Table 17.6).

The total number of lymphocytes in the body is about 10^{12} but when all immune cells are included the total number is probably greater than in many organs (e.g. liver,

kidney, neurones in the brain). Even resting immune cells (i.e. those not yet responding to an immune stimulation) require a considerable rate of ATP generation, which increases in response to an invasion by pathogens or in response to trauma.

To provide fuels and building materials for all these cells, in a severe infection or trauma (including surgery), requires large changes in the metabolism, which may be such as to impose a significant metabolic stress upon the patient (see Chapter 18).

Table 17.6 Some processes in proliferation that require ATP hydrolysis

DNA and RNA synthesis	
De novo formation of purine and pyrimidine nucleotide	
Nucleoside diphosphate reductase	
Thymidylate synthase	
Polymerase reactions	Chapter 20
Cell cycle	
Synthesis and degradation of cyclins	
Peptide and protein synthesis	
Protein kinase reactions	
Mitosis	
Complex carbohydrate synthesis	Appendix 6.2
Lipid requirements	
Cholesterol synthesis	Chapter 11
Phospholipid synthesis	

Pages in parenthesis indicate where the processes are discussed in the book. See also Chapter 9.

Table 17.5 Sources of precursors for formation of polymers required for proliferation of cells

Original precursor and source of precursor	Immediate precursor	Product
[a]Glucose (from food or synthesised within the liver from non-carbohydrates)	acetyl-CoA	cholesterol[d]
	NADPH	RNA, DNA
	ribose 5-phosphate	essential fatty acids elongation and desaturation to produce
	non-essential amino acids	peptides, proteins
Glutamine, which is synthesised in muscle or adipose tissue	nucleotides	RNA, DNA
[b]Polyunsaturated (essential) fatty acids (both omega-6 and omega-3) from food or adipose tissue	phospholipids	membranes
[c]Muscle protein	amino acids	peptides, proteins (including cytokines, acute phase proteins)

[a]Glucose must be synthesised if trauma or infection results in anorexia: the process is gluconeogenesis.

[b]A normal diet may not provide sufficient essential fatty acids to satisfy proliferation in trauma or infection, so that they are mobilised from adipose tissue.

[c]Muscle protein is broken down in trauma even on a normal diet (Chapter 18).

[d]Other essential compounds are synthesised from acetyl-CoA, see Appendix 9.4.

Fuels used by the immune cells

The question arises: what fuels are used by immune cells to maintain the rate of ATP generation, not only for the essential processes in the cell but for supporting the cell cycle and secretions, such as antibodies or cytokines (Table 17.6)? There is not a satisfactory method for obtaining quantitative or even qualitative information in vivo. The problem is the same for other dispersed cells, e.g. those cells in the bone marrow, different muscle fibres in a single muscle, Küpffer cells in the liver, neurones in the brain and erythrocytes in circulating blood. In contrast, the fuels and their rates of utilisation by heart, kidney, brain and some large muscles can be measured in vivo from arteriovenous differences across the tissue or organ. For the immune cells (and other cells, e.g. tumour cells, see Chapters 9 and 21), two indirect methods have been used.

(i) Measurement of maximal activities of key enzymes in extracts of immune cells (see Chapters 3 and 9);

(ii) Measurement of the rates of fuel utilisation by immune cells in culture or during short-term incubation. Details of the methods and results are presented in Appendix 17.4.

Glutamine and glucose as fuels

The two fuels used by all immune cells are glucose and glutamine. Most of the glucose utilised is converted to lactic acid and very little is oxidised: i.e. ATP is generated from *anaerobic* glycolysis (see Chapter 6). Most of the glutamine utilised is converted to aspartate and very little is completely oxidised. This pathway is termed glutaminolysis (see Chapters 8, 9 and 21).

Not only are glucose and glutamine key fuels for immune cells but they also provide the carbon and nitrogen for formation of the nucleotides that are required for synthesis of DNA and RNA (see Chapter 20). This dual role of glutamine is summarised in Figure 17.38. Since glutamine is so important for immune cells, the extent and rate of proliferation of, for example, lymphocytes is reduced if the concentration in the culture medium is decreased below the physiological level (Figure 17.39, Table 17.7). Other processes in immune cells are also dependent on the concentration of glutamine (Table 17.8).

Sources of glutamine for immune cells

Although glutamine is present in protein and therefore made available in the lumen of the intestine by digestion, most of this is metabolised after absorption by the enterocytes in the intestine. Glutamine, therefore, has to be synthesised in the body, the precursors for which are glucose and branched-chain amino acids. It is synthesised in muscle, adipose tissue and the lung (see Figure 8.23). Furthermore, muscle

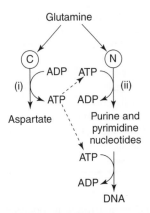

Figure 17.38 *A simple diagram illustrating the two roles of glutamine: (i) generation of ATP, via glutaminolysis, (ii) formation of purine and pyrimidine nucleotides, for the synthesis of nucleic acids in proliferating cells (Chapter 20).* Ⓒ represents the carbon atoms of glutamine, one of which is released as CO_2 and the others are converted to aspartate, via part of the Krebs cycle (Chapter 9). Ⓝ represents the amide nitrogen of glutamine.

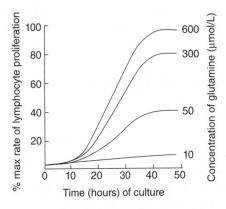

Figure 17.39 *Effect of changes in glutamine levels in culture media on proliferation of human lymphocytes.* The numbers represent the concentration of glutamine, in µmol/L, in culture media. Note that both the maximum rate and the rate of response are decreased as the glutamine concentration in culture medium is decreased. The plasma glutamine level in normal humans is 600 µmol/L.

Table 17.7 Approximate half-times for maximal rates of proliferation of T-lymphocytes from rats and humans in culture

Concn of glutamine (mmol/L)	Approx. half-times for maximum proliferation (hour)	
	Rat	**Human**
0.01	>60	>60
0.05	50	>48
0.30	39	30
0.60	–	27
1.00	27	–

The normal plasma concentration in the rat is 0.9 mmol/L and in the human is 0.6 mmol/L. It is likely to be lower in other body fluids, e.g. lymph, interstitial fluid.

Table 17.8 Processes in immune cells dependent on glutamine

ATP generation
DNA synthesis
RNA synthesis
Protein synthesis
Cytokine production
Glutathione production
Proliferation
Phagocytosis

stores glutamine in a similar manner to the storage of glucose in the liver. It can be considered that there is a 'division of labour' between these three tissues, so that glutamine is provided from the 'production' tissues to the specific sites where it is required. The lung provides glutamine for the many alveolar macrophages (it is estimated that there are 70 macrophages for each alveolus in the lung and possible one billion alveoli are present in the body); adipose tissue provides it for lymphocytes and other cells in the lymph node embedded in the adipose tissue; muscle provides it for other cells (e.g. endothelial cells) (Figure 17.40).

Essential fatty acids and proliferation

In a similar manner to the requirement of nucleotides for synthesis of DNA in proliferating cells, essential fatty acids are required to synthesise phospholipids for new membranes. The synthesis of phospholipids for membranes requires omega-3 and omega-6 polyunsaturated fatty acids (see Chapters 11 and 20). Although these can be obtained from food, a store of these fatty acids is maintained in the triacylglycerol in adipose tissue, which can be made available if they are absent from the diet or during starvation (including patients in hospital who are anorexic) (Chapter 18). Consequently, under these conditions, mobilisation of these fatty acids may be essential during an immune response in which lymphocytes are proliferating in the lymph nodes, and precursor immune cells are proliferating in the bone marrow. It is therefore not surprising that almost all lymph nodes in the body, where proliferation takes place, are embedded within adipose tissue depots for local provision of essential fatty acids and glutamine. For the same reason, proliferating cells in the bone marrow are surrounded by adipocytes (Figure 17.41). The adipocytes provide both the essential fuel (glutamine) and precursor molecules for formation of new membranes. Note that in addition to the hornzone sensitive lipase, adipocytes contain a lipase which is continuously active to provide fatty acids even it the hornzone–sensitine lipase is inacitve.

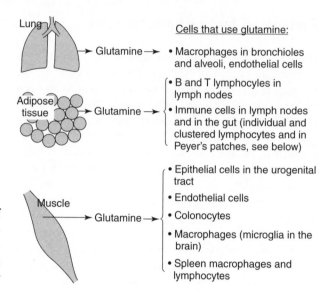

Figure 17.40 *A suggested division of labour between the tissues providing glutamine and the immune and other cells using the glutamine.* Note that all the cells in this figure also use glucose as a fuel (Chapter 9). The lung provides glutamine for the many macrophages present in the bronchioles and in the alveoli, where they are essential for dealing with pathogens, environmental particles and pollutants entering the body with each breath. In addition, the lung contains the largest number of endothelial cells in the body. Adipose tissue provides glutamine for the immune cells in the lymph nodes (i.e. >90% of all lymphocytes); most nodes are present in adipose tissue depots. Muscle provides glutamine for most other cells.

The lymph nodes

A lymph node consists of a cortex and an inner medulla. The cortex is composed of an outer cortex, which contains B-lymphocytes, within lymphoid follicles, and paracortical areas, which contain mainly T-lymphocytes and dendritic cells. The proliferation of B-cells occurs in central areas, called germinal centres. The medulla consists of 'strings' of macrophages and the B-cells that secrete the antibodies (i.e. the effector cells): these are the medullary cords (Figure 17.42). Lymph carries immune cells (e.g. lymphocytes, antigen-presenting cells) and pathogens from the tissues to the lymph nodes, via the afferent lymphatics.

A single artery enters the lymph node at the hilum. After branching, the vessels pass through the medulla within the medullary cords and into the inner and outer cortex. The vessels branch into capillaries around the follicles which provide oxygen, fuels and precursor molecules for the proliferating cells.

Lymphatic fluid enters the node through several afferent lymphatics, percolates through most areas of the node and

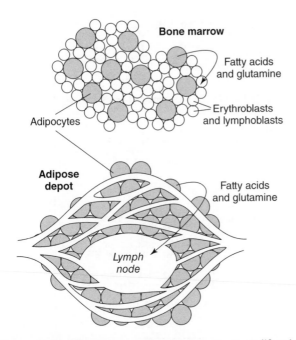

Figure 17.41 *The close proximity of adipocytes to proliferating and developing cells in the bone marrow and the presence of a lymph node in an adipose tissue depot.* Note adipocytes in both locations provide essential nutrients, i.e. fatty acids and glutamine. Proliferation of both B and T lymphocytes takes place in the lymph nodes (see below).

leaves in the efferent lymphatics. The cells in the node can therefore 'examine' the lymph and blood for the presence of pathogens. Clusters of lymph nodes are strategically placed in areas such as the neck, axillae, groin and particularly in the abdominal cavity: areas of the body that are most at risk of invasion. Furthermore, the immune cells can be transported from one lymphoid organ to another, either via the lymph or via the blood, so that all can be 'primed' to deal with the infecting pathogens.

The roles of lymph nodes are as follows:

(i) Particulate matter, including pathogens in the lymph, are 'filtered' and phagocytosed by macrophages, which are both free roaming and fixed within the node.

(ii) The various compartments slow the flow of lymph, providing a better opportunity for examination of its contents.

(iii) In response to an infection, the number of germinal centres in the nodes covering the area of infection increases since it is here where the massive proliferation of lymphocytes occurs, so that the nodes become swollen, tender and more easily palpable.

Tonsils are effectively a small group of lymph nodes in the pharynx which opens to the surface, so that microbes entering via inspired air or food can be dealt with at an early stage.

Lymphocyte recirculation Lymphocytes continuously circulate throughout the body in both the lymph and the blood, which increases the likelihood that they will meet an antigen. Lymphocytes enter the lymph node from the blood through post-capillary venules which contain endothelial cells that are specialised to facilitate the entry of the circulating lymphocytes. These are known as high endothelial cells and possess on their luminal surface adhesion molecules which have a high affinity for the proteins present on the surface of lymphocytes. After adhesion, the lymphocytes traverse the junction between the endothelial cells to enter the node. This is similar to the passage of neutrophils through the spaces between the endothelial cells (see Figure 17.3). The high endothelial cells are located in the post-capillary venules, which are so arranged

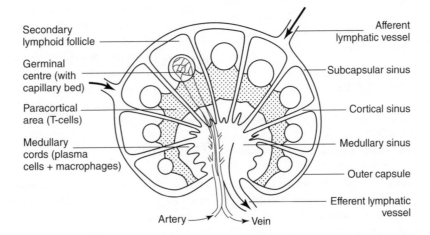

Figure 17.42 *Structure of a lymph node.* See text for details.

It has been claimed that the magnitude of this stress in some blood vessels is similar to the shear stress at the nozzle of the famous water fountain in Lake Geneva. The fountain spouts 500 litres of water per second with a mean velocity of 200 km/hr which reaches a height of about 140 m; the shear stress at the nozzle is about 40 dynes/cm³.

that the flow of blood through them is reduced, which decreases the shear stress to ensure that the lymphocytes bind to the endothelial cells with sufficient affinity not to be swept away by the current. Hence, they have sufficient time to enter the node. In other parts of the vasculature, it is estimated that the flow of blood is such as to produce a much higher shear stress.

In addition, the binding of lymphocytes to endothelial cells in the capillaries that surround the follicles, which contain proliferating B-cells, could restrict the supply of oxygen, fuels and building materials to the proliferating cells. Thus the binding of a large number of lymphocytes occurs at a later position in the blood supply to the node.

The spleen

The spleen functions as a large lymph node and is supplied with blood via a single artery which progressively divides into smaller branches. Small arterioles are surrounded by areas of lymphocytes, which are known as the 'white pulp' of the spleen. The arterioles ultimately end in vascular sinusoids, which contain different types of cells but especially erythrocytes, hence forming the 'red pulp' of the spleen.

The functions of the spleen are similar to those of the lymph nodes but the spleen is the major site of immune responses to blood-borne antigens, whereas lymph nodes are mainly involved in responses to antigens in lymph. The spleen contains large numbers of lymphocytes, where they also proliferate in response to an infection. It also acts as a store of erythrocytes and stores about 30% of the platelets in the body. These are released when required. The spleen also contains macrophages in the red pulp; these are responsible not only for the phagocytosis of pathogens but also of senescent erythrocytes. These macrophages are very important in the salvage of iron present in the erythrocyte (Chapter 15: Figure 15.16).

Activation of lymphocytes

Activation is the term used to describe processes involved in the secretion of antibodies by B-cells and secretions of cytokines by T-cells, but it can also describe the process of proliferation of both types of cells.

The binding of the antigen to its receptor on either the B- or T-cell is an essential stimulus but the binding of other molecules (i.e. co-stimulators) is also necessary. In addition, the metabolic condition (e.g. energy status) within the

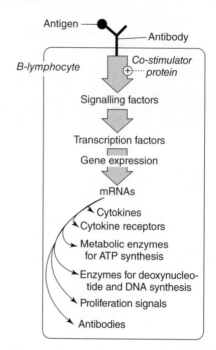

Figure 17.43 *Some intracellular signals in the activation of lymphocytes.* The sequence of factors involved in the signalling process is similar if not identical to such sequences in tumour cells (Chapter 21). The costimular protein can respond within or outside the lymphocyte.

cells and the physiological conditions within the body are also essential for proliferation to proceed. Hence, proliferation will not proceed unless sufficient deoxyribonucleotides, essential fatty acids and fuels are available in the cell. Proliferation may not proceed very effectively under stressful conditions, when fuels or 'building materials' are in short supply, or if pathogens are competing effectively for essential micronutrients (e.g. iron, see Chapter 18: Figure 18.4).

The stimulus provided by the binding of the antigen and the co-stimulator initiates the proliferation process. The sequence of signalling mechanisms that results is shown in Figure 17.43. The process of proliferation is the process of the cell cycle which is described in Chapter 20. Some of the factors that are important for stimulation of the cell cycle in immune cells are indicated in Figure 17.44.

Tolerance

During and after the Second World War, attempts were made to graft tissues from one individual to another, especially skin grafts in order to reduce disfigurement caused by burns. Unfortunately, the grafts were always rapidly

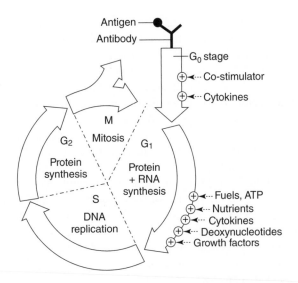

Figure 17.44 *Some factors that are required for completion of the cell cycle in lymphocytes.* The cell cycle is described in detail in chapter 20. The initial signal is binding of the antigen to the receptor immunoglobulin in the membrane of the B cell.

rejected. The rejection process is an immune reaction mounted by the immune cells of the recipient of the graft to the cells of the tissue from the donor. Progress in understanding this problem was made in 1953 when Peter Medawar showed that rejection of skin grafts in mice could be prevented if cells from a donor animal were injected into the prospective recipient either in utero or shortly after birth, i.e. the immune system would accept as 'self' any antigen to which it was exposed in foetal or early neonatal life. This state of unresponsiveness of the recipient to the graft was termed *immune tolerance.*

This is explained by the fact that the lymphocytes that respond to antigens during development are eliminated by apoptosis, a process known as clonal deletion. The lymphocytes subsequently present in the foetus will respond only to foreign proteins, for example, those in a graft from another person which then leads to its rejection. Not surprisingly, the system is more complicated than this. The MHC proteins are required to help the lymphocytes differentiate between self and non-self proteins and much of the antigenic difference between individuals lies in the MHC-class II protein.

Chronic inflammation and autoimmunity

Approximately 5% of all Europeans and North Americans suffer autoimmune diseases, two-thirds of them are women;

it is not clear why females have higher risk. There are two major types of autoimmune disease: organ-specific and systemic. Examples of organ-specific disease are insulin-dependent type 1 diabetes and rheumatoid arthritis; an example of a systemic autoimmune disease is lupus erythematosus. How do these diseases arise?

For the immune system to act effectively but only against invading organisms, a mechanism exists by which it distinguishes between the individual's own proteins and those which are foreign; that is, it can distinguish between 'self' and 'non-self' proteins. As indicated above, during the perinatal period the immune system learns to recognise the structure of the self-proteins to which it is exposed at that time, so that it never responds to these proteins.

Three mechanisms are involved in failure of this system:

(i) Any proteins that are not normally exposed to the immune system during the perinatal period, for example, those locked within the cell by an impermeable membrane or hidden within a membrane will not be learned as 'self'. If subsequent cell damage, for example as a result of viral attack or exposure to a toxic chemical, exposes such proteins, the immune system may treat them as non-self. In this case, activation of B-cells produces antibodies that interact with the self protein; and activation of T_H cells leads to activation of macrophages and consequent, damage to the cells that possess this protein.

(ii) A defect in the mechanism of clonal deletion, that is, relevant lymphocytes that fail to be eliminated could cause an immune attack on those self-proteins that possess complementary antigenic sites to the antibody receptor on the B-cell.

(iii) If the structure of an antigen on a pathogen closely resembles the structure of a self-protein, the antibodies produced against the antigen may not distinguish between the two proteins. Consequently, immune cells attack the self-protein in the tissues, as well as that on the pathogen, so that the tissue is damaged. This mechanism is termed *antigenic mimicry.*

> The bacteria *Campylobacter jejuni* possesses an antigen similar or identical to a ganglioside sugar that is present on peripheral motor nerves. Infection with these bacteria results in the production of antibodies to bacterial antigens which also recognise the ganglioside sugar. This results in immune attack on motor neurones, which damages the myelin membrane. Hence failure to transmit the action potential results in paralysis which can affect all four limbs and the respiratory muscle, requiring assisted ventilation. It is known as Guillain–Barré syndrome. It is named after the French neurologists G. Guillain and A. Barré. Guillain–Barré syndrome is, therefore, an example of molecular mimicry.

Immunosuppressive agents

Autoimmunity is, in many cases, severely debilitating. Consequently, there is considerable interest in the development of immunosuppressive agents that will control or remove this problem. Current immunosuppressive agents have been very successful in suppressing rejection of transplanted organs including bone marrow transplants but, unfortunately, much less success has been achieved in treating diseases caused by autoimmunity. There are currently three well-established immunosuppressive agents: cytotoxic compounds, cyclosporin-A and similar compounds, and steroids. However, polyunsaturated fatty acids may have an immunosuppressive action in some conditions (see below).

Cytotoxic agents   Since proliferation of cells is essential for the immune response, agents that inhibit DNA synthesis have been used as immunosuppressive agents for many years. The first were used in the 1960s, particularly to prevent rejection of a transplanted organ, for example purine and pyrimidine analogues. These agents are not now used in autoimmune diseases but are still used in cancer chemotherapy (Chapter 21).

Cyclosporin-A   In 1972, the immunosuppressive effect of cyclosporin-A, an antibiotic secreted by some fungi, was discovered. Other antibiotics also have immunosuppressive effects. They interfere with proliferation of T-helper cells by preventing the entry of a transcription factor into the nucleus. This prevents transcription of the genes involved in the proliferative process. Their use is restricted to patients after transplantation, since there are serious side-effects, for example, toxic effects on the tubules of the kidney. This precludes their use for treatment of non-life-threatening autoimmune diseases.

Steroids   It was noted in the 1950s that, during pregnancy, there was sometimes a remission of inflammatory disorders. This was attributed to the increase in the level of steroid hormones. Since then, glucocorticoids have been used therapeutically to reduce inflammation and autoimmunity. Their mechanism of action is not understood but it is likely to result from inhibition of phospholipase activity, decreasing formation of arachidonic acid and hence a decrease in production of prostaglandins and leucotrienes, which are involved in control of proliferation of cells of the immune system. Steroids, however, have serious long-term side-effects, glucose intolerance, weight gain and osteoporosis, which restrict their chronic use (See Chapter 12).

Polyunsaturated fatty acids   Polyunsaturated fatty acids, especially omega-3 fatty acids, in vitro, inhibit proliferation of lymphocytes, phagocytosis by macrophages and killing by NK cells (Chapter 11). Supplementation of the diet with fish oils has been shown to have beneficial effects in a number of autoimmune/chronic inflammatory diseases, e.g. lupus erythematosus, rheumatoid arthritis and inflammatory bowel diseases. The mechanism of action of these fatty acids is unclear. At least two suggestions have been made. (i) They change the balance of prostaglandins and leucotrienes that are produced during the stimulation of the phospholipase. (ii) They change the structure of phospholipids and hence the structure of membranes which interface with the binding of some components of signalling processes, which are involved in control of proliferation, e.g. those involved in the MAP kinase cascade. Failure of some components of the signalling cascade to bind to the cell membrane reduces thin signalling activity in the cascade (Chapter 21).

Conditions that reduce the effectiveness of the immune system

There are a number of conditions that are considered to reduce the effectiveness of the immune system and hence increase the risk of infections. These include malnutrition, stress and intense prolonged physical activity.

Malnutrition

It is generally accepted that malnutrition decreases the effectiveness of the immune system so that it increases the incidence of infections. Thus famine and infections are always considered together since it is assumed that the former exacerbate the latter. The influenza pandemic in 1918–19 in Europe led to the deaths of more than 20 million people. Poor nutrition caused by the First World War may well have impaired the immune system in many people, thus contributing to the large number of deaths. This topic is discussed in Chapter 18.

Stress

There is evidence that stressful conditions decrease the effectiveness of the immune system: the loss of a job, a divorce or a bereavement increases the risk of development of cancer. This is due to impairment of the process of immune surveillance carried out by the neutrophils and other immune cells. They kill tumour cells that are migrating from a primary tumour to establish another tumour in a different tissue. The impairment may be due to: chronic activation of the sympathetic system, which increases the local concentrations of catecholamines, which inhibit proliferation of immune cells; an increase in the blood level of glucocorticoids; a decrease in the blood level of glutamine, which is the important fuel for immune cells. There is, as yet, no firm evidence to support any of these suggestions.

Physical activity

Prolonged and regular physical activity of high intensity decreases the effectiveness of the immune system. For example, there is an increased incidence of infections in athletes during intense periods of training and in young recruits in the armed forces, especially in the first few months of training. In contrast, regular, mild activity can increase the effectiveness of the immune system. (For discussion, see Chapter 13.)

Factors that increase the effectiveness of the immune system

Factors that can increase or are claimed to increase the effectiveness of the immune system include: healthy nutrition (see Chapter 15); low stress levels; mild physical activity; some nutritional supplements; and vaccination. Many factors that promote the effectiveness of the immune system are opposite to those that decrease the effectiveness. Unfortunately, it is unclear what is meant by healthy nutrition but several nutritional supplements (neutraceuticals or functional foods) have been suggested to increase the effectiveness of the system. These are discussed in Chapters 15 and 18.

Immunisation

The terms vaccination, inoculation and immunisation are used synonymously; they are used to describe the processes for the production of immunity by artificial means. Vaccination has a long history and is a fascinating topic (Box 17.5). The term 'vaccine' is defined as an infectious agent that is either dead or attenuated, and which is introduced into the body with the object of increasing the ability of the immune system to resist or get rid of a disease.

The definition of vaccination, in the strict use of the term, stems from the work of Edward Jenner with cowpox and smallpox. Jenner invented the Latin name *Variolae vaccinea* for cowpox. From this, the term vaccination arose.

Box 17.5 Smallpox, cowpox and Edward Jenner

Smallpox is a very ancient disease that has been mentioned in medical writing dating back to the third century AD. It originated in Egypt or India and became endemic in both of these countries. In the seventeenth and eighteenth centuries, it is estimated to have killed almost half a million people every year in Europe. The Spanish conquerors (conquistadors) of Mexico and Peru took smallpox with them. It is estimated to have killed over three million Aztec Indians, who had no immunity against this disease because, until then, it was unknown in the New World.

Well before the work of Edward Jenner in the 1790s, a means of immunisation was practised in China and India. The procedure was to lesion a small vein of a healthy child and transfer pus from a smallpox pustule (vesicle) to the open wound. After a few days, the child became ill with a fever due to a minor smallpox infection but, after about a week, the child recovered. Not surprisingly, since just the right amount of pus needed to be used, it was not completely safe and those who became infected spread the disease. Salvation came via an obscure country doctor, Edward Jenner (1749–1823), who was practising in the Cotswolds in England. He was familiar with the 'old wives' tale' that the unsightly but harmless disease of cowpox, which milkmaids caught from cows' udders, gave protection against smallpox. He hit on the idea that cowpox might be artificially induced and provide protection against smallpox. He carried out an experiment. On 14th May 1796 he took pus from the hand of the dairymaid (Sarah Niemes) who was suffering from cowpox, and inserted it into the arm of an eight-year-old boy (James Phipps) by means of two superficial incisions. On 1st July, pus obtained from a pustule of a patient with smallpox was administered to James Phipps: he did not develop smallpox. Several months later he was again inoculated with pus from a patient suffering from smallpox; once again, smallpox did not develop!

Jenner attempted to publish the results of his experiment in *Transactions of Royal Society* but the paper was rejected. He then repeated the experiment with 13 volunteers and published the results in his treatise, '*An inquiry into the causes and effects of the variolae vaccinae, a disease discovered in some parts of the western counties of England, particularly Gloucestershire, and known by the name of the Cow Pox*', published in 1798. Following its publication, vaccination quickly spread through Europe and America, earning Jenner worldwide fame.

In the 1960s, the WHO initiated a 10-year mass vaccination campaign which has eradicated the disease. In 2005, large-scale production of the smallpox vaccine was undertaken in the USA for storage in case of a biological attack by terrorists.

There are two types of immunisation: passive and active. Passive immunisation involves transfer of antibodies formed in response to an antigen in one individual to another. Such antibodies were first produced in animals but now most antibodies used for passive immunisation are of human origin, which minimises allergic reactions. This form of immunisation gives immediate protection but it does not last very long, since the antibodies are soon degraded in the body. It is used, for example, to protect against tetanus, rabies and the toxins in snake venom.

Active immunisation uses killed or live microorganisms. The immunisation is effective because of the production of memory cells. Unfortunately, the immune response to dead pathogens is poor since T-cells are poorly activated, so that live pathogens are preferred. These are treated to make them innocuous, whilst retaining the ability to elicit a response from the immune cells: it is then known as an attenuated vaccine. One way to achieve this is to grow the causal agent in large amounts and then treat it with a chemical (e.g. formalin).

Current knowledge in the fields of biochemistry and molecular biology is being harnessed to produce antigens more simply and in greater amounts. For example, surface proteins, bearing antigenic sites from the pathogen, can be sequenced and the corresponding DNA synthesised. This 'artificial gene' can be transfected into a bacterium which can then be grown to produce large quantities of antigen for use in immunisation.

DNA vaccines

Once the virus has achieved an intracellular location, it is protected from antibodies and only cell-mediated immunity is effective. Difficulties in producing effective vaccines against some viral infections, such as HIV, herpes simplex and hepatitis, have spurred the development of new strategies that use DNA rather than protein.

The process is as follows: a piece of DNA, which encodes a gene for a surface protein of the virus, is incorporated into a plasmid. From a suitable vector, DNA is taken up by host cells and incorporated into their DNA. A large amount of viral protein is produced within the host cell, hydrolysed and the resultant peptide complexed with MHC class I molecules presented on the cell surface. This is then seen and responded to by the T_H cells which proliferate and form memory cells that will result rapidly in the death of host cells infected by the virus in subsequent infections. DNA vaccines are safer than 'live-virus' vaccines and, furthermore, several genes that produce different viral antigens can be constructed on the same piece of DNA.

Return of the 'old' infectious diseases

Despite the remarkable natural defence mechanisms that have evolved, they have not always been successful and infections have, on occasions, produced catastrophic effects on the human population (Box 17.6). Two examples are the plague pandemic of the mid-1300s (Box 17.7), and the influenza pandemic of the early 1900s. This is due, in part, to the ability of the pathogen to avoid detection or killing by the immune system.

Infectious diseases are probably the oldest recognised diseases of humans. Evidence of leprosy is found on Egyptian mummies dating from the second century BC and the Black Death is mentioned throughout history. So successful were immunisation and antibiotics in the fight against infectious diseases that in 1969 the US Surgeon General, William H. Stewart, reported to Congress that, 'it was time to close the book on infectious diseases'. This statement was somewhat premature! Unfortunately, 'old' infectious diseases, such as tuberculosis, cholera, typhus, leprosy, syphilis, and even the Black Death, are now returning.

Some of the reasons for the return are as follows: (i) new breeding grounds for the insects that are vectors for some pathogens; (ii) antigenic drift in viruses and bacteria; (iii) resistance to antibiotics; (iv) a decrease in the effectiveness of the immune system due to the presence of other more chronic infections, poor nutrition or stress; (v) expansion of air travel.

New breeding grounds for insects

One example is the mosquito. Conversion of forests into farmland in many developing countries produces places that accumulate water which are ideal breeding grounds for mosquitoes. Global warming may increase the size of areas that are warm enough for mosquitoes to breed (Box 17.8).

Antigenic drift

The continual mutation of pathogens results in a phenomenon known as antigenic drift. Such mutations can change the structure of the protein components of the membrane, which may then not be recognised by the memory lymphocytes, so that the pathogen now readily infects an unprotected population. A good example is the influenza virus. It regularly undergoes so much antigenic drift that the modified proteins of the new strain are not recognised by memory cells of most or all members of a population. Such

Box 17.6 Strategies employed by pathogens to evade detection or killing by the immune system

Microorganisms have evolved a wide range of compounds that interfere with defence mechanisms provided by the immune system.

Host cell mimicry Sialic acid is part of the glycocalyx of many host cells. The coat of some bacteria contain sialic acid which may reduce the effectiveness of the response of the immune system to the 'foreign' nature of the coat.

Complement Some microorganisms produce proteins that bind to and inactivate components of the complement system and hence decrease activation of the cascade, e.g. the vaccinia virus secretes a protein that inhibits activation of both the classical and alternative pathways. Some bacteria produce a protein that mimics the action of an acceleration factor, which increases the rate of destruction of the active convertase; this factor is normally produced by the host when the complement response is no longer required.

Cytokines Some microorganisms produce proteins that mimic receptors for the cytokines: these sequester the cytokines and prevent them from reaching their normal targets,

e.g. poxviruses produce proteins that bind interleukin-1, tumour necrosis factor or interferon; the hepatitis B virus expresses, on its surface, a protein that binds interleukin-6. The secretion of some cytokines by host cells requires intracellular proteolytic digestion of a pro-cytokine: cowpox virus produces a protein that inhibits a serine protease in macrophages inhibiting the cleavage of the pro-interleukin 1-β. The synthesis of interferon and interleukin-2 is normally decreased by IL-10; some viruses produce a homologue of the latter.

MHC-class I proteins Some microorganisms have developed a considerable number of tricks to interfere with the presentation of the MHC-class I protein peptide complex on the surface of the infected host cell: for example, production of a protein that binds the MHC protein and causes it to be retained within the endoplasmic reticulum.

Synthesis of steroid hormones Some microorganisms synthesise and secrete steroids that have an immunosuppressive effect.

Box 17.7 The plague: Black Death

Surprisingly, the term 'plague' is difficult to define. *The Oxford English Dictionary* (2nd edn) describes plague as 'an affliction, calamity, evil "scourge" *esp.* a visitation of divine anger or justice, a divine punishment; often with reference to the ten plagues of Egypt "Egypt was smyten with *x* plages and diseases"'. The term has been used in descriptions of outbreaks of many diseases including leprosy, smallpox, tuberculosis and yellow fever but it is now more generally associated with Black Death, the popular name for bubonic or pneumonic plagues, which devastated Europe in the middle of the fourteenth century. It is estimated that it killed 25 million people. It recurred at intervals until the final pandemic of 1664–1665, after which it disappeared from much of Europe.

The cause of Black Death is a gram-negative bacterium, *Yersinia pestis*, formerly known as *Pasteurella pestis*. The name was changed after Alexandre Yersin, a student of Pasteur, first identified it in buboes of corpses in Hong Kong. It lives in the blood of rodents and is spread between rodents by their fleas. It spread to humans when domestic rats become infected and their fleas bite humans, allowing the bacteria to enter the blood and lymph. The organisms enter lymph nodes and elicit a major inflammatory response so that the nodes swell markedly (bubo). Haemorrhagic lesions occur in many

organs and those under the skin result in loss of blood which turns black due to degradation of haemoglobin, one reason for the name Black Death.

During the fourteenth century traders travelled from the Mediterranean and the Black Sea to China to bring back silks and furs. Returning from such a trip in 1343 a group of merchants from Genoa were attacked by a band of Tartars (Cossacks) and they took refuge in the walled town of Caffa in the Crimea. The Tartars besieged the town for three years without success until, instead of catapulting rocks over the wall, they catapulted the corpses of their own men who had died of the plague. The whole town became infected, as did most of the Tartars. Some Genoans survived and made it back to Europe but presumably many were infected and they then transferred it to their families so that the disease spread within Europe. Between the mid-fourteenth and the nineteenth century there were at least 10 plague pandemics (see *The Great Mortality: an intimate history of the Black Death*, by John Kelly (2005)).

The organism still thrives in many parts of the world in wild rodents. The total number of cases reported to the WHO in 1998 by various countries was almost 2500. Of further concern, a multidrug resistant strain of *Y. pestis* has been found.

Box 17.8 Malaria

Malaria is the most important parasitic infection in humans: it accounts for more than two million deaths each year and many of these are children. Since it is most prevalent in areas containing swamps, it was once believed to be caused by poisonous gases emanating from the swamps, hence the name given to the disease, *mal aria*. The distinguished Australian immunologist, Macfarlane Burnet, stated, 'there is no doubt that malaria has caused the greatest harm to the greatest number, not through cataclysms, as with bubonic plague, but through its continual winnowing effect'. Malaria has caused problems for humans for thousands of years. The fevers characteristic of malaria were known to the Greeks, and X-ray analysis of Egyptian mummies confirms that early Egyptians suffered from malaria. Unfortunately, it remains out of control across much of the world.

The protozoan parasite *Plasmodium* is responsible for the disease. It lives within stomach and salivary glands of the *Anopheles* mosquito and is transmitted to humans through mosquito bites. The parasites move through the bloodstream to the liver, where they proliferate during an incubation stage of a couple of weeks. Returning to the blood, they infect red blood cells, where they break down haemoglobin to release amino acids. This breakdown occurs within what are called 'food vacuoles' in the parasite, which are similar to lysosomes in that the pH is low. The breakdown of the protein to provide amino acids releases the haem which is poisonous to the parasite. It converts the haem to haemozoin which is not toxic. This conversion is catalysed by an enzyme, and some of the antimalarial drugs inhibit this enzyme. The discovery that the parasite resided in the mosquito was made by Donald Ross in 1897, for which he was awarded the Nobel Prize for Medicine in 1902.

The drug chloroquine was introduced in the late 1940s: it was cheap, non-toxic, and soon became the mainstay of therapy and prevention. By now, however, widespread resistance has developed to the drug. So far there has been no success in producing a vaccine against the parasite, but in June 2005, US president George Bush pledged 'an official commitment' by the USA to provide US$1.2 billion over eight years to fight malaria.

---

antigenic drift accounts for the influenza pandemics that sweep across the world from time to time: the pandemic in 1918–19 resulted in the death of 20–40 million people, more than were killed in the First World War. Another shift resulted in the 'Hong Kong' influenza pandemic in 1968 and yet another gave rise to 'Singapore influenza', between 1957 and 1967. Since the influenza virus is transmitted directly from one human to another, the disease is highly contagious, but since it has a short latent period it needs a large population in order to establish itself. The massive size of many cities throughout the world, in which many of the population may be malnourished, explains why there is considerable concern about emergence of new strains and accounts for the fact that the WHO has over 100 laboratories reporting on the influenza virus worldwide, so that vaccines for a new strain can be produced as quickly as possible in an attempt to prevent another pandemic.

Resistance to antibiotics

Resistance to antibiotics is usually due to the acquisition of genes that express enzymes that can inactivate the antibiotics (e.g. β-lactamase degrades the lactam ring in penicillin (see below) or that can modify the structure of proteins that are necessary for the antibiotic to enter the bacterial cell. Acquisition of the genes for resistance can occur quickly since transfer of genes from one microorganism to another can readily occur. Indeed, resistance to penicillin developed within the first two years of its introduction in the 1940s (Box 17.9).

Resistance is now common in several organisms, for example *Yersinia pestis* and *Vibrio cholerae*. In addition, the parasite that causes malaria (*Plasmodium*) is resistant to most drugs (Box 17.8).

The bacterium *Staphylococcus aureus*, which is a major cause of infection in the developed countries, is now resistant to most antibiotics. It is usually present on the skin, where it causes no problems, but it can invade the body through cuts and wounds, including those caused by surgery. These bacteria are now prevalent in many hospitals, so that infection is a major problem for the medical staff in hospitals. The resistant bacterium is known as methicillin-resistant *Staphylococcus aureus* (MRSA). It is also known in the mass media as the 'super bug'. Penicillin kills bacteria because the β-lactam group in the antibiotic inhibits a reaction that is essential for bacterial cell wall production. Consequently, the bacteria cannot proliferate. Resistance to penicillin in many bacteria is due to production of an enzyme, β-lactamase, that degrades β-lactams. The antibiotic methicillin is one of a group of semisynthetic penicillins in which the β-lactam group is not

Box 17.9 A short history of penicillin

In 1928 Alexander Fleming discovered that under certain circumstances the mould *Penicillium notatum* produced a diffusible substance that inhibited the growth of some species of bacteria. He named it penicillin. Very little was done with this substance in the ensuing years, since it was very unstable. In 1939, Ernst Chain, a refugee from Eastern Europe, and Howard Florey who was Head of the Dunn School of Pathology in Oxford, began work on penicillin as part of a comprehensive programme of research on antibacterial substances. Norman Heatley joined Chain and Florey and made a major contribution to the programme. He devised the culture conditions for growth of the mould, designed a method for purification of penicillin from the culture broth, and set up a sensitive assay for penicillin. All these techniques were essential for progress of the study. Eventually enough stable penicillin was accumulated for an animal experiment and ultimately enough to permit trial of the drug on a patient suffering from a severe bacterial infection. The patient began to recover but unfortunately the amount of penicillin available was not sufficient to complete the trial and the patient died. This work may be regarded as the true dawn of the age of antibiotics. Chain and Florey, along with Fleming, shared the Nobel Prize for Physiology and Medicine in 1945.

broken down by β-lactamase. However, a form of β-lactamase has evolved in *Staphylococcus aureus* that is resistant to this antibiotic. Hence it is a significant cause of death in many hospitals.

Malnutrition

Even when a disease has been effectively treated in a country, retention of the disease in small isolated pockets can act as a source for future infection. This can readily occur when movement of even a small number of infected people into a population in which the immune system is impaired can lead to the spread of this new disease to areas that were previously free of it. Chronic illness due to malnutrition can weaken the immune system which then facilitates the spread of an infection. This is a particular problem if malnutrition is accompanied by a chronic illness. Although malnutrition and accompanying diseases are associated with developing countries, the phenomenon also occurs in developed countries, for example, in the very poor, the homeless, drug abusers and the elderly. It is now considered to be a major factor in the increased incidence of tuberculosis in these groups.

Air travel

The jet plane has made it easy to travel to and from previously inaccessible parts of the world, for business or vacation. Infectious diseases or agents that transmit these diseases can, therefore, be rapidly transferred to countries in which the population has never encountered the diseases, so that the infection rapidly spreads throughout the population. Such transport of pathogens is reminiscent of the transmission of diseases that were previously unknown in North America but were transported from Africa in the slave ships. For example, the mosquito *Aedes aegypti*, which transmits the virus that causes yellow fever, was probably transported in water barrels on these ships.

New infectious diseases

Several new diseases have emerged in the last 20–30 years including AIDS, SARS, Lyme disease and Legionnaire's disease. Several explanations have been given for this emergence, including changes in human behaviour and new technologies or new industries. The recent increase in the use of air conditioning systems provides a warm, stationary aqueous environment and the bacterium, *Legionella*, thrives in it. When the bacteria infect humans they cause Legionnaire's disease. The increasing pursuit of walking in areas frequented by deer or horses has increased the risk of a bite from ticks that normally attack these animals and this bite can transfer a parasite, *Borrelia burgdorferi*, to humans, which results in Lyme disease. The immune deficiency virus (HIV) is resident in some types of monkey without causing serious problems but it was transferred to humans by the bite of a monkey with the development of the acquired immune deficiency syndrome (AIDS). A whole section in this text is devoted to AIDS, since it is having a great impact on the human population in some parts of the world, and there is an intense research effort for both treatment and prevention of the disease.

Lyme disease and AIDS are two of a number of diseases that have transferred from animals to humans. Diseases that have been transmitted from domestic animals to humans are known as zoonoses (some examples are given in Table 17.9).

Acquired immune deficiency syndrome (AIDS)

A quite remarkable story began in Los Angeles in 1981. A pharmacist noted a sudden increase in the number of prescriptions for pentamidine, which is a drug used to treat

Table 17.9 Common diseases for which the microorganisms transferred from animals to humans, or diseases that can be transmitted from animals

Animal	Disease	Comments
Cattle	tuberculosis	an increasing problem in developing and developed countries;
	smallpox	probably evolved from cowpox;
Pigs and birds	influenza	epidemics arise as new strains evolve;
Horses	common cold	the rhinovirus is responsible for the common cold;
Dogs, cattle	measles	a serious illness that can result in death;
Many domesticated animals	Salmonella infection	common cause of food poisoning, which is increasing due to poor hygiene;
Monkey	AIDS	the most serious disease since the bubonic plague;
Water polluted by animal faeces	poliomyelitis, cholera, typhoid	pathogens that cause these diseases are developing resistance to antibiotics.

a type of pneumonia found in immunosuppressed patients (e.g. those given immunosuppressive agents after transplantation). However, these prescriptions were being presented mostly by homosexual men, intravenous drug users or haemophiliacs. This distribution suggested the existence of an infectious agent that was transmitted via blood or other body fluid. A search began for a virus. In 1983, a new virus, the human immunodeficiency virus (or HIV) was identified independently by two research groups. This virus, which is a retrovirus, is responsible for the disease that is known as acquired immune deficiency syndrome (AIDS), in which the function of both the cellular and humoral arms of the immune system is impaired.

The human immunodeficiency virus (HIV)

The virus is transmitted from one subject to another by transfer of body fluids and the most common means of transfer is sexual activity. It is also transferred between drug abusers due to use of the same syringe and needle. It can be transferred from mother to offspring via breastfeeding. The disease caused by HIV is the most devastating outbreak of a sexually transmitted disease since the emergence of syphilis in Europe 500 years ago. In the year 2003, it was estimated that 45 million people worldwide were infected with AIDS, most of them in the sub-Saharan Africa. In 2006, it was estimated that almost 5 million people in India, alone, were infected.

> In view of the high level of infection amongst male homosexuals, it is of interest that lymphoid follicles, equivalent to Peyer's patches, are abundant in the human rectal mucosa.

Retroviruses

A retrovirus consists of double-stranded RNA (the genome) and a few proteins within a lipid envelope bilayer and inside the envelope. The virus enters the body and then binds to a protein on the surface of a host cell. Once the virus has bound, it fuses with the cell membrane, releasing its RNA and proteins into the cell. Within the cell, the viral enzyme, known as reverse transcriptase, catalyses three separate reactions: RNA-directed DNA synthesis; DNA-directed DNA synthesis; and hydrolysis of viral RNA.

The result is that viral RNA is used to synthesise DNA which acts as a template for the synthesis of complementary DNA so that viral RNA is no longer required and it is hydrolysed. The double-stranded viral DNA then inserts into the DNA of the host cell; this involves two integration proteins which are also contained within the virus. The insertion of the viral DNA into host DNA requires cell division of the host cell. One reason for this is that the levels of deoxyribonucleotides are not high enough for sufficient DNA synthesis to occur until the S phase of the cell cycle in the host cell takes place, when deoxyribonucleotides are synthesised within the cell (Chapter 20) (Figure 17.45). Thus, replication of the virus is enhanced after prophylactic administration of influenza or hepatitis vaccine to HIV-infected patients. In contrast, in the macrophage, the levels of deoxyribonucleotides are always high to allow DNA repair to occur whenever necessary, so that the virus readily infects macrophages, which include the microglia in the brain.

Several glycoproteins, which are present in the lipid bilayer of the virus, are necessary for infection. One is known as GP120. It binds to the CD4 protein on the surface of the T_H lymphocyte (i.e. the CD4+ cell). This initiates fusion with the plasma membrane of the CD4+ cell so that the viral RNA and its proteins enter the cell (i.e. it infects the CD4+ cell). The original infection probably occurs in the peripheral circulation but the lymphocytes will be transported by the blood to the spleen, other lymph nodes and the brain, where the microglia become infected (Figure 17.46).

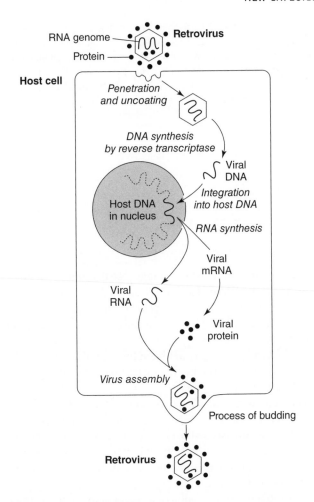

Figure 17.45 *Summary of the life history of a retrovirus. See text for details.*

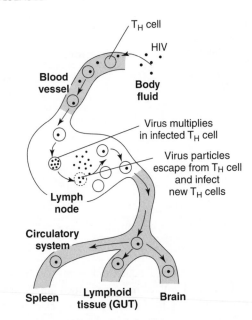

Figure 17.46 *Transfer of HIV from body fluid to lymph node and thence, via blood and lymph, to lymphoid tissue and brain.*

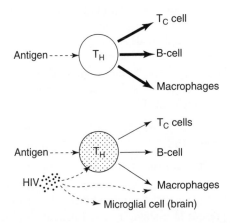

Figure 17.47 *Comparison of a response to an antigen from a normal T_H cell with a T_H cell infected with HIV. Thickness of lines indicates extent of the cytokine stimulus. Note that the virus infects peripheral and resident macrophages as well as T_H cells and microglial cells in the brain.*

As long as an infected cell is quiescent, the virus remains latent and the patient shows no obvious signs of infection, except for the presence of the antibodies against the surface proteins of the virus as a result of the initial infection. However, when a CD4+ cell encounters an antigen to which it responds, host and viral DNA are transcribed and the cell is compelled to produce mRNA for the viral proteins. New viruses are then assembled, the host cell dies and the viruses are released. They then infect and kill other cells. As a result, there is a marked decrease in the number of CD4+ cells, so that they fail to produce sufficient cytokines in response to an infection, to activate macrophages, cytotoxic T-cells and B-cells (Figure 17.47). The severity of the disease is accounted for by the fact that the role of the CD4+ cells is to provide the central coordinating system in the adaptive immune response. Failure to activate B-cells and macrophages in response to a bacterial, fungal or viral infection means that the patient is open to opportunistic infections, particularly herpes, oral candida,

tuberculosis, pneumonia and a particular form of cancer, Kaposi's sarcoma. Patients also suffer from dementia, due to infection of the microglial cells in the brain. More general symptoms include fatigue, diarrhoea, fever and wasting (cachexia). Since the CD8+ cells are not affected by the infection, an indication of the extent of the infection is provided by the ratio of the number of CD4+ cells to the number of CD8+ cells (i.e. CD4+/CD8+).

There is considerable variation in progress of the disease but the usual clinical course is shown in Figure 17.48. In

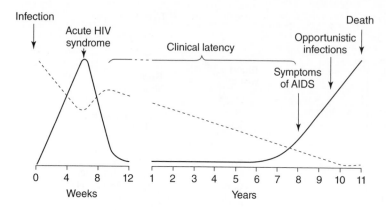

Figure 17.48 *Sequential changes in the number of CD4+ lymphocytes and clinical signs and symptoms of AIDS. The dashed line represents the number of CD4+ lymphocytes; the solid line represents the number of virus particles. The timing of the changes is approximate. (Adapted from Rang et al. (2000).)*

brief, there is an initial acute influenza-like infection, which is associated with an increase in the number of viral particles in the blood and is known as the *acute HIV syndrome*. Then there is a progressive decrease in the number of virus particles probably due to effects of antibodies, complement and phagocytosis. After this, there is a symptom-free period when, presumably, the virus distributes to the lymphoid tissue and organs such as the brain. This latent period may be as long as 10 years. Eventually, when all the lymph nodes and lymphocytes have been infected, the organisation and coordination of the immune system by the CD4+ cells fails, simply due to a serious depletion of the cells. The signs and symptoms of AIDS then become apparent (Figure 17.48).

Treatment of AIDS

A vaccine   The search for a vaccine against the virus is intense. Unfortunately, progress is difficult since the virus frequently mutates. It is estimated that 10^6 to 10^8 variants may be present in a single infected patient and different variants may be present in patients in different parts of the world. Such a mixed population of a virus has been called a 'quasi species'.

Drugs   Since viruses use much of the machinery of the host cell for proliferation, only enzymes catalysing reactions that are specific to the virus offer points of attack. Antiretroviral drugs inhibit one of two such enzymes:

• HIV peptidase (HIV protease) which hydrolyses precursors of retroviral proteins to produce active proteins;

• reverse transcriptase.

Since one of the catalytic actions of reverse transcriptase is DNA synthesis, analogues of pyrimidine and purine nucleotides inhibit this process. A drug, zidovudine, azi-

dothymidine, AZT, is an analogue of thymidine, whereas didanosine (dideoxyinosine) is an analogue of a purine. These compounds are phosphorylated by host enzymes to produce triphosphates that inhibit the transcription process.

The viral protease is inhibited by several drugs, including indinavir. Surprisingly, such protease inhibitors not only affect the virus but they also affect fat metabolism. They cause loss of adipose tissue in the limbs and face, which is balanced by its accumulation in other areas, intra-abdominal, breasts and over the cervical vertebrae, the last giving rise to the term 'buffalo hump', a change which is especially distressing for the patient. The cause of this effect is not known.

Although administration of a combination of these drugs does delay progress of the disease, they are not totally effective. Thus in health care workers who had been infected in work-related accidents, administration of the drugs did not prevent the progression to AIDS. The use of a combination of drugs at the optimal doses for their action is known as *highly active antiretroviral therapy* (HAART).

Transmissible spongiform encephalopathies

Encephalopathies have been known in animals for many years, e.g. scrapie, which occurs in sheep. Others include spongiform encephalopathy in domestic cats and, more recently, bovine spongiform encephalopathy. The latter developed in the UK in the 1980s: symptoms include unco-ordinated movement and frenzy (hence *mad cow disease*). There is some evidence that it occurred in cows after they were fed offal that was prepared from tissues of other cattle.

Two human forms of spongiform encephalopathies are kuru and Creutzfeldt–Jacob disease (CJD).

Kuru

Kuru occurred in a small isolated population in the highlands of New Guinea. Presenting signs and symptoms were ataxia, shivering and inability to perform normal movements. The disease was due to the custom of eating the brain of a relative who had recently died. The disease has now almost disappeared with the end of the custom.

Creutzfeldt–Jacob disease (CJD)

This disease is the most common form of spongiform encephalopathy in humans. It falls into several categories: sporadic, familial and a new form, known as variant CJD. The disease was first reported in patients who had received growth hormone prepared from the pituitaries of cadavers, but new variant CJD was recognised in the UK in the 1990s. Although it is similar to the other forms of the disease, there are differences, hence the name. The patients are generally young (median age is 29 years) and presenting features are behavioural changes, ataxia, peripheral sensory disturbance and progressive dementia. Death results in about one year after the first signs of the disease appear. It is possible that this variant form arose from eating meat that contained nervous tissue that had been prepared from cows suffering from mad cow disease.

The cause of spongiform encephalopathies

One possible agent that may be the cause of encephalopothies is known as a prion, which is a peptide (mol.mass 27–30 kDa) that is resistant to high temperature, ultraviolet light, formalin and proteolytic enzymes. For this reason, it has been termed *proteinase resistant protein* or prion protein (abbreviated to PrP). Prions have been isolated from the brains of animals with these diseases, which suggested that infection with this peptide causes the disease. However, a gene for PrP has been isolated from brains of normal animals. Furthermore, the genes from both infected and non-infected animals are identical, so that the proteins are identical, although they can exist in two different conformational forms or isoforms. One isoform is present in the normal brain and is known as Pr^C, and the other in the infected brain, which is known as Pr^{SC}. It is, therefore, suggested that it is not the prion itself that is the transmissible factor but some unknown agent that converts the normal isoform into the disease-producing isoform (i.e. it converts Pr^C into Pr^{SC}). As yet, there is no indication as to the identity of the factor. It is, probable, that Pr^{SC} more readily polymerises to produce aggregates, known as plaques, that interfere with brain function.

The concept that a small molecule, when bound to a site, could cause a conformational change that markedly affects the properties of a protein was proposed in 1963: the allosteric model (Chapter 3). A molecule binds to a specific site, changes the activity of an enzyme that regulates metabolism. This is achieved by a conformational change in the structure of the enzyme, which lowers the activity. It is suggested that a similar situation might exist with the prion. A small molecule that binds to the Pr^C isomer results in a conformational change that converts it to the Pr^{SC} isomer, which then aggregates to form plaques. The identity or, indeed, the existence of such a small molecule is not known.

Defence in the intestine

The mucous surface of the gastrointestinal tract is chronically exposed to pathogens. There are several lines of defence in the intestine to protect against the entry of pathogens from the lumen into the blood: the physical barrier of the mucous membrane of the epithelial cells of the intestine; and the mucus which covers the membranes prevents pathogens from adhering to the epithelial cells which is essential for them to enter the cell or enter the body. Mucus traps pathogens which are, therefore, localised for attack by the IgA antibody in the lumen. Acid in the stomach and also help in the small intestine alkaline conditions. The immune system is particularly well developed within the gastrointestinal tract. It is called the gut-associated lymphoid system, abbreviated to GALT.

It is usually considered that pathogens that cause problems in the intestine have entered via the mouth, but more than one million pathogens are present in the large intestine. These can enter the bloodstream when the intestinal barriers are not sufficient to prevent translocation (Chapter 18). Conditions that increase translocation and can, therefore, lead to peritoneal sepsis are as follows.

- Disruption of the 'balance' between non-pathogenic and pathogenic bacteria can occur when broad-spectrum antibiotics are used to control an infection in another part of the body.

- Conditions such as malnutrition, starvation, infection or trauma can reduce the fuels available for the colonocytes (glutamine, short chain fatty acids) so that the barrier is less effective.

- Proinflammatory cytokines can result in damage to the cells which can result in damage to the barrier (Chapter 18).

Gut-associated lymphoid tissue (GALT)

Collectively, GALT is the largest lymphoid organ in the body and contains a high proportion of all of the B-cells in the body. Both B- and T-lymphocytes are found individually within the epithelial layer and the lamina propria of the intestinal wall (Chapter 4). However, most of the immune cells are organised into anatomically and functionally distinct compartments, which are Peyer's patches (aggregated lymphoid follicles), isolated follicles and mesenteric lymph nodes. The fact that a large number of T-lymphocytes and antibody-producing plasma cells are always present in the intestine has led to the suggestion that it is in a continual state of inflammation, which has been termed '*controlled inflammation*'.

A number of features distinguish the immune responses of the gut from those of the general immune system. IgA is the predominant class of the immunoglobulins released by the B-cells in the gut. Moreover, there is considerable communication within GALT, so that an antigenic response at one site can 'prepare' another site for encounter with the same organism, and this communication can also protect other mucosal systems, such as those of the respiratory and urogenital tracts, from an infection by the same pathogen. (Another example of an early warning system.)

Specialised epithelial cells, M-cells, overlie Peyer's patches and isolated follicles, to facilitate the transport of antigens from the lumen of the intestine to the lymphoid cells below. These are the 'front men' of the intestinal immune system. They do not possess microvilli or a glycocalyx, so that they can more readily bind proteins and pathogens from the lumen and convey them through the epithelium to the antigen-presenting cells beneath. This results in activation of T_H cells and therefore secretion of cytokines to stimulate the B-cell cascade in follicles within the patch. This results in production of antibodies against the introduced antigen; some of the B-cells migrate along the lymphatic vessels to the mesenteric lymph nodes where further proliferation, maturation and effector cell production occurs. Some of these cells can continue to migrate until they enter the thoracic duct from where they enter the blood and are carried to other parts of the intestine, where they enter the lamina propria to provide the individual lymphocytes that are present in the epithelium of most of the gut. They can also be transported to mucosal surfaces of the respiratory system and urogenital tracts to provide defence against the entry of these pathogens through those surfaces (Chapter 4).

18
Survival After Trauma: Metabolic Changes and Response of the Immune System

There is a circumstance attending accidental injury which does not belong to disease – namely that the injury done has in all cases a tendency to produce both the disposition and means of cure.

(Hunter, 1794)

Unfortunately, even in the face of accumulated knowledge, standard hospital care knowingly produces energy and nutrient deficits while otherwise attempting to provide acceptable care . . . Disregard for the effect of malnutrition on care seems primarily due to both undergraduate and postgraduate medical training programmes that present only limited aspects of nutrition and in a fashion that makes unclear any relevance to medical practice.

(Tucker, 1997)

The muscle wasting associated with critical illness cannot be prevented/reversed by provision of calories and nitrogen alone. The need to overcome this catabolic drive has led to a number of trials of supplementary feeding regimens with, for example, either glutamine or growth factors. Whilst both have shown some benefit, there is no unequivocal evidence to support their use and the quest continues for other strategies.

(Griffiths *et al.*, 1999)

'Trauma' is defined as physical injury to the body, usually resulting from an external source and, therefore, includes surgery or burns. 'Sepsis' is defined as the clinical responses of pyrexia, anorexia and lethargy in association with bacterial, protozoal, fungal or viral infections. These symptoms also occur in patients when there is no obvious infection so that the term 'sepsis syndrome' is sometimes used to describe this condition. Since the physiological and biochemical responses to infection, sepsis and trauma are similar they have been termed 'systemic inflammatory response syndrome' (SIRS). In this chapter, they are discussed together under the umbrella term *trauma*. In the UK approximately 15 000 people die every year from trauma, 5000 in road traffic accidents, making this the most common cause of death in adults under 35. It is estimated that the cost to the nation is about £7.5 billion. In the USA, it is estimated to cause more than 1.5 million deaths each year.

The extract from Hunter's treatise above raises a challenge in the 21st century: to provide a biochemical explanation for this tendency. The extract from Tucker emphasises the importance of nutrition for the wellbeing and recovery of the patient. This chapter considers the functional biochemistry not only of the immune system but of the whole body, to provide a greater understanding of the biochemistry of, and the recovery from, trauma.

Physiological and metabolic responses: the ebb & flow phases

The response to trauma can be divided into two phases: the ebb and the flow. The ebb phase lasts for about 12 to 24 hours after the insult, after which there is a smooth

Functional Biochemistry in Health and Disease by Eric Newsholme and Tony Leech
© 2010 John Wiley & Sons Ltd

Table 18.1 Biochemical and physiological changes during the ebb and flow phases of trauma

Parameter	Changes during:	
	Ebb phase	Flow phase
Blood glucose	+	+
Plasma fatty acids	+	+
Blood catecholamines	+	+
Plasma insulin	−	−
Cardiac output	−	+
Oxygen consumption	−	+
Body temperature	−	+

+ Indicates an increase.

− Indicates a decrease.

transition to the flow phase. The ebb phase is characterised by decreased cardiac output and oxygen consumption with a low temperature, whereas the flow phase is characterised by increased cardiac output and oxygen consumption with an elevated temperature (Table 18.1).

In the *ebb* phase, there is increased activity of the sympathetic nervous system and increased plasma levels of adrenaline and glucocorticoids but a decreased level of insulin. This results in mobilisation of glycogen in the liver and triacylglycerol in adipose tissue, so that the levels of two major fuels in the blood, glucose and long-chain fatty acids, are increased. This is, effectively, the stress response to trauma. These changes continue and are extended into the flow phase as the immune cells are activated and secrete the proinflammatory cytokines that further stimulate the mobilisation of fuel stores (Table 18.2). Thus the sequence is trauma → increased endocrine hormone levels → increased immune response → increased levels of cytokines → metabolic responses.

During the *flow* phase there is a marked increase in energy expenditure and oxygen consumption. To satisfy this, the cardiac output is increased, from a resting value of about 5 litres per minute, by two fold during mild trauma, or as much as threefold in severe trauma or sepsis. Resistance to blood flow in peripheral arteries decreases, which can lead to hypotension, with poor perfusion of some organs and hence hypoxia in these organs which, if severe, can give rise to acidosis. Other physiological effects include anorexia, myalgia, somnolence and fatigue; perhaps due to the increased plasma levels of cytokines.

> The method used for measuring energy expenditure of patients is indirect calorimetry, although it can be calculated from the Harris & Benedict or Schofield equations (Chapter 2).

> A rate of 15 litres per minute is remarkably high; it is not very different from that achieved during intense sustained exercise by an athlete.

Metabolic changes

The integrated response to trauma depends on increases in the rates of metabolic processes, which are summarised as follows:

(i) mobilisation of fatty acid from adipose tissue (Chapter 7);

(ii) fat oxidation and ketone body production (Chapter 7);

(iii) release of lactate from muscle and from immune cells (Chapter 6);

(iv) gluconeogenesis (Chapter 6);

(v) Cori cycle activity (Chapter 6);

Table 18.2 Endocrine hormones and cytokines responsible for metabolic changes in trauma

Metabolic changes	Hormones responsible for changes	
	Endocrine	Cytokine
Increased hepatic glycogen breakdown and gluconeogenesis	{ Glucagon, Catecholamines	Tumour necrosis factor
Stimulation of muscle protein breakdown and gluconeogenesis	Glucocorticoids[a]	Interleukin-1, Tumour necrosis factor, Interleukin-6
Increased fatty acid mobilisation from adipose tissue	Catecholamines (sympathetic activity)	Interleukin-2, Tumour necrosis factor
Insulin resistance	Growth hormone	Interleukin-6

Initially the level of insulin decreases, favouring increased rates of lipolysis, fatty acid oxidation, muscle protein degradation, glycogenolysis and gluconeogenesis. It soon increases, however, as a result of insulin resistance, when the stimulation of the above processes will depend on the cytokine levels. For details of endocrine hormone effects, see Chapter 12. For details of cytokines see Chapter 17.

Cytokines stimulate the anterior pituitary to secrete growth hormone, antidiuretic hormone and ACTH.

[a] In response to increased levels of ACTH.

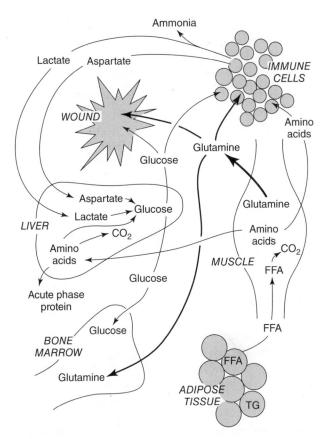

Figure 18.1 *A summary of the metabolic response to trauma.* A major fuel in trauma is glutamine: it is synthesised, stored and released by muscle and taken up particularly by tissues in which cell proliferation is occurring. These are cells in the wound, which include lymphocytes, macrophages and fibroblasts. Glutamine is also used by proliferating immune cells in the lymph nodes and the cells in the bone marrow. These cells also use glucose. They produce aspartate from glutamine and lactate from glucose. The muscle releases amino acids, all of which (except for lysine and leucine) are precursors for glucose in the liver (via gluconeogenesis) and the branched-chain amino acids are precursors for glutamine in muscle (Chapter 8). Fatty acids are mobilised from adipose tissues, to be oxidised by muscle. Glucose, glutamine and fatty acid metabolism generates ATP for these cells (Chapter 9). Almost all of the immune cells are present in the lymph nodes.

(vi) intra- and inter-tissue triacylglycerol/fatty acid substrate cycling (Chapters 11 and 21);

(vii) degradation of muscle proteins, primarily actin and myosin, and consequent release from the muscle of amino acids, particularly alanine and glutamine (Chapter 8).

In addition, rates of glucose oxidation fall and the blood glucose level rises.

The importance of these changes in response to trauma all as follows.

- The increased mobilisation of fatty acids from adipose tissue raises the plasma concentration, which increases the rate of fat oxidation by muscle. It also releases some essential fatty acids from the store in the triacylglycerol in adipose tissue. These are required for formation of new membranes in proliferating cells and those involved in repairing the wound (e.g. fibroblasts) (Chapters 11 and 21: Figure 21.22).

- The increased oxidation of fatty acids decreases the rate of glucose utilisation and oxidation by muscle, via the glucose/fatty acid cycle, which accounts for some of the insulin resistance in trauma. An additional factor may be the effect of cytokines on the insulin-signalling pathway in muscle. An increased rate of fatty acid oxidation in the liver increases the rate of ketone body production; the ketones will be oxidised by the heart and skeletal muscle, which will further reduce glucose utilisation. This helps to conserve glucose for the immune and other cells.

- Any pyruvate produced from glucose in muscle, which is not oxidised, will be converted to lactate and released into the blood.

- The increased production of lactate by muscle and by the growing number of immune cells, fibroblasts and proliferating cells in bone marrow results in an elevation of the plasma lactate concentration. This stimulates glucose formation in the liver (i.e. gluconeogenesis). The conversion of glucose to lactate in muscle and the conversion of lactate to glucose in the liver is known as the Cori cycle but, as more tissues than muscle are involved during trauma, it is known as an extended Cori cycle (see Figure 6.10).

- Protein degradation in muscle provides amino acids for protein synthesis in proliferating cells, synthesis of peptides in the liver and immune cells (e.g. acute-phase proteins, cytokines, glutathione) and formation of some conditionally essential amino acids (arginine, cysteine and glutamine). The fate of other amino acids is catabolism in the liver: the nitrogen is converted to urea and the carbon is converted to glucose via gluconeogenesis. The glucose produced from the amino acids is used by the brain (which is why provision of glucose is particularly important in those patients who are not eating or not being fed), by the immune cells and by fibroblasts in the tissues that are being repaired after the trauma (see Figure 8.14).

- Stimulation of the intra- and inter-tissue triacylglycerol/fatty substrate cycles results in better control of the intra- and extracellular concentrations of fatty acids, since high concentrations of fatty acids can result in tissue damage (Chapter 7). The increased rates of these cycles, plus that of the Cori cycle, increase the turnover of ATP and hence account, at least in part, for the increased energy expenditure and consequently increased oxygen consumption. These changes can also account for the increase in body temperature. It is important to note that gluconeogenesis not only removes lactate from the blood but also removes protons, i.e. it removes lactic acid, which helps to minimise or prevent acidosis.

Fuels in trauma

The major fuels used in trauma are glucose, glutamine and fatty acids. The source of the fuels will obviously depend on whether the patient is eating or being fed but, if not, they will be produced from endogenous sources, i.e. liver glycogen, muscle protein and adipose tissue triacylglycerol (Figures 18.1 and 18.2).

Glucose is the fuel for brain and other glucose-dependent cells and tissues (Chapters 6 and 9).

Fatty acids are used as a fuel for a number of tissues, especially muscle, liver and kidney.

Ketone bodies It is unclear how important ketone bodies are in trauma. The plasma concentration is not greatly

increased but this may be explained by an increased rate of utilisation by muscle or other tissues, such as the kidney. If ketone bodies are a significant fuel they will complement the effects of fatty acids, as in other conditions.

Glutamine is used as a fuel especially for endothelial cells, lymphocytes, macrophages (in brain, lung and other organs) and for proliferating cells in the bone marrow. Glutamine is also a nitrogen donor for synthesis of purine and pyrimidine nucleotides in proliferating cells and is a source of aspartate, which is important for provision of nitrogen in synthesis of pyrimidine nucleotides. These roles are especially important in the immune and other proliferating cells that are involved not only in 'fighting' pathogens but for repair of the damage (See Figure 17.38 and Table 17.7).

Nutrition

The basic human need to be fed extends to patients in hospital. Despite this, many patients may not be adequately fed: indeed, it has been claimed that 25–50% of patients in intensive care units (ICUs) are malnourished. The invasion by feeding of patients is discussed in detail in Chapter 15.

Adequate nutrition improves wound healing, decreases the risk of infection, reduces muscle wasting and maintains muscle function after trauma. The following extract emphasises this:

> *Function begins to return as soon as nutritional support is introduced, even before any tissue gain, suggesting an immediate and direct effect on cellular function. Skeletal and respiratory muscle function deteriorates steeply after a loss of 20% of body protein stores (equivalent to 15% total body weight loss) and improves by 10–20% within the first few days of nutritional support.*
>
> (Lobo & Allison, 2000)

Undernutrition decreases the levels of ATP and phosphocreatine in muscle which affects function, but they return to normal levels with adequate energy intake (Table 18.3). The store of glutamine in muscle is also decreased.

Most patients who cannot feed normally can be fed either by infusion of a nutrient solution into a vein (the parenteral route) or by a tube into the stomach or the jejunum (the enteral route). These solutions deliver all the essential nutrients. Indeed, many patients who cannot eat normally live productive lives while being nourished exclusively by one or both of these routes.

There is considerable discussion in the medical literature on the relative merits of enteral or parenteral feeding. For enteral feeding, the choice is between elemental (or monomeric) preparations, which contain glucose and amino acids, or polymeric preparations which contain protein and

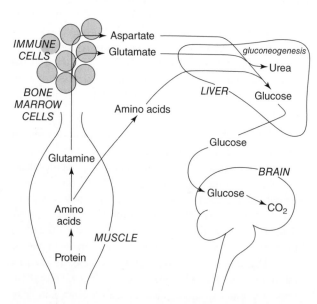

Figure 18.2 *Source of glutamine for use by immune cells and cells in the bone and fate of its metabolic products.* Glutamine, which is synthesised and stored in muscle, is released and used by the immune cells in lymph nodes and in the wound, and by the cells in the bone marrow. For other tissues that provide glutamine and those that use it, see chapter 8 and see Figure 17.40.

Table 18.3 Percentage changes in concentrations of some amino acids, ATP and phosphocreatine in muscle of patients suffering from severe trauma

Metabolite		Percentage change
Energy metabolites	ATP	−12
	Phosphocreatine	−22
	Creatine[a]	+80
Amino acids	Aromatic amino acids[b]	+88
	Basic amino acids	−49
	Branched-chain amino acids[c]	+39
	Glutamine	−72
	Total amino acids	−59

These compounds are measured in biopsy samples of muscle (vastus lateralis).

[a]The increased creatine concentration will restrict the breakdown of phosphocreatine which converts ADP to ATP.

[b]Aromatic acids include tyrosine and phenylalanine: these are not metabolised by muscle so increased levels in muscle are consistent with increased rate of protein degradation.

[c]Branched-chain amino acids are leucine, isoleucine and valine: the increased concentrations are also consistent with an increased rate of degradation, as muscle protein contains a high proportion of these amino acids. The extent of the decrease in ATP concentration is even greater than in exaustive physical activity. Note the very large fall is glutamine concentration.

> Small peptides rather than amino acids may be more beneficial in enteral diets, because there are transporters for peptides in the luminal wall of the intestine and this may result in a greater increase in the plasma level of amino acids.

starch. In general, the latter are favoured, despite the need for digestion, because they do not pose osmotic problems. Fibre is sometimes added for fermentation by the bacteria in the colon, which form short chain fatty acids that are used by the colonocytes (Chapter 4).

High concentrations of nutrients in parenteral feeds can damage the vein but this is overcome by supplying the nutrients into a major vein, either the subclavian or internal jugular, where the rate of blood flow is sufficient to dilute the nutrients rapidly. Alternatively, an isotonic, energy-rich solution can be administered via a peripheral vein; it usually contains a fat emulsion together with glucose and amino acids. The fat consists of triacylglycerol containing omega-3, omega-6, medium-chain and some saturated fatty acids. Phospholipid is included to aid the formation of an emulsion. The lipid content can provide a significant proportion of the energy requirement. This is important, because the resting energy expenditure is increased in patients suffering from trauma.

Several commercial companies prepare solutions for both enteral and parenteral nutrition: an example of commercial feeds that are designed to maintain function of the immune system is presented in Table 18.4.

Table 18.4 Composition of two parenteral feeds designed to stimulate the immune response

	Content (g/L)	
	Impact	Immun-Aid
Carbohydrate	132	120
Fat	28	22
Omega-3 fatty acids	2	1.1
Amino acids	59	80
Arginine	14	15
Glutamine	–	12
Nucleic acid (RNA)	1.3	1.0

For example, a standard feed for one day provides approximately 8 MJ (2000 kCal) from a mixture of fat and carbohydrate plus amino acids (12–14 g nitrogen). Both feeds are currently in clinical use.

Nutritional supplements: nutraceuticals

Some cells require specific nutrients to maintain their function. These nutrients are included in some feeds for the patient to help fight disease and aid recovery. These are sometimes called 'nutraceuticals', a term which can be loosely defined as a fuel or substrate that is provided by the physician or surgeon as part of the treatment for a specific condition. They include nucleotides, arginine, cysteine, glutamine and polyunsaturated fatty acids. The metabolic bases for the beneficial effects are described in Chapter 15. A summary is given below.

> Differences between nutraceuticals and functional foods are explained in Chapter 15.

Nucleotides (RNA) Proliferation of cells requires synthesis of the nucleic acids, DNA and RNA. The content of DNA is doubled during the cell cycle. The nucleotides required for formation of the nucleic acids are synthesised from glucose and glutamine via the *de novo* pathways or via the salvage pathways from nucleosides (Chapter 20). The best means of providing nucleosides is by using RNA directly as a supplement. It is very stable in the commercial feeds and it is degraded by intestinal enzymes to nucleosides, which are absorbed into the blood, transported into cells and then converted to nucleotides in the proliferating cells.

$$\text{RNA} \xrightarrow[\text{(i)}]{gut} \text{nucleosides} \xrightarrow[\text{(ii)}]{salvage\ in\ cells} \text{nucleotides}$$

$$\xrightarrow[\text{(iii)}]{polymerases} \text{RNA and/or DNA}$$

For details of the pathways (i, ii and iii) see Chapter 10: Appendix 10.2 and Chapter 20.

Cysteine   Cysteine is a conditionally essential amino acid but is not included in commercial feeds because it is not stable in such solutions. It is, however, provided as *N*-acetylcysteine, which is slowly hydrolysed *in vivo* to release cysteine:

$$N\text{-acetylcysteine} + H_2O \rightarrow acetate + cysteine$$

Polyunsaturated fatty acids   Polyunsaturated fatty acids are provided in the triacylglycerols on phospholipids in the feeds. They are required for synthesis of phospholipids, which are required for formation of new membranes in proliferating cells, and as precursors for fat signalling molecules that are important in control of proliferation (see below).

> A very high intake of polyunsaturated fatty acids should be avoided in trauma since they have an immunosuppressive effect. They are used in the treatment of chronic inflammation.

Arginine   Arginine is a precursor for a number of compounds, including proline and nitric oxide. Nitric oxide is a signalling molecule, proline is needed for formation of collagen, some of which is needed for provision of new tissue. Arginine has been shown to be beneficial for the healing of wounds and for this reason is included in some parenteral feeds (Table 18.4).

> Glutamine is not very soluble in solution and it may not be stable for long periods or stable to heat, so that dipeptides containing glutamine have been preferred by some clinicians (e.g. alanylglutamine). Glutamine administration also increases body weight in malnourished patients, for example those with AIDS or those receiving chemotherapy for cancer.

Glutamine   Glutamine is the most frequently discussed nutraceutical in the clinical literature as it is an essential fuel for immune and other proliferating cells and the plasma concentration decreases in a number of clinical conditions (Table 18.5). Provision of glutamine (or a dipeptide that contains glutamine), either enterally or parenterally, has

Table 18.5 Plasma glutamine levels in normal subjects and patients

Condition	Approximate glutamine concentration (µmol/L)
Normal	540
Acute pancreatitis	380
Acute pancreatitis and multi-organ failure	266

Data from Professor Eric Roth. Chirurgisches Forschungslaboratorium, Klinik Chirurgie Waehringer, Guertel 18–20, 1090 Wien, Austria. (Personal communication.)

Table 18.6 The effect of enteral glutamine after major trauma on complications caused by infections

| Type of complication | Percentage of patients with complications | |
	Control group	Glutamine group
Pneumonia	45	17
Emphysema	6	0
Abdominal abscess	13	3
Urinary tract infection	23	17
Bacteraemia	42	7
Sepsis	26	3
Central line infection	16	3
Wound infections	13	3

From Wilmore (2001) based on original paper by Houdjik *et al.* (1998). (Glutamine was provided for at least five days. Number of patients in each group was 30.) Another study showed that glutamine-enriched total parenteral nutrition (TPN) resulted in an overall 15% reduction in the total cost of hospital care which, when expressed as cost per survivor, is 50% less than standard TPN at 6 months. The benefits of glutamine provision by either enteral or parenteral routes in the critically ill patient are discussed in detail in the following references: Wilmore (1998); Griffiths *et al.* (2002); Novak *et al.* (2002); Wischmeyer (2001).

at least five beneficial effects in patients suffering from trauma: (i) It maintains the size of the glutamine store in skeletal muscle; (ii) It prevents or reduces the breakdown of the intestinal barriers, limiting bacterial translocation from the lumen of the colon into the blood or peritoneal cavity and hence decreases the risk of infection; (iii) It reduces the incidence of infection in the intensive care unit after trauma or major surgery (Table 18.6); (iv) It is essential for maintenance of the function of the immune cells (macrophages, lymphocytes and neutrophils); (v) Administration of glutamine after elective surgery reduces the length of stay in hospital.

Mobilisation of triacylglycerol and protein in trauma

One obvious symptom of a patient suffering from trauma is loss of body weight. This is due to increased mobilisation of both triacylglycerol in adipose tissue and degradation of protein in skeletal muscle.

Loss of triacylglycerol

Mobilisation of triacylglycerol is due to the increased rate of lipolysis. Adipose tissue triacylglycerol contains some essential fatty acids (linoleic and linolenic acids) and these are mobilised along with the non-essential fatty acids. The latter are oxidised in various tissues, especially skeletal muscle, to offset the decreased energy intake due to anorexia and to provide energy to support the hypermetabolism. Essential fatty acids have a different fate: they are taken up by proliferating cells including immune cells, damaged endothelial cells and cells in the bone marrow for the following processes:

- formation of phospholipids for new membranes;

- production of local hormones (e.g. prostaglandins, leukotrienes);

- production of intracellular messengers (e.g. diacylglycerol, arachidonic acid).

In addition to the increased mobilisation of fatty acids, there is an increase in the rates of cycling in the intra- and inter-cellular triacylglycerol/fatty acid cycles that contribute to increased energy expenditure in trauma.

Control of lipolysis

The increased levels of catecholamines and glucocorticoids, the increased sympathetic activity and the decreased level of insulin increase the activity of hormone-sensitive lipase in adipose tissue that is responsible for the increased rate of lipolysis. Also important are the proinflammatory cytokines, TNFα, interleukins 1 and 6. The significance of fat as a fuel in trauma explains why the cytokines have a marked lipolytic effect on adipose tissue: the lipolytic effect of TNFα is greater than that of the endocrine hormones, particularly in adipose tissue depots in which lymph nodes are embedded. This is significant because it provides a local supply of fatty acids for immune cells within the nodes, where the rate of proliferation is very high (Chapter 17: Figure 17.41).

The adipocytes nearest to the lymph node in these depots contain more essential fatty acids than adipocytes further removed from lymph node. Thus the essential fatty acids necessary for proliferation are located as close as possible to the proliferating lymphocytes.

Loss of protein

The amount of protein lost depends on the severity of the injury, in the following order: burns > multiple trauma > simple trauma > major elective surgery.

The response in sepsis depends on the extent of infection. If severe, the loss of skeletal muscle can be large enough to impair physiological function and this increases mortality, for reasons given in Chapter 16. The loss can be as much as 150–200 g per day, despite adequate enteral or parenteral nutrition. Because this is detrimental to the recovery of the patient, there is considerable emphasis on development of treatments that could reduce this loss: provision of

glutamine has some effect but the mechanism is not known.

The breakdown of protein provides amino acids that are beneficial since:

- They are precursors for the synthesis of glucose via gluconeogenesis that is required for the increased number and activity of cells of the immune system, for those involved in repair (e.g. fibroblasts) and for the brain, especially if the patient cannot feed.

- They are required for protein synthesis for proliferating cells.

- They are required for synthesis of peptides or proteins that have a specific role in trauma: e.g. antibodies, acute-phase proteins, cytokines, glutathione.

- They are required for the synthesis of conditionally essential amino acids, e.g. arginine, glycine, cysteine and glutamine.

Control of proteolysis

Loss of muscle protein in trauma is caused by increased degradation rather than decreased synthesis. The degradation is controlled by changes in the levels of glucocorticoids, insulin and the proinflammatory cytokines TNFα and IL-1. The proteolytic enzyme complex that degrades the protein is the proteasome (Chapter 8). The mechanism by which the enzyme is activated is not known, but increased activities of the enzymes involved in ubiquitination of proteins and an increase concentration of ubiquitin may play a role (Chapter 8).

As early as 1929, David Cuthbertson reported a loss of 856 g of protein over a period of 10 days after a leg fracture. This is a similar value to those currently reported. Sadly, the work of Cuthbertson is rarely referred to or even acknowledged.

Metabolic changes in trauma and in starvation

Some metabolic changes in trauma are similar to those that occur in starvation but there are quantitative differences (Table 18.7). A comparison of the two conditions provides insight into the possible metabolic significance of such changes. (Some of the changes also occur in cancer patients; these are discussed in Chapter 21.) In both conditions, the changes are necessary for survival: in starvation, they maintain the blood glucose level; in trauma they provide fuels for the cells involved in combating an infection or in repair and recovery. The rate of gluconeogenesis is higher in trauma to provide this glucose (Figures 18.1 and 18.2).

One important difference between the two is that the loss of muscle protein continues at a high rate throughout

Table 18.7 Extent of metabolic changes in trauma and starvation

	Direction and extent of change in	
Process	Trauma	Prolonged starvation[a]
Energy		
Oxygen consumption	+++	–
Resting energy expenditure	+++	–
Fat and carbohydrate metabolism		
Primary fuels	mixed	fat
R value[b]	0.85	0.75
Gluconeogenesis	++	++
Lipolysis	+++	++
Ketone body production	++	++++
Insulin resistance	+++	+
Nitrogen metabolism		
Muscle protein loss	+++	+
Synthesis of body protein[c]	+	+
Degradation of protein in muscle	++	+
Ureagenesis[d]	+++	+
Concn of glutamine in muscle	–	–
Concn of total amino acids in muscle	–	–

– decrease; + increase.

Data presented are obtained from several sources.

[a] For details of these changes, see Chapter 16.

[b] R is the respiratory quotient (originally abbreviated to RQ). It is the ratio of the CO_2 produced to the O_2 consumed: it is unity for carbohydrate oxidation and 0.7 for fat oxidation.

[c] In the arduous crossing of Antarctica, during which both men lost more than 20 kg in body weight, high rates of protein synthesis were maintained (Streud *et al.* 1996)

[d] In prolonged starvation more ammonia than urea is excreted in order to batter the urine from the acidic ketone bodies that are present in the urine.

trauma whereas in starvation it decreases over time. In early starvation, the rate of loss of protein is similar to that in trauma (i.e. about 150 g/day), but it decreases over several weeks to about 20 g per day. This is possible because the primary role of protein degradation, in early starvation, is to provide glucose, via gluconeogenesis, from the amino acids derived from skeletal muscle, for oxidation by the brain. As starvation proceeds, the brain gradually increases its use of ketone bodies as an energy source, so that less glucose and less gluconeogenesis is required and the degradation of muscle protein can decrease. In trauma, by contrast, protein degradation provides amino acids for protein synthesis, for the synthesis of key peptides and proteins and for formation of the conditionally essential amino acids, especially glutamine.

Fever

Hippocrates wrote about fever as follows:

The worst, most protracted diseases were the continued fevers. These showed no real intermissions although they did show paroxysms in the fashion of tertian fevers, one day remitting slightly and becoming worse the next. They began mildly but continually increased, each paroxysm carrying the disease a stage further. Shivering fits and sweats were least frequent and most irregular in these patients. The extremities were chilled and could be warmed with difficulty, and insomnia was followed by coma.

And Sir William Osler commented that 'Humanity has but three great enemies: fever, famine and war. Of these, the greatest by far, the most terrible, is fever.'

The term 'hyperthermia' denotes an elevation in body temperature, regardless of its cause. The heat produced by prolonged intense exercise is due to an increased metabolic rate which raises the body temperature: a condition which is easily reversed. In contrast, the hyperthermia in trauma or infection is not readily reversed and is known as fever. The source of the heat is not known but there are four possible sources and all may contribute:

- Uncoupling of oxidative phosphorylation from respiration in brown adipose tissue and possibly other tissues (e.g. in muscle) (Chapter 9).

- Substrate cycling (e.g. the Cori cycle and the intra- and inter-cellular triacylglycerol/fatty acid cycles (Chapter 3)) in which there is no net metabolic change so that the energy from ATP hydrolysis is released as heat.

- Shivering.

- Increased work of the heart.

Body temperature is maintained by means of a balance between heat loss and heat production. These processes are controlled via the sympathetic nervous system and the effects of cytokines.

If these processes produce too much heat, the body attempts to lose heat by vasodilation within the skin (via convection) and sweating (via evaporation of the water in the sweat). Both are well-known characteristics of fever. The patient's experience of alternate shivering and sweating (so well described by Hippocrates) probably represents an impairment of the thermoregulatory centre in the hypothalamus that regulates the balance between heat loss and heat production, resulting in fluctuations in body temperature.

A scientific understanding of fever only came about once it was realised that heat is generated from metabolism. The first work on body temperature and disease was carried out by Carl Wunderlich, Professor of medicine at

Leipzig. In 1866 he published his book *The Temperature in Disease* in which he reported the body temperature in no fewer than 25 000 patients. Wunderlich commented on the 'unparalleled value of the method in giving an accurate and reliable insight into the condition of the sick': a statement that is still valid today, as body temperature is routinely used as an indicator of infection.

Causes of fever

In 1888 William Welch speculated that microbial agents produced fever indirectly by release of 'ferment', possibly

> William Welch (1850–1934) was an American pathologist and microbiologist who first demonstrated that the bacteria that live on the skin, *Staphylococcus epidermis*, could cause wound infections. He was also the first to describe the anaerobic organism which causes gas gangrene, which was named *Clostridium welchii* after him.

> An elegant study identified interleukin-1β as one cytokine that stimulates the pyrogenic response. Poxviruses possess genes for proteins that interfere with the host's immune response. One gene encodes a protein that binds IL-1β with high affinity and specificity, thus inactivating the cytokine and increasing the virulence of the virus. When mice are injected with wild type smallpox virus, which possesses this gene, the infection does not elicit a fever. However, injection with a mutant strain of the smallpox virus which lacks only this gene results in fever (Alcami & Smith, 1996).

from white cells, which then acted upon the brain. It was shown much later that white cells produce a fever-inducing factor, which was termed endogenous pyrogen. The cytokines, interleukins 1 and 6 and tumour necrosis factor, are such pyrogens and are produced by monocytes and macrophages in response to an invasion by pathogens.

The cells in the hypothalamus that control body temperature respond to the cytokines by stimulating the activity of the membrane bound phospholipase, which results in the formation of arachidonic acid, the substrate for the enzyme cyclooxygenase-2 (COX-2) which is the rate-limiting step in the pathway for synthesis of prostaglandins. Prostaglandins influence cells in the hypothalamus that are responsible for temperature regulation.

Role of fever

The role of fever is to help defend the body against bacterial infection. Evidence for the benefit of raised body temperature in helping to fight infection is as follows:

• Lizards infected with a gram-negative bacteria consistently choose 'hot spots' under a moving light, in order to achieve an increase in body temperature.

• Animals that were infected with bacteria and were maintained for a week at a higher than normal temperature showed a marked increase in survival compared with animals maintained at the normal temperature.

• In children suffering from gastroenteritis due to bacterial infection, there was a negative correlation between the magnitude of fever and the duration of the infection.

One mechanism by which fever has an antibacterial effect is that it decreases the blood concentration of iron, which is necessary for bacterial proliferation (Chapter 17). However, iron is also necessary for the proliferation of immune cells in the lymph nodes and in the bone marrow (for formation of the iron-containing proteins, haemoglobin and mitochondrial proteins). This leads to competition for iron in the

> Body temperature is lowered by drugs that inhibit the activity of the cyclooxygenase (e.g. aspirin, paracetamol) and hence decrease the formation of prostaglandins. In view of the benefit that fever may provide, it is questionable whether this is beneficial for patients in recovery from infection.

blood, as indicated in Figure 18.3. It is the cytokines that lower the Fe^{2+} concentration in blood by stimulating uptake by the liver and also by proliferating lymphocytes (Figure 18.4).

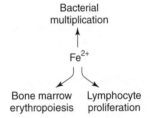

Figure 18.3 *Competition for iron between proliferating bacteria, erythropoiesis in the bone marrow and proliferating lymphocytes in the lymph nodes.* The iron ion is required for synthesis of haemoglobin, cytochromes and iron-sulphur proteins, and for maintenance of the structure of DNA.

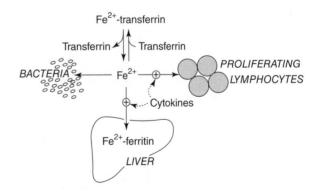

Figure 18.4 *Effects of cytokines on fate of iron during an infection.* Transferrin is the form in which iron is transported in the blood (Chapter 15). Cytokines increase the number of ferritin receptors in proliferating lymphocytes, to facilitate the uptake of iron by lymphocytes. They also stimulate synthesis of apoferritin in the liver, which removes iron from the blood to reduce that available for bacteria. (See Chapter 15 for discussion of iron metabolism)

Acute-phase proteins (heat-shock proteins)

Acute-phase proteins have long been associated with fever, as their original name, heat shock proteins, suggests. The cytokines released by macrophages in response to infection, and/or tissue damage, stimulate the synthesis and secretion of these proteins by the liver. They include:

- C-reactive protein, which binds to bacteria and activates the complement cascades. It also binds and removes host proteins which are released from damaged tissues which, if allowed to persist, could induce an autoimmune response.

> The rise in fibrinogen causes erythrocytes to form stacks (rouleaux) that sediment more rapidly than individual erythrocytes. Hence, the erythrocyte sedimentation rate is a simple test for prolonged low-level infection.

- Complement proteins, which increases the concentrations of these proteins in blood.

- Fibrinogen, which is a substrate for the clotting cascade in blood, an essential response to tissue damage.

- Caeruloplasmin, which binds Cu^{2+} ions in the circulation and at the sites of damage, to prevent them from reaching a concentration that would catalyse the formation of the very dangerous hydroxide radical, OH^{-} (Appendix 9.6).

- A protease inhibitor to prevent formation of intravascular blood clots due to activity of thrombin.

Summary of the effects of trauma on the immune system and the whole body

The role of the immune system in trauma or an invasion by pathogens is described and discussed in Chapter 17, but the response of the whole body, in addition to that of the immune system, is essential in recovery from trauma or fighting an invasion. A summary of the integration achieved via cytokines is described in Figure 18.5.

Multiple organ failure

Some patients fail to recover from trauma because they develop multiple organ failure (MOF), also known as multiple systems failure (MSF). It is estimated to kill almost 200 000 patients each year in the USA. The name arises from the fact that several organs are involved and they are distant from the site of trauma. Since it can arise in septic patients, it is also known as 'septic shock syndrome'. It can occur in any condition that induces a major inflammatory response. The time between the insult and the malfunction of the organs is variable and may run into weeks. As expected, the larger the number of organs involved, the greater the risk of death (Box 18.1).

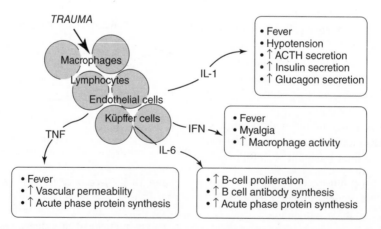

Figure 18.5 *A summary of the biochemical, physiological and immunological changes brought about by cytokines in response to trauma.* Cytokines can be produced in trauma from macrophages, lymphocytes, endothelial cells in the tissue that is damaged, and also by Küpffer cells if the liver is damaged. IL-1, IL-6 – interleukins 1 and 6; TNF – tumour necrosis factor, IFN – interferon.

Box 18.1 Wars and trauma

The treatment of casualties in major wars has provided unparalleled opportunities to study the clinical problems of major trauma and its treatment. Indeed, the history of trauma management is mainly the history of military surgery. As one problem was identified, and successfully treated, further problems were identified and subsequently overcome. In the First World War many casualties died from irreversible 'wound shock', because doctors at the time did not appreciate the relationship between shock and the loss of a large amount of blood. Between the two world wars, this cause of shock was identified, so that in the Second World War infusion of blood to replace that lost was a basic treatment to reduce deaths from shock caused by wounds. Despite this, many of the injured went on to die of acute renal failure, until subsequent studies established the importance of maintaining the extra-vascular volume for adequate perfusion of the kidney. Patients were therefore infused with solutions of electrolytes and glucose to prevent renal failure. However, as the patients survived longer, a new problem arose: post-traumatic pulmonary insufficiency, identical to adult respiratory distress syndrome (ARDS), and impaired gas exchange. Thus, in the 1970s the lungs became the organ limiting survival and assisted ventilation was introduced. However, infection can still lead to multiple organ failure and death.

A description of the trauma caused by war wounds suffered by a soldier injured in the American Civil War is to be found in *Walt Whitman's Civil War* (Lowenfels, 1989):

I dress a wound in the side, deep, deep,
But a day or two more, for see the frame all wasted
and sinking,
And the yellow-blue countenance see.
I dress the perforated shoulder, the foot with the
bullet-wound,
Cleanse the one with a gnawing and putrid gangrene,
so sickening,
so offensive, . . .

In the American Civil War and the First World War estimates of mortality from penetrating wounds were more than 80%. Many deaths were as the result of infection, with gangrene being a particular problem. The penicillin mould was identified in 1928 but it was only in the early 1940s that optimal culture conditions and a purification procedure to produce pure penicillin were established. Large-scale production then commenced in the USA, where adequate facilities and financial investment were available. The objective was to produce as much penicillin as possible to treat injured troops with open wounds so that they could recover as quickly as possible during the invasion of France and subsequent liberation of Europe in 1944–45.

It is claimed that his experience in treating infected soldiers in World War I led Alexander Fleming to follow up his chance observation that a mould restricted growth of bacteria in a culture dish in his laboratory.

Causes of damage to organs

There are several hypotheses to explain MOF: all are based on an acute overactivity of the immune system. In trauma or infection, the macrophage plays a central role in coordinating the responses not only of the immune system but of the whole body. It does this by secretion of the proinflammatory cytokines (e.g. tumour necrosis factor, various interleukins). The concentration of these cytokines in the blood is regulated not only by the rate of their secretion (and degradation) but also by adsorption onto cytokine-binding proteins or soluble cytokine receptors, which are present in blood and are known as anti-inflammatory cytokines (Chapter 17). If the levels of the proinflammatory cytokines overwhelm the anti-inflammatory cytokines, then damage can occur to organs that are far removed from the site of trauma or infection. Once the damage has occurred, treatment of the initial cause (e.g. recovery from the trauma or infection) does not halt or reverse MOF. An excessive inflammatory response to trauma, therefore, can result in permanent impairment of organ function. It is not known how cytokines damage the organs, but three proposals have been put forward:

(i) Damage to vascular endothelium Excessive levels of cytokines in the circulation cause activation of endothelial cells in some organs. This results in the expression of adhesion molecules on their surface, to which monocytes, macrophages and lymphocytes adhere. Here they release growth factors, chemotactic factors and free radicals that damage the endothelial cells, one result of which is platelet adhesion followed by local thrombosis. If this obstructs blood flow in the organ, further damage results and acts as a focus for inflammation, which attracts more phagocytic cells. The process is then a vicious circle, similar to that which occurs in the development of atherosclerosis (Chapter 22). (A rather cumbersome medical term is used to describe this phenomenon: disseminated intravascular coagulation, abbreviated to DIC.)

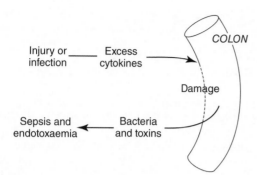

Figure 18.6 *Damage to the colon caused by excessive levels of* proinflammatory *cytokines and its consequences.* The excessive levels of proinflammatory cytokines that result from severe trauma can damage the physical barrier of the colon, so that bacteria and toxins in the colon leak into the peritoneal cavity, which can cause sepsis and/or endotoxaemia.

(ii) Reperfusion damage to vascular endothelium   Trauma or sepsis results in shock with hypotension, leading to poor perfusion of one or more organs. Alternatively, distribution of the blood may be affected, so that some organs are overperfused but others are underperfused. Poor perfusion results in hypoxia which itself may not be severe enough to cause permanent damage but, when perfusion improves, free radicals are formed that damage endothelial cells and the microvascular system. This phenomenon is known as reperfusion damage, that is, marked reperfusion damage is the cause of MOF.

(iii) Bacterial translocation from colon due to hypoxic damage   Reperfusion damage to the colonocytes can impair the physical barrier between the contents of the colon and the blood. This leads to translocation of bacteria (some of which are pathogens (e.g. *E. coli*)), or lipopolysaccharide (endotoxin) into the peritoneal cavity and eventually into the bloodstream. Activation of the gut-associated immune system results in local inflammation which can also damage the barrier and hence increase bacterial translocation. This sets up a vicious circle, with the development of peritonitis and possible systemic sepsis (Figure 18.6).

Treatment

So far there are no satisfactory treatments for multiple organ failure, although steroids are sometimes used to inhibit the activity of the immune system. Provision of antibodies to one or more of the proinflammatory cytokines is one novel approach.

19
Sexual Reproduction

The problems and the pathology arising from failure of the biochemistry, endocrinology or physiology in reproduction are not always given the same emphasis in medical schools as that of, for example, cancer or coronary artery disease; yet in the number affected, it is perhaps of greater pathological significance and the misery it causes can be as severe and debilitating.

(Masters *et al.*, 1995)

Reshuffling the genetic message is at the heart of sexual reproduction.

(Jones, 1993)

The essential feature of sexual reproduction is that genetic information from two individuals is combined to determine the characteristics of their offspring. In fertilisation, one cell (the ovum), which contains only one half of the normal number of chromosomes, receives chromosomes from another cell, the sperm, that also contains half the normal number of chromosomes. The result is a cell that then contains the normal number of chromosomes. The process that makes this possible is meiosis (Chapter 20) which not only results in a halving of the number of chromosomes, but almost randomly determines which characteristics are passed on to the offspring.

Much of the behaviour of humans and society as a whole is designed to provide an opportunity for a sperm and an ovum to meet in the uterus. However, sexual reproduction is not the only means of reproduction; asexual reproduction occurs frequently in the animal kingdom and in many tissues in order to increase the number of cells during growth or to maintain the number due to cell death. The question arises, therefore, why sexual reproduction has arisen at all since it is very expensive in the use of resources: for example, in the process of selecting a suitable mate, and in the requirement for males, whose only function in some species is fertilisation of the females, yet they use resources that could be used by the females and by the offspring. Perhaps because of this, a few species of vertebrates (e.g. some lizards) have retained the option for asexual reproduction. This process is known as parthenogenesis. Mammals, however, have lost this option. This enables them to 'shuffle the genes' to provide genetic variability in a population. This results in sufficient variability in the immune system in a population that some individuals will be able to produce an immune response to any genetic change in a pathogen or any new pathogen that may appear (Chapter 17). Hence, the population as a whole will survive.

Male reproductive system

The male reproductive system includes the two testes, ducts that store and transport the sperm to the exterior, and the glands that secrete into these ducts and the penis; it also includes the hormones that integrate the reproductive activities. Basic details of the anatomy of the male reproductive system are given in Figures 19.1 and 19.2.

Testes

The testes are suspended outside the abdomen, in the scrotum, which is divided into two sacs, one for each testis. This location outside the abdominal cavity helps to maintain the testes at a temperature below normal (about 6 °C lower) which is essential for satisfactory spermatogenesis. Indeed, the arterial and venous supply is so organised that it provides a heat exchange mechanism in which the arterial blood is cooled and the venous blood is warmed. The

Functional Biochemistry in Health and Disease by Eric Newsholme and Tony Leech
© 2010 John Wiley & Sons Ltd

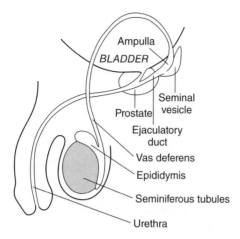

Figure 19.1 *A diagrammatic representation of the male reproductive tract.* Much of the volume of the testes consists of convoluted seminiferous tubules in which the spermatozoa form. In the interstitial tissue that surrounds the seminiferous tubules are the Leydig cells which produce and secrete androgens, oestradiol and the peptides inhibin and activin. The epididymis is a single but convoluted tube. Sperm from the epididymis enter the vas deferens and pass through the ejaculatory duct into the urethra, mainly at the time of ejaculation. Just at the transition of the vas deferens to ejaculatory duct, two large glands, the seminal vesicles, drain into the two vasa deferentia. Prior to joining the urethra, the ejaculatory ducts pass through the prostate gland which lies below the bladder and surrounds the upper part of the urethra, into which prostatic fluid is secreted.

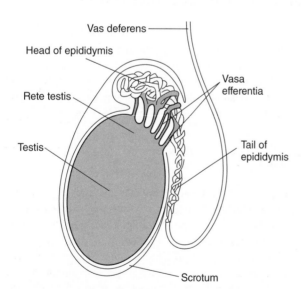

Figure 19.2 *A diagram of a section through a human testis to show the general structure.* The rete testis is a small structured component (not shown) that joins the seminiferous tubules to the vasa efferentia and hence to the epididymis, where the sperm are stored.

testes have two major products, spermatozoa and hormones. These are produced in two discrete compartments: spermatozoa are produced and develop, in the seminiferous tubules, in close association with Sertoli cells, while the hormones are synthesised and secreted by Leydig cells. These are also known as interstitial cells, which are situated between the seminiferous tubules.

Spermatogenesis

During embryonic development, the cells destined to become spermatozoa, i.e. the germ cells, divide to form spermatogonia, of which there are many thousands in each testis. Spermatogonia are, in fact, stem cells. Spermatogenesis begins at puberty and continues throughout life. It is divided, arbitrarily, into three parts:

- Proliferation of the spermatogonia by mitosis to produce large numbers of primary spermatocytes.

- Proliferation of spermatocytes by meiosis, in which the number of chromosomes is halved. The process of meiosis is discussed in Chapter 20, where it can be compared with mitosis (Figure 20.29). The cells are now called spermatids.

> Vitamin A is essential for spermatogenesis: the retinoic acid may be required in maturation (i.e. differentiation) of spermatid into spermatozoa. This is important in male fertility.

- The spermatids develop to produce spermatozoa (a process described as maturation).

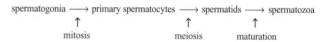

This process is complete in about 60 days. With the large number of mitotic divisions, each primary spermatozyte is one sibling in a large clonal family. During meiosis a process known as 'crossing over' occurs which is an almost random physical exchange of alleles between homologous chromosomes (i.e. the genes are shuffled). The result is that each spermatozoon is genetically unique although they all have a common ancestral cell. Hence any offspring resulting from fertilisation are also genetically unique.

> Adjacent alleles are more likely to end up in the same chromosome, during meiosis, than those which are far apart. This is known as linkage, which, therefore, can be used as a means of estimating the distance between them.

Sertoli cells

The seminiferous tubules are lined by large cells, the Sertoli cells, which surround and almost enclose the developing spermatozoa: in fact, it can be considered that much of the development of a spermatozoon occurs within a Sertoli cell. The functions of the Sertoli cells are:

- To provide a barrier between the tubules and the blood: the blood–testes barrier.

- To provide the nutrients for proliferation of the spermatogonia and spermatocytes and maturation of the sperm. These are essential amino acids, glutamine and glucose as fuels and also as precursors for formation of nucleotides; essential fatty acids for formation of phospholipids (for discussion of these topics, see Chapters 20 and 21). It is presumed that Sertoli cells extract these compounds from the blood.

- To convert testosterone into the more active androgen, dihydrotestosterone (5α-dihydrotestosterone), in a reaction catalysed by the enzyme 5α-reductase.

- To secrete a sex-hormone binding protein (SHBP) which binds the sex hormones in the blood to protect them from metabolism and hence inactivation.

- To respond to follicle-stimulating hormone (FSH) and testosterone, which stimulate the production of local hormones that increase maturation of the sperm.

- To phagocytose defective sperm.

The blood–testes barrier separates two parts of the seminiferous tubules, the part in which the spermatozoa are produced from the outer part which provides the blood supply. It has two functions: (i) it prevents spermatozoa leaking into the blood or lymph, since proteins on the surface could act as antigens; (ii) it maintains the distinct composition of fluid inside the tubules, which is necessary for spermatogenesis.

The tubes in which the spermatozoa are produced, stored and eventually released are depicted in Figures 19.1 and 19.2.

The journey for spermatozoa from seminal vesicles to ejaculation, and then to the oviduct for fertilisation

The journey that each spermatozoon must take prior to ejaculation is presented in a linear manner in Figure 19.3. This emphasises that the journey is long. It is approximately 20 cm, but the spermatozoon contributes little, if at all, to the transport, which is achieved by pressure of fluid and actions of cilia lining, some of the tubes through which it passes. A cell about 4 μm long must be transported in a fluid environment containing many other identical cells, for a distance, almost 50,000 times longer than itself. Yet it must be structurally and biochemically able to travel a

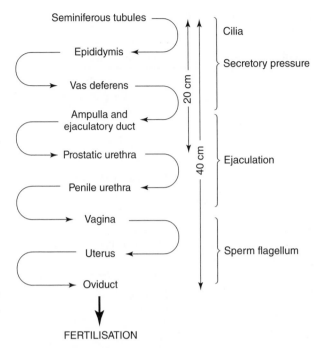

Figure 19.3 *The lifetime journey of a spermatozoon.* The journey of the spermatozoa starts in the seminiferous tubules of the male and finishes in the oviduct of the female. For ejaculation they must travel from storage in the epididymis and vas deferens to the penile urethra, a distance of about 20 cm. The distance travelled in the female reproductive system is also about 20 cm.

similar distance though its own effort after ejaculation (Figure 19.3) and in a relatively hostile enviroment (see below).

Semen

Semen is a suspension of spermatozoa in seminal fluid, the volume of which in a normal human ejaculation varies between 2 and 10 mL. There are about 20 million sperm per mL of semen, although they comprise only a small percentage of the volume. Seminal fluid arises from secretions of accessory glands. The prostate gland and the seminal vesicles secrete the bulk of the fluid. The sources, contents and functions of the secretions from the accessory glands are given in Table 19.1. It is only during ejaculation that sperm are mixed with seminal fluid. The pH of vaginal fluid is normally about

Antonie van Leeuwenhoek, a Dutch lens maker, devoted his life to studies with the microscope that he had constructed. In this work, he discovered, in about 1680, motile sperm in semen. He believed, because of their motility and presumably their shape, that they were animal parasites within the semen. Consequently, under this misapprehension, he called them *animals in semen*, i.e. spermatozoa (another example of a name which, albeit incorrect, is still used).

Table 19.1 Compounds present in the secretions from various glands associated with the male reproductive system and their functions

Gland	Secretion	Some suggested functions
Testis	Testosterone/androstenediol	Male steroid hormones
	Lactate dehydrogenase	Not known
	Oestradiol	Possibly encourages contraction of smooth muscle in walls of the vagina and uterus
Epididymis	Carnitine	Facilitates acetyl-CoA oxidation by spermatozoa (Chapter 9)
	Inositol	Precursor for formation of phosphatidylinositol bisphosphate
	Phosphatidylcholine	Buffer to maintain pH and a source of choline
	Cholesterol	Stabilises membranes
	Glycoproteins	They coat the surface of the sperm to protect against IgA
Seminal vesicles	Fructose	Fuel for sperm
	Prostaglandins	Not known but may stimulate contractions of smotth muscle in walls of vagina and uterus
	Potassium ions	Required by glycolytic enzymes
Prostate	Citric acid	Oxidation generates ATP and removes protons (Chapter 6)
	Acid phosphatase	Hydrolyses phosphatidylcholine
	Spermidine	Not known
	Zinc ions	Helps to stabilise structure of DNA (see Table 15.4)
	Calcium ions	Stimulate coagulation of semen in vagina
	Proteolytic enzymes	Gradual breakdown of coagulum to release sperm
Bulbourethral gland	Mucoproteins	Lubrication
	IgA antibodies	Bacteriocidal

Acid phosphatase retains its activity for a long period and hence is useful in forensic science to detect semen but has now been superseded by DNA fingerprinting. The activity in blood was used in the diagnosis of prostatic cancer but was superseded by PSA (prostate specific antigen).

LH – luteinising hormone; FSH – follicle-stimulating hormone.

The concentration of fructose in semen is about 10 mMol/L.

It is not known if any of these constituents of semen are important not for spermatozoa but for proliferation of the zygote after fertilisation (e.g. cholesterol, inositol, zinc ions).

4.5, which was considered to be largely due to lactic acid, produced by the mucosal cells of the vagina, presumably from blood glucose or stored glycogen. However, short chain fatty acids may also contribute. This low pH protects against infection of the vagina by pathogens. It is, however, one of the several hazards faced by spermatozoa in their journey along the vagina, since at this low pH, motility of spermatozoa is low, so that semen contains buffers for the hydrogen ions (Table 19.2).

Fructose is the fuel provided in the semen. The advantage of this is that it is transported across the plasma membrane of the sperm by the transporter protein that is specific for fructose, the GLUT-5 transporter. Since the cells of the vagina and the uterus do not possess this transporter, fructose uptake and oxidation is restricted to the spermatozoa.

Fructose is converted to fructose 6-phosphate by hexokinase, which phosphorylates both glucose and fructose. It then is converted to pyruvate via glycolysis, which is converted to acetyl-coenzyme-A in the mitochondria for oxidation by the Krebs cycle. The enzyme that is specific for fructose metabolism, fructokinase, has not been found in spermatozoa. However, there are no reports of studies on the maximum activities of hexokinase and oxoglutarate dehydrogenase to indicate the maximum capacities of glycolysis and the Krebs cycle in human spermatozoa.

Spermatozoa

The spermatozoon consists of a head, a midpiece and a tail or flagellum. The head contains the nucleus and a vesicle known as the acrosome (Figure 19.4). It contains hydrolytic enzymes that are required during fertilisation of the ovum.

The midpiece contains the mitochondria which are wrapped around the proximal part of the flagellum. The beating of the flagellum, and hence the swimming of the sperm involves the motor protein known as dynein, which requires ATP hydrolysis. In some species, the diffusion of 'energy' in the spermatozoa is increased by the presence of the creatine/phosphocreatine shuttle (Chapter 9); that is, phosphocreatine and creatine diffuse throughout the cytosol

Table 19.2 Some biochemical hazards faced by spermatozoa during their journey from the vagina to the oviduct

1 The spermatozoa have to swim approximately 20 cm to reach the site of fertilisation.

2 They must remain mobile for a considerable period of time, many hours, in the uterus and oviduct to increase the chance of meeting an ovum after ovulation. Spermatozoa can survive in the female genital tract 30 to 50 hours.

3 Sufficient fructose must be present in the semen to generate enough ATP to keep the cell alive, and power the swimming. On the basis of the relative size of a spermatozoon and a human, this distance is equivalent to 0 km (approximately a full marathon).

4 Sufficient oxygen must be available within the three tubes for complete oxidation of fructose, to provide for oxidative ATP generation, the most efficient process (Chapter 9).

5 The spermatozoa must swim through a viscous fluid (mucus) so that the 'drag' will be significant.

6 Antibodies are present in the vagina and the sperm is a foreign body. Since spermatozoa are not present at the period of clonal deletion, they possess proteins that are antigenic (Chapter 11). They must either survive an immunological attack or are antigenically 'silent'. Coating with glycoprotein may protect against the antibodies. Whether any of these factors may account for infertility of men is not known: no studies have been reported.

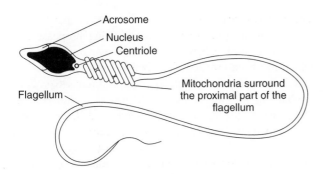

Figure 19.4 *The structure of a spermatozoon.*

from the mitochondria to the flagellum and back (instead of ATP and ADP, respectively). The shuttle provides a rapid supply of ATP for the dynein, along the whole length of the flagellum. Phosphocreatine may also provide a greater burst of energy for the final action of the spermatozoon, that is, penetration of the membrane of the ovum at fertilisation (an 'activity' similar to sprinting in human skeletal muscle (Chapter 9).

Female reproductive system

The female reproductive system includes the two ovaries, two oviducts (also known as Fallopian tubes) a uterus and a vagina (Figure 19.5).

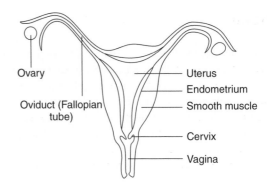

Figure 19.5 *A diagrammatic representation of the female genital tract.*

Ovary

The ovary has two functions:

(i) the production of ova (oogenesis) and their release (ovulation);

(ii) the synthesis and secretion of the steroid hormones, oestradiol and progesterone.

Provided these two processes are in phase, the conditions are then optimal for fertilisation, proliferation of the zygote to form the blastocyst, implantation of the latter and maintenance of pregnancy. Before ovulation, the development of the ovum and hormone secretion take place in a single structure, the follicle.

> The term 'zygote' refers to the immediate result of fertilisation, i.e. one cell, whereas the 'embryo' contains more than one cell. The term 'foetus' refers specifically to mammalian development but has more of a legal meaning than a biological one. It is defined as the embryo from the end of the eighth week after fertilisation. Nonetheless, the term is still used in biology and biochemistry and is used in this text to describe the developing embryo (also known as the conceptus).

Oviducts

The oviducts are also known as the Fallopian tubes or uterine tubes. The wall of the oviduct consists of three layers: a mucosa, a muscular layer and an outer layer of connective tissue. When the ovum is released by the ovary, it is taken up by the oviduct. The ends of the oviducts are not directly attached to the ovaries but open into the abdominal cavity close to them. The opening of each is funnel-shaped and surrounded by long finger-like projections, fimbriae, with ciliated epithelium which 'catch' the ovum as it is released. The other ends open directly into the uterus. The lower end of the latter is known as the cervix and opens into the vagina.

Uterus

In the human, the uterus lies like an inverted pear in the pelvis. The wall of the uterus consists of three layers: a thin outer layer, the perimetrium; a middle muscular layer, the myometrium; and an inner glandular mucosa, the endometrium, which lines the uterine cavity. Once the ovum is fertilised, the zygote undergoes several cell cycles (Chapter 20) to form a rounded mass of cells, the blastocyst, which passes into the uterus where it implants in the wall (see below).

Vagina

The vagina is a fibromuscular tube that connects the uterus to the exterior of the body. The wall of the vagina is similar to that of the other tubes in the female genital tract, consisting of three layers: a mucosal layer consisting of epithelial cells that line the cavity of the vagina; a muscular layer; and the outer adventitia, a connective tissue layer. The latter contains elastic fibres, and blood vessels that provide nutrients and oxygen to the inner layers. The responses of all three layers during coitus play an important role in providing conditions for motility of the sperm (see below).

The menstrual cycle

Unlike the production and release of spermatozoa, only one ovum develops and is released from the ovary approximately every 28 days. Hence, in the female, the reproductive system has regular cyclic changes. This is necessary since the endometrium can only provide optimal conditions for implantation of the blastocyst for a short period (no more than seven days). After this period, the blood supply is shut off and the result is necrosis (Chapter 20). This leads to shedding of the endometrium, with loss of fragments of tissue, some cells and blood. This is known as menstruation, with the cycle known as the menstrual cycle (Figure 19.6). By common usage, the cycle starts with the first day of menstruation and the length in women is about 28 days. The cycle can be divided into two approximately equal halves, which are separated by ovulation. The half before ovulation is known as the follicular phase, whereas the half after is known as the luteal phase. The hormones that control the changes that occur in the former phase are the oestrogens (oestradiol, oestriol and oestrone) and that in the luteal phase is progesterone. Hence the cycle can be summarised as oestrogen–ovulation–progesterone–menstruation.

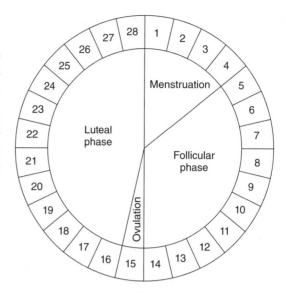

Figure 19.6 *A 28-day menstrual cycle.* The approximate number of days for menstruation, ovulation and the follicular and luteal phases are shown.

Meiosis in the oocyte

The first cell division of meiosis occurs in the primary oocyte but the process is arrested during prophase and remains so until puberty. Just before ovulation, meiosis, which has been arrested since before birth, resumes. The first division halves the number of chromosomes to produce the haploid secondary oocyte. The process is the same as that in spermatozoa (Chapter 20: see Figure 20.29) except that the two resulting haploid daughter cells are unequal in size. One is the large functional secondary oocyte whereas the other is much smaller and is known as the first polar body. The second meiotic division is arrested at metaphase. It is completed only at fertilisation. Once again, the division is unequal. One cell is large, the secondary oocyte. The other is small, a second polar body, which is discarded.

The follicular phase

A follicle is a sac that contains extracellular fluid, cells and the ovum. The name comes from the Latin for a little bag, *folliculus*. At birth an ovary contains several million follicles, each of which contains an immature ovum (the primary oocytes). These are known as

At the end of the reproductive life of the female, very few immature follicles are left. As this stage is approached, the frequency of cycles is reduced and they become irregular. This state, known as the climacteric, may last for several years until, at the menopause, the cycles, and hence ovulation, cease completely.

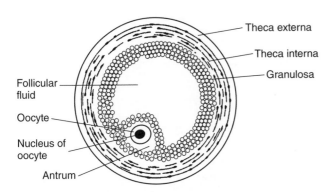

Figure 19.7 *A diagram of a Graafian follicle.* In the rim of cytosol around the nucleus are spherical mitochondria, ribosomes and the Golgi. A mature follicle is relatively large, about 1 cm in diameter.

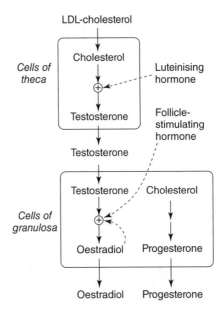

Figure 19.8 *A brief summary of the pathways for formation and secretion of oestradiol and progesterone within the cells of the follicle.* Cholesterol is taken up by thecal cells in a complex with low density lipoprotein. In the thecal cells, cholesterol is converted to testosterone which is released to be taken up by granulosa cells where it is converted into oestradiol. For synthesis of progesterone in the granulosa cells, cholesterol is synthesised *de novo* within the cells from acetyl-CoA. In the follicle the enzyme aromatase, which produces the aromatic ring in the female sex hormones, is restricted to the granulosa cells. The reactions that are stimulated by LH and FSH increase synthesis and, therefore, secretion of testosterone and increased synthesis of oestrogens and progesterone.

primordial follicles. Of these, only about 400 develop into follicles that can ovulate. Of those that ovulate, only a few will be fertilised, and of those only a few will implant and even fewer will develop into a foetus.

During each cycle one, and only one, follicle develops and at mid-cycle releases its ovum (about day 14). This mature follicle is called the Graafian follicle (Figure 19.7).

> After the 17th-century Dutch microscopist, Reynier de Graaf, who first described them.

The follicular phase covers the development of the follicle, which involves an increase in its size due to an increase in follicular fluid, growth of the ovum (i.e. an increase in the contents of RNA and protein) and an increase in the number of cells that surround the ovum. These cells are of two types, the granulosa and the thecal cells. The role of these cells is to synthesise and secrete the steroid hormones oestrogens (mainly oestradiol). The precursor molecule for their synthesis is cholesterol. There is a 'division of labour' between these cells: the thecal cells convert cholesterol into the male sex hormones androstenedione and testosterone, which are released into the blood to be taken up by the granulosa cells where they are converted to the oestrogens (Figure 19.8). For details of pathways, see Appendix 19.1.

The advantage of this division of labour is not obvious but one possibility is that secretion of testosterone by the cells not only provides a precursor for oestradiol but provides a hormone that is secreted into the bloodstream to influence, perhaps surprisingly, sexual activity in women, particularly sexual arousal and interest.

Ovulation

Just prior to ovulation, the cells surrounding the ovum die, probably due to apoptosis, and the ovum bulges out from the follicle wall. The remaining granulosa cells secrete proteolytic enzymes that digest the follicle and peritoneal walls and the ovum 'escapes' into the peritoneal cavity where it is collected by the oviduct.

The luteal phase

After ovulation, the follicle left behind 'collapses' and is transformed into the corpus luteum. In the latter, the granulosa cells hypertrophy and there is a marked increase in the blood supply to these cells. The hormone that is secreted by the corpus luteum is progesterone. It is synthesised from cholesterol within the granulosa cells. On about day 24 of the cycle the corpus luteum begins to regress which after several days results in menstruation. However, if fertilisation occurs, the corpus luteum does not regress. It persists to maintain the correct endocrine balance to support the initial period of pregnancy prior to the longer-term changes in hormones that maintain this condition (see below).

Chemical communication in male and female reproduction

Different chemicals provide communication between tissues/organs of the reproductive system. They are classified into three groups: steroids, peptides and fatty messengers. They can also be classified as endocrine or local hormones, some of which are discussed in Chapter 12.

Steroid hormones These are the oestrogens (mainly oestradiol in women) progesterone, testosterone and 5α-dihydrotestosterone.

Peptide hormones and growth factor Endocrine peptide hormones that are involved in reproduction are gonadotrophin-releasing hormones, gonadotrophins (follicle-stimulating hormone and luteinising hormone) and chorionic gonadotrophin. There are also many growth factors that play a role in reproduction but all except activin and inhibin are discussed in other chapters.

Fatty messengers Fatty messengers are discussed in Chapter 11. Those involved in sexual reproduction are the eicosanoids. The prostaglandins that are involved in reproduction are particularly PGE_2 and $PGF_{2\alpha}$.

Chemical communication provides for a widespread interaction between many cells, tissues or organs of the reproductive system. A general overview is provided in Figure 19.9.

The effects of the reproductive steroid hormones are discussed below.

These hormones play the major role in regulation of all the processes described above. Although it is complex, the student may find it easier to understand the endocrine control when it is appreciated that the endocrinology of the male and female is similar (Figures 19.10 and 19.11). The endocrine organs that play the central role in both sexes are the hypothalamus and the pituitary: they control the secretions of the steroid hormones by the ovary and testes. The hypothalamus secretes gonadotrophin-releasing hormone (GnRH) which is transported to the anterior pituitary. This transport is direct via a portal blood system which is a specific link between the two glands (Chapter 12). This ensures that the main hormone that controls and integrates the hormonal control of reproduction is not diluted in the peripheral circulation prior to its stimulatory effect on the pituitary. As its name suggests, GnRH controls the secretion of the gonadotrophins, which are the follicle-stimulatory hormone (FSH) and luteinising hormone (LH) (Figures 19.10 and 19.11).

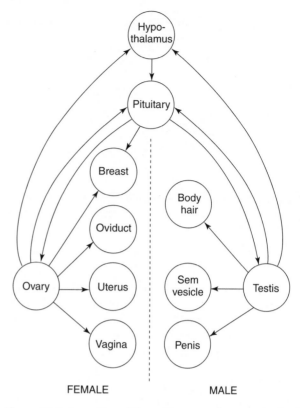

Figure 19.9 *An outline of the main routes of communication by endocrine and local hormones within the reproductive system.* Note the communication can occur in both directions. The hypothalamus secretes gonadotrophin releasing hormone (GnRH). This stimulates secretions of LH and FSH by the anterior pituitary. The secretions of GnRH and FSH and LH occur in pulses (i.e. oscillations in secretion). In the male, LH stimulates secretion of testosterone by the Leydig cells. FSH and testosterone both act on the Sertoli cells to maintain the process of spermatogenesis. The androgens stimulate development of the external genitalia, the distribution of body hair and an increase in muscle bulk in the male. Oestrogens are involved in development of breasts at puberty and have marked effects on the oviduct and vagina, changes which facilitate movement of spermatozoa. They affect development of the ovum and cause changes in the uterus during the menstrual cycle. Oestrogen and testosterone also act as feedback regulators of hormone secretions by the hypothalamus and pituitary. Oestrogen is additionally important in stimulating the LH surge during the cycle which stimulates ovulation. Two regulatory peptides, inhibin A and B, are secreted by the ovary. Inhibin B 'assists' oestrogen and progesterone in feedback inhibition of secretions of gonadotrophin by the pituitary, whereas inhibin A 'assists' oestrogen in follicular development during the menstrual cycle.

Feedback regulation

The control of the levels of the sex hormones, oestrogen, progesterone and testosterone, is achieved by feedback inhibition of the secretions of both the pituitary and hypo-

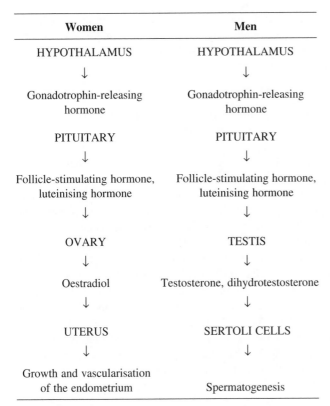

Women	Men
HYPOTHALAMUS	HYPOTHALAMUS
↓	↓
Gonadotrophin-releasing hormone	Gonadotrophin-releasing hormone
PITUITARY	PITUITARY
↓	↓
Follicle-stimulating hormone, luteinising hormone	Follicle-stimulating hormone, luteinising hormone
↓	↓
OVARY	TESTIS
↓	↓
Oestradiol	Testosterone, dihydrotestosterone
↓	↓
UTERUS	SERTOLI CELLS
↓	↓
Growth and vascularisation of the endometrium	Spermatogenesis

Figure 19.10 *Summary of the endocrine glands, the hormones that they secrete and the tissues they affect in men and women.*

thalamus. This is an additional means of communication. The secretion of GnRH by the hypothalamus and the secretions of FSH and LH by the anterior pituitary are inhibited by androgens in men and by oestradiol and progesterone in women. It is the plasma levels of these hormones which provide for the feedback.

Hormonal control of ovulation

Ovulation is essential for fertilisation and hence for reproduction. Not surprisingly, therefore, control of ovulation involves a complex endocrine interplay.

As the blood concentrations of oestradiol increase towards the end of the follicular phase, the effect of it on the pituitary changes completely. Normally, as indicated above, oestradiol has a negative feedback effect on the secretion of the gonadotrophins but, at a particular concentration, the effect changes to positive feedback. In combination with increased secretion of GnRH from the hypothalamus and increased sensitivity of the pituitary to GnRH, this positive feedback effect of oestradiol results in a marked increase in the rate of secretion of LH, and therefore in the plasma level. This is known as the mid-cycle surge of LH (Figure 19.12). The latter initiates a series of events in the follicle, which results in rupture, release of the ovum and formation of the corpus luteum. The biochemical mechanism by which a negative feedback can be transformed into positive feedback is not known.

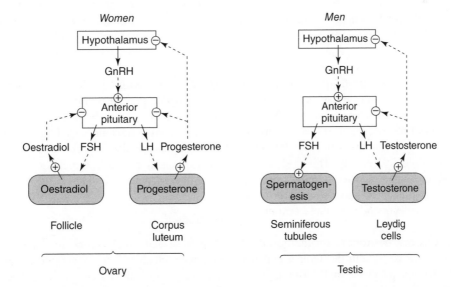

Figure 19.11 *Hormones secreted by the hypothalamus, anterior pituitary, ovary and testis and feedback regulation.* GnRH is gonadotrophin-releasing hormone; the gonadotrophins are follicle-stimulating hormone (FSH) and luteinising hormone (LH). The effect of these hormones on activities in the ovary and testes is shown. FSH stimulates synthesis and secretion of oestradiol from follicle, and spermatogenesis in testis. LH stimulates synthesis and secretion of progesterone from corpus luteum and synthesis and secretion of testosterone by the Leydig cells.

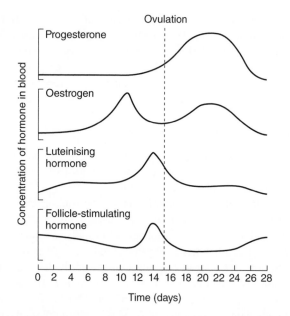

Figure 19.12 *Representation of changes in hormone levels during the menstrual cycle.* Note that LH peaks about one day before ovulation. Oestrogen has two peaks: one in the luteal phase prior to LH surge and a smaller one in the follicular phase. Progesterone peaks in the follicular phase to stimulate development of endometrium.

Effects of the reproductive steroid hormones

An account of the principles which help to understand how hormones achieve their roles in the body is given in Chapter 12. The understanding is based on separation of the effects of hormones into three components: the action, the effects (biochemical and physiological) and the function. A steroid hormone binds to a cytosolic intracellular receptor, which then moves into the nucleus where it binds to DNA at a specific site (the steroid response element) and activates genes which result in the formation of proteins that elicit biochemical and physiological effects. This is discussed for cortisol in Chapter 12 and aldosterone in Chapter 22. Much of the interest in the reproductive steroid hormones is in the physiological effects and how these account for their functions.

A long-held view is that oestrogens are the hormones that control reproductive activity in the female and that androgens control the activity in the male. Although this is broadly the case, such a complete separation is too simple. For example, testosterone is secreted by the thecal cells in the ovary and although some is taken up by the granulosa cells for conversion to an oestrogen, some remains in the blood, where it has several roles (e.g. stimu-lation and maintenance of sexual drive in the female). In men, oestrogens play an essential role in transport of spermatozoa within the reproductive tract.

Androgens: testosterone and 5α-dihydrotestosterone

Effects in the embryo Testosterone makes an early appearance in embryonic life. A gene on the Y-chromosome, called the SRY gene, forms a transcription factor that activates genes that are involved in the development of a testis which, even at this early stage, contains both Leydig and Sertoli cells. The Leydig cells secrete testosterone whereas the Sertoli cells secrete a protein, the Müllerian inhibiting hormone (MIH), also known as anti-Müllerian hormone (AMH). This hormone inhibits the formation of the female internal and external genitalia so determines that the individual become male. In the absence of MIH, the Müllerian ducts develop into the Fallopian tubes, uterus and the vagina. Testosterone maintains the Wolffian ducts which develop into the epididymis, vas deterens and the seminal vesicles. If testosterone secretion fails, the Wolffian duct system regresses and the organs fail to develop. Testosterone, however, has no influence on the Müllerian duct system; this is affected only by MIH.

Effects at puberty Androgens are responsible for the increase in size of the external genitalia, the increased secretory capacity of the seminal vesicles, prostate and bulbo-urethral glands, development of sexual drive, stimulation of spermatogenesis, deepening of the voice by increasing the size of the vocal cords, changes in the distribution and amount of body hair, increased secretion of the sebaceous glands, an increase in the amount of muscle and bone and more aggressive behaviour (Boxes 19.1 and 19.2).

Effects in the adult In the adult, testosterone maintains spermatogenesis. It also influences sexual interest, arousal and behaviour. Nonetheless, although testosterone plays a role in sexual behaviour, social, environmental and emotional factors are also important. Indeed, neither testosterone in the male nor oestradiol in the female is essential for sexual interactions in humans.

Oestrogens and progesterone

Oestradiol and progesterone regulate the structural and functional changes in oviducts, uterus, cervix and vagina that occur during the menstrual cycle. They provide conditions in the oviduct for the upward motility of sperm, and the downward movement of ova, and also conditions favourable for fertilisation in the oviduct and implantation in the uterus. Another effect is to stimulate vaginal secretions.

Box 19.1 Is there a biochemical or genetic cause of homosexuality?

Homosexuality is sexual inversion in which orientation is towards individuals of the same gender. The term 'sexual orientation' refers to a personal potential to respond with sexual excitement to persons of the same sex, the opposite sex or both.

The sexual histories of several thousand North Americans were collected by Kinsey and collaborators more than 50 years ago. From these studies, they estimated that 8% of men and 4% of women were exclusively homosexual for a period of at least three years during adulthood. However, subsequent studies have yielded lower estimates.

Studies on homosexuality in homosexual male and female monozygotic twins compared with dizygotic twins and familial aggregates of homosexuality suggest a genetic influence. Hence, there is a debate about whether a gene for homosexuality exists; if it does, it is probably located on the X-chromosome. It is also suggested that prenatal levels of androgens may influence development of the homosexuality. A current view is that the level of androgens in utero is responsible for sexual orientation: a low level results in male homosexuality whereas a high level results in female homosexuality. The mechanism for this could be the effect of the steroid hormones on the brain at this particular time of development.

Box 19.2 Imperato-McGinley syndrome (guevedoces)

The enzyme steroid 5α-reductase converts testosterone to dihydrotestosterone, which has more androgenic activity than testosterone and is responsible for development of the external genitalia. A deficiency of this enzyme gives rise to a type of male pseudohermaphroditism which was first identified in a number of families in the village of Salinas in the Dominican Republic but has also been identified in Europe. The affected males are born with a marked ambiguity of the external genitalia to such an extent that they were raised by their families as girls (until the inhabitants became aware of the condition). However, at puberty, their voices deepen and they develop male phenotype characteristics, with substantial increase in muscle mass. At the same time, the penis and scrotum enlarges. Hence the descriptive term used by the Dominicans, guevedoces, which means 'penis at twelve'.

The explanation for this phenomenon is that in the pre-pubertal phase the amount of testosterone that is secreted is not sufficient to cause development of the male genitalia, which is normally stimulated by dihydrotestosterone. However, at puberty the secretion of testosterone increases. The increase is large in this syndrome, because lack of dihydrotestosterone decreases the extent of the feedback inhibition of the hormone secretions by the pituitary so that more gonadotrophins are released which stimulate, markedly, the rate of testosterone secretion and hence its concentration in the blood. At these high levels, testosterone has sufficient androgenic effects to stimulate development of the external genitalia.

Vaginal secretions   It is well established that vaginal secretions and their odours are important in many animals for stimulating sexual interest by the males. They also play a role in human sexual activity. Of considerable biochemical interest, a contribution to the odours is provided by short-chain fatty acids, including acetic, propionic, isobutyric and isovaleric acids. The concentration of these in the vagina varies during the menstrual cycle; and they disappear after ovariectomy but are restored after oestrogen treatment. The immediate source of these acids appears to be secretion by the vaginal mucosa but they may also arise from the action of microorganisms in the vagina upon the contents of vaginal secretions. These odours can be considered as pheromones, which have effects on the male via sensors in the cilia in the nasal passages (Chapter 12; Figure 12.15).

Growth of the endometrium during the menstrual cycle

The endometrium consists of an epithelium, the tunica propria, which contains glands that produce mucus, and the endometrial stroma. The latter is divided into the lamina basalis and the lamina functionalis. It is much of the latter that is lost during menstruation. The stem cells present in the lamina basalis must replenish those lost during menstruation and regenerate this functional component. The cell cycle must, therefore, be activated in these stem cells by the effects of the hormones, particularly oestrogen and progesterone (Figure 19.13). This regeneration lasts from about day five to day 14 or 15 of the cycle, i.e. during the follicular phase. During this time, the endometrium increases markedly in thickness. Consequently, cell cycling must be increased, that is, many stem cells in the G_0 phase of the cell cycle must be activated. For such a large amount of tissue that must be replenished during these 10 days, a considerable quantity of essential fuels and building materials must be provided to the stem cells in the lamina basalis: these include glutamine, glucose, essential amino acids and essential fatty acids (Chapters 8, 11 and 20). A considerable quantity of fuel will also be required to generate sufficient ATP. In addition, erythrocytes that are lost must be replaced by erythropoiesis in the bone marrow. This requires iron and vitamins B_{12} and

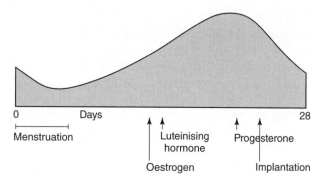

Figure 19.13 *Relative thickness of the endometrium during the menstrual cycle.* Arrows show peak concentrations of hormones involved in control of regeneration of the endometrium.

folic acid (Chapter 15). Hence, the anaemia due to menor-rhagia, which is usually explained by a lack of iron, may also be caused by a deficiency of essential and condition-ally essential amino acids or essential fatty acids. It is possible that failure to provide such materials could lead to poor regeneration of the endothelium after menstruation and may be one cause of infertility in women.

Coitus and the sexual response in the male and female

Coitus (sexual intercourse or copulation) is sexual contact between men and women during which the penis is inserted into the vagina and the resulting ejaculation deposits semen there. Considering the enormous amount of time, financial and social investment that is put into this activity by humans, it is surprising that it was only in the 1960s that significant physiological research into this topic was carried out and, even now, by comparison with many other fields, biochemical research in this field is modest. The studies by Masters & Johnson (Masters *et al.*, 1995) were pioneer-ing in the field of human sexual activity. As part of this study, they proposed a model for the sexual responses in both men and women during coitus, known as the EPOR model. According to this there is:

- an initial phase of excitement (E);

- a plateau phase in which sexual arousal is intensified (P);

- the orgasmic phase, which entails a few seconds of involuntary climax and an explosion of intense pleasure (O);

- the resolution phase (R) in which sexual arousal is dis-sipated and the increased flow of blood to the penis and vagina returns to normal.

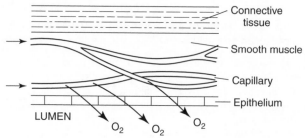

Figure 19.14 *Diffusion of oxygen from the vaginal smooth muscle into the lumen of the vagina.* Increased contractions of the smooth muscle during coitus and especially during an orgasm increase the flow of blood through the muscle and hence increase the diffusion of oxygen into the lumen, where it is required by spermatozoa for mitochondrial ATP generation (see below).

Little is known about the biochemistry that is involved in these various stages. Two biochemical topics discussed here are the role of orgasms and sexual arousal in men, which results in penile errection.

Role of the orgasm in both the male and the female

An obvious role of the orgasm is the period of intense pleasure in both men and women which, encourages sexual intercourse. In men, the coincidence of ejaculation and orgasm ensures deposition of spermatozoa in the vagina.

In women, the orgasm results in contraction of the smooth muscle of the vagina which facilitates movement of sperm along the vagina. In addition, it increases the blood flow to the muscle and the vaginal epithelium, which facilitates the diffusion of oxygen from the capillaries in the epithelium into the lumen. This provides the oxygen for the spermatozoa, which is essential for the complete oxidation of fructose and hence for mitochondrial genera-tion of ATP (Figure 19.14). This may explain why multiple orgasms can occur in women but not in men. Multiple orgasms will provide a more sustained increase in the flow of blood through the muscle and hence the epithelium, which provides for a more sustained diffusion of oxygen into the lumen of the vagina. Multiple orgasms in the male are of no advantage since the first ejaculation deposits a massive number of sperm into the vagina.

Penile erection

Penile erection is essential for insertion of the penis into the vagina and therefore is essential for reproduction. This process depends upon relaxation of smooth muscle in the

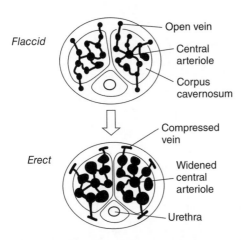

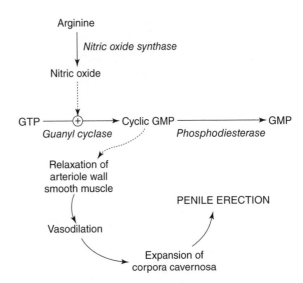

Figure 19.15 *The hydraulic mechanism for penile erection.* Erection begins when the penile arterioles dilate, under the influence of nitric oxide, allowing more blood to enter the corpora cavernosa of the penis. These contain a spongy tissue which now expands but at the same time compresses the veins so that the pressure within the cavernosa increases further, producing more expansion of the cavernosa and maintaining the erection.

Figure 19.16 *Role of nitric oxide synthase in control of penile erection.* Nitric oxide synthase catalyses conversion of arginine to nitric oxide, which then acts to activate guanyl cyclase which results in an increase in the concentration of cyclic GMP. The latter relaxes smooth muscle in the arterioles that supply blood to the corpora cavernosa in the penis so that blood flow increases and erection results.

arterioles that control the flow of blood into the corpus cavernosa in the penis. Erection is caused by the increased flow of blood so that the cavernosa enlarge, which compresses the venules that control the outflow of blood from the cavernosa, leading to more engorgement, that is a positive feedback process which ensures erection (Figure 19.15). Consequently, failure of relaxation of these smooth muscles results in impairment of erection, a problem known as erectile dysfunction.

Relaxation of smooth muscles is controlled by the concentration of cyclic GMP in the muscle. This is regulated by the activities of the enzyme that forms cyclic GMP (i.e. guanyl cyclase) and the enzyme that degrades cyclic GMP, that is, cyclic GMP phosphodiesterase (see Box 12.2). This is analogous to the enzyme system that regulates the concentration of cyclic AMP, by the activities of adenyl cyclase and phosphodiesterase:

$$GTP \rightarrow cyclic\ GMP \rightarrow GMP$$

$$ATP \rightarrow cyclic\ AMP \rightarrow AMP$$

As discussed in Chapter 22, the activity of guanyl cyclase is regulated by the messenger nitric oxide: this gas stimulates the activity of this enzyme. Nitric oxide is generated from arginine catalysed by the enzyme nitric oxide synthase (NOS) (Figure 19.16).

The biochemical cause(s) of erectile dysfunction is not known but one possibility is a low concentration of cyclic GMP in the smooth muscle. On the basis of the principle of the regulation of second messengers see Chapter 12 and Box 12.2, the following might be responsible for a low

level cyclic GMP, a low activity of guanyl cyclase or too high an activity of the phosphodiesterase. This principle predicts that inhibition of the activity of the phosphodiesterase should lead to an increase in the level of cyclic GMP. Indeed drugs have been developed that inhibit the phosphodiesterase and facilitate erection of the penis: the first was Viagra (sildenafil). In addition, anxiety or stress, presumably acting via the sympathetic system and noradrenaline, and biochemically via cyclic AMP, restricts blood flow to the corpora cavernosa and hence can interfere with the control of erection (Figure 19.17).

Ejaculation

Ejaculation is the result of contractions of smooth muscles in the urethra and the striated muscles in the bulbocavernosus, both of which are controlled by Ca^{2+} ions. Semen coagulates within the vagina upon ejaculation: the coagulum protects the sperm from the low pH of the vaginal fluid, until it is neutralised by the semen. It also minimises loss of semen from the vagina. The coagulum liquifies slowly, aided by proteolytic enzymes in the semen which also break down the mucus that surrounds

> Viagra has been reported to improve arousal, orgasm and enjoyment of sexual activity in women, which suggests that nitric oxide and cyclic GMP via vasodilation, affect the vagina.

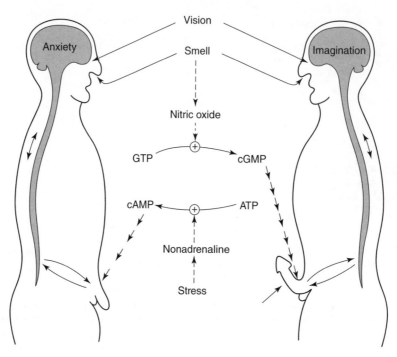

Figure 19.17 *The biochemistry and physiology responsible for penile erection.* Sexual activity itself begins with a state of arousal that leads to erection. Arousal results in part from stimulation of the sense organs. The hypothalamus coordinates the sensations and activates the autonomic nervous system. Sensory nerves from the skin of the penis and other erogenous zones stimulate the parasympathetic system. This activates nitric oxide synthase and the resultant nitric oxide, via cyclic GMP, causes vasodilation of the arterioles. This increases blood flow through the corpora cavernosa which then expands producing an erection. Pheromones secreted by the female can stimulate the odour detecting system in the nasal cavity of the male (Chapter 12 and see above). Stress, however, activates the sympathetic system releases cyclic AMP which can result in vasoconstriction of the arterioles. Other factors that can interfere with an erection are physical fatigue and alcohol.

the cervix in order to facilitate the entry of sperm into the uterus. These biochemical actions reduce some of the hazards for the movement of the sperm in the vagina.

Vasodilation, therefore, has two key effects that are vital for fertilisation: erection of the penis and provision of oxygen in the lumen of the vagina for generation of mitochondrial ATP to support swimming of the spermatozoa (Figure 19.18).

Fertilisation

Fertilisation can be defined as the fusion of the nucleus of a sperm with that of the ovum but many biochemical events must occur before this process can take place.

The ovum possesses two membranes, an outer protective coat (the zona pellucida) and the plasma membrane. The sperm has to traverse both of these membranes to gain access to the nucleus of the ovum. As the sperm approaches the ovum, three biochemical changes occur:

(i) The motility of the sperm increases which provides the momentum that increases the chance of the sperm crossing the outer protective coat of the ovum.

(ii) A protein is lost from the surface of the sperm which exposes binding sites on the head of the sperm for attachment to specific receptors on the membrane of the ovum.

(iii) The binding initiates a process in which proteolytic enzymes are released from the acrosome (Figure 19.4) and hydrolyse proteins in the zona pellucida. This is known as the acrosomal reaction.

These three changes take several hours, and the whole process is termed capacitation. These changes allow a sperm to reach the plasma membrane of the ovum. Then the plasma membrane of the sperm fuses with that of the ovum, the head is engulfed within the membrane of the ovum and the head, and only the head, enters the ovum. Within the ovum, the nucleus of the sperm is released and the two nuclei fuse. A diploid zygote is the result.

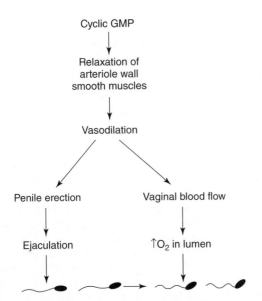

Figure 19.18 *The role of cyclic GMP and vasodilation in provision and preparation of spermatozoa for fertilisation.* Vasodilation is regulated by the concentration of cyclic GMP by relaxation of smooth muscle. The resultant increase in blood flow to the corpora cavernosa results in erection of the penis for the ejaculation of spermatozoa into the vagina. The increase in blood flow to the vaginal smooth muscle provides more oxygen for diffusion into the lumen. Here it provides for oxidative phosphorylation in the mitochondria of the-mid section of the spermatozoa, which provides the ATP for the beating of the flagellum and hence for swimming to the oviduct for fertilisation.

It is important to emphasise that only the head of the sperm enters, so that it is only the nuclear genetic material that enters the ovum, that is, the genetic material in the mitochondria of the sperm does not enter the ovum and hence it does not appear in the zygote. Consequently, mitochondrial genes are inherited only from the mother. This has implications for some mitochondrial diseases (Chapter 9).

Fertilisation and phospholipase

The binding of the sperm to a receptor on the membrane of the oocyte either activates a membrane-bound phospholipase or releases a phospholipase into the oocyte. The phospholipase hydrolyses phosphatidylinositol bisphosphate to produce the two intracellular signals, inositol trisphosphate (IP$_3$) and diacylglycerol within the ovum. As in other cells, the IP$_3$ signal increases the level of cytosolic Ca^{2+} ions and the diacylglycerol (DAG) signal activates protein kinase C.

$$PiP_2 \rightarrow IP_3 + DAG$$

The increase in the Ca^{2+} ion concentration activates calmodulin-dependent protein kinase. Thus two protein kinases are activated which initiate two essential processes in fertilization:

The effects can be mimicked in vitro by a Ca^{2+} ionophore, an agent that increases Ca^{2+} ion entry into a cell, plus artificial activation of the phospholipase.

(i) Activation of enzymes that change the structure of the zona pellucida to prevent binding of other spermatozoa and prevent penetration if they do bind. If an additional sperm entered, this could cause triploidy (i.e. the zygote would possess three sets of chromosomes). Changes in membrane potential on fertilisation also prevent entry of sperm.

(ii) The second meiotic division of the oocyte was arrested at the metaphase stage of the process prior to ovulation. The high level of Ca^{2+} ions and the kinase activities result in completion of the second cell division, to produce a haploid ovum (Chapter 20). The ovum can now be fertilised to produce the zygote containing 46 chromosomes.

Implantation

The zygote floats free for several days in the oviduct, during which further cell divisions occur, until it reaches a stage known as the blastocyst, which consists of an inner and an outer layer of cells: the inner gives rise to the embryo, whereas the outer gives rise to the trophoblast. The trophoblast surrounds the embryo throughout development and provides nutrition and secretes hormones. By about day 21 of the cycle the endometrium is ready to accept the blastocyst and implantation can begin (see Figure 19.13). The trophoblast cells adhere, proliferate and then penetrate the endometrium. Proteolytic enzymes are secreted by the trophoblast cells to break down some of the intercellular matrix of the endometrium to aid penetration of the trophoblast cells. (This is similar to invasion by tumour cells of a new tissue for development of a secondary tumour, Chapter 21, Figure 21.18.) During this period the endometrium undergoes changes, including increased vascularisation and increased vascular permeability which provides fuel, 'building materials' and oxygen for the developing zygote, that is, nutrition of the early embryo is provided by the endometrial cells. Soon, however, trophoblast and maternal tissue combine to produce the placenta. This organ exchanges materials between the blood of the mother and that of the embryo during the remainder of the pregnancy, but the structure of the placenta is such that it prevents direct contact between the two bloods. This is a most effective system for transfer of fuels and other nutrients ('building materials') for the developing embryo.

Fuels used by the early embryo

The major fuel used by the early embryo is lactate, not glucose. The lactate is provided, via glucose and glycolysis, by cells in the endometrium. Thus the transporter for lactate but not glucose is present in the plasma membrane of the cells in the early embryo. (This relationship between the endometrium and the embryonic cells is similar to that between the neurones and the glial cells in the brain, whereby the glial cells take up glucose and convert it to lactate, which is released for uptake by neurons, Chapter 9.) As the number of cells increases, glucose transporters replace lactate transporters so that glucose becomes the dominant fuel which will be provided, from mother, via the developing placenta. This allows much more rapid growth of the foetus.

Assisted fertilisation

The sequence of events culminating in the fusion of the gametes to form a zygote is long and complex, involving behaviour, physiology, endocrinology and biochemistry. If any one of these goes wrong, infertility can occur. Even for fertile couples, the average conception rate is only about 20% for each cycle. To overcome the problem of infertility in either one of a couple, assisted conception has been developed. The ova are extracted from Graafian follicles, followed by in vitro fertilisation of the ovum with donor sperm and the resultant embryo(s) is placed into the uterus. The process is helped by the fact that spermatozoa, ova and embryos, up to the eight-cell stage of development, can be stored frozen in liquid nitrogen where they will remain viable for many years. Knowledge of endocrine control of the reproductive cycle permits artificial control by administering natural hormones or hormone analogues. A history of assisted fertilisation is presented in Box 19.3.

Pregnancy

If fertilisation does not occur, the corpus luteum regresses and the levels of oestrogen and progesterone fall. However, if fertilisation does occur, the corpus luteum is maintained. Regression of the corpus luteum is prevented by the implanting blastocyst which secretes a hormone, human chorionic gonadotrophin (hCG). This occurs about a week after fertilisation. The structure of this hormone is similar to that of LH and it takes on the role of LH in maintenance of the corpus luteum. To do this, it binds to the LH receptor

---

> **Box 19.3 A brief history of assisted conception**
>
> The first 'test-tube' baby was born in July 1978. An oocyte was aspirated from a Graafian follicle which was developing in a natural cycle in the mother. It was then mixed with the father's sperm in a Petri dish and fertilisation took place. The zygote was cultured in vitro and allowed to divide several times until it was at a stage when it could be transferred into the uterus of the mother. An initial problem with the technique was the short luteal phase, induced by the GnRH, which was not sufficient for the endometrium to develop to a satisfactory stage for implantation. Progesterone normally promotes this development but neither injections of progesterone, at levels high enough to sustain the endometrium, nor a 'long-lasting' analogue, were successful. The finding that zygotes could be preserved in liquid nitrogen until a suitable time in the mother's normal menstrual cycle for implantation overcame this problem.
>
> Initially, the idea that sperm had to be capacitated in the female tract before they could be used for in vitro fertilisation seemed to create a complication for IVF because of the need to recover spermatozoa from the vagina. However, this was shown to be invalid: ejaculated spermatozoa, provided they are washed, were found to be effective.
>
> Assisted fertilisation is not always successful and it can be an expensive procedure. Discriptions of some aspects of the biochemistry of reproduction in this chapter lead to suggestions for simple nutritional supplements is overcome some conditions that may lead to infertility. These are discribed in Appendix 19.2.

---

on the granulosa cells which maintain the activity of adenyl cyclase and hence the production of cyclic AMP and the activity of protein kinase-A. The latter regulates the synthesis of oestrogen and progesterone from cholesterol in the corpus luteum.

The production of progesterone and oestrogen is important for the first 9–10 weeks of pregnancy. After this, the placenta and also the adrenal cortex of the foetus gradually take over the production of these hormones.

The interaction between the adrenal cortex of the foetus and the placenta in production of steroid hormones is complex. In outline, the placenta produces progesterone from cholesterol (which is available from the maternal blood) whereas the foetal adrenal cortex produces corticosteroids and androgens from the progesterone produced in the placenta. The placenta then converts some of these androgens into oestrogens. The interplay between the placenta and the foetal adrenal cortex is acknowledged by the use of the term 'foeto-placental unit' to describe steroidogenesis during pregnancy. Especially during the second half of pregnancy, very large quantities of the steroids are secreted by the foeto-placental unit and this is reflected in the excretion of large amounts of pregnanediol (from progesterone) and oestradiol in the urine. The rate of excretion of oestradiol is 1000-fold greater than during the follicular stage of the menstrual cycle, and can be used, along with maternal blood concentration of the same hormone, to indicate foeto-placental wellbeing.

Progesterone is the hormone of pregnancy, relevant effects of which include:

- Maintenance of proliferation of cells of endometrium.

- A decrease in the contractile activity of the myometrium which is necessary for implantation. This is achieved by inhibition of prostaglandin formation.

- Development of the endometrium to form the placenta after implantation.

- Prevention of development of another follicle in the ovary.

Immunology and pregnancy

It would be expected that, since the foetus is a foreign body, it should be recognised as such by the immune system and rejected. Furthermore, foetal tissue is part of the placenta, so that maternal and foetal tissue are intimately structurally associated. This is no evidence that the maternal immune system is suppressed during pregnancy since the mothers do not suffer from an abnormally high number of infections.

Four suggestions have been made as to why foetal cells are not rejected, although there is no firm evidence for any:

- Trophoblast cells do not produce MHC proteins, so that foetal peptides cannot be expressed on their surface to direct an attack by cytotoxic T-cells (Chapter 17).

- The trophoblast cells may possess many transporters for glutamine, so that the local extracellular concentration is maintained very low, which may restrict the activity of immune cells locally (Chapter 17).

- The endometrium contains only a few B-lymphocytes so that few antibodies to foetal antigens are produced.

- There is little or no direct contact of blood between mother and the foetus, so that transmission of antibodies or immune cells from mother to foetus does not occur.

Just as it is important to 'protect' the foetus from the maternal immune system, it is also important to provide the foetus with immunological protection. The most susceptible time for infection by pathogens is soon after birth. To provide protection, before the milk is secreted, a yellow fluid known as colostrum is secreted. It contains antibodies which can enter the blood of the foetus since the gut in the newborn is permeable to the antibodies and other proteins. Milk contains the IgA antibodies which are derived from mother since IgA can cross membranes (Chapter 15). These protect, in particular, the intestine of the baby from pathogens.

Parturition

Parturition is the process by which the mother expels the foetus from the uterus. The process involves a considerable amount of biochemistry. The maintenance of pregnancy requires many conditions, two of which are particularly relevant here: (i) inactivity of the muscles of the uterus (myometrium), and (ii) closure of the cervix to provide physical support for the growing foetus. Consequently, prior to parturition there must be a restoration of physical activity of the myometrium, so that contractions can increase intrauterine pressure, which is essential for expulsion. Contractions must, however, be periodic, not continuous, so that blood supply to the foetus is not occluded. There must also be a 'relaxation' of the cervix.

The cervix consists of the connective tissue, collagen, in the form of fibres within a matrix consisting of proteins and proteoglycans. The chemical messengers, prostaglandins and nitric oxide, cause release of metalloproteinase and other enzymes which break down the matrix and loosen the fibres. This is similar to the action of enzymes released by tumour cells to enable escape from a large tumour (Chapter 21). The production of nitric oxide and prostaglandins is controlled by the hormone oxytocin, which is secreted by the pituitary. A peptide, relaxin, which is secreted by the corpus luteum, loosens ligaments in the pelvic area. Both oxytocin and prostaglandins are involved in parturition.

These messengers also play a role in regulating contraction of myometrium, which consists of smooth muscle fibres. Contraction is controlled by increases in the concentration of cytosolic Ca^{2+} ions. Prostaglandins activate Ca^{2+} ion channels in the plasma membrane of the fibres: oxytocin activates release of Ca^{2+} from intracellular stores. The increase in concentration of Ca^{2+} ions leads to activation of myosin light-chain kinase which leads to cross-bridge cycling and contraction (as described in Chapter 22: Figure 22.12).

The increase in the size of the foetus causes stretch of the uterus and the cervix which, via sensory nerves, together with steroid secretion from the adrenal cortex of the foetus, lead to secretion of oxytocin. This hormone increases formation of prostaglandins by stimulation of the enzyme cyclooxygenase, the key enzyme in prostaglandin synthesis (Chapter 11). (Note that aspirin, which inhibits cyclooxygenase, delays parturition). A decrease in the number of receptors for the hormone progesterone, on the membrane of the myometrial fibres, increases the sensitivity of contraction of the myometrium to the various stimuli.

Unfortunately, it is not known which fuels are used by the myometrium to generate the ATP to 'power' the contractions. Since glycogen levels in the myometrium increase prior to parturition, it is likely that this is a significant fuel, but whether it is converted to lactic acid or fully oxidised is not known. It is also not known if blood glucose or fatty acids are significant fuels for this muscle. The activity of the muscle must be sustained over several hours, which suggests that ATP must be generated by oxidative phosphorylation in mitochondria, a process that requires not only carbohydrate, fat or amino acids as fuels but also oxygen (Chapter 9). Since there are at least 600,000 births in the UK every year, the subject of fuel supply is very relevant to many mothers each year, particularly since parturition can cause considerable fatigue. It is not known if this is peripheral or central fatigue. It is interesting to note that, although it is known which fuel is involved in the generation of ATP in the muscles of athletes in every Olympic athletic event, as well as the rate of utilisation of each of these fuels, the fuels that are used by uterine muscle during a significant physiological and psychological event in the life of most women are not known.

Energy costs of pregnancy

This is discussed in Chapter 15. In brief, the total energy cost for development of the foetus from conception to full term is about 200 MJ in developed countries. This is the cost incurred from the activities of the mother in relation to pregnancy and the development of the foetus. However, it is much less in undernourished mothers. For example, it is about 45 MJ in mothers in Gambia. There is considerable 'energy' sparing (see below). Adaptations that can occur to reduce energy expenditure are a decrease in physical activity and a reduction in diet-induced thermogenesis. (These are normal adaptations to an energy restricted diet, Chapter 16.) Despite this, in undernourished mothers there is a 'struggle' to achieve energy balance in pregnancy, so that a satisfactory outcome is not always possible.

Foetal growth

During the early life of the embryo, up to about eight weeks after fertilisation, the basic form of the human body is laid down, in miniature, without a great increase in size.

In contrast, from about nine weeks, growth is rapid and continues until after birth. Early growth is achieved by cell division, i.e. the cell cycle is rapid (see Chapter 20). This growth activity is characterised by critical periods, during which selected tissues/organs engage in rapid bursts of cell division. Since proliferation requires considerable ATP generation and therefore considerable amounts of fuel plus building materials, including amino acids, essential fatty acids and specific nutrients (e.g. glutamine, aspartate), it is not surprising that proliferation is restricted to a few tissues/organs at a time. In particular, the omega-3 fatty acids are important for growth of the neurones in the brain because of the large number of axons, dendrites and synapses, the membranes of which contain phospholipids comprising a high proportion of the omega-3 fatty acids. Indeed it is interesting to speculate that a deficiency of these fatty acids during development in the uterus might be responsible for development of some degenerative diseases of the brain in later life (Chapters 14 and 15).

The diversity of size and form of babies born even after normal pregnancy is remarkable. However, this is not only due to genetic variation but to the behaviour and nutrition of the mother: adequate nutrition, macro- plus micronutrients, plus the essential fatty acids, essential amino acids and the conditionally essential amino acids. Too high an intake of alcohol and smoking tobacco are known to affect the size of the baby with, in some cases, a reduction in the number of cells in particular tissues/organs.

In relatively recent years, it has become clear that undernutrition of mother leads to low birth weight of the baby and this can increase the risk of development of degenerative disease in later life, e.g. hypertension, obesity, type 2 diabetes. One hypothesis is that the foetus adapts metabolically to deficiencies by increasing the number of cells in organs that perform specific functions that can overcome the deficiency, e.g. an increase in the number of liver cells that carry out gluconeogenesis, an increase in cells in the adrenal cortex to produce more of the 'chronic stress' hormone, cortisol. These changes are 'carried over' into adulthood which can lead to an inadequate response of the liver to insulin so that insulin resistance develops. So far, however, it is unclear whether deficiencies in specific nutrients or undernutrition per se are responsible for such changes (Chapter 15).

Contraception

Contraception is the prevention of conception by a variety of means. Only humans actively seek to avoid fertilisation and/or implantation of the embryo (except for animals in which a male attempts to achieve paternity by preventing subsequent mating).

Rhythm method

This consists of avoidance of coitus around the time of ovulation (i.e. about day 15 of the cycle).

Vasectomy

In a quick and simple operation, the vasa deferentia are ligated. The spermatozoa behind the ligature either break down within the epididymis or leak through the epididymal wall where, because of the antigen present on the surface of sperm, they are destroyed by the immune system.

Condoms and diaphragms

Barriers: condoms are the commonest form of mechanical contraception in use. Their use is important not only in contraception but in protection against infection by HIV and other sexually transmitted diseases (see below). It should be noted that diaphragms do not protect against such diseases.

Spermicides

Spermicides kill sperm. They are not completely effective as contraceptives when used alone but they improve the effectiveness of barrier methods. For centuries, women have used a variety of substances in conjunction with pads or sponges to prevent conception. Acids, such as vinegar, lemon juice and even Coca-Cola, restrict the motility of sperm, so that movement through the vagina is significantly restricted.

Intrauterine devices (IUD)

For centuries, camel traders prevented their animals from becoming pregnant during the long journeys in the desert by inserting a pebble into the uterus. The use of plastic IUDs in women follows the same principle. These devices stimulate prostaglandin production which increases motility of the myometrium thus preventing implantation of the blastocyst. For this reason, the intrauterine device can also be used for postcoital contraception (see below). Side-effects include pain, menorrhagia, increased risk of infection and possible perforation of the uterus.

Oral steroids

Since the development of the oral contraceptive steroids in the 1960s, various steroids and procedures have been used. Synthetic steroids are now used since their half-lives in the body are longer than the natural steroids. Combined oestrogen–progesterone contraception involves taking one pill each day which contains a synthetic progestogen plus an oestrogen. The pill is taken for 21 dyas, with a break for seven days when no pill is taken, which stimulates menstruation, so providing a more normal cycle. Progestogen-only contraception involves taking a pill containing only a progestogen on a daily basis, or a subcutaneous depot of the progestogen which releases it continually: injection of sufficient to produce a large depot can last for months or much longer. Despite their effectiveness, it is still unclear how the oral steroids act as contraceptives. It is likely that several effects are involved:

- Interference with the LH surge by loss of the preise feedback mechanism on the pituitary and hypothalamus.

- Progestogens change the composition of the secretions of the endometrium so that sperm motility is decreased and/or implantation of a fertilised ovum made more difficult.

- Inhibition of FSH, secretion so that a follicle fails to develop (Box 19.4).

There is still debate about the side-effects of the pill. The general consensus is that, although the risk of thrombosis and breast cancer is increased, the risk of endometrial or ovarian cancer is decreased.

Coitus interruptus

The withdrawal of the penis from the vagina prior to ejaculation has been, and remains, one of the most frequently used forms of contraception. The technique is credited (together with abortion) as being responsible for much of the decline in the birth rate at the time of the industrial revolution in Europe.

Postcoital contraception ('emergency' contraception)

Postcoital contraception can be defined as the use of a drug or device to prevent pregnancy soon after intercourse. It has been estimated that the use of emergency contraception in the USA has prevented over 2 million unwanted pregnancies.

Several methods of emergency contraception have been, or are being, used:

- Administration of a high dose of oestrogen was originally used but is now discontinued since it caused nausea and vomiting.

- A pill containing a combination of high dose oestrogen is taken followed by one containing a progestogen about 12 hours later.

- An anti-progesterone (mifepristone, RU 486) pill is used to inhibit ovulation and implantation.

- Introduction of a device into the uterus.

Box 19.4 The development of the steroid oral contraceptive

The history of the development of the oral contraceptives has been reviewed by Petrow (1966). In 1921, Häberlandt showed that if ovaries of pregnant rabbits were transplanted into non-pregnant animals, temporary sterility resulted. He suggested that this 'sterilisation' method might be applied to fertile women. In 1936, it was shown that daily injections of progesterone inhibited the oestrus cycle of the rat, and that the inhibition of ovulation after mating in the rabbit and guinea pig was due to progesterone. In 1938, Kurzrok noted that, during treatment of dysmenorrhoea with oestrogen, the normal menstrual rhythm was upset and ovulation was delayed. Thus oestrogen could inhibit ovulation and this offered an approach to fertility control. The idea of attempting to mimic the normal cycle pattern of steroid hormone secretion by oral administration of oestrogen followed by oestrogen plus progesterone was proposed by Albright in 1945. In the early 1950s it was shown that the daily administration of progesterone to women from the fifth to the 25th day of the cycle caused a large reduction in the frequency of ovulation. Partially synthetic progestogens became commercially available in the mid-1950s. In 1960 the Food and Drug Administration (FDA) in the USA approved the use of the steroid oral contraceptive (the 'pill'). Within the first five years, it is reported that over five million women in the USA were using this form of contraception. At this time the pill contained a high level of oestrogen plus a synthetic progestogen. However, by the late 1960s it was evident that it posed significant health problems, particularly thrombosis, cardiovascular disease and strokes. Consequently, the content of the steroids in the pill was decreased. This made it safer without compromising its effectiveness as a contraceptive. Later, the so-called 'mini pill' was introduced which contains no oestrogen and only a small amount of a progestogen.

The hormones probably decrease the magnitude of the midcycle LH surge and, therefore, effectively prevent ovulation. The anti-progesterone drug delays endometrial maturation so that implantation does not take place. It may be effective up to five days after intercourse, since implantation occurs seven days after ovulation.

Surgical termination of pregnancy

An unwanted pregnancy can be terminated by suction curettage during the first trimester. The procedure is simple, relatively inexpensive, safe and completes the termination at the time of the procedure. It is estimated that approximately 50 million pregnancies worldwide are legally terminated each year by this surgery.

The menopause

The menopause is the period after the final menstrual period which occurs normally at about the age of 50. It is estimated that by the year 2030, more than 20% of the population in the USA will be over 65 and a high proportion will be women who will be menopausal. Clinical problems associated with the menopause will then be of even greater concern for the health services.

Menstrual cycling is dependent upon the hormones secreted by the Graafian follicles. Once the primordial follicles are depleted, the secretion of oestrogen and progesterone progressively fails and menstrual cycle activity gradually decreases, a period which is characterised by irregular cycles. The differences in steroid hormone production or plasma levels between pre- and post-menopausal women are as follows:

- The rate of oestrogen production decreases.

- The cyclical production of oestrogen and progesterone is lost.

- The plasma androgen level is increased.

- Oestrone rather than oestradiol is the major oestrogen present in the plasma.

- The plasma levels of FSH and LH increase.

The clinical problems that arise in the menopause are hot flushes, sweating, depression, decreased libido, increased risk of cardiovascular disease and osteoporosis. The latter results in increased incidence of hip, radial and vertebral fractures. Oestrogen is one factor controlling synthesis of active vitamin D and osteoporosis is in part due to a deficiency of vitamin D. Not surprisingly, to reduce these problems, administration of oestrogen is recommended (known as hormone replacement therapy or HRT). HRT reduces some of the risk factors for coronary artery disease since it reduces blood pressure and decreases the blood level of LDL-cholesterol and increases that of HDL-cholesterol. However, there is considerable debate about whether HRT increases the risk of breast or endometrial cancer.

Sexually transmitted diseases

In recent years it seems that AIDS has somewhat eclipsed other sexually transmitted diseases. A problem, however, is that sexually transmitted diseases not only cause ill health in their own right, but can have long-term consequences including infertility, ectopic pregnancy and genital cancers. Consequently, there is increasing concern about the increasing incidence of sexually transmitted diseases other than AIDS.

There are five main sexually transmitted diseases: AIDS, gonorrhoea, syphilis, genital herpes and genital candidiasis. AIDS is discussed in Chapter 17.

Syphilis is a chronic disease caused by the bacterium *Treponema pallidum*, which results in the formation of lesions throughout the body. The bacteria usually enter the body during sexual intercourse through the mucous membranes of the vagina or urethra.

Gonorrhoea is caused by the bacterium *Neisseria gonorrhoeae*. It affects the mucous membranes of the genital tract of either sex. Symptoms include pain on urination and discharge of pus from the penis or vagina.

Genital herpes is caused by the herpes simplex virus. Symptoms are variable, starting with a small blister (vesicle) leading on to recurrent small genital sores.

Genital *candidiasis* is a common disorder caused by a yeast, *Candida albicans* infects the mucous membrane of the vagina.

20
Growth and Death of Cells and Humans: The Cell Cycle, Apoptosis and Necrosis

Borsook's review article (1952) is a good starting point for a consideration of changes in thought which have occurred with regard to biosynthesis of protein ... the possibility that synthesis could occur by reverse of hydrolysis of proteins ... is discussed (and disagreed with) and so is protein synthesis by transpeptidation. Both theories were current at the time ... Borsook needs only half a page to discuss the role of nucleic units in protein synthesis.

(Korner, 1964)

In human society, suicide often seems an irrational and impulsive act. Not so in the society of cells in an organism. Like obedient soldiers making a personal sacrifice for the common good, excess cells, or those that pose a threat to the well-being of the organism often commit suicide on command, via an orderly process of programmed cell death, or apoptosis.

(Marcia Barinaga, 1966)

Growth is difficult to define. It implies an increase in size but this can be measured in a number of ways; for example, an increase in the number of cells, an increase in size of a tissue or organ, an increase in body mass or an increase in height. Although an increase in the number of cells must occur in order to grow, it is unusual to use this as a measure of growth: even the growth of a tumour is measured by the increase in size. At the other end of the measurement spectrum, an increase in weight or an increase in height is used. The latter is due to an increase in the size of the skeleton, particularly the increase in length of the long bones which is usually used as a measure of growth in children or adolescents (see Appendix 15.2).

The increase in the number of cells in a tissue or organ is a balance between the rate of cell division, also known as proliferation, and the rate of cell death. The process of proliferation is known as the cell cycle. Cell death occurs either by necrosis or apoptosis (programmed cell death).

Introduction to cell proliferation

There are two types of cell division, mitosis and meiosis. Mitosis is the division of the cells that produces two genetically identical daughter cells. It is the process by which new cells are produced for growth or for replacement of lost or damaged cells. Meiosis is the division of cells that is essential for sexual reproduction.

Meiosis

Meiosis is cell division that produces two daughter cells with only half the number of chromosomes of the parent cell; these are haploid cells. It occurs during the formation of spermatozoa and ova. The normal number of chromosomes is then achieved at fertilisation of an ovum by a spermatozoon. Meiosis also results in 'crossing over', which is the physical reciprocal exchange of parts

---

Functional Biochemistry in Health and Disease by Eric Newsholme and Tony Leech
© 2010 John Wiley & Sons Ltd

of homologous chromosomes, during which an almost infinite number of chromosomal combinations can occur. This results in genetic variation during formation of spermatozoa or ova, so that the uniqueness of each individual is assured. It is considered to be a major advantage of sexual reproduction as opposed to asexual reproduction (see below for details of meiosis).

Cell division: mitosis

Mitosis was first characterised through use of the light microscope, and was divided into four phases: prophase, metaphase, anaphase and telophase according to the behaviour of the chromosomes. Subsequently, partition of the cytosol and the formation of a furrow that constricts the diameter of the plasma membrane and pinches the cell into two cells occurs and is known as cytokinesis. The period between mitoses is known as the interphase, so that cell division consists of two parts, interphase and mitosis. Whereas mitosis could be studied with the light microscope, interphase can only be studied by the techniques of biochemistry. The whole process of cell division, interphase plus mitosis, is now called the cell cycle. The cycle, therefore, comprises both continuous and discontinuous processes: the latter are synthesis (i.e. duplication) of DNA and the splitting of one cell into two daughter cells (i.e. mitosis). The synthesis of other macromolecules (proteins, lipids, carbohydrates) occurs in the continuous stages. Both the continuous and discontinuous processes and the regulation of the cell cycle are described and discussed in this chapter.

The cell cycle

The biochemical studies on the interphase established that it could be divided into three phases so that, with mitosis, the whole process can be separated into four phases known as G_1, S, G_2, M (Figure 20.1(a)). Although it is usually called a cycle, it can also be considered as a linear process (Figure 20.1(b)). The S phase is that in which DNA is duplicated and the M phase is the process of splitting, i.e. mitosis. The G_1 and G_2 phases represent gaps or breaks in the cycle, during which preparation for the next phase takes place. Regulation of the cycle also takes place during these phases. It is essential that: (i) one process is completed before the next process starts (for example, if mitosis took place before DNA duplication had been completed, mitosis would fail and the cell would die during division); (ii) sufficient material is present to complete the next stage of the cycle (for example, if a sufficient concentration of even one deoxyribonucleotide is low, DNA duplication may fail

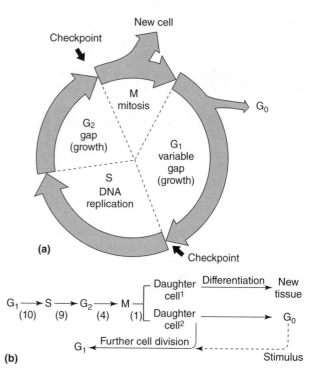

Figure 20.1 **(a)** *The cell cycle.* For details see text. **(b)** *A 'linear' cell cycle.* The process need not be cyclical: stem cells may proliferate to produce daughter cells that differentiate and daughter cells that enter the G_0 phase where they remain for some time, indeed some may remain in G_0 until death (e.g. nerve cells). A stimulus, however, can cause a cell in go to enter the cycle (broken arrow). The numbers in parentheses indicate the time (in hours) that a human liver cell, in culture, takes to complete each phase.

or mutations may occur). Consequently, prior to the S phase (i.e. during the G_1 gap) and also prior to the M phase (i.e. during the G_2 gap) 'decisions' have to be taken whether to stop progression into the next phase, i.e. to arrest the cycle at that stage. For this reason, these have been termed *checkpoints*, which are points when the cell has to check whether it is 'safe' to proceed into the next phase. However, these checkpoints perform the same function as regulatory steps in control of other biochemical processes (Chapters 2 and 3). Regulation at checkpoints can stop the process completely, whereas regulation in metabolism is usually concerned with regulation of the rate of the process. Nonetheless, some metabolic processes must be arrested, i.e. completely inhibited, under some conditions. For example, glycogenolysis in muscle and liver must be completely inhibited in order for glycogen to be synthesised since the capacity for glycogen breakdown is close to 100-fold greater than that for synthesis. Not surprisingly, the same mechanism of regulation applies to the cycle and glycogenolysis, i.e. phosphorylation by protein kinases and/or dephosphorylation by protein phosphatases.

In addition to the four stages outlined above, one further stage exists. After the M phase, a resultant daughter cell can enter a state in which, for a period, no further change occurs: the cell is in a quiescent stage, which is known as G_0. It can, however, at a later stage re-enter the cycle, for example, after stimulation of the cell by a hormone, growth factor or other signal (stimulation of a lymphocyte to proliferate in response to binding an antigen).

Once the cycle has begun, the sequence of events is almost always completed in a time which is approximately constant for a given cell: about 24 hours for a typical human cell. The largest variation in time occurs in the G_1 phase. Very short cell cycles, 8 to 60 minutes, occur in early embryonic cells, during which cell division results in the formation of many smaller cells. In these cells, both the G_1 and G_2 phases are massively shortened, so that most of the time of cycling is spent in the S and M phases.

Biochemical requirements for the cell cycle

Although the mechanisms that regulate the entry into both the S and M phases of the cell cycle are of considerable biochemical interest and importance, so that they are intensively investigated, the cycle cannot proceed unless the precursors for synthesis of all macromolecules and the pathways for their synthesis are present in the cell (Figure 20.2).

A summary of some of these processes is as follows: synthesis of phospholipids and cholesterol; *de novo* synthesis of ribonucleotides; synthesis of RNA; *de novo* synthesis of deoxyribonucleotides; regulation of synthesis of deoxyribonucleotides; salvage pathways; duplication of DNA; transcription and translation (polypeptide synthesis). After this series of topics, those of fuels and ATP generation, mitosis and, finally, regulation of the cycle, are described and discussed.

Synthesis of phospholipids and cholesterol

The pathways for the synthesis of phosphoglycerides and sphingolipids are described in detail in Chapter 11. They are, therefore, described only in brief here in order to emphasise the importance of the essential fatty acids in proliferation and how the cell cycle could be impaired by failure to provide these acids.

The initial reactions produce phosphatidate, in which the two hydroxyl groups of glycerol 3-phosphate are esterified with long-chain fatty acids, catalysed by enzymes known as acyltransferases. An important point is that, due to the difference in specificities of the acyltransferase enzymes,

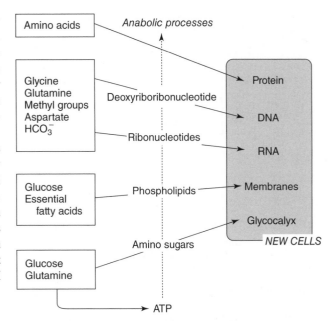

Figure 20.2 *A summary of the pathways for synthesis of the major macromolecules or macrostructures that are required to complete a cell cycle.* These processes, except for formation of amino sugars and ATP generation, are described and discussed in this chapter. The synthesis of amino sugars is described in Appendix 6.2. The pathways for the generations of ATP from both glucose and glutamine are described in Chapters 6, 9 and 11. The role of the glycocalyx in the cell is discussed in Chapters 4 and 5.

different fatty acids are esterified into the two positions of the glycerol phosphate: a saturated fatty acid initially occupies the first and a monounsaturated occupies the second. However, in most phosphoglycerides in membranes, the second position is utimately occupied by a polyunsaturated fatty acid, usually either arachidonic or eicosapentaenoic acid. This is important since it can modify not only the structure of the membrane but also the activities of membrane-bound enzymes, transport molecules and receptors. The more unsaturated the fatty acid in the phosphoglyceride, the more fluid is the membrane. In addition, the hydrolysis of the second ester bond, by a specific membrane-bound phospholipase, releases the polyunsaturated fatty acid, which is then a precursor for formation of eicosanoids, which are fatty messengers that play key roles in the regulation of proliferation. The two essential fatty acids result in synthesis of different eicosanoids, and the different eicosanoids may play different roles in proliferation. The importance of this in health is that the proportion of these fatty acids in the phosphoglyceride can be influenced by the source of fat in the diet (Figure 20.3). Such changes may interfere the cell cycle and may be important in tumour growth (Chapter 21).

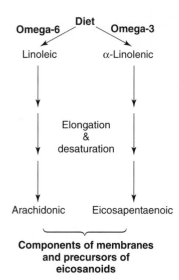

Figure 20.3 *Essential fatty acids in the diet, production of 'physiological' essential acids and their roles in the cell cycle. Essential fatty acids in the diet are mainly linoleic and α-linolenic but they are converted by desaturation and elongation reactions to the essential acids that are used in phospholipid formation and synthesis of eicosanoids. (For details of the elongation and desaturation reactions and eicosanoid formation, see Chapter 11.).*

The incorporation of the polyunsaturated fatty acid in position two depends upon removal of the monounsaturated fatty acid and replacement by the polyunsaturated acid. This is achieved by the action of two enzymes, (i) a deacylase and (ii) an acyltransferase.

(i)   phosphatidylcholine + H_2O →
        lysophosphatidylcholine + fatty acid

(iia) lysophosphatidylcholine + arachidonyl-CoA →
        phosphatidylcholine + CoASH

(iib) lysophosphatidylcholine + eicosapentaenoyl-CoA →
        phosphatidylcholine + CoASH

Thus, the three enzymes must be present in proliferating cells to ensure satisfactory phospholipid synthesis. These enzymes are also important in repairing damaged polyunsaturated fatty acids (e.g. caused by free radicals, see Chapter 9). In fact, it is possible that repairing damaged fatty acids during the cell cycle may be as important as repairing damaged nucleotides in DNA.

It is possible that some of the disorders caused by essential fatty acid deficiency are due to failure of the cell cycle in proliferating tissues: alopecia and skin problems (e.g. scaly skin) due to poor proliferation of epithelial cells;

increased susceptibility to infection could be due to poor proliferation of immune cells; increased capillary fragility could be due to poor proliferation of endothelial cells; slow wound healing could be due to poor proliferation of fibroblasts. The inadequate proliferation in all cases could be due, not to a defect in a key step in the cycle but a supply of essential fatty acids.

Sphingolipids   The structure and synthesis of sphingolipids are darcibed in Chapter 11.

Cholesterol   The pathway for synthesis of cholesterol is described in Appendix 11.9. Cholesterol is important in the structure of membranes since it can occupy the space that is available between the polyunsaturated fatty acids in the phospholipid (Chapter 4). In this position, cholesterol restricts movement of the fatty acids that are components of the phosphoglycerides and hence reduces membrane fluidity. Cholesterol can be synthesised *de novo* in proliferating cells but it can also be derived from uptake of LDL by the cells, which will depend on the presence of receptors for the relevant apolipoproteins on the membranes of these cells (Appendix 11.3).

Bases, sugars, nucleosides: the components of ribonucleotides and deoxyribonucleotides

The four bases in DNA are the two purines, adenine and guanine, and the two pyrimidines, cytosine and thymine. The same bases are present in RNA, except that uracil replaces thymine (Table 20.1). All these bases, plus the Watson–Crick base pairs that hold the bases together in DNA and RNA, are shown in Figure 20.4. The corresponding nucleosides (i.e. base plus sugar) are adenosine, guanosine, cytidine, thymidine and uridine. The corresponding nucleotides (base plus sugar plus phosphate), in the form of monophosphates, are adenylate (AMP), guanylate (GMP), cytidylate (CMP), thymidylate (TMP) and uridylate (UMP) (Table 20).

Ribonucleotides are required to synthesise RNA whereas deoxyribonucleotides (dA, dG, dC, dT) are required to synthesise DNA (Figure 20.5). The difference in the structure between the two types of nucleotide is relatively little: the sugar in ribonucleotides is ribose, whereas that in deoxyribonucleotides is deoxyribose, in which the hydroxyl group at position 2 of the ribose ring is replaced by a hydrogen atom (Figure 20.6). In fact, ribonucleotides are the precursors for formation of deoxyribonucleotides (see below); reduction of position 2 of the ribose ring to remove the hydroxyl group occurs within the ribonucleotide molecule. Only two enzymes are required for this reaction. These are ribonucleoside diphosphate reductase and thioredoxin reductase (see below).

Table 20.1 Classification and nomenclature of bases, nucleosides and nucleotides

Base[a]	Abbreviation	Type	Nucleoside[b]	Nucleotide	Occurrence
Adenine	A	purine	adenosine	adenylate (AMP)	DNA and RNA
Cytosine	C	pyrimidine	cytidine	cytidylate (CMP)	DNA and RNA
Guanine	G	purine	guanosine	guanylate (GMP)	DNA and RNA
Thymine	T	pyrimidine	thymidine	thymidylate (TMP)	DNA only
Uracil	U	pyrimidine	uridine	uridylate (UMP)	RNA only

[a]Methylated bases occur in small amounts in deoxyribonucleotides. The role of methylation, especially of cytosine in DNA, is discussed below and in Chapter 15.

The structure of the bases and the base pairs formed via hydrogen bonds are shown in Figure 20.4.

[b]Nucleotides are named by adding a term denoting the number of phasphate groups (and their position) to the name of the nucleoside e.g. adenosine 5-monophosphate, AMP (or adenylate). In general, nucleotides are assumed to be ribonucleotides unless the prefix deoxy (or d- as abbreviation) is used. (See Figures 20.5 and 20.6)

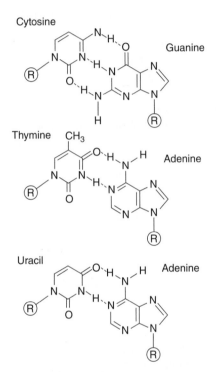

Figure 20.4 *The bases present in RNA and DNA and the Watson–Crick base pairing relationships.* Uracil is present in RNA but is replaced by thymine in DNA: that is, the pairs C–G and T–A are found in DNA: the pairs C–G but U–A are found in RNA. The pairing is brought about by hydrogen bonding, indicated by a broken line.

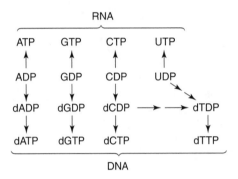

Figure 20.5 *Nucleotides that are required for RNA or DNA synthesis.* Note that the ribonucleotide diphosphates are the precursors for the formation of deoxyribonucleotides. It is the triphosphates that are required for polymerisation to form either RNA or DNA (see text).

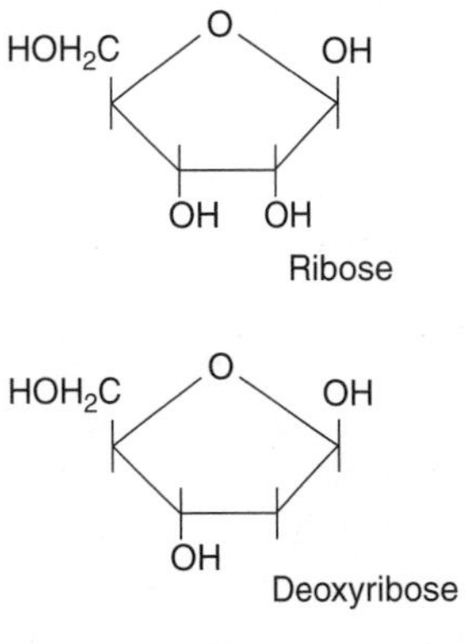

Figure 20.6 *The structures of ribose and deoxyribose.*

De novo synthesis of ribonucleotides

The two classes of nucleotide that must be synthesised are the pyrimidine and purine ribonucleotides for RNA synthesis and the deoxyribonucleotides for DNA synthesis. For the original sources of the nitrogen atoms in the bases of the pyrimidine and purine nucleotides, see Figure 20.7. The pathway for the synthesis of the pyrimidine nucleotides is relatively straightforward since only two compounds are involved. Furthermore the synthesis of the deoxy nucleotides is also relatively straightforward, since they are synthesised directly from the ribonucleotides (see below).

All proliferating cells have the capacity to synthesise nucleotides *de novo*. An outline of the pathways is provided in Figure 20.8.

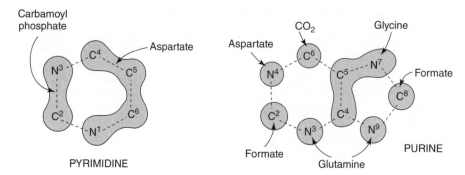

Figure 20.7 *Sources of carbon and nitrogen atoms in pyrimidine and purine bases.* The dotted lines represent the bonds between the atoms. The shaded areas represent atoms or groups of atoms from precursor molecules.

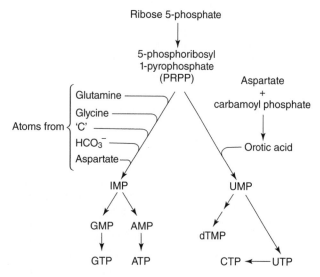

Figure 20.8 *Summary of pathways for* de novo *synthesis of purine and pyrimidine nucleotides.* 'C' represents transfer of a single carbon atom (a one-carbon transfer). Details are provided in Appendix 20.1. IMP – inosine monophosphate. For thymidylate synthesis, see Figure 20.12a.

In the purine nucleotide pathway, the purine nucleotide is synthesised upon the phosphoribose using several small molecules. The first purine nucleotide formed is inosine monophosphate (IMP): it is an intermediate on the pathway for the synthesis of both adenine and guanine nucleotides (Figure 20.8).

A different, 'simpler', pathway is involved in the synthesis of pyrimidine nucleotides. A pyrimidine base (orotate), is synthesised first. Then the ribose is added from 5-phosphoribosyl 1-pyrophosphate. The two precursors for the formation of orotate are carbamoylphosphate and aspartate, which form carbamoyl aspartate, catalysed by aspartate carbamoyltransferase.

In various chapters in this book, emphasis is placed on the roles of glucose and glutamine in provision of compo-

nent atoms or component structures for pyrimidine and purine nucleotide synthesis (e.g. Chapters 6, 8, 15, 17, 18 and 21). The reactions in both these synthetic pathways, where glucose or glutamine are involved, directly or indirectly, are presented in Figures 20.9 and 20.10 (Box 20.1).

To convert purine and pyrimidine nucleoside monophosphates to triphosphates, two kinase enzymes are required: mono- and di-nucleotide kinases. For example,

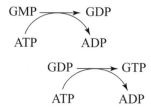

Synthesis of RNA

The enzymes that synthesise RNA and DNA are known as nucleic acid polymerases. They are classified as nucleotidyl transferases (Chapter 3). The basic reaction can be represented as follows:

$$nNTP \longrightarrow \text{nucleic acid} + PPi$$

The enzymes catalysing the synthesis of RNA, using DNA as a template, are known as DNA-dependent RNA polymerases. The term dependency indicates a requirement for a template. The RNA polymerases synthesise RNA in the 5′ to 3′ direction in formation of the phosphodiester linkage. This requires that the DNA template is read in the 5′ to 3′ direction (see below for explanation).

In eukaryotic cells, there are three classes of RNA polymerases (I, II and III) which synthesise different classes of RNA, as follows:

- RNA polymerase I catalyses the synthesis of the RNA molecules that are present in ribosomes (Chapter 1 and see below).

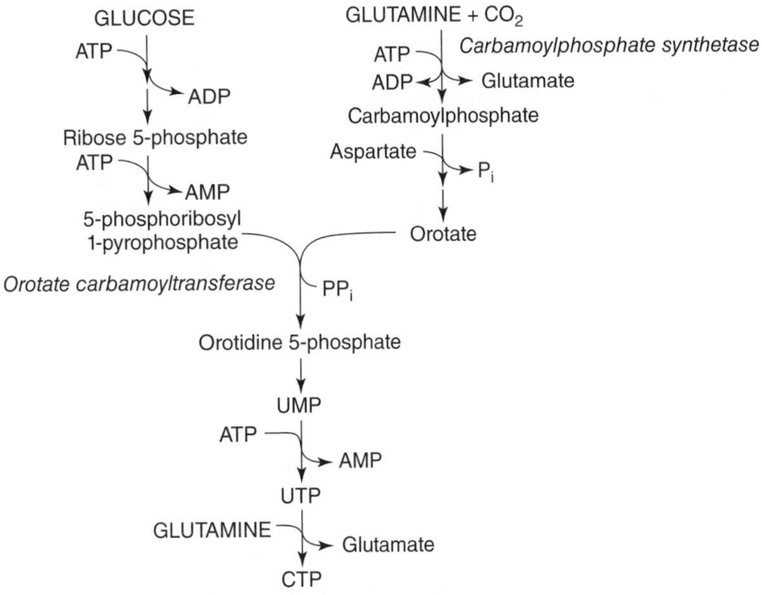

GLUCOSE
ATP
→ ADP
Ribose 5-phosphate
ATP
→ AMP
5-phosphoribosyl
1-pyrophosphate

GLUTAMINE + CO₂
ATP   *Carbamoylphosphate synthetase*
ADP → Glutamate
Carbamoylphosphate
Aspartate → P$_i$
Orotate

Orotate carbamoyltransferase → PP$_i$

Orotidine 5-phosphate

UMP
ATP
→ AMP
UTP
GLUTAMINE
→ Glutamate
CTP

Figure 20.9 *The positions in the pathway for* de novo *pyrimidine nucleotide synthesis where GLUCOSE provides the ribose molecule and GLUTA-MINE provides nitrogen atoms.* Glucose forms ribose 5-phosphate, via the pentose phosphate pathway (see chapter 6), which enters the pathway, after phosphorylation, as 5-phosphoribosyl 1-pyrophosphate. Glutamine provides the nitrogen atom to synthesise carbamoylphosphate (with formation of glutamate), and also to form cytidine triphosphate (CTP) from uridine triphosphate (UTP), catalysed by the enzyme CTP synthetase. It is the amide nitrogen of glutamine that is the nitrogen atom that is provided in these reactions.

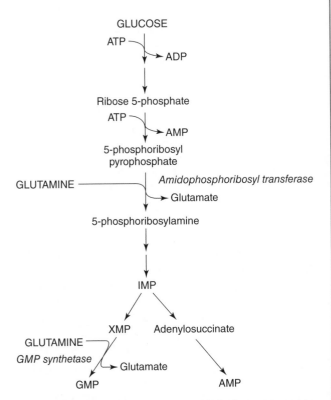

Figure 20.10 *The positions in the pathway for* de novo *purine nucleotide synthesis where GLUCOSE provides the ribose molecule and GLUTAMINE provides nitrogen atoms.* The pathway begins with glucose which provides ribose 5-phosphate, via the pentose phosphate pathway (Chapter 6). Glutamine provides its amide nitrogen in two reactions: formation of 5-phosphoribosylamine and formation of guanosine monophosphate (GMP) from xanthosine 5-phosphate (XMP).

Box 20.1 Purine and pyrimidine antimetabolites

In the 1950s, George Hitchings and his colleagues at the Wellcome Research Laboratories in Tuckahoe, New York, prepared purine and pyrimidine analogues. Two of the purine derivatives, 2,6-aminopurine and 6-mercaptopurine, were synthesised as antibacterial drugs, but were much more effective in inhibiting the proliferation of tumour cells and hence for use in cancer chemotherapy. 6-Mercaptopurine, together with 6-thioguanine, is still used, especially for treating acute leukaemia. These are the 6-thiol analogues of hypoxanthine and guanine respectively and in the cell are metabolised to their corresponding ribonucleotides which inhibit reactions in the *de novo* pathway of purine synthesis so denying the cell a supply of precursors for nucleic acid synthesis. Pyrimidine analogues also, for example cytarabine (cytosine arabinoside) and 5-fluorouracil (see Figure 21.2) are metabolised to their respective nucleotides, which inhibit DNA polymerase and thymidylate synthase, respectively. Until the discovery of cyclosporin and similar agents, they were used as immunosuppressant drugs. Another pharmaceutical application of purine and pyrimidine analogues has been as antiviral agents. The first to prove clinically useful was acycloguanosine (acyclovir) used against the herpes virus and zidovudine (AZT) against HIV. Acyclovir works because the enzyme which phosphorylates it, and ultimately allows it to become incorporated into DNA (where it causes chain termination) is coded for by a part of the viral genome. Zidovudine is phosphorylated by a host enzyme but is a specific inhibitor of RNA-dependent DNA-polymerase (reverse transcriptase), an enzyme required only for viral reproduction (see Figure 17.45).

- RNA polymerase II catalyses the synthesis of messenger RNA (see below).

- RNA polymerase III catalyses the synthesis of small RNA molecules, such as the transfer RNAs (see below).

These enzymes use DNA as a template and the ribonucleotide substrates must be present in the nucleus, i.e. ATP, GTP, CTP and UTP. Similarly, for the synthesis of DNA, the deoxyribonucleotides dATP, dGTP, dCTP and dTTP must be present in the nucleus. In addition, since the ribonucleoside diphosphates are required for synthesis of deoxyribonucleotides, these diphosphates must also be present. The concentrations of these various nucleotides have not been measured in the nucleus but it may be assumed that the concentrations of the ribonucleotides will be similar in the nucleus to those in the cytosol.

The concentrations of the deoxynucleotide triphosphates have been estimated from information in the scientific literature and are presented below. The concentrations are presented on the assumption that they are distributed equally between the nucleus and cytosol.

De novo synthesis of deoxyribonucleotides

The deoxyribonucleotides, except for deoxythymidine nucleotide, are formed from the ribonucleotides by the action of an enzyme complex, which comprises two enzymes, ribonucleoside diphosphate reductase and thioredoxin reductase (Figure 20.11). The removal of a hydroxyl group in the ribose part of the molecule is a reduction reaction, which requires NADPH. This is generated in the pentose phosphate pathway. (Note, this pathway is important in proliferating cells not only for generation

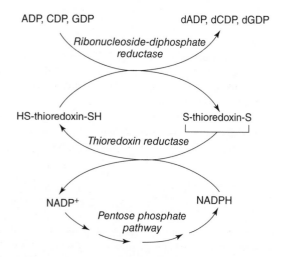

Figure 20.11 *Enzyme complex that converts ribonucleotides to deoxyribonucleotides and role of the pentose phosphate pathway. For details of the pentose phosphate pathway, see Chapter 6, (Appendix 6.8).*

of NADPH but in generation of ribose 5-phosphate for the synthesis of 5-phosphoribosyl 1-pyrophosphate.)

Since thymidine ribonucleotides do not exist, deoxythymidine ribonucleotides cannot be formed by the processes described in Figure 20.11. However, the chemical difference between the two bases, uracil and thymine, is relatively small. The latter has a methyl group attached to position 5 of the ring. Hence, deoxyuridine in the monophosphate form is converted to deoxythymidine monophosphate by methylation in a reaction catalysed by the methylating enzyme, deoxythymidine synthase, also known as thymidylate synthase (Figure 20.12(a)). The methyl group 'provider' is N^5, N^{10}-methylene tetrahydrofolate (see Figure 15.2). The deoxyuridine monophosphate can be formed either from deoxyUTP or deoxyCDP (see Figure 20.12(b)). The formation of all the four deoxynucleotides required for DNA synthesis is summarised in Figure 12(c).

Regulation of the synthesis of the deoxynucleotides to ensure equal concentrations of all four nucleotides

An extract from a paper by P. Reichard (1988) illustrates the importance of ensuring the normal balance of all the deoxynucleotides prior to DNA synthesis.

The cell requires each of the four deoxyribonucleotides (dNTPs) to replicate and repair its DNA. This is a trivial statement. More interesting is to say that the cell needs a balanced supply of the four NTPs to replicate and repair its DNA properly. The importance of the regulation of the NTP pools becomes apparent from disturbances in cell function when a bias in the 'normal' balance occurs. Two extensively studied cases, leading to cell death, are the effects of severe dTTP deprivation ('thymineless' death) and certain immune diseases accompanied by the accumulation of dATP and dGTP.

Regulation of the balance of the concentrations of the four deoxyribonucleotides depends on the properties of only two enzymes, the ribonucleotide reductase complex and deoxy-CMP deaminase. The balance between pyrimidine deoxynucleotides is brought about by the properties of the deoxy-CMP deaminase, which is inhibited by deoxy-TTP and stimulated by deoxy-CTP. The ribonucleotide reductase also possesses allosteric sites which bind all four deoxynucleotide triphosphates, the effect of which is to maintain approximately similar concentrations of all the triphosphates.

Estimates of the concentrations of all four deoxynucleotide triphosphates have been collated from the small number of studies so far published (on proliferating lymphocytes). As expected from the properties of the reductase, the concentrations of three nucleotides are similar but that for dCTP is considerably higher: dATP, 0.02 mM/L, dGTP 0.02 mM/L, dTTP 0.04 mM/L and

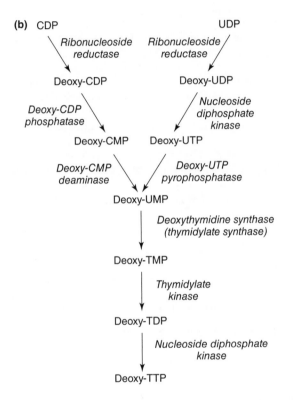

(a)

Deoxyuridine
monophosphate
(dUMP)

Methylene
FH₄

Thymidylate
synthase

Deoxythymidine
monophosphate
(dTMP)

(b) CDP UDP

Ribonucleoside
reductase

Ribonucleoside
reductase

Deoxy-CDP Deoxy-UDP

Deoxy-CDP
phosphatase

Nucleoside
diphosphate
kinase

Deoxy-CMP Deoxy-UTP

Deoxy-CMP
deaminase

Deoxy-UTP
pyrophosphatase

Deoxy-UMP

Deoxythymidine synthase
(thymidylate synthase)

Deoxy-TMP

Thymidylate
kinase

Deoxy-TDP

Nucleoside diphosphate
kinase

Deoxy-TTP

(c) ADP CDP GDP → dADP dCDP dGDP → dATP dCTP dGTP → DNA

CDP → → → → dUMP → → → dTTP

Figure 20.12 (a) *Details of reaction catalysed by thymidylate synthase. Methylene FH₄ represents N⁵,N¹⁰ methylene tetrahydrofolate (see Figure 15.2).* **(b)** *Reactions in the pathways in which either CDP or UDP gives rise to deoxythymidine monophosphate. Note that two processes can be involved in synthesis of deoxyuridine monophosphate. It is not known if one process dominates, but in (c) it is assumed that the pathway from CDP dominates formation of dTTP.* **(c)** *A summary of the reactions required for synthesis of deoxyribonucleotides required for DNA replication.*

dCTP 0.2 mM/L. The concentrations of the latter two are, however, consistent with the properties of deoxy-CMP deaminase, which is stimulated by dCTP. These concentrations are relatively low; for comparison it should be noted that the concentration of ATP is approximately 100-fold higher. This suggests either that the concentrations of the deoxynucleotides are higher in the nucleus or that the DNA polymerase possesses low K_m values for these nucleotides.

Problems caused by an imbalance of the concentrations of deoxynucleotides In experiments with replicating cells in culture, if the culture medium contains an imbalance in the concentrations of the four deoxynucleotides, especially an imbalance between the pyrimidine deoxynucleotides, the mutation rate increases markedly. Deficiencies or low activities of at least three enzymes can result in an imbalance of the nucleotides and do, in fact, give rise to disorders (Box 20.2).

Salvage pathways

The purine and pyrimidine bases can be converted to their respective nucleotides by reaction with 5-phosphoribosyl 1-pyrophosphate. Since these bases are not very soluble, they are not transported in the blood, so that the reactions are only of quantitative significance in the intestine, where the bases are produced by degradation of nucleotides. In contrast, in some cells, nucleosides are converted back to nucleotides by the activity of kinase enzymes. In particular, adenosine is converted to AMP, by the action of adenosine kinase, and uridine is converted to UMP by a uridine kinase

$$\text{adenosine} + \text{ATP} \rightarrow \text{AMP} + \text{ADP}$$

$$\text{uridine} + \text{ATP} \rightarrow \text{UMP} + \text{ADP}$$

These are known as salvage pathways.

Indeed, the biochemical interconversions are such that these two salvage reactions can supply all the nucleotides necessary for DNA synthesis. This is demonstrated by the results of the following experiment. If the culture medium for proliferating cells lacks glutamine, the *de novo* pathway for synthesis of nucleotide phosphates cannot take place, DNA synthesis fails and proliferation does not take place (see above and Chapter 17). However, glutamine can be replaced by adenosine plus uridine in the culture medium, when the proliferation is fully restored. This is explained as follows: when both these nucleosides are phosphorylated by the salvage pathways, adenosine gives rise to AMP and uridine gives rise to UMP. Then AMP can be converted to GMP and UMP can be converted to TMP and CMP (Figure 20.13). Consequently all deoxynucleotide diphosphates can be formed, so that DNA can be duplicated and cells can proliferate (Table 20.2).

Box 20.2 Deficiencies or low activities of enzymes involved in purine nucleotide metabolism

Deficiencies of adenosine deaminase and hypoxanthine-guanine phosphoribosyltransferase or a low activity of thymidylate synthase are responsible for clinical disorders.

Severe combined immunodeficiency disease The enzyme adenosine deaminase degrades deoxyadenosine which is produced during DNA degradation (Chapter 10). Deficiency of the enzyme results in accumulation of deoxyadenosine which is a substrate for adenosine kinase and leads to production of deoxyadenosine and deoxyguanosine triphosphates, which reach high concentrations. This disturbs the balance of deoxy nucleotides which results in failure of DNA replication. This enzyme is normally present in lymphocytes so that a deficiency prevents proliferation of the lymphocytes, which is essential in combatting an infection. Consequently, patients are very susceptible to infections. This is one disease that is effectively treated by gene therapy.

Lesch–Nyhan syndrome A deficiency of hypoxanthine-guanine phosphoribosyltransferase results in accumulation of purine bases (Chapter 10). This causes a marked increase in the plasma level of uric acid, and hence can give rise to gout, but it also causes a severe neurological disorder, known as Lesch–Nyhan syndrome, the symptoms of which include

mental retardation, spasticity and a compulsive form of self-mutilation (e.g. finger and lip biting). The neurological disturbance may be caused by high concentrations of the purine bases, especially hypoxanthine, in the brain, which probably interferes with the binding of adenosine to its receptor in the brain (Chapter 14). It is tempting to speculate that a high level of uric acid or other purine metabolites may be one factor involved in the problem of 'self-harm' which is not uncommon in teenagers in developed countries.

Thymineless death of cells (Megaloblastic anaemia and reverse conditions) The synthesis of thymidine monophosphate from uridine monophosphate requires the transfer of a methyl group (see above). The methyl group is supplied from tetrahydrofolate but the availability of methyl groups depends upon methionine which is formed by methionine synthase. This enzyme requires the vitamins folic acid and B_{12} (Chapter 15). Lack of these vitamins in the diet leads to a deficiency of methyl groups and hence a deficiency of thymidine deoxy-nucleotides, so that DNA cannot be synthesised during a cell cycle. This rapidly results in problems with production of erythrocytes and gives rise to megaloblastic anaemia (Chapter 15). Lack of tetrahydrofolate also leads to death of cells in culture, when the effect is known as thymineless death.

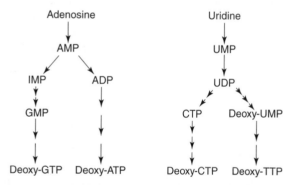

Figure 20.13 *Summary of the reactions by which all four deoxyribonucleoside triphosphates can be synthesised from the nucleosides, adenosine and uridine.* The reactions are summaries of the processes presented in Figures 20.8, 20.9 and 20.12. AMP is converted to IMP by a deaminase (Chapter 6). The conversion of UTP to CTP is catalysed by CTP synthetase.

Table 20.2 Effect of glutamine or adenosine plus uridine on proliferation of lymphocytes in culture

Contents of culture medium	Magnitude of proliferation
(i) Complete plus glutamine	+++++++++
(ii) Complete but minus glutamine	+
(iii) Complete minus glutamine but plus adenosine	+
(iv) Complete minus glutamine but plus uridine	+
(v) Complete minus glutamine but plus adenosine and uridine	+++++++++

Condition (i) is a complete culture medium, and proliferation is high. (ii) Glutamine is omitted from medium and proliferation is very low (Chapter 17). (iii) Addition of adenosine to condition (ii) makes no difference to proliferation. (iv) Addition of uridine to condition (ii) makes no difference. (v) Addition of uridine plus adenosine to condition (ii) restores proliferation to a rate identical to that in complete culture medium.

DNA replication

It is essential that, in mitolic cell division, genetic information in the parent cell, which is encoded in the base sequence of its DNA, is copied exactly so that each daughter cell receives genetic information identical to its parent cell. Consequently, before a cell divides, its DNA must have been replicated completely and accurately. The

latter is achieved by the base-pairing relationships, which were originally recognised by Watson & Crick; that is adenine (A) hydrogen-bonds with thymine (T) and cytosine (C) hydrogen-bonds with guanine (G) (Figure 20.4). The DNA is present in the cell nucleus as a double-stranded molecule (duplex) in which the sense strand (which determines the amino acid sequence in proteins) is hydrogen-bonded to a 'nonsense' strand according to these base-pairing relationships.

In principle, DNA replication is straightforward and involves the following:

- separation of the two strands;

- alignment of the new nucleotide molecules against the other strand according to the base-pairing 'rules';

- condensation reactions catalysed by DNA polymerase to link the new nucleotides together.

Preparations for DNA replication

Duplication of DNA is initiated at different positions along the DNA duplex, at sites which are defined by specific DNA sequences. This means that separation of the two strands, at any one time, occurs along short stretches of the DNA. Initially this is achieved by two helicase enzymes, moving away from each other, which disrupt the hydrogen bonds between the bases. This facilitates unwinding of the DNA that results in separation of the two strands. Since it occurs over only a short region of the duplex, it results in the production of what is known as a replication bubble. However, separation of the two strands would create a local tightening of each individual strand but this is prevented by action of the enzyme, DNA topoisomerase. This enzyme catalyses the breaking and then reforming of a diester bond between two nucleotides in a single strand. (That is, once unwound, the enzyme re-establishes the link). Consequently, the enzyme is also known as a nicking–closing enzyme, which aptly describes its action. In order to prevent the two strands from, once again, interacting to produce the duplex, another protein binds to the strands, a single strand-stabilising protein (Figure 20.14). Within the replication bubble, the enzymes involved in replication have the opportunity to carry out their catalytic activities in a localised environment which results in duplication of DNA (Table 20.3).

Although this description summarises, albeit briefly, the processes involved in DNA duplication, it conceals a considerable amount of biochemistry. Nonetheless, the result is two identical double-stranded DNA molecules. This means of replication is known as semi-conservative since, in each new DNA duplex, one strand is identical to that from the parent molecule, i.e. it is conserved.

The simplest way of organising DNA duplication would be to start polymerisation at one end of the strand and proceed to the other end. There are, however, a number of constraints which preclude this simple procedure.

- Complete separation of the strands would create long, fragile and unwieldy single strands that could readily

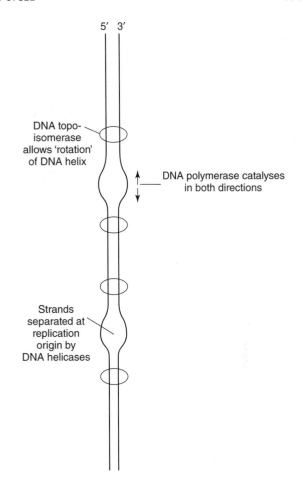

Figure 20.14 *Preparation for DNA replication.* At least three proteins are involved: DNA helicases disrupt hydrogen bonds between bases to allow the two strands to unwind; single-strand DNA-binding proteins stabilise the unbound strands in preparation for base pairings; DNA topoisomerase relaxes strain in the strands to facilitate polymerisation.

break or become damaged. This would restrict, if not prevent, accurate or complete duplication. To overcome this problem, only short stretches of the DNA duplex separate at any one time and duplication then takes place along these stretches.

- A property of the DNA polymerase enzyme is that it only proceeds in one direction, that is, from the 5′ hydroxyl to the 3′ hydroxyl end of the strand. However, in double-stranded DNA, the strands are orientated in opposite directions, that is, they are antiparallel (Figure 20.15). Polymerisation of nucleotides along the strand that runs in the 5′ to the 3′ hydroxyl group direction, from the beginning to the end of the strand, poses no problems. However, there is a problem for polymerisation in the opposite direction. The solution to this problem is quite ingenious, if a little complicated. In this case, the polymerase catalyses polymerisation only in short stretches,

Table 20.3 Some enzymes and binding proteins involved in DNA replication

Enzymes or binding proteins	Function
DNA helicases	Unwinds the two strands of a DNA duplex in preparation for polymerisation.
Single-strand DNA-binding proteins	Stabilise the unwound single strands of DNA in preparation for base pairing and polymerisation.
DNA topoisomerases	Relaxes strain in the strand to facilitate polymerisation.
Origin-binding proteins	Bind to DNA and unwind at the sites of the origins of replication.
Primases	Synthesise the short RNA primers that initiate DNA synthesis.
DNA polymerases	Polymerise the deoxy ribonucleoside triphosphates.
Ligases	Link fragments of the DNA once they are synthesised

It should be noted that all these enzymes are targets for inhibitors that act as antitumour (anticancer) drugs (Chapter 21).

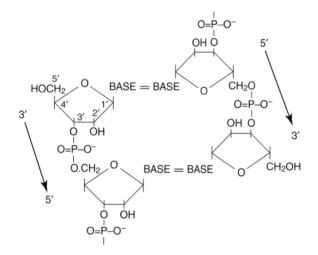

Figure 20.15 *Diagram of a small portion of the DNA duplex to illustrate antiparallel nature of the two DNA strands.* The bonds = between bases are the hydrogen bonds. The prime (') is necessary since the atoms in the base are numbered as are the atoms in the sugar. Numbers in the latter are therefore distinguished by a superscript prime on the pertinent carbon atom. The numbers in this diagram are indicated on only one ribose to keep its diagram in the simplest form.

so that it produces fragments of DNA, known as Okazaki fragments. To do this, the enzyme catalyses 'backwards', so that, over these short stretches, it polymerises in the 'correct' (i.e. the 5' to 3') direction. These fragments are then linked together by another enzyme, a DNA ligase. Since the former strand can be polymerised more rapidly, it is known as the leading strand, whereas the latter is called the lagging strand (Figure 20.16).

• A further problem is the lack of a mechanism for initiating polymerisation. DNA duplication must always start from an RNA primer, i.e. a short sequence of polymerised ribonucleotides. The sequence of the primer is achieved through base pairing with the short sequence

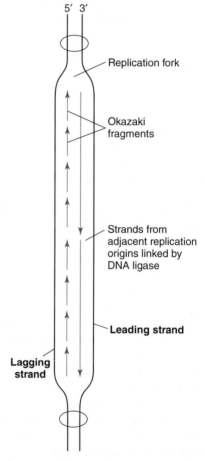

Figure 20.16 *An overview diagram to illustrate polymerisation of the deoxyribonucleoside triphosphates in both the leading and lagging DNA strands, by DNA polymerase.* The diagram illustrates the processes that take place in the replication bubble. In the leading strand, the adjacent sequences that are polymerised in short stretches between origins of replication (i.e. initiation points) have to be linked together by a DNA ligase. Similarly the much shorter sequences produced by the 'backward' movement of the polymerase in the lagging strand, i.e. the Okazaki fragments, are linked together by a DNA ligase.

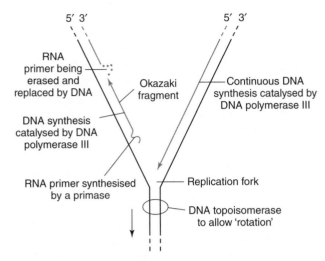

5′ 3′   5′ 3′

RNA
primer being
erased and   Okazaki
replaced by DNA   fragment   Continuous DNA
                                      synthesis catalysed by
                                      DNA polymerase III
DNA synthesis
catalysed by DNA
polymerase III

RNA primer synthesised   Replication fork
by a primase

                                DNA topoisomerase
                                to allow 'rotation'

Figure 20.17 *Diagram to illustrate the processes involved in DNA duplication of the leading and lagging strands.* The major DNA polymerase (polymerase III) is present as a complex with other proteins which assist in chain elongation. The RNA primers that initiate chain elongation are ten nucleotides long and are synthesised by a primase enzyme. The primase is a DNA-dependent RNA polymerase. Synthesis of the leading strand requires only one primer but, in the lagging strand, each Okazaki fragment (typically 100–200 nucleotides long) requires a new primer. The elongation of the chain is catalysed by DNA polymerase III which catalyses the condensation of adjacent deoxyribonucleoside triphosphates with the elimination of pyrophosphate. In the lagging strand, once an Okazaki fragment is synthesised, the polymerase must move backwards to a new RNA primer to produce another fragment. The primer must play an important role in directing the polymerase to the correct site for the next stage of polymerisation. Finally, the RNA primer must be removed. This is catalysed by another DNA polymerase that removes the ribonucleotides and replaces them with deoxyribonucleotides according to the base-pairing rules. A ligase enzyme then joins the fragments together.

of a single DNA strand at the initiation point. The RNA primer is subsequently hydrolysed and hence removed from the DNA. This sequence of ribonucleotides is replaced with a DNA sequence filler, also achieved by base pairing (Figures 20.16 and 20.17).

DNA polymerase III

DNA polymerase III is the main DNA replicating enzyme and consequently is involved in duplication of both the leading and the lagging strands. Besides the deoxyribonucleoside triphosphates as substrates, it requires a single-stranded DNA template and an RNA primer. The RNA primer is synthesised according to the DNA sequence at each replication fork. It is DNA polymerase I that is

involved in the removal of RNA primers from the Okazaki fragments in the lagging strand (Figure 20.17). The DNA polymerase not only catalyses formation of the diester bond and hence polymerisation but it also proof reads the immediate sequence that has been produced, i.e. the enzyme checks whether the newly incorporated base can form a base pair according to the Watson–Crick rules (Figure 20.4). If the rule has been broken, the last added nucleotide is removed, and then the enzyme resumes the polymerisation.

Comparison between DNA repair and phospholipid repair The processes that can lead to DNA damage and the type of damage are described in Chapter 9 and Appendix 9.6. The repair processes involve removal of the specific nucleotide(s) by an exonuclease and replacement of the nucleotide by a DNA polymerase. Since the strand must be broken to remove the damage (by an endonuclease) these parts of the strand must be repaired by a ligase. The process is known as excision–repair. Of interest, there is a degree of similarity between the removal of damaged polyunsaturated fatty acids from phospholipids in membranes and replacement with a new fatty acid by two enzymes, a deacylase and an acyltransferase (see above and Chapter 11), and excision-repair of DNA.

Excision

(i) Membrane phospholipid-(damaged FA) → damaged fatty acid + membrane-lysophospholipid

(ii) DNA-(damaged nucleotides) → (DNA minus nucleotide) + damaged-nucleotide

Repair

(i) membrane-lysophospholipid + acyl-CoA → membrane phospholipid + CoASH

(ii) (DNA minus nucleotide) + deoxyribonucleoside triphosphate → DNA + PPi

It is the damage to DNA in the epithelial cells of the skin that is usually considered to be the cause of the development of melanoma due to excessive exposure to sunlight (Chapter 21). However, an alternative or additional mechanism could be the damage to polyunsaturated fatty acids in membrane phospholipid in the epithelial cells. This could be due, as in the case of DNA damage, to the local production of free radicals (Appendix 9.6). The damaged polyunsaturated fatty acids (e.g. peroxidised or hydroperoxide fatty acids) will disrupt the membrane which might facilitate the binding of key proteins of proliferation to these membranes or result in the production of abnormal eicosanoids either of which could facilitate inappropriate proliferation.

Transcription and translation

Transcription is the term used to describe the synthesis of RNA from a DNA template. Translation is the process by which information in RNA is used to synthesise a polypeptide chain. In a little more detail, the genetic information encoded in DNA is first transcribed into a complementary copy of RNA (a primary RNA transcript) which is then processed to form messenger RNA (mRNA). This leaves the nucleus and is translated into a polypeptide in the cytosol. This then folds into a three-dimensional structure and may be further biochemically modified (post-translational modification) to produce a protein (Figure 20.18).

The relationship between the base sequence in DNA and the amino acid sequence in the protein is known as the genetic code. With four bases (A, C, G and T) 64 three-base combinations are possible to provide the code for the amino acids (e.g. GTA, CCG). All but three of these are used to code for the polymerisation of the 20 different amino acids (in fact, 21, see Chapter 8) to form a polypeptide chain that can then form a protein. Most amino acids are, therefore, coded for by more than one three-base combination (Appendix 20.2). The link between the three-base sequence (codon) on mRNA and the relevant amino acid is achieved through another RNA molecule, transfer RNA (tRNA), so that a different tRNA molecule exists for each of the amino acids. One part of the tRNA molecule binds covalently to its specific amino acid while another region forms complementary hydrogen bonds with the codon on the mRNA. As the tRNA molecules transfer their amino acids in sequence along the mRNA, the amino acids are linked to form the polypeptide (i.e. translation, see below for details).

A stretch of DNA that is transcribed as a single continuous RNA strand is called a transcription unit. A unit of transcription may contain one or more sequences encoding different polypeptide chains (translational open reading frames, ORF) or cistrons. The transcription unit is sometimes termed the primary transcript, pre-messenger RNA or heterogeneous nuclear RNA (hnRNA). The primary transcript is further processed to produce mRNA in a form that is relatively stable and readily participates in translation. In order to understand the primary need for processing of this RNA, the biochemical definition of a gene must be discussed.

For some years, it was considered that a gene was simply a contiguous sequence of bases within the DNA molecule (i.e. within the sense strand of DNA). In 1977, however, it was shown that this assumption, i.e. that there is a strict one-to-one relationship between the nucleotide sequence of a gene and the amino acid sequence of a polypeptide that it encodes, was not necessarily valid.

The stretch of DNA, i.e. the DNA template, that is described as a gene contains both coding and non-coding regions. The latter are known as introns (i.e. interfering sequences in the DNA). The coding regions are known as exons (expressed regions of the DNA). Nonetheless, the entire sequence of nucleotides in the template is transcribed to form the complementary sequence. To produce mRNA, the introns must be removed and the exons ligated, a process known as RNA splicing. That is, the introns are removed and the exons are then joined together, which is carried out by a particle known as the spliceosome (Figure 20.19). Splicing occurs in the nucleus so that, once formed, mRNA must leave the nucleus and enter the cytosol, where translation takes place. An advantage of the presence of introns is that it permits alternative splicing of the primary transcript which, if regulated, can lead to the production of different mRNAs that will produce slightly different peptides and hence proteins with different properties, and possibly different functions. Thus regulated variation in splicing could be of considerable significance in producing slightly different proteins in different tissues to match the function of the protein to the role of that particular tissue.

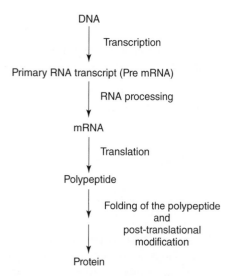

DNA

↓ Transcription

Primary RNA transcript (Pre mRNA)

↓ RNA processing

mRNA

↓ Translation

Polypeptide

↓ Folding of the polypeptide and post-translational modification

Protein

Figure 20.18 *The central dogma of molecular biology: a summary of processes involved in flow of genetic information from DNA to protein.* The diagram is a summary of the biochemical processes involved in the flow of genetic information from DNA to protein via RNA intermediates. This concept had to be revised following the discovery of the enzyme, reverse transcriptase, which catalyses information transfer from RNA to DNA (see Chapter 18). It may have to be modified in the future since changes in the fatty acid composition of phospholipids in membranes can modify the properties of proteins, and possibly their functions, independent of the genetic information within the amino acid sequence of the protein (See Chapters 7, 11 and 14).

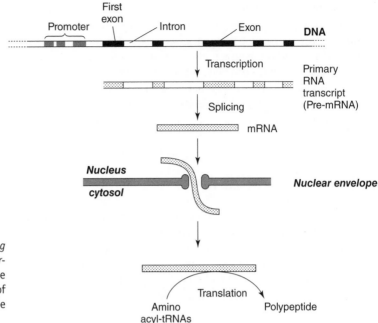

Figure 20.19 *Summary of transcription, RNA splicing entry of mRNA into the cytosol and polypeptide formation.* The difference in shading is to indicate the change from DNA to RNA. Splicing is just one of the four processes involved in the processing of the primary RNA transcript (Figure 20.20).

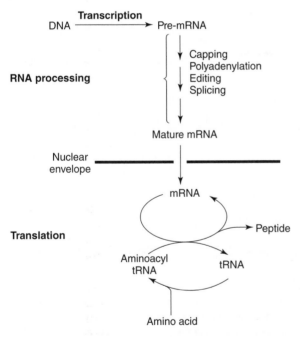

Figure 20.20 *Summary of transcription, RNA processing and polypeptide synthesis.* Polymerisation of the DNA template by RNA polymerase produces pre-mRNA (the primary transcript): this is transcription. The pre-mRNA is now processed, which involves capping, polyadenylation, editing and splicing (see text). The resultant mRNA transfers from the nucleus to the cytosol, where amino acids are polymerised to produce a polypeptide using the instructions present in the codons of the mRNA.

RNA processing

Splicing is just one of the four processes that can modify the mRNA molecule prior to transfer into the cytosol. The other three are cap formation, polyadenylation and editing of RNA. These are colectively known as RNA processing (Figure 20.20).

Cap formation This process is the addition of a single guanine base to the 5′ end of the RNA molecule. The guanine is attached to the terminal nucleotide via a triphosphate link. The guanine is methylated in position 7 of the base, catalysed by a methyltransferase. The cap plays a role in translation by facilitation of the binding of mRNA to the ribosome (see below).

Polyadenylation This process is the addition of many AMP molecules to the 3′ end of the RNA molecule. Once bound it is known as a poly A tail, since it is comprised of 100 or more such adenylate (AMP) residues. The tail is thought to stabilise the mRNA molecule in the cytosol, probably by preventing hydrolysis by an RNAase.

RNA editing This process involves insertion, deletion or modification of nucleotides in the pre-mRNA. The result of one or more of these modifications is that the mRNA is sufficiently modified that the properties of the proteins that are produced are slightly different in different tissues. A cytidine base can be converted to uracil in mRNA in a reaction catalysed by a deaminase enzyme. This can result in the formation of a stop codon. For example, apoB-100 is produced in the liver, where it is used as a structural protein in the lipoprotein particle VLDL, which is produced and secreted by the liver. However in the intestine, the messenger for apoB-100 is modified by a change from cytidine to uracil. The result is that a smaller version of apoB-100, that is apoB-48, is generated which is involved in stabilising the chylomicrons produced in the intestine (Chapter 4).

Initiation of transcription

RNA polymerase II is responsible for transcription of genes to produce the primary transcript that will, eventually, be converted to mRNA. The polymerase, in order to initiate transcription at the start point for transcription of the first exon in the gene, has to bind to specific short sequences of DNA. They act as binding sites for the RNA polymerase and also additional proteins that aid the binding of the polymerase and the initiation of transcription. These sequences are upstream of the start site for transcription of the first exon. They are collectively known as the promoter which, unfortunately, can give rise to the impression that it is a protein or a group of proteins. However, as indicated above, it is, in fact, a series of short nucleotide sequences which are not contiguous. Furthermore, not all the sequences in the promoter for any gene are known, and probably vary from one gene to another, but one that is very common is rich in adenine- and thymine-containing nucleotides (e.g. TATAAA) known as a TATA box. It is a general consensus sequence for transcription of genes. The combination of the RNA polymerase, the promoter and the essential accessory proteins is known as the transcription–initiator complex (Figure 20.21(a)). As an example of transcriptional control the regulation of the expression of enzymes necessary for metabolism of a nutrient is presented in Figure 20.21(b).

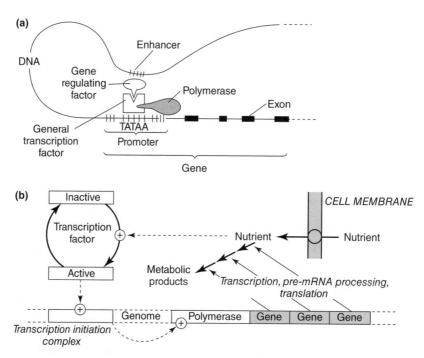

Figure 20.21 **(a)** *Summary of the factors that activate RNA polymerase and hence can regulate transcription of a gene.* The promoter comprises three sequences, including the TATA box. A general transcription factor assists the binding of the polymerase to the promoter to form the transcription–initiator complex. The gene comprises exons, introns and the promoter. The enhancer can activate (or inhibit) the activity of the promoter via a gene regulating factor, the identity of which is not known. The enhancer can be affected by specific transcription factors. The enhancer is separated from the promoter within the DNA sequence by many nucleotides but the DNA loops back so that the enhancer is physically close to the promoter. **(b)** *Regulation of gene expression by a nutrient.* This is a diagram illustrating the general principles by which a particular compound – in this case, a nutrient – can influence the expression of genes that produce enzymes necessary to metabolise the nutrient. The nutrient activates a transcription factor that binds to and leads to activation of the transcription–initiator complex which initiates transcription. This leads to formation of processing to mRNA and then to the expression of enzymes that can metabolise the nutrient. A specific example of such a process, glucose stimulates gene expression of enzymes involved in fatty acid synthesis from glucose so that glucose is converted to fat, is described in Chapter 11.

Transcription factors

In several chapters in this book, the regulation of biochemical processes by activation of gene expression by changes in the level, distribution or phosphorylation state of transcription factors is described and their significance stressed (e.g. see below for role of transcription factors in control of the cell cycle). Transcription factors have their effects by binding to promoter and/or enhancer sites. The region of the gene that initiates transcription is the promoter, which is the position on DNA where the RNA polymerase binds to start the polymerisation process. It also is the site for binding proteins that regulate the process, i.e. transcription factors. Another DNA sequence that regulates transcription, via transcription factors, is the enhancer sequence. It can increase the rate by up to 1000-fold. However, enhancers can be located well away from the initiation site, even more than 3 kilobases away. This implies that the three-dimensional shape of DNA in the nucleus positions the enhancer close to the transcription–initiator complex (Figure 20.21(a)). Transcription factors may be present in the cell as inactive factors and then activated by a specific process (e.g. phosphorylation) or may be expressed by activation of another gene (see p 493, where there is a description of how a protein appressed by a tumour suppressor gene acts as a transcription factor).

To influence the activity of RNA polymerase, transcription factors bind to DNA at enhancer sites and/or promoter sites. Four different structures of these factors are known. They possess structures that allow specific binding to DNA. They are given names that indicate the key protein domains that encourage this binding, as follows:

(i) Zinc finger. In the zinc finger protein, the DNA-binding region contains many repeats of 30 amino acid sequences, which contain a zinc atom, and which protrube from the protein and appear as 'fingers'.

(ii) A leucine zipper protein is a protein in which a leucine residue occurs every seventh amino acid within an α-helix, such that the leucines occurs every two turns on the same side of the helix, as in a zip.

(iii) A helix-turn-helix factor consists of two α-helices that are separated by a β-turn.

(iv) The helix-loop-helix is a protein in which two α-helices are linked by a loop of amino acids.

The mechanism by which transcription factors increase (or decrease) the rate of transcription is unclear. It is likely that, bound to the enhancer or promoter DNA sequences, they interact with other proteins to stimulate on inhibit formation of the transcription–initiator complex.

Translation: polypeptide synthesis

Translation involves three stages: initiation, elongation and termination. A brief summary of these processes is provided below. However, the first step in polypeptide synthesis, from intracellular amino acids, is the formation of aminoacyl-tRNA. This reaction is particularly important so that the biochemistry is discussed in some detail. In addition, it is also important in the regulation of the rate of translation (see below).

The reactions that form aminoacyl-tRNAs and their significance in translation

Transfer RNAs are small, single-stranded polynucleotides (70–90 bases long). The tRNA molecule is linked to its specific amino acid in a reaction that is catalysed by the enzyme tRNA-aminoacyl synthetase. It occurs in two stages:

(i) The amino acid reacts with ATP to produce an aminoacyl adenylate (with the release of pyrophosphate).

(ii) The enzyme transfers the adenylate to the terminal hydroxyl of the tRNA.

Once the amino acid is bound to tRNA, the complex is known as *charged tRNA*. The sequence of reactions is:

$$\text{amino acid} + \text{ATP} \rightarrow \text{aminoacyl adenylate} + \text{PP}_i$$

$$\text{tRNA} + \text{aminoacyl adenylate} \rightarrow \text{aminoacyl-tRNA} + \text{AMP}$$

There is one separate tRNA for each amino acid and one separate specific synthetase. The enzyme must bind not only the correct amino acid but also the correct tRNA, so that each synthetase has specific recognition sites for both. Transfer RNAs contain a three-base sequence that is an anticodon, which binds to its complementary codon on messenger RNA. The importance of the synthetase in relation to fidelity of translating the information in messenger RNA is indicated by the fact that, once an amino acid is bound to tRNA, its identity as an amino acid is dictated by the anticodon site on the transfer RNA and not by the amino acid itself. (The enzyme can be considered as a dictionary, since it provides a cross-reference between the nucleic acid and amino acid languages.)

The three processes are now described, but in the order elongation, initiation and termination, since, in this order, the overall process is perhaps more readily understood.

Elongation The process takes place on ribosomes which bring together the many components required in the correct sequence. Each ribosome consists of a large and a small subunit, the latter bearing two binding sites, the P-site, which binds the growing peptide chain, and the A-site which binds the incoming amino acid in the form of the aminoacyl-tRNA. The basic process in elongation is the transfer of the peptide chain from site P to condense with the amino acid at the A-site, which is catalysed by peptidyltransferase. This creates a peptide chain longer by one amino acid and linked to site A through the tRNA of the 'last' amino acid. To continue the process, the peptidyl-tRNA must be transferred from the A-site back to the P-site, with the release of the tRNA. The next aminoacyl-tRNA can now bind at the vacant A-site, allowing the process to continue and hence the peptide is elongated.

Initiation The process outlined above begins, not at the physical 5'-hydroxyl end of the mRNA, but at an AUG initiating codon a little distance from it. AUG is the initiating codon for methionine, which initiates the whole process of translation. At the beginning, both the A and P sites on the ribosome are free but peptide synthesis can only occur if both sites are occupied. The problem is overcome by the binding of an initiating factor (IF) to the initiator methionine-tRNA. This complex, i.e. the initiating factor and methionine-tRNA, binds to the P-site on the small ribosome. When this ribosome unites with the large ribosome, the initiating factor is displaced allowing an aminoacyl tRNA to bind to the A-site, so that both the A and P sites are occupied and polymerisation can begin. It can be considered that the complex, methionine tRNA–initiating factor, is a molecular 'disguise' for a peptidyl tRNA, so that both sites are occupied. The result of this 'biochemical trick' is that the eventual resultant peptide has an extra methionine at the N-terminus. At the end of the process, this extra methionine is removed.

The same mRNA molecule can be read many times, so that further initiations can begin before the whole mRNA has been translated. Therefore, several ribosomes (small and large) may be attached to a single mRNA molecule to form a polysome. The number of polysomes in a cell is sometimes used as a rough indication of the extent of peptide synthesis in that cell.

Termination Three codons (UAA, UAG and UGA) are stop codons which do not code for any amino acid but, instead of attaching to a tRNA molecule, they bind a protein release factor. When one of these factors is encountered by the ribosome, peptidyl transfer is aborted, the completed polypeptide chain released by hydrolysis and the ribosome subunits separate. The N-terminal methionine unit is then removed from the polypeptide chain.

A description of the individual reactions in the overall process for peptide synthesis

The sections presented above provide an account of the separate topics into which translation can be divided. These act as an introduction to the current section, in which a description of the individual reactions in peptide synthesis is presented in a single diagram, i.e. a diagram that encapsulates the whole process (Figure 20.22). An analysis of each separate reaction provides a simple explanation of the interactions that are required in a sequential manner to form the various complexes in the pathway, the activities of which result in the synthesis of, initially, a dipeptide but then a growing peptide. The repetition of the formation of the complexes for each amino acid results in the synthesis of the final peptide, as dictated by the base sequence in the mRNA.

This informative diagram not only presents details of the individual reactions but shows how it is possible to consider peptide synthesis as one large biochemical pathway. Although the pathway is long and appears complex, the biochemistry within each reaction is, as in most other pathways, straightforward, the information present in Figures 20.14, 20.16, 20.17 and 20.22 provides a description of the essential processes that constitute the central dogma that is summarised in Figure 20.18, i.e. from DNA to protein which includes the process of peptide synthesis.

Polypeptide synthesis as a metabolic pathway

Polypeptide synthesis (i.e. translation) can be depicted as a sequence of reactions involving co-substrates that are recycled during flux through the pathway (Figure 20.23(a)). When viewed in this way, it can be considered as a metabolic pathway, similar to the pathway of glycolysis (Figure 20.23(b)). This comparison might remove some of the concern of complexity when studying the process of translation.

Mechanisms for the regulation of rate of peptide synthesis

The discussion provided above explains how different peptides can be synthesised, at any one time during the cell cycle, via activation of specific genes to produce different mRNA molecules. However, in addition to the regulation of the type of protein produced, regulation of the rate of synthesis is important to ensure that a sufficient amount of protein is available at each stage in the cycle.

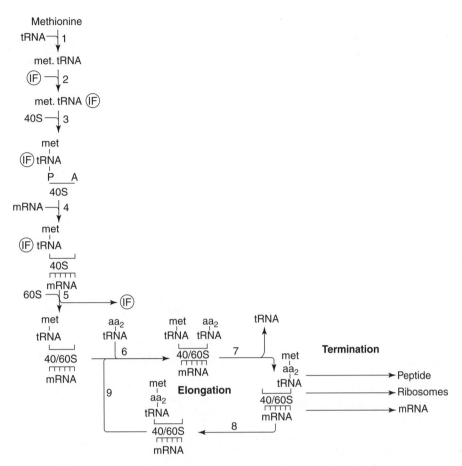

Figure 20.22 *Details of the sequence of reactions involved in the process of translation: from amino acids to a peptide.* Each step in the overall process is identified by a number. *Step 1*, in this diagram, involves the amino acid methionine, which is the initiating amino acid that starts the whole process of peptide synthesis for poly-peptide. However, *Step 1* must occur repetitively, converting all the amino acids that are required for the peptide synthesis to their respective amino acyl-tRNA complexes, which then bind to the A-site on the ribosome prior to condensation with the extending peptide chain. *Step 2.* An initiation factor (IF) forms a complex with methionine-tRNA. *Step 3.* The small ribosome (40S ribosome) joins to form a complex, in which met-tRNA-IF binds to the P-site on the ribosome. *Step 4.* Messenger RNA joins the complex. *Step 5.* The large ribosome (60S) joins the complex and the initiation factor dissociates from the complex. *Step 6.* The next amino acid (aa_2) (i.e. the first amino acid in the actual peptide, see below for explanation) joins the complex, as tRNAaa_2, as directed by the codon on the messenger RNA. It should be noted that several (many) initiation factors are involved but the function of all of them is not understood. *Step 7.* The two amino acids are linked together by a peptide bond to form a peptide. This is catalysed by the enzyme peptidyltransferase, which is a component of the ribosome. This is the beginning of the process of elongation, which requires an additional protein known as an elongation factor (there are also several elongation factors). At this stage, also, the first tRNA molecule now dissociates to return to reaction 1. (The importance of this in regulation of the rate of the process is described below.) There are now *two* choices:

(i) If the complete peptide has not yet been synthesised, the process of elongation continues, via *Steps 8 and 9*. The ribosome moves one codon along the mRNA, and the growing peptide chain is moved from the A-site to the P-site, so that the whole process can begin once more, and another aminoacyl-tRNA (i.e. tRNAaa_3) binds to the A-site on the ribosome, directed by the codon on the mRNA. *Step 6* continues the process of peptide bond formation, since another amino acid is now available at the correct position for catalysis by peptidyltransferase.

(ii) However, if a termination codon on the mRNA is reached by the ribosome, the peptide is released: this is the *termination process*. At this stage, the ribosomes and the mRNA dissociate. The mRNA rejoins to continue the process at *Step 4* and the ribosomes rejoin at *Steps 3 and 5*. (Note, the S in 40S and 60S ribosomes indicates the size as detected, originally, in the ultracentrifuge. S refers to T. Svedberg, who developed the techniques of ultracentrifugation.)

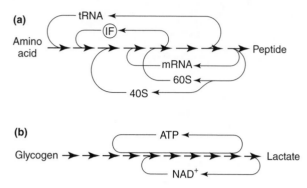

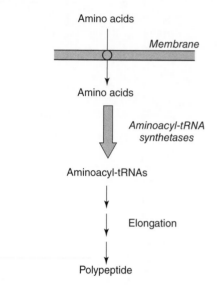

Figure 20.23 (a) *Polypeptide synthesis illustrating recycling of mRNA, ribosomes and tRNA.* The process of polypeptide synthesis (translation) can be considered as a linear biochemical pathway with requirements for what are effectively co-substrates (Chapter 3). These include transfer tRNA, 40S and 60S ribosomes, and mRNA: this is similar, in principle, to a 'classical' metabolic pathway (e.g. glycolysis). **(b)** *Glycolytic pathway illustrating recycling of ATP and NAD⁺.* ATP is utilised in initial reactions and regenerated in the later reaction (Chapter 6).

Figure 20.24 *The physiological pathway of polypeptide synthesis.* The flux-generating step is that catalysed by the aminoacyl-tRNA synthetases, indicated by the broad arrow. The assumption implicit in this interpretation is that the physiological pathway starts with the intracellular amino acids and ends with the peptide that is formed in the elongation and termination processes. For the majority of enzymes, the concentration of intracellular amino acids is higher than the K_m for the synthetase (Chapter 3).

The procedures for studying mechanisms of regulation of the rates of biochemical processes are described in Chapter 3. These procedures are applied to peptide synthesis in this discussion. There are three questions that must be answered prior to discussion of mechanisms. (i) Which reactions are non-equilibrium? (ii) Which is the flux-generating reaction and how is it regulated? (iii) What reactions other than the flux-generating steps are regulated by external factors, i.e. external to the intermediates in peptide synthesis?

The answers to these questions are:

(i)   It is likely that most if not all the reactions in the process are non-equilibrium, but a study of the pathway to identify non-equilibrium steps, as described in Chapter 2, has not been carried out.

(ii) The formation of the tRNA–amino acid complex is the flux-generating step, which is catalysed by the aminoacyl-tRNA synthetases. That is, the enzymes are saturated with their specific amino acid substrate, since the intracellular concentrations of amino acids are in excess of the K_m of the synthetases. (Note that the intracellular concentrations of amino acids are much higher than those of the extracellular amino acids – Chapter 8.) Not only is the flux-generating step an important step in the regulation of the rate of synthesis but it defines the physiological pathway of peptide synthesis, which is presented in Figure 20.24. One mechanism for regulation of the synthetases probably the primary mechanism, is changes in the

concentrations of free tRNAs for each amino acid (Figure 20.25).

(iii) The two reactions in translation that are regulated by external factors are (a) the initiation step (step 2 in Figure 20.22), which is regulated by initiation factors (see below); and (b) the translation of mRNA (steps 6 and 7) which is regulated by a protein kinase. (The mechanism of regulation by the initiation factors is described below.)

A fourth question is how regulation at these steps can communicate with the flux-generating step. An increase in the rates of steps 2, 6 and 7 would not result in an increase in peptide synthesis unless they communicate with the flux-generating step. This is usually achieved by feedback mechanisms. The mechanism, in this case, is feedback communication rather than inhibition. This is achieved via changes in the concentrations of free (i.e. uncharged) tRNAs. Initially, tRNAs are bound to the ribosomes in the process but are released as free tRNAs in step 7, after 6. Once released, the free tRNAs return to the flux-generating step, as the co-substrate for another synthetase reaction. That is, changes in the concentration of uncharged tRNAs regulate the activity of the individual synthetases (Figure 20.25).

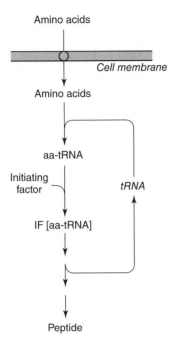

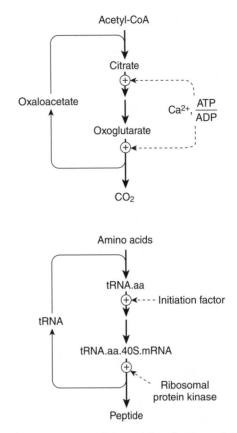

Figure 20.25 *Regulation of the activities of the aminoacyl-tRNA synthetases by the concentrations of free tRNAs (i.e. uncharged tRNA).* Changes in the concentrations of free tRNAs provide the mechanism for communication between control via the initiation factor (Figure 20.20) and ribosomal protein kinase (steps 6 and 7) and the flux-generating step.

Figure 20.26 *Regulation of the activity of aminoacyl-tRNA synthetases and hence the rate of peptide synthesis by the concentration of free tRNAs, and regulation of activity of citrate synthase and hence the rate of the Krebs cycle by the concentration of oxaloacetate.* Note that there are two positions in which external regulation occurs in the peptide sythetase pathway and two positions in the Krebs cycle (Chapter 9).

Despite the complexity of translation it is helpful to note that the mechanism of regulation of the rate is similar to regulation of flux through the Krebs cycle. The flux-generating step in the cycle is that catalysed by citrate synthase which is saturated with the compound that is oxidised in the cycle, that is, acetyl-CoA. The activity of citrate synthase is regulated by changes in concentration of oxaloacetate, the co-substrate for the process (Chapter 9). The concentration of the latter is controlled by other steps in the cycle, particularly isocitrated dehydrogenase and oxoglutarate dehydrogenase (Figure 20.26).

Simplicity in the regulation of peptide synthesis

During peptide synthesis, the reactions catalysed by the various aminoacyl-tRNA synthetases must occur repetitively for each amino acid. Simplicity in regulation is provided by the fact that all these different synthetases are saturated with their respective amino acids, that is, the K_m values of all the synthetases for their respective amino acids must be much lower than the concentrations of the amino acids, so that the enzyme activity is regulated solely by the changes in concentration of the co-substrates, i.e. the free tRNAs. This property, together with external regulation at steps 2 and 7, provides a relatively simple mecha-

nism for regulation of the rate of the whole process of peptide synthesis. This emphasises that regulation of the initiation step in translation must be of key importance in ensuring that sufficient proteins have been synthesised for the essential processes in the cycle to take place during the G_1 and G_2 phases (see below).

Regulation of initiation by the initiating factors

The initiation factor proteins regulate the formation of the complex between a tRNA-amino acid, the 40S ribosome and mRNA. There are at least ten initiation factors but the roles of all these factors are not known. One key factor is usually abbreviated to eIF2 but is abbreviated to IF in the present discussion. The mechanism of regulation of the

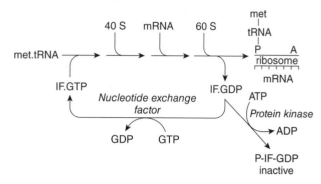

Figure 20.27 *A simplified description of the mechanism of regulation of the activity of the initiation factor for initiation of peptide synthesis.*

activity of this initiation factor involves a protein kinase and a G-protein (see Figure 20.27).

In this figure, the initiation factor is presented as IF. In other publications, the initiation factor is sometimes presented as eIF2: in this abbreviation, e indicates that it is an initiation factor in eukaryotic cells, since the factor in prokaryotic cells is quite distinct, and the number 2 is to indicate that, although there are many initiation factors, it is number 2 that appears to be directly involved in regulation of initiation, via a G-protein.

The initiation factor, IF, forms a complex with GTP and methionine-tRNA which promotes binding to the 40S small ribosome, which increases the rate of the reaction. As the process proceeds to form a complex with the 60S ribosome, GTP is hydrolysed to GDP and the initiation factor is now inactive. It is re-activated by nucleotide exchange, i.e. GDP is exchanged for GTP. The nucleotide exchange factor is a different initiation factor to eIF2. (The role of GTP/GDP exchanges in the regulation of the activity of many proteins is discussed in Chapter 12.) In an additional reaction, phosphorylation of the IF–GDP complex, by a protein kinase, increases the affinity of the factor for GDP, so that nucleotide exchange is inhibited and the initiation factor cannot be activated. Regulation at this step must involve the nucleotide exchange factor and the protein kinase. However, the mechanism(s) is not known for regulation of peptide synthesis in the cell cycle.

Note that the protein kinase, which phosphorylates the IF–GDP complex is structurally similar to the protein kinase that is activated by double-stranded RNA, i.e. the genome of some viruses. This protein kinase phosphorylates the IF–GDP complex in an infected host cell, so that viral peptide synthesis is inhibited and the virus cannot multiply. Synthesis of this kinase is stimulated by the cytokine, interferon, which is released by virus-infected cells as an early-warning system to adjacent cells not yet infected (Chapter 17: see Figure 17.32).

Mitosis

Mitosis is defined strictly as the process of chromosome segregation that occurs during the division of eukaryotic cells, but the term usually includes the division of the cell into the two daughter cells, the process called cytokinesis. The name, mitosis, derives from the Greek word for a thread, since, in the early stages of mitosis, the chromosomes become visible as thread-like structures. Mitosis was discovered with the light microscope, when microscopists identified four stages, which are described in Figure 20.28. Mitosis is the terminal stage of the cell cycle, when the cycle is presented as a linear pathway (see Figure 20.1(b)). During interphase, the mass and volume of the cell, including the macromolecules, double, due to completion of all the synthetic processes (including the preliminary events that prepare for the syntheses). All these processes are described or are referred to in the above discussions.

Meiosis

The process of meiosis is introduced at the beginning of this chapter and its significance in sexual reproduction is described in Chapter 19, since it occurs during formation of ova and spermatozoa. It is presented in this chapter since it is best understood by direct comparison with mitosis.

All nucleated cells except eggs and sperm (gametes) carry two sets of genetic material in their nuclei, one from the individual's mother and one from the father. In human cells there are 22 pairs of homologous chromosomes plus an X chromosome and a Y chromosome. In the formation of gametes, this must be reduced to a single set, that is, the diploid number of chromosomes (46) must be reduced to the haploid number (23).

This reduction is achieved through meiosis, a process in which two successive cell divisions occur without intervening DNA duplication. As with the cell cycle, DNA duplication will have already taken place prior to the cell division, so that during the initial prophase each pair of identical DNA molecules forms a pair of chromatids attached to each other at a centrosome. In contrast to mitosis, however, homologous chromosomes now pair up (metaphase I). Note that in a pair of homologous chromosomes each will contain the same genes (one of the pair having come from the mother and one from the father) but many of the alleles will be different.

The first division of meiosis results in two daughter cells, each of which contains one complete (haploid) set of chromosomes. The second division of meiosis (which follows the same course as mitosis) results in these two cells becoming four, each of which can develop into a gamete (Figure 20.29).

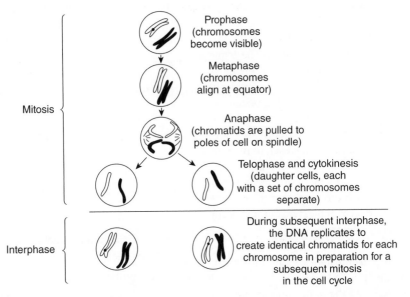

Mitosis
- Prophase (chromosomes become visible)
- Metaphase (chromosomes align at equator)
- Anaphase (chromatids are pulled to poles of cell on spindle)
- Telophase and cytokinesis (daughter cells, each with a set of chromosomes separate)

Interphase
- During subsequent interphase, the DNA replicates to create identical chromatids for each chromosome in preparation for a subsequent mitosis in the cell cycle

Figure 20.28 *Diagrammatic representation of mitosis in a cell with a single pair of homologous chromosomes.* In prophase, the chromatin condenses into chromosomes, each of which consists of a pair of chromatids that have been formed by replication during interphase, and the nuclear envelope disappears. In metaphase, each chromatid attaches to the spindle fibres (microtubules) at a centre point, the centromere. In anaphase, the two chromatids of each chromosome become detached from each other and move to opposite poles of the cell along the microtubules. In telophase, the chromatids have reached the poles. Two nuclear envelopes then form and enclose each new set of chromatids, now once again called chromosomes. The microtubules disappear and the chromosomes uncoil and re-form into the long chromatin threads. Finally the cell membrane is drawn inward by a band of microfilaments to form a complete constriction between the newly formed nuclei, and two new cells are formed. The process is called cytokinesis.

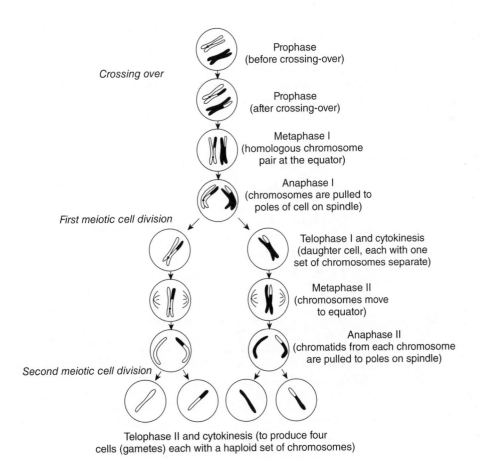

Crossing over
- Prophase (before crossing-over)
- Prophase (after crossing-over)
- Metaphase I (homologous chromosome pair at the equator)
- Anaphase I (chromosomes are pulled to poles of cell on spindle)

First meiotic cell division
- Telophase I and cytokinesis (daughter cell, each with one set of chromosomes separate)
- Metaphase II (chromosomes move to equator)
- Anaphase II (chromatids from each chromosome are pulled to poles on spindle)

Second meiotic cell division

Telophase II and cytokinesis (to produce four cells (gametes) each with a haploid set of chromosomes)

Figure 20.29 *Diagrammatic representation of meiosis in a cell with a single pair of homologous chromosomes.* The nuclear membrane is not shown. In reality the chromosomes would not be visible in early prophase.

If this process occurred simply as described above, two out of the four daughter cells would contain the same alleles as the original father and two the same alleles as the original mother, that is, genetic variation would not occur. To ensure that variation does occur, the phenomenon of crossing-over takes place during the initial prophase. Sections of chromatids exchange with corresponding sections in homologous chromatids so that each of the four chromatids, which subsequently separate, contains combinations of alleles derived from both the individual's father and mother. The points at which crossing-over takes place are known as chiasmata and, on average, two or three occur in each chromosome during meiosis.

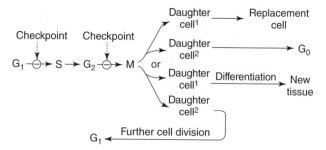

Figure 20.30 *Checkpoints in the cell cycle and fate of cells after mitosis.*

Velocity of movement of chromosomes   The movement of chromosomes along the spindle to the poles is achieved via the microtubule system and the motor protein, kinesin or dynein. Movement is 'powered' by ATP hydrolysis. This process is similar to the movement of vesicles along the axons of nerves, which is known as axonal flow (Chapter 14). Chromosomal movement is, however, very much faster than that achieved by axonal flow. The high speed is necessary because the DNA in chromosomes is not so well protected from damage, caused by, for example, free radicals, radiation or pollutants, as it is when complexed with protein in chromatin. In this form, they are protected from such factors. Speed is therefore essential in the separation to each pole since, once at the pole, the nuclear envelope can form to protect the DNA from chemicals within the cytosol and allow the formation of chromatin in which the DNA is protected by the associated proteins (e.g. histones). It is not known if this 'high speed' requires a greater rate of ATP utilisation i.e. a greater requirement for ATP generation.

Regulation of the cell cycle

The regulation of biochemical processes, described so far, in this book, is primarily the regulation of flux; that is, the amount of biochemistry that is achieved in a certain period of time. In contrast, in the cycle, at least other than in the embryo, time is much less important. The important decision that a cell has to take is whether there are sufficient materials and fuels available in the preceding stage to allow the next step to proceed satisfactorily. If not, then the cycle must be arrested to prevent entry into the next phase. The cell, therefore, has to check if all the conditions are satisfactory, so that these positions of regulation are known as checkpoints. There are at least two; one is present in the G_1 phase, the other in the G_2 phase (Figure 20.30). For example, DNA duplication can only proceed if:

(i)   There is sufficient activity of enzymes that synthesise the ribonucleotides and deoxyribonucleotides to produce sufficient amounts of each deoxynucleotide for complete duplication of DNA.

(ii)   The template DNA is not damaged.

(iii)   The concentration ratio ATP/ADP is sufficient that, upon hydrolysis, enough energy is available to 'power' the reactions to which it is coupled. (This is of key importance since there are many reactions or processes that require ATP hydrolysis during the cell cycle). The maintenance of a high ATP/ADP concentration ratio in the cytosol and presumably in the nucleus is discussed in Chapter 9.

(iv)   That sufficient fuel is available to generate sufficient ATP to carry out all the reactions involved in DNA duplication.

As might be expected from other mechanisms of regulation described in this text, phosphorylation and dephosphorylation of key proteins is the main mechanism for regulating the cycle, i.e. reversible phosphorylation, also known as interconversion cycles (discussed in Chapter 3). In the cell cycle, several of these interconversion cycles play a role in control at the checkpoints. Two important terms must be appreciated to help understand the mechanism of regulation of the cycle: the phosphorylation of proteins is catalysed by specific protein kinases, known as cell-division kinases (cdck) or cell cycle kinases (cck) and these enzymes are activated by specific proteins, known as cyclins.

It is an increase in the concentration of cyclins that stimulates the activity of the cycle kinases. The concentration of the cyclins is controlled by the balance between the rates of synthesis and degradation.

$$\text{amino acids} \xrightarrow[\text{translation}]{\text{transcription}} \text{cyclin} \xrightarrow{\text{proteolysis}} \text{amino acids}$$

There are several cyclins, but they fall into two main classes:

(i)   Cyclins which are produced during the G_1 phase and bind to and activate kinases that regulate entry into the S phase of the cycle.

(ii) Cyclins that are produced during the G_2 phase that bind to and activate kinases that control the entry into the M phase:

$$G_1 \longrightarrow S \longrightarrow G_2 \longrightarrow M$$

$$\begin{array}{cc} \uparrow & \uparrow \\ G_1 \text{ cyclins} & G_2 \text{ cyclins} \end{array}$$

The principle underlying regulation of the concentration of a cyclin

The synthesis of cyclin proteins is achieved at the transcription/translational levels by activation of genes, and degradation is achieved by proteolysis. Hence, the concentration of the cyclins, and, therefore, the activity of the cell cycle kinases, and the whole cycle, depends upon the balance of protein synthesis and proteolysis. The principle underlying this mechanism of regulation of the concentration of specific proteins is identical to that underlying the regulation of the activity of many enzymes (or other proteins) by covalent modification. For example, in reversible phosphorylation, a protein kinase catalyses phosphorylation of another enzyme which increases the activity from zero to a high activity, whereas dephosphorylation converts an enzyme that has a high catalytic activity to an enzyme that has zero catalytic activity. In the case of a cyclin, the process of translation increases the concentration of the cyclin from zero to a high concentration, whereas proteolysis leads to a decrease from a high concentration to a zero concentration of the cyclin (Figure 20.31).

The remarkable magnitude of these changes is due to the fact that translation and proteolysis are non-equilibrium

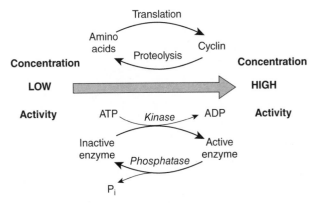

Figure 20.31 *The principle of interconversion cycles in regulation of protein activity or changes in protein concentration as exemplified by translation/proteolysis or protein kinase/protein phosphatase. They result in very marked relative changes in regulator concentration or enzyme activity. The significance of the relative changes (or sensitivity in regulation) is discussed in Chapter 3. The principle of regulation by covalent modification is also described in Chapter 3. The modifications in cyclin concentration are achieved via translation and proteolysis, which, in effect, is an interconversion cycle. For the enzyme, they are achieved via phosphorylation and dephosphorylation reactions. In both cases, the relative change in concentration/activity by the covalent modification is enormous. This ensures, for example, that a sufficient increase in cyclin can occur so that an inactive cell cycle kinase can be converted to an active cell cycle kinase, or that a cell cycle kinase can be completely inactivated. Appreciation of the common principles in biochemistry helps in the understanding of what otherwise can appear to be complex phenomena.*

reactions, as are phosphorylation and dephosphorylation reactions. That is, the relative change (percentage change) in the concentration of cyclin is enormous, which results in an 'all or nothing' effect, which is so essential for satisfactory regulation of each stage of the cell cycle. The effect of this is that, when the cyclin concentration is zero, the cycle is arrested, but, when the concentration is high, the cycle proceeds through that checkpoint (Figure 20.31). A discussion of this principle in biochemical regulation is provided in Chapter 3. Its biochemical significance is underlined by the fact that almost one thousand processes are regulated by reversible phosphorylation, and many are regulated by protein synthesis and degradation (i.e. interconversion cycles).

Mechanism of regulation of cell cycle kinases and activation of checkpoints

The mechanisms are best explained by provision of answers to two questions:

(i) What is the mechanism by which cyclins regulate the activity of cyclic-dependent cell cycle kinases and how do the latter control the process of proliferation?

Box 20.3 Cell cycle terminology

Unfortunately, this field is made difficult for a student by the terminology used by biochemists. For example, there is a protein kinase that regulates the cycle between G_2 and M, i.e. the second restriction point. This protein is known as maturation (or mitosis) promoting factor, since it promotes entry into mitosis. It phosphorylates a protein, probably a transcription factor, in the nucleus. The kinase has a molecular mass of 34 kDa. Hence it is known as p34 cell division kinase, abbreviated to p34-cdc and, since it is regulated at restriction point 2, it is known as p34-cdc-2 protein, which is sometimes written as p34^{cdc-2}. This kinase is normally inactive until it binds a cyclin. Hence the active maturation-promoting factor is, in fact, a protein kinase–cyclin complex, which is referred to as p34^{cdc}–cyclin complex. It is hoped that this piece of information may help a student (or lecturer from another field) to understand one part of a review article that contains the abbreviation p34^{cdc}–cyclin complex, without explanation, or other similar pieces of biochemical shorthand.

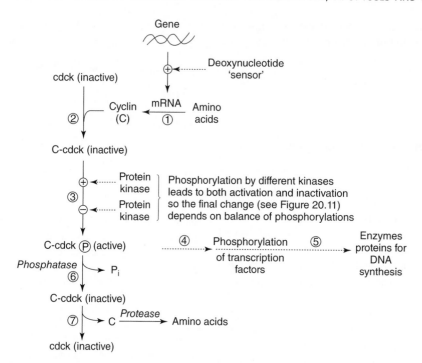

Figure 20.32 *A hypothesis for regulation of cyclin-dependent cell cycle kinase by changes in the concentration of a cyclin and how this accounts for the synthesis of enzymes required for DNA duplication.* The sequence of events that leads to changes in the concentration of a cyclin and the effects on enzymes involved in DNA synthesis are as follows:

1. A hypothetical sensor assesses the concentrations of deoxyribonucleotides. The sensor will, if the concentrations are sufficient for DNA synthesis, activate a transcription factor that leads to transcription of a gene that expresses a cyclin (C).

2. The cyclin binds to a cyclin-dependent cell division cycle kinase (cdck). (In the absence of cyclin, this enzyme is sometimes known as a *naked* protein kinase.)

3. The kinase can now be phosphorylated by several protein kinases at different positions in the protein which activates the enzyme but kinases are also present that inactivate the enzyme, so that the final state of activation depends upon the balance of different kinases.

4. Once activated, the kinase phosphorylates transcription factors that result in activation of these factors by, for example, translocation of these factors into the nucleus.

5. These transcription factors activate genes that express the enzymes for purine and pyrimidine deoxyribonucleotide synthesis and DNA duplication.

6. The active cyclin-dependent cell cycle kinase is dephosphorylated by a protein phosphatase which inactivates the enzyme.

7. Finally, the cyclin is degraded by a proteolytic enzyme so that the cyclin concentration decreases to zero and the kinase is now completely inactive and cannot be activated again until the cyclin is, once again, synthesised and binds to the naked kinase, i.e. degradation of the cyclin arrests the cycle.

This system ensures that the cdck is either completely active, to facilitate sufficient enzyme synthesis that the deoxynucleotides reach a satisfactory level for DNA synthesis, or completely inactive, so that DNA synthesis cannot occur.

(ii) What is the mechanism by which the cell 'knows' that there are sufficient material and fuels available for the cycle to proceed through each particular discontinuous phase (i.e. DNA duplication and mitosis)?

An answer to question (i) is presented in Figure 20.32. In summary, a cyclin is synthesised after activation of a gene and it then binds to a cell cycle kinase, and the kinase is then susceptible to phosphorylation and/or dephosphorylation: in this case, both effects lead to activation of the enzyme. This enzyme activates key factors in the nucleus. The phosphorylated proteins stimulate transcription factors that will lead to activation of genes that express the enzymes and proteins that are necessary for DNA duplication (Table

20.3). Once these enzymes and proteins are available, they will carry out the process described in Figures 20.11 and 20.12, that is, the production of ribonucleotides, deoxyribonucleotides and specific proteins and those involved in the generation of ATP. Once the proteins and enzymes necessary for the processes that are described in Figures 20.16 and 20.17 are available, DNA will be replicated.

This is similar to the situation when, if all the glycolytic enzymes, plus activators and co-factors are placed together, in a flask or tube, glucose will be converted to lactic acid or ethanol, i.e. a mixture of enzymes plus co-factors can carry out a process that occurs in a living cell (e.g. muscle or yeast). Such experiments on the enzymes of glycolysis, in a flask, were first performed by Th. Büchner over 100 years ago.

An overall summary of the regulation of the cell cycle is provided in Figure 20.32. Although the description is oversimplified it provides a basis for discussion of the mechanisms.

A tumour suppression gene and a cell cycle kinase An example of how a cell cycle kinase can activate three genes that express proteins essential for DNA duplication is provided in Chapter 21 (see Figure 21.17). The tumour suppressor gene (*Rb*) forms a protein that produces a complex with a transcription factor, which normally activates genes that express three proteins that are essential for DNA duplication. Within the complex, the transcription factor is inactive so that these proteins are not produced so that DNA synthesis and hence proliferation of tumour cells cannot take place. A cell cycle kinase, however, can phosphorylate the Rb protein, which results in dissociation of the transcription factor from the complex, and activation of three genes that express the three proteins that are essential for DNA synthesis and hence progression through the cell cycle. The three proteins are DNA polymerase, tetrahydrofolate reductase which produces a necessary cosubstrate for thymidylate synthesis) and a cyclin (p. 494).

Cell cycle sensors

To provide an answer to question (ii) that is raised above, the concept of biochemical sensors for the essential factors required for each stage of the cell cycle, particularly the S and M stages, is put forward (Figure 20.33). This is a hypothetical system that is proposed primarily to promote thought and discussion of a subject that is of significance in biochemical regulation: e.g. in understanding development and growth of tumour cells and the proliferation of lymphocytes during an infection. Although the subject of sensors may be novel in relation to the cell cycle, sensors that are involved in other areas of biochemistry are well established (e.g. AMP as a sensor of the ATP/ADP concentration ratio).

Death

Death can be considered at two different levels: death of cells, and death of humans due to old age (which may be described as physiological) or disease.

Death of cells

Cells die in two ways that are fundamentally different: necrosis and apoptosis (also known as programmed cell death). Necrosis is usually caused by a decrease in the ATP/ADP concentration ratio in the cell, which can be due, for example, to non-specific external damage, intracellular damage (e.g. uncoupling of oxidative phosphorylation, Chapter 9) or lack of oxygen. Apoptosis, on the other hand, is controlled by specific extracellular or intracellular signals (i.e. it is programmed by specific factors). There are two important distinctions between the two: the cellular changes during death are different and the effect of the death upon the whole tissue or organ is different (Figure 20.34).

Necrosis

It is likely that the fall in the ATP/ADP concentration ratio, sometimes described as 'energy stress', is gradual since, when the ratio begins to fall, processes that generate ATP are stimulated and processes that utilise ATP are inhibited. As the ratio falls, the energy that can be transferred to various processes will fall and, eventually, these processes will fail. The processes likely to fail first are those involved in ion transport (e.g. Na^+/K^+ ATPase, Ca^{2+} ion pumps) so that uncontrolled entry of Na^+ and Ca^{2+} ions into the cell occurs. This increases the entry of water into the cell due to osmosis, and Ca^{2+} ions are taken up by the mitochondria which causes mitochondrial damage and further restricts ATP generation (see Chapter 9). As a result, the cell swells, the plasma membrane is further damaged and lysis results. The cells can die within two or three hours after the initial injury. Although the structural debris is eventually cleared by the macrophages, some of it and the contents of the cells can damage local cells, which initiates an inflammatory process. This can cause considerable pain and chronic damage.

In some cases, death due to disease is caused by necrosis. For example, failure of oxygen supply to some cardiomyocytes, after occlusion of an arteriole or artery, will decrease the ATP/ADP concentration ratio in these cells, which will

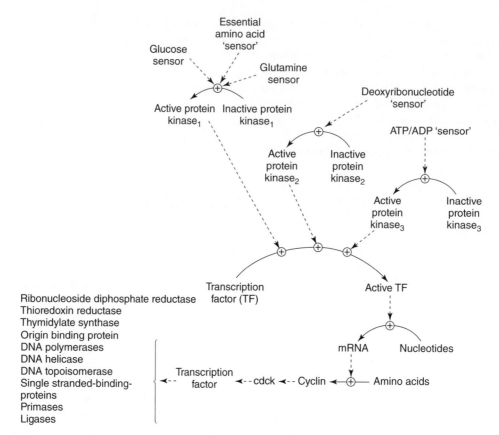

Figure 20.33 *Speculative mechanism by which sensors lead to activation of enzymes that result in DNA duplication.* Several sensors are required as shown in this figure, for example, to assess the concentration of all the deoxynucleotides, to assess the ATP/ADP concentration ratio and to assess the concentrations of plasma or intracellular glucose and glutamine. Specific sensors activate their specific protein kinases (1, 2 or 3). Each kinase activates either several transcription factors or all the kinases contribute to the activation of one key transcription factor. The latter then leads to expression of several proteins from several genes. If all the sensors are required to be active before the transcriptions sector can be fully activated, this would ensure that all precursor molecules, together with a suitable ATP/ADP concentration ratio and fuel levels, are adequate prior to expression of all enzymes required for the S and M phases of the cycle. Examples of such sensors are already discussed in this book. For example, the AMP concentration is a sensor for the ATP/ADP ratio (Chapter 6). The intracellular glucose concentration is a sensor for glycogen synthesis in the liver (Chapter 6: see Figure 6.33). Adenosine may be a sensor for recovery from hypoxia in cardiomyocytes (Chapter 22: see Figure 22.17) and 5-hydroxytryptamine may be a sensor for central fatigue originating in the brain (Chapter 13).

die due to necrosis. If a sufficient number of cells are affected, contraction of the left ventricle may be inadequate to supply blood to essential organs (the heart itself and the brain) and death is the probable result (Chapter 22).

Necrosis of cells may be the cause of other conditions that may be less obvious and less severe than the heart attack, but can still be the cause of considerable damage and pain. These include frostbite, in which blood supply to part of the lower limbs (especially fingers and toes) is restricted by extreme cold (vasoconstriction) so that oxygen supply is not sufficient to maintain normal rate of ATP generation in the cells in the skin and, in severe cases, the

underlying tissues. It can occur relatively rapidly in subjects in whom blood supply to the extremities is already compromised by peripheral vascular problems. Hence, diabetic patients, for example, should take particular care of their toes and fingers in cold environments. A pressure sore (e.g. a bedsore), known as a decubitus ulcer, occurs when there is constant pressure on one area of the skin and chronic vasoconstriction occurs so that the oxygen supply to this area of skin is decreased and ATP generation is impaired. Long-term pressure from confinement to a bed, wheelchair, cast or splint is a common cause of decubitus ulcers.

Apoptosis

The existence of the process of apoptosis indicates that it has benefit to the organism. Three examples of the benefits of apoptosis are provided:

1   When a cell is infected with a virus, the latter utilises the 'metabolic machinery' within the host cell to generate viral proteins, RNA and DNA to produce more virus particles which then escape to infect other cells. The process is stopped by death of the host cells so that generation of new viruses is halted. The major mechanism that results in death of the host cell is apoptosis. The cells that are responsible for the death of the infected cells are either cytotoxic lymphocytes or natural killer cells. Death is caused either by release of toxic biochemicals and/or proteolytic enzymes or by binding to a death receptor, which is present on many cells. The entry of proteolytic enzymes or binding to the death receptor results in activation of initiator caspases. These activate effector caspases that cause damage to the cell which results in death due to apoptosis (Chapter 17: Figures 17.28, 29 and 30).

2   When a tissue is invaded by a bacterium, neutrophils are attracted to the site in large numbers to attack and destroy the invader. Once the infection has been dealt with, the neutrophils are no longer required. They undergo apoptosis (Chapter 17: Figure 17.3).

3   If DNA in a cell is damaged, the concentration of a protein known as P53, increases. This protein can delay proliferation, indirectly, by arresting the cell cycle (Chapter 21: Figure 21.14). This delay gives time for the DNA to be repaired. However, if the damage cannot be repaired, P53 can initiate apoptosis, the ultimate means of removing damaged DNA from the body. It is presumed that P53 initiates apoptosis by activating the caspase cascade (see below). P53 is the product of a tumour suppressor gene.

Death by apoptosis is much less spectacular and is usually quicker than necrosis. Cells shrink, possibly to a quarter of their previous size, and vesicles appear in the cytosol and in the nucleus. As the cytosol contracts, the chromatin fragments into a number of distinct particles and the endoplasmic reticulum fuses with the plasma membrane. The dying cell breaks up into small vesicles (apoptotic bodies) which are removed by phagocytosis (Figure 20.34). Since little if any of the cytosolic contents escapes, there is little or no inflammation. That is, death by apotosis rather than reduces the risk of local or general inflammation. In fact, death by apoptosis is so inconspicuous that it was not discovered for many years.

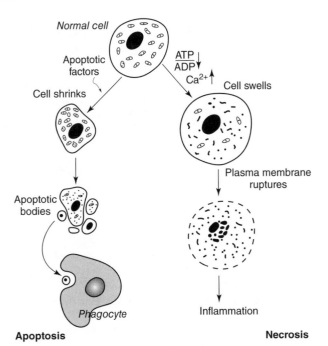

Figure 20.34 *A simple representation of the processes of necrosis and apoptosis leading to the death of cells.* Necrosis is initiated by a decrease in the ATP/ADP concentration ratio, which slows ion pumps, which leads to cation imbalance (e.g. ca^{2+} ion entry into the cytosol) and hence intracellular damage, entry of water and lysis which can lead to local inflammation. Apoptosis is initiated by specific extracellular or intracellular factors, which lead to cell shrinkage and disruption into apoptotic bodies which are removed by phagocytes.

There are two signalling systems that are responsible for apoptosis: the extrinsic and intrinsic systems (Figure 20.35).

Extrinsic system   This system is specific for cell death via apoptosis. Many cells possess receptors on their surface, known as death receptors, to which specific death ligands bind. These ligands include the cytokine, TNFα (see Chapter 17), and the fibroblast associated cell surface ligand (abbreviated to Fas). The intracellular event that results from binding one of the ligands is activation of a cascade of proteolytic enzymes, known as caspases. These have the amino acid cysteine in their active site and they cleave their substrates at peptide bonds made up of one aspartate amino acid: hence their name cysteine-aspartatyl specific proteases (caspases). They are present in the cell as inactive procaspases, which are activated by hydrolytic cleavage by another caspase. (The activation of inactive proenzymes by hydrolytic cleavage is described in Appendix 4.1.) They are divided into two classes, initiator and

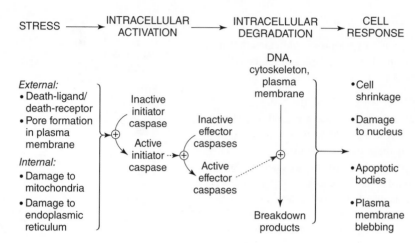

Figure 20.35 *Mechanisms by which external or internal stress leads to cell damage resulting in apoptosis.* The stress leads to activation of initiator proteolytic enzymes (caspases) that initiate activation of effector caspases. These enzymes cause proteolytic damage to the cytoskeleton, plasma membrane and DNA. The activation of DNAases in the nucleus results in cleavage of DNA chains between histones that produces a specific pattern of DNA damage which, upon electrophoresis, gives a specific pattern of DNA fragments. The major endproduct of apoptosis are the apoptolic bodies which are removed by the phagocytes.

effector caspases. The first class are activated by the death receptor, and they cleave and activate the effector enzymes. The effector caspases then activate several enzymes that degrade the cytoskeleton, weaken the plasma membrane and activate an endonuclease that breaks down DNA in the nucleus.

Intrinsic system   Apoptosis is initiated due to changes to the mitochondria during which the inner membrane becomes permeable to large molecules, probably as a result of a decrease in the membrane potential. This can result from intracellular damage (e.g. accumulation of Ca^{2+} ions), lack of oxygen or fuel. This results in release of cytochrome *c* and other proteins from the mitochondria which stimulate apoptosis. In fact, these apoptotic proteins plus cytochrome *c* form a complex, the apoptosome, which activates an initiator caspase.

There is a family of proteins (the Bcl-2 family) that can regulate the potential across the inner mitochondrial membrane, and its permeability. Hence, they can influence apoptosis. Some of this family stabilise the inner membrane and maintain the potential, so that they suppress apoptosis. They are known as cell survival molecules (e.g. the proteins, Bcl-2, Bcl-X). Other members of the family destabilise the mitochondria and decrease the membrane potential which facilitates apoptosis. They are known as cell killer proteins (e.g. the proteins, Bax, Bak). The mechanisms by which they have their effects on mitochondria are not known. They are, of course, of considerable interest for the development and control of growth of tumour cells (See Chapter 21).

Death of humans

The cause of death in a human is usually considered to be inadequate cardiac output or inadequate exchange of gases in the lung. Either of these can be caused by failure of function of vital centres in the brain that control the beat of the heart or the act of breathing. Questions as to how a particular disease or trauma leads to failure of these activities are discussed throughout this book.

The advent of transplant surgery has, however, led not only to an answer as to the cause of death but to the need for a definition of death, since organs can only be removed from a patient who has been certified as dead. The need for a definition is important, since artificial ventilation, the use of drugs to maintain blood pressure, the development of techniques for intravenous or intestinal feeding and the availability of satisfactory feeds to maintain fuel supply for months or even years, and removal of toxic metabolic products by dialysis has led to the ability to maintain almost if not all the basic functions of the body, except for function of the brain. The brain stem controls not only the part of the brain responsible for consciousness but also that for maintenance of breathing and contractions of the heart. Consequently, death is defined as occurring when the function of the brain stem fails and the failure is irreversible. This is known as brain death or, more specifically, death of the brain stem. In fact, as early as 1983, a Working Party of the Department of Health in the UK published the statement that 'Death can be diagnosed by the irreversible cessation of brain stem function' and drew up a series of criteria to identify brain stem death. It is only when these criteria can be satisfied that organ(s) may be removed.

Physiological death

In view of the advances in understanding the cause of death in cells, it is somewhat ironic that we understand so little as to what causes death in otherwise healthy humans as they become old (i.e. death due to senescence). In fact, in the UK it is now not permissible to write 'old age' or a similar phrase on a death certificate. The most likely cause of death in humans under most circumstances, is, as indicated above, failure of the vital centres in the brain, e.g. the cells in the brain stem. A suggestion for the cause of death in old age is the progressive decrease in cardiac output with age. Nonetheless, under normal conditions, even a lower cardiac output will not compromise the function of the brain. However, a mild trauma, mild stress or mild infection requires an increase in cardiac output (to support the biochemical changes described in Chapter 18). If this required increase can no longer be attained by the heart in an elderly person, the provision of blood to the brain could be insufficient for this organ. The part of the brain that maintains the essential functions of the body, that is, the brain stem, may be the first to be affected, so that the control of the contraction of the heart or breathing are so significantly impaired that death results. If the trauma or the infection is particularly mild, the person may exhibit no or few clinical signs of ill health and, therefore, the cause of death may be difficult to diagnose. One claimed advantage of exercise in the elderly is to maintain an effective cardiovascular system so that a mild trauma or infection does not overload the system and the brain receives its required supply of oxygen and fuel under such conditions: i.e. physical fitness allows the elderly person to respond adequately to a mild stress not normally considered to be a clinical problem (Chapters 9 and 13). Perhaps the best definition of physiological death was provided by Shakespeare. The problem for the medical profession is identification of the *little pin* and the techanism by which it bores through the *castle wall*, as described by Shakespeare.

Shakespeare's description of physiological death in *Richard II*:

... For within the hollow crown
That rounds the mortal temples of a king
Keeps Death his court and there the antic sits,
Scoffing his state and grinning at his pomp,
Allowing him a breath, a little scene,
To monarchize, be fear'd and kill with looks,
Infusing him with self and vain conceit,
As if this flesh which walls about our life,
Were brass impregnable, and humour'd thus
Comes at the last and with a little pin
Bores through his castle wall, and farewell king!

V   SERIOUS DISEASES

21
Cancer: Genes, Cachexia and Death

25 years ago, then President Nixon "declared" War on Cancer. In this personal commentary, the war is reviewed. There have been obvious triumphs, for instance in cure of acute lymphocytic leukaemia and other childhood cancers, Hodgkin's disease, and testicular cancer. However, substantial advances in molecular oncology have yet to impinge on mortality statistics. Too many adults still die from common epithelial cancers. Failure to appreciate that local invasion and distant metastasis rather than cell proliferation itself are lethal, obsession with cure of advanced disease rather than prevention of early disease, and neglect of the need to arrest preneoplastic lesions may all have served to make victory elusive.

(Sporn, 1996)

In 1848, Rupert Willis, a British pathologist, defined a tumour as '*a mass of tissue the growth of which exceeds and is uncoordinated with that of normal tissues and persists in the same excessive manner after cessation of the stimuli which evoked the change*'.

> *Malnutrition, and its ultimate form cachexia, are encountered every day in cancer and haematology wards. Malnutrition results from the 'parasitic' metabolism of the tumour at the expense of the host, from the impact of the tumour on the metabolism of the host . . . The major consequence is an increased risk of complications and death during the course of chemotherapy, radiation therapy and major surgery. It is thus important to offer nutritional support, in order to stop or reverse the process of malnutrition. Nutritional intervention should be founded on the abundant literature devoted to cancer cachexia, including the pathophysiology of the disease . . .*
>
> (Nitenberg & Raynard, 2000)

Along with coronary heart disease, cancer is one of the major causes of death in developed countries and increasingly so in developing countries. At least one person in three in developed countries will develop cancer and one in four men and one in five women will die from it. The incidence of cancer increases with age, particularly above 45 years. Although the number of cases of cancer is expected to be more than 14 million worldwide by 2020, it is claimed that most cancers are avoidable because they are caused by unhealthy food, an unhealthy lifestyle and/or an occupation that involves contact with carcinogens. The study of cancer is called oncology, derived from the Greek word for a lump (*onkos*). The term 'cancer' derives from the Greek for a crab, *karcinos*. Cancer is a disease characterised by uncontrolled proliferation of cells that produces a tumour, from which the cells can invade local tissues and then spread to give rise to metastases. Cancers are classified into three groups: carcinomas, sarcomas and leukaemias/lymphomas. Carcinomas, the most common, arise in epithelial cells, sarcomas in connective tissue (e.g. bone, cartilage) and leukaemias and lymphomas from white blood cells.

In 1867, Heinrich Wilhelm Waldeyer-Hartz proposed, with remarkable foresight, that cancer cells developed from normal cells during cell division and then spread by the blood or lymph to other sites in the body. He was an anatomist who not only realised the cause of cancer but also identified the basic unit of the nervous system, to which he gave the name *neuron*.

A tumour cell develops from a normal cell when the processes that control proliferation fail. In fact, this is the

> Most cancers in humans can be divided into three groups according to age: (i) embryonic tumours (neuroblastoma, Wilms' tumour); (ii) tumours found mainly in the young (leukaemias, bone, testis) and (iii) tumours that appear frequently in middle and old age (prostate, colon, skin, bladder and breast).

> Hippocrates coined the word 'karcinos' after observing that distended veins radiating from a breast tumour resembled the legs of a crab.

Functional Biochemistry in Health and Disease by Eric Newsholme and Tony Leech
© 2010 John Wiley & Sons Ltd

best definition of a tumour cell since it is not possible to define a tumour cell in absolute terms. The result is a lump or a mass of tissue. Both benign and malignant tumours develop from uncontrolled proliferation but a benign tumour remains confined to the site of origin. A malignant tumour cell invades and destroys adjacent tissues and then escapes into the circulatory system, within which it travels to distant tissues where it produces secondary tumours (metastases). Benign tumours can be removed by surgery or destroyed by radiation but this is not the case for malignant tumours once the cells have spread to other tissues. Uncontrolled growth i.e. a benign tumour, therefore, is not necessarily a major problem unless it interferes with the function of the tissue or organ.

Basic information

Understanding cancer involves not only knowledge of the genetic changes that cause a normal cell to develop into a tumour cell but also the response of the whole body to a tumour, factors that increase or decrease the risk of development of a tumour and the current therapies that arrest growth of, or kill, tumour cells. These topics are discussed in this chapter, but some basic information is required first: the cell cycle, the growth of a tissue, the fuels used by tumour cells and, finally, the role of genes.

The cell cycle

The sequence of events that occurs in the proliferation of individual cells is known as the cell cycle, which is discussed in detail in Chapter 20. The cycle has four phases (Figure 21.1).

* G_1, which controls progression through the cycle so that, once the cell passes G_1, it is committed to divide.

* The S phase in which DNA is synthesised and the original DNA duplicated.

* G_2, in which the chromosomes separate and move to the two ends of the cell.

* The M phase, in which the cell divides to produce two daughter cells.

When regulation of the cycle fails, a mass of tissue is produced.

Growth of a tissue

Normal adult tissue can be separated into three classes:

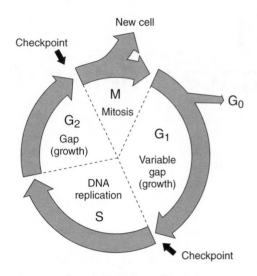

Figure 21.1 *The cell cycle.*

* *Continually renewing*: the tissue contains cells that proliferate continually to replace those that die or are lost (e.g. the skin, intestine, bone marrow). These cells are known as stem cells.

* *Conditionally renewing*: the tissue contains cells (stem cells) that can be stimulated to proliferate in response to an appropriate signal (e.g. liver after injury, lymphocytes during an infection).

* *Non-renewing*: the cells in the tissue never proliferate (e.g. nerve, muscle).

Most primary tumours arise in continually renewing tissues (often epithelium). When cells divide they do not normally produce two identical daughter cells: one is another daughter stem cell, but the other differentiates to replace the cell that was originally lost. In contrast, when a tumour cell divides it produces two cells that continue to proliferate. Hence, the increase in the number of cells is exponential. Tumour cells do not always proliferate more rapidly than normal cells and the cell cycle is not of shorter duration. In a wound, fibroblasts and lymphocytes proliferate many times faster than some tumour cells but this ceases when the wound has healed. A tumour can be considered as a wound that does not heal. The growth fraction of the cell population is the proportion of the cells that are proliferating at any one time. In the initial stages the number of proliferating cells in a tumour is small, possibly only one, but the increase is soon exponential. Eventually, however, growth is limited by lack of sufficient fuel and oxygen. The tumour then develops new blood vessels to provide a better supply of blood, which allows further enlargement.

Normally, proliferation is inhibited when a tissue reaches its optimal size, due to contact between cells which is dependent on the binding proteins present on the cell surface – a phenomenon known as contact inhibition. This, however, is lost in tumour cells, since the adhesive proteins are either ineffective or absent. The number of cells in a tissue is also restricted by cell death, which occurs by apoptosis or necrosis. Both processes occur in a tumour but they do not usually stop growth.

Fuels used by tumour cells

Measurements of which fuels are used and the rates of utilisation by tumours *in vivo* are difficult to make, due to the lack of a self-contained blood supply, which precludes measurement of arteriovenous differences. Measurements have been done either on human tumour cell lines or on tumours removed from patients during surgery. The methods that have been used and results obtained are presented in Appendix 21.1. These studies indicate that tumour cells generate ATP primarily from the conversion of glucose to lactate and glutamine to aspartate; details of these pathways are presented in Chapters 6 and 8. Furthermore, the rates of utilisation of these fuels are high. In these properties, tumour cells resemble other proliferating cells such as lymphocytes (Table 21.1). Even a small tumour (2 cm in diameter) will contain about 10^9 cells, larger tumours, especially if metastases are present, will contain many more. This compares with just 10^{11} cells in a human liver and 10^{11} neurons in the brain. Not surprisingly, therefore, marked changes in metabolism occur in patients suffering from cancer, which can place considerable metabolic stress on a patient. Excessive mobilisation of body fat and protein which may be necessary to feed the tumour can produce a marked fall in body weight, which is often the initial indication of the presence of a tumour that alerts the patient to seek medical advice.

The role of genes

In 1944, Oswald Avery discovered that DNA isolated from tumour cells could transform normal cells in culture. Hence, DNA (i.e. genetic factors) were responsible for tumour development, so that mutations or damage to DNA can bring about tumour formation with the change in DNA being passed on to daughter cells during proliferation of the original cell. This increases the risk of development of a tumour cell. There is considerable evidence to support the view that tumourigenesis is a multi-step process. Thus a cell accumulates a series of mutational events, all of which contribute to development of a tumour. A cell is considered to be transformed when changes in a cell have occured that increase the risk of a tumour cell. Some changes that characterise transformation are presented in Table 21.2.

Two types of gene are associated with development of a tumour cell: oncogenes and tumour suppressor genes. Oncogenes encode proteins involved in the process of proliferation; that is, those involved in the cell cycle. These proteins are synthesized in amounts or with an activity that accelerates progression through the cycle. In contrast, tumour suppressor genes encode proteins that can decelerate or arrest progression through the cycle. A mutation in a tumour suppressor gene can, therefore, result in uncontrolled proliferation (Figure 21.2).

An oncogene has been compared with a car with a stuck accelerator: the car moves even if you take your foot off the accelerator. However, the car will stop if the brake is pressed (this is the effect of the tumour suppressor gene). A mutation in or the loss of a tumour suppressor gene is analogous to a faulty brake, i.e. it no longer halts proliferation of the abnormal cell (Vogelstein & Kinzer, 2004).

There are two classes of tumour suppressor genes: gatekeeper genes and caretaker genes. If the activity of a

Table 21.1 Metabolic characteristics of normal proliferating cells and tumour cells

Metabolic characteristic	Maximum activity or capacity	
	Normal proliferating cell[a]	Tumour cell
Hexokinase activity (required for glycolysis);	moderate	high
Growth dependence upon glucose;	present	present
Capacity of glycolysis (glucose conversion to lactate);	moderate	high
Glutaminase activity (required for glutaminolysis);	high	high
Capacity of glutaminolysis (glutamine conversion to aspartate);	high	high
Growth dependence on glutamine;	present	present[b]
Oxidation via complete Krebs cycle (i.e. from acetyl-CoA)	low	low

[a] A lymphocyte is taken as an example of a normal proliferating cell.

[b] Some tumour cells possess the enzyme glutamine synthetase and so are able to synthesise glutamine from glutamate: as expected, proliferation of these tumour cells is not dependent upon the presence of glutamine in the culture medium.

Data on which this table is based were obtained from Board (1991) and Colquhoun (1990).

Table 21.2 Some changes in cell biological characteristics that indicate transformation of a cell

Characteristics	Change
Density-dependent inhibition of growth	reduced or lost
Growth factor requirements	reduced or lost
Anchorage dependence	lost
Proliferative lifespan	indefinite
Contact inhibition of movement	lost
Adhesiveness	lost
Microscopic appearance	less differentiated than normal cells

Transformation is something of a loose term, since it is unclear whether one or more than one of these change must occur to define a transformation.

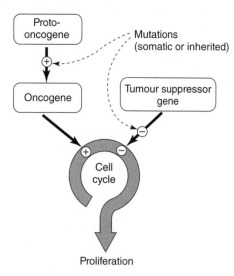

Figure 21.2 *A diagram representing a simplified account of how oncogenes and tumour suppressor genes affect progrssion through the cell cycle and the possible effects of mutations.* An oncogene increases progression through the cell cycle (+) and a tumour suppressor gene decreases or arrests progression (−). The dotted line from mutations to the conversion of a proto-oncogene to an oncogene indicates activation of the process, whereas that to the tumour suppressor gene indicates inactivation of the gene.

gatekeeper gene that has been mutated is regained, the tumour suppressor effect is restored. Caretaker genes have a more general role in that they encode proteins that maintain the integrity of the genome; for example, enabling DNA damage to be repaired, preventing loss of genes or preventing transfer of genes within the genome. An example of a gatekeeper gene is the *rb* gene, inactivation of which leads to retinoblastoma. An example of a caretaker gene is the *p53* gene that inhibits duplication of damaged DNA, which gives time for repair of the damage,

if this is possible. The effects of the proteins encoded by these genes are described below.

Oncogenes and proto-oncogenes

Several genes are involved in the stimulation of proliferation in a normal cell; that is, they express proteins that directly or indirectly regulate this process. These proteins include growth factors, growth factor receptors, signalling proteins and transcription factors, which are organised into a sequence that links a growth factor to the process of proliferation (Figure 21.3). If a mutation occurs in any one of these genes which results in an increase in the activity of one of the proteins in this sequence, the risk of development of a tumour increases. Such a mutated gene is termed an oncogene: it is defined by its ability to transform cells in culture. The precursor gene for the oncogene is termed a proto-oncogene. It should be clear, therefore, that any of the genes that express proteins in the sequence shown in Figure 21.3 are potential proto-oncogenes. The proteins expressed by oncogenes are similar to those expressed by proto-oncogenes but are more active or less well controlled than the normal protein.

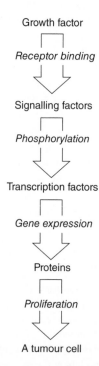

Figure 21.3 *Sequence of intracellular factors that link a growth factor receptor to proliferation.* Phosphorylation is one of several means by which signalling factors communicate with transcription factors. This sequence is similar to the mechanism by which some hormones affect the biochemistry within a cell (Chapter 12). It could be entitled, A Carcinogenic Pathway.

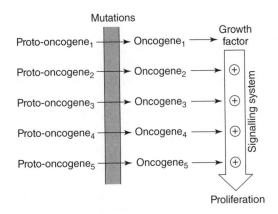

Figure 21.4 *Representations of a sequence of conversions of proto-oncogenes to oncogenes and effects on the signalling sequence leading to stimulation of proliferation. It is very likely that more than five oncogenes are involved. Note that they affect different components of the sequence, so that all five conversions would be required to enhance proliferation. A single conversion would have no effect.*

Consequently, identification of a proto-oncogene depends on identifying an oncogene. Increases in the activities of all of the proteins in the signalling sequence will result in excessive stimulation of proliferation and, therefore, development of a tumour cell (Figure 21.4).

Discovery of oncogenes

Oncogenes were not originally discovered in tumour cells but, surprisingly, as an integral part of the genome of some retroviruses. When RNA, isolated from a retrovirus, was transfected into cells in culture, the cells were transformed. However, the amount of the genetic material that caused this change was much smaller than the whole genome. This was isolated and transcribed to produce DNA which, when transfected into cells, resulted in transformation. The sequence of this DNA was similar to that of a gene that was present in normal cells and expresses a protein that is involved in stimulation of proliferation. Furthermore, such genes were found in tumour cells but in a mutated form. These genes were then termed oncogenes, so that the normal genes in the normal cells that gave rise to oncogenes were termed proto-oncogenes.

The life history of a retrovirus is described in chapter 17 (see Figure 17.45). A summary is presented here. The genome of a retrovirus is composed of RNA not DNA but, when a retrovirus infects a host cell its RNA is transcribed into DNA, catalysed by the enzyme, reverse transcriptase. This DNA is then incorporated into the genome of the host. On transcription of the host DNA, during cell division, viral mRNA and viral genomic RNA are produced. The

mRNA produces new viral proteins and, together with the genomic RNA, new viruses, which escape from and kill the host cell.

During the process of transcription and production of viral genomic RNA, a portion of the host DNA sequence is sometimes incorporated into the viral genome. This host genetic sequence may contain a proto-oncogene, which is then subject to a mutation that accompanies viral replication. During a subsequent round of viral infection, the mutated proto-oncogene sequence (now an oncogene) will be inserted into the DNA of the host cells. Not only may infection result in tumorigenesis but new viral particles are produced, all of which carry an oncogene (see below).

Proteins expressed by oncogenes

The naming of oncogenes is often abstruse. For example, although oncogenes originate from normal genes that express proteins involved in cell proliferation, this is not reflected in the name of the oncogene. Nonetheless, there are a few guiding principles: a viral oncogene is abbreviated to *v-onc* and its cellular equivalent to *c-onc* and the names often reflect the species of origin of the virus. For example, the *simian sarcoma* virus is termed the *Sis* oncogene. Some clarification is provided in Table 21.3 and Figure 21.5.

Oncogenes that express growth factors and growth factor receptors

Growth factors are paracrine or endocrine agents, peptides or proteins, released by one cell to regulate proliferation in adjacent cells (chapter 12) (Table 21.4). Normal cell proliferation requires the presence of growth factors but proliferation of tumour cells requires either no growth factors or much lower concentrations. There are several different characteristics of tumor cells that enable them to respond to very low concentrations of growth factors:

• The tumour cells produce and respond to their own growth factors (an autocrine effect).

• The growth factor receptor(s) that are expressed by oncogenes are always active, even in the absence of the growth factor.

• An increase in the number of receptors in the membrane of the tumour cells (i.e. over-expression of the gene) allows the receptors to respond to low concentrations of growth factor. (See chapter 12 for discussion of hormone receptors and response to hormones.)

Table 21.3 Some oncogenes, their protein products and functions

Oncogene	Products of oncogene	Function of the product
Sis	Growth factor	Binds to receptor to initiate a signal transduction process.
Erb-B	Growth factor receptor	Binds with a growth factor to produce an overactive growth-factor-receptor complex.
Ras	Signal transduction factors	Monomeric G-protein with very low GTPase activity, which maintains the protein in the active state.
Jun *Fos*	Transcription factor	Both associate to produce a complex that acts as a transcription factor the activity of which is greater than each individual factor.
Myc	Transcription factor	A transcription factor that increases concentrations of the enzymes required for S-phase of cell cycle, so that DNA synthesis can proceed without any limitation in precursor concentrations (chapter 20).

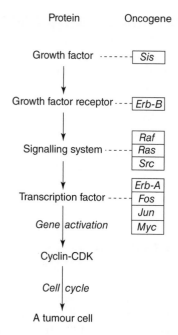

Figure 21.5 *The proteins encoded by some specific oncogenes and their roles in proliferation.* This figure provides examples of oncogenes represented diagrammatically in Figure 21.4. For origin of oncogene names, see Appendix 21.2.

Oncogenes that express signals in the signal transduction pathway

When the growth factor binds to its receptor, the first component of the intracellular signalling process is activated and this then stimulates the activity of all the other components in sequence (see Figure 21.3). A signal transduction process that has particular significance in proliferation is the membrane-bound phospholipase system, which produces the messengers diacylglycerol and inositol trisphosphate from phosphotidylinositol phosphoinositide bisphosphate (Figure 21.6).

G-proteins and signal transduction in tumour cells

G-proteins are a family of guanine nucleotide-binding proteins and represent an initial stage of signal transduction. Many G-proteins are trimers and mediate a wide range of responses to hormonal action, which are discussed in Chapter 12. However, G-proteins involved in cell proliferation are monomers. The significant property of all G-proteins is that they are active (and transmit a signal) when GTP is bound but are inactive when GDP is bound.

Monomeric G-proteins: the Ras protein

The Ras protein (the name derives from *rat* sarcoma virus) is present in some human tumour cells. It is the of the *Ras* oncogene and product is a component of the signalling pathway for cell proliferation. It possesses many of the usual properties of G-proteins; that is, it is active when GTP is bound and inactive when GDP is bound. Activation occurs when GDP is exchanged for GTP in an ATP-hydrolytic reaction catalysed by the guanine nucleotide exchange factor. Inactivation occurs when GTP is hydrolysed to GDP by a GTPase activity: an intrinsic component of *Ras* (Figure 21.7).

Control of Ras The oncogenic form of *Ras* has lost its intrinsic GTPase activity, so that the bound GTP cannot be converted to GDP and Ras remains permanently active (Figure 21.8). When Ras is active, it stimulates the MAP kinase (mitogen-activated kinase) cascade which leads to activation of transcription factors in the nucleus (Figure 21.9). The activated transcription factors stimulate genes that express proteins concerned with progression through the cell cycle. Thus, proliferation of cells is stimulated, even in the absence of a growth factor or a low density of a growth factor receptor.

Table 21.4 Some growth factors secreted by tumour cells. Tumour cells can secrete several growth factors, all of which can stimulate excessive proliferation.

Growth factor	Abbreviation	Normally produced by	Normally stimulates proliferation of
Erythropoetin	EPO	kidney;	erythroid cells (erythrocyte precursors);
Epidermal growth factor	EGF	salivary glands: intestine;	cells of epidermal origin (e.g. skin cells);
Platelet derived growth factor	PDGF	platelets;	fibroblasts and smooth muscle cells;
Vascular endothelial growth factor	VEGF	cells suffering from hypoxia;	endothelial cells of small blood vessels;
Fibroblast growth factor	FGF	?	fibroblasts; endothelial cells of large vessels;
Transforming growth factor alpha	TGFα	platelets;	variety of adult and foetal cells;
Transforming growth factor beta	TGFβ	many cells;	stimulates deposition of extracellular matrix, by increasing synthesis of proteins such as fibronectin and the integrins;
Insulin-like growth factors	IGF-I; IGF-II	ICF-I is secreted by the liver and is stimulated by growth hormone.	ICF-II is most active in the foetus. They stimulate growth of tissue throughout body.

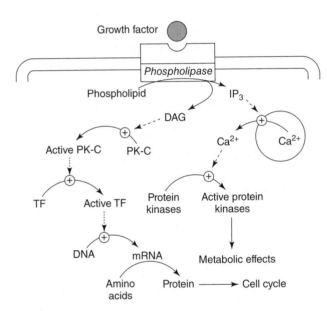

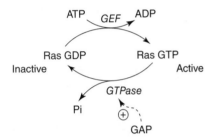

Figure 21.7 *Control of the activity of* Ras *by a balance of the activities of guanine nucleotide exchange factor and GTPase. GAP is the abbreviation for GTPase-activating factor and GEF for guanine nucleotide exchange factor. Both are enzymes. Both the activities are controlled by stimuli from various cell surface receptors.* Ras *oncogenes are present in about 30% of all human tumours.*

Figure 21.6 *One mechanism of activation of the cell cycle by a growth factor.* Binding of growth factor to its receptor activates membrane-bound phospholipase-C. This hydrolyses phosphatidylinositol bisphosphate in the membrane to produce the messengers, inositol trisphosphate (IP_3) and diacylglycerol (DAG). IP_3 results in release of Ca^{2+} from an intracellular store. The increased Ca^{2+} ion concentration activates protein kinases including protein kinase-C (PK-C). DAG remains membrane-bound and also activates protein kinase-C (PK-C) which remains in the activated form as it travels through the cell where it phosphorylates and activates transcription factors. This results in activation of genes that express enzymes involved in nucleotide synthesis, DNA polymerases and cyclins, which are all required for the cell cycle (See Chapter 20 for provision of nucleotides and cyclins for the cell cycle).

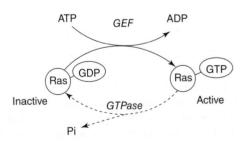

Figure 21.8 *Maintenance of Ras in the activated from by loss of GTPase activity.* A point mutation in the *Ras* proto-oncogene leads to a very low activity of the GTPase so that Ras protein remains in the active form. The broken line indicates low activity of GTPase.

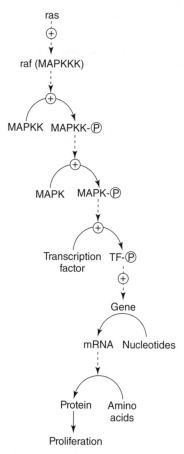

Figure 21.9 *The mitogen-activated protein kinase cascade (MAP kinase cascade).* The active protein *Ras* activates *Raf* by promoting its recruitment to a cell membrane. Through a series of phosphorylations MAP kinase is activated as follows: MAP kinase kinase kinase (*Raf*) phosphorylates MAP kinase kinase which, in turn, phosphorylates MAP kinase, the final target enzyme. MAP kinase phosphorylates transcription factors for genes that express proteins involved in proliferation. Another nomenclature for the enzymes is also used: *raf* is MEKK; MAPKK is MEK and finally ERK is MAP kinase (ERK is the abbreviation for extracellular-signal-related kinase): For comparison, the reader is referred to the metabolic phosphorylase cascade, which is discussed in Chapter 12 (Figure 12.12).

Role of membrane attachments The Ras-stimulated cascade depends on the binding of components of the cascade to the cell membrane. Ras itself is attached to the plasma membrane and it activates Raf (the first enzyme in the MAP kinase cascade) by causing it to bind to the membrane.

It is tempting to suggest that the binding of Ras and Raf to membranes may provide a link between diet and tumour development. There is evidence that a chronic alteration in the phospholipid composition of membranes, which can be caused by modifying the fat content of the diet (e.g. by increasing the proportion of the polyunsaturated omega-3 fatty acids), can change the structure of a membrane. Such a modification in the diet reduces the proliferation of immune cells and, therefore, has an immunosuppressive effect (Chapter 17). It is speculated that this is due to interference in the binding of components of the MAP kinase cascade to cell membranes. It is possible that such changes could have a similar effect on tumour cells (see below).

Transcription factors

The effect of transcription factors is enhanced by processes which increase their concentration in the nucleus or their activity. Such processes include stimulation of transcription of the gene encoding the transcription factor; increasing the rate of its translocation from the cytosol to the nucleus; activation of the transcription factor by phosphorylation or by increasing its affinity for DNA. An oncogenic transcription factor could be more effective after phosphorylation i.e. when phosphorylated, it has a greater effect on its target gene.

Processes by which proto-oncogenes can be activated or converted to oncogenes

Several modifications can convert a proto-oncogene into an oncogene or can activate the proto-oncogene (see Figure 21.10):

(i)   The proto-oncogene suffers a somatic mutation to produce an oncogene.

(ii) The proto-oncogene is duplicated, so that the protein is over-expressed.

(iii) A chromosomal translocation moves a promoter to a new position where it affects a proto-oncogene, so that its protein is over-expressed.

(iv) A virus acquires a proto-oncogene from a host cell which mutates into an oncogene.

(v)   A virus inserts its DNA into the host genome next to a proto-oncogene. This could result in enhanced transcription of the proto-oncogene, when the host cell next proliferates.

A viral infection that results in insertion of viral DNA adjacent to a proto-oncogene will be a rare event, so that the risk of tumour development will be very low. In contrast, infection by a virus that which has actually acquired an oncogene will increase the risk markedly.

Cell survival molecules: apoptosis

Apoptosis, or programmed cell death, is a sequence of events beginning with the appearance of condensed chromatin within the nucleus and extensive cytoplasmic protrusions or 'blebs' on the cell surface and ending with the death of the cell. Events proceed rapidly and cell death occurs within an hour of the first signs of cytoplasmic blebbing (See Figure 20.34). Cell debris is then phagocytosed by macrophages. The occurrence of apoptosis depends on the balance between proteins which stimulate

it (cell killer proteins) and those which suppress it (cell survival proteins). Both types of proteins belong to a protein family, the *Bcl-2* (Chapter 20). Inappropriate formation of cell-killer protein will have a similar effect to activation of a proto-oncogene.

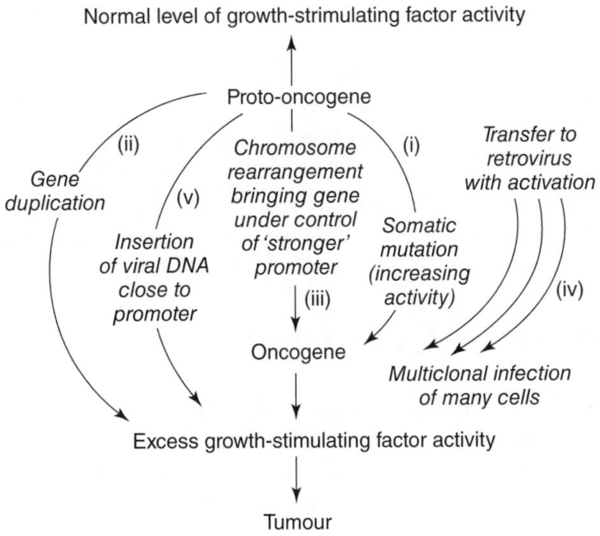

Tumour suppressor genes

Loss of tumour suppressor activity leads to production of a tumour in a homozygous recessive manner. This means that both alleles must be mutated for loss of control of the cell cycle to be lost and uncontrolled proliferation to occur. At least 30 different tumour suppressor genes have been identified but the mechanisms of suppression will probably be different for each gene. The mechanisms for two are presented.

The *p53* gene

The *p53* gene encodes a protein of molecular weight 53 kDa. The presence of damaged DNA in the genome increases the concentration of the p53 protein, which acts as a transcription factor to activate another gene, the *p21* gene, which produces a 21 kDa protein and arrests the cell cycle between the G_1 and S phases. This gives time for the damaged DNA to be repaired before the next cycle is initiated. Hence, it prevents transmission of damaged DNA to the next generation of cells. In this way, it prevents accumulation of damage during each cell cycle and therefore reduces the risk of development of a tumour (Figures 21.11 & 21.12). If p53 does not stop the cycle it initiates apoptosis (i.e. it results in death of the cell).

Consequently, the concentration of the p53 protein, via p21, controls the rate of the cycle. The concentration of p53 is controlled by its rate of degradation, which is achieved by proteolysis, catalysed by the proteasome. The rate of proteolysis depends upon the phosphorylation state of p53 protein, i.e. phosphorylated-p53 is resistant to degradation and, therefore, phosphorylation results in an

Figure 21.10 *Modifications by which proto-oncogenes can be activated.* Conversion to an oncogene or excess activation of proto-oncogene results in excess growth-promoting activity, and consequently increases the risk of tumour development (see text). Process (iv) can result in many cells in the host being 'infected' with an oncogene that has been formed from a proto-oncogene within the virus.

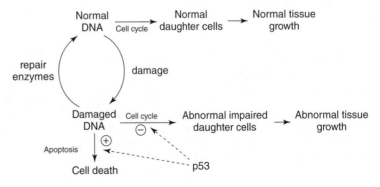

Figure 21.11 *Role of p53 as a tumour suppressor gene. p53* induces a cell with damaged DNA either to initiate apoptosis, or arrest the cell cycle, to give time for damaged DNA to be repaired. Damage can be, for example, a mutation, DNA strand breakage or chromosomal rearrangement.

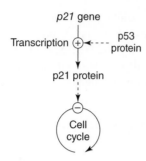

Figure 21.12 *Effect of p53 on transcription of* p21 *gene.* The p53 protein is a transcription factor for *p21*, which arrests the cell cycle.

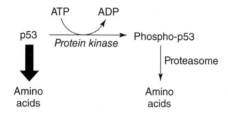

Figure 21.13 *Regulation of the rate of proteolysis of p53 by phosphorylation.* The thick arrow indicates a high rate of proteolysis by the intracellular proteolytic enzyme, the proteasome. Phosphorylation of *p53* reduces the rate of proteolysis (thin arrow): phosphorylation is catalysed by a DNA damage-sensitive protein kinase.

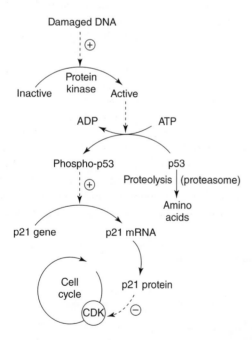

Figure 21.14 *A summary diagram of how damaged DNA leads to arrest of the cell cycle.* Damaged DNA activates a protein kinase which phosphorylates and hence stabilises *p53*, which stimulates transcription of the *p21* gene which expresses a protein that arrests the cycle. CDK, cell cycle division kinase.

increased concentration of p53. The phosphorylation is catalysed by a protein kinase, which is activated by damaged DNA, that is, by a damaged-DNA-dependent kinase. There are several such enzymes, which respond to different types of DNA damage, such as those caused by free radicals or chemical carcinogens (Figures 21.13 & 21.14).

> Such damage is more likely to occur when DNA is stripped of the proteins that stabilise it (e.g. histones). Consequently, the concentration of p53 is high in rapidly dividing cells but low in quiescent cells.

Although this sequence seems complicated (Figure 21.14), it provides a necessary degree of complexity to ensure the effectiveness of the system. The sequence is, in fact, a cascade, so that a small degree of damage to DNA produces a large amount of p21 protein to inhibit the cycle effectively. It also provides additional steps within a cascade for opportunities where extra control can be exerted (e.g. additional transcription factors or other proteins).

> The protein p21 can be induced by p53-independent factors, including drugs known as statins. These drugs are used to lower blood cholesterol by inhibition of its synthesis, via the enzyme HMG-CoA reductase. It is not known how statins affect the p21 protein.

> In any signalling process, it is essential that the signal travels only in one direction (e.g. action potential in a nerve, signalling in hormone action). To do this, non-equilibrium reactions must be included in the sequence.

The retinoblastoma gene

Retinoblastoma is a retinal tumour that occurs in either an inherited or sporadic form. The sporadic form is rare since both alleles must be subject to somatic mutations for this form of the disease to develop.

The *Rb* locus that is affected by the mutation, and that causes retinoblastoma, is a tumour suppressor gene. It encodes the *Rb* protein which is an inhibitor of transcription. The *Rb* protein controls the expression of three genes, which are essential for cell proliferation by forming a complex with, and thus inactivating, a transcription factor for these genes. These are:

- The gene for a cyclin protein that is involved in control of the cell cycle (Chapter 20, see Figure 20.32).

- The enzyme tetrahydrofolate reductase, which is essential for the synthesises deoxythymidine monophosphate (dTMP) from deoxyuridine monophosphate, a process essential for DNA synthesis. This enzyme catalyses formation of methylene tetrahydrofolate (CH₃-FH₄) a necessary co-substrate for synthesis of d-TMP catalysed by thymidylate synthase (See Figure 20.12(a) and p. 477).

- The enzyme DNA-dependent DNA polymerase, which is also essential for DNA replication (Figure 21.15).

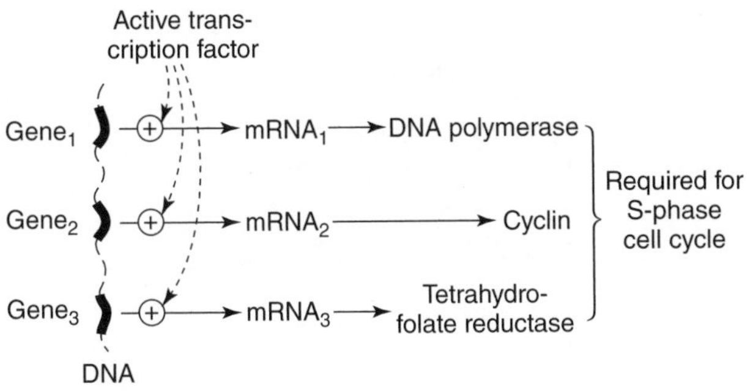

Figure 21.15 *Regulation of three genes that express proteins essential for cell cycle.* The same transcription factor activates three genes that encode for (1) DNA polymerase (2) a cell cycle cyclin (3) tetrahydrofolate reductase.

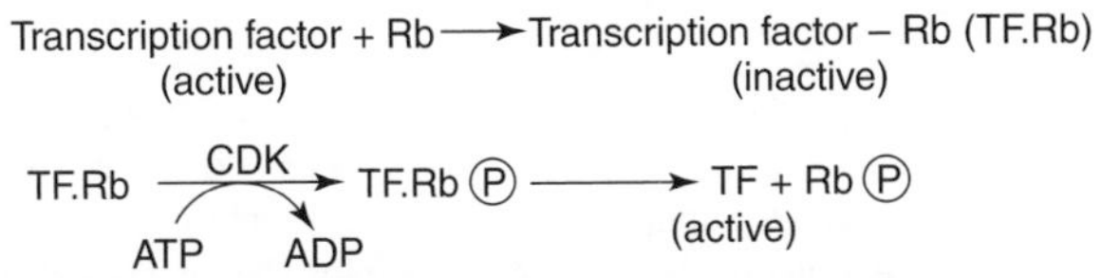

Figure 21.16 *A diagram of the mechanism by which retinoblastoma protein* (Rb) *regulates transcription factor activity.* The rb protein binds to the transcription factor, which forms a complex in which the transcription factor for three genes is inactive. Phosphorylation of Rb by a cell division cycle kinase results in dissociation of transcription factor from the complex and hence activation.

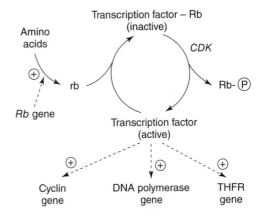

Figure 21.17 *Overview of the regulation of the genes that express three proteins essential for DNA synthesis.* The *Rb* gene expresses Rb which inactivates the transcription factor by forming a complex. Phosphorylation of the Rb protein by a cell cycle kinase causes dissociation of complex and release of transcription factor, which is now active and stimulates expression of the three genes. THFR, tetrahydrofolate reductase. See chapter 20 for details of the actions of cyclins, DNA polymerase and THFR in the cell cycle.

The *Rb* protein binds to the transcription factor, forming a complex in which the factor is inactive. However, when the *Rb* protein is phosphorylated by a cell cycle kinase, the complex dissociates, so that the transcription factor is released and is then available to activate the three genes allowing cell proliferation to occur (Figures 21.16 & 21.17). Mutations in the *Rb* gene that result in the synthesis of a protein with a reduced capacity to bind the transcription factor, results in uncontrolled cell proliferation.

Telomeres and telomerase in tumour cells

Telomeres are sequences of six-nucleotide repeats found at the ends of the chromosomal DNA strands. Many thousands of repeat units $(TTAGGG)_n$ may be present at the end of the 3′ strand and $(AATCCC)_n$ at the end of the 5′ strand. These are present at the ends of the strands to overcome a problem posed by the semi-conservative mechanism of DNA replication, known as 'the end replication problem'. Replication of the ends of the chromosomes presents particular difficulties, since DNA polymerase can only elongate a pre-existing DNA strand (it cannot initiate DNA synthesis) and can only polymerise DNA in one direction (5′ to 3′). Since there is no complementary strand from which to synthesise primers at the ends of the DNA molecule, up to 200 base pairs of DNA would be lost at each cell division. To overcome this problem, a special mechanism exists to prevent chromosomal shortening with each round of DNA replication. The mechanism involves the activity of an enzyme, telomerase, which adds on repetitive nucleotides to maintain the length of the telomeres on the newly synthesised daughter DNA strands. In a normal cell, telomere length is gradually lost during cell replication and, when it is completely lost and then coding sequences are lost, the cell dies via apoptosis. This process is implicated in ageing and normal death of cells with age. Tumour cells, however, have an enhanced activity of telomerase which maintains the length of the telomeres which contributes to their evasion of normal apoptotic mechanisms.

Metastasis

Benign tumours are usually only a problem when they impair the function of organs or cause metabolic stress. They can be removed by surgery or radiation therapy. Malignant tumours are much more of a problem, since the cells can escape from the primary tumour to other sites in the body, where they settle and develop into secondary tumours (metastases). Then, chemotherapy is the only treatment available. The process is known as metastasis.

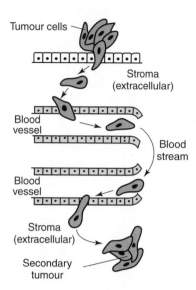

Figure 21.18 *Exit of tumour cells from a tumour into the blood-stream and entry into a new tissue for formation of a new tumour.* Exit depends on lack of adhesion between tumour cells and release of proteolytic enzymes to destroy extracellular stroma. Cells travel to new tissue via the blood on lymph. They secrete proteolytic enzymes and 'bore a hole' in the extracellular stroma of new tissue and then infiltrate and proliferate.

The loss of contact inhibition between cells, which characterises tumour growth, arises from inactivation of genes encoding cell surface adhesion molecules, such as fibronectin. The absence of such proteins allows the cell to detach from the primary tumour. In addition, some genes are activated which express metalloproteases and, when these are secreted from the cell, they break down the basement membrane and extracellular matrix of the tissue, which provides an escape route into the circulatory system. The cells are then transported to new tissues/organs by blood or lymph. At the new site, proteolytic enzymes are expressed on the surface of the tumour cell, allowing invasion of tissue and establishment of a secondary tumour (Figure 21.18). The proteolytic enzymes contain metal ions, so that they are known as metalloproteases (abbreviated to MMP). These changes are the result of a sequence of inactivation and activation of particular genes that result eventually in formation of malignant cells. An example of the development of a colon carcinoma from an epithelial cell is shown in Figure 21.19.

Angiogenesis and enlargement of the tumour

As the primary tumour enlarges, fuels and oxygen become limiting for further growth. Consequently, for growth to continue, the blood supply must expand. At least two proteins are secreted by tumour cells that stimulate angio-

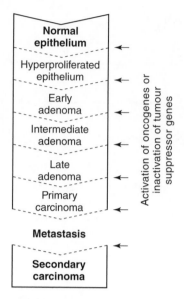

Figure 21.19 *Development of a secondary carcinoma from a normal epithelium by effects of activated genes,* i.e. oncogenes, and inactive tumour suppressor genes. It is somatic mutations in four or five genes in a given cell plus hypomethylation changes in histones and chromatin stracture that are involved. It is the accumulation of these genetic alterations, not the sequence, that determines the progression to a tumour cell.

genesis: *vascular endothelial growth factor* (VEGF) and *fibroblast growth factor* (FGF). The genes that encode these factors are activated by the hypoxic conditions towards the centre of the tumour. VEGF stimulates proliferation and migration of endothelial cells and increases vascular permeability, with the result that endothelial cells and blood proteins enter the tumour. This is followed by activation of clotting and conversion of fibrinogen to fibrin, which forms fibres that act as scaffolding to support endothelial cells involved in the production of new capillaries. Moreover, tumour cells secrete cytokines which stimulate proteolysis in muscle, to release amino acids, and stimulate lipolysis in adipose tissue to release essential fatty acids, both of which are required for the growth of endothelial cells for formation of new vessels and for the further growth of the tumour.

Metabolic changes in cancer patients

Intense research has revealed much about the molecular mechanisms involved in tumour development but much less is known about the metabolic changes in cancer. Nonetheless, most deaths from cancer are caused not directly by the tumour but by infections due to an immune system that has been impaired by metabolic disturbances in the whole body. These can also result in a marked loss of weight which is very distressing for the patient. Unfortunately, there is considerable variation in the data due to the type of cancer, the extent of metastases, anorexia or cachexia, marked loss of weight due to illness (Chapter 16), suffered by the patient. Consequently, it is only the most important and most marked metabolic changes that are analysed in the following section.

> An involuntary weight loss of more than 10% is a major risk factor for survival. This is due to a high rate of secretion of cytokines, especially from aggressive tumours.

Carbohydrates

Glucose turnover is increased due to high rates of utilisation and conversion to lactate by the tumour cells. The lactate is released into the blood, taken up by the liver and reconverted to glucose via gluconeogenesis, for re-use by the tumour. This is equivalent to the Cori cycle between muscle and liver (Figure 21.20). In addition, glucose tolerance by the cancer patient is reduced. This may be partly explained by high rates of fatty acid oxidation by muscle, which reduces glucose utilisation, and the fact that the tumour cell secretes cytokines that stimulate lipolysis and fatty acid release from adipose tissue (as discussed in Chapter 18).

Fat

Hydrolysis of triacylglycerol and release of the resultant fatty acids from adipose tissue is increased and accounts, in part, for loss of body weight. Despite this, the plasma fatty acid level is not always increased, which suggests enhanced uptake of fatty acids by muscle and liver. The fatty acids taken up by muscle will be oxidised to provide energy, particularly in the event of anorexia. The fatty acids taken up by the liver will be oxidised or esterified. Oxidation generates ATP for gluconeogenesis to convert lactic acid into glucose. Esterification produces triacylglycerol which is incorporated into very low density lipoproteins (VLDL) and secreted into the blood. The VLDL is hydrolysed by lipoprotein lipase in adipose tissue with some of the resultant fatty acids being released into the blood and taken up once again by the liver for further esterification and secretion as VLDL. These processes constitute the extracellular fatty acid/triacylglycerol substrate cycle (Figure 21.21) and contribute to increased energy expenditure and therefore the hypermetabolism characteristic of many cancer patients.

The cycle makes available essential fatty acids that are required for the phospholipid synthesis necessary for new membrane formation in the proliferating tumour cells (Figure 21.22) and for synthesis of eicosanoids for regulation of proliferation. Such factors are also required by proliferating immune cells that may attack tumour cells, so that there will be competition for essential fatty acids between immune and tumour cells.

Protein

Degradation of muscle protein contributes to weight loss and wasting in patients. It supplies all amino acids for protein synthesis in tumour cells and some specific amino acids that have the following key roles:

- Branched-chain amino acids that are metabolised in muscle transfer nitrogen to glutamate to form glutamine (See Figure 8.23).

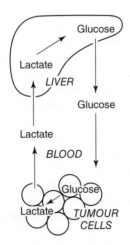

Figure 21.20 *Diagram of a 'Cori cycle' in a patient with a tumour.* Lactate produced from glucose by tumour cells is converted back to glucose in the liver (gluconeogenesis) and released into the blood for re-uptake by tumour cell, an ATP-requiring process. Note that muscle, immune cells and red blood cells will also contribute to the cycle (see, Chapter 6; Figure 6.10).

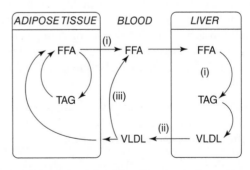

Figure 21.21 *Diagram to illustrate the intertissue triacylglycerol/ fatty acid cycle.* (i) Fatty acids released from adipose tissue are esterified in the liver, (ii) The triacylglyceral is released in the form of VLDL. (iii) The triacylglycerol in the latter is hydrolysed in the capillaries in the adipose tissue. Some fatty acids are taken up by adipose tissue, but about 30% are release in the circulation that give life to the extracellular cycle. The intracellular cycle exists in the adipocytes.

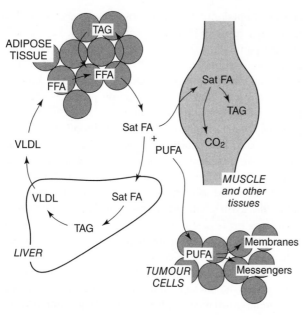

Figure 21.22 *Provision of polyunsaturated fatty acids (PUFA) for tumour cells.*

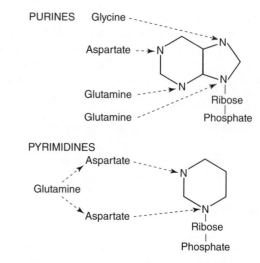

Figure 21.23 *The positions where the nitrogen atoms of glutamine glycine and aspartate are required for nucleotide synthesis. Glutamine also gives rise to aspartate via glutaminolysis (Chapter 8). (See Appendix 20.1 and Figure 20.8).*

- Methionine, which is involved in methyl group donation for nucleotide synthesis, methylation of bases in DNA (Chapter 20) and conversion of choline to ethanolamine for membrane synthesis (Chapter 11).

- Degradation of amino acids in the liver produces oxoacids (for glucose synthesis) and ammonia for synthesis of non-essential amino acids, conditionally essential amino acids (glutamine, cysteine, glycine and arginine) and other nitrogen-containing compounds (e.g. cytokines, growth factors) (Chapter 8). Glutamine and glucose are both precursors for synthesis of purine and pyrimidine nucleotides for DNA and RNA synthesis. The positions where nitrogen from the amide nitrogen in glutamine is used in purine and pyrimidine nucleotide synthesis are shown in Figure 21.23 (See also Figure 20.7, 20.9 and 20.10).

- The glucose produced in the liver is important as a fuel for the brain and other tissues, especially if the patient is suffering from anorexia (Chapter 6).

Weight loss

Weight loss in cancer patients is usually a characteristic symptom and often the one that forces them to seek medical advice. It occurs even when the tumour load is small.

Several processes contribute to the loss of weight: loss of triacylglycerol in adipose tissue; loss of protein in muscle hypermetabolism (increased energy expenditure, due to the high metabolic rate of tumour cells and also increased rate of metabolic cycles) and anorexia. The extent of these processes varies with type of cancer, the patient, and whether cachexia is present. However, when hypermetabolism and anorexia coincide, the prognosis is poor.

Anorexia

The anorexia suffered by cancer patients is likely to arise from a combination of psychological stress, altered senses of taste and smell and increased levels of cytokines, which influence the appetite and satiety centres in the hypothalamus. There are several consequences: micronutrient intake will be diminished and this may contribute to the signs and symptoms of the disease. Plasma amino acid levels will fall, as in starvation (Chapter 16). Synthesis of glutamine (by muscle, adipose and lung), aspartate (by liver), glutathione (by the intestine) and arginine (by the kidney) will all be compromised. The metabolic significance of all of these is discussed in Chapter 18.

The following extract illustrates the current clinical concern surrounding nutritional provision for the cancer patient:

Research is currently directed towards a better understanding of the metabolic alterations of cancer patients, the definition of nutritional regimens that can efficiently support the host without promoting tumour growth, and on the impact of nutritional pharmacology on the host-tumour relationship. Glutamine, arginine, ornithine-

alpha ketoglutarate, omega-3 fatty acids, nucleotides, antioxidants and growth factors are presently under extensive investigation.

(Nitenberg & Raynard, 2000)

Mediators of metabolic changes

Increased plasma concentrations of glucocorticoids and proinflammatory cytokines (tumour necrosis factor and some interleukins) produced by both tumour cells and macrophages within the tumour stimulate lipolysis and protein degradation. This is accompanied by low levels of insulin which also encourage lipolysis and proteolysis.

Cancer cachexia

Cachexia accompanies many diseases, which include sepsis, diabetes and AIDS, as well as cancer (Chapter 16). The metabolic changes in cachexia are an extension of those presented above, but are more severe. An overview of the metabolic changes in cancer is given in Figure 21.24. The metabolic changes that occur in patients, after trauma and during cancer can be compared by reference to Table 21.5.

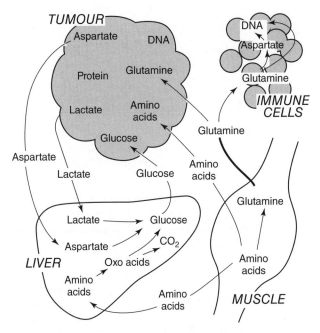

Figure 21.24 *An overview of amino acid metabolism, particularly amino acid metabolism, in a patient suffering from cancer.* The tumour acts as a sink for glucose, amino acids and glutamine. As tumour grows in size, the 'sink' is exaggerated and cachexia develops. This diagram can be considered with that in Figure 21.22 in order to include fatty acids in tumour metabolism. Note the thicker line to indicate magnitude of release of glutamine by muscle.

Table 21.5 Metabolic changes in trauma patients and in cancer patients

| | Direction and extent of change in | |
Process	Trauma	Cancer[a]
Energy		
Oxygen consumption	+++	−
Resting energy expenditure	+++	+
Fat and carbohydrate metabolism		
Primary fuels	mixed	fat
R value (respiratory exchange ratio)	0.85	?
Gluconeogenesis	+++	++
Lipolysis	+++	++
Ketone body production	++	±
Insulin resistance	+++	+
Nitrogen metabolism		
Muscle protein loss	+++	++
Synthesis of protein in muscle	+	?
Degradation of protein in muscle	++	++
Ureagenesis	+++	?
Concentration of glutamine in muscle	−	?
Concentration of total amino acids in muscle	−	?

−, decrease; +, increase; ? not known.

Data presented for both conditions are obtained from several sources (see also Chapter 18).

[a]These are interpretations by the authors from current available information in the literature (Schattner & Schike, 2006). There is a large variation in the data from different patients.

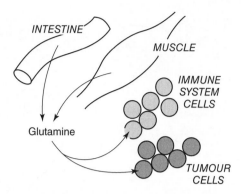

Figure 21.25 *Competition for plasma glutamine by immune and tumour cells.* Glutamine is synthesised and released by muscle. However, if diet is supplemented with glutamine it is also released from the intestine.

Glutamine and cancer

High rates of glutamine consumption by tumour cells can deplete the level of this amino acid in blood. This may impair the function of immune cells which depend upon glutamine. Immunosurveillance, is particularly importance in cancer patients. It involves phagocytosis of tumour cells that have detached from the primary tumour during their transit to develop metastases. The phagocytes are the natural killer and cytotoxic T-cells. These cells also provide defence against pathogens. Consequently, many cancer patients succumb to opportunistic infection, particularly pneumonia. The lung relies on defence by macrophages, and phagocytosis by macrophages depends upon glutamine (Chapter 8). Consequently, in cancer patients there may be competition between immune cells and tumour cells for glutamine (Figure 21.25). Supplementation of the diet with glutamine has been suggested for cancer patients. Although, it carries with it the danger of encouraging further tumour growth, but some studies indicate that glutamine provision is beneficial and does not accelerate tumour growth. Moreover, glutamine supplementation may ameliorate some of the side-effects of chemotherapy or radiotherapy, such as damage to proliferating cells in the intestine. This may be due to the glutamine requirement for cell proliferation in all tissues but especially in the small intestine where it is very high.

Overview of cancer

The sequence of events from mutations or damage to proto-oncogenes and leads to tumour suppressor genes, loss of development of cancer, with its metabolic disturbances and cachexia. Finally these changes can lead to

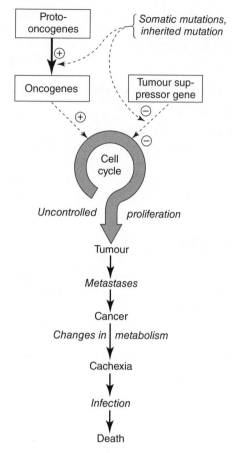

Figure 21.26 *From development of a tumour to death of the patient.*

death, summarised in Figure 21.26. The diagram emphasises that development of the tumour cell is only one component of the disease cancer. Many other steps and processes are involved.

Cancer-causing agents or conditions

A cancer-causing agent or carcinogen may be physical, chemical or viral (Table 21.6). Carcinogens damage DNA or change the position of genes in the genome. Each carcinogen probably works in a different way.

Viruses

Infection with a retrovirus is the most common way in which viruses cause cancer (see above). The first work that led to the discovery that a virus could cause cancer was that of Dennis Burkitt which was published in 1958 (Box 21.1). As

Table 21.6 Some risk factors for cancer

Sites of cancer	Major established factors	Minor established risk factors	Other possible factors
Lung	Smoking[a]	Radon, Asbestos, Arsenic, Chromates, Bischolormethyl ether, Polycyclic aromatic hydrocarbons.	Low intake of β-carotene or vitamin E
Colon		*Polyposis coli*[b] Sedentary lifestyle, Obesity	Low fibre intake, Low vegetable intake, High fat intake, High meat intake
Breast	Early menarche, Late menopause, Late first birth	Postmenopausal obesity	Steroid hormones (endogenous and exogenous oestrogens)
Bladder	Smoking[a]	Aromatic amines	Oral contraceptives
Cervix	Sexual intercourse introducing infectious agents (e.g. viral herpes, human papilloma virus)		
Stomach	Infection with a bacterium, *Helicobacter pylori*		
Pancreas	Smoking[a]		
Mouth, pharynx, oesophagus	Smoking[a], alcohol intake		
Liver	Infection with hepatitis virus		

[a] Smoking refers to tobacco.

See also Willett & Giovannucci (2006).

[b] A hereditary disease in which multiple polyps develop in the colon at puberty.

[c] Controversial.

Box 21.1 Viruses and cancer: the first link

In Kampala, Uganda, after the Second World War, a surgeon, Dennis Burkitt was treating children with a type of facial cancer (lymphoma). He noted a strong geographical demarcation for its occurrence: it was restricted to the North and East of Uganda. Tony Epstein speculated that a virus was responsible and identified the Epstein-Barr virus, which normally causes glandular fever. Since the cancer coincided with the distribution of malaria, it was suggested that immune stress allowed the Epstein-Barr virus to invade. Burkitt treated these children with a cytotoxic drug which completely cured the tumour. This was the first evidence of a link between viral infections and cancer. Burkitt had found not only the cause but the cure for a cancer, despite having no scientific training. Not satisfied, he went on, along with T.L. Cleave, to link the lack of dietary fibre in the Western diet to disease (Chapter 4).

early as 1911, Francis Peyton Rous (1879–1970) had published evidence that a chicken sarcoma could be transmitted by cell-free filtrates, but this was before the discovery of viruses (Box 21.2).

Radiation

Strongly ionising radiation (α-particles, neutrons and protons) is highly carcinogenic, because it causes breaks in both strands of DNA, which are difficult to repair. The radiation stimulates formation of free radicals which are responsible for the damage. Weakly ionising radiation (X-rays, gamma rays) causes breaks in single strands of DNA, which can often be repaired when the cell cycle is

Two accidents of vastly differing severity have occurred at nuclear power plants. On 28 March 1979, an accident occurred in the nuclear power plant at Three Mile Island, Pennsylvania, USA. The radiation was contained and the small amount released had negligible effects on the health of individuals at the plant. On 26 April 1986 an accident occurred in the nuclear power plant 10 miles from the city of Chernobyl, then part of the Soviet Union. The chain reaction in the radioactive core of one of the four reactors became uncontrolled. Steam pressure rose to dangerous levels; there were several explosions and a subsequent fire took several hours to extinguish. Large amounts of radioactive material were scattered over a wide area and into the atmosphere (later descending in a dilute form in rain all over the world).

Box 21.2 The discovery of the Rous sarcoma virus

The following is from the front page plus part of the discussion of the historic paper by Peyton Rous:

A SARCOMA OF THE FOWL TRANSMISSIBLE BY AN AGENT SEPARABLE FROM THE TUMOUR CELLS BY PEYTON ROUS, M.D.

(From the Laboratories of the Rockefeller Institute for Medical Research, New York)

A transmissible sarcoma of the chicken has been under observation in this laboratory for the past fourteen months, and it has assumed of late a special interest because of its extreme malignancy and a tendency to widespread metastasis. In a careful study of the growth, tests have been made to determine whether it can be transmitted by a filtrate free of the tumour cells. Attempts to so transmit rat, mouse, and dog tumours have never succeeded; and it was supposed that the sarcoma of the fowl would not differ from them in this regard, since it is a typical neoplasm. On the contrary, small quantities of a cell-free filtrate have sufficed to transmit the growth to susceptible fowls.

For the first experiments on this point, ordinary filter paper was used, and the ground tumour was suspended in Ringer's solution. It was supposed that the slight

paper barrier, which allows the passage of a few red blood cells and lymphocytes, would suffice to hold back the tumour and render the filtrate innocuous. Such has been the experience of other workers, with rat, mouse, and dog tumours. But in the present instance characteristic growths followed the inoculation of small amounts of the watery filtrate, or of the fluid supernatant after centrifugalization of a tumour emulsion . . .

. . . The first tendency will be to regard the self-perpetuating agent active in this sarcoma of the fowl as a minute parasitic organism. Analogy with several infectious diseases of man and the lower animals, caused by ultramicroscopic organisms, give support to this view of the findings, and at present work is being directed to its experimental verification. But an agency of another sort is not out of the question. It is conceivable that a chemical stimulant, elaborated by the neoplastic cells, might cause the tumour in another host and bring about in consequence a further production of the same stimulant. For the moment we have not adopted either hypothesis . . .

This paper was written in 1911, well before the discovery of viruses.

---

halted by p53 (see above). Ultraviolet light from the sun, cosmic rays from outer space and radon gas (released from granite) are all sources of environmental radiation. Radioactive waste and accidents at nuclear power stations produce larger and more harmful doses of radiation.

Chemicals

Most chemical carcinogens have been identified from epidemiological studies, for which incidence of cancer is linked to a particular occupation or lifestyle (Table 21.6). Such studies are enormously beneficial, because identification of a risk factor can lead to changes in behaviour which decrease the incidence of cancer. This is illustrated by a quotation from Roy Porter's book, *Greatest Benefit to Mankind* (1999):

The best break we have had with cancer has come from epidemiology. Concerned with rising British mortality from lung cancer, the Medical Research Council, in 1947, commissioned Austin Bradford Hill (1897–1991) and Richard Doll (1912–) to analyse the possible causes. Their meticulous statistical survey (published in 1951) of patients from twenty London hospitals showed that 'smoking is a factor, and an important factor, in the production of cancer of the lung.'

A review on smoking and disease by Richard Doll (Box 21.3) was instrumental in the adoption of health measures to reduce lung cancer. In terms of occupational carcinogens, progress in identifying the precise mechanisms is often hampered by the long latency periods between exposure and development of symptoms.

Occupation

The first description of an occupational cancer was made by a British surgeon, Percival Potts in 1775. This was cancer of the scrotum in young chimney sweeps in the 18th century in Britain. Many young boys, who were small enough to climb up chimneys, were employed to remove the soot. They probably worked whilst clad in very few clothes so that there would be frequent damage to the scrotum which, with direct exposure to soot, could lead to the development of cancer, which was known as chimney sweep's cancer or epithelioma of the scrotum.

> Potts was also the first to identify tuberculosis of the spine, a disease which is named after him (Pott's disease).

Diet

In 1981 Doll & Peto estimated that diet accounts for 30% of all cancers, a claim strongly endorsed by a report published by the World Cancer Research Fund in 1997:

We estimate that recommended diets, together with maintenance of physical activity and appropriate body mass,

---

Box 21.3 Tobacco and cancer: a short history

Taken from Richard Doll's review 'Uncovering the effects of smoking: historical perspective', in *Statistical Methods in Medical Research* 1998; **7**: 87–117.

Tobacco grows naturally in central America and the custom of burning the leaves and inhaling the smoke was adopted by the Mayans at least 2,500 years ago. At first, the leaves were burnt in religious ceremonies and the priests, who were also the physicians, credited the plant with powers of healing. Later, tobacco came also to be burnt and the smoke inhaled for pleasure. The use of tobacco for these purposes spread north and south in America and east to the Caribbean islands, where leaves were presented to Spaniards when they invaded the continent at the end of the fifteenth century. Within a few years, tobacco was brought to Spain and Portugal and was claimed to have medicinal value. Its use for medical purposes spread through Europe, where it was chewed, taken nasally as a powder, or applied locally in the treatment of cough, asthma, headaches, stomach cramps, gout, diseases of women, intestinal worms, open wounds, and malignant tumours. Although the plant was named after Jean Nicot, he did not encounter it until 1559 in Lisbon, where he had been sent on a diplomatic mission. While there he became enthused by the reports of its healing powers, wrote about it to the Cardinal of Lorraine, and gave some seeds to a visiting dignitary from the French Court.

Smoking tobacco in pipes for pleasure . . . was made socially acceptable by Sir Walter Raleigh. Many, however, thought it disgusting and the use of tobacco in this way was virulently attacked. The opposition was led by James I in 1603, and he published a pamphlet against it in Latin in the same year and in English anonymously, under the title of A counterblaste to tobacco, a year later. The pamphlet was read widely, dutifully praised, and generally ignored. His attempt to persuade Parliament to increase taxation on tobacco failed and the main effect of his opposition was to diminish imports from Virginia and increase the amount grown at home.

Cigarette consumption increased rapidly in the First World War, particularly in Britain, and by the end of the Second World War, cigarettes had largely replaced other tobacco products in most developed countries. By this time, smoking had become so much the norm for men, that in Britain 80% were regular smokers and some doctors offered a cigarette to patients, when they came to consult them, to put them at ease. . . .

Cigarette smoking has now been found to be positively associated with nearly 40 diseases or causes of death and to be negatively associated with eight or nine more. In some instances the positive associations are largely or wholly due to confounding, but the great majority have been shown to be causal in character. The few diseases negatively associated with smoking are for the most part rare or non-fatal and their impact on disease incidence and mortality as a result of smoking is less than 1% of the excess of other diseases that are caused by smoking. The most recent observations show that continued cigarette smoking throughout adult life doubles age-specific mortality rates, nearly trebling them in late middle age.

It is not surprising, therefore, that in many developed countries smoking has been banned in public places.

---

can in time reduce cancer incidence by 30–40%. At current rates, on a global basis, this represents 3–4 million cases of cancer per year that could be prevented by dietary and associated means.

We believe that action to prevent cancer is rational, timely and important, and should be, therefore, a major priority and responsibility for international agencies, government, industry, non-governmental organisations, medical and health authorities, and for all working in the public interest at international, national and community level.

(World Cancer Research Fund, 1997)

Some dietary practices are thought to contribute to cancer development (Table 21.7). For example, high-temperature cooking produces carcinogens such as heterocyclic amines and polycyclic aromatic hydrocarbons, and storing food can allow microbial carcinogens, such as aflatoxin, to be produced (see Figure 21.27).

Moreover, unfavourable changes in intestinal bacteria can reduce their ability to inactivate carcinogens and, together with deficiencies of folic acid or methionine, predispose to colorectal cancer.

By contrast, other compounds in food may decrease cancer risk (Table 21.7). Free radical scavengers such as the antioxidants, vitamins E and C, carotenoids and flavenoids have anti-cancer activity, while vitamins A and D and other retinoids may encourage a cell to differentiate rather than proliferate (Box 21.4). Plant oestrogens in soya products may be protective since they compete with human oestrogens for the oestrogen receptors in breast and ovary but elicit no response.

Dietary recommendations designed to reduce cancer risk are to eat a balanced diet containing fruit, vegetables and fibre (to provide colonic bacteria with fuel to produce short-chain fatty acids which may regulate colonocyte proliferation) with limited alcohol and fat (in particular to avoid obesity).

Table 21.7 Dietary risk and protective factors for common cancers

Type of cancer	Risk factors	Protective factors
Mouth, pharynx, oesophagus	Alcohol	Adequate intake of fruit, vegetables and micronutrients
Stomach	High intake of salt	Fruit and vegetables
Colon	Red and processed meat; obesity[a]	Plant foods rich in fibre
Liver	Alcohol; foods contaminated with aflatoxin	None established
Pancreas	Not known	None established
Lung	Not known	Possibly fruit and vegetables
Breast	Alcohol; obesity[a] after menopause	None established
Endometrium	Obesity[a]	None established
Prostate	None established	None established

[a] and therefore excessive energy intake.

Data from Key (2003).

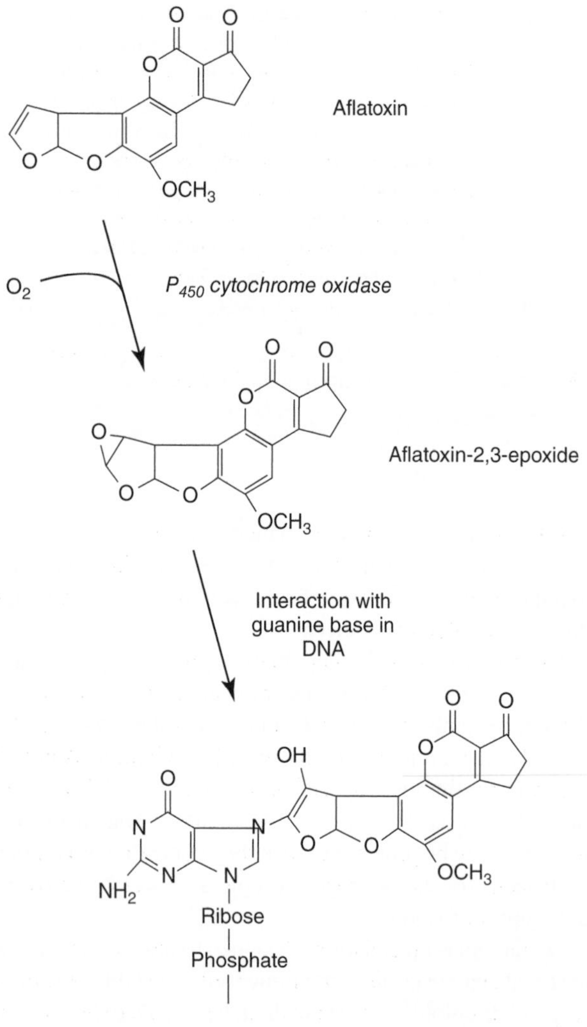

Figure 21.27 *Formation of a carcinogen.*

Box 21.4 Latitude, cancer and rickets

Rickets has been recognised as a disease for almost 500 years but it was not until 1890 that Theodore Palm showed it was probably associated with a deficiency of sunlight, as it occurred mostly at latitudes of 37° or higher. The disease was also prevalent in northern cities such as Manchester and Glasgow, which were polluted with chemicals that absorb UV light. A similar association of breast and colon cancer with latitude has been observed in North America. Deaths are highest in the Northeastern and New England states and in Canada where the intensity of UV light is relatively low during most of the year. Mortality rates are much lower in sunnier states such as Hawaii, Arizona and New Mexico. As with rickets, this suggests that vitamin D or its metabolites may have a protective effect, although the mechanism is not known.

Lifestyle

Several lifestyle factors predispose to cancer development, including smoking tobacco and exposure to sunlight (especially for children and the fair-skinned). It should be noted that low levels of continuous sun exposure may protect against breast and colon cancer, perhaps as a result of raising vitamin D levels which has already been discussed described.

High levels of oestrogen, such as in the non-pregnant state, increase breast cancer risk in nulliparous women, whereas exposure to the human papilloma virus (during sexual intercourse with multiple partners) increases the risk of cervical cancer. Physical activity (especially in young women) is considered to be beneficial. This may be due to shortening of the luteal phase of the menstrual cycle (when oestrogen levels are high) or to a reduction in quantity of adipose tissue (with its aromatase activity for production of oestrogen, see Appendix 19.1).

Chemotherapy

A wide range of anti-cancer drugs are in current use (see Figure 21.28; Table 21.8), including cytotoxic drugs to inhibit cell division, alkylating agents to cross-link bases in DNA, inhibiting its synthesis, and antimetabolites to prevent synthesis of purine and pyrimidine nucleotides (e.g. methotrexate, nucleotide analogues). Hormone analogues, which occupy receptors and prevent the binding of hormones that stimulate proliferation (e.g. in prostate, breast and endometrium) are also used. For example, tamoxifen, a synthetic oestrogen, binds to oestrogen receptors and oestrogens are used in the treatment of prostate cancer.

A related therapy for prostate cancer involves administration of gonadotrophin-releasing factor antagonists (GnRH), which reduce the synthesis of GnRH and hence that of testosterone. Molecular targets in chemotherapy include aromatase, which normally converts the male sex hormones, androstenedione and testosterone, into the female sex hormones, oestrone and oestradiol. Aromatase is present in adipose, liver, muscle, normal breast and breast tumour cells as well as follicular granulosa cells, where it is involved in the production of the female sex hormones during the menstrual cycle.

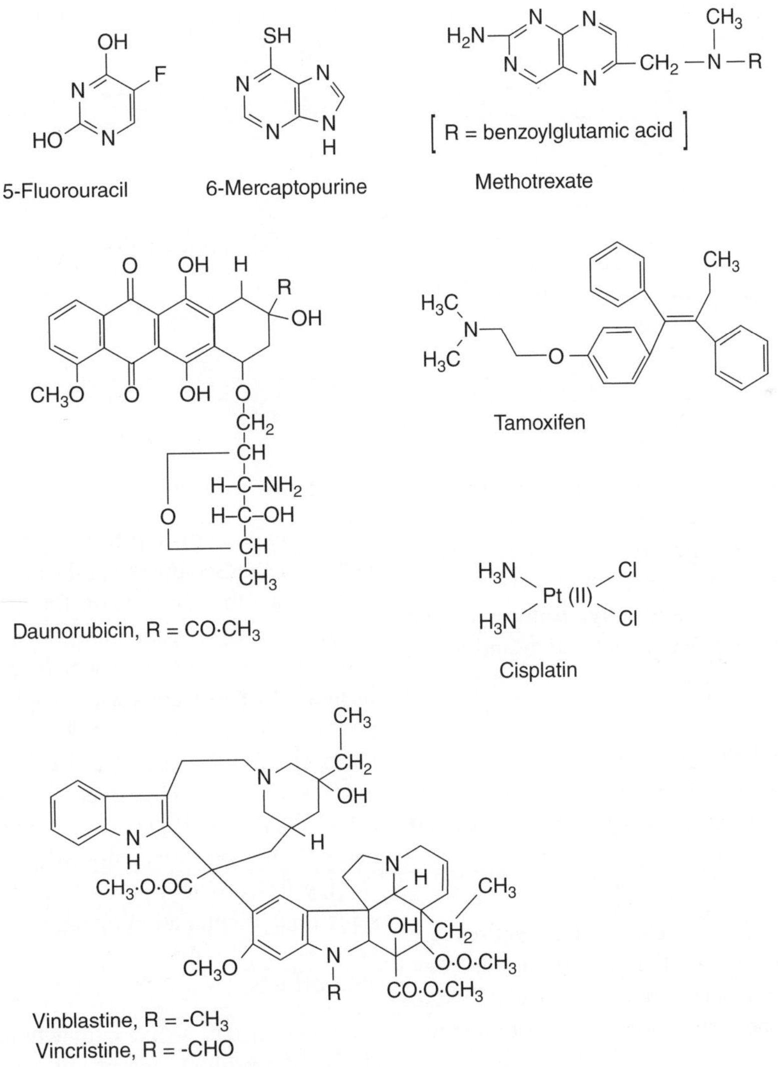

Figure 21.28 *Drugs used in chemotherapy for cancer.*

Table 21.8 Some chemotherapeutic agents and their mechanisms of action

Agent	Mechanism
Antimetabolites (inhibition of purine and pyrimidine nucleotide synthesis)	
Methotrexate	Folic acid antagonist, inhibits tetrahydrofolate reductase and therefore dTMP synthesis
6-Mercaptopurine	Interferes with purine synthesis
5-Fluorouracil	Inhibits dTMP synthesis
Alkylating agents	
Nitrogen mustards (cyclophosphamide)	
Busulphan	Crosslinks DNA strands
Platinum coordination complexes (cisplatin)	
Nitrosoureas	
Antibiotics	
Daunorubicin	Inhibit topoisomerase, stimulate free radical formation
Doxorubicin	
Plant alkaloids	
Taxol (paclitaxel)[a]	Inhibits mitosis
Vinblastine	Interfere with mitosis, block β tubulin polymerisation
Vincristine	
Molecular targets	
Anti-oestrogens	
Aromatase inhibitors	
Androgens	
Hormone therapy	
Tamoxifen	Anti-oestrogen, forms complex with oestrogen receptor

[a] isolated from bark of western yew tree

Antibody directed targets   Techniques for fusing myeloma cells with B-lymphocytes and production of monoclonal antibodies have allowed specific targeting of drugs by recognition of cell-surface antigens present only on the tumour cell. Thus, cytotoxic drugs or radioactive isotopes can be conjugated with the tumour-specific antibodies and delivered directly to the tumour cell.

Antibiotics   Two antibiotics have been used in cancer therapy: actomyosin D which binds to DNA and prevents proliferation, and bleomycins which cause single-strand breaks in DNA.

Plant extracts   Some plant extracts may have anti-cancer activity, such as epipodo phyllotoxins from the May apple plant (mandrake plant). They were used by the American Indians and the early colonists against infection by helminths.

Resistance to drugs

Some tumours develop resistance to drugs due to several factors: decreased uptake of the drug by the cell; mutations in enzymes that are targets for the drugs or increase in activity of a membrane transporter (e.g. the multiple drug resistant (MDR) protein) which expels toxic compounds from a cell. Those cells with a higher activity of the transporter or lower activity of the uptake process are more likely to survive a course of treatment and therefore have a competitive advantage over the cells with a lower activity of the transporter. Hence, the proportion of cells with resistance to the drug gradually increases. One way of overcoming this problem is to use more than one drug at the same time: combination therapy.

Side effects

Cytotoxic drugs, which inhibit tumour cell proliferation, also inhibit proliferation of normal cells. This produces a range of side effects, including decreased production of immune cells and erythrocytes in the bone marrow and impaired resistance to infection and slow wound healing; stunted growth in children; sterility; loss of hair; damage to gastrointestinal epithelium leading to vomiting and diarrhoea, mucositosis and teratogenicity.

Radiotherapy

Radiotherapy is the use of ionising radiation to damage and kill tumour cells. *Radical* radiotherapy involves using radiotherapy to cure the tumour. It may be combined with chemotherapy before, during or after radiotherapy. *Adjuvant* radiotherapy, for example following surgery to remove a tumour, is given to eradicate residual tumour cells. *Palliative* radiotherapy is administered in short courses, for example for bone metastases to treat pain, to shrink tumour masses or to relieve symptoms such as bleeding.

Radiotherapy is a local treatment aiming to achieve local control or cure of locally confined tumours. It cannot treat metastases. Radiotherapy may be administered as external beam radiotherapy with X-rays or gamma rays, in sealed radioactive sources (e.g. prostate brachytherapy), or unsealed sources (e.g. orally administered radioiodine for thyroid cancer, intravenous strontium-89 for bone metastases). In external beam radiotherapy, the X-ray or gamma ray beams are targeted at the tumour to damage and kill the tumour cells. Inevitably, surrounding normal tissues are also affected resulting in the early and late side effects of radiotherapy.

22
Atherosclerosis, Hypertension and Heart Attack

These three subjects are discussed together in one chapter since they are all related and have a number of features in common. They are also involved in other medical problems that are discussed in other chapters.

Atherosclerosis

The advanced lesions of atherosclerosis represent the culmination of a series of cellular and molecular events, involving replication of both smooth muscle cells and macrophages which have previously entered the artery wall. The interactions of these cells with T lymphocytes also in the lesion and the overlying endothelium can lead to a massive fibroproliferative response over which connective tissue from smooth muscle cells form a fibrous cap. This covers the advanced lesion or fibrous plaque of atherosclerosis, deeper portions of which consist of macrophages, T lymphocytes, smooth muscle cells, connective tissue, necrotic debris and varying amounts of lipids and lipoproteins.

(Ross, 1999)

Atherosclerosis is a degenerative condition of the arteries in which fat, yellowish plaque, known as atheroma, is present in medium and large arteries. This results in thickening, lessening of elasticity and narrowing of the lumen. It was once believed that atheroma developed and then grew steadily until it eventually obstructed the artery and seriously impeded blood flow (Figure 22.1). This view has now been abandoned. The current view is that atheroma starts with injury to endothelial cells in an artery and that this affects the immune system in a similar manner to the development of chronic inflammation. This is the *injury–inflammation hypothesis*. It is, however, not new: a similar account was put forward by Rudolf Virchow over 150 years ago (Box 22.1).

Atherosclerosis and coronary heart disease are amongst the most intensively investigated subjects in medical science. Despite this, it must be pointed out that atherosclerosis is not itself a disease but a process that contributes to diseases, such as an intermittent claudication, coronary arterial disease and stroke.

> In most developed countries coronary disease is the commonest single cause of death. In the USA almost half of all deaths are related to atherosclerosis.

Development of atherosclerosis: the response to injury and inflammation hypothesis

The atherosclerotic plaque consists, on the lumen side, of a layer of connective tissue containing smooth muscle cells and macrophages covering a deeper layer of macrophages containing so much lipid that they are known as foam cells due to their microscopic appearance. This layer also contains a varying amount of cell debris and extracellular lipid. Outside it, there is often a region of proliferated smooth muscle cells.

Even during the first or second decade of life, small deposits of lipid, fatty streaks, are often detectable in arterial walls. In a study by R. Ross over half of the children (age 10–14) examined at autopsy had fatty streaks in their arteries. These are the first indications of the entry of fat and cholesterol into macrophages in the subendothelial space of an artery. This initiates a sequence of processes that eventually produces a plaque. A prerequisite for the development of fatty streaks, and hence atherosclerosis, is injury to the endothelial cells lining the arterial wall. Many factors are suspected of causing this, including pollutants,

Functional Biochemistry in Health and Disease by Eric Newsholme and Tony Leech
© 2010 John Wiley & Sons Ltd

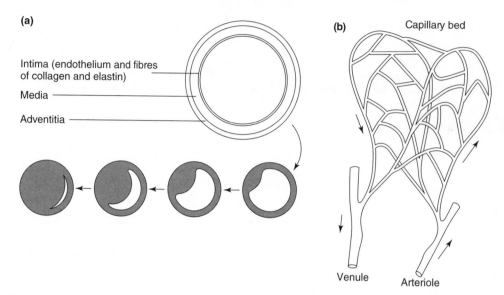

Figure 22.1 (a) *Structure of an artery.* The innermost layer, the intima, is composed of a single layer of endothelial cells resting on a thin layer of connective tissue and elastic fibres. Outside this is an elastic lamina which is surrounded by the media, largely composed of smooth muscle fibres. Finally, outside this is the adventitia which is composed mostly of collagen fibres. Contraction or relaxation of the smooth muscle regulates the diameter of the vessel and hence the flow of blood. The space between the intima and the media is known as the subendothelial space. Beneath is a diagram representing the older view of the development of atheroma: i.e. it expands gradually and finally occludes the vessel. **(b)** *Diagram of a capillary bed.* Blood is supplied by an arteriole, which is a small artery. Arterioles are of prime importance in regulating blood flow through the capillary bed. This is controlled by the diameter of the arteriole, which is controlled by contraction/relaxation of the smooth muscle (see below).

Box 22.1 The first description of inflammation in the development of atherosclerosis

Rudolf Virchow (1821–1902), considered by some to be the greatest pathologist of all time, delivered 20 lectures at the Pathological Institute of Berlin during the months of February, March and April, 1858. These lectures were later translated into English and published. The following is an extract from this translation which demonstrates that Virchow was aware of the possibility that inflammation could be involved in the development of atheroma more than 150 year ago.

Today you see very extensive patches in the aorta, in which the atheromatous change has taken place. But, as is wont to be the case in changes of this kind, in addition to the specific transformation attendant upon the chronic inflammatory processes going on in the deeper parts, you find on the surface also a simply fatty change, so that we have the two processes occurring together. If now we examine atheromasia a little more minutely, for example in the aorta, where the process is the most common, the first thing we see present itself at the spot where the irritation has taken place, is a swelling of larger or smaller size . . . I have therefore

felt no hesitation in siding with the old view in this matter, and in admitting an inflammation of the inner arterial coat to be the starting point of the so-called atheromatous degeneration; and I have moreover endeavoured to shew that this kind of inflammatory affection of the arterial coat, is in point of fact exactly the same as what is universally termed endocarditis, when it occurs in the parietes of the heart. . . . Now if the mass be examined which is present at the close of this process, numerous plates of cholestearine are seen, which display themselves even to the naked eye as glistening lamellae; large rhombic tablets, which lie together in large numbers, side by side, or covering one another, and altogether produce a glittering reflection. In addition to these plates, we find under the microscope black-looking granule-globules, in which the individual fat-granules are at first very minute. These globules are often present in very large quantity; some of them are seen, breaking up, and falling to pieces, particles of them swimming about . . .

Virchow (1858)

chemicals in tobacco smoke, various lipids (e.g. cholesterol, long-chain fatty acids, lysophospholipids), proinflammatory cytokines and bacteria. Disturbance in the pattern of blood flow at bends and branchpoints in the major arteries can also lead to endothelial injury, particularly if the blood pressure is raised. Such injuries result in the appearance of adhesion molecules on the surface of the endothelial cells to which monocytes adhere before entering the subendothelial space, where they develop into macrophages. Factors that are released from these cells attract more monocytes and other immune cells, and the process of inflammation begins (Figure 22.2).

> Infection with cytomegalovirus (a common virus of the herpes virus group) is associated with atherosclerosis.

A central event in the generation of plaque is the uptake of low density lipoproteins (LDL) by macrophages in the subendothelial space. LDL enters this space through the damaged endothelial cells. The uptake occurs via endocytosis, after the binding of LDL to one of three receptors on the macrophage:

- The normal LDL-receptor, i.e. that which is present on most other cells.

- A non-specific receptor, which binds and takes up LDL when the plasma concentration is above normal.

- A scavenger receptor, which binds and takes up damaged LDL.

The normal LDL-receptor is used to transport cholesterol into cells when it is required, for example, in the formation of membranes or synthesis of steroid hormones. The number of these receptors is regulated by the intracellular level of cholesterol; the higher the level, the lower is the receptor number. However, no such regulation exists for the non-specific or scavenger receptors, so that both are responsible for high and uncontrolled rates of entry of LDL into the macrophage. Damage to LDL is caused by high blood glucose levels, which cause glycosylation of a protein in the LDL (apolipoprotein B), and free radicals, which oxidise unsaturated fatty acids in the phospholipids of the LDL.

Within the macrophage, the LDL is degraded and the resultant free cholesterol is esterified to form cholesterol ester. This accumulates and then damages and eventually kills the macrophages to produce what are known as foam cells. The dead and dying macrophages secrete cytokines and chemotactic agents, which encourage the entry of more monocytes and lymphocytes into the developing plaque, and growth factors that stimulate proliferation of smooth muscle cells to further increase the size of the plaque. The process, therefore, has the characteristics of a vicious circle (Figure 22.3).

Atherosclerosis and arterial occlusion

Although the diameters of the major, and particularly the coronary, arteries of many adults in developed countries are decreased due to plaque, the plaque is normally quite stable and impinges little on lifestyle. Despite this, changes can occur in the artery or in the plaque that result in occlusion of the artery. These events include:

- Acute vasoconstriction (vasospasm) of smooth muscles in the artery in the vicinity of the plaque, which can severely reduce the diameter of the artery.

- Formation of one or more blood clots at the site of the plaque.

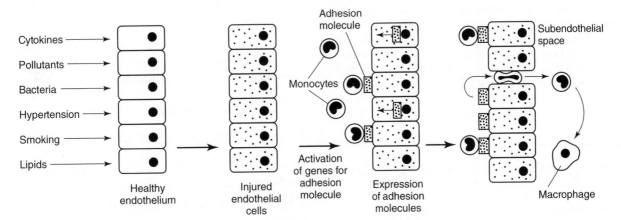

Figure 22.2 *Injury to endothelial cells, adhesion of monocytes and entry into subendothelial space.* Injury to the cells by various factors activates genes for expression of adhesion molecules on the luminal surface of the cells. Monocytes attach to these molecules and then enter the subendothelial space. Here they are activated to form macrophages.

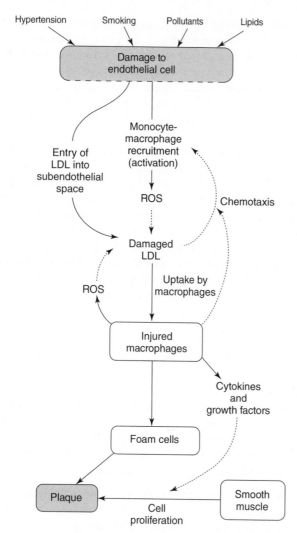

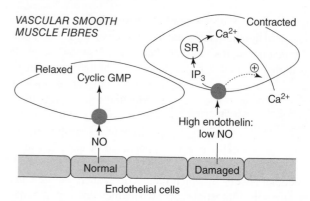

Figure 22.4 *Injury to endothelial cells can lead to vasospasm.* Normal endothelial cells release nitric oxide (NO) which relaxes smooth muscle: this is achieved by nitric oxide increasing the concentration of cyclic GMP within smooth muscle fibres and cyclic GMP relaxing the smooth muscle. Injured endothelial cells secrete very little nitric oxide but secrete more endothelin. The latter increases the formation of inositol trisphosphate (IP_3), which binds to the sarcoplasmic reticulum (SR) where it stimulates the Ca^{2+} ion channel. The Ca^{2+} ion channel in the plasma membrane is also activated. Both effects result in an increase in cytosolic Ca^{2+} ion concentration, which then stimulates contraction (vasospasm). This reduces the diameter of the lumen of the artery.

occlusion occurs in one of the coronary arteries, normal pumping by the heart can fail.

Vasospasm

Both endothelial cells and platelets contribute to contraction of smooth muscle, as follows:

- Normal endothelial cells produce nitric oxide, which relaxes smooth muscle, whereas damaged cells release less nitric oxide but more of a local hormone, endothelin, which stimulates contraction of smooth muscle (Figure 22.4).

- Platelets in the vicinity of, or within the plaque, release factors that stimulate contraction of smooth muscle (e.g. thromboxanes, platelet activating factor).

Formation of a blood clot

The normal endothelial cells possess characteristics that prevent thrombus formation: indeed, their surface is one of the few that has such properties. Three of these characteristics are:

- The presence of heparin on their surface which exerts an anticoagulant effect.

- The capacity to produce prostaglandins, which inhibit platelet aggregation.

Figure 22.3 *Sequence of events leading from injured endothelial cells to the formation of plaque.* Injury to endothelial cells leads to entry of monocytes and LDL into the subendothelial space. Production of reactive oxygen species (ROS) by macrophages leads to damage to the structure of LDL. This damaged LDL is taken up by the macrophages. This leads to accumulation of cholesterol within the macrophages which is then converted to cholesterol ester, which also accumulates and leads to death of the macrophage. The cholesterol ester and fat within the dead macrophages are released into the subendothelial space where, together with foam cells and the larger number of smooth muscle fibres, they form the plaque (see Figure 22.5(b)).

- Disruption of the plaque which then breaks up and blocks the artery.

Whatever the mechanism, the occlusion results in failure of blood supply and, hence, oxygen supply to the organ, which leads to a decrease in the rate of ATP generation and then failure of the normal function of the organ or tissue. If the

- The capacity to produce or activate enzymes (e.g. plasmin, urokinase) that break down clots once they have begun to form.

Damage to or loss of endothelial cells before or after the formation of the plaque interferes with these properties, thus increasing the risk of thrombosis. The presence of a large number of platelets producing thromboxanes, in comparison with the decreased number of endothelial cells producing prostacyclins, results in an increase in the local concentration ratio, thromboxane/prostacyclin, which favours platelet aggregation and hence clot formation (Chapter 11). Finally, cells within the plaque also release thrombotic factors which, for example, can stimulate the extrinsic clotting cascade. Any one or all of these changes can lead to local thrombosis (Figures 22.5(a)&(b)). Furthermore, if all of these changes occurred simultaneously, a clot could suddenly form, precipitating an infarct and possibly sudden death.

Rupture of the plaque

A number of factors increase the risk of disruption of the plaque: the presence of a considerable amount of fat in the plaque; the effects of physical stress on the vessel wall, particularly in the hypertensive patient; and the activity of macrophages. Macrophages release proteases (e.g. collagenase, elastase) which lead to breakdown of the plaque. Rupture of the plaque produces fissures that are sites for aggregation of platelets. Fragments from the rupture can occlude blood vessels.

Atherosclerosis in peripheral, cerebral and coronary arteries

Atherosclerosis can affect the function of any organ or tissue, muscles in the legs, the brain or the heart.

- Atherosclerosis in the arteries supplying the muscles in the legs results in pain in the calves and buttocks upon walking, which is known as *intermittent claudication*. A restricted supply of oxygen to these muscles decreases the generation of ATP, which leads to a decrease in the ATP/ADP concentration ratio and to pain and fatigue, as described in Chapter 13. In severe cases of atherosclerosis, a bacterial infection in the lower leg may not be readily controlled, since sufficient immune cells cannot be delivered to the site of infection. Furthermore, in the hypoxic environment of the muscle in this condition, anaerobic organisms such as *Clostridium* species can survive and proliferate, leading to gangrene.

- Atherosclerosis in a cerebral artery can result in an occlusion, which could be caused by any of the mecha-

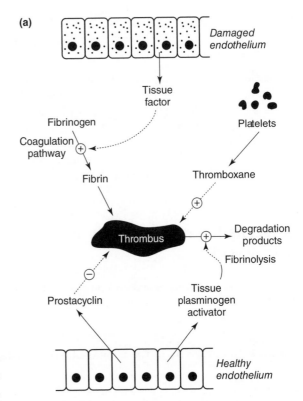

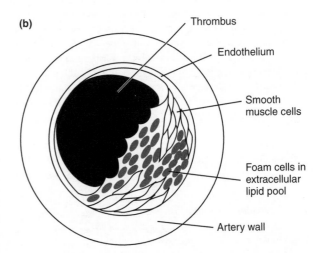

Figure 22.5 (a) *Injury to endothelial cells can lead to thrombosis.* Healthy endothelial cells secrete two factors that inhibit thrombus formation: (i) prostacyclins, which inhibit aggregation of platelets; and (ii) a factor that facilitates conversion of plasminogen to plasmin. Injury to endothelial cells can facilitate thrombosis since (i) they do not secrete prostacyclins; (ii) they do not secrete plasminogen activator; and (iii) they secrete a factor that stimulates thrombosis. **(b)** *Diagram of an atherosclerosed artery containing a thrombus.* A thrombus is blocking the lumen of the artery that is not totally blocked by plaque.

nisms described above. The resultant lack of oxygen causes damage or death to neurones, a condition known as an ischaemic stroke.

- Atherosclerosis in a coronary artery can lead to an infarct (see below). If the atherosclerosis is less severe, the reduction in the blood flow cannot provide enough oxygen in some conditions to support the work of the heart, for example during exercise or acute anxiety, and the resulting pain, angina pectoris, if frequent reduces the quality of life of the patient (Box 22.2).

Risk factors for atherosclerosis and how they work

As early as 1980, Ancel Keys and his colleagues studied over 12 000 men in seven countries and identified three major risk factors for coronary heart disease: smoking tobacco, hypertension and a high serum level of cholesterol (Table 22.1). Since these have been known for some time, they are described as traditional factors, whereas those that have been identified more recently are described as novel factors. What are the mechanisms by which these factors increase the risk? There are several answers, including the following:

> As a rule of thumb for the physician, the presence of one of these factors doubles the risk, two factors increase the risk fourfold, whereas all three increase the risk sevenfold.

- Smoking tobacco causes damage to endothelial cells due to free radicals present in tobacco smoke. It is estimated that each puff of a cigarette produces 10^{14} free radicals. In addition, the resultant lack of oxygen causes damage or death to neurones, and nicotine and carbon monoxide, both present in tobacco smoke, cause an increase in blood pressure.

Box 22.2 The discovery of the benefit of amyl nitrite for angina pectoris

Angina pectoris is the pain or spasm felt in the centre of the chest and sometimes in the arms and jaw (angina from the Greek *agkhone* for strangling) which occurs upon exercise or sometimes during acute anxiety. It is due to the inability of the coronary arteries to supply sufficient blood and therefore oxygen to maintain the necessary rate of mitochondrial ATP generation required for normal contractile activity of the myocardium. Organic nitrates (e.g. glyceryl trinitrate) are beneficial since they rapidly vasodilate the coronary circulation. The first drug used to treat angina was amyl nitrate, the effect of which was discovered by Lauder Brunton in 1867. Part of the paper published by Brunton in 1867 is presented below.

On the USE OF NITRITE OF AMYL IN ANGINA PECTORIS. By T. Lauder Brunton, B.Sc., M.B., Senior President of the Royal Medical Society and Resident Physician to the clinical wards of the Royal Infirmary, Edinburgh.

> *Few things are more distressing to a physician than to stand beside a suffering patient who is anxiously looking to him for that relief from pain which he feels himself utterly unable to afford. His sympathy for the sufferer, and the regret he feels for the impotence of his art, engrave the picture indelibly on his mind, and serve as a constant and urgent stimulus in his search after the causes of the pain, and the means by which it may be alleviated.*
>
> *Perhaps there is no class of cases in which such occurrences appear so frequently as in kinds of cardiac disease, and serve as a constant and urgent stimulus . . . angina pectoris forms at once the most prominent and the most painful and distressing symptom . . . I am now publishing a statement of the results which I*

have obtained in the treatment of angina pectoris by nitrite of amyl . . . On application to my friend Dr. Gamgee, he kindly furnished me with a supply of pure nitrite which he himself had made; and on proceeding to try it in the wards, with the sanction of the visiting physician, Dr. J. Hughes Bennett, my hopes were completely fulfilled. On pouring from five to ten drops of the nitrite on cloth and giving it to the patient to inhale, the physiological action took place in from thirty to sixty seconds; and simultaneously with the flushing of the face the pain completely disappeared, and generally did not return till its wonted time next night. Occasionally it began to return about five minutes after its first disappearance; but on giving a few drops more it again disappeared, and did not return.

The mechanism by which organic nitrates relieve the pain of angina pectoris was not discovered until nitric oxide was identified as the agent which was responsible for vasodilation of arteries. It was known for many years that endothelial cells released a factor that resulted in vasodilation; a factor appropriately called *endothelial relaxing factor* (EDRF). It was, however, some time before the factor was identified, probably because it turned out to be a gas – nitric oxide – which was totally unexpected. Nitric oxide is now known to be a very important messenger molecule involved in the regulation of many other systems. The mechanism by which it causes vasodilation is described in Chapter 13.

The factor responsible for the pain is not known but the increased concentrations of adenosine and/or potassium ions are considered to be involved. These factors are also considered to be a cause of the pain that occurs in the skeletal muscle during peripheral fatigue (Chapter 13).

Table 22.1 Risk factors for development of atherosclerosis and coronary heart disease

Traditional risk factors	Novel risk factors
Increasing age;	Antioxidants, low intake;
Family history of hypertension;	Fibrinogen, high serum level;
Smoking tobacco;	Homocysteine, high plasma level;
Total cholesterol, high plasma level;	Dimethylarginine, high plasma level;
LDL cholesterol, high serum level;	Lipoprotein (a), high serum level;
HDL cholesterol, low serum level;	Physical inactivity;
Family history of myocardial infarction;	*Trans* fatty acids, high intake.
Triacylglycerol, high serum level.	

- Hypertension causes damage to endothelial cells due to the shear stress as the blood flows through the arteries, especially at positions where arteries branch.

> It should be noted that the shear stress caused by the flow of blood through arteries and capillaries can be very high.

- High levels of serum cholesterol can damage endothelial cells, although a better indication of the risk is provided by the serum LDL-cholesterol level. Desirable levels of serum cholesterol and LDL-cholesterol level are given in Table 22.2.

- A high serum triacylglycerol level, especially VLDL, increases the activity of some clotting factors, particularly factor VIII, and therefore increases the risk of intravascular clots (Figure 22.6).

- Cholesterol ester transfer protein catalyses the transfer of triacylglycerol from VLDL or chylomicrons to LDL and to HDL. However, it is the removal of this triacylglycerol from LDL and HDL, which occurs in the liver, via hepatic lipase, that causes problems: small and dense LDL particles, which are more atherogenic than normal LDL particles, are produced. The removal from the HDL produces a form that is less effective in reverse cholesterol transfer. The latter is a beneficial process since it transfers cholesterol from cells in various tissues to the liver (see below).

- *Trans* fatty acids are produced during the commercial hydrogenation of plant oils (Chapter 11). Some margarines contain these fatty acids, as do some commercially prepared snack foods (e.g. biscuits, cookies, cakes, crisps, chips). In addition, bacteria in the rumen of ruminants produce *trans* fatty acids, which are therefore present in dairy produce and meat. *Trans* fatty acids can be incorporated into the phospholipids of the plasma membrane of endothelial and other cells, resulting in damage to the membranes. Furthermore, these abnormal fatty acids can interfere in the production of thromboxanes, prostacyclins or leucotrienes and hence interfere in control of blood clotting, immune cell activity and inflammation (Chapter 11).

> The concern over a high intake of *trans* fatty acids is indicated by the fact that food which contains a high proportion of *trans* fatty acids has been banned from restaurants in New York and California.

- Lipoprotein (a) (abbreviated to Lp(a)) is a complex between LDL and apoprotein (a) that forms spontaneously in blood. Lp(a) is secreted by the liver but its function is unknown. A high plasma level of Lp(a) interferes with the conversion of plasminogen to plasmin, the role of which is to break down blood clots and even disperse small clots.

- Homocysteine is an intermediate in the metabolism of methionine and disturbances in metabolism of this amino acid can lead to an elevated plasma level of homocysteine. The risk is due not to homocysteine itself but to a thiolactone, to which homocysteine is converted spontaneously (Figure 22.7; Box 22.3).

> An elevated plasma level of homocysteine is also a risk factor for Alzheimer's disease (Chapter 14).

Table 22.2 Optimal and desirable levels of serum total cholesterol and LDL-cholesterol levels

Classification of level	Total cholesterol		LDL-cholesterol	
	mg/dL	mmol/L	mg/dL	mmol/L
Optimal	<150	<3.9	<100	<2.6
Desirable	150–199	3.9–5.1	100–129	2.6–3.3
Borderline high	200–239	5.2–6.2	130–159	3.4–4.1
High	>240	>6.2	>160	>4.1

Information from US National Cholesterol Education Program (1993, 1994).

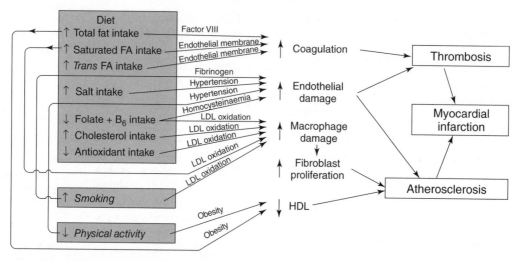

Figure 22.6 *How various factors increase the risk of atherosclerosis, thrombosis and myocardial infarction.* The diagram provides suggestions as to how various factors increase the risk of development of the trio of cardiovascular problems. The factors include an excessive intake of total fat, which increases activity of clotting factors, especially factor VIII; an excessive intake of saturated or *trans* fatty acids that change the structure of the plasma membrane of cells, such as endothelial cells, which increases the risk of platelet aggregation or susceptibility of the membrane to injury; excessive intake of salt – which increases blood pressure, as does smoking and low physical activity; a high intake of fat or cholesterol or a low intake of antioxidants, vitamin B_{12} and folic acid, which can lead either to direct chemical 'damage' (e.g. oxidation) to the structure of LDL or an increase in the serum level of LDL, which also increases the risk of chemical 'damage' to LDL. A low intake of folate and vitamin B_{12} also decreases metabolism of homocysteine, so that the plasma concentration increases, which can damage the endothelial membrane due to formation of thiolactone.

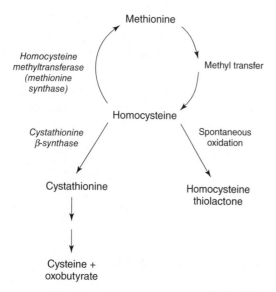

Figure 22.7 *Homocysteine formation from methionine and formation of thiolactone from homocysteine.* The homocysteine concentration depends upon a balance between the activities of homocysteine methyltransferase (methionine synthase) and cystathionine β-synthase. Both these enzymes require vitamin B_{12}, so a deficiency can lead to an increase in the plasma level of homocysteine. (For details of these reactions, see Chapter 15.) Homocysteine oxidises spontaneously to form thiolactone, which can damage cell membrane.

• As with some other amino acids in proteins, arginine is methylated, i.e. a methyl group ($-CH_3$) is attached to the amino acid, in a process called post-translational modification. Indeed, the amino acid is methylated in two positions to produce asymmetrical dimethylarginine or symmetrical dimethylarginine (Figure 22.8(a)). When the protein is hydrolysed, as part of the normal turnover of proteins, the dimethylarginine is released into the cell and, since it is not reincorporated into the protein, it enters the bloodstream. It is a risk factor for atherosclerosis since it inhibits the enzyme nitric oxide synthase, which catalyses the formation of nitric oxide, which relaxes smooth muscles in arterioles (it is a vasodilator). Failure to produce nitric oxide, therefore, leads to vasoconstriction which increases blood pressure, and, for similar reasons, increases the risk of vasospasm. Normally, dimethylarginine is degraded by an enzyme, dimethylarginine aminohydrolase, which is present in endothelial cells (Figures 22.8(b)&(c)). In view of the danger of dimethylarginine, this enzyme can be considered to be a detoxifying enzyme. Consequently, when endothelial cells are damaged, their capacity to remove dimethylarginine is decreased and its concentration in the blood rises, which restricts vasodilation or increases vasoconstriction, either of which increases the risk of hypertension or vasospasm.

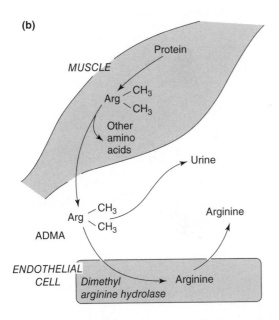

(a)

Arginine

Asymmetrical dimethyl arginine (ADMA)

Symmetrical dimethyl arginine (SDMA)

(b)

MUSCLE

Protein

Arg — CH₃ / CH₃

Other amino acids

Urine

Arg — CH₃ / CH₃

ADMA

Arginine

ENDOTHELIAL CELL

Dimethyl arginine hydrolase → Arginine

(c)

Protein
↓
Dimethyl arginine — *Dimethyl arginine aminohydrolase* → Arginine
↓
Arginine ⊖→ Nitric oxide + Citrulline
Nitric oxide synthase
↓
Relaxation of smooth muscle
↓
Vasoconstriction

Figure 22.8 (a) *Structures of arginine, asymmetrical dimethylarginine and symmetrical dimethylarginine.* It is the asymmetrical form that is a particularly strong inhibitor of nitric oxide synthase. **(b)** *Formation of dimethylarginine from the turnover of muscle protein and degradation of methylarginine in endothelial cells.* Methylation of arginine occurs when arginine is a component of body protein; i.e. it is a post-translational process. The methylarginine is released during protein degradation that is part of the natural turnover process. The released methylarginine has two fates: (i) excretion in the urine; (ii) hydrolysis to arginine catalysed by the enzyme, dimethylarginine hydrolase, which is present in endothelial cells. **(c)** *Inhibition of nitric oxide synthase by dimethylarginine and its physiological relevance.* The inhibition decreases the level of nitric oxide, which, especially in vascular smooth muscle, reduces relaxation, which can lead to vasoconstriction.

Box 22.3 The discovery of homocysteine as a risk factor for coronary heart disease

A simple observation led to the identification of homocysteine as a risk factor for coronary heart disease. Homocysteine is an intermediate in metabolism of the amino acid methionine. Indeed, the first reaction in the catabolism of methionine involves the formation of homocysteine but it can be converted back to methionine in a reaction that is catalysed by methionine synthase (see Figure 22.7).

$$\text{Methionine} \xleftarrow{(1)} \text{Homocysteine} \xrightarrow{(2)} \text{Cystathionine} \longrightarrow \longrightarrow CO_2$$

Enzyme (1) is methionine synthase: enzyme (2) is cystathionine β-synthase

In a totally different field, studies were being carried out on children who had a deficiency of methionine synthase and an impaired ability to convert homocysteine to methionine, so that they had increased blood levels of homocysteine. It was noted that these children had an increased incidence of thrombosis in cerebral and coronary arteries. This led to a study which eventually showed that an increased level of homocysteine was a risk factor for coronary artery disease in adults. Since methionine synthase requires the vitamins, folic acid and B₁₂, for its catalytic activity, it has been suggested that an increased intake of these vitamins could encourage the conversion of homocysteine to methionine and hence decrease the plasma level of homocysteine. This is particularly the case for the elderly who are undernourished (see Chapter 15 for a discussion of nutrition in the elderly).

Further to this, the enzyme cystathionine β-synthase is involved in the catabolism of homocysteine, so that a deficiency of this enzyme also results in an elevated level of homocysteine in the blood. Consequently, patients with a deficiency, even a partial deficiency, could also suffer an increased risk of coronary artery disease.

- Rosenman and Friedman suggested in 1974 that psycho-social factors play an important role in development of cardiovascular disorders. Subjects, who were described as exhibiting type A personality behaviour, that is 'ambition, competitive drive, aggressiveness, a strong sense of time urgency', were much more likely to suffer from atherosclerosis than type B personalities, who are more relaxed and unhurried. This suggestion was and still is controversial, but one component of type A behaviour, aggression, is now less controversial as a risk factor.

Protective factors

Protective factors for atherosclerosis are fish oils, antioxidants, physical activity and garlic. These are discussed below.

Fish oils   These contain high levels of omega-3 fatty acids, which have a number of properties that could explain why fish oils or a diet high in oily fish have a protective effect:

- They lower the blood level of VLDL and therefore they lower plasma level of LDL.

- They decrease the formation of blood clots.

- They decrease the formation of thromboxane A_2 and prostacyclin I_2 in favour of thromboxane A_3 and prostacyclin I_3, changes which protect against thrombosis (see below).

- They have a hypotensive effect.

The major omega-3 fatty acid in fish oil is eicosapentaenoic acid, which contains five double bonds compared with only four present in the omega-6 fatty acid, arachidonic acid. When eicosapentaenoic acid is substrate for eicosanoid production, it gives rise to prostacyclins and thromboxanes of the three series (Figure 22.9(a)) whereas when arachidonic acid is substrate, it gives rise to the two series, thromboxane A_2 and prostacyclin I_2. Thromboxane A_3 has much less of a thrombolytic effect than thromboxane A_2 whereas prostacyclin I_3 has more of an antithrombotic activity than prostacyclin I_2. Hence, the risk of formation of a thrombus is decreased when omega-3 fatty acids are the substrate for the cyclooxygenase (Figure 22.9(b)). There is considerable epidemiological evidence that fish oils are protective against atherosclerosis.

A study in Norway initiated after the end of the Second World War noted that within 12 months of the German occupation in 1940 there was a substantial fall in heart disease. The occupation resulted in an abrupt change in the Norwegian diet: meat was replaced by fish.

In a multi-centre trial involving more than 11 000 patients, supplementation of diet with fish oil reduced mortality after a myocardial infarction (EISSI-Prevenzione Trial 1999).

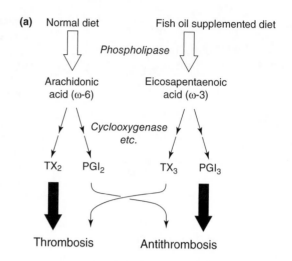

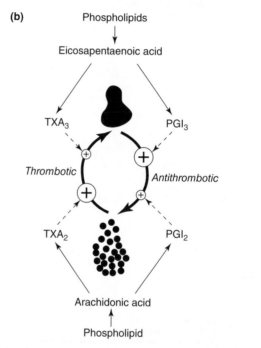

Figure 22.9 (a) *Proposed effects of a normal diet and one supplemented with fish oils on the type of thromboxanes or prostacylins produced in the reaction catalysed by cyclooxygenase and their effects on thrombosis. TX and PGI are abbreviations for thromboxane and prostacyclin, respectively. Subscripts 2 or 3 indicate the two or three series (i.e. containing two or three unsaturated carbon–carbon bonds) (see Chapter 11 for discussion). The thickness of the lines is an attempt to demonstrate the magnitude of the effects.* **(b)** *Formation of thromboxanes and prostacyclins of the two and three series from arachidonic or eicosapentaenoic acids: their effects on platelet aggregation. The unsaturated fatty acids, eicosapentaenoic and arachidonic acids, are produced from their respective phospholipids in the cell membrane by phospholipase C. Formation of PGIs and TXs from these acids involves cyclooxygenase and prostacyclin synthase or thromboxane synthase (Chapter 11).*

- One role of high density lipoprotein (HDL) is to collect unesterified cholesterol from cells, including endothelial cells of the artery walls, and return it to the liver where it can not only inhibit cholesterol synthesis but also provide the precursor for bile acid formation. The process is known as 'reverse cholesterol transfer' and its overall effect is to lower the amount of cholesterol in cells and in the blood. Even an excessive intracellular level of cholesterol can be lowered by this reverse transfer process (Figure 22.10). Unfortunately, the level of HDL in the subendothelial space of the arteries is very low, so that this safety valve is not available and all the cholesterol in this space is taken up by the macrophage to form cholesteryl ester. This is then locked within the macrophage (i.e. not available to HDL) and causes damage and then death of the cells, as described above.

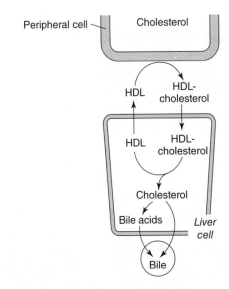

Figure 22.10 *Reverse cholesterol transfer.* High density lipoprotein (HDL) 'collects' cholesterol from cells in various tissues/organs; the complex is then transported in the blood to the liver where it binds to a receptor on the hepatocyte, is internalised and the cholesterol is released into the hepatocyte. This increases the concentration in the liver cells which then decreases the synthesis of cholesterol by inhibition of the rate-limiting enzyme in cholesterol synthesis, HMG-CoA synthase. The cholesterol is also secreted into the bile or converted to bile acids which are also secreted into the bile, some of which is lost in the faeces (Chapter 4).

Antioxidants These are naturally occurring compounds that have the ability to lower the levels of free radicals: they include vitamins C and E, the carotenoids and the flavonoids. Vitamin E and the carotenoids are particularly important in preventing oxidation of the unsaturated fatty acids within the LDL particle and within membranes of cells.

An extension of the Seven Countries Study, by Ancel Keys and colleagues, indicated the particular importance of the flavonoids. During the study, which was reported in 1980, random samples of the food eaten by the participants were taken and stored for future analysis. When this was done, it was found that the average intakes of the antioxidants vitamin E, β-carotene and vitamin C were not related to the 25-year coronary heart disease mortality rates. In contrast, the intake of flavonoids was inversely related to the mortality.

Physical activity Evidence for the beneficial effects of physical activity on the development of atherosclerosis first arose from a series of epidemiological studies (Box 22.4). This activity is now known to cause several changes, all of which are beneficial in decreasing the risk of development of atherosclerosis. These are:

- a fall in the total serum level of cholesterol;

- an increase in the serum HDL-cholesterol level;

- a fall in the serum LDL-cholesterol level;

- a fall in the plasma triacylglycerol level;

- loss of weight;

- reduction of blood pressure.

It also increases the sensitivity of tissues to insulin, which may provide better control of the blood glucose level to

minimise the risk of damage to LDL by glycosylation (see above).

Garlic Ingestion of garlic, onion or other tubers, protects against coronary artery disease. The active agents in these tubers are the allyl sulphides. (e.g. allylcysteine sulphoxide, $CH_2=CH \cdot SO \cdot CH_2CH(NH_2)COOH$). Protection by the sulphides against coronary heart disease is due to a decrease in the blood levels of triacylglycerol and cholesterol and decrease in platelet aggregation. However, the major effect is considered to be an increase in the NADH/NAD$^+$ and NADPH/NADP$^+$ concentration ratios particularly in endothelial cells. This results in an increased level of *reduced* glutathione (See Figure 6.21). This is important for a number of reasons but particularly since it can remove oxygen free radicals and hydrogen peroxide and can reduce the formation of the dangerous hydroxyl radical. In addition, it can repair damage caused by these radicals (especially damage to unsaturated fatty acids in membranes and to proteins) (Figure 22.11).

For references and discussion of other health benefits of exercise see Chapter 13.

Box 22.4 Early studies on physical activity and cardiovascular disease

The possibility that physical activity is protective against the development of coronary artery disease was first suggested from studies carried out by Jeremy Morris in 1953. The studies showed that coronary artery disease occurred earlier, more frequently and with greater severity in those men with a sedentary job than those with a more active one, e.g. sedentary clerical workers compared with those who delivered the mail. Although differences in employment-related physical activity were advanced as the most attractive explanation for the findings, Morris was well aware of the alternative explanation: that is one of self-selection where the relatively unfit men chose the more sedentary occupation. In a similar study in the USA, Ralph Paffenbarger studied more than 6000 San Francisco longshoremen who were followed for 22 years (1951–1972), and similar findings were observed: for example, men in jobs expending more than 34 MJ (8500 kcal) per week had much lower risk of fatal heart disease than those with less strenuous jobs.

Other epidemiological studies investigated the relationship between leisure time activities and the risk of coronary heart disease: Jeremy Morris used recall recording of activities outside working hours of almost 17 000 male civil servants. There were no significant associations until he used 'participation in vigorous physical exercise' as a major criterion for his analysis. The analysis then revealed that those who engaged in vigorous exercise during their leisure time suffered a heart attack with only one-third the frequency of matched controls who were inactive in their leisure time. Vigorous exercise was defined as that which cost more than 31 kJ (i.e. 7.5 kcal) per hour. Paffenbarger collected information on the amount of exercise in former students of Harvard: men who reported climbing 50 steps per day appeared to have a 20% lower risk of coronary heart disease than men who climbed fewer. Those who walked more than half a mile each day had a 21% lower risk. Some benefit was observed in those who had an energy expenditure of no more than 8 MJ (2000 kcal) per week; these activities are quite modest and represent a level which can be achieved with only small changes in customary lifestyle. For references and discussion of other health benefits of exercise see Chapter 13.

Treatment

At least three treatments are used to decrease the development of atherosclerosis: diet, drugs and surgery.

Diet   In primary prevention and in treatment, changes in diet are recommended.

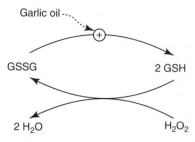

Figure 22.11 *Effect of garlic oil on the conversion of oxidised to reduced glutathione and degradation of nudrogen peroxide.* GSSG represents oxidised glutathione, GSH represents reduced glutathione (Chapter 6). The allyl sulphide in garlic oil reduces glutathione (i.e. GSSG → 2GSH). The latter is able to convert hydrogen peroxide to water in a reaction catalysed by glutathione peroxidase. Hydrogen peroxide can produce the dangerous hydroxyl free radical so that its removal is vitally important. Reduced glutathione is also important in restoring the normal structure to molecules damaged by free radicals (e.g. proteins; See Figure 6,21).

Drugs   Drugs that lower the blood levels of cholesterol are frequently used as part of the treatment; these include: (i) *Oral bile acid binding exchange resins.* Resins such as cholestyramine are effective because, when taken by mouth, they prevent the reabsorption of bile acids in the lower small intestine, so that they are excreted in the faeces. Since bile acids are formed in the liver from cholesterol, synthesis of more acids requires more cholesterol uptake by the liver from the blood, which occurs via LDL-cholesterol, so that the concentration of the latter is decreased.

(ii) *HMG-CoA reductase inhibitors.* The enzyme HMG-CoA reductase is the rate-limiting enzyme in synthesis of cholesterol and hence drugs that inhibit this enzyme are particularly effective in lowering the blood cholesterol level. These drugs are known as statins (e.g. simvastatin, provastatin).

Aggressive treatment with drugs and changes in diet in patients who have a high risk for coronary heart disease are particularly beneficial, especially if the treatment leads to a lowering of the level of LDL-cholesterol (Table 22.2). Furthermore, in the long term it can be cost effective for the medical services: the cost of treatment with drugs is considerably less than that of subsequent surgery for coronary artery disease (e.g. coronary artery bypass surgery).

There is some evidence that statins can reduce the risk of stroke and even senile dementia. Consequently, there is current clinical discussion as to whether statins should be recommended for all patients with an increased risk of stroke and possibly for protection against senile dementia.

Surgery Surgical intervention, e.g. coronary artery bypass or balloon angioplasty, is frequently used. The latter surgery is known as percutaneous transluminal coronary angioplasty. A catheter with a small balloon at the end is inserted into a coronary artery: when the balloon is expanded, the vessel dilates which allows a greater flow of blood. Although the technique appears crude, it produces clinical benefit. An extension is the use of stents: a small metallic mesh is crimped onto an angioplasty balloon and then inflated within the artery, so that the tube embeds permanently into the vessel wall, holding the artery open.

promotes interaction between myosin and actin, that is cross-bridge cycling increases, which increases the force of contraction.

When the Ca^{2+} ion concentration falls, the activity of the kinase falls, and a protein phosphatase now dephosphorylates the light chain (smooth muscle myosin light chain phosphatase) (Figure 22.12).

The simultaneous activities of the kinase and phosphatase are yet another example of regulation by reversible protein phosphorylation (i.e. an interconversion cycle – Chapter 3). An increased force of contraction could be caused either by inhibition of the phosphatase or by activation of the kinase. However, physiologically relevant inhibitors of the phosphatase have not yet been discovered.

Although this is the primary mechanism for regulation of contraction in smooth muscle, there are at least two additional mechanisms.

Hypertension

Hypertension is a common and progressive disorder which, if not effectively treated, results in an increased risk of atherosclerosis (see above), haemorrhagic stroke and damage to the kidney. For most cases of hypertension, there is no obvious cause, hence it is known as *essential hypertension*, so called because it was originally thought to be essential to maintain tissue perfusion. In order to better understand the regulation of blood pressure, a brief description of the regulation of contraction of smooth muscle is provided.

Regulation of contraction of smooth muscle

As in skeletal muscle, Ca^{2+} ions play a central role in the regulation of contraction in smooth muscle, but the mechanisms for controlling the Ca^{2+} ion concentration are quite different. The Ca^{2+} ions enter the cytosol from both the endoplasmic reticulum and the extracellular fluid: the importance of each depends on the amount of reticulum in the muscle. Depolarisation leads to activation of a Ca^{2+} ion channel in the plasma membrane and Ca^{2+} ions enter the cell down the large concentration gradient (about 1000-fold). This increase in the Ca^{2+} ion concentration activates the Ca^{2+} ion channels in the reticulum, leading to a further increase in the concentration in the cytosol. This results in an increase in the binding of Ca^{2+} ions to calmodulin which binds to a protein kinase, known as myosin light chain kinase (MLCK), which increases the activity of this enzyme. The kinase catalyses phosphorylation of the regulatory light chain of the myosin molecule which

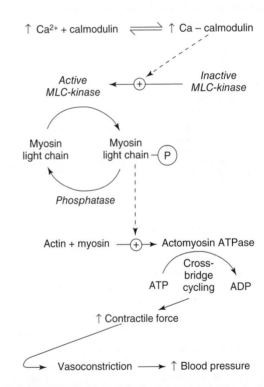

Figure 22.12 *Regulation of actin–myosin interaction in smooth muscle via the light-chain kinase and phosphatase and effect on blood pressure.* Ca^{2+} *ions bind to calmodulin and the complex stimulates the conversion of inactive myosin light chain kinase (MLCK) to active MLCK which then phosphorylates the light chain. This results in activation of the cross-bridge cycle. The overall effect is vasoconstriction of the arteriole, which increases blood pressure.*

(i) The reticulum in some smooth muscles possesses receptors for inositol trisphosphate (IP_3) which activates a Ca^{2+} ion channel in the reticulum and hence increases the release of ions and raises the Ca^{2+} ion concentration in the cytosol. The IP_3 concentration is raised due to activation of a phospholipase C which hydrolyses the phospholipid, phosphatidylinositol bisphosphate, in the plasma membrane.

$$\text{Phosphatidyhinositol bisphosphate} \rightarrow \text{inositol trisphosphate} + \text{diacylglycerol}$$

The phospholipase is activated via the adrenergic α-receptor (Figure 22.13) so that the force of contraction is increased.

(ii) Relaxation of the vascular smooth muscle is achieved through an increase in the level of nitric oxide, which is produced from arginine. Nitric oxide activates the enzyme guanyl cyclase which increases the level of cyclic GMP which, in turn, inhibits the activity of light chain kinase, thereby causing relaxation of the smooth muscle leading to vasodilation (see Figure 22.12). For details of nitric oxide formation and its effect on guanyl cyclase, see Chapter 19.

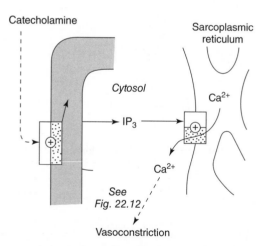

Figure 22.13 *α-Adrenergic receptor control of contraction of smooth muscle.* IP_3 represents inositol trisphosphate. Binding of a catecholamine to an α-receptor activates a membrane-bound phospholipase which hydrolyses phosphatidylinositol bisphosphate within the membrane to produce IP_3 and diacylglycerol (DAG). IP_3 binds a receptor on the sarcoplasmic reticulum in smooth muscle, which activates a Ca^{2+} ion channel and the cytosolic Ca^{2+} ion concentration increases, which results in contraction of smooth muscle in arterioles. This results in vasoconstriction and hence decreases blood flow which can leading to an increase in blood pressure.

Regulation of blood pressure

Both physiological and metabolic factors play a role in regulation of blood pressure. The former involves the vasomotor centre at the base of the brain and the basoreceptor located in the carotid sinus and the aortic arch. These mechanisms provide short-term regulation of the blood pressure, which is not discussed in this book.

The renin–angiotensin system

The renin–angiotensin system is the most important biochemical process in regulation of blood pressure. It consists of a group of proteolytic enzymes which together provide a sequence of reactions that produce and metabolise a messenger molecule, angiotensin-II. The first enzyme in the sequence is renin. It catalyses the hydrolysis, in the bloodstream, of one peptide bond in a protein known as angiotensinogen to produce a small peptide, angiotensin-I. Another proteolytic enzyme converts angiotensin-I to angiotensin-II; this enzyme is known as angiotensin-converting enzyme (ACE). Finally, angiotensin II is degraded to angiotensin III by an aminopeptidase:

$$\text{Angiotensinogen} \xrightarrow{E_1} \text{angiotensin-I} \xrightarrow{E_2} \text{angiotensin-II} \xrightarrow{E_3} \text{angiotensin-III} \rightarrow\rightarrow \text{amino acids}$$

where E_1 is renin, E_2 is angiotensin-converting enzyme and E_3 is an aminopeptidase: E_2 is the flux-generating step and E_3 is the inactivating step.

ACE is present in endothelial cells of several organs, but lung is quantitatively the most important, since it possesses the most endothelial cells. This is another example of a sequence of reactions that produce and then inactivate (metabolise) a messenger molecule. Angiotensin-II can be considered as a messenger, as discussed in Chapter 3, so that the principles governing the changes in the concentration of a messenger also apply to angiotensin-II. Thus there is a regulatory step (a zero order process), the conversion of angiotensin-I to angiotensin-II, and there is an inactivating step, the conversion of angiotensin-II to angiotensin-III, catalysed by an aminopeptidase. As with other messengers, for this process to be effective, angiotensin-II must have a short half-life; it is in fact about one minute.

> Somewhat surprisingly, all the enzymes in the angiotensin sequence are present in adipose tissue, from where angiotensin-II can be released and contribute to vasoconstriction, further complicating the regulation of blood pressure. The significance of this is unknown.

The positions of hydrolysis of the peptide bond of angiotensinogen by renin and angiotensin-I by the converting enzyme are shown in Figure 22.14.

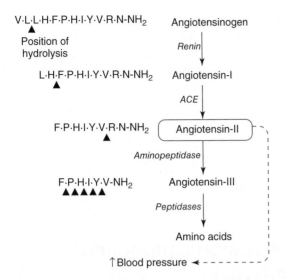

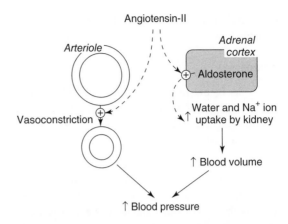

Figure 22.15 *How angiotensin-II increases blood pressure.* Angiotensin-II causes vasoconstriction of smooth muscle in arterioles; in addition, it stimulates synthesis and hence secretion of aldosterone from the adrenal cortex. Both of these effects increase blood pressure.

Figure 22.14 *Formation and inactivation of angiotensin II and positions of hydrolysis of the peptide bonds in angiotensinogen, angiotensin-I and angiotensin-II by the sequence of peptidases.* The amino acids in the peptide are identified by the standard single letter abbreviation (see Chapter 8). Renin hydrolyses the peptide bond between two leucine amino acids, indicated by ▲. The enzyme ACE hydrolyses the bond between histidine and phenylalanine, indicated by ▲, and the aminopeptidase hydrolyses between valine and arginine. Note that angiotensin-II is an octapeptide, and that the peptidase that hydrolyses the peptide bond in angiotensin-I, ACE, is in fact identical to the enzyme that degrades bradykinin, known as kininase-II. Consequently ACE inhibitors also inhibit the degradation of bradykinin. Consequently, the inhibitors of ACE have two effects: they lower the concentration of angiotensin-II but increase that of bradykinin (Chapter 17). Both these effects result in a decrease in blood pressure. This is an advantage of ACE inhibitors over angiotensin-II blockers in the control of blood pressure (see Figure 22.16).

Renin is secreted into the bloodstream by specialised cells in a portion of the distal tubules in the kidney (known as the macula densa). The rate of secretion is stimulated by several factors, including a low Na^+ ion concentration in the blood and an increase in sympathetic activity (i.e. an increase in the local level of noradrenaline).

How angiotensin-II increases blood pressure

Angiotensin-II raises the blood pressure by two mechanisms:

- Vasoconstriction of arterioles.

- Stimulation of aldosterone secretion from the adrenal cortex, which increases Na^+ ion reabsorption in the tubules of the kidney. Via an osmotic effect, this increases water uptake from the glomerular filtrate, which increases blood volume, and hence increases cardiac output, and therefore blood pressure in order to pump the extra fluid around the body.

Angiotensin-II binds to receptors on the plasma membrane of smooth muscle cells within the arterioles. This results in contraction and so reduces the diameter of these vessels, thus causing more resistance to blood flow. It also binds to a receptor on the cells in the adrenal cortex that secrete aldosterone and this increases the synthesis and secretion of aldosterone (Figure 22.15).

> A reduction in diameter of arterioles is of major importance in control of blood pressure, since resistance is directly proportional to the fourth power of the radius (Poiseuille's law), i.e. a change in radius, therefore, produces a massive change in blood flow.

The same basic biochemical control mechanism causes contraction of the smooth muscle as well as secretion of aldosterone. The binding of angiotensin to its receptor activates a membrane phospholipase-C. It catalyses the hydrolysis of phosphoinositide phosphatidylinositol bisphosphate to produce the two intracellular messengers, inositol trisphosphate (IP_3) and diacylglycerol (DAG).

(i) In the smooth muscle cells, IP_3 binds to its receptor on the sarcophasmic reticulum which stimulates the release of Ca^{2+} ions so that the cytosolic concentration increases which results in an increased force of contraction (see above) and hence vasoconstriction.

(ii) In the aldosterone-secreting cells, diacylglycerol activates protein kinase C resulting in phosphorylation of transcription factors that increase the transcription of a gene which expresses the key enzyme controlling the synthesis of aldosterone in the steroid-secreting cells of the adrenal cortex. This results in an increased rate of secretion of aldosterone and hence an increase in the plasma level.

Treatment

On the basis of the mechanisms that control blood pressure, it is not surprising that the drugs used in the treatment of hypertension include diuretics, ACE inhibitors, angiotensin-II receptor blockers, and Ca^{2+} ion channel inhibitors.

- *Diuretics* These inhibit reabsorption of Na^+ ions from the glomerular filtrate so that the concentration of Na^+ ions in the filtrate increases which, via an osmotic effect, increases the loss of water and hence decreases the blood volume.

- *Angiotensin-converting enzyme inhibitors* These lower the plasma concentration of angiotensin-II.

- *Angiotensin-II-receptor antagonists* These block the binding of this messenger to its receptors on the two target tissues, i.e. smooth muscle in the arterioles and the aldosterone-secreting cells in the adrenal cortex.

- *Ca^{2+} ion channel inhibitors* These inhibit the Ca^{2+} ion channels in the plasma membrane of the smooth muscle cells and result in a decrease in the cytosolic level of Ca^{2+} ions in smooth muscle of the arterioles, which reduces contraction of these muscles.

Treatment with a diuretic and an ACE inhibitor is usually very effective. ACE inhibition may be preferable to the use of a receptor antagonist since the inhibition has a dual effect on vasodilation: it decreases the concentration of angiotensin II, which is a vasoconstrictor, but increases that of bradykinin, which is a vasodilator (see Figure 22.16).

Heart attack (myocardial infarction)

Just like the brain, the mammalian heart has a high requirement for oxygen and fuels, and for the same reason, to provide for mitochondrial generation of ATP. The fuel and

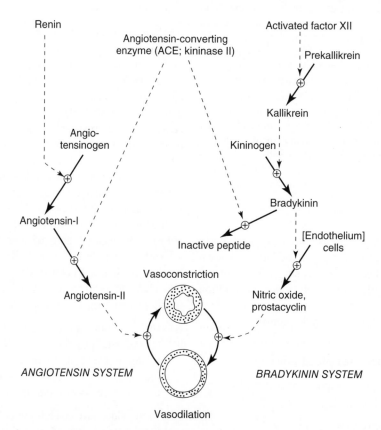

Figure 22.16 *Regulation of vasoconstriction/vasodilation by angiotensin-II and bradykinin.* The mechanism by which angiotensin-II stimulates vasoconstriction is shown in Figure 22.15. Angiotensin-converting enzyme is also responsible for bradykinin inactivation. Bradykinin stimulates endothelial cells to produce and secrete nitric oxide and prostacyclin, both of which are vasodilators. Consequently the effect of an ACE inhibitor is to decrease the concentration of angiotensin-II, which lowers blood pressure, and to increase the concentration of bradykinin, which also lowers blood pressure.

oxygen required by the heart is provided from the coronary circulation, which is the special circulatory system found only in the mammalian heart. It is essential for supplying fuels and oxygen to the thick wall of the left ventricle. Of oxygen consumed by the heart, about 70% is used by the left ventricle. This is not surprising, since its contractions provide blood for all the organs of the body, including the brain. A high blood pressure is required especially in bipedal animals to ensure an adequate supply of blood to the brain.

An occlusion in one of the coronary arteries, produced by one of the mechanisms described above, diminishes or blocks the blood flow to a portion of the myocardium, which decreases the supply of oxygen. This results in a marked decrease in the rate of mitochondrial generation of ATP, so that those cells that are affected by the occlusion cannot provide sufficient ATP to satisfy the energy demands of both contraction and ion pumps in the cardiomyocytes in that part of the heart. As a consequence, the ATP/ADP concentration ratio progressively decreases, so that the heartbeats become feebler and feebler and then become irregular. If the latter effect is severe, arrhythmias and ventricular fibrillation develop, during which the heart fails to function as a pump. In order to understand the processes underlying the reduction in force of the myocardium after an occlusion, knowledge of the regulation of contraction in the cardiomyocytes is required.

Regulation of contraction of cardiac muscle

Depolarisation of the membrane of the cardiomyocyte, resulting from the action potential, initiates contraction in cardiac as in skeletal muscle. This depolarisation arises in the sinoatrial node, a small group of cells in the right atrium, and then spreads through the heart causing, first, the muscles in the atria to contract and then the muscles in the ventricles to contract.

As in skeletal muscle, the action potential is transmitted into the cardiomyocytes by a T-system. The depolarisation opens Ca^{2+} ion channels in the T-system (i.e. the plasma membrane) which leads to a small increase in the cytosolic Ca^{2+} ion concentration. This activates Ca^{2+} ion channels in the sarcoplasmic reticulum (a process known as a Ca^{2+}-induced Ca^{2+} ion release) which results in the movement of more Ca^{2+} ions into the cytosol. The amount of Ca^{2+} ion released from the reticulum varies according to the stimulation of the ion channel, i.e. the amount of Ca^{2+} ions released from the T-tubule. This provides for variations in the cytosolic Ca^{2+} ion concentration and hence in the force of contraction.

An increase in the cytosolic Ca^{2+} ion concentration increases the activity of myosin ATPase. This increases the rate of cross-bridge cycling and hence the force of contraction (i.e. greater cardiac output). Relaxation occurs due to a lowering of the cytosolic Ca^{2+} concentration. As in skeletal muscle, the uptake of Ca^{2+} by the sarcoplasmic reticulum is mainly responsible for reducing the Ca^{2+} ion concentration, but the Na^+/Ca^{2+} exchange process in the plasma membrane plays an additional role, i.e. as Na^+ ions enter, Ca^{2+} ions are extruded. The decrease in intracellular Ca^{2+} ion concentration decreases the force of contraction.

Sympathetic activity increases both the rate and force of contraction of the heart through the increase in the local concentration of noradrendine which binds to β-adrenergic receptors on the plasma membrane. This leads to activation of adenylate cyclase, which increases the cytosolic level of cyclic AMP. This, in turn, increases the activity of protein kinase A which phosphorylates the following:

- *A slow Ca^{2+} ion channel* in the plasma membrane. This increases its activity and hence cytosolic Ca^{2+} ion concentration increases. This stimulates the Ca^{2+} ion channel in the reticulum, leading to a further increase in the flux of Ca^{2+} ions into the cytoplasm.

- *Myosin*, which increases the rate of the cross-bridge cycling.

- *A sarcoplasmic reticulum* protein, known as phospholambam, phosphorylation of which increases the activity of the Ca^{2+} ion uptake process in the reticulum and hence the concentration of Ca^{2+} ions in the cytoplasm decreases more rapidly. This results in a decrease in the relaxation time of the muscle and hence an increase in heart rate.

- *Troponin-I*, which decreases the sensitivity of myosin ATPase to Ca^{2+} ions so that, *in contrast to the above*, the result is a decrease in the effect of adrenaline on the force of contraction. This is, therefore, a change that prevents too marked an increase in contractility, i.e. a feedback control mechanism.

These effects explain how blockers of the β-adrenergic receptors, or blockers of the calcium ion channels in the heart, decrease the force of contraction of the heart and therefore lower blood pressure.

Stimulation of the parasympathetic system releases acetylcholine at the neuromuscular junction in the sino-atrial node. The binding of acetylcholine to its receptor inhibits adenylate cyclase activity and hence decreases the cyclic AMP level. This reduces the heart rate and hence reduces cardiac output. This explains why jumping into very cold water can sometimes stop the heart for a short period of time: intense stimulation of the vagus nerve (a parasympathetic nerve) markedly increases the level of acetylcholine in the sinoatrial node, which halts nervous activity in the node and the heart stops beating. Fortunately, the enzyme acetylcholinesterase rapidly lowers the acetylcholine level so that contractile activity of the heart is soon restored.

The cause of the decrease in contractile activity after an occlusion

The cause of the decreased contractile activity is likely to be due to a decrease in ATP/ADP concentration ratio in the cardiomyocyte, since hydrolysis of ATP now results in less energy being transfered for each molecule of ATP that is hydrolysed. That is, sufficient energy is not available to power the maximum rate of the cross-bridge cycle or power the transport of ions across the plasma membrane (e.g. the Na^+/K^+ pump or the Ca^{2+} ion channels). Changes in concentrations of such ions can lead to disturbance of electrical activity that controls the contractions of the fibres.

A further cause of disturbance in electrical activity is the psychological stress of the infarction, which increases the plasma level of fatty acids, which can lead to increased intracellular levels of fatty acids and fatty acyl CoA. These can damage the plasma membrane and hence disturb electrical activity (Chapter 7).

Biochemical mechanisms to maintain the ATP/ADP concentration ratio in a cardiomyocyte when oxygen becomes limiting

In an attempt to maintain the ATP/ADP concentration ratio close to the normal value, the myocardium responds in two ways. The rate of ATP generation is increased and the rate of ATP utilisation is decreased. Indeed it is the decrease in the ATP/ADP concentration ratio that results in metabolic changes (i.e. produces metabolic signals) that increase ATP generation and decrease ATP utilisation in the cardiomyocyte. In fact, one signal has both effects, adenosine (see below).

Mechanisms to increase ATP generation   The two processes that increase ATP generation are glycolysis and oxidative phosphorylation. The latter is due to increased oxygen supply due to vasodilation (see below). The signals that stimulate glycolysis are increases in the concentrations of AMP and phosphate and a decrease in the level of phosphocreatine. These changes occur as follows. The three nucleotides, ATP, ADP and AMP are in equilibrium, catalysed by the enzyme adenylate kinase, as follows:

$$ATP + AMP \rightleftarrows 2ADP$$

The equilibrium is such that, as the ATP/ADP concentration ratio decreases, the AMP concentration increases. The

increase in the phosphate concentration and the decrease in phosphocreatine occur as follows. In an attempt to maintain the normal ATP/ADP concentration ratio, phosphocreatine transfers its phosphate to ADP in a reaction catalysed by creatine kinase

$$PCr + ADP \rightarrow Cr + ATP$$

Since this change occurs as ATP is hydrolysed to ADP, by myosin ATPase

$$ATP + H_2O \rightarrow ADP + P_i$$

the net result is an increase in the concentration of phosphate as follows: i.e. the net reaction is

$$PCr \rightarrow Cr + P_i$$

The key regulatory enzymes in glycolysis are phosphorylase, hexokinase, phosphofructokinase and pyruvate kinase, the activities of which are stimulated by the increase in the concentrations of AMP and phosphate and the decrease in that of phosphocreatine. These mechanisms are discussed in Chapters 6 and 9; Figures 6.16 and 9.27.

Vasodilation is stimulated by an increase in the concentration of adenosine. The enzyme AMP deaminase catalyses the formation of adenosine from AMP as follows:

$$AMP + H_2O \rightarrow adenosine + ammonia$$

The activity of this enzyme is regulated by changes in the concentrations of ATP and phosphate: the former inhibits whereas the latter activates it. These are the signals that increase the concentration of adenosine. It is transported out of the cell, so that the extracellular concentration also increases. This then stimulates relaxation of the smooth muscle in the arterioles which results in vasodilation and increased blood flow and consequently a greater supply of oxygen to the cardiomyocytes. The blood flow will be increased to those parts of the myocardium that are not totally occluded by the clot, so that more mitochondrial generation of ATP can occur. Provided the portion of the myocardium that is totally occluded is not too large, the heart can then continue to function as a pump.

Mechanism that decreases the rate of ATP utilisation
The increase in the extracellular concentration of adenosine has two effects that result in a decrease in the force of contraction and hence a decrease in the rate of utilisation of ATP:

- It opens a K^+ ion channel in the plasma membrane which increases the loss of K^+ ion from the cell which then increases the membrane potential, so that the ability to initiate an action potential is decreased. For discussion of a similar phenomenon in skeletal muscle see Chapter 13.

- It opens a K⁺ ion channel in the sinoatrial node, which slows the initiation of the electric signals that control the heartbeat (this is known as sinus bradycardia).

The roles of adenosine in these safety events have led to the term 'adenosine revival' emphasising the important role of this nucleoside.

   A summary of the overall mechanism by which a decrease in the ATP/ADP concentration increases ATP generation and decreases ATP utilisation is provided in Figure 22.17.

Treatment for a myocardial infarct

Three forms of treatment are used: drugs, enzymes and metabolic interventions.

Drugs

The stress of an infarction increases the blood levels of catecholamines which increase the force of contraction of

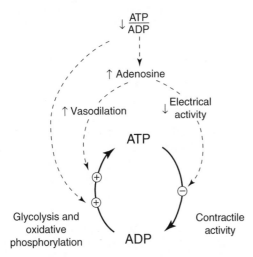

Figure 22.17 *Summary of mechanisms to maintain the ATP/ADP concentration ratio in hypoxic myocardium.* A decrease in the ATP/ADP concentration ratio increases the concentrations of AMP and phosphate, which stimulate conversion of glycogen/glucose to lactic acid and hence ATP generation from glycolysis. The changes also increase the activity of AMP deaminase, which increases the formation and hence the concentration of adenosine. The latter has two major effects. (i) It relaxes smooth muscle in the arterioles, which results in vasodilation that provides more oxygen for aerobic ATP generation (oxidative phosphorylation). (ii) It results in decreased work by the heart (i.e. decrease in contractile activity), (mechanisms given in the text) which decreases ATP utilisation.

the heart. This effect is reduced by β-adrenergic receptor blockers, which reduce the rate of ATP utilisation and help to maintain the ATP/ADP concentration ratio which helps the function of the heart. The force of contraction is controlled by Ca²⁺ ion concentration in the cytosol. Reduction in Ca²⁺ ion transport into the cell decreases cytosolic Ca²⁺ ion concentration and hence the force of contraction. Hina, drugs that block (close) Ca²⁺ ion channels help to maintain the ATP/ADP concentration ratio.

Enzymes

Three proteolytic enzymes, streptokinase, urokinase and tissue plasminogen activator, hydrolyse peptide bonds in fibrin which loosens the structure of the clot and can results in its dispersal. This can restore flow of blood to the part of the myocardium affected by the clot. These are known as 'clot-busting' enzymes. One or more of these enzymes is introduced into the circulation, and provided that this is done very soon after an infarct, damage to that part of the myocardium can be minimised.

Metabolic interventions

One metabolic intervention has been shown to be beneficial. This is the intravenous provision of a solution of glucose, insulin and K⁺ ions (known as GIK therapy). This procedure raises the blood levels of glucose and insulin, which result in three changes:

- The increased plasma concentration of glucose increases the rate of glucose transport into the cardiomyocytes which increases the rate of glycolysis.

- The increase in the plasma concentration of insulin causes a decrease in fatty acid oxidation by reducing mobilisation of fatty acids from adipose tissue and by increasing the malonyl-CoA (see Chapter 7). This encourages the oxidation of glucose rather than that of fatty acid, which is beneficial when oxygen supply is limiting, since glucose oxidation requires less oxygen to generate the same amount of ATP. Moreover, a decrease in the plasma level of fatty acids may reduce the risk of arrhythmias

- Since K⁺ ions are required by several enzymes in glycolysis and since K⁺ ions may be lost from an hypoxic muscle fibre, administration of a small amount of K⁺ ion may help to maintain the intracellular concentration, particularly since K⁺ transport into the cell is stimulated by insulin.

Index

Note: Boxes are indicated by **emboldened page numbers**, Figures and Tables by *italic numbers*, and Appendices (on website) by *App.*; abbreviations: CNS = central nervous system; GI = gastrointestinal